Logic Design of Digital Systems

Logic Design of
Digital Systems

Second Edition

Donald L. Dietmeyer
University of Wisconsin

Allyn and Bacon, Inc.
Boston   **London**   **Sydney**   **Toronto**

Copyright © 1978, 1971 by Allyn and Bacon, Inc.
470 Atlantic Avenue, Boston, Massachusetts 02210

Library of Congress Cataloging in Publication Data
Dietmeyer, Donald Leo, 1932-
  Logic design of digital systems.

  Bibliography: p.
  Includes index.
  1. Electronic digital computers – Design and
construction. 2. Logic design. 3. Logic circuits.
I. Title.
TK7888.3.D52   1978   621.3819'58'2   77-15056

ISBN 0-205-05960-0
ISBN 0-205-06122-2 (International)

Contents

9 *Computer Manipulation of Switching Functions*   *535*

10 *Minimization of Switching Functions*   *609*

11 *Multiple Level Synthesis*   *683*

Preface

Digital technology has changed substantially since the first edition of this book was published. This edition reflects suggestions of users of the first edition as well as technological advances. The most obvious change is the reorganization of material into shorter chapters sequenced for use in a comprehensive introductory course (Chapters 1-8) followed by an advanced course emphasizing logic design automation. The introduction and substitution of new material and methods of presentation is less obvious, but no less extensive. For example, basic concepts and components of digital systems are introduced immediately in Chapter 1 together with a minimal treatment of binary codes and number systems to motivate students and prepare them to accept and appreciate the mathematics of switching functions. A survey of digital electronics now opens Chapter 2 so that the important relation between digital electronics and logic design may be established or emphasized as the instructor prefers. The bulk of Chapter 2 concentrates on switching function representation and logic network analysis. The combinational logic syntax and semantics of design language DDL are collected in a final section of this chapter where they may be easily referenced.

Chapter 3 concentrates on the synthesis of combinational networks with relatively informal, manual techniques. The use of PLAs, ROMs and multiplexers as well as traditional gates to realize switching functions is shown. Arithmetic systems are used extensively to illustrate such techniques, and the relations between algorithms, coding theory and hardware. Sequential network analysis and design begins with Chapter 4. Latches and edge-triggered and master-slave flip-flops are considered in detail. Sources of delay in electronic amplifiers are reviewed for those who wish to emphasize the relation between logic design and digital electronics. Chapter 5 emphasizes the analysis and synthesis of synchronous sequential networks. The sequential syntax of DDL is collected in one section for easy reference. Chapter 6 dwells on the design of sequential structures commonly found in digital systems. The use of both ROMs and PLAs in data processing and control applications is illustrated.

Chapters 7 and 8 are concerned with information processing systems. An elementary digital computer is rather completely specified, designed and programmed, not because the system is of any significance in itself, but to show the considerations that lie behind the architecture of such a system and to indicate clearly the relevance of all of the previous chapters to digital systems design. This system is then expanded in detail to illustrate common features and organizations of digital computers such as register files and microprogramming and, more important, to point out the possibilities, problems, and limitations of digital-systems architecture. Extensive use of the digital design language makes expansion in detail possible and meaningful.

Material on the "cube" representation and manipulation of switching functions has been collected in Chapters 9-11. First the Boolean algebra of cubes is developed and applied in introductory, but practical algorithms. Then Chapter 10 treats minimization algorithms. The first section of this chapter presents the Quine-McCluskey Algorithm which some may wish to study in conjunction with map minimization as presented in Chapter 2. Chapter 11 covers multiple level network design via algorithms. New material on the use of arrays to describe such networks and feed-forward network design has been introduced.

Most if not all types of hazards in combinational and asynchronous sequential switching circuits are encountered by analysis of various models of such circuits in Chapter 12. The definitions of "state variable," "race," etc. are general so that they may be applied to the very detailed models of switching circuits previously encountered only in speed-independent switching theory, as well as to the more traditional models of asynchronous switching circuits. The treatment of synthesis in Chapter 13 emphasizes single-transition-time state assignments and the use of the delay which logic gates introduce rather than that of additional delay elements to eliminate sources of malfunction in asynchronous circuits.

A one semester course in logic design may be slanted toward either the practical or the mathematical by appropriate selection of topics from Chapters 1-6. A one year course would more completely cover these chapters as well as Chapters 7 and 8. Chapters 9 through 13 are used at the University of Wisconsin-Madison in a graduate course in algorithmic logic design, but selected sections might well be included in more traditional logic design courses.

Professor Charles R. Kime has been most helpful in preparing this revision. The countless suggestions and corrections provided by students and other users of the first edition are also very much appreciated.

1

An Introduction to Digital Systems

Digital systems are assemblages of interacting parts that are capable of storing, communicating, and processing information expressed in discrete form. *Logic design* is the process of solving problems of subsystem and system organization so as to achieve desired information processing, communication, and storage. Other facets of digital system design such as the solution of mechanical, chemical, thermal, electrical, economic, and computer science problems impact logic design. The logic design of digital systems is really much easier than these general definitions suggest, as we can see by examining the words used in the definitions.

1.1 BASIC CONCEPTS OF DIGITAL SYSTEMS

Information—recorded or communicated facts or data—takes a variety of physical forms when being stored, communicated, or manipulated. Printed symbols store and convey information; behind the retina of the eye they take the form of chemical and electrical activity. In telephone and radio communications, information takes the form of voltages and currents that vary with time. Variations in air pressure encode information when people converse. The Indians used smoke. The list of examples can go on and on; for any parameter of nature that can be controlled by man can be, and probably has been, used to express information.

All forms that information may take can be classified as being either (1) discrete (composed of distinct parts) or (2) continuous. Numbers printed on paper record information in a discrete manner. A plot based on those numbers and printed on that same paper may record the same information in a continuous form. Usually the hands of a clock move continuously to display the time of day. But some clocks show a sequence of digits, which changes abruptly from time to time to give the same information, but in discrete form.

"Digit" originally meant a finger or toe. These were probably the first tools used by people to assist in counting. As a result "digit" also refers to symbols commonly used to express numeric information. The symbols 0, 1, 2, 3, 4, 5, 6, 7, 8, and 9 are familiar digits. In electronic computation and data-processing, letters, punctuation marks, and mathematical symbols are stored, communicated, and processed in much the same fashion as these familiar digits. These other symbols are used on occasion to express numeric information. The definition of "digit" has been generalized as a result. *Digital* has become synonymous with "discrete"; *digit* has become synonymous with "symbol." A set of digits is called an *alphabet*.

Communicating information is moving it through space. In electronic networks, digital information is usually expressed by the value of electrical variables, voltage or current. A different value or range of values of such a parameter is used to express each digit of an alphabet. Wires then provide a means of moving information from one point to another. Each of several wires may convey a digit of a digit sequence (parallel communication), or a single wire may convey the digits of a message sequentially in time (serial communication). Such movement does take time: fundamental laws of physics limit the speed with which information can be moved. Electromagnetic waves can travel no faster than 3×10^8 meters per second. It thus takes light and electronic signals at least 3.3×10^{-9} seconds (3.3 nanoseconds) to travel one meter. Mechanical and acoustic waves travel much more slowly.

Storing information is carrying that information through time. Since communicating takes time, communication mechanisms are sometimes used to provide memory. An electrical signal that is applied at one end of a circuit one meter long arrives at the other end approximately 3.3 nanoseconds later. The applied signal may have changed during that interval. The wire *delays* the supplied signal for a brief, but nonzero interval. *Dynamic* memory units carry information through long periods of time by repeatedly passing that information through a delay line. *Static* memory elements store by using special electronic circuits or by recording the information in a preservative fashion.

Processing information consists of forming new information by altering given information according to specified rules. Arithmetic operations are familiar ways of processing given information to generate new information; we know the rules. We commonly perform many other operations on information without specifying rules in detail. Selecting one of many units of information is an example.

Digital systems consist of processing and memory elements that are interconnected by communication mechanisms (see Fig. 1.1). Communica-

Digital System

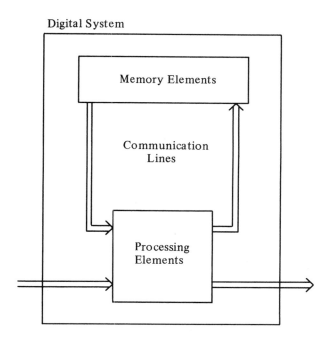

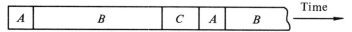

A: Memory Elements Capture New Information
B: Memory Elements Hold Information
    Communication and Processing Are Performed
C: New Results Are Available to the Memory Elements

FIG. 1.1   Structure and Activity Cycle of Digital Systems

ting and processing both take time; units that communicate and process require that the same information be continuously supplied for greater than some minimum interval of time. Memory elements hold and continuously supply information while it is carried to the processing elements and processed, and the results are conveyed back to the memory elements. Activity is cyclic: memory elements acquire new information; it is propagated to the processing elements and processed, which usually involves communication between elements; and the new information is propagated to the memory elements. Since useful systems must be able to accept information from their environment and supply information to it, the environment is also pictured in Fig. 1.1 as a source of information to be processed and a sink for results.

Coordination of activity is essential in digital systems. Processing elements must be told which set of rules to use; memory elements must be told when to "forget" old information by capturing new information. It is natural to distinguish *data*, i.e., information to be processed to answers, from control information, which governs the processing steps taken to arrive at answers. Figure 1.1 does not distinguish between these two kinds of information; in a digital system both data and control information are stored, communicated, and processed by identical means. Thus, distinctions between types of information are external to the actual system. They may be difficult to make and are not always clear. When adding two numbers, their signs and magnitudes influence the steps taken as well as serve as data.

While the model of digital systems presented in Fig. 1.1 is very general, it is also very vague. The remainder of the book refines this model through detail and examples of specific systems. When considering the details, keep Fig. 1.1 in mind. Digital systems, like forests, consist of huge numbers of leaves, twigs, branches, etc. By placing details in perspective, one can often avoid becoming lost or entwined in brambles.

1.2 BASIC COMPONENTS OF DIGITAL SYSTEMS

In practice, electronic circuits that can reliably determine what digit a wire conveys to it and in turn generate the electrical representation of the same or another digit deal only with two different digits. Reliability is the key. Digital systems today consist of tens of thousands to millions of amplifying circuits constructed of components with characteristics over which only limited control can be exercised. Transistor characteristics vary from unit to unit, and vary for a given unit with time and temperature, for example. While a few circuits might be made to deal with more than two digits by carefully selecting components and constantly "tuning" the circuits, large assemblages of circuits can not be economically manufactured and maintained by such means. Circuit operation must be made as nearly independent of component characteristics as possible. To date, the use of vacuum tubes, diodes, and transistors as switches, i.e., either turned fully on or fully off, has proved to be the most economical method of achieving such independence and hence the required reliability of circuit performance.

Thus, today at least, electronic digital information-processing systems deal with information expressed in terms of two digits, commonly symbolized by 0 and 1. The word "binary" expresses this "twoness." A digit from a binary alphabet is a "binary digit" or *bit*. Information theory defines the smallest unit of information as the *bit*. We will not distinguish between these two definitions, and consider one binary digit to convey one unit of information.

When we wish to express more than one unit of information, more than one binary digit must be used. These digits are linked together (*concatenated*) to form a binary sequence. The small circle ∘ will be used to express the concatenation operation. Thus linking a 1 on the right of a 0, 0 ∘ 1, gives one binary sequence. Linking a 0 on the right of a 1, 1 ∘ 0, gives a different sequence. Usually we omit the concatenation symbol and simply write the binary digits next to each other in the desired order. Thus, 01, 10, 101101, and so forth, are binary sequences.

Communicating Elements

In a machine different digits take the form of different levels of some physical quantity such as voltage, current, pressure, or light intensity. A *terminal* is a communication mechanism that is capable of conveying binary expressed information. In electronic systems a terminal is often just a wire (a return path for current is assumed). When the rate with which information is to be conveyed is high, or the distance is long, the wire may take the form of a coaxial cable, perhaps with special driving and receiving amplifiers. Regardless of their physical construction or length, we will usually ignore the delay provided by terminals and think of communication as instantaneous. Even when delay is not ignored, we will assume that terminals have no ability to store information. We will take the delay as introducing a limit on the rate at which information may be presented to a terminal.

A *signal* is the level of the conveyed quantity (voltage/current) at a specific point in a circuit. The value of a signal changes with time as that level changes, i.e., as different digits are conveyed in turn. Since terminals were defined to be conveyors of bits, the signals on terminals take values from the binary set $\{0, 1\}$.

It is very useful to assign names to terminals and signals. Confusion is greatly reduced if the same name is used for a terminal and the signal it bears (and later for the generator of that signal). A name to which any of a number of values may be assigned is known as a *variable*. The names assigned to signals are thus mathematical variables. If terminal X bears a binary signal, then signal X takes values from the set $\{0, 1\}$. At one instant variable X may be assigned value 0,

$$X = 0$$

at other instants it may be assigned the value 1,

$$X = 1$$

If terminal X is to bear the signal carried by terminal C, then variable X is to take the value of variable C.

$$X = C$$

In terms of signals, terminal X is to be *connected* to terminal C to insure that they carry the same signal. Thus, the equal sign ($=$), which says that variables are assigned the same values, may be thought of as demanding a connection of terminals. Touching two bare wires connects them; cutting a wire disconnects the two halves. We will usually place electronic circuits between so that connections may be made and broken at high speeds.

An ordered collection of n items is known as an *n-tuple*. An *n*-tuple is usually written as a list of the n items enclosed in parentheses, the items being separated by commas. Thus, $(1,2.5)$ is a 2-tuple and $(0, 1, 0, 0, 1)$ is a 5-tuple. If the items are numeric values, as in these examples, or variables representing values, the *n*-tuple is called a *vector*. Vectors are valuable to us when a terminal of many wires is used to achieve parallel communication. Terminal X might consist of five wires that can be individually identified with subscripts.

$$X = (X_1, X_2, X_3, X_4, X_5)$$

If terminal C consists of the same number of wires, the connection of the C wires to the X wires may be more concisely expressed with vector notation,

$$X = C$$

than with the detailed but verbose collection of 5 connections.

$$X_1 = C_1$$
$$X_2 = C_2$$
$$X_3 = C_3$$
$$X_4 = C_4$$
$$X_5 = C_5$$

Figure 1.2 shows how a 5-pole switch may be used to make and break this connection.

When a single wire, say A, is to be connected to many wires of a set, vector notation may again be used. Let $B = (B_1, B_2, B_3)$, then

$$B = A$$

illustrates this type of connection. Figure 1.2 also shows how this connection might be accomplished, and justifies calling the distribution of the single signal a *fan-out*.

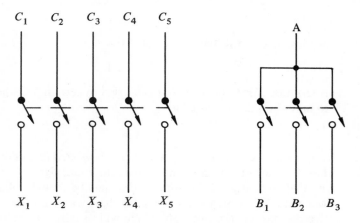

FIG. 1.2   Terminal Sets and Connections $X = C$ and $B = A$

CHAPTER 1   An Introduction to Digital Systems

Memory Elements

A *memory element* is a system component that is able to store one unit of information until commanded to replace it with another. A variety of electronic circuits are able to hold one bit until commanded to replace it with another. Chapter 4 explores this variety; we will introduce an idealized version of one such circuit as a specific example here. The *D flip-flop* has two input terminals, *D* and *C*, and two output terminals as shown in Fig. 1.3. The signal on the named output terminal (Q in Fig. 1.3) indicates the bit being stored; when the flip-flop is storing a 1, variable Q has value 1. The output terminal with the small circle bears the *complementary* signal value. Hence it is labeled with a bar over the name. When Q has value 1, $\bar{Q}$ has value 0; when Q has value 0, $\bar{Q} = 1$. Table 1.1 summarizes this relation between the two output signals. Tables that completely enumerate a relation between binary variables are known as *truth tables*.

TABLE 1.1. THE
NOT RELATION
BETWEEN BINARY
VARIABLES

Q	$\bar{Q}$
0	1
1	0

The *C* input terminal of the *D* flip-flop (DFF, for short) is the point at which control information is supplied. Often a periodic signal called a *clock* is supplied to this terminal. The *C* may therefore stand for either control or clock. We will defer practical details here by assuming that a clock signal with very brief pulses as shown in Fig. 1.3 is supplied to all *C* terminals of all flip-flops of a digital system. Data to be captured and held are supplied at the *D* terminal at the times of the clock pulses. Captured values are displayed on the *Q* line shortly after the clock pulses.

Each time that the *C* signal goes to logic *l* and back to 0, the flip-flop captures the value of the *D* signal. It holds that value until the next *C* pulse. Only the values of the *D* signal at the instants of the clock pulses are important. What goes into the DFF at one clock pulse is held until the next clock pulse. If time is treated as a discrete variable and the values of signals examined only at the times of the clock pulses, *Q* has at the $i + 1st$ instant of time the value that *D* had at the *i*th instant.

$$Q(i + 1) = D(i) \tag{1.1}$$

This memory element delays its data signal by one unit of time—the time between clock pulses. Thus, this memory element is a "delay" flip-flop.[1]

---

[1] Real *D* flip-flops are more complex than suggested here. Practical details are deferred until Chapter 4.

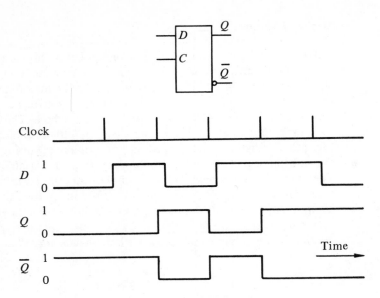

FIG. 1.3   The D Flip-flop

A number of memory elements that collectively hold the digits of a sequence is known as a *register*. A collection of flip-flops can hold a binary sequence. As with terminals, there is an advantage in naming the collection and using subscripts to identify individual members of the collection. Thus, in a system we might find register A to consist of one flip-flop named A (no subscripts are necessary), register B consisting of 5 flip-flops, B_1, B_2, B_3, B_4, and B_5, and register C, which consists of 25 flip-flops. The name assigned to a flip-flop is also assigned to the output terminal of that flip-flop and the signal it bears. In addition, the same name will be used to refer to the (generally fictitious) data input terminal of the flip-flop.[2] Whether the input or output terminal is being referenced is determined by the way in which the name is used.

The activity in which each flip-flop of a register captures a new bit of data is known as a *register transfer*. A left-pointing arrow will be used to denote this activity. The register named on the left of the arrow is known as the information *sink*. The register or terminal named on the right is the information *source*; thus,

$$B \leftarrow C_{1:5}$$

indicates that information is to be transferred to the B register, and the reference is to the data input terminals of the B flip-flops. The subscript notation "$i:j$" will be used to reference the range of components i, $i + 1$, ..., j. Thus, $C_{1:5}$, or equivalently, $C[1:5]$, indicates that the left five flip-flops of the C register are the source of information; the output terminals of these flip-flops are being referenced.

[2]One terminal of the D flip-flop can be identified as the "data" input terminal. This is not the case with most other types of memory elements.

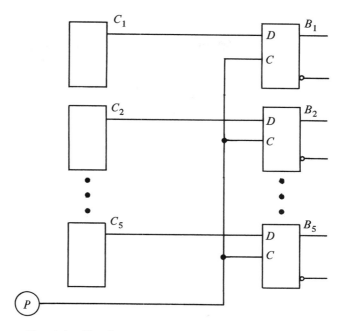

FIG. 1.4 Circuitry to Accomplish the Register Transfer
If P Then $B \leftarrow C_{1 : 5}$.

The controlling clock signal is not mentioned above so the register transfer description is not complete. In English we might say, "If the clock signal goes to 1, then load B from $C[1 : 5]$." "*If ... then*" enclosing the condition that must be satisfied, i.e., take the value of logic 1, is a very natural way of describing control. Let P be the name of a clock signal.

If P then $B \leftarrow C_{1 : 5}$.

The period terminates the "if ... then" condition just as it terminates the English sentence. Figure 1.4 suggests how DFFs must be interconnected to accomplish the register transfer described.

The following declarations describe a simple digital system.

System *SHIFT*:
Clock P.
Terminal X.
Register $A[8]$.
If P then $A \leftarrow X \circ A[1 : 7]$.

The statement "Register $A[8]$." tells us that register A exists in the system and consists of eight flip-flops. With each clock pulse the eight flip-flops of the A register are to capture and hold new information. The leftmost flip-flop, A_1, gets its information from terminal X. Flip-flop A_2 gets its information from A_1. Flip-flop A_8 has A_7 as its source. Thus, with each clock pulse the binary sequence held by A is moved one place to the right, and a new bit supplied by terminal X is introduced at the left. Register A is an example of a *shift register*. Figure 1.5 shows how A may be constructed of DFFs and how the binary sequence that it holds changes with clock pulses.

Shift registers find many applications in digital communication and computing systems. For example, system SHIFT above may be used to convert the form of information. Eight-bit binary sequences conveyed in serial fashion over terminal X are available in parallel form on the eight output terminals of the flip-flops after every eight clock pulses. This is called serial-to-parallel conversion. Conversely, the eight-bit sequence held by register A is available in serial form at terminal A_8—parallel-to-serial conversion. The shift register of Fig. 1.5 also can be thought of as delaying a signal for eight units of time; i.e., the bit sequence at terminal A_8 is the same as that at terminal X, but delayed with respect to signal X. The first zero entering the shift register in Fig. 1.5 is emphasized so that these conversions may be more easily visualized.

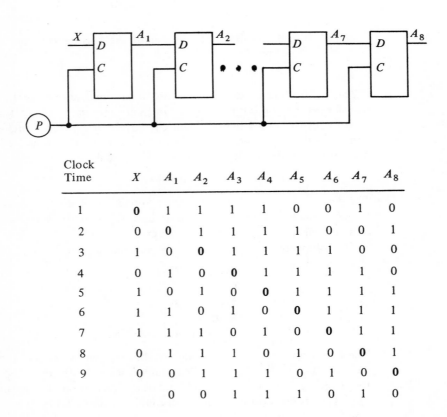

Clock Time	X	A_1	A_2	A_3	A_4	A_5	A_6	A_7	A_8
1	0	1	1	1	1	0	0	1	0
2	0	0	1	1	1	1	0	0	1
3	1	0	0	1	1	1	1	0	0
4	0	1	0	0	1	1	1	1	0
5	1	0	1	0	0	1	1	1	1
6	1	1	0	1	0	0	1	1	1
7	1	1	1	0	1	0	0	1	1
8	0	1	1	1	0	1	0	0	1
9	0	0	1	1	1	0	1	0	0
		0	0	1	1	1	0	1	0

FIG. 1.5   A Shift Register

CHAPTER 1   An Introduction to Digital Systems

Processing Elements

Simple binary processing elements are known as *gates*. Subsequent chapters will show how networks of gates can be designed that do complex processing. Now we will examine gates of a very basic set to illustrate the processing aspect of digital systems and to define notation for describing binary processing.

The complement relation between binary signals was defined in the truth table of Table 1.1. A single input circuit whose output signal is the complement of its input signal is called a NOT gate or *inverter*. Figure 1.6(a) shows one American National Standards Institute (ANSI) symbol for the inverter. The triangle symbolizes an amplifying type circuit; the small circle indicates that the output signal is the complement of the input signal.

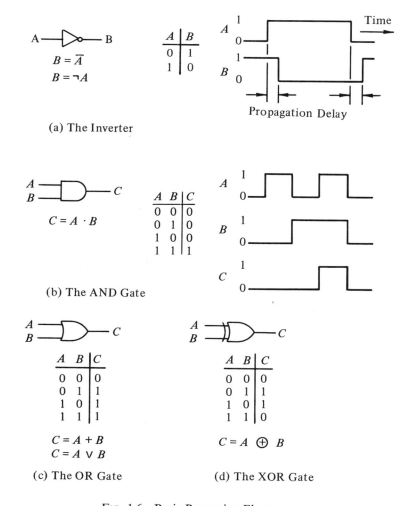

(a) The Inverter

(b) The AND Gate

(c) The OR Gate

(d) The XOR Gate

FIG. 1.6   Basic Processing Elements

If A and B label the input and output terminals of the inverter, as shown in Fig. 1.6(a), then $\bar{A}$ (or $\neg A$) describes the complement of signal A. The connection

$$B = \bar{A}$$

dictates that terminal B be connected to a source of that complementary signal. If it does not exist, then an inverter must be used to generate it. Thus " $B = \bar{A}$ " describes the manner in which terminals are to be connected. If A and B are thought of as binary variables, then " $B = \bar{A}$ " is referred to as a Boolean[3] equation that says variable B always has the value that variable A does not have.

The inverting amplifier does not provide the complement of its input signal instantaneously. If the input signal value changes, the output signal changes value at a later time. The interval between changes is known as the *propagation delay* of the gate. All of the processing elements introduce nonzero propagation delays. It is essential that the period of the clock of a digital system be greater than several gate propagation delays; usually it is very much greater than the gate delay. Thus, gates can be thought to act instantaneously, since their action is complete long before the flip-flops "look at" the gate output signals. Boolean equations do not usually include time as a variable; gates are thought to act instantly when Boolean equations are being used to model them.

Figure 1.6(b) shows the symbol and truth table for the AND processing element, or *AND gate*. Two input signals are combined to form one output signal. Output signal C has the value 1 only when both input signal A and input signal B have value 1. The processing performed is much like multiplication. As a result, the multiplication symbol ($\cdot$) is used to denote AND combining. The statement

$$C = A \cdot B$$

may be viewed as a Boolean equation relating dependent variable C to independent variables A and B or as a connection of the output terminals of an AND gate to labeled terminal C.

From another point of view, if A is thought of as a control signal and B as a data signal, then data B is passed to the output terminal ($C = B$, the gate is open) when $A = 1$. When $A = 0$, the gate is closed, and B is not connected to C. Thus, AND gates constitute a means of electronically making and breaking connections. They would replace the switches in Fig. 1.2.

Figure 1.6(c) presents the *OR gate*. Two input signals are combined such that the output signal has value 1 wherever either or both input signals have value 1. This is then "inclusive or" processing in that the "both" case is included. The *exclusive-or gate* is defined in Fig. 1.6(d). Note that its output signal takes the value 1 when one or the other but not both

---

[3]George Boole first studied logic variables and their mathematics in *An Investigation of the Laws of Thought, on Which Are Founded the Mathematical Theories of Logic and Probability*, 1849.

input signals have value 1. The OR processing table is somewhat like an addition table; thus, engineers often use the $+$ symbol to denote OR combining. Others use ∨ as the OR operator in Boolean equations to avoid the ambiguity that $+$ introduces (Does it denote a logic or an arithmetic operation?). We will use the $+$ to symbolize the OR operation in Chapter 2 where it is very clear that logic, not arithmetic, is being discussed. This should assure that the reader is also comfortable with such usage.

The XOR gate can be thought of as a selective complementer. When $A = 0$, data B appears on output terminal C, $C = B$; but when $A = 1$, the complement of B is present on that terminal, $C = \bar{B}$.

$C = A + B$ and $C = A \oplus B$ can be interpreted as Boolean equations or connection statements. We will not continue to point out such dual interpretations. Boolean mathematics is especially valuable in that it describes the hardware to accomplish processing as well as that processing. Truth tables describe processing by enumeration. At times their detail is also very valuable.

Networks of interconnected AND, OR, XOR, and NOT gates perform complex processing. The wires that interconnect them are communication elements. Figure 1.7 provides an example. Signals provided by the environment of a system under study are called *primary input* signals. The signal or signals supplied to the environment are *output* signals. Signals entirely within the network are intermediate or *secondary* signals. In Fig. 1.7, A, B, C, and D are input signals. Perhaps flip-flops are the actual sources of these signals; the actual physical source is of no real concern here. Interconnecting terminals that bear secondary signals have been named α, β, and γ in Fig. 1.7 so that we can easily refer to them. The following set of Boolean equations can be written from Fig. 1.7, or it could have been presented in place of Fig. 1.7 as a definition of the example network.

$$\alpha = \bar{B}$$
$$\beta = A \cdot \alpha$$
$$\gamma = C \oplus D$$
$$F = \beta \vee \gamma$$

The gate definitions of Fig. 1.6 enable us to calculate what value F has for any assumed values of A, B, C, and D. Given values for primary input variables, we evaluate secondary variables that depend only on the primary input variables; then we evaluate secondary variables that depend on evaluated secondary as well as primary variables; etc. Finally the output variables are evaluated. For example, if we assume that all primary input variables have value 0, then

$$\alpha = \bar{B} = \bar{0} = 1$$
$$\gamma = C \oplus D = 0 \oplus 0 = 0$$
$$\beta = A \cdot \alpha = 0 \cdot 1 = 0$$
$$F = \beta \vee \gamma = 0 \vee 0 = 0$$

If we assume $A = B = C = 0$ while $D = 1$, then

$$\alpha = \bar{B} = \bar{0} = 1$$
$$\gamma = C \oplus D = 0 \oplus 1 = 1$$
$$\beta = A \cdot \alpha = 0 \cdot 1 = 0$$
$$F = \beta \vee \gamma = 0 \vee 1 = 1$$

We must be very familiar with the defining truth tables of Fig. 1.6. These enable us to carry out these calculations efficiently.

The dependency of an output variable on primary input variables can be described with a single complex Boolean equation. While this equation does not necessarily reveal the detailed structure of a processing network, it may suggest that structure, and does summarize the logic function of the network. The network of Fig. 1.7 is suggested by:

$$F = (A \cdot \bar{B}) \vee (C \oplus D)$$

Beta takes the value of 1 whenever A has value 1 and B does not $(A \cdot \bar{B})$. Gamma has value 1 whenever C and D have different values $(C \oplus D)$. The OR gate produces $F = 1$ whenever either or both β and γ have value 1. Thus the summary Boolean equation says, "$F = 1$ whenever $A = 1$ and $B = 0$ or $C \neq D$." Here we are converting the summary Boolean equation to English in a very literal fashion. In some cases the English language provides words that permit briefer, clearer statements. Saying "C not equal to D", rather than "C exclusive-or D equals 1", is an example of this possibility.

Parallel processing usually accompanies parallel communication. System *SELECT* illustrates how we extend the logic operators when operands are vectors.

System *SELECT*:

Terminals A, $B[10]$, $C[10]$, $D[10]$, $E[10]$, $F[10]$.

$$D = A \cdot B,$$
$$E = \bar{A} \cdot C,$$
$$F = D \vee E$$

The ten F terminals are to be connected to the output terminals of ten OR gates, each of which combines one D signal with the corresponding E signal.

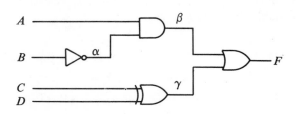

FIG. 1.7   A Processing Network

$$(F_1, F_2, \ldots, F_{10}) = (D_1, D_2, \ldots, D_{10}) \vee (E_1, E_2, \ldots, E_{10})$$
$$F_1 = D_1 \vee E_1$$
$$F_2 = D_2 \vee E_2$$
$$\vdots$$
$$F_{10} = D_{10} \vee E_{10}$$

Since A (and hence $\bar{A}$) has unit dimension, and B and C have greater dimensions, a fan-out of the A and $\bar{A}$ signals is implied. The D terminals are connected to the output terminals of ten AND gates, each of which combine A with one B signal.

$$(D_1, D_2, \ldots, D_{10}) = A \cdot (B_1, B_2, \ldots, B_{10})$$
$$D_1 = A \cdot B_1$$
$$\vdots$$
$$D_{10} = A \cdot B_{10}$$

The E terminals are connected to ten AND gates (label the output terminals of those gates), which each combine $\bar{A}$ with a C signal.

$$(E_1, E_2, \ldots, E_{10}) = \bar{A} \cdot (C_1, C_2, \ldots, C_{10})$$
$$E_1 = \bar{A} \cdot C_1$$
$$\vdots$$
$$E_{10} = \bar{A} \cdot C_{10}$$

Thus $^-$, $\cdot$, $\vee$, and $\oplus$ are applied on a bit-by-bit basis across the elements of terminals. Figure 1.8 shows a diagram of one of the ten gate networks of this processing system. To draw the full system would require ten times as much time and space.

It is not necessary to explicitly name all terminals of a processing network; in fact, it may be confusing to do so. The system above is also described by:

$$F = (A \cdot B) \vee (\bar{A} \cdot C)$$

or

$$F = \text{If } A \text{ then } B \text{ else } C.$$

This last form most clearly reveals the processing performed. If $A = 1$, then the F terminals are connected to the B terminals or else (i.e., $A = 0$)

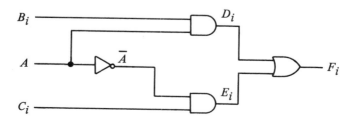

FIG. 1.8   A Typical Cell of Processing System SELECT

they are connected to the C terminals. Selection is accomplished. Either the information carried by B or that carried by C is carried by F. Signal A controls the processing. A processor that selects information from one of several sources is commonly known as a *multiplexer*.

To summarize this section, we put processing, memory, and communication together in the following digital system:

> System *COUNT*:
> Clock P.
> > Register $A[4]$.
> > Terminals $C[4]$, IN.
> > $\quad C = A \cdot (C_{2:4} \circ IN)$,
> > $\quad$ If P then $A \leftarrow A \oplus (C_{2:4} \circ IN)$.

Figure 1.9 shows the logic of this system, DFF's are used to realize register A. Very soon after a clock pulse, information captured by the A flip-flops is presented and held on the A output terminals. It propagates through a network of four AND gates and four XOR gates. The AND gates each combine an A signal with the C signal from the processing cell on their right. The rightmost AND gate combines A_4 with input signal IN; $C_4 = A_4 \cdot IN$. The XOR gates combine the same signals as their corresponding AND gates to form the flip-flop input signals. Then a clock pulse is supplied and the A register is loaded with the data presented to it.

If $IN = 0$, then the register holds the same contents for another clock period. To see this we start at the right and calculate the C signals.

$$C_4 = A_4 \cdot IN = A_4 \cdot 0 = 0$$
$$C_3 = A_3 \cdot C_4 = A_3 \cdot 0 = 0$$
$$C_2 = A_2 \cdot C_3 = A_2 \cdot 0 = 0$$
$$C_1 = A_1 \cdot C_2 = A_1 \cdot 0 = 0$$

If $IN = 0$, then $C_4 = 0$. $C_3 = C_2 = C_1 = 0$ follows, regardless of the values of the A signals. The input signals to the flip-flops are identical to the output

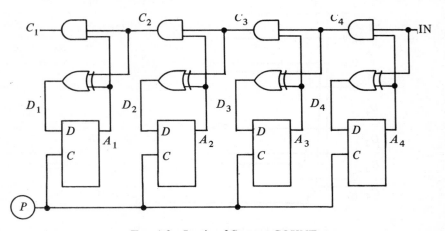

FIG. 1.9 Logic of System *COUNT*

signals; the exclusive-or logic of Fig. 1.6 must be examined carefully to see this.

$$D_4 = A_4 \oplus IN = A_4 \oplus 0 = A_4$$
$$D_3 = A_3 \oplus C_4 = A_3 \oplus 0 = A_3$$
$$D_2 = A_2 \oplus C_3 = A_2 \oplus 0 = A_2$$
$$D_1 = A_1 \oplus C_2 = A_1 \oplus 0 = A_1$$

In summary, when $IN = 0$ the register contents remain unchanged. The processing circuitry simply routes the signal out of each flip-flop to its input terminal—where it is captured with the next clock pulse.

With $IN = 1$, the register contents will change. Thus we will assume that register A holds a specific bit pattern in order to reveal the nature of the change. Suppose that all flip-flops hold 0 after a clock pulse and $IN = 1$. The C and D signals are calculated as above.

$$C_4 = A_4 \cdot IN = 0 \cdot 1 = 0 \qquad D_4 = A_4 \oplus IN = 0 \oplus 1 = 1$$
$$C_3 = A_3 \cdot C_4 = 0 \cdot 0 = 0 \qquad D_3 = A_3 \oplus C_4 = 0 \oplus 0 = 0_{\rightarrow}$$
$$C_2 = A_2 \cdot C_3 = 0 \cdot 0 = 0 \qquad D_2 = A_2 \oplus C_3 = 0 \oplus 0 = 0$$
$$C_1 = A_1 \cdot C_2 = 0 \cdot 0 = 0 \qquad D_1 = A_1 \oplus C_2 = 0 \oplus 0 = 0$$

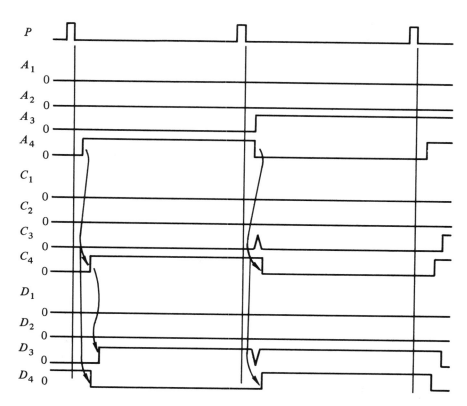

FIG. 1.10   Detailed Activity Within System $COUNT$ (IN = 1)

Only D_4 has a different value. With the next clock pulse the contents of the register will change from $(0, 0, 0, 0)$ to $(0, 0, 0, 1)$.

If the significance of these calculations is not clear, perhaps Fig. 1.10 will be helpful. Each of the 14 signals in system $COUNT$ is shown for two clock periods. At the left, register A holds $(0, 0, 0, 0)$. Since $D_4 = 1$ at the first clock pulse shown, flip-flop A_4 is set by that pulse. It does not switch state instantaneously, but when it does switch the register content becomes $(0, 0, 0, 1)$. $A_4 = 1$ causes the AND gate that provides C_4 to calculate $C_4 = 1$. Again, it does take some time for this gate to act. At the same time the XOR gate that generates signal D_4 calculates and provides the new result, $D_4 = 0$. Following the change in C_4, the XOR gate that provides D_3 gives $D_3 = 1$. No further action takes place until the next clock pulse when the D signals are examined and their values are transferred to the A lines.

During the second clock period shown in Fig. 1.10, the resetting of flip-flop A_4 causes C_4 and D_4 to take new values. Temporary false values are shown on the C_3 and D_3 lines. Since $C_3 = A_3 \cdot C_4$, if the A_3 flip-flop switches before the C_4 signal drops $C_3 = 1$ for a short time. Temporary false values cause no ill-effects as long as the correct values are reinstated before the next clock pulse. What conditions cause D_3 to take the value 0 for a short time?

If $IN = 1$ is held for many clock periods, we will find the contents of the A register at successive clock times to be:

C_1	A_1	A_2	A_3	A_4	Possible Decimal Interpretation
0	0	0	0	0	0
0	0	0	0	1	1
0	0	0	1	0	2
0	0	0	1	1	3
0	0	1	0	0	4
0	0	1	0	1	5
0	0	1	1	0	6
0	0	1	1	1	7
0	1	0	0	0	8
0	1	0	0	1	9
0	1	0	1	0	10
0	1	0	1	1	11
0	1	1	0	0	12
0	1	1	0	1	13
0	1	1	1	0	14
1	1	1	1	1	carry 15
0	0	0	0	0	0

These results are obtained by repeating the example above 15 times. Carry C_1 takes the value 1 only when all flip-flops hold 1's. The next contents will start the sequence over again. In the following sections we will begin to assign meaning to binary tuples and sequences of tuples. One meaning is shown by the table above: The successive contents of A represent successive integers of the natural binary number system. In a restricted sense this system "counts" the number of clock pulses during which $IN = 1$.

1.3 THE CODING PROBLEM

An *alphabet* is a finite collection of digits. Previous civilizations have introduced many. Today we use a variety of these alphabets for different purposes. An alphabet of ten symbols is commonly employed to express numeric information. In the printing of information a rather large alphabet is used, which includes the numeric alphabet; English alphabet (both upper- and lowercase); perhaps Greek, Roman, and Hebrew characters; punctuation marks; mathematical symbols; and possibly new symbols defined as part of the recorded information. In speaking, an alphabet of syllables is used.

Because finite alphabets are used, a single digit can convey only a finite amount of information. Oriental civilizations use very large alphabets. They convey as much information with one symbol as we convey with several English letters. Digits of a binary alphabet (0, 1; yes, no; true, false; and so forth) each convey a minimum of information. A yes or no answer to a question provides the minimum nonzero amount of information to the questioner. The field of intellectual activity known as *information theory* provides a unit for measuring information and the mathematical means of computing the information content of a message. A yes or no answer to a question when either answer is equally probable to the person asking the question is defined as providing one unit of information—one *bit* of information. A single digit from a binary alphabet can then convey as much as one bit of information. Ignoring the "equal probability" requirement, we will say that each digit from a binary alphabet conveys exactly one bit of information and will refer to a binary digit as a bit. Generalizing, each digit of an alphabet of R digits can convey as much as $\log_2 R$ bits of information.

Sequences of digits are used to convey a greater amount of information than that provided by a single digit. One letter follows another to form words, sentences, and paragraphs. One syllable follows another when we speak. A speaker or writer who commands a small number of words requires longer strings of words than one with a richer vocabulary. To express n bits of information requires a binary string of n or more digits. If an alphabet of R digits is used, a sequence of $n/\log_2 R$ of those digits is (theoretically) capable of expressing the same information. Each of the R digits expresses as much information as $\log_2 R$ binary digits. Thus, each decimal digit conveys the same information as 3.32 binary digits; binary sequences will usually be more than three times as long as corresponding decimal sequences.

Formally, a digit sequence is formed by linking the digits together in a sequence—concatenating the digits. The order in which digits appear in a digit sequence is usually important. The sequence \$531 appearing on a paycheck does not convey the same information as the sequence \$153 would; yet both are sequences of the same symbols. Concatenation is not a commutative operator; that is, the order of digits can not be altered without affecting the results.

Let n be the number of binary digits in a binary sequence. How many different binary sequences exist for a specific value of n? We can find out by constructing all of the sequences, at least for small values of n. If $n = 1$, then two sequences 0 and 1 obviously exist. The binary sequences of length two can be constructed by concatenating first a 0 and then a 1 to each of the sequences of length 1.

$$
\begin{array}{cc}
0 & 00 \\
  & 01 \\
1 & 10 \\
  & 11
\end{array}
$$

The sequences of length 3 can be constructed from those of length 2 in a similar manner. In each case the length of the list doubles (obviously?). The number of binary sequences of length n is then 2^n. Table 1.2 lists the sequences for $n \le 4$.

If an electronic digital system must be capable of storing, processing, and communicating any one of m different messages, then a unique binary sequence must be assigned to represent each message. Usually all such sequences are of the same length. Thus, we must find a value of n such that

$$
m \le 2^n
$$

so that a sufficient number of sequences are available. A lower bound is placed on n; n must be an integer greater than or equal to $\log_2 m$. The ten decimal digits can be represented with sequences of length 4 or more. The 26 English letters can be represented with sequences of length 5 or more. Members of an alphabet containing both the ten decimal digits and the 26 English letters are representable by sequences of length 6 or more.

Having accepted a specific, suitable value of n, we now must assign a specific binary sequence to each message. $N = 2^n$ choices are available for the first message; $N - 1$ binary sequences are available for assignment to the second message. In all

$$
N \times (N - 1) \times (N - 2) \times \cdots \times (N - m + 1) = \frac{N!}{(N - m)!}
$$

different assignments can be made. If we have only two messages (yes and no, for example) and elect to use 1-bit sequences, then

$$
\frac{2!}{(2 - 2)!} = 2
$$

assignments can be made.

$$
\begin{array}{cc}
\text{yes—0} & \text{yes—1} \\
\text{no—1} & \text{no—0}
\end{array}
$$

TABLE 1.2. BINARY SEQUENCES OF LENGTH $n \leq 4$

Decimal Index	1	2	$n =$ 3	4
0	0	00	000	0000
1	1	01	001	0001
2		10	010	0010
3		11	011	0011
4			100	0100
5			101	0101
6			110	0110
7			111	0111
8				1000
9				1001
10				1010
11				1011
12				1100
13				1101
14				1110
15				1111

If the ten decimal digits are represented by 4-bit sequences, approximately 2.9×10^{10} different assignments can be made.

An assignment of a sequence to each message is a *code*. Each assigned sequence is called a *code word*. Codes may be specified in many ways. One way is to tabulate the messages and their assigned code words. Table 1.3 lists portions of two widely used codes for an extended set of characters. The Hollerith punched card code uses 12 bits to encode each character. Most of the $2^{12} = 4096$, 12-bit sequences are not assigned to any character. The American Standard Code for Information Interchange (ASCII) is widely used in the digital communication and computing industries. It is capable of encoding alphabets of up to $2^8 = 256$ characters. The hexadecimal expression of the listed code words will be discussed in the next section.

The messages that we desire to communicate and process are seldom simple characters from Table 1.3. However, sequences of such characters constitute useful messages. The 80 column punched card provides transportation and storage of 80-character sequences. If any of 60 characters can be punched in each column, 60^{80} different sequences can be punched in principle; we do not desire to begin to list these 1.8×10^{142} sequences. Enumerating the binary sequence to be assigned to each message is out of the question. A very practical approach consists of encoding the characters and then forming the code word for a sequence of characters by concatenating the code words for the characters. Using the ASCII code, the six character message (blank is a character) "$Z = 1$." is represented by the 48-bit sequence

01011010 ∘ 00100000 ∘ 00111101 ∘ 00100000 ∘ 00110001 ∘ 00101110

$\quad$ Z $\qquad$ blank $\qquad$ = $\qquad$ blank $\qquad$ 1 $\qquad$ •

This message can be communicated in parallel using 48 terminals or in serial using one terminal and 48 clock periods, or using 8 terminals and six clock periods, and so forth. Forty-eight memory elements are required to store the message. They may be organized as one 48-bit register, six 8-bit registers, etc.

We usually express quantity by a sequence of decimal digits and a selected few punctuation marks. Such digit sequences can be encoded using the ASCII code, of course; the decimal number 736.2 is expressed

00110111 ∘ 00110011 ∘ 00110110 ∘ 00101110 ∘ 00110010

$\quad$ 7 $\qquad$ 3 $\qquad$ 6 $\qquad$ • $\qquad$ 2

A total of 40 bits are used in this case. Numbers are processed so frequently by digital systems that we seek and often use codes for numbers that are more efficient; i.e., fewer bits are used and code words may be easily processed according to arithmetic rules to find code words for arithmetic results. While discussion of the most common numeric codes and rules for processing them is deferred until Chapter 3 (where we will be in a position to design suitable processing hardware), we can now examine the foundation upon which these codes are based—the binary number system. In fact, we will review the decimal number system at the same time because both the binary and decimal systems are special cases of positional number systems.

1.4 POSITIONAL NUMBER SYSTEMS

Archaeology indicates that early man was able to conceive only of quantities of one, two, and many. Knots in rope, stones in a pile, notches in a stick, and marks in sand and stone were used to indicate and record larger quantities. We still use this *unitary* number system on occasion. With the frequent need to express and record a larger range of quantities, new words and symbols and methods of combining them to express still other quantities were introduced. We use these inventions when we say " twenty three " and write " 23 " rather than recording 23 explicit marks. We use the symbols " 2 " and " 3 " to denote specific quantities and have rules whereby the sequence " 23 " denotes a specific quantity.

Digit sequences as used today to express numeric information differ from those of earlier number systems in that: (1) the digits are, for the most part, used exclusively to express numeric information (the Greeks at one time used the same alphabet for expressing numeric and alphabetic information) and (2) the relative positions of the digits in a sequence are extremely important. The Egyptians first introduced the concept of *positional value* in their hierographic number system (1850–1650 B.C.). We are more familiar with the Roman number system in which position

TABLE 1.3. TWO WIDELY USED ALPHAMERIC CODES

Character	Hollerith Punched Cards (Rows Punched)	ASCII Binary	Hexadecimal
(blank)	none	00100000	20
$	11, 3, 8	00100100	24
(	0, 4, 8	00101000	28
)	12, 4, 8	00101001	29
*	11, 4, 8	00101010	2a
+	12	00101011	2b
,	0, 3, 8	00101100	2c
—	11	00101101	2d
.	12, 3, 8	00101110	2e
/	0, 1	00101111	2f
0	0	00110000	30
1	1	00110001	31
2	2	00110010	32
3	3	00110011	33
4	4	00110100	34
5	5	00110101	35
6	6	00110110	36
7	7	00110111	37
8	8	00111000	38
9	9	00111001	39
=	3, 8	00111101	3d
A	12, 1	01000001	41
B	12, 2	01000010	42
C	12, 3	01000011	43
...			
I	12, 9	01001001	49
J	11, 1	01001010	4a
K	11, 2	01001011	4b
...			
R	11, 9	01010010	52
S	0, 2	01010011	53
T	0, 3	01010100	54
...			
Y	0, 8	01011001	59
Z	0, 9	01011010	5a

affects value to a relatively small extent; i.e., VI does not express the same quantity as IV, but in the digit sequence XX the X on the left has neither greater nor less significance than the one on the right.

Today we use positional number systems that consist of: (1) a *radix* or *base R* and (2) an ordered alphabet of R digits. These systems associate a weight with each position in a sequence. To ensure that every quantity is expressed by a unique sequence of digits, each of the R digits of the alphabet must express a different number, less than the radix R. One digit expresses the quantity of none; we use the symbol "0" for this digit. One digit expresses unit quantity; here we use "1." The remaining digits of the ordered alphabet express quantities of two, three ..., $(R - 1)$ units. The ordering of the alphabet ranks the digits according to the quantity they express. A hierarchy of digits is established. Zero is the least of the digits; 1 is greater than 0 and less than the digits that express greater quantity. The digit that expresses the greatest quantity $(R - 1)$ is greater than all other digits.

With the understanding that a positional number system of a given radix R is being employed, we express numeric information by a sequence of digits, $\ldots d_2 d_1 d_0 d_{-1} \ldots$, each of which is understood to multiply the weight associated with its position in the sequence.

$$N_R = \sum_{i=+\infty}^{-\infty} d_i R^i = \cdots + d_2 R^2 + d_1 R^1 + d_0 R^0 + d_{-1} R^{-1} + \cdots \quad \textbf{(1.2)}$$

The digit sequence 793 is interpreted as 7 hundreds (10×10) of units plus 9 tens of units plus 3 units. The binary ($R = 2$) sequence 1101 expresses 1 eights of units ($2^3 = 8$) plus 1 fours of units plus no pair of units plus 1 unit. The decimal sequence 13 expresses the same quantity.

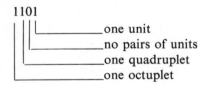

Equation (1.2) suggests that all numbers must be expressed with digit sequences of infinite length. Such sequences are unmanageable, so we commonly collapse infinite sequences to finite ones by implying, rather than writing, leftmost and rightmost zeros, and terminating the expression of fractional information.

Advancing a digit of an ordered alphabet consists of replacing that digit with the next digit in the hierarchy. The greatest digit is replaced by the least digit in the hierarchy. Thus in the decimal system, 1 is advanced by replacing it with 2, and 9 is advanced by replacing it with 0. In the binary number system, which utilizes a binary alphabet of 0 and 1, 0 is advanced to 1, and 1 is advanced to 0.

The *integers* of a positional number system may be generated by means of the following algorithm, which assumes 0 to be the least digit in the ordered alphabet.

1. The first integer is 0. (An infinite sequence of 0's)
2. The next integer in the list is obtained from the preceding integer in the list by:
   (a) advancing the rightmost digit one place in the list and

(b) if any digit is advanced to zero (0), advancing the digit to its left.

We recognize the formation of a list of integers as equivalent to *counting*. Counting was probably the first numeric operation performed by men. We actually have not made much progress; addition, multiplication, and other arithmetic operations are only sophisticated methods for counting, as we will see. As a result, digital machines are sometimes said to operate by counting, and we will see that a great deal of counting does in fact go on within such machines.

A partial list of integers of several positional number systems is given in Table 1.4. Observe the pattern in the binary table and in Table 1.2. The digital system COUNT of Fig. 1.9 works according to the counting algorithm above (for the first 16 integers). Hence, it generates the first 16 integers in binary. If in that system a carry signal C_i has value 1, what message does that 1 encode? What does $C_i = 0$ say?

Since 8 (and 16) is a power of two, there are exactly the same number of octal (hexadecimal) digits as there are binary sequences of length 3 (4). Table 1.4 thus suggests a very natural binary code for the octal (hexadecimal) digits. The octal integer 237_8 is expressed $010 \circ 011 \circ 111 = 010011111$ with this code, and the octal digit sequence 237 expresses the same quantity as 010011111 expresses in the binary number system. Conversely, the long binary sequence 101100011 may be more compactly expressed in octal as 543_8. Table 1.3 expressed each 8-bit ASCII code word as a two-hexadecimal digit sequence; break each 8-bit code word into two 4-bit sequences. Each 4-bit binary sequence has a unique, equivalent hexadecimal digit as listed in Table 1.4.

When we express rational numbers in writing we add another symbol known as the *radix mark* (decimal point) to the alphabet we use, but do not include this digit in the hierarchy of our ordered alphabet. This digit establishes the position coordinate system for a reader in its placement to the immediate right of the digit in the 0 position, i.e., the digit to multiply weight R^0. Thus the decimal sequence 13.72 expresses 13 units plus 7 tenths of a unit plus 2 hundredths of a unit. The radix mark is placed between the units and tenths positions in a sequence to orient the reader and not to express additional numeric information, and hence is not used as a multiplier of any weight. And we often include commas in long digit sequences to assist a reader further in determining the position of a particular digit. Again, these commas express other than numeric information.

Radix Conversion

All positional number systems are equally interesting to the mathematically inclined person. The practical person argues that only one is needed and decimal is the only realistic choice today. Electronic digital systems force binary and hence octal and/or hexadecimal on the practical person; one must be interested in methods of converting a number from

TABLE 1.4. INTEGERS OF SEVERAL POSITIONAL NUMBER SYSTEMS

Radix	Decimal 10	Binary 2	Ternary 3	Octal 8	Hexadecimal 16
Digits	0, 1, 2, 3, 4 5, 6, 7, 8, 9	0, 1	0, 1, 2	0, 1, 2, 3, 4, 5, 6, 7	0, 1, 2, 3, 4, 5, 6, 7, 8, 9, a, b, c, d, e, f
Integers	0	0	0	0	0
	1	1	1	1	1
	2	10	2	2	2
	3	11	10	3	3
	4	100	11	4	4
	5	101	12	5	5
	6	110	20	6	6
	7	111	21	7	7
	8	1000	22	10	8
	9	1001	100	11	9
	10	1010	101	12	a
	11	1011	102	13	b
	12	1100	110	14	c
	13	1101	111	15	d
	14	1110	112	16	e
	15	1111	120	17	f
	16	10000	121	20	10
	17	10001	122	21	11
	18	10010	200	22	12
	19	10011	201	23	13
	20	10100	202	24	14

one system to another. We have already seen that if one radix is a power of another, then conversion may be carried out by simple algorithms. In general, conversion is more involved. Because of the practical importance and wide knowledge of the decimal number system, general conversion algorithms are specialized below as means of converting to and from decimal.

The simplest and most natural means of obtaining the decimal equivalent of an *integer* is to:

1. Convert the radix and each digit of the given integer into the equivalent decimal integer, and
2. Evaluate, using decimal arithmetic, the polynomial defined by the given integer.

$$I_R = \sum_{i=0}^{\infty} d_i R^i \qquad (1.3)$$

Thus by the definition of a positional number system and assuming that 0, 1, ..., 9, a, b, c, d, e is used as the alphabet for base 15 numbers

$$10b5_{15} = (0 \times 15^4) + (1 \times 15^3) + (0 \times 15^2) + (11 \times 15^1) + (5 \times 15^0) = 3545_{10}$$

Conversion of an integer from decimal to a system of another radix is best accomplished by an algorithm called *repeated division*. Decimal integer I_{10} is repeatedly divided by R (expressed in decimal) with all arithmetic performed in the decimal system. At each division the remainder is expressed as the corresponding digit of the base R alphabet and not used in subsequent divisions. Repeated division is terminated when the integer quotient has been reduced to zero. The remainders of each step of division taken in reverse order, i.e., last to first, then express I_R.

$$I_{10} = 3545, \ R_{10} = 15$$

$$
\begin{array}{r|l l}
15 & 3545 & \text{Remainders} \\
15 & 236 & 5 \\
15 & 15 & 11_{10} = b_{15} \\
15 & 1 & 0 \\
& 0 & 1 \quad 3545_{10} = 10b5_{15}
\end{array}
$$

The preceding algorithms have emphasized conversion to and from the decimal system. Actually both are entirely general and can be used to convert from any system to any other system. Arithmetic in other than the decimal system must be performed, however, and such generality must consequently wait until we are able to perform such arithmetic.

Conversion of integers from decimal to binary is frequently accomplished by a variation of the repeated division algorithm that makes use of our familiarity with powers of two. Successively smaller powers of 2 are subtracted from the decimal integer, and a 1 is recorded when the subtraction leaves a positive remainder. 0 is recorded when the subtraction would not leave a positive remainder and hence is not accomplished.

$$43_{10}$$

$$\frac{-32}{11} = 2^5$$

$$\frac{-8}{3} = 2^3 \qquad \text{hence } 101011_2 = 43_{10}$$

$$\frac{-2}{1} = 2^1$$

$$\frac{-1}{0} = 2^0 \qquad 1 \times 2^5 + 0 \times 2^4 + 1 \times 2^3 + 0 \times 2^2 + 1 \times 2^1 + 1 \times 2^0 = 43_{10}$$

A *fraction* may be expressed by using the defining polynomial:

$$F_R = \sum_{i=-1}^{-\infty} d_i R^i \tag{1.4}$$

As with the integral part, the fractional part of a number is best converted to decimal by expanding the defining polynomial.

$$N_4 = 123.321_4 = (1 \times 4^2) + (2 \times 4^1) + (3 \times 4^0) + (3 \times 4^{-1})$$
$$+ (2 \times 4^{-2}) + (1 \times 4^{-3}) = 27.8906_{10}$$

$$N_2 = .11011_2 = (1 \times 2^{-1}) + (1 \times 2^{-2}) + (0 \times 2^{-3}) + (1 \times 2^{-4})$$
$$+ (1 \times 2^{-5}) = 1/2 + 1/4 + 1/16 + 1/32 = .84375_{10}$$

The dual of repeated division may be used to convert a fraction from decimal to another base. *Repeated multiplication*, with the integers at each step taken in forward order as the digits of the result, constitutes the algorithm. For example to convert 27.8906_{10} to base 4:

Convert Integer	Convert Fraction
4 $\lfloor 27$	.8906
	$\times 4$
4 $\lfloor 6$ 3	3.5624
	$\times 4$
4 $\lfloor 1$ 2	2.2496
	$\times 4$
0 1	0.9984
	$\times 4$
	3.9936
	$\times 4$
	3.9744
	$\vdots$

Put integer and fraction together $123.32033\ldots_4$

Converting to binary is particularly easy: we are good at multiplying by 2.

$$
\begin{array}{l}
.8575_{10} \\
\underline{\times 2} \\
1.7150 \quad .8575_{10} = .110110\ldots_2 \\
1.430 \\
0.86 \\
1.72 \\
1.44 \\
0.88
\end{array}
$$

Arithmetic

The arithmetic operations of addition, subtraction, multiplication, and division, which are so familiar in the decimal system, can be performed by

TABLE 1.5. BINARY AND OCTAL ARITHMETIC TABLES

+	0	1
0	00	01
1	01	10

Sum bit (right digit), Carry bit (left digit)

Binary

$b_i - a_i$	b_i	
	0	1
a_i 0	00	01
1	11	00

Difference bit, Borrow bit

+	0	1	2	3	4	5	6	7
0	0	1	2	3	4	5	6	7
1		2	3	4	5	6	7	10
2			4	5	6	7	10	11
3				6	7	10	11	12
4					10	11	12	13
5						12	13	14
6							14	15
7								16

Octal

identical procedures in any positive radix number system. But those procedures are so familiar that we perform them automatically, and perhaps no longer recognize the individual steps of the algorithms.

ADDITION of integer I to integer J, $I + J$, is the process of advancing integer I, J times. Normally we do not have a list of integers available, and do not wish to create such a list. We do not perform addition by counting. Memorizing an ADDITION TABLE that summarizes the sums of all combinations of decimal digits and rules which govern the use of that table enable us (and some machines) to add two or more decimal numbers with relative ease. To add in other number systems we now must memorize the tables of those systems. But we do not need new rules: the same rules apply for all positive base systems.

The binary addition table shown in Table 1.5 is most easily memorized because of its simplicity and brevity. It and the accompanying octal table are quickly generated from the list of integers of the binary and octal number systems. Note that in some cases we must use a 2-digit sequence to express the sum of two digits. The left digit of such a pair is called the *carry* digit. Addition of numbers in general is then accomplished by aligning coefficients of equal powers of the radix and performing digit-by-digit addition beginning with the rightmost (least significant) digits of the numbers and including in the addition any carries propagated from lower order positions. Leading and trailing zeros may be written so that each number is expressed by the same number of digits.

Binary	Octal
1111.1010	325.71_8
+ 0110.0011	014.50_8
10101.1101_2	342.41_8

Note that when we add two numbers having p digits to the left of the radix point, it is possible to have a 1 (but no more than 1) propagate into the $(p + 1)$st position.

SUBTRACTION of integers, $I - J$, is formally the process of beginning with integer I and retreating J integers, but, as with addition, we usually employ a digit DIFFERENCE TABLE when performing subtraction by hand. Table 1.5 includes a difference table for the binary number system. Borrow bits arise in the subtractive process and propagate to the left as carry bits did in adding. Common practice in the handling of borrow bits is to decrease the minuend (upper) digit in the next more significant position and proceed with the indicated digit subtraction in that position, or to advance the subtrahend (lower) digit in the next higher order position, and then subtract the result from the original minuend digit. Note that a borrow (carry) bit is always generated. When the borrow (carry) bit is a zero, we ignore it when subtracting (adding) in the next more significant position. Machines do not have the "intelligence" to ignore; they must be constructed to introduce the borrow or carry bit at each digit position regardless of the value of that bit.

Aligning the radix points and beginning at the right are familiar steps when subtracting two rational numbers $I - J$ by hand. Also, we have been taught to compare the two numbers before beginning to perform the subtraction algorithm. If I is a smaller number than J, we actually change the given problem to another problem which we prefer to solve: we subtract I from J, i.e., compute $J - I$, and affix a minus sign to the result to denote that it is negative. Signs of numbers are enclosed in parentheses to distinguish them from arithmetic operators in the following examples.

$$
\begin{array}{ll}
1101.1010 & 0110.0011 \\
-0110.0011 & -1101.1010 \\
\hline
0111.0111_2 & \text{change the}
\end{array}
\qquad
\begin{array}{l}
1101.1010 \\
-0110.0011 \\
\hline
(-)0111.0111_2
\end{array}
$$

problem to
remembering to affix

If this change is not made we get strange results in decimal or binary or any other number system: borrows propagate ad infinitum indicating a negative result.

$$
\begin{array}{ll}
\text{Decimal} & \text{Binary} \\
017 & 0010001 \\
-043 & -0101011 \\
\hline
\ldots 99974 & \ldots 1111100110
\end{array}
$$

When numbers are signed, we also change given problems to equivalent problems that we prefer to solve. Thus to subtract a negative number from a positive one, we actually add the two magnitudes and affix a plus sign to the sum.

$$
\begin{array}{ll}
(+)\ 17 & (+)\ 17 \\
-(-)\ 43 & +(+)\ 43 \\
\hline
\text{change problem to} & (+)\ 60
\end{array}
$$

TABLE 1.6. TRANSFORMATION OF ARITHMETIC PROBLEMS

Command	Sign of I	Sign of J	Action	Sign of Result
$I + J$	+	+	Add	+
$I + J$	+	−	Compare Subtract	Can not predict
$I + J$	−	+	Compare Subtract	Can not predict
$I + J$	−	−	Add	−
$I - J$	+	+	Compare Subtract	Can not predict
$I - J$	+	−	Add	+
$I - J$	−	+	Add	−
$I - J$	−	−	Compare Subtract	Can not predict

Table 1.6 summarizes the changes we make to arithmetic problems we face. These same changes can be made when working in any number system.

MULTIPLICATION is defined as repeated addition. Thus:

$$I \times J = \underbrace{I + I + I + \ldots I}_{J \text{ terms}} = \sum_{i=1}^{J} I \qquad (1.5)$$

Again we find a digit table to be most useful. MULTIPLICATION TABLES for the binary and octal systems are shown in Table 1.7. Where a product is shown as having two digits, the carry digit is propagated one digit position to the left and added to the product of digits in that position.

In contrast to the situation in addition and subtraction, the positions of the radix points do not affect the actual operations carried out for multiplication and division, provided a point is inserted in the proper location in the result. Hence the multiplication algorithms described below are applicable to integers, fractions, or mixed numbers.

TABLE 1.7. MULTIPLICATION TABLES

×	0	1
0	0	0
1	0	1

Binary

×	0	1	2	3	4	5	6	7
0	0	0	0	0	0	0	0	0
1		1	2	3	4	5	6	7
2			4	6	10	12	14	16
3				11	14	17	22	25
4					20	24	30	34
5						31	36	43
6							44	52
7								61

Octal

We often use a special multiplication algorithm without being aware that we are doing so. Multiplication of number N_R by the radix of the system in which N_R is expressed is accomplished by *shifting* all digits of N_R one position to the left. Thus with $R = 2$:

$$R \times 101.1 = 1011.0$$

which is the same as adding N_R to itself.

$$
\begin{array}{r}
101.1 \\
+ \quad 101.1 \\
\hline
1011.0
\end{array}
$$

In the decimal system

$$123.45 \times 10 = 1234.5, \quad \text{and} \quad 123.45 \times 100 = 12345,$$

which points out that when multiplying by a power of the radix R^m we need only shift all digits m places (with respect to the radix mark) to the left or right, as the exponent is positive or negative, respectively.

We use this special multiplication technique when performing multiplication by hand. Each digit of the multiplier J must multiply every digit of I. We prefer to record the partial product of each digit of J and all digits of I and then add all partial products at one time. But each partial product we record is shifted one place to the left with respect to the previous partial product. This shift accomplishes multiplication by a power of the radix: the power is increased by one for each partial product.

$$
\begin{array}{lll}
I & 101.1 & \\
J & \times 11.0 & \\
\hline
& 00.00 & I \times 0 \ (2)^{-1} \\
& 101.1o & I \times 1 \ (2)^{0} \\
& 1011.00 & I \times 1 \ (2)^{1} \\
\hline
& 10000.10 & \text{sum}
\end{array}
$$

DIVISION is repeated subtraction and hence another counting operation. We use multiplication, subtraction, and shifting to accomplish division, efficiently. A "trial divisor" serves to reduce the number of actual multiplications and subtractions that we perform. But this trial divisor is actually an educated guess arrived at by performing a rough multiplication and subtraction in our heads. In binary the only possible divisors are 0 and 1 and multiplying by either of these is very easy. Thus we can decide upon a trial division and perform binary division more easily than decimal division.

$$
\begin{array}{r}
0.1101 \ldots \\
1101 \overline{\smash{\big)}\ 1011.0000} \\
-1101 \\
\hline
\end{array}
$$

$$10010$$
$$-1101$$
$$\overline{010100}$$
$$-1101$$
$$\overline{111}$$

With the ability to perform arithmetic in any positive radix number system, the algorithms for radix conversion can be generalized. To convert an integer expressed in the R_1 number system to the R_2 number system:

1. Convert R_1 and each digit of the given integer into the equivalent R_2 integer by table look-up, and
2. Evaluate the polynomial defined by the given integer using R_2-arithmetic.

Table 1.4 and the octal arithmetic tables are valuable when converting 201_3 directly to octal using this algorithm.

$$201_3 = (2 \times 3^2 + 0 \times 3^1 + 1 \times 3^0)_8$$
$$= (2 \times 11 + 0 + 1)_8$$
$$= 23_8$$

Ternary arithmetic tables are needed to convert 23_8 to base 3; from those tables

$$23_8 = (2 \times 22 + 10 \times 1)_3$$
$$= (121 + 10)_3$$
$$= 201_3$$

1.5 SUMMARY

Digital systems consist of memory elements called flip-flops, which store binary digits (bits), processing elements called gates that combine bits to form new results, and interconnecting wires that communicate bits between memory and processing elements. While the use of bits rather than decimal digits or other symbols is forced on us by available technology, it is not considered to be a serious disadvantage. The meaning assigned to a bit or collection of bits is arbitrary. If we assign meaning to tuples of bits, we generate a code. Widely used codes for alphabetic symbols were examined. The binary number system provides the most widely used codes for numbers. It, like the decimal number system, is a positional number system. Thus, the rules of binary arithmetic are essentially the same as the familiar rules for performing decimal arithmetic. In the following chapters we will look at all of these topics in greater detail.

1.2-1. Draw and label blocks and lines that represent wires declared by each of the following statements.
(a) Terminals *STOP, START, SPEED* [10].
(b) Terminal *COLOR*[7].
(c) Register *A*[15], *B*[5].

1.2-2. The single-pole, double-throw (SPDT) switch (also known as a "transfer contact") connects one terminal to either of two others, but never both simultaneously.

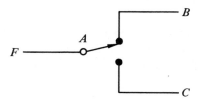

A names a five-pole, double-throw (5PDT) switch, and the following terminals are defined.

Terminal *B*[5], *C*[5], *F*[5]

Show a wiring diagram of a system described by:

$$F = \text{If } A \text{ then } B \text{ else } C$$

In what ways do the 5PDT switch and network of Fig. 1.8 differ? How are they the same?

1.2-3. Draw a logic diagram to show how AND gates may be used to make and break the following connections.

Terminals *A, B*[5], *C*[5]

(a) If *A* then *B* = *C*.
(b) If *Ā* then *C* = *B*.

1.2-4. Terminals *A, B*[4], *C*[4], *D*[4].
(a) Show how AND gates may be used to establish the following connections. What is on terminal *B* when the connection is broken, i.e., *A* = 0?

If *A* then *B* = *C*

(b) Repeat (a) for:

If *Ā* then *B* = *D*

(c) With the above results in mind, draw a diagram of circuitry that makes the following connections.

$$B = \text{If } A \text{ then } C \text{ else } D$$

(Output terminals of gates may not be wired together! Gate output signals are combined with other gates.)

1.2-5. Terminals *A, B, X*[5], *Y*[5], *Z*[5].
A network is needed that meets the following specifications:

If *A* then *X* = *Y*.
If *B* then *X* = *Z*.

Are the given specifications complete? ambiguous? contradictory?

(a) Express these requirements in a Boolean equation.

(b) Draw a logic diagram for circuitry that satisfies all requirements.

(c) At different times the A, B, Y, and Z terminals bear the binary values shown. What appears on the X terminal of your network in each case?

	A	B	Y	Z	X
(i)	0	0	00111	10110	
(ii)	0	1	11111	00001	
(iii)	1	0	10000	01111	
(iv)	1	1	10001	00011	

1.2-6. (a) The clock signal shown is supplied to the DFF. Complete the drawing of its output signal waveform.

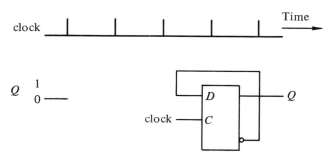

(b) Describe this digital system.

1.2-7. The digital circuit shown is known as a "T flip-flop."

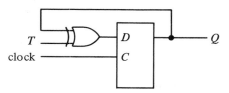

(a) Relate this circuit to the general model of Fig. 1.1.

(b) Complete the output waveform for the given clock and T waveform.

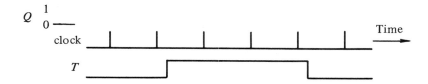

(c) Equation 1.1 characterizes the DFF; write the characteristic equation for the TFF that shows how $Q(i+1)$ depends on $Q(i)$ and $T(i)$.

(d) What does the *T* in the name of this flip-flop stand for?

(e) How is this problem related to Prob. 1.2-6?

1.2-8. The 0 or 1 held by a flip-flop determines its *state*. When it holds a 1, it is in its 1-state or is "set"; when it holds a 0, it is in its 0-state or is "reset." A *state transition table* shows in tabular form how its *next state* $[Q(i+1)]$ depends on its present state $[Q(i)]$ and present input signal values.

(a) From Eq. (1.1), fill in the state transition table for the DFF.

$D(i)$	$Q(i)$	$Q(i+1)$
0	0	
0	1	
1	0	
1	1	

(b) From the results of Prob. 1.2-7(c), prepare the state transition table for a TFF.

1.2-9. The triangle in the DFF symbol below indicates that the 0 to 1 transition of the clock signal causes the flip-flop to capture the value of the *D* input signal. The 1 to 0 transition of the clock signal has no effect.

(a) Draw the waveform for *Q* given the clock signal shown.

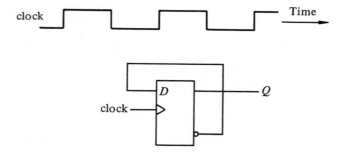

(b) Draw the waveform for *Q1* and *Q2*.

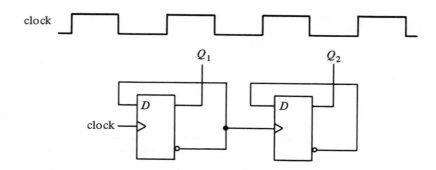

(c) How are the frequencies (periods) of *Q1* and *Q2* related to that of the clock signal?

(d) In what sense are these circuits "counters"?

1.2-10. In the network below, assume each gate and inverter introduces 5 nano-seconds of propagation delay and input signal I is varying as shown. Draw the waveforms of all gate output signals. What is the total propagation delay of each network?

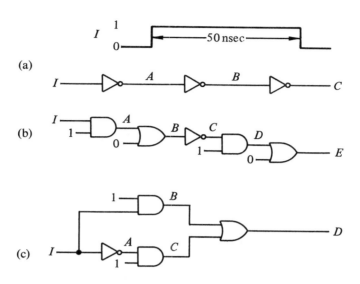

(a)

(b)

(c)

1.2-11. Write the set of gate equations for each network. Evaluate the output variable(s) for each set of values of input variables; record your findings in the table to complete the truth table for the network.

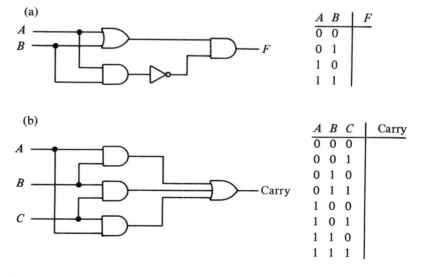

(a)

A	B	F
0	0	
0	1	
1	0	
1	1	

(b)

A	B	C	Carry
0	0	0	
0	0	1	
0	1	0	
0	1	1	
1	0	0	
1	0	1	
1	1	0	
1	1	1	

1.2-12. The "half-adder", or HA block, is a gate circuit that acts according to the binary add table shown in Table 1.5. The S output gives the sum bit; the C output gives the carry bit. Complete the truth table for the network shown.

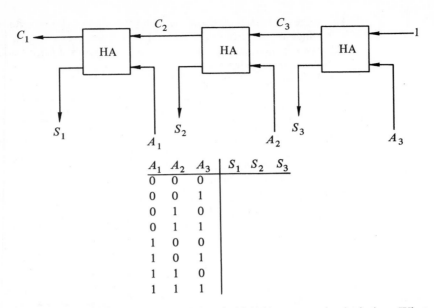

A_1	A_2	A_3	S_1	S_2	S_3
0	0	0			
0	0	1			
0	1	0			
0	1	1			
1	0	0			
1	0	1			
1	1	0			
1	1	1			

1.2-13. Each of the shift registers below hold 1111 at a certain clock time. What do they hold at successive clock times? What happens after n clock times? What value of n is important for each circuit?

(a)

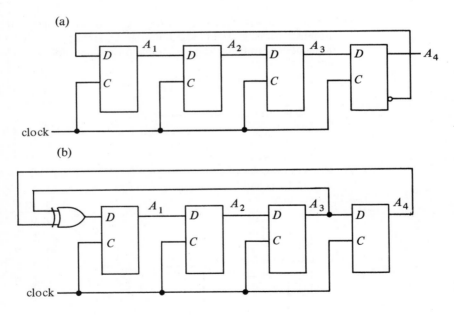

(b)

1.3-1. How many bits of information are conveyed by one digit of each of the following alphabets? (Assume all digits of an alphabet are equally probable.)

(a) $\{\alpha, \beta, \gamma, \delta, \varepsilon\}$

(b) English letters, both upper- and lowercase

(c) FORTRAN alphabet (uppercase English, decimal digits, 12 punctuation marks)

1.3-2. What is the minimum length of code words for each of the alphabets of Problem 1.3-1?

1.3-3. How many binary sequences are there of length:
(a) 10 bits?
(b) 15 bits?
(c) 20 bits?

1.3-4. A pocket calculator displays 8 decimal digits. Each digit is represented internally with a 4-bit code word. How many flip-flops must exist behind the digit display devices?

1.3-5. How many bits of information can be stored on one punch card (80 columns, 12 rows)?

1.3-6. Characters are encoded with the ASCII code given in part by Table 1.3. Write out the bit sequence for the message "CAT = DOG".

1.4-1. Give one or more examples of the use of positional number systems with the following radices. (Example: 2 pints to the quart.)

$$2, \; 3, \; 4, \; 5, \; 12, \; 60$$

1.4-2. Generate the first 15 integers of positional number systems with radix (a) 5, (b) 11.

1.4-3. The symbols α, a, and A are to be used as digits of a number system. Select an ordering of these digits and write the first ten integers of the number system.

1.4-4. The first expedition to Mars found only the ruins of a civilization. From the artifacts and pictures, the explorers deduced that the creatures who produced this civilization were four-legged beings with a tentacle that branched out at the end with a number of grasping "fingers." After much study, the explorers were able to translate the Martian mathematics. They found the following equation:

$$5x^2 - 50x + 125 = 0$$

with the indicated solutions $x = 5$ and $x = 8$.
This was strange mathematics. The value $x = 5$ seemed legitimate enough, but $x = 8$ required some explanation. Then the explorers reflected on the way in which our number system developed, and found evidence that the Martian system had a similar history. How many fingers would you say the Martians had? (from "The Bent" of Tau Beta Pi, Feb., 1956)

1.4-5. The Romans insist that

$$XVII = 17_{10}$$

If I, V, and X are really digits in a positional number system, what is the radix of the system and what is the order of the digits?

1.4-6. Convert each integer to each of the bases given in the following table. Indicate what alphabet was used, especially for the base 11 and hexadecimal integers.

	base			
	10	2	3	7
10_{10}	10	1010		
111_4				
1010101_2				
$1t9_{11}$				
$9d_{16}$				

1.4-7. Convert each of the given sequences directly to a binary sequence of the indicated number of bits.
- (a) 377_8 (8)
- (b) 125_8 (8)
- (c) 144007_8 (16)
- (d) 79_{16} (8)
- (e) $1af_{16}$ (12)
- (f) $cdb6_{16}$ (16)

1.4-8. Prove algebraically that the repeated division algorithm for converting integers is valid.

1.4-9. Convert each fraction to each base.

	10	2	3	7
$.5_{10}$	.5	.1		
$.123_4$				
$.10101_2$				

1.4-10. In positional number systems of what base(s) can each of the following repeating decimal fractions be expressed exactly with a finite number of digits?
- (a) .33333 ... (3 is obvious, find some others)
- (b) .1428571428 ...
- (c) .27272727 ...

1.4-11. Express π and $e = 2.718$ ... in binary to at least ten bits.

1.4-12. For each of the pairs of decimal numbers given below: (a) Convert both to binary; (b) In binary form $A + B$; (c) In binary form $A(A + 2B)$; (d) Convert the answers to decimal and check.
- (i) $(A, B) = (21, 10)$
- (ii) $(A, B) = (7.1, 24.9)$

1.4-13. Perform the indicated subtractions in binary. Convert the answer to decimal and check.
- (a) $17 - 3 = 14$
- (b) $14.625 - 4.0624$

1.4-14. Develop base 5 addition and multiplication tables. Use these tables to perform the following arithmetic in the quinary (base 5) number system. Check your answers.
- (a) $25.4 + 12.5$
- (b) 11×9

1.4-15. Perform the indicated division in binary. Convert your answer to decimal and compare. Discuss any discrepancies.

$$\frac{.1000}{.1010} = ?$$

1.4-16. Convert 10010_2 directly to base 5 (without going to decimal as an intermediate step) by repeated division doing all division in binary. Check.

ANSWERS

Answers are offered to some problems, particularly the more routine problems, so that students may gain confidence in their computational abilities. Solutions are sketched for problems that are more challenging and for which the student may have difficulty in determining how to proceed to obtain an answer.

1.2-2.

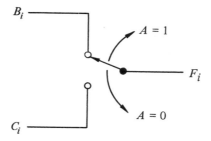

Switches are mechanical and therefore much slower than electronics. Switches are bilateral; F is fed to B as well as B being fed to F. Both perform the same logic processing. B is connected to F when $A = 1$; otherwise, C is connected to F.

1.2-4. (a)

$B_i = 0$ when $A = 0$.

(b)

$B_i = D_i$ when $A = 0$; $B_i = 0$ when $A = 1$.

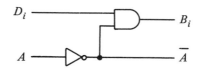

(c) See Fig. 1.8.

1.2-5. (a) $X = A \cdot Y \vee B \cdot Z$

(c) All $X_i = 0$ when both $A = 0$ and $B = 0$.

All $X_i = Y_i \vee Z_i$ when both $A = 1$ and $B = 1$.

If the specifications are interpreted as requiring that two things be placed simultaneously on the X terminal, they call for the impossible.

1.2-7. (a) The flip-flop is the memory; the XOR gate is the processing network of the system.

(b)

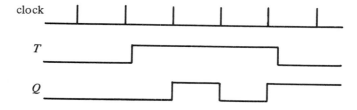

(c) The value of Q is changed (complemented) with each clock pulse for which $T=1$. The XOR gate provides the selective complementing.

$$Q(i+1) = T(i) \oplus Q(i)$$

(d) T for "toggle" or "trigger".

1.2-9. (a)

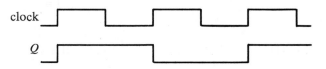

1.2-10. (c)

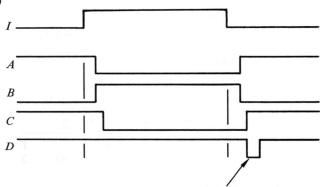

Temporary False Value, "Glitch"

1.2-12.

A_1	A_2	A_3	S_1	S_2	S_3	C_1	C_2	C_3
0	0	0	0	0	1	0	0	0
0	0	1	0	1	0	0	0	1
0	1	0	0	1	1	0	0	0
0	1	1	1	0	0	0	1	1
...								

1.3-1. (a) 5 digits. $\log_2 5 = \log_{10} 5 \,/\, \log_{10} 2 = .699 \,/\, .301 = 2.33$.

1.3-2. (a) Integer greater than 2.33 is 3.

1.4-4. The coefficients 5, 50, and 125 are not necessarily decimal integers. Express them in decimal

$$5_R = 5_{10}$$

$$50_R = 5R$$

$$125_R = R^2 + 2R + 5 \qquad \text{where } R \text{ is the unknown radix.}$$

Now substitute the given solution $x = 5$

$$5(5)^2 - 5R(5) + R^2 + 2R + 5 = 0$$

$$R^2 - 23R + 130 = 0 \qquad R = 10, \; 13$$

Substitute the other solution $x = 8$

$$5(8)^2 - 5R(8) + R^2 + 2R + 5 = 0$$

$$R^2 - 38R + 325 = 0 \qquad R = \overline{13}, \; 25$$

1.4-5. Three symbols appear. $R \geq 3$.
Try 3

$$
\begin{array}{r|l}
3 & 17 \quad 2 \\
3 & 5 \quad\ \ 2 \\
3 & 1 \quad\ \ 1 \\
& 0 \quad\ \ 0
\end{array}
$$

$01\ 22_3 = 17_{10}$ $X \equiv 0$

$|\ |\ |\ |$

XVII $V \equiv 1$

   $I \equiv 2$

1.4-6. $111_4 =$ $1 \times 4^2 + 1 \times 4^1 + 1 = 21_{10}$

$$
\begin{array}{r|l}
2 & 21 \\
2 & 10 \quad 1 \\
2 & 5 \quad\ \ 0 \\
2 & 2 \quad\ \ 1 \\
& 1 \quad\ \ 0 \\
& 1
\end{array}
$$

$21_{10} = 10101_2 = 111_4$

	10	2	3	7
1010101_2	85		10011	151
$1t9_{11}$	240	11110000	22220	462

1.4-8. Integer I_{10} is to be converted to base R.

$$
I = \left(\sum_{i=0}^{\infty} d_i R^i \right)_{10} \qquad \text{where the } d_i\text{'s are unknown.}
$$

Divide by R_{10}

$$
\frac{I_{10}}{R_{10}} = \frac{(\sum d_i R^i)_{10}}{R_{10}} = \underbrace{\left(\sum_{i=1}^{n} d_i R^{i-1} \right)_{10}}_{\text{integral part}} + \frac{(d_0)_{10}}{R_{10}}
$$

$\therefore$ the first remainder is $(d_0)_{10}$; convert this to base R. Complete proof.

1.4-9. $.123_4 = \dfrac{1}{4} + \dfrac{2}{4^2} + \dfrac{3}{4^3} = \dfrac{16 + 8 + 3}{64} = \dfrac{27}{64} = .421875_{10}$

$$
.421875_{10} = .102\ldots_3 = .123_4
$$

$$
\begin{array}{r}
.421875 \\
\times 3 \\
\hline
1.265625 \\
\times 3 \\
\hline
0.796875 \\
\times 3 \\
\hline
2.390625
\end{array}
$$

1.4-12. $A = 21_{10} = 10101_2$ $B = 10_{10} = 1010_2$

                                          $2B = 10100_2$

$$A + B = 10101$$
$$01010$$
$$\overline{11111}$$

$$A + 2B = \qquad 10101$$
$$10100$$
$$\overline{101001} = 32 + 8 + 1 = 41_{10}$$
$$\times 10101$$
$$\overline{101001}$$
$$1010010$$
$$1010010$$
$$\overline{1101011101}_2 = 861_{10}$$

2

Combinational Logic Fundamentals

Logic networks may be analyzed and designed to a large extent with little or no knowledge of electronics, hydraulics, or other technologies. However, if design is to lead to optimized operational hardware, then the logic designer should—must—respect the technology to be used to realize a paper design. Limitations and imperfections of technology must be overcome by logic design. Unusual characteristics may be taken advantage of. This chapter therefore begins with a review of digital electronics that emphasizes characteristics of concern to the logic designer.

Then the chapter turns to *combinational* logic networks, networks that only process information. Their output signal values depend entirely on input signal values at each instant of time; we usually think of them as acting instantaneously. *Sequential* logic networks store as well as process information. Output signal values of such networks depend upon past as well as current input signal values. Time is a very important variable in sequential network analysis and synthesis, as will be shown in Chapter 4.

As with many other aspects of engineering, substantial advances in the engineering of switching circuits and systems were made only after appropriate mathematical tools were provided. Claude Shannon [6] set forth the mathematics of switching in 1938, when he recognized that the algebra developed by George Boole [1] in 1854 to facilitate the study of the logic of the English language could be applied directly to relay switching circuits. Subsequently, that algebra was shown to be applicable to vacuum tube, diode, and transistor switching circuits, and now it is the language most commonly used to specify such circuits. Thus it is essential that this mathematics be studied before considering methods of designing switching systems.

Finally, this chapter introduces several other mathematical models of combinational logic networks and offers first methods for transforming from one model to another. We have already seen that equations, tables, logic diagrams, and vector notations are all valuable ways of representing digital systems. Each has advantages and disadvantages. Hence these and others to be developed are best suited for different analysis and design tasks.

2.1 DIGITAL ELECTRONICS

Digital circuits are also called "switching circuits" because they were first constructed using switches and relays. Switches are binary devices. When the two pieces of metal that form Switch A of Fig. 2.1 are brought together, low resistance is offered to electric current and the lamp lights. When the switch is opened, its high resistance prevents currents and the lamp is dark. These two conditions of the switch and of the lamp may be encoded with the logic 0 and 1 as Fig. 2.1 suggests.

Since the position of a switch will vary with time, its name may be considered to be a binary variable. Variable L may be assigned to express the condition of the lamp. For the circuit of Fig. 2.1

$$L = A \tag{2.1}$$

Dependent variable L always takes the value of independent variable A.

$$L = 1 \quad \text{whenever } A = 1$$

The lamp is on whenever the switch is closed.

$$L = 0 \quad \text{whenever } A = 0$$

The lamp is off whenever the switch is open. This then is the meaning of "equality" (symbolized by =) in logic equations.

Switch positions may be functions of the positions of other switches. The relay of Fig. 2.2 provides simple examples. With switch A open and no current in the relay coil, the contacts of switch B are closed while those of C

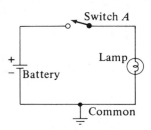

Position of Switch A	Light	Resistance	Conductance	Ideal
open	off	very large	very small	0
closed	on	very small	very large	1

FIG. 2.1 A Switch is Binary

CHAPTER 2   Combinational Logic Fundamentals

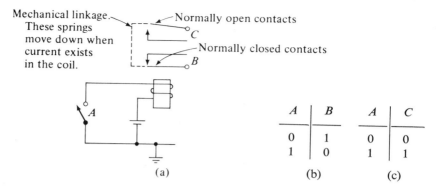

Mechanical linkage. These springs move down when current exists in the coil.

Normally open contacts
Normally closed contacts

A	B		A	C
0	1		0	0
1	0		1	1

(a)

(b)

(c)

Fig. 2.2   Relay Switching

are open. When A is closed, coil current causes B to open and C to close. In summary the table of Fig. 2.2(c) indicates

$$C = A \qquad\qquad (2.2)$$

while the table of Fig. 2.2(b) indicates that B has the value that switch A does not. This dependency is expressed

$$B = \bar{A} \qquad\qquad (2.3)$$

where the overbar expresses the logic complement or NOT operation.

Two imperfections of switches may be noted at this point. First, mechanical or electromechanical switches in which two pieces of metal are forced together or apart, usually with a spring action, have less than ideal transient characteristics. After mechanical contacts collide, for example, they bounce apart and then are forced together again. The bounce may be repeated. If appreciable current is carried by the closed contacts an electrical arc may exist. Rather than providing an ideal transition from 0 to 1, the conductance of the contact pair may vary in the irregular fashion suggested by Fig. 2.3.

Second, Fig. 2.3 indicates that contacts do not open or close at the instant they are commanded to do so. Mechanical parts have mass and hence inertia. Forces must act through time before motion causes contacts to open or close. For relays this delay is measured in milliseconds. Equations (2.2) and (2.3) are not valid immediately following a change in value

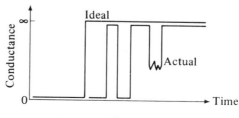

Fig. 2.3   Contact Bounce

of A. This imperfection takes a subtle secondary form when two or more contacts are mechanically linked to open and close simultaneously. Linkage imperfections result in one contact reaching a final value after bouncing, before all others. For a short time the positions of the contacts are not related in the expected manner. Or, we may say that different delays exist between the independent (input) variable and the dependent (output) variables. Electronic switching circuits have these same imperfections.

When two switches connected in series control a lamp as shown in Fig. 2.4(a), the lamp is lit only when both switches are closed. Both the truth table of Fig. 2.4(c) and the equation

$$L = A \cdot B \qquad (2.4)$$

express this relation of binary variables. Parallel switches shown in Fig. 2.4(b) turn the lamp on whenever either or both are closed. The OR truth table of Fig. 2.4(d) and the equation

$$L = A \vee B \qquad (2.5)$$

describe the inclusive-OR combination of binary variables. Thus the AND, OR, and NOT gates introduced in Fig. 1.6 may be realized with switch and relay contacts. The exclusive-OR gate of that figure can be most easily built using single-pole, double-throw switches or "transfer" contacts on relays. One movable contact always touches one of two fixed contacts as suggested by the schematic diagram of Fig. 2.5. The lamp may be turned off or on by changing the position of either switch.

$$L = A \oplus B \qquad (2.6)$$

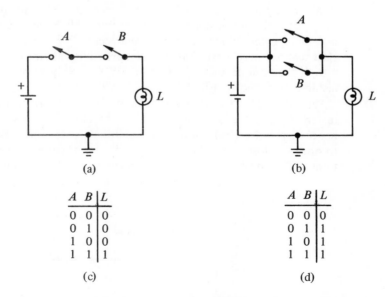

A	B	L
0	0	0
0	1	0
1	0	0
1	1	1

(c)

A	B	L
0	0	0
0	1	1
1	0	1
1	1	1

(d)

FIG. 2.4   Switches in Series Give AND Logic; Switches in Parallel Give OR Logic

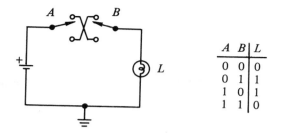

A	B	L
0	0	0
0	1	1
1	0	1
1	1	0

FIG. 2.5   The XOR Gate Using Transfer Contacts;
A = 1 When the Switch Is Up

Diode Logic

In electronic digital systems, two voltage levels are used to represent the logic 0 and 1. Two codes are available, and while both are used, one predominates. In the " positive logic " code, the lower voltage level represents logic 0 and the higher level expresses the 1. Quite often the ground or common level is the lower level and a positive supply voltage of $+V$ is the higher level. In the less common " negative logic " code, the lower voltage encodes the logic 1 while the higher voltage represents the 0.

Diodes are devices that conduct current from their anode to their cathode, but prohibit current in the other direction. Ideal diodes have the characteristics shown in Fig. 2.6. Zero volts exist across ideal diodes when they are conducting positive current. If a negative voltage is placed across an ideal diode, no current can exist. Silicon junction diodes conduct when the positive voltage placed across them exceeds a breakpoint voltage, V_D, of about 0.7 volts, and hence are slightly less than ideal.

If two diodes and a resistor connected as shown in Fig 2.7 are driven by sources of binary signals A and B that take ideal values of 0 and $+V$ volts, then output signal C will take values close if not equal to the minimum of the input signals. C is at the high level only when both A and B are at the high level; otherwise, one or both input signals pull C low via its diode. The AND truth table is observed for ideal diodes. With real diodes, the lower level of C is not the ideal value of V volts. If this circuit were driving a load, then the high voltage level would be less than $+V$ volts. The two

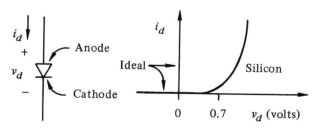

FIG. 2.6   The Symbol and Characteristics of the Ideal and Silicon Diode

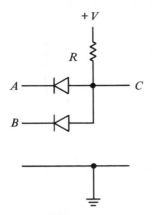

		Ideal Diodes	Real Diodes
A	B	C	C
0	0	0	0.7 volts
0	V	0	0.7
V	0	0	0.7
V	V	V	V

FIG. 2.7   Diode AND Gate

voltage levels provided by this circuit are closer together than the ideal levels supplied to it. If a number of such circuits were cascaded, the two levels would come together and information would be lost.

If one of the diodes in Fig. 2.7 is conducting and thereby holding C at a low voltage, it must pass a current of approximately V/R amperes. The generator of the low input signal must also pass this current to ground, or " sink " it, while holding the input signal at or near the ideal low level. If it cannot do this, then it cannot successfully drive this gate. It is not automatically true that any gate can drive any other gate.

The diode AND gate can sink current supplied by a load circuit connected to its output terminal, but with a 0.7 volt increase in the low level. It can supply or " source " current i to a load through resistor R, but at a high level of

$$v_c = V - iR \qquad (2.7)$$

volts. Thus only modest currents can be sourced without affecting the high level substantially.

Diode OR gates as shown in Fig. 2.8 have characteristics that are the dual of AND gate characteristics. If either or both input signals are at the high ideal level, the output signal is 0.7 volts less positive. The drivers must source current at the high level. Diode OR gates source current to load circuits with a 0.7 volt loss of level, and sink current through resistor R at a not necessarily low level of iR volts.

Both the deterioration of signal amplitude when passing through diode gates and their limited sinking and sourcing abilities restrict their application. After every two levels of diode gating, power amplifiers that restore signal levels to nearly ideal values are required.

The Common Emitter Amplifier

Enlarging (amplifying) a signal requires " active " devices that can convert energy supplied by a power source to signal energy. Transistors

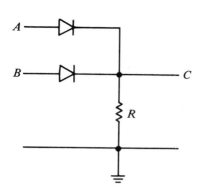

		Ideal Diodes	Real Diodes
A	B	C	C
0	0	0	0 volts
0	V	V	$V-0.7$
V	0	V	$V-0.7$
V	V	V	$V-0.7$

FIG. 2.8   Diode OR Gate

are the most often used active devices. When the base terminal of an NPN
silicon transistor is less positive than a breakpoint voltage V_{BEO} of approximately 0.7 volts with respect to the emitter terminal [see Fig. 2.9(a)], the
transistor is *cut off* and no collector current is permitted. With the base-to-
emitter voltage, v_{BE}, greater than V_{BEO} and the collector more positive
than the base, the transistor is *active*. Then its collector current, i_C, is a
large multiple of its base current. The " current gain ", h_{fe}, of the transistor
relates these currents.

$$i_C = h_{fe} \times i_B \tag{2.8}$$

When $v_{BE} > V_{BEO}$ and the collector is less positive than the base, the
transistor is *saturated* and the collector-to-emitter voltage is a small, fixed
value $V_{CE(SAT)}$ of 0.05 to 0.15 volts. This highly simplified description of the
low frequency operation of a transistor is summarized in the model of
Fig. 2.9(b). Ideal diodes prohibit negative base current and limit v_{CE} to a
minimum value of $V_{CE(SAT)}$ volts.

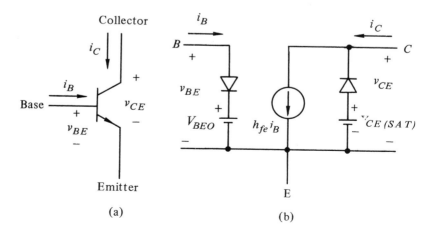

FIG. 2.9   The NPN Transistor Symbol and Model

Combining a transistor with two resistors in the manner of Fig. 2.10(a) results in the *common emitter* amplifier. When input signal v_{IN} is less than V_{BEO} volts, the transistor is cut off; that is, $i_B = 0$ and therefore $i_C = 0$ and $v_{OUT} = +V_{CC}$, the power supply voltage. When v_{IN} exceeds V_{BEO},

$$i_B = \frac{v_{IN} - V_{BEO}}{R_B} \tag{2.9}$$

Collector current is h_{fe} times greater and v_{OUT} has a value below V_{CC}.

$$
\begin{aligned}
v_{OUT} &= V_{CC} - i_C R_C \\
&= V_{CC} - h_{fe} i_B R_C \\
&= \left(V_{CC} + h_{fe} \frac{R_C}{R_B} V_{BEO} \right) - h_{fe} \frac{R_C}{R_B} v_{IN} \tag{2.10} \\
&= K + A v_{IN} \tag{2.11}
\end{aligned}
$$

where K is a constant and A is the amplification of the circuit. Small variations, v_{in}, in the input signal result in larger variations, v_{out}, in the output signal.

$$A = \frac{dv_{OUT}}{dv_{IN}} = \frac{v_{out}}{v_{in}} = -h_{fe} \frac{R_C}{R_B} \tag{2.12}$$

The output voltage must not drop below $V_{CE(SAT)}$, of course; when it reaches that value the transistor saturates and the collector current is as large as the power supply can provide through the collector resistor.

$$I_{C(SAT)} = \frac{V_{CC}}{R_C} \tag{2.13}$$

The smallest input voltage that causes saturation is

$$v_{IN(SAT)} = \frac{(V_{CC} - V_{CE(SAT)})}{-A} + V_{BEO} \tag{2.14}$$

Figure 2.10(b) summarizes some of these findings in a voltage transfer characteristic for a common emitter amplifier with $A = -10$ and a 10 volt power supply. Typical values of V_{BEO} and $V_{CE(SAT)}$ for silicon transistors are shown. We see that any voltage in the range 0 to 0.7 volts results in an ideal high output voltage of $+10$ volts. Any voltage in the range 1.7 to 10 volts results in an ideal low voltage of 0.1 volts. All input voltages in a range are interpreted as logic 0; all voltage in another range are interpreted as logic 1 by this amplifier. The intermediate range 0.7 to 1.7 volts is the *threshold* range of this amplifier. Input voltages in this range do not result in ideal output voltages.

The generator supplying v_{IN} must source current to the common emitter amplifier when v_{IN} is high and sink essentially zero current when v_{IN} is low. Input I-V characteristics such as that of Fig. 2.11(b) are used to indicate the demands that a common emitter amplifier places on a driving circuit.

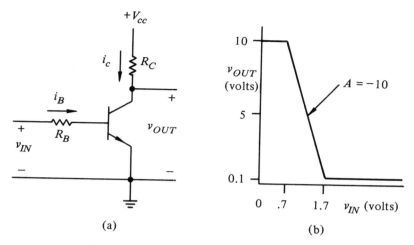

(a)   (b)

FIG. 2.10   The Common Emitter Amplifier

How able is this amplifier to supply a signal to a load such as one or more other amplifiers or diode gates? Two cases must be considered. With the transistor cut off, the amplifier can source current to a load, but via the collector resistor R_C. Thus, as the current demanded by the load increases, the output voltage drops. With the transistor saturated, the amplifier can sink current to ground with almost no change in v_{OUT}. If the supplied base current is I_B, then the sinking capacity of the amplifier is $h_{fe} \times I_B$ amperes without pulling the transistor from saturation, and hence without substantial change in v_{OUT}. Figure 2.12 summarizes these output characteristics.

Suppose one common emitter amplifier must drive n identical amplifiers. With the driving transistor cut off, base current must be supplied to n others. The driving amplifier must source

$$i_{OUT} = \frac{V_{CC} - V_{BEO}}{R_C + R_B / n} \text{ amperes} \tag{2.15}$$

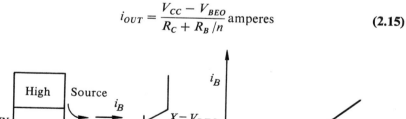

(a)   (b)

FIG. 2.11   The Input Characteristics of the Common Emitter Amplifier

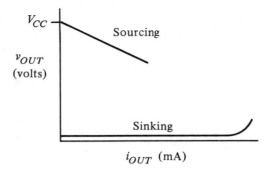

FIG. 2.12   Output Characteristics of the Common Emitter Amplifier

to saturate those transistors. The output voltage of the driving amplifier
will drop to:

$$v_{OUT} = V_{CC} - \frac{V_{CC} - V_{BEO}}{R_C + R_B/n} R_C \text{ volts} \tag{2.16}$$

This voltage must not fall in or below the threshold range for correct logic
operation. With the driving transistor saturated, all driven transistors will
be cut off and only very small reverse leakage currents must be returned to
ground.

The transfer characteristic of the driving amplifier varies with n as shown
in Fig. 2.13. With increasing values of n, the high level of v_{OUT} drops from
the ideal toward the threshold region. For some value of n, it will enter the
threshold region and the driven amplifiers will be supplied a voltage that is
not a clear representation of logic 1. A limit called the *fan-out* limit is
imposed upon the logic designer to prevent this possibility.

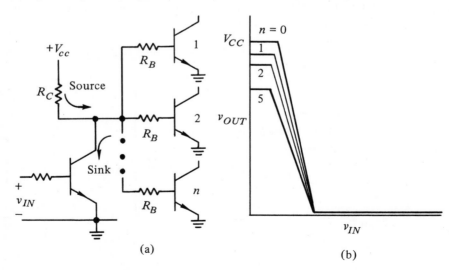

FIG. 2.13   Fan-out and Its Effect on the Transfer Characteristic of the Driving
Amplifier

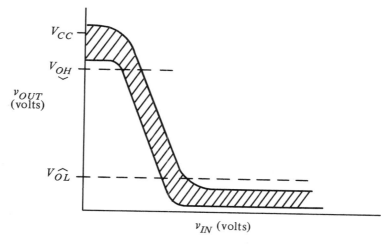

FIG. 2.14 Transfer Characteristics of Many Common Emitter Amplifiers Operated at All Permitted Power Supply Voltages, Temperatures, and Fan-out

Digital systems consist of hundreds to hundreds of thousands of amplifiers, each driven by different sources and driving different loads, powered by a variety of supplies, and all expected to operate flawlessly over a range of environmental conditions. Furthermore, the resistors, diodes, transistors, and other components of the amplifiers cannot be manufactured with exact parameter values. The characteristics of one amplifier operating at room temperature with ideal drive, load, and power supply are of limited value to systems designers.

Suppose that we build a large number of "identical" amplifiers using components with parameters that fall within established ranges. We operate these amplifiers under all combinations of conditions that fall within established limits. If the transfer, input, output, and other characteristics are collected for each amplifier tested, we will obtain *group characteristics* from which best and worst cases can be identified. Figure 2.14 shows a group transfer characteristic for common emitter amplifiers.

Any amplifier or amplifiers in a system might consist of components with parameters that lead to worst case characteristics. If the system is to function properly, the design must be such that worst case amplifier performance is acceptable. Two worst case values can be determined from Fig. 2.14. All amplifiers provide a low output voltage less than $V_{\widehat{OL}}$ (Output Low, maximum). We may even include a safety factor in selecting $V_{\widehat{OL}}$. All amplifiers provide a high output voltage greater than $V_{\underset{\smile}{OH}}$ (Output High, minimum). These worst case limits can be guaranteed if our testing is complete; they establish output voltage ranges for logic 0 and logic 1 that can be assured.

Figure 2.15 shows that $V_{\widehat{IL}}$ is the greatest low input voltage for which we can be sure of obtaining an output voltage greater than $V_{\underset{\smile}{OH}}$. $V_{\underset{\smile}{IH}}$ is the smallest high input voltage for which the output voltage will be less than $V_{\widehat{OL}}$. If amplifiers are to be used in systems, then it is essential that $V_{\widehat{IL}}$

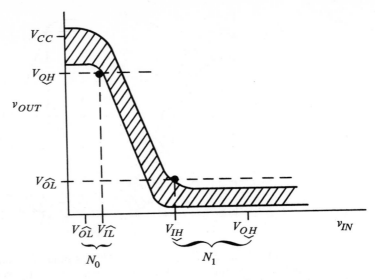

Fig. 2.15   Noise Margins

exceed $V_{\widehat{OL}}$. The difference between these two worst case values is called the dc *noise margin* with output low, N_0.

$$N_0 = V_{\widehat{IL}} - V_{\widehat{OL}} \tag{2.17}$$

Noise to this magnitude (maximum) may be added to the low output signal of a driving amplifier without the driven amplifiers entering their threshold region of operation. Similarly the dc noise margin with output high is

$$N_1 = V_{\underset{\smile}{QH}} - V_{\underset{\smile}{IH}} \tag{2.18}$$

A large part of digital systems' success is due to the ability of digital amplifiers to reject noise.

Dynamic behavior of digital amplifiers is characterized with several parameters. The time taken by a signal to change from low to high value is its *rise time*, t_r. Designers usually use the interval between the 10 percent and 90 percent points of the transition. Figure 2.16(a) shows that *fall time* t_f is similarly defined. With the output signal of an amplifier changing from high to low, t_{PHL} measures the interval between the input and output signals in passing their 50 percent points. Figure 2.16(b) illustrates this definition for an inverting amplifier and the similar definition of t_{PLH}, propagation delay with the output signal changing from a low to a high value. The average of t_{PHL} and t_{PLH} is usually quoted as the average *propagation delay* of the amplifier.

$$t_P = \frac{t_{PHL} + t_{PLH}}{2} \tag{2.19}$$

Propagation delay is of great concern to the logic designer. Activity in logic networks is affected by it. The rate at which information can be

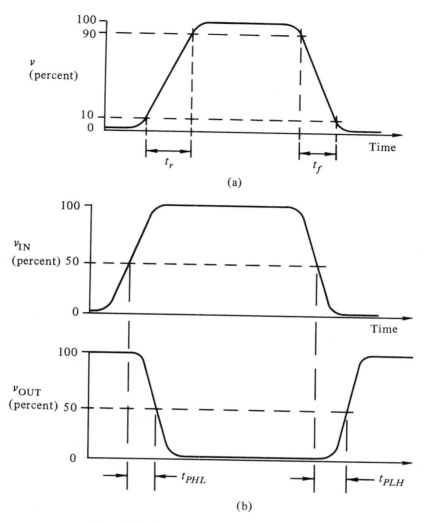

FIG. 2.16　Parameters of Dynamic Behavior

processed is limited by it. At times a logic design must be redone because of undesired activity that is the direct result of propagation delay; a circuit may do the wrong thing, or do the right thing, but too slowly. Factors such as temperature, which causes propagation delay to change, must also be taken into account. For example, if load capacitance is one such factor, fan-out may be limited by it rather than static electrical considerations. This will keep propagation delay within bounds and thereby achieve high processing rates.

Dynamic response of the common emitter amplifier is only partially determined by the transistor model of Fig. 2.9. When v_{IN} of Fig. 2.10 is zero, the transistor is cut off and $v_{OUT} = +V_{CC}$ volts. If v_{IN} suddenly rises to $+V_{CC}$ volts, the transistor conducts, but does not immediately saturate

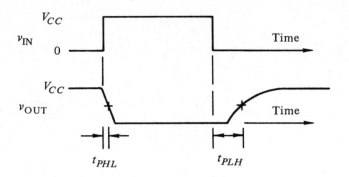

FIG. 2.17   Dynamic Response of the Common Emitter Amplifier

with $v_{OUT} = V_{CE(SAT)}$. Capacitance between collector and emitter must be discharged by the almost constant collector current, $h_{fe} \times i_B$. During this discharge, v_{OUT} drops to $V_{CE(SAT)}$ in an almost linear fashion, as shown in Fig. 2.17. The transistor itself, wiring between the amplifier and its load, and that load all contribute to the capacitance that must be discharged. Greater capacitance results in larger values of t_f and t_{PHL} for this amplifier.

When the transistor is turned off by dropping v_{IN} to zero, v_{OUT} remains at the low value for a time while charge stored in the base region of the transistor is removed via a reverse base current. Then the transistor turns off suddenly, but v_{OUT} rises slowly with an exponential waveform as the load capacitance is charged through collector resistor R_C. Both t_r and t_{PLH} depend on R_C as well as the capacitance. Propagation delay thus depends on many things.

Designers of digital systems must also be concerned with power required by logic amplifiers. The cost of power supplies and cooling equipment is not trivial. When the transistor of the common emitter amplifier is saturated, energy is converted to heat at the rate V_{CC}^2 / R_C watts. With the transistor off, no power is dissipated. Assuming that in a large system transistors are on as often as they are off, an average power level of $V_{CC}^2 / 2R_C$ watts may be quoted.

RTL, I²L, DTL and TTL Logic Circuits

Two common emitter amplifiers that share a single collector resistor [Fig. 2.18(a)] form a Resistor Transistor Logic (RTL) NOR gate. This is symbolized in Fig. 2.18(b). When either or both input signals are high, the corresponding transistor(s) are saturated and the output signal is low. Output variable C takes the logic 1 value only when neither input variable A nor input variable B have that value. While the NOR gate can be thought of as an OR gate followed by an inverter, as its symbol suggests, no distinct OR gate and inverter exist. Other circuits consisting of resistors and transistors only are also known as RTL.

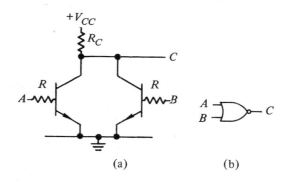

FIG. 2.18   The RTL NOR Gate

If the base resistors are removed from the circuit of Fig. 2.18(a), the circuit becomes a Direct Coupled Transistor Logic (DCTL) NOR gate. Collectors are connected directly to the bases of driven amplifiers. Fewer parts are required and other logic functions can be realized, for example, by stacking transistors in series between ground and a collector resistor. It is essential that $V_{CE(SAT)}$ be less than V_{BEO} for the transistors used. Signal amplitudes and noise margins are small. The low voltage level is $V_{CE(SAT)}$, while the high level is V_{BEO}.

The first digital integrated circuits provided RTL gates and flip-flops with propagation delays in the 10 to 30 nsec range and power dissipations of 10 mW per gate. While RTL is no longer being developed, one idea found here continues to be employed. If the final stage of a gate amplifier is a common emitter amplifier without the collector "pull-up" resistor, then these output terminals may be wired together to form a NOR gate at the cost of interconnecting wires only. This is called *dot* or *wired* logic; the dot refers to the symbol used in electrical drawings to denote connection of wires. "Open collector" gates are available in both the DTL and TTL families of circuits.

When circuits are fabricated in integrated form, additional considerations impact the logic designer. With the entire circuit formed from one crystal of silicon, the surface area taken by the components and their interconnections is important. Usually several hundred circuits are manufactured simultaneously on the surface of one large crystal. Then they are separated and tested. A greater number of circuits that require a smaller area can be made from one large crystal, but in addition, the probability of a crystal defect is less with a smaller area. Thus, a greater fraction of the manufactured circuits will work. The silicon "chips" that function must be packaged to protect them from the environment and provide ease of handling. Connections are made to the packaged integrated circuit via "pins", which must be large and strong so that the IC can be handled and soldered to a printed circuit board. Desires to have small packages and large numbers of pins are contradictory.

Logic designers who develop circuits to be manufactured must minimize the number of gates and flip-flops specified. This is to reduce the area,

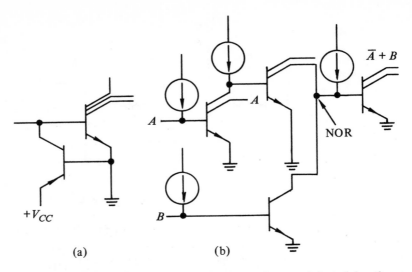

$\overline{A} + B$

A

NOR

A

B

$+V_{CC}$

(a)   (b)

FIG. 2.19 The I²L Inverter and NOR Gate. Current injected by the pnp transistor is passed to ground via the driving transistor or the base-to-emitter of the driven transistor, turning it on.

power, and propagation delay of the circuit. They may have to modify designs so that the components can be formed and interconnected with a restricted sequence of manufacturing steps. The logic designer who uses fabricated ICs must attempt to use a minimum number of packages for each one costs money, takes space, and requires power. Within each package the designer finds a fixed number of gates, each with a fixed number of input terminals called the *fan-in* of the gate. When a 14-pin standard package is adopted, it may contain either four 2-input gates, three 3-input gates, two 4-input gates, or one 8-input gate. If the designer requires some other fan-in, the system must be reworked to eliminate that requirement, let some input terminals go unused or use more than one gate to build a network that acts as the required gate.

The area and power required by the resistors of RTL and its slow speed have caused RTL to fall in favor from other logic families. One of the emerging families is a variation of DCTL. Integrated Injection Logic (I²L), also known as Merger Transistor Logic (MTL), uses a pnp transistor as a current source. If the transistor driving the amplifier of Fig. 2.19(a) is saturated, the constant collector current of the pnp transistor is diverted to ground. If the driving transistor is off, it serves as the base current for the multicollector transistor. Dot logic provides NOR gates; the multiple collectors facilitate forming electrically independent NOR gates. Signals vary from a low value of $V_{CE(SAT)} = 50$ mV to a high value of $V_{BEO} = 750$ mV. Power dissipation depends on the transistors used, but ranges from nanowatts to microwatts. Propagation delay varies inversely with power dissipation and varies over the 10 to 100 nsec range.

Interest in I²L logic arises from the small area required by each amplifier. Since the pnp and npn transistors are highly interconnected, they may be

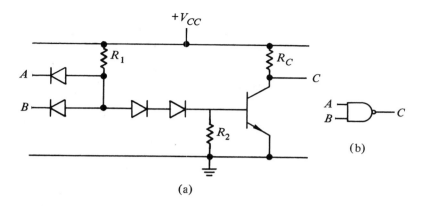

FIG. 2.20   A DTL NAND Gate and Its Symbol

fabricated as one, eliminating any isolation space that usually surrounds transistors. No resistors are required so an entire I^2L amplifier takes less area than a single multi-emitter transistor of a TTL gate and less area than MOS amplifiers (yet to be discussed). Fewer processing steps are required than for most other logic circuits. And finally, I^2L can be manufactured on the same silicon chip at the same time as other bipolar transistor logic circuits. Thus, I^2L integrated circuits can be made compatible with logic circuits of other families by using buffers of those technologies between the pins and the I^2L circuits. The small signal magnitudes are then not brought to pins and exposed to noise.

When a diode AND gate is combined with a common emitter amplifier, a Diode Transistor Logic (DTL) NAND gate is formed. Figure 2.20(a) shows one of a variety of DTL NAND circuits. An identifiable AND gate is followed by an inverting amplifier. The two interconnecting diodes move the threshold region of the transfer characteristic to higher voltages so that the low level noise margin N_0 is more nearly equal to N_1. The current sourcing and sinking characteristics of the common emitter amplifier are well matched to the current requirements of the input side of the DTL circuit. Fan-out limits of 10, propagation delays of 15 nsec, and power dissipations of 5 mW per gate are typical of DTL.

While development of the DTL circuit family has ceased and the use of DTL has declined, the newest form of TTL, low-power Schottky, uses the diode AND gate again. In more conventional Transistor Transistor Logic (TTL), the diodes are replaced with a multi-emitter transistor. The substitution is not a trivial replacement.

With all of the input signals of Fig. 2.21 at the high level, transistor T_1 operates in its reverse mode (emitter serves as collector, collector serves as emitter) to supply base current to T_2. The generators supplying the input signals must source current in this case. To provide generators that can source as well as sink current, and to avoid the long rise time of the common emitter amplifier, most TTL circuits include a "totem pole" output amplifier. In Fig. 2.21, transistor T_3 pulls the output terminal low and sinks current when it is active and saturated. When T_4 is turned on,

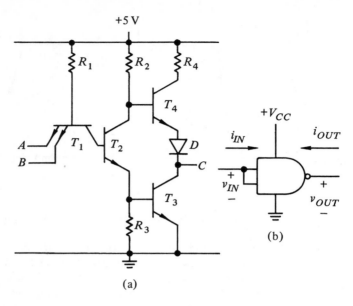

+5 V

A
B

R_1 R_2 R_4

T_4

T_1 T_2 D

C

T_3

R_3

(a)

$+V_{CC}$

i_{IN} i_{OUT}

$+$
v_{IN}
$-$

$+$
v_{OUT}
$-$

(b)

FIG. 2.21   Basic TTL NAND Gate

it pulls the output terminal to a high voltage. While the phase splitting circuit of T_2 minimizes the time when both T_3 and T_4 are simultaneously conducting, R_4 is included to limit the "spiking" current demanded of the power supply when this does happen.

Figure 2.22(a) shows a typical voltage transfer characteristic of a single input TTL amplifier and values of $V_{\widehat{IL}}$, V_{IH}, $V_{\widehat{OL}}$, and V_{OH} that are guaranteed for the SN54/74xx series of TTL gates. With $v_{IN} < V_a$ (about 0.6 V), transistor T_1 is saturated. Its low collector potential ensures that T_2 is cut off. As a result, T_4 is saturated with base current supplied via R_2, and the output voltage is V_{BEO} (for T_4) plus V_D (for the diode) below the power supply level of +5 volts. The driver must sink the emitter current of T_1. Since $V_{\widehat{OL}} = 0.4$ V is guaranteed, the worst case input low current, $I_{\widehat{IL}}$, is guaranteed at that voltage to be less than 1.6 mA in magnitude. This point is shown in Fig. 2.22(b). Figure 2.22(c) shows that the output amplifier is able to source many milliamperes while maintaining $v_{OUT} > V_{OH} = 2.4$ V. The amplifier is guaranteed to source at least 400 μA at a voltage greater than $V_{OH} = 2.4$ V.

With the input voltage in the range $V_a \leq v_{IN} \leq V_b$, transistor T_2 operates in the active mode with low gain determined by R_2/R_3; T_3 is still cut off. When v_{IN} reaches V_b, T_3 begins to conduct and R_3 is paralleled by its constant base to emitter voltage, V_{BEO}. Transistor T_2 acts as a common emitter amplifier with high gain, and v_{OUT} drops rapidly as v_{IN} increases through the range $V_b \leq v_{IN} \leq V_c$. Since both T_3 and T_4 are active and R_4 is a small resistance, a great deal of current is demanded of the power supply; hence V_{CC} will drop and provide noise to other gates, unless a

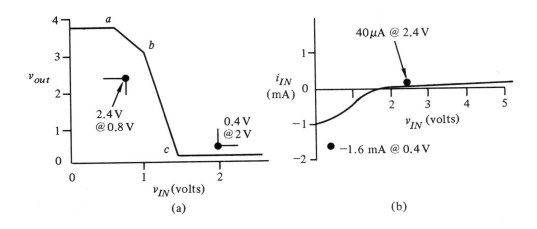

(a)

(b)

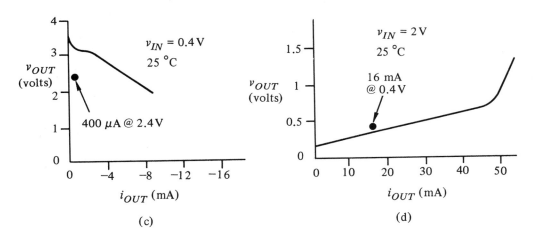

(c)

(d)

FIG. 2.22   Typical TTL (a) Transfer, (b) Input, (c) Output High, and (d) Output
Low Characteristics

capacitor with short leads is connected between the supply and ground
terminals of the amplifier.

When v_{IN} exceeds V_c, transistor T_4 is cut off and T_3 is saturated. T_1 is
acting in a reverse mode to supply base current to T_2. The ability of the
output amplifier to sink load current varies widely with temperature; a
value of 16 mA at $v_{OUT} = V_{\widehat{O}L} = 0.4$ V is guaranteed, as shown in Fig.
2.22(d). Load currents greater than 50 mA pull T_3 out of saturation.

The guaranteed values previously discussed ensure noise margins of
400 mV and a fan-out of 10. Figure 2.23 shows a way of recording worst
case numbers so as to make these facts obvious. Propagation delays vary
with temperature, supply level, and capacitance load. The effect of capacitive load is quite linear. A typical gate exhibits delay

$$t_P = 0.06\ C + 9 \text{ nsec} \tag{2.20}$$

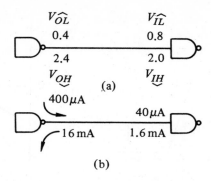

(a)

(b)

FIG. 2.23   Worst Case Limits for Determining (a) Noise Margin and (b) Fan-out Limit

where C is given in picofarads. Thus a 100 pF load increases propagation delay to 15 nsec.

If the output terminals of two totem pole amplifiers are wired together, the amplifiers may be destroyed whenever one attempts to pull the terminal up while the other is pulling down. The loss of wired-logic is particularly severe in bus oriented systems where a *bus* is a long wire or bundle of wires

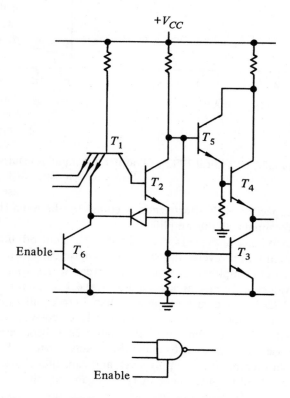

FIG. 2.24   A Tri-state NAND Gate

upon which any one of many transmitting amplifiers can place information at any instant and from which any of many receiving amplifiers can accept that information. Wired-logic permits the physical separation of the transmitters. While open-collector output amplifiers are available in the TTL family, they are substantially slower and less able to source current to the receiving amplifiers. *Tri-state* TTL circuits retain the advantages of the totem pole amplifier while serving as bus transmitters.

In the amplifier of Fig. 2.24, transistor T_6 turns other transistors off when it conducts. In particular, both T_3 and T_4 are cut off; this leaves the output terminal floating at a high impedance and drawing a maximum of 50 μA of leakage current. This then is the third state. When T_6 is cut off, the gate functions normally and provides either a low or high output voltage with current sinking and sourcing capability. The addition of T_5 increases drastically the current sourcing capability of the amplifier to supply the leakage current of other wired transmitters as well as the current required by receiving amplifiers. Only one transmitter on a bus may be enabled at any time, of course.

The propagation delay of TTL gates may be decreased by using Schottky transistors, which include a Schottky diode between the base and collector as shown in Fig. 2.25. The Schottky diode has a breakpoint voltage of about 0.3 volts. Thus the collector is prevented from going more than 0.3 V negative with respect to the base. The transistor is prohibited from becoming heavily saturated, and the turn-off delay due to stored charge is avoided. The various lines of TTL logic circuits are compared in Table 2.1.

Emitter-Follower Logic

The emitter follower amplifier shown in Fig. 2.26(a) is a current amplifier with a voltage gain of less than one. Thus, while voltage signal amplitude cannot be restored, current amplitudes can. If $v_{IN} \leq 0.7$ V ($i_B = 0$), the transistor is cut off; $v_{OUT} = 0$ V and $i_{OUT} = 0$. With $v_{IN} = V_{CC}$, $v_{OUT} = V_{CC} - V_{BEO}$ with the transistor operating in its active mode. Hence i_{OUT} is approximately V_{CC}/R_L. Since saturation is avoided, propagation delays are smaller. No time is required to remove stored base charge.

If two emitter followers share a load resistor [Fig. 2.26(b)], either or both may pull the output voltage to the high level. The circuit forms an OR gate, but one which does not restore the high voltage level to an ideal value.

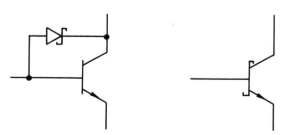

FIG. 2.25   The Schottky transistor

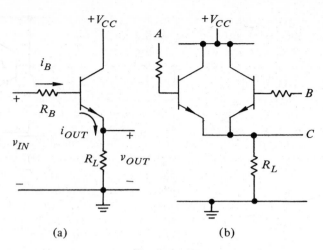

(a)   (b)

FIG. 2.26   The Emitter Follower

This structure will be found in a number of the logic amplifiers to be discussed.

Emitter Coupled Logic (ECL) achieves high switching speeds by avoiding saturation. In Fig. 2.27, transistors T_1 and T_2 form a differential amplifier that draws constant current regardless of which transistor is conducting. The negative supply and resistor R_E form a constant current source. Either T_1 or T_2 must pass this current. If input signal A is less positive than the reference voltage $V_{BB} = -1.175$ V, T_1 is active and T_2 is cut off. The collector of T_2 is at ground potential and output terminal C is at -0.7 volts. The collector of T_1 is at a negative voltage because T_1 is conducting, and terminal B will be at approximately -1.7 volts. When input signal A is more positive than V_{BB}, T_2 conducts, T_1 is cut off, and the output voltages at B and C will be high $(-0.7$ V$)$ and low $(-1.7$ V$)$, respectively. While voltage changes are small, very definite current switching takes place; the current in T_2 is either zero or at a reference value. The emitter followers shift voltage levels by 0.7 V so that the output signals can be used directly as input signals to other amplifiers. They also provide the possibility of dot logic OR gates via the circuit of Fig. 2.26(b) and low impedance for driving

TABLE 2.1   TTL FAMILIES

Name	Designation	Propagation Delay	Gate Power
Standard	54/74xx	10 nsec	10 mW
Low Power	54L/74Lxx	33	1
High Speed	54H/74Hxx	6	20
Schottky	54S/74Sxx	3	20
Low Power Schottky	54LS/74LSxx	5	2

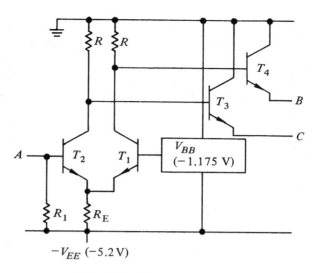

FIG. 2.27   The ECL Inverter

the lines connecting gates. With switching speeds approaching one nano-second, connecting wires longer than a few inches must be treated as transmission lines.

ECL gates are formed by placing one or more transistors in parallel with T_2 (see Fig. 2.28). Additional emitter follower transistors may also be included as shown in Fig. 2.28. These provide logically identical, electri-cally isolated output signals that can be combined via dot logic with different signals. If either A or B are at the high level, then T_1 is off and C is at the high level. Both D and E are at the low level. This circuit then provides the OR function and its complement, the NOR of the input variables. While noise immunity is only about 100 mV for ECL circuits, impedance levels are low. Propagation delays and switching times are in the 1 to 5 nsec range. At these high speeds interconnecting wires must be treated as transmission lines. Emitter resistors are not built into the transmitting circuits, but are a part of the receiving circuit where they act to terminate the transmission line (to eliminate reflections).

Several lines of ECL logic circuits have been developed. While ECL offers a small propagation delay of 2 nsec, this speed is paid for with a power dissipation of 25 mW per gate. Thus husky power supplies and forced cooling are required in digital systems constructed of ECL circuits.

MOS Logic Circuits

Metal-over-Oxide-over-Semiconductor, field-effect transistors are used in logic circuits because they occupy very little area on the semiconductor crystal, require fewer manufacturing steps to fabricate than bipolar transis-tors, and use very little power. They also have a very high input impedance

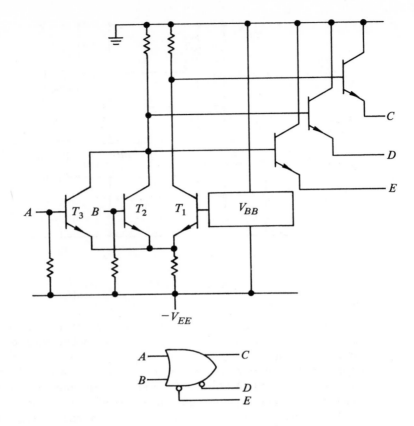

FIG. 2.28   ECL NOR/OR Gate

and excellent isolation of the input from the output terminal. The metal "gate" of the device is insulated from the semiconductor "channel" beneath it by the oxide layer; the "source" and "drain" terminals form the ends of the channel. If the channel conducts electrons, the device is an n-channel or NMOS transistor. P-channel or PMOS devices utilize "hole" conduction. While the source and drain are physically indistinguishable, in p-channel devices the substrate must be connected to the most positive voltage, as the source usually is. In n-channel devices, the substrate must be tied to the most negative voltage, again as the source usually is. Thus, in circuits using insulated gate field effect transistors (IGFET) with symbols of Fig. 2.29, the source may be identified as the terminal to which the substrate is tied.

The equivalent circuit for an FET is shown in Fig. 2.30. Resistance between the gate and source is on the order of 10^{12} ohms, hence is not shown. The gate-to-source and gate-to-drain capacitances are approximately 2 pF, which, while small, cannot be ignored. In fact, these capacitances are utilized to store information in some MOS circuits. The drain-to-sources resistance, r_{DS}, is sensitive to the gate-to-source voltage, V_{GS}, and varies from 10^9 ohms or greater down to 50 to 200 ohms with V_{GS}.

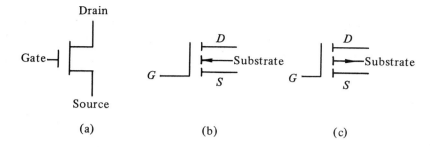

(a)   (b)   (c)

FIG. 2.29   Symbols for (a) Junction FET, (b) n-Channel IGFET, and
(c) p-Channel IGFET

Figure 2.30 shows how r_{DS} varies with V_{GS} for a n-channel IGFET. When V_{GS} is a high voltage, r_{DS} is a low resistance and the transistor will conduct current. When V_{GS} is close to zero volts, r_{DS} is very high and the transistor is cut off.

FETs can be used in common-source amplifiers, but in integrated circuits the load resistor is replaced with a second MOS device designed and biased to provide the desired load resistance. This second device requires much less area than a resistor, and temperature effects tend to cancel since both devices are at the same temperature and fabricated from the same semiconductor material. Paralleling two FETs leads to the NOR gate; connecting two in series as in Fig. 2.31 gives a NAND gate. Series–parallel combinations can give circuits that realize more complex switching functions.

When both n-channel and p-channel devices are fabricated on the same chip, complementary MOS (CMOS) logic emerges. The basic inverter circuit is shown in Fig. 2.32 together with its nearly ideal transfer characteristics. When the input voltage is low, the n-channel (lower) device presents very high resistance. The p-channel device charges the load capacitance through its low resistance to the supply level. Once the capacitance is charged, no further current and hence power is required of the

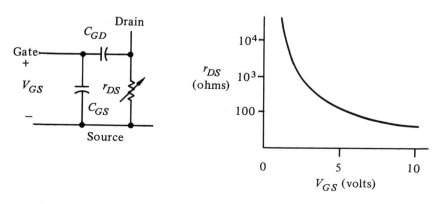

FIG. 2.30   An Equivalent Circuit of the Field Effect Transistor

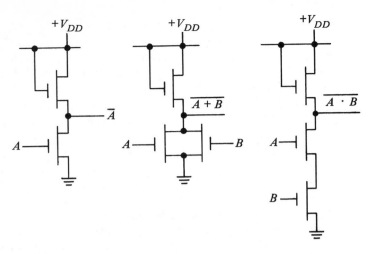

FIG. 2.31   Basic MOS Circuits

V_{DD} supply. When the input voltage is high, the n-channel device offers low resistance while the p-channel transistor is cut off. Again, currents exist only while the capacitances are charging and discharging; power consumption increases with switching frequency. The signal source must supply power to charge and discharge the gate-to-source capacitances only.

CMOS NOR and NAND circuits are shown in Fig. 2.33. In the NOR circuit, if either input signal is at the high level, its p-channel device is off and its n-channel device is on, which pulls the output terminal to a low voltage. Both p-channel devices are on and both n-channel devices are off when both input signals are at a low level. In the NAND circuit the output is pulled high by conducting p-channel transistors when either or both input signals are at the low level. The series connected, n-channel devices can pull the output to a low voltage through a low resistance, only when both input signals are high. Series–parallel networks of n-channel devices together with the dual parallel–serial network of p-channel devices may be used to realize more complex switching functions.

As a result of leakage currents, these CMOS circuits consume a few nanowatts of power under static conditions. If they are switched at a megahertz rate, however, their power dissipation exceeds 1 mW. At higher frequencies it exceeds the power requirement of TTL or even ECL. CMOS is somewhat more flexible than these families in that it will accept any supply voltage in the range of 3 to 18 volts. Supply level influences power dissipation, of course, but it also effects switching speed. Higher supply voltages lead to greater dissipation, and smaller propagation delays. With a 5 V supply, typical propagation delay is 35 nsec, whereas with $V_{DD} = 10$ V propagation delay is 25 nsec. Propagation delay increases linearly with load capacitance. Thus while fan-out in CMOS circuits is unlimited at low frequencies, a fan-out limit of 50 or less may be placed to limit delays to acceptable limits. The low-frequency noise immunity of CMOS amplifiers is approximately 40 percent of the supply voltage. Unfortunately, the high

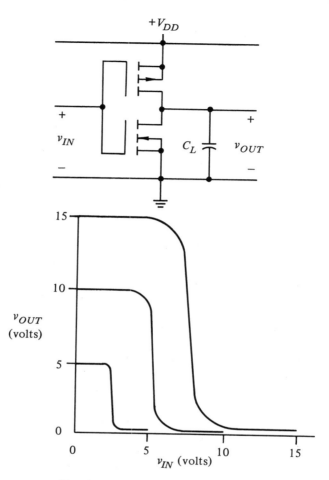

Fig. 2.32   The CMOS Amplifier

impedance levels of CMOS circuits results in poor rejection of high frequency capacitively coupled noise.

Buffered CMOS consists of a logic circuit of Fig. 2.33 followed by one or two of the amplifiers shown in Fig. 2.32. Only two transistors of the final amplifier need to be physically large to obtain required output drive capability. The area required may be less than that of a basic CMOS circuit, while amplification is increased. Buffering results very nearly in an ideal static transfer characteristic. The response of buffered CMOS to slowly varying input signals is far superior to basic CMOS circuits. Unfortunately, the extra amplifier stages increase propagation delay.

The symmetry of MOS transistors and the availability of both p- and n-channel devices in CMOS technology permit a circuit that is very much like a mechanical switch in that it offers low or high resistance between two points and it permits currents in either direction when it is conducting. The *transmission gate* consists of a p- and an n-channel transistor connected

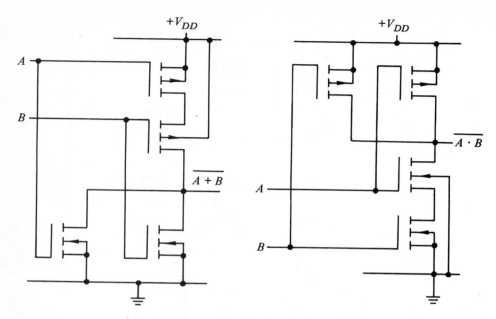

FIG. 2.33   CMOS NOR and NAND Circuits

as shown in Fig. 2.34. When control signal C is low, both devices place very high resistance between points A and B. When C is high, both devices place low resistance between A and B. Also, either end of each device may serve as its source. The resistance may be switched from 10^9 ohms to a few hundred ohms in approximately 30 nsec.

Semiconductor technology has, and apparently, will continue to advance rapidly. This section is not a substitute for continuing surveillance of the literature for reports of new devices and circuits with lower delays and power requirements. Hopefully this section will make it possible to understand and appreciate such reports.

2.2   BOOLEAN ALGEBRA: A FORMAL DEFINITION

Boolean algebra, like set theory, is an axiomatic branch of mathematics: axioms are selected and theorems are developed from these axioms and other already proven theorems. Such mathematical systems often have a bearing on a subset of real world situations and hence are of importance in solving problems associated with those situations. We must never forget that results predicted by Boolean algebra, and other algebras for that matter, will be observed in the real world only if the axioms upon which the algebra is based are satisfied in that real situation.

Boolean algebra is based upon: (1) a set P of two or more distinct elements α, β, ..., (2) two rules denoted $+$ and $\cdot$ here for combining elements of P to form other elements of P, (3) an equivalence relation denoted $=$ here on the members of P, and (4) a set of axioms that define the

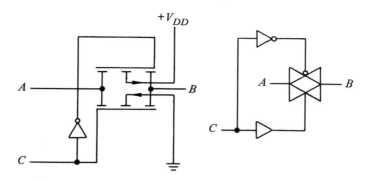

FIG. 2.34   The CMOS Transmission Gate

combining operators and place limits on the membership of P. Set P must have at least two members. In this section a two-member set will be emphasized even though nothing in the section is restricted to this small set; the next section will illustrate sets with larger memberships. The items in the binary set and the symbols used to represent those items are arbitrary so we will use "0" and "1" as the two items.

$$P_2 = \{0, 1\} \tag{2.21}$$

The members of set P will also be referred to as *values* and *constants*.

$A, B, C, \ldots$ will be used as *variables* to express the axioms and theorems of Boolean algebra. As such, each represents any element of P at each instant of time. We say that the variable has "the value of" the element it is representing. Variables are important because we may think of their assuming different values at different times, or we may think of them as representing any (an arbitrary) member of P, as we prefer or find convenient. Thus variable A may be assigned the value α at one instant and the value β at another, or we may use A to symbolize any element of P without being specific. We will write $A \in P$ meaning "A takes the value of some (or any) member of P."

If two or more members of set P are to be treated as being "the same," the equivalence relation is recorded by writing a selected symbol between them. A set of all equivalent elements is an *equivalence* class. Each member of P belongs to one equivalence class. Since in most useful sets, each element is equivalent only to itself, $\alpha = \alpha$, $\beta = \beta, \ldots$, we will write "unique element" as an abbreviation of "unique equivalence class". Variables may be related by the same equivalence relation. If variables A and B have the same value at all times, we record this by writing

$$A = B \tag{2.22}$$

Then, since the equivalence is known, we can replace appearances of A with B, and vice versa if we like. This is the mathematical *law of substitution*.

We will find each member of set P to be related to a unique other member of P in Boolean algebra. This *complement* relation is denoted with an overbar. Thus, if constant α is the complement of β we write

$$\alpha = \bar{\beta} \tag{2.23}$$

α is the value complementary to β where "complementary value" is yet to be defined. Similarly, one variable A may always take the value complementary to that of another variable B.

$$A = \bar{B} \tag{2.24}$$

At times we wish to speak of either a variable or its complement, not both simultaneously, without being specific. We then speak of a *literal*, which is a symbol representing a variable or its complement. It can be whichever we like but not both simultaneously.

A literal or two or more literals combined with the $\cdot$ operator constitute a *product term*. A, $A \cdot B$, and $1 \cdot \bar{A} \cdot B$ are examples. A *sum term* consists of a literal, or two or more literals combined with the $+$ operator. A, $A + \bar{B}$, and $\bar{A} + B + C$ are examples. Clearly the "product" is a "logic product" and the "sum" is a "logic sum;" we are not discussing arithmetic here.

We begin to define *Boolean expressions* by stating that:

a constant is a Boolean expression, and
a variable is a Boolean expression.

Now we generalize. If Be_1 and Be_2 are Boolean expressions, then

$\overline{Be_1}$ is a Boolean expression,
(Be_1) is a Boolean expression,
$Be_1 + Be_2$ is a Boolean expression, and
$Be_1 \cdot Be_2$ is a Boolean expression.

For now we will conclude the definition by saying that no other expressions are valid Boolean expressions. Soon we will accept other symbols such as the exclusive-or $\oplus$, and not insist that the $\cdot$ symbol always appear.

Since "A" is a Boolean expression, then so is "$\bar{A}$". "B" is a Be and hence "$\bar{A} \cdot B$" is a Be. $\overline{\bar{A} \cdot B}$ and $\overline{\bar{A} \cdot B} + C$ are Boolean expressions constructed in turn. We can construct Bes from variables and constants, or given a proposed Be we can verify its validity using the definition above. $A + \cdot B$ and $A - B$ are not Boolean expressions. The axioms of Boolean algebra tell us some Boolean expressions to be treated as equivalent. Recording such an equivalence gives a *Boolean equation*.

$$Be_1 = Be_2 \tag{2.25}$$

The law of substitution permits us to replace appearances of Be_1 with Be_2, or vice versa, if we choose to do so. We use this approach to derive theorems—other equations that indicate the equivalence of Boolean expressions. We also write Boolean equations when we wish to introduce new variables with desired properties. Variable Z may be defined to be equivalent to the Boolean expression "$A \cdot B$." If $Z = A \cdot B$ is declared, then $A \cdot B + C$ may also be written $Z + C$, and $X + Y \cdot Z$ may be written $X + Y \cdot (A \cdot B)$. Substitution is a very fundamental and powerful aspect of algebra.

A Boolean expression is said to be in *sum-of-products* form if it consists of product terms combined with the $+$ operator. $A + (B \cdot C)$ and $(A \cdot \bar{B}) + (B \cdot C) + \bar{D}$ are examples. The form is *product-of-sums* if sum terms are combined with the $\cdot$ operator. $A \cdot (B + \bar{C})$ is an example. We begin to sense a dualism: sum versus product terms, *sop* versus *pos* Boolean expressions. Where one has " $\cdot$ "s, the other has " $+$ "s, and so forth. More completely, the *dual* of a Boolean expression is another *Be* obtained by:

> replacing all 0 with 1
> replacing all 1 with 0
> replacing all $+$ with $\cdot$
> replacing all $\cdot$ with $+$

It is advisable to enclose all sum and product terms in parentheses before beginning to replace so that the resulting expression is not misinterpreted. Thus the dual of

$$(A \cdot B) + (\bar{B} \cdot (C + D))$$

is

$$(A + B) \cdot (\bar{B} + (C \cdot D))$$

In general, dual Boolean expressions are not equivalent or complements of each other. The value of recognizing duals appears as we turn to the axioms and theorems of Boolean algebra. Notice that the axioms and theorems appear in pairs, each member of the pair being the dual of the other member.

Many different sets of axioms may be employed in the development of Boolean algebra. The axioms of a set should be (1) consistent, so that no contradictions arise in the logical development of the algebra by deduction; (2) mutually independent and therefore free of redundancy; and (3) simple, to facilitate development of the algebra. The set presented below was selected with priority given to this third point; its members should appear reasonable and even familiar to us.

Definition: A *Boolean Algebra* is a closed mathematical system consisting of a set P of two or more distinct elements, and two binary operators, denoted $+$ and $\cdot$, which satisfy the following axioms. ["Closed" means that the result of operating (with $\cdot$ or $+$) on any members of P is itself a member of P.)

AXIOM 1. *The operators are* **commutative**.
For every $A \in P$ and $B \in P$: $\quad A + B = B + A$
$$A \cdot B = B \cdot A$$

AXIOM 2. *The operators are* **associative**.
For every $A, B, C \in P$: $\quad (A + B) + C = A + (B + C)$
$$(A \cdot B) \cdot C = A \cdot (B \cdot C)$$

These two axioms are familiar because they are valid in the algebra of real numbers.

AXIOM 3. *The operators are **distributive** over each other.*
For every $A, B, C \in P$: $\quad A + (B \cdot C) = (A + B) \cdot (A + C)$
$$A \cdot (B + C) = A \cdot B + A \cdot C$$

The second of this pair of distributive rules is again familiar from the algebra of real numbers. The first is not and hence should receive special attention!

Note also that a hierarchy of operators is proposed in the second part of this axiom by the omission of parentheses around the terms $A \cdot B$ and $A \cdot C$. Thus when no parentheses are present to indicate otherwise all $\cdot$ combinations are to be performed before any $+$ combinations. This is a carry-over from arithmetic, where multiplication is taken as a higher operation than addition, i.e. $4 \cdot 5 + 2 = 22$, not $4 \cdot 5 + 2 = 4(5 + 2) = 28$. Another carry-over will be noted subsequently: when the meaning is clear, the symbol $\cdot$ will be deleted and $A \cdot B$ written simply AB.

AXIOM 4. *Unique elements **0** and **1** exist.*
For every $A \in P$ there exists a unique element of P named 0 and another named 1 such that:

$$A + 0 = A$$
$$A \cdot 1 = A$$

This axiom asserts that while set P may contain other elements, it must contain two elements that are unique in their relation to other elements of P. The special names 0 and 1 are universally given to these elements to emphasize their unique relationship with other elements.

AXIOM 5. *The **complement** of each element exists.*
For every $A \in P$ there exists a related element in P named $\bar{A}$ such that:

$$A + \bar{A} = 1$$
$$A \cdot \bar{A} = 0$$

A second element denoted $\bar{A}$ is said to exist for every element A; it is related to A and the unique elements 0 and 1. Set P must therefore consist of pairs of elements.

While a profusion of theorems can be developed from these axioms, only a sample will be presented here to illustrate deductive development. Other theorems will be presented in subsequent sections that expand upon their use.

Duality plays an important role in the following development. Only the proof for one part of each theorem will be presented; the proof of the second part may be obtained by replacing each Boolean expression of the given proof with its dual.

THEOREM 1. *Operations with 0 and 1.*

$$0 + 0 = 0 \qquad 1 + 0 = 1$$
$$1 \cdot 1 = 1 \qquad 0 \cdot 1 = 0$$

Proof: In Axiom 4, A represents any member of P and hence represents elements 0 and 1. Direct substitution in Axiom 4 gives these sets of relationships.

If the second of these sets is examined with Axiom 5 in mind, the 1 can be observed to satisfy all requirements as the complement of 0, and vice versa.

COROLLARY:

$$\bar{0} = 1$$
$$\bar{1} = 0$$

THEOREM 2.

$$A + 1 = 1$$
$$A \cdot 0 = 0$$

Proof:

$A + 1 = (A + 1) \cdot 1$	Axiom 4
$= (A + 1) \cdot (A + \bar{A})$	Axiom 5
$= A + 1 \cdot \bar{A}$	Axiom 3
$= A + \bar{A}$	Axiom 4
$= 1$	Axiom 5

COROLLARY: *More operations with 0 and 1.*

$$0 + 1 = 1 \qquad 1 + 1 = 1$$
$$1 \cdot 0 = 0 \qquad 0 \cdot 0 = 0$$

Proof: Substitution of 0 and 1 for A in Th. 2 gives these results.

Exponents and multiplicative constants found in arithmetic have no place in Boolean algebra.

THEOREM 3. *The operators are **idempotent**.*

$$A + A = A$$
$$A \cdot A = A$$

Proof:

$A + A = (A + A) \cdot 1$	Axiom 4
$= (A + A) \cdot (A + \bar{A})$	Axiom 5

$$= A + (A \cdot \bar{A}) \qquad \text{Axiom 3}$$
$$= A + 0 \qquad \text{Axiom 5}$$
$$= A \qquad \text{Axiom 4}$$

Some of the results of Theorems 1 and 2 can be obtained by substituting 0 and 1 for A in Th. 3. We can develop theorems in many ways.

THEOREM 4. *Absorption.*

$$A + A \cdot B = A$$
$$A \cdot (A + B) = A$$

Proof:

$$A + AB = A \cdot 1 + AB \qquad \text{Axiom 4}$$
$$= A \cdot (1 + B) \qquad \text{Axiom 3}$$
$$= A \cdot (B + 1) \qquad \text{Axiom 1}$$
$$= A \cdot 1 \qquad \text{Th. 2}$$
$$= A \qquad \text{Axiom 4}$$

THEOREM 5. *Simplification.*

$$A + \bar{A}B = A + B$$
$$A(\bar{A} + B) = AB$$

Proof:

$$A + \bar{A}B = (A + \bar{A})(A + B) \qquad \text{Axiom 3}$$
$$= 1 \cdot (A + B) \qquad \text{Axiom 5}$$
$$= A + B \qquad \text{Axiom 4}$$

THEOREM 6. *The complement of A is unique.*

Proof: Assume that A has two complements, $\bar{A}_1$ and $\bar{A}_2$. Then:

$$A + \bar{A}_1 = 1 \qquad \qquad A + \bar{A}_2 = 1$$
$$A \cdot \bar{A}_1 = 0 \qquad \qquad A \cdot \bar{A}_2 = 0 \quad \text{all by} \quad \text{Axiom 5}$$
$$\bar{A}_1 = 1 \cdot \bar{A}_1 \qquad\qquad\qquad\qquad\quad \text{Axiom 4}$$
$$= (A + \bar{A}_2)\bar{A}_1 \qquad\qquad\qquad \text{Axiom 5}$$
$$= A\bar{A}_1 + \bar{A}_2\bar{A}_1 \qquad\qquad\qquad \text{Axiom 3}$$
$$= 0 + \bar{A}_2\bar{A}_1 \qquad\qquad\qquad\quad \text{Axiom 5}$$
$$= A\bar{A}_2 + \bar{A}_2\bar{A}_1 \qquad\qquad\qquad \text{Axiom 5}$$
$$= \bar{A}_2 A + \bar{A}_2\bar{A}_1 \qquad\qquad\qquad \text{Axiom 1}$$
$$= \bar{A}_2(A + \bar{A}_1) \qquad\qquad\qquad \text{Axiom 3}$$
$$= \bar{A}_2 \cdot 1 \qquad\qquad\qquad\qquad\quad \text{Axiom 5}$$
$$= \bar{A}_2 \qquad\qquad\qquad\qquad\qquad \text{Axiom 4}$$

THEOREM 7. *Involution*.

$$\bar{\bar{A}} = A$$

Proof: What is the complement of $\bar{A}$?
By Axiom 5, A is one complement of $\bar{A}$.
By Th. 6 the complement is unique, so A is the only complement of $\bar{A}$.

THEOREM 8. *De Morgan's laws*.

$$\overline{(A + B)} = \bar{A} \cdot \bar{B}$$

$$\overline{(A \cdot B)} = \bar{A} + \bar{B}$$

Proof: Since

$$
\begin{aligned}
(\bar{A} \cdot \bar{B}) + (A + B) &= (\bar{A} \cdot \bar{B} + A) + B & &\text{Axiom 2} \\
&= (\bar{B} + A) + B & &\text{Th. 5} \\
&= A + (\bar{B} + B) & &\text{Axioms 1, 2} \\
&= 1 & &\text{Axiom 5, Th. 2}
\end{aligned}
$$

and

$$
\begin{aligned}
(\bar{A} \cdot \bar{B}) \cdot (A + B) &= \bar{A} \cdot \bar{B} \cdot A + \bar{A} \cdot \bar{B} \cdot B & &\text{Axiom 3} \\
&= 0 + 0 & &\text{Axiom 5} \\
&= 0 & &\text{Th. 1}
\end{aligned}
$$

$(A + B)$ satisfies Axiom 5 as the complement of $\bar{A} \cdot \bar{B}$. By Th. 6, $A + B$ is the only complement of $\bar{A} \cdot \bar{B}$. Theorem 7 completes the proof.

Looking back we can see how an algebra is formally developed. First we assume a set of elements and relationships (axioms) between those elements. Then we propose other relationships (theorems) and prove, if we can, that these relationships are valid under the assumption of the axioms. Proof is achieved by deduction—usually converting one side of a proposed equality to the other in a step-by-step manner using only axioms and theorems already proven—or by contradiction where we assume something to be true and then show that it can not be so. Other methods of proving theorems are also available, and we will see them, but at this point we may question the value of all of this formal mathematics.

2.3 EXAMPLES OF BOOLEAN ALGEBRA

The algebra with set $P_2 = \{0, 1\}$ is by far the most widely known and used Boolean algebra. Variables are binary, just as we found to be the case in the electronic circuits of Section 2.1. The truth tables of Figs. 1.6, 2.2, and 2.4 fully define the two combining operators, and the complement of each member of P_2, just as do Theorems 1 and 2 of the previous section. Our interest in Boolean algebra arises from this fact and its ramifications; that is, Boolean equations express the logic behavior of binary switching circuits.

Every combinational logic network *realizes* a switching function. The logic diagram of that network and its corresponding Boolean equation are mathematical models of the network as they express that switching function. Often they are taken to model more than just the function. Gate symbols in a logic diagram are taken to symbolize gate circuits in a network. Each AND, OR, XOR, etc., operator in a Boolean equation is taken to symbolize a gate circuit. Actual networks, logic diagrams, and Boolean equations are considered to be very closely related; and the simplicity of going from one to another is one of the things that make Boolean algebra so useful. Boolean algebra provides means of easily altering a Boolean equation without modifying the function expressed. Each new Boolean equation describes a new network; all networks so described realize the same switching function.

Algebra P_2 is so powerful that many practicing logic designers are unaware that other Boolean algebras exist, even when they use them. In Chapter 1, we found binary sequences to be very useful in describing networks of parallel terminals and logic elements. Boolean algebras may be based upon any complete set P_2^n of binary n-tuples.

$$P_2^n = \{all\ binary\ n\text{-}tuples\} \tag{2.26}$$

In Section 1.3, we found that there are 2^n different binary n-tuples. Thus for $n = 2$

$$P_2^2 = \{(0, 0), (0, 1), (1, 0), (1, 1)\} \tag{2.27}$$

or abbreviating,

$$P_2^2 = \{00, 01, 10, 11\} \tag{2.28}$$

Given a set P_2^n, we must find **0** and **1** elements, a unique complement for each element, and **·** and **+** operators (note bold face); all of which simultaneously satisfy Axioms 1 through 5 of the previous section. This is not hard to do, especially with algebra P_2 in mind.

Let $A = (A_1, \ldots, A_n)$, $B = (B_1, \ldots, B_n)$, etc., be variables in this 2^n-value algebra. Then examine the following definitions that utilize the operators of the P_2 algebra in the bit-by-bit fashion we discussed in Chapter 1.

$$\begin{aligned}
\mathbf{0} &= (0, \ldots, 0) \\
\mathbf{1} &= (1, \ldots, 1) \\
\bar{A} &= (\bar{A}_1, \ldots, \bar{A}_n) \\
A \cdot B &= (A_1 \cdot B_1, \ldots, A_n \cdot B_n) \\
A + B &= (A_1 + B_1, \ldots, A_n + B_n)
\end{aligned} \tag{2.29}$$

For P_2^2, these definitions yield the operator tables of Fig. 2.35. Note the symmetry of the **·** and **+** tables about their major diagonals—the Commutative axiom is seen. Notice in the first column of the $+$ table and the last column of the $\cdot$ table that Axiom 4 is satisfied. The $+$ combination of each value and its complement yields the **1** value; their $\cdot$ combination is the **0** value. Thus Axiom 5 is satisfied. That the Associative and Distributive axioms are satisfied may be shown by working all possible examples, but this extensive effort can be saved. If we are willing to argue that since the P_2 operators are associative and distributive, then the $\cdot$ and $+$ operators are also in each coordinate position and, hence, in general. All of the theorems are also satisfied; try a few for practice.

A	$\bar{A}$	$+$	B 00	01	10	11	$\cdot$	B 00	01	10	11
00	11	00	00	01	10	11	00	00	00	00	00
01	10	A 01	01	01	11	11	A 01	00	01	00	01
10	01	10	10	11	10	11	10	00	00	10	10
11	00	11	11	11	11	11	11	00	01	10	11

FIG. 2.35   Operators for Algebra P_2^2

We used these extended operators with vector notation in Chapter 1. To see another specific, practical example of the use of P_2^n algebra, let us consider truth tables for a moment. In Chapter 1, we used truth tables to define the operation of gates. In the next section we will indicate in more detail that a truth table can be used to record any Boolean function. On the left we list, usually as ascending binary integers, all of the possible combinations of values that the variables may assume. On the right we list the corresponding values of a variable or variables that depends on the left variables. The left variables are *independent* while the right variables are *dependent*. Each dependent variable is a function of the independent variables. If we have n independent variables, there will be 2^n combinations of their values and 2^n rows in the truth table. Each dependent variable is described by a column of 0's and 1's that can be thought of as a binary 2^n-tuple.

Let us take an example with $n = 2$. Let A and B be the independent variables. Truth tables for dependent variables C, D, and E are then

A	B	C	D	E
0	0	0	1	0
0	1	0	0	0
1	0	1	1	1
1	1	1	0	0

A quick comparison indicates that $C = A$ and $D = \bar{B}$ in this example. Or, thinking in terms of tuples, $A = 0011$, $B = 0101$, $C = 0011$, and $D = 1010$. D is the complement of B in the P_2^4 algebra with its operators of Eq. (2.29). $E = 0010$ can be seen to be the $\cdot$ combination of C and D in the same algebra. Or, by substitution, $E = A \cdot \bar{B}$, and we see that E takes the value 1 only when $A = 1$ and $B = 0$—as the P_2 algebra would have it.

Some logic designers prefer to construct truth tables for Boolean functions by ANDing, ORing, and complementing independent variable columns in a "row-by-row" fashion. They are using the operators of a P_2^n algebra when they do so, just as we did in Chapter 1 when we operated on vector variables.

We should not be led to believe that only the operators of Eq. (2.29) may be used to form a Boolean algebra on a set P_2^n. Any tuple can be used as the **0** element; any other one can serve as the **1**. Suitable operators can be defined although the definition may not be easy to record or use. In fact, any set of 2^n unique items can be used. One way to find suitable operators consists of associating each item with a binary n-tuple and then of using the operators from Eq. (2.29) to derive operator tables expressed in terms of the new elements. To clarify this technique, let us use set $G_4 = \{\alpha, \beta, \gamma, \delta\}$ and the association

α	01
β	00
γ	10
δ	11

We see that β will be the **0** element and δ the **1**. The complement of each item is:

A	$\bar{A}$
α	γ
β	δ
γ	α
δ	β

The AND and OR tables are formed by rewriting those tables of Fig. 2.35.

One final point should be observed. The equivalence relation on set P need not be the common one with each element equivalent only to itself. Suppose that we have set $P_3 = \{0, 1, 2\}$ with the equivalence relation ($\equiv$) defined as follows:

$$0 \equiv 0 \quad 1 \equiv 1 \quad 2 \equiv 2 \quad 1 \equiv 2$$

Since $1 \equiv 2$, we can replace 1 with 2, and vice versa, whenever we like. We can define suitable AND and OR operators that may not appear to obey the Commutative axiom and Idempotent theorem.

$\cdot$	0	1	2
0	0	0	0
1	0	1	2
2	0	2	2

$+$	0	1	2
0	0	1	2
1	1	2	1
2	2	2	1

They do, however, under the given equivalence relationship. If we replace all 2's with 1's, these operators are found to be the operators of P_2 algebra.

When the equivalence relationship on a set is not the common one, why do we not eliminate all but one of the equivalent items? The undersirability of doing so can be illustrated with the following set.

$$Z = \{all\ Boolean\ expressions\} \qquad (2.30)$$

The items of Z are Boolean expressions; all Boolean expressions are members of Z. Clearly Z is an infinite set. The subset with members A, $A \cdot A \ldots$ is infinite and suggests many other infinite subsets. Z remains infinite even if we restrict the number of variables used to form members of Z. $0, 0 \cdot 0, 0 \cdot 0 \cdot 0, \ldots$ are all Boolean expressions, and no variables are used in writing them.

We desire a Boolean algebra based upon set Z. Complement, $\cdot$ and $+$ operators that give other members of Z from given members are needed. While we could attempt to describe such operators in terms of tables that indicate the results of combining each pair of members of Z, it is very practical to think of the operators as being identical to those used to describe the members of Z. We complement, AND, and OR Boolean expressions to find other Boolean expressions. The equivalence relation that we insist on using equates all Boolean expressions that the axioms and theorems of Boolean algebra relate as equivalent.

If we eliminated the redundancy and insisted upon having only one Boolean expression for each switching function, we would have a relatively uninteresting algebra. It would not show us the variety of combinational logic circuits that realize a switching function. The algebra could not be the powerful design tool that we saw at the beginning of this section.

2.4 SWITCHING FUNCTIONS

A *function* (*mapping* and *transformation* are other names) is an indication of the values of dependent variables for all combinations of values of independent variables. Our interest in functions is restricted here. We are interested in mathematical descriptions of what combinational logic circuits do. Since such circuits are most often considered to have binary input and output signals, we will defer generalizing and talk in terms of binary variables.

A *switching function*[1] of n variables, $x_1, x_2, \ldots, x_n$, each taking values from the set $P_2 = \{0, 1\}$, is a rule that associates every n-tuple of valued variables $(x_1, x_2, \ldots, x_n)$ with an m-tuple of valued variables $(z_1, z_2, \ldots, z_m)$, where each z_i takes values only from P_2. The mathematical notation used to express this definition is:

$$f: P_2^n \longrightarrow P_2^m \qquad (2.31)$$

---

[1]The variables of *Boolean functions* take values from any set P upon which a Boolean algebra may be based. It is not unusual to overlook the small difference between switching and Boolean functions.

A switching function mathematically describes the logic performance of a combinational switching circuit with n input terminals $x_1, x_2, \ldots, x_n$ and m output terminals $z_1, z_2, \ldots, z_m$. We assume that there exist n binary signal sources $x_1, x_2, \ldots, x_n$ and their connections to the input terminals of the circuit, which generates m binary signals $z_1, z_2, \ldots, z_m$. We say that the circuit *realizes* or is a realization of the switching function. To deal efficiently with switching circuits, a switching function and the circuit which realizes it must become "the same" to us. We must be able to draw the block diagram of a circuit which realizes a given switching function (synthesis), and write the function which describes a given switching circuit (analysis).

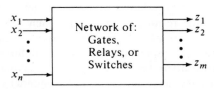

FIG. 2.36   Model of a Combinational Switching Circuit

An *input symbol*, denoted x^i in general, is an n-tuple of valued input variables $(x_1, x_2, \ldots, x_n)$. One orderly way to name input symbols is to view the n-tuple as the binary expression of an integer and let i be the decimal expression of that integer. Thus if $n = 5$,

$$x^0 = (x_1 = 0, x_2 = 0, x_3 = 0, x_4 = 0, x_5 = 0)$$
$$x^1 = (0, 0, 0, 0, 1)$$
$$\vdots$$
$$x^{31} = (1, 1, 1, 1, 1)$$

In general, any of the 2^n possible input symbols may be presented to a combinational circuit. But in some cases it is known that some input symbols will never be presented. The *input alphabet*, X, of a combinational circuit is the set of all input symbols that may appear at the input terminals of the circuit.

$$X = \{\xi^1, \xi^2, \ldots, \xi^p\} \subseteq \{x^0, x^1, \ldots, x^{2^n-1}\} = P_2{}^n \qquad \textbf{(2.32)}$$

Similarly, an *output symbol*, z^i, of a circuit is an m-tuple of valued output variables $(z_1, z_2, \ldots, z_m)$, and the *output alphabet*, Z, is the set of all output symbols that the circuit can generate.

$$Z = \{\zeta^1, \zeta^2, \ldots, \zeta^q\} \subseteq \{z^0, z^1, \ldots, z^{2^m-1}\} = P_2{}^m \qquad \textbf{(2.33)}$$

When $m = 1$, the output alphabet is very small: $Z = \{0, 1\}$.

The use of subscripted variables, x_i and z_i, is very desirable when treating a general case. But when dealing with a specific circuit, more concise names such as A, B, ..., or names which describe a signal such as $START$, $LOAD$, ADD, etc. may be preferred. In still other cases many terminals may bear signals of identical significance, and names such as $A_1, A_2, ..., A_{15}$ may be appropriate. The terminals which bear the bits of a binary sequence A constitute such a case. We will use all such names to identify terminals. Further, when the use of traditional subscripts is awkward, square brackets [ ] will be used to enclose subscripts. For example, $A[1], A[2], ...$. Such notation is generally necessary today when we attempt to present Boolean equations to a digital computer, and we will eventually consider doing just that.

Methods of Expressing Switching Functions

The specification of a particular switching function may take any of a number of forms, several of which we have already seen. Boolean algebra provides one form. Defining a (dependent) variable to be equivalent to a Boolean expression by writing a Boolean equation specifies a switching function. Such an equation does indicate the value of the dependent variable for each combination of values of the independent variables that appear in the Boolean expression. For example, a specific rule that relates the binary variables, A, B, and C (or x_1, x_2, and x_3, if you prefer), to the binary variable F is

$$F = A + B \cdot \overline{C}$$

Given specific values for A, B, and C and the axioms and theorems of Boolean algebra, a specific value of F can be ascertained. The process of doing so is called *evaluation*. Equivalent Boolean expressions must always evaluate to the same value. We use this fact to "find" the value of the dependent variable. If we choose to assign value 0 to all independent variables, $A = B = C = 0$, then we may substitute that value for the variables.

$$F = 0 + 0 \cdot \overline{0} \qquad \text{NOT combination first}$$
$$= 0 + 0 \cdot 1 \qquad \text{Then AND combinations}$$
$$= 0 \qquad \text{Finally, OR combinations}$$

An exhaustive evaluation of F for all possible combinations of values A, B, and C can assume provides the necessary data for a second representation of a switching function—the truth table. But first, how many different n-tuples of valued binary variables exist? Knowledge of this number is important. It tells us how many unique n-tuples exist and thus how much work is to be done. The first variable can assume two values; the second can assume two values. Thus the first two variables offer $2 + 2 = 4$ possible combinations. The remaining variables are also binary, so in general

TABLE 2.2 TRUTH TABLE SPECIFICATION OF SWITCHING FUNCTION F.

A	B	C	F
0	0	0	0
0	0	1	0
0	1	0	1
0	1	1	0
1	0	0	1
1	0	1	1
1	1	0	1
1	1	1	1

$$2 \times 2 \times \cdots \times 2 = 2^n \tag{2.34}$$

unique combinations exist. For example, $n = 3$; thus 8 input symbols exist. Table 2.2 shows these input symbols and the value of F for each, and thus is a specification of the mapping of A, B, C to F. Each row of the table presents a unique 3-tuple and an associated 1-tuple.

The next section presents a method for deriving a Boolean expression of a function from a truth table representation.

Circuit diagrams offer a third means of specifying a switching function. Figure 2.37(a) utilizes the symbolism introduced in Sec. 1.2 to specify the function F. Note the direct correspondence between the Boolean equation expression of F and the circuit diagram shown. One of the values of the Boolean equation is the ease with which a corresponding circuit may be derived. When designing logic, Boolean expressions may be manipulated to a desired form, and then the circuit diagram drawn as a final effort. Each product term suggests an AND gate; each sum term suggests an OR gate. Each complemented variable or sum or product term suggests an inverter. For the most part, each of the suggested logic blocks can be drawn. The parentheses or hierarchy of operation determine how the gates are interconnected. Here, for example, the OR of A and the logical product of B

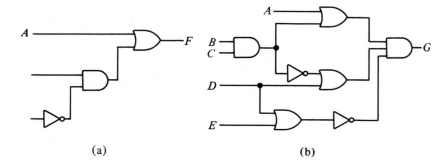

(a)   (b)

FIG. 2.37 Circuit Specification of Switching Function $F = A + B\bar{C}$.

and $\bar{C}$ is specified; so the output line of the AND gate must be connected to an input terminal of the OR gate. Because the OR operator is commutative we need not be concerned about the specific input terminal of the OR gate to which the connection is made.

When a sum or product term, or complemented variable or term, appears more than once in a Boolean expression, it is usually not necessary to show its corresponding gate more than once in a network. Thus the diagram of Fig. 2.37(b) might be drawn from the following Boolean expression, in which duplicated terms are underlined for emphasis.

$$G = (A + \underline{B \cdot C}) \cdot (D + \overline{\underline{B \cdot C}}) \cdot (\overline{D + E})$$

An n-tuple of valued variables $(x_1, x_2, \ldots, x_n)$ can be used to determine a point in an n-dimensional space: the value of each variable x_i determines the distance of the point from the origin along the ith axis of the n-dimensional space. $(A, B, C) = (0, 0, 0)$ locates the origin of a 3-dimensional space. $(A, B, C) = 011$ locates the point 0 units from the origin along the first axis, unit distance along the second axis, and unit distance from the origin parallel to the third axis. Figure 2.38 shows the eight vertices of a 3-dimensional *unit cube*, the vertices of a 1-dimensional unit cube, and the mapping of the vertices of the input variable 3-dimensional space to the vertices of the output variable space dictated by the example function. Compare this mapping with Table 2.2: it presents exactly the same information but in a graphic form. J. P. Roth [4] first introduced and explored such topological representations of switching functions.

Figure 2.38 is a rather confusing way to represent a switching function and is therefore usually simplified in the manner of Fig. 2.39. Vertices of the n-dimensional cube or n-cube which are mapped to the 1 are emphasized. Alternatively, there is often advantage in deleting those vertices which are mapped to 0 and showing only the adjacencies which exist between vertices mapped to the 1.

We will eventually come to realize that the main value of the topological representation lies in the terminology and means of visualizing switching functions that it provides. Actually, a modification (generalization) of the truth table suggested by the topological representation is a most valuable representation of switching functions. From Fig. 2.39(b) we clearly see that

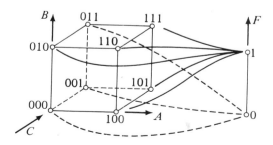

FIG. 2.38 N-space Mapping F: $\{0, 1\}^3 \rightarrow \{0, 1\}^1$

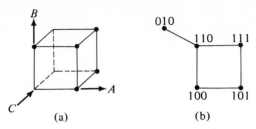

FIG. 2.39   Cube Representations of a Switching Function

five different input n-tuples cause the output variable F to have a value of 1. These five conditions appear in the truth table of Table 2.2 also, but the topology emphasizes the distances between these elements, and that they form certain geometric figures. The set of these 5 n-tuples is referred to as an *ON-array* of function F in Fig. 2.40, i.e., a set of input symbols that cause the function to have the value of 1.

The vertices not shown in Fig. 2.39(b) correspond to input symbols for which $F = 0$. The set of these input symbols is an *OFF-array* for F. While the ON- and OFF-arrays of a function express the same information as a truth table, we will find it is not necessary to work with both arrays, and that they can be expressed more compactly. And as we work with more and more complex switching functions, we will find this n-tuple notation to be more concise and easy to work with than Boolean algebra.

Rather than having an input symbol designate a point in space, we may let it determine an area in which we place the corresponding value of the dependent variable(s). The collection of input symbol cells is called a *map*. We will organize the cells in the fashion of a *Karnaugh* map [2]. Given a large square or rectangular area, we begin to form the Karnaugh map by dividing the area in half. One half will contain all cells for which $x_1 = 0$; the other half then contains all cells for which $x_1 = 1$. Now we divide the overall area in half again, but in a manner that divides each of the x_1 areas in half, labelling one half the $x_2 = 0$ area and the other half $x_2 = 1$. We have divided the overall area into quarters, each quarter being labelled with a unique (x_1, x_2) 2-tuple. If we have more variables, the process continues. For each variable up to 4 (5 and 6 variable maps will be considered later), we divide the area in half so as to divide each prior cell in half.

Figure 2.41 shows Karnaugh maps for $n = 2$, 3, and 4. Several representative input symbols have been placed in the cells they designate. Notice

$$
\begin{array}{ccc}
A & B & C \\
\end{array}
\left\{
\begin{array}{ccc}
0 & 1 & 0 \\
1 & 1 & 0 \\
1 & 0 & 0 \\
1 & 0 & 1 \\
1 & 1 & 1 \\
\end{array}
\right\}
\qquad
\begin{array}{ccc}
A & B & C \\
\end{array}
\left\{
\begin{array}{ccc}
0 & 0 & 0 \\
0 & 0 & 1 \\
0 & 1 & 1 \\
\end{array}
\right\}
$$

(a)   (b)

FIG. 2.40   ON- and OFF-arrays of F

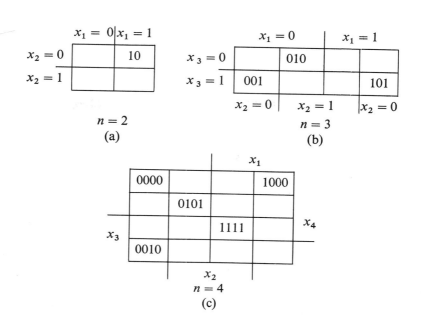

FIG. 2.41   Karnaugh Map Organization

that the labelling of map halves has been abbreviated on the $n = 4$ map, but is no less clear or precise than the more detailed labellings.

Input symbols that differ in one variable value only are said to be *adjacent* or unit distance apart. In Fig. 2.38, adjacent input symbols designate points connected by a line, an edge of the n-cube. These lines remain in Fig. 2.39 and clearly reveal a geometric structure to us. This same "distance" idea is to be preserved in the Karnaugh map. Cells that have a side in common are identified by adjacent input symbols in Fig. 2.41. But in the $n = 3$ map, the cells on the left, such as 001, are not next to those on the right, such as 101, while their input symbols are adjacent. The $n = 3$ map should be rolled into a cylinder, which is not a convenient surface to work on. Therefore, we must remember that left is next to right when we cut the cylinder and flatten it. The $n = 4$ map must be rolled into a cylinder and its ends brought together as forming a toroid. This is even less convenient so we remember left is next to right and top is next to bottom. Then in Fig. 2.41, cell 0000 is next to cells 0010 and 1000. M. Karnaugh first showed the importance of arranging cells so that adjacent input symbols identify cells that share a side and therefore are geometrically adjacent.

More generally, the Hamming distance between two input symbols is the number of positions in which they differ. Both Karnaugh maps and n-cube drawings display Hamming distance as geometric distance. In Fig. 2.41(c), 0000 and 0101 are two steps apart; 0010 and 1111 are three steps apart, on the surface of a toroid.

We place the value of a switching function in the cells of a Karnaugh map. The Karnaugh map can be thought of as a truth table in which the single long column of function values has been rearranged into a square or

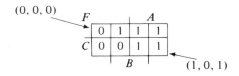

FIG. 2.42   Karnaugh Map Representation of the Switching Function
$$F = A + B \cdot \bar{C}$$

rectangular table. It is more, however, in that the order of the rows and columns of function values must be such that the geometric distances are correct. Thus the columns of the maps of Fig. 2.41(b) and (c) are ordered $x_1 x_2 = 00, 01, 11, 10$, which is not the order of ascending binary integers. In the Karnaugh map of example function F shown in Fig. 2.42, we see geometric patterns of cells containing 1's that are identical to the geometric structures (line and square) in the cube representation of F shown in Fig. 2.39(b). The importance of such geometric structures will be seen later.

To summarize, we can express Boolean functions as follows:

(1) truth tables,
(2) Boolean equations,
(3) logic diagrams,
(4) n- to m- space mappings,
(5) cubes in n-space,
(6) ON and OFF arrays, and
(7) Karnaugh maps.

Truth tables, n- to m-space mappings, and Karnaugh maps completely enumerate a switching function. Every possible input symbol and the corresponding output symbol are listed. Each switching function then has a unique truth table, space mapping, and Karnaugh map. (Permuting the rows in a truth table, axes in an n-space, and rows and columns of a Karnaugh map do not yield a different table, mapping, and map, respectively.) Logic diagrams provide logic network structures as well as switching functions. Output symbols may be calculated for given or assumed input symbols from logic diagrams. Boolean equations and arrays bridge the gap between enumeration and networks. We will see that they can enumerate a function. We have already seen that they can also be written to suggest network structures.

Are all of these representations really necessary? No, but each has advantages in certain situations and thus should be available when those situations arise. A mechanic could get along with one adjustable wrench, but can do better work more efficiently if he has a complete set of tools.

The Number of Switching Functions

How many switching functions are there? Since each has a unique truth table, we need only determine how many different truth tables exist. A complete truth table has 2^n rows, one for each of the 2^n input symbols of

n independent variables. Let α_d represent the value of the dependent variable for input symbol d.

x_1	x_2	$\cdots$	x_n	z
0	0	$\cdots$	0	α_0
0	0	$\cdots$	1	α_1
		$\vdots$		
1	1	$\cdots$	1	α_{2^n-1}

Each switching function is specified with a different dependent variable column, a different 2^n-tuple $(\alpha_0, \alpha_1, \ldots, \alpha_{2^n-1})$. How many binary 2^n-tuples are there? Equation (2.34) indicates that this count is $2^{(2^n)}$, where n is the number of independent variables. Thus, 2^{2^n} Boolean functions of n variables exist.

Table 2.3 lists the number of Boolean functions of n variables for small values of n only. The number of functions increases very rapidly with n. For n less than 3 the work involved in listing and studying all functions is modest and most informative, but we do not care to write down and study each of the functions of 4 or more variables.

In the previous section we examined truth table algebra. This is now seen to be an algebra in which the values are switching functions. We operate on functions to obtain other functions. Each value (switching function) has a complement (switching function with the complementary truth table), for example. Table 2.3 indicates the number of values available in this algebra. We also found an algebra based upon set Z, the set of all Boolean expressions. Since each Boolean expression describes a switching function when it is equated to a dependent variable, we now see that Z can be divided into subsets such that the members of each subset give the same switching function. There will be 2^{2^n} such subsets, of course. Each subset has an infinite number of members all of which are taken to be equivalent.

Functions of Two Variables

Table 2.4 suggests a systematic way of finding all of the Boolean functions of n variables (use each binary sequence of length 2^n as a truth table), and

TABLE 2.3   NUMBER OF
SWITCHING FUNCTIONS OF n
VARIABLES

n	2^n	2^{2^n}
0	1	2
1	2	4
2	4	16
3	8	256
4	16	65536
5	32	$\sim 4.295 \times 10^9$
6	64	$\sim 1.845 \times 10^{19}$

TABLE 2.4 BOOLEAN FUNCTIONS OF TWO VARIABLES

f_d	α_0	α_1	α_2	α_3	Sum-of-product Expression	Function name	Operator notation
f_0	0	0	0	0	0		
f_1	0	0	0	1	$A \cdot B$	AND	$A \cdot B$
f_2	0	0	1	0	$A \cdot \bar{B}$	INHIBIT	$A\,|\,B$
f_3	0	0	1	1	A		
f_4	0	1	0	0	$\bar{A} \cdot B$	INHIBIT	$B\,|\,A$
f_5	0	1	0	1	B		
f_6	0	1	1	0	$\bar{A} \cdot B + A \cdot \bar{B}$	EXCLUSIVE OR	$A \oplus B$
f_7	0	1	1	1	$A + B$	OR	$A + B$
f_8	1	0	0	0	$\bar{A} \cdot \bar{B} = \overline{A + B}$	NOR	$A \downarrow B$
f_9	1	0	0	1	$\bar{A}\bar{B} + AB$	EQUIVALENCE or EXCLUSIVE NOR	$A \equiv B$ or $A \odot B$
f_{10}	1	0	1	0	$\bar{B}$	NOT	$\bar{B}$
f_{11}	1	0	1	1	$A + \bar{B}$	IMPLY	$B \rightarrow A$
f_{12}	1	1	0	0	$\bar{A}$	NOT	$\bar{A}$
f_{13}	1	1	0	1	$\bar{A} + B$	IMPLY	$A \rightarrow B$
f_{14}	1	1	1	0	$\bar{A} + \bar{B} = \overline{A \cdot B}$	NAND	$A \uparrow B$
f_{15}	1	1	1	1	1		

lists the 16 functions of two variables. Each function is the complement of another. This can be seen by examining either the algebraic expressions or the truth tables.

The familiar AND (f_1), OR (f_7) and NOT (f_{10}) functions appear in Table 2.4 as three of many functions which have been given names. While we have been successful with only these three Boolean operators, and could proceed using only these operators, it is to our advantage to expand the set of switching operators with which we are conversant. The NAND (f_{14}) and NOR (f_8) functions are especially important because practical transistor switching circuits generate these functions rather than the AND and OR functions. These then are the operators which we actually will be using to describe transistor networks. We should recognize that by definition the NAND (for NOT-AND) and NOR (for NOT-OR) functions are complements of the AND and OR functions, respectively.

The EXCLUSIVE-OR (f_6) and EQUIVALENCE (f_9) functions are complements of each other. While electronic circuits to realize these functions directly are seldom employed, these functions do occur frequently in digital computer designs; expression of those designs in Boolean equations can be substantially abbreviated through use of symbols for these functions. Note that the truth table for the EXCLUSIVE-OR operator is equivalent to the sum bit component of the binary addition table. Thus the $\oplus$ operator may be expected to make an appearance when circuits to perform binary addition are being designed.

TABLE 2.5 FORMS OF 2-VARIABLE
FUNCTIONS

Form	Examples
0	0
$l_1 \cdot l_2$	$A \cdot B, \bar{A} \cdot B$
l_1	$A, \bar{B}$
$l_1 \cdot l_2 + l_1 \cdot l_2$	$A \oplus B$
$l_1 + l_2$	$A + B, \bar{A} + \bar{B}$
1	1

The INHIBIT (f_2 and f_4) and IMPLY (f_{11} and f_{13}) complementary functions are seldom expressed with the operator notation of Table 2.4 because reliable, high-speed electronic circuits that realize these functions are not available. This situation may change as new switching devices and circuits are developed.

The sum-of-products expressions in Table 2.4 suggest that some functions are similar to others. Functions $f_1(A \cdot B), f_2(A \cdot \bar{B}), f_4(\bar{A} \cdot B),$ and $f_8(\bar{A} \cdot \bar{B})$ are all products of two literals. All of these functions also have a single 1 in their truth tables. Let l_1 and l_2 be two literals. Then each of the 2-variable functions of Table 2.4 is of one of the six functional forms of Table 2.5.

Functional forms do have practical significance. If we have a gate network that realizes a switching function, then it can be used to realize all other functions of the same form. We only have to permute the terminals to which the input signals are supplied and/or provide complements of input signals. A circuit that realizes $A + B \cdot C$ realizes $B + AC$ when we interchange the A and B wires. It realizes $\bar{C} + A \cdot \bar{B}$ when $A, \bar{B},$ and $\bar{C}$ are supplied to the correct input terminals. When designing, we may concentrate on finding the best circuit for a given functional form and use it whenever we must realize a function of that form, rather than redesigning "best" circuits over and over again.

2.5 REPRESENTATION TRANSFORMS

Since truth tables, space mappings, Karnaugh maps, and the arrays we have looked at so far are all complete enumerations of a switching function, we can change from one form to another with ease. Just draw the lines and move the bits to the right place on the new form of expressing the function. Only practice is required to be able to make such changes quickly. We have seen how Boolean equations suggest logic diagrams, and diagrams give equations. The next section will explore this last change in greater detail. We saw in the last section how evaluating equations for each input symbol gives the truth table expression of the switching function. In this section we will find how to obtain truth tables and Karnaugh maps from Boolean equations more efficiently, and how to obtain Boolean equations from truth tables and Karnaugh maps.

Equations to Tables

We can quickly find the truth table or Karnaugh map from the equation of a switching function if we are willing to place the equation in sum-of-products or product-of-sums form. For this we use the axioms and theorems, primarily the distributive axiom. Let us examine the sum-of-products form equations first. For purposes of example we will use

$$F = \overline{A}\overline{B}\overline{C}\overline{D} + \overline{A}BD + AC$$

Each product term in a *sop* equation is called an *implicant* of the switching function; that is, knowing that a product term evaluates to 1 implies that the function has value 1 by Th. 2 ($A + 1 = 1$). Thus if $\overline{A}\overline{B}\overline{C}\overline{D}$ has value 1, then $F = 1$, and we do not have to examine the other product terms to reach this conclusion.

A product term has the value 1 when all of the variables that appear in true form have value 1, and all of the variables that appear complemented have value 0. In our example:

$$\overline{A}\overline{B}\overline{C}\overline{D} = 1 \text{ when } A = 0, B = 0, C = 0, D = 0$$
$$\overline{A}BD = 1 \text{ when } A = 0, B = 1, D = 1$$
$$AC = 1 \text{ when } A = 1, C = 1$$

Another way of recording this information is with tuples in a compressed ON-array.

A	B	C	D	
0	0	0	0	$\overline{A}\overline{B}\overline{C}\overline{D} = 1$
0	1	x	1	$\overline{A}BD = 1$
1	x	1	x	$AC = 1$

The x's indicate that the variable does not appear in the product term and that its value does not influence the value of the product term. Notice the pattern in this array. One tuple is present for each product term. If a variable appears true in the product term, a 1 appears in the tuple. Complements give rise to 0's. If a variable is missing, an x is placed in its column. The x is a "place marker" so that we do not have to write the heading down if we know which variable is associated with each column. This is especially important when we write the tuples with element values from the set $\{0, 1, x\}$ by themselves. Tuple "11" does not clearly indicate which variables are missing; "$1x1x$" does clearly indicate that the second and fourth variables are missing while the first and third appear in true form in the product term. While the use of "x" as a place marker is quite traditional, any symbol other than 0 and 1 may be used; $-$, *, ϕ, ?, etc., appear in the literature.

To summarize, product terms can be represented by tuples, and vice versa. Each independent variable of a switching function is assigned a position in tuples. In going from tuples to product terms, or vice versa, we use the code of Table 2.6.

Tuples enable us to place entries in truth tables quickly. If no x's appear in the tuple, then the tuple is the input symbol for which the product term

TABLE 2.6. TUPLE–PRODUCT–TERM CODE

Tuple Component	Product-Term Component
0	Complement
1	True Variable
x	Missing Variable

has value 1. We place a 1 in the dependent variable truth table column opposite that input symbol. Such tuples also identify a vertex of an n-dimensional cube. If we are preparing an n to m-space mapping of the function, such vertices are connected to the $F = 1$ point in the output space. In a cube representation such as that of Fig. 2.39(a), we emphasize such vertices.

Tuples with a single x, such as $01x1$ in our example, lead us to two input symbols (vertices). First we replace the x with a 0, then with a 1 to obtain two tuples that are free of x's. Since C does not appear in the product term $\bar{A}BD$, the term has value 1 for either value of C if the other variables have the correct value. Finding the two input symbols by replacing the x is equivalent to performing the algebra

$$\bar{A}BD = \bar{A}B \cdot 1 \cdot D$$
$$= \bar{A}B(C + \bar{C})D$$
$$= \bar{A}BCD + \bar{A}B\bar{C}D$$

which converts a product term with a missing variable into the sum of two product terms in which all variables appear. We just saw that each such product term corresponds to a row in the truth table. Hence, a product term with one variable missing gives two input symbols, opposite which we may place 1's in the truth table.

If more than one x appears in a tuple (more than one variable is missing), then to find input symbols we must replace all x's with 0's and 1's in all possible ways. Product term AC illustrates this.

$$1x1x \rightarrow \begin{array}{cccc} 1 & 0 & 1 & 0 \\ 1 & 0 & 1 & 1 \\ 1 & 1 & 1 & 0 \\ 1 & 1 & 1 & 1 \end{array}$$

If 3 x's had appeared, we would have obtained 8 input symbols. In general, r x's in a tuple give rise to 2^r input symbols.

The same results can be obtained with algebra using the technique illustrated above to introduce missing variables.

$$AC = A(\bar{B} + B)C(\bar{D} + D)$$
$$= A(\bar{B} + B)C\bar{D} + A(\bar{B} + B)CD$$
$$= A\bar{B}C\bar{D} + ABC\bar{D} + A\bar{B}CD + ABCD$$

The four product terms produced here correspond to the four input symbols generated just above. We have just done the same thing two different ways; however, we will not continue to do everything twice. Tuples and product terms will be used interchangeably because it is so easy to convert one to the other.

Each product term with r variables absent in a *sop* equation produces 2^r 1's in the truth table. Examining the product terms then permits us to place quickly the 1's in the truth table. If no product terms have value 1, the dependent variable has value 0. Any rows in the truth table that do not contain 1 must therefore contain 0. Figure 2.43 illustrates this for the example function.

Filling the Karnaugh map from a *sop* equation is not substantially different. Each product term designates a collection of 2^r adjacent cells where r is the number of variables absent from the product term. If no variables are absent, the product term designates a single cell. In Fig. 2.44, $\overline{A}\overline{B}\overline{C}\overline{D}$ designates a single cell into which a 1 is to be placed. $\overline{A}BD$ designates an area of two cells into which 1's are to be placed. This area is outside of A (inside the $\overline{A}$ half of the map), and inside B and inside the D half of the map.

A	B	C	D	F	
0	0	0	0	1	⟵ $\overline{A}\overline{B}\overline{C}\overline{D}$
0	0	0	1	0	
0	0	1	0	0	
0	0	1	1	0	
0	1	0	0	0	
0	1	0	1	1	⟵
0	1	1	0	0	} $\overline{A}BD$
0	1	1	1	1	⟵
1	0	0	0	0	
1	0	0	1	0	
1	0	1	0	1	⟵
1	0	1	1	1	⟵
1	1	0	0	0	} AC
1	1	0	1	0	
1	1	1	0	1	⟵
1	1	1	1	1	⟵

FIG. 2.43. The Truth Table of $F = \overline{A}\overline{B}\overline{C}\overline{D} + \overline{A}BD + AC$

Two cells satisfy all of these requirements. The area that is both inside the A and inside the C half map consists of 4 cells. Empty cells are to contain 0's, although we often neglect to fill them in.

Truth tables can be obtained from Boolean equations in product-of-sums form in a similar but dual manner. Rather than asking when a product term takes the value of 1, we now must ask when a sum term takes the value 0; for knowing that a sum term has value 0 permits the conclusion that the dependent variable has that value ($A \cdot 0 = 0$). Sum terms of a *pos*

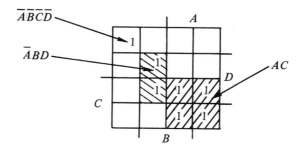

FIG. 2.44. The Karnaugh Map of $F = \bar{A}\bar{B}\bar{C}\bar{D} + \bar{A}BD + AC$

expression of a switching function are *implicates* of that function. To illustrate we will use the dual of function F.

$$F_d = (\bar{A} + \bar{B} + \bar{C} + \bar{D}) \cdot (\bar{A} + B + D) \cdot (A + C)$$

If we know that one or more implicates have value 0, then we know that $F_d = 0$. Knowing that one implicate has value 1 does not permit the conclusion that $F_d = 1$.

A sum term has value 0 when the variables that appear in true form have value 0, and those variables that appear complemented have value 1; thus

$$\bar{A} + \bar{B} + \bar{C} + \bar{D} = 0 \quad \text{when } A = 1, B = 1, C = 1, \text{ and } D = 1$$
$$\bar{A} + B + D = 0 \quad \text{when } A = 1, B = 0, \text{ and } D = 0$$
$$A + C = 0 \quad \text{when } A = 0 \text{ and } C = 0.$$

The conditions listed on the right are phrased in terms of "ands", and hence give product terms. They describe tuples that are easily turned into input symbols.

A	B	C	D	
1	1	1	1	$\bar{A} + \bar{B} + \bar{C} + \bar{D} = 0$
1	0	x	0	$\bar{A} + B + D = 0$
0	x	0	x	$A + C = 0$

The code for forming these tuples directly from the sum terms is given in Table 2.7.

TABLE 2.7. TUPLE–SUM–TERM CODE

Tuple Component	Sum Term Component
0	True variable
1	Complement
x	Missing variable

Since these tuples are derived from sum terms, they indicate input symbols for which the dependent variable takes value 0. For the remaining input symbols, no sum term evaluates to 0 so all must have value 1, and the dependent variable must also have that value. Figure 2.45 gives the truth table for F_d, derived in this manner. Comparing Figs. 2.43 and 2.45 reveals that F and F_d have different truth tables that are not complements. Dual Boolean expressions do not necessarily describe the same or complementary functions.

Maps can be filled from product-of-sums equations in the same manner. Each sum term describes a group of r adjacent cells into which 0's must be placed. Tuples formed with the code of Table 2.7 describe the area of this group, or if you prefer to use algebra, complement each sum term forming a product term that describes the area of cells into which 0's are placed.

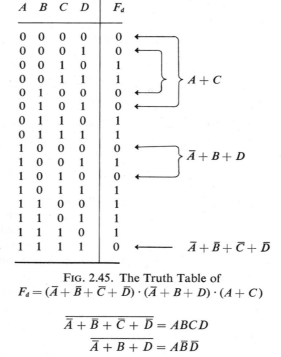

A	B	C	D	F_d
0	0	0	0	0
0	0	0	1	0
0	0	1	0	1
0	0	1	1	1
0	1	0	0	0
0	1	0	1	0
0	1	1	0	1
0	1	1	1	1
1	0	0	0	0
1	0	0	1	1
1	0	1	0	0
1	0	1	1	1
1	1	0	0	1
1	1	0	1	1
1	1	1	0	1
1	1	1	1	0

$A + C$

$\bar{A} + B + D$

$\bar{A} + \bar{B} + \bar{C} + \bar{D}$

FIG. 2.45. The Truth Table of
$$F_d = (\bar{A} + \bar{B} + \bar{C} + \bar{D}) \cdot (\bar{A} + B + D) \cdot (A + C)$$

$$\overline{\bar{A} + \bar{B} + \bar{C} + \bar{D}} = ABCD$$

$$\overline{\bar{A} + B + D} = A\bar{B}\bar{D}$$

$$\overline{A + C} = \bar{A}\bar{C}$$

Figure 2.46 identifies the areas in this manner.

One other form of expression arises out of networks in which AND gates drive an OR gate. If the OR gate in turn drives an inverter that produces the output signal, the equation derived is in AND-OR-INVERT, or "AOI" form. De Morgan's theorem can be used to convert such expressions to *pos* form. The Distributive axiom can be used to convert to pure *sop* form. But all of this work can be avoided by finding the truth table or Karnaugh

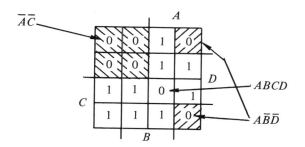

FIG. 2.46. The Karnaugh Map of
$$F_d = (\bar{A} + \bar{B} + \bar{C} + \bar{D})(\bar{A} + B + D)(A + C)$$

map for the complementary function, and then by complementing the truth or map.

Let

$$G = \overline{\bar{A}\bar{B}\bar{C}\bar{D}} + \bar{A}BD + AC$$

then,

$$\bar{G} = \bar{A}\bar{B}\bar{C}\bar{D} + \bar{A}BD + AC$$

$$= F$$

Complementing the dependent variable values in Figs. 2.43 and 2.44 then gives the truth table and Karnaugh map of G.

Truth Tables to Equations

A truth table may be converted to an equation by reversing these processes. The dependent 1's identify input symbols (tuples) that specify product terms via the code of Table 2.6. Dependent 0's identify input symbols (tuples) that specify sum terms via the code of Table 2.7. We can obtain either *sop* or *pos* equations with ease, but the product and sum terms so obtained are unusual and worthy of deeper investigation.

Since no x's appear in an input symbol, all variables appear in their product terms (sum terms). Each variable appears once, either in true or complement form. Such product terms are *minterms*. To see their "minimum" nature, we need only look at their Karnaugh maps or truth tables.

Since there are 2^n binary n-tuples, there are 2^n minterms of n independent variables. Table 2.8 lists the minterms for $n = 3$ and suggests that each minterm be named m_d, where d is the decimal equivalent of the corresponding binary input symbol. If each minterm in turn is thought to specify a function and its truth table found, Fig. 2.47 results. A single 1 appears in each such truth table. This is the minimum number of 1's that can appear if we exclude the trivial truth table of all 0's.

Recalling the Boolean algebra of truth tables as discussed in Sec. 2.3, we see now that any truth table can be formed by ORing the truth tables of

TABLE 2.8. MINTERMS OF 3 VARIABLES

Input Symbol	Decimal Equivalent	Minterm	Name
000	0	$\bar{x}_1 \bar{x}_2 \bar{x}_3$	m_0
001	1	$\bar{x}_1 \bar{x}_2 x_3$	m_1
010	2	$\bar{x}_1 x_2 \bar{x}_3$	m_2
011	3	$\bar{x}_1 x_2 x_3$	m_3
100	4	$x_1 \bar{x}_2 \bar{x}_3$	m_4
101	5	$x_1 \bar{x}_2 x_3$	m_5
110	6	$x_1 x_2 \bar{x}_3$	m_6
111	7	$x_1 x_2 x_3$	m_7

selected minterms. Conversely, any truth table can be decomposed into a set of minterm truth tables. We may conclude from this that:

THEOREM 9a. *Any Boolean function can be written as a unique sum of minterms.*

The dependent 1's in the truth table of a function identify the minterms that appear in its simplified sum-of-minterms expression. These then are the minterms " of" the function.

To present this in another way, let α_d be the dependent entry opposite input symbol d in the truth table for function f. Then f is expressed in sum of minterm form by

$$f = \sum_{d=0}^{2^n-1} \alpha_d \cdot m_d \tag{2.35}$$

The truth table of Fig. 2.48 leads to

$$H = 0 \cdot \bar{A}\bar{B}\bar{C} + 0 \cdot \bar{A}\bar{B}C + 0 \cdot \bar{A}B\bar{C} + 1 \cdot \bar{A}BC + 1 \cdot A\bar{B}\bar{C}$$
$$+ 0 \cdot A\bar{B}C + 1 \cdot AB\bar{C} + 1 \cdot ABC$$
$$= \bar{A}BC + A\bar{B}\bar{C} + AB\bar{C} + ABC$$

A	B	C	m_0	m_1	m_2	m_3	m_4	m_5	m_6	m_7
0	0	0	1	0	0	0	0	0	0	0
0	0	1	0	1	0	0	0	0	0	0
0	1	0	0	0	1	0	0	0	0	0
0	1	1	0	0	0	1	0	0	0	0
1	0	0	0	0	0	0	1	0	0	0
1	0	1	0	0	0	0	0	1	0	0
1	1	0	0	0	0	0	0	0	1	0
1	1	1	0	0	0	0	0	0	0	1

FIG. 2.47. The Truth Table of a Minterm Contains a Single 1

A	B	C	H
0	0	0	$0 = \alpha_0$
0	0	1	$0 = \alpha_1$
0	1	0	$0 = \alpha_2$
0	1	1	$1 = \alpha_3$
1	0	0	$1 = \alpha_4$
1	0	1	$0 = \alpha_5$
1	1	0	$1 = \alpha_6$
1	1	1	$1 = \alpha_7$

FIG. 2.48.   Truth Table in Which the α_d Are
Identified

Four 1's appear in that figure; four minterms remain in the sum after simplification.

Earlier in this section we saw how to introduce missing variables into product terms. This technique may also be used to find the minterms of a function from an *sop* expression.

$$
\begin{aligned}
H &= A\bar{C} + BC \\
&= A(\bar{B} + B)\bar{C} + (\bar{A} + A)BC \\
&= A\bar{B}\bar{C} + AB\bar{C} + \bar{A}BC + ABC \\
&= m_4 + m_6 + m_3 + m_7 \\
&= \Sigma\ 3,\ 4,\ 6,\ 7
\end{aligned}
$$

Maxterms are sum terms in which all variables appear, each once in either true or complement form. Truth tables and Karnaugh maps of maxterms contain $2^n - 1$ ones, the maximum less than the trivial case of all 1's. Table 2.9 lists the maxterms of 3 variables and suggests names M_d for each. Note that these names have been chosen so that maxterm M_d is the complement of minterm m_d for all d.

TABLE 2.9. MAXTERMS OF 3
VARIABLES

d	M_d
0	$x_1 + x_2 + x_3$
1	$x_1 + x_2 + \bar{x}_3$
2	$x_1 + \bar{x}_2 + x_3$
3	$x_1 + \bar{x}_2 + \bar{x}_3$
4	$\bar{x}_1 + x_2 + x_3$
5	$\bar{x}_1 + x_2 + \bar{x}_3$
6	$\bar{x}_1 + \bar{x}_2 + x_3$
7	$\bar{x}_1 + \bar{x}_2 + \bar{x}_3$

Any truth table can be constructed by ANDing selected maxterm truth tables. Any truth table can be decomposed to maxterm truth tables. The 0's determine which maxterms are needed. Corresponding input symbols translate to d values that can be used in Table 2.9.

THEOREM 9b. *Any Boolean function can be expressed as a unique product of maxterms.*

$$f = \prod_{d=0}^{2^n - 1} (\alpha_d + M_d) \tag{2.36}$$

Four 0's appear in Fig. 2.48. We should expect four maxterms to survive simplification, those maxterms associated with 0's in the truth table.

$$H = (0 + A + B + C)(0 + A + B + \bar{C})(0 + A + \bar{B} + C)(1 + A + \bar{B} + \bar{C})$$
$$(1 + \bar{A} + B + C)(0 + \bar{A} + B + \bar{C})(1 + \bar{A} + \bar{B} + C)(1 + \bar{A} + \bar{B} + \bar{C})$$
$$= (A + B + C)(A + B + \bar{C})(A + \bar{B} + C)(\bar{A} + B + \bar{C})$$

Maxterms can be found algebraically by using the dual of the technique used to find minterms.

$$H = (A + C)(B + \bar{C})$$
$$= (A + \bar{B} \cdot B + C)(\bar{A} \cdot A + B + \bar{C})$$
$$= (A + \bar{B} + C)(A + B + C)(\bar{A} + B + \bar{C})(A + B + \bar{C})$$
$$= M_2 \cdot M_0 \cdot M_5 \cdot M_1$$
$$= \Pi\ 0, 1, 2, 5$$

While we have found equations from truth tables, these equations may not describe " best " networks. Sum-of-minterm expressions describe a large number of AND gates, each with maximum fan-in. Product-of-maxterm expressions suggest the need for a large number of OR gates. To find *sop* expressions with a minimum number of product terms, each with a minimum number of variables, is the difficult task considered in Chapter 10. For now we will see how to find " good ", if not best, *sop* and *pos* expressions from the Karnaugh map.

Karnaugh Map to Equations

If two adjacent cells of a Karnaugh map contain 1's, the minterms corresponding to those cells are of the function and differ in one variable position only. The variable appears true in one and complemented in the other. The Karnaugh map is constructed so that this is the case. Now the sum of those two minterms can be replaced by the Minimization theorem with a single product term from which the differing variable is absent.

THEOREM 10. *Minimization*

$$A \cdot B + A \cdot \bar{B} = A$$
$$(A + B)(A + \bar{B}) = A$$

Proof:

$$A \cdot B + A \cdot \bar{B} = A(B + \bar{B})$$
$$= A \cdot 1 = A$$

To see how we can employ this theorem, suppose we face the function:

$$f = \sum 5, 6, 7 = A\bar{B}C + AB\bar{C} + ABC$$

Theorem 10 allows us to replace the sum of minterms m_5 and m_7 with a single product term.

$$A\bar{B}C + ABC = AC(\bar{B} + B) = AC$$

And we could express f as $f = AC + AB\bar{C}$. Alternatively, we may replace the sum of m_6 and m_7 by a single, simpler term.

$$AB\bar{C} + ABC = AB$$

Then f may be expressed $f = A\bar{B}C + AB$.

Can we make both replacements? Yes, of course we can, if we remember the idempotent theorem. In effect, minterm m_7 may be repeated in the original expression of f above, both combinations made, and f written:

$$f = AB + AC$$

Some groups of four minterms permit further simplification. If we add minterm m_4 to the three used above, m_4 may be combined with m_5, and m_6 may be combined with m_7.

$$g = \Sigma\, 4, 5, 6, 7$$
$$= A\bar{B}\bar{C} + A\bar{B}C + AB\bar{C} + ABC$$
$$= A\bar{B} + AB$$

Theorem 10 can clearly be applied again to give

$$g = A$$

We could have combined m_4 with m_6, and m_5 with m_7, to obtain the same result.

One way to simplify a switching function then might consist of expressing that function in sum-of-minterms form. All pairs of adjacent minterms are combined to form product terms with one variable absent, all adjacent product terms are combined via Th. 10 to form product terms with two variables absent, and so forth. We will explore this procedure later. For now, let us see how Karnaugh maps serve as a tool to accomplish the same thing.

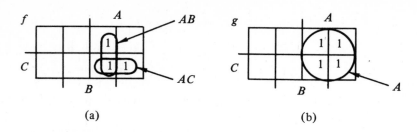

FIG. 2.49   Karnaugh Maps of (a) $f = \Sigma\, 5, 6, 7$ and
(b) $g = \Sigma\, 4, 5, 6, 7$ Facilitate Their Immediate Simplifications

The Karnaugh map of a function pictures adjacent minterms, groups of four minterms, etc., that describe product terms with missing variables. Such groups consist of 2^r cells, each of which is adjacent to r other cells of the group. A group of 2^r such cells is described with a product term from which r variables are absent. It is easy for us to recognize patterns of cells that permit simplification of a switching function. In fact, having recognized all such patterns, we can easily select those that we wish to use. In the Karnaugh maps of Fig. 2.49, we spot and encircle the groups of 2 and 4 cells containing 1's that permit the algebraic reduction of f and g illustrated above. Having encircled groups, we write a product term for each and have the minimum *sop* expression for f and g found with the algebra.

The patterns of cells holding 1's are not always so obvious. In Fig. 2.50(a), we are reminded that the left is next to the right column on a Karnaugh map.

$$\Sigma\, 0, 8 = \bar{A}\bar{B}\bar{C}\bar{D} + A\bar{B}\bar{C}\bar{D}$$
$$= \bar{B}\bar{C}\bar{D}$$

Any row filled with 1's constitutes a group of four minterms that can be replaced with a single product term.

$$\Sigma\, 3, 7, 11, 15 = \bar{A}\bar{B}CD + \bar{A}BCD + A\bar{B}CD + ABCD$$
$$= \qquad \bar{A}CD \qquad + \qquad ACD$$
$$= CD$$

In Fig. 2.50(b), a filled column gives a similar result.

$$\Sigma\, 12, 13, 14, 15 = AB\bar{C}\bar{D} + AB\bar{C}D + ABC\bar{D} + ABCD$$
$$= \qquad AB\bar{C} \qquad + \qquad ABC$$
$$= AB$$

So do the four corners (which form a square when we roll the Karnaugh map into a toroid).

$$\Sigma\, 0, 2, 8, 10 = \bar{A}\bar{B}\bar{C}\bar{D} + \bar{A}\bar{B}C\bar{D} + A\bar{B}\bar{C}\bar{D} + A\bar{B}C\bar{D}$$
$$= \qquad \bar{A}\bar{B}\bar{D} \qquad + \qquad A\bar{B}\bar{D}$$
$$= \bar{B}\bar{D}$$

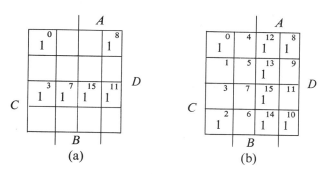

FIG. 2.50   Patterns to be Recognized on the Karnaugh Map

One other group exists in Fig. 2.50(b). The upper right two cells and the lower right two cells form a square that permits simplification.

$$\Sigma\ 8,\ 10,\ 12,\ 14 = A\bar{B}\bar{C}\bar{D} + A\bar{B}C\bar{D} + AB\bar{C}\bar{D} + ABC\bar{D}$$
$$= \qquad A\bar{B}\bar{D} \qquad + \qquad AB\bar{D}$$
$$= A\bar{D}$$

If Fig. 2.50(b) were the Karnaugh map for dependent variable h, which way would we express it?

$$h = AB + \bar{B}\bar{D}$$

or

$$h = AB + \bar{B}\bar{D} + A\bar{D}$$

The first expression is simpler and suggests a simpler network of fewer gates than the second expression. Since all minterms are represented by both expressions, they both express h. The $A\bar{D}$ term is *redundant*, and we might well choose to ignore it. The Karnaugh map form of expressing functions makes this easy to do.

For now let us assume that gates cost more than connections between gates. To reduce the cost of a network, we eliminate redundant gates and interconnections. A minimum cost network, then, has as few gates as possible, and each gate in the network has as few input terminals as possible. To find the minimum *sop* expression of a Boolean function from the Karnaugh map requires the following two steps: (1) Circle the minimum number of the largest possible groups of cells such that each cell containing a 1 is enclosed in at least one circle. (2) Write a product term for each encircled group.

Be careful when executing this procedure. In Fig. 2.51(a), the center group of four cells is obvious, but it is not so obvious that this grouping is redundant, and therefore that the switching function is best expressed without it. Four groups of two cells each contain all cells holding 1's.

$$F = \bar{A}B\bar{C} + \bar{A}CD + ABC + A\bar{C}D$$

Four groups of four cells each must be used to encircle all cells containing 1's in Fig. 2.51(b) with the minimum number of the largest possible groups.

$$G = \bar{B}\bar{C} + \bar{C}\bar{D} + \bar{A}B + \bar{A}\bar{D}$$

Some cells are circled more than once. Including a minterm in more than one product term is not wrong, or undesirable. If we fail to do so, then we cannot obtain a *sop* expression that is as minimum as the one given above.

Finding minimum AOI expressions for switching functions is just as easy. With the map for G given in Fig. 2.51(b), imagine what the map for $\bar{G}$ would be—0's where there are 1's and 1's where there are 0's. We are not going to prepare the map for $\bar{G}$ for our imagination permits us to minimize $\bar{G}$ without it. Alternatively, we can encircle groups of cells containing 0's to find a minimum *sop* expression for $\bar{G}$. From Fig. 2.51(c):

$$\bar{G} = AC + BD$$

We can quickly write a minimum AOI expression from the minimum *sop* expression of the complementary function.

$$G = \overline{AC + BD}$$

In this example the minimum AOI expression specifies less circuitry than the minimum *sop* expression.

Using De Morgan's theorem, a minimum AOI expression yields a minimum product-of-sums expression.

$$G = \overline{AC + BD}$$
$$= (\bar{A} + \bar{C})(\bar{B} + \bar{D})$$

Such expressions can also be found with the dual of the procedure used to find minimum *sop* expressions. The maxterms of a switching function are combined with the Minimization Theorem $[(A + B)(A + \bar{B}) = A]$ to simpler sum terms. If two simpler sum terms differ only in that a variable appears true in one and complemented in the other, they may be combined to a single term in which the differing variable is absent. From Fig. 2.51(b):

$$G = \Pi\ 5,\ 7,\ 10,\ 11,\ 13,\ 14,\ 15$$
$$= (A + \bar{B} + C + \bar{D})(A + \bar{B} + \bar{C} + \bar{D})(\bar{A} + B + \bar{C} + D)$$
$$\cdot (\bar{A} + B + \bar{C} + \bar{D})(\bar{A} + \bar{B} + C + \bar{D})$$
$$\cdot (\bar{A} + \bar{B} + \bar{C} + D)(\bar{A} + \bar{B} + \bar{C} + \bar{D})$$

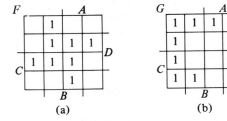
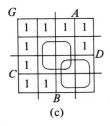

(a)   (b)   (c)

FIG. 2.51   Maps to be Minimized

$$= (A + \bar{B} + \bar{D})(\bar{A} + \bar{B} + \bar{C})(\bar{A} + \bar{B} + \bar{D})(\bar{A} + \bar{B} + \bar{C})$$
$$= (\bar{B} + \bar{D})(\bar{A} + \bar{C})$$

Not all of the possible maxterm combinations are shown.

The 0's in a truth table or Karnaugh map dictate the maxterms of a function. Adjacent cells holding 0's in a Karnaugh map describe maxterms that can be combined via the Minimization Theorem. Then, encircling groups of cells containing 0's, as in Fig. 2.51(c), is a means of recording simplified sum terms. If 2^s cells form a group, then s variables are absent from the associated sum term. The literals that appear in the sum term are most easily found by complementing the literals used in the product term that describes the encircled area. Thus the center cell in Fig. 2.51(c) is described by BD; the sum term of this area is then $(\bar{B} + \bar{D})$. Alternatively, we can complement (mentally) the labels on the map when we write sum terms. The lower right square in Fig. 2.51(c) is then entirely in the $\bar{A}$ area and entirely in the $\bar{C}$ area. The B and D boundaries are crossed by this area, and B and D do not appear in the sum term $(\bar{A} + \bar{C})$ as a result.

A substantial number of very fundamental ideas and procedures were introduced in this section. It is important that you fully understand all of them.

Product terms in *sop* expressions are implicants.
Product terms may be expressed with tuples, and vice versa.
Product terms (tuples) may be converted to minterms (input symbols).
Product terms determine the 1's in a truth table and Karnaugh map.
Sum terms in *pos* expressions are implicates.
Sum terms may be expressed with tuples, and vice versa.
Sum terms determine the 0's in a truth table and Karnaugh map.
AOI means And Or Invert.
Minterms (maxterms) are product (sum) terms in which all variables appear once.
Each switching function is a unique sum(product) of minterms(maxterms).
The 1's (0's) of a truth table dictate the minterms (maxterms) of the function.
Sum (product) terms may be combined with the Minimization Theorem.
Recognizing groups of cells on the Karnaugh map graphically accomplishes algebraic simplification.
Groups of 2^r Karnaugh map cells containing 1's (0's)—each adjacent to r others—describe product (sum) terms with r variables absent.
Selecting the minimum number of maximum sized groups on a Karnaugh map gives minimum Boolean equations.

2.6 ANALYSIS OF COMBINATIONAL LOGIC NETWORKS

Because logic diagrams are so close to actual networks, often reflecting their actual construction as well as the switching function they realize, to find a truth table, Karnaugh map, or Boolean equation from a logic diagram is the *analysis problem*. The reverse process of *synthesis* or *design* begins with one of these other forms of expressing switching functions and produces a logic diagram from which an actual circuit can be constructed. Design is more difficult than analysis, but the steps taken are those of analysis to an extent, only their order is reversed. Thus analysis is important to design as well as to maintaining and evaluating digital systems.

Simulation

Determining a truth table from a logic diagram is so routine that we often use a computer to carry it out. Input symbols are assumed present on the input terminals in turn. For each the input variable values are driven through the diagram much as signals propagate in the actual network. We *simulate* the network. Signals are modified by gates according to their defining truth tables. If input symbol $(A, B, C, D, E) = 10001$ is applied to the network of Fig. 2.52, OR gate 7 combines B and C to produce 0. Inverter 5 produces a 1 in response. Gate 6 combines the values of D and E to logic 1, which inverter 4 converts to 0. AND gates 2 and 3 both have a 0 on an input terminal and hence deliver a 0 to OR 1, which produces $F = 0$. One of 32 input symbols has been simulated. The remaining 31 may be treated in the same way to obtain a complete truth table.

Walk-back

If we are analyzing by hand, then finding a Boolean equation from the logic diagram—and simplifying or just changing that equation to one of the forms discussed in the last section—can lead us to a truth table more quickly. Equations can be written " by inspection " from logic diagrams of AND, OR, and NOT networks. We begin at the output terminal and see what type of gate provides the output signal. In Fig. 2.52, F is generated by a 3-input OR gate. This finding may be recorded.

$$F = (\qquad) + (\qquad) + (\qquad)$$

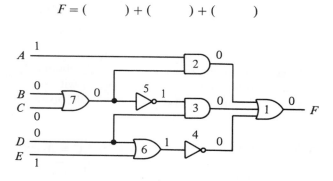

FIG. 2.52   A Network to be Simulated and Analyzed

The three signals that gate 1 combines are determined by scanning each of the lines driving gate 1 until a gate or named input signal is encountered. Scanning from gate 1 to gate 2 in Fig. 2.52 gives

$$F = ([\quad\quad] \cdot [\quad\quad]) + (\quad\quad) + (\quad\quad)$$

Gate 2 combines A with the OR of B and C.

$$F = A \cdot (B + C) + (\quad\quad) + (\quad\quad)$$

Gate 3 combines the NOT of the signal from gate 7 with D.

$$F = A(B + C) + D \cdot (\overline{\quad\quad}) + (\quad\quad)$$
$$= A(B + C) + D \cdot (\overline{B + C}) + (\quad\quad)$$

Finally, scanning from gate 1 to NOT 4 and then gate 6 completes the equation.

$$F = A(B + C) + D(\overline{B + C}) + (\overline{D + E})$$

We could convert this equation to *sop* form and quickly fill in a truth table or Karnaugh map, using the techniques of the previous section. The important point here, however, is the technique for inspecting a logic diagram so that an equation can be written directly. All paths of a network are scanned from the output terminal to all input terminals. For each gate encountered, we write a logic operator; for each input variable terminating a path we record that input variable. If we number gates as they are encountered in tracing a path, the numbers indicate the *level* of the gate with respect to the path traced.

This technique is general and usable even if other types of gates appear. However, substantial algebraic manipulation may be desirable. In networks of NAND and/or NOR gates, shortcuts are possible to avoid this manipulation. Before we use the networks of Fig. 2.53 to find these shortcuts, we will analyze by the method above and by using the NAND ($\uparrow$) and NOR ($\downarrow$) operator symbols introduced in Table 2.4.

$$F = (\quad\quad) \uparrow (\quad\quad) \uparrow (\quad\quad)$$
$$= (A \uparrow B) \uparrow (C \downarrow D) \uparrow E$$
$$G = (\quad\quad) \downarrow (\quad\quad) \downarrow (\quad\quad)$$
$$= (A \uparrow B) \downarrow (C \downarrow D) \downarrow E$$

While these Boolean equations are easily obtained, they do not describe the network in terms of the logic operators with which we are most familiar. We can eliminate the NAND and NOR operators by substituting their AND-NOT and OR-NOT definitions.

$$A \uparrow B \uparrow C = \overline{A \cdot B \cdot C} = \overline{A} + \overline{B} + \overline{C}$$
$$A \downarrow B \downarrow C = \overline{A + B + C} = \overline{A} \cdot \overline{B} \cdot \overline{C} \tag{2.37}$$

Substituting and simplifying give more revealing expressions for F and G.

$$F = (A \uparrow B) \uparrow (C \downarrow D) \uparrow E$$

$$= \overline{(A \cdot B) \cdot (\overline{C + D}) \cdot E}$$

$$= AB + C + D + \bar{E}$$

$$G = (A \uparrow B) \downarrow (C \downarrow D) \downarrow E$$

$$= \overline{(\overline{A \cdot B}) + (\overline{C + D}) + E}$$

$$= AB(C + D)\bar{E}$$

De Morgan's theorem is seen to be of great importance.

Symbolic Replacement

In fact, De Morgan's theorem suggests that either of two symbols are valid representations of the NAND (NOR) gate. From Eq. (2.37), we see the validity of the alternative symbols shown in Fig. 2.54. Some logic designers prefer to use only one symbol for a gate; others find an advantage in using both symbols in the same diagram.

If we use the alternative symbols for the output level gates of Fig. 2.53, we find internal lines with circles on both ends. The Involution theorem permits deleting double inversions, and the diagram of Fig. 2.55(b) may be drawn. The circles on the E line must remain since they have no counterpart at the source end of the line. Figure 2.55(b) shows an AND-OR network from which Boolean equations for F and G can be written by inspection. As far as F is concerned, gate NAND 1 of Fig. 2.53 acts as an OR gate, while NAND 3 acts as an AND gate; and NOR 4 is an OR gate. As far as G is concerned, NOR 2 of Fig. 2.53 is an AND gate, NAND 3 is an AND gate, and NOR 4 is an OR gate. We see these facts from the simplified Boolean equations for F and G, as well as from Fig. 2.55(b).

We have not yet seen examples of all the rules for efficiently analyzing NAND and NOR networks so we turn to the network of Fig. 2.56(a). Basic analysis is complicated by the 15 paths from the output to the input terminals. Gate 1 is at the output or first level. If we replace it with the basic-OR symbol for the NAND gate, we see AND gates 2, 3, and 4 driving OR gate 1. These are second level gates. Gate 5 is at the third level from the output terminal and can be drawn in basic-OR form. Gate 6

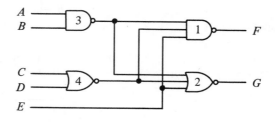

FIG. 2.53   A NAND-NOR Network That Reveals
Analysis Rules

FIG. 2.54   Alternative Symbols for NAND and NOR Gates

causes problems with the symbolic replacement method of analysis. Along some paths it drives second level gates and is best represented in basic-OR form. It drives third level gate 5 along other paths and is best drawn with the basic-AND symbol for these paths. If we insist upon using the symbol replacement method of analysis, gates that drive both odd and even level gates must be shown twice, once in basic-AND form for those paths in which they appear at an even level and once in basic-OR form for those paths in which they appear at an odd level. Figure 2.56(b) shows the AND-OR network that results after the Involution theorem is applied. From it we can write

$$f = A(\bar{A} + \bar{B} + BC) + (\bar{A} + \bar{B} + BC)B(\bar{B} + \bar{C}) + (\bar{B} + \bar{C})C$$

by very careful inspection. A four level NAND network leads to a "sum-of-products-of-sum-of-products" form equation. Simplification with Boolean algebra is easily performed.

$$f = A\bar{B} + AC + \bar{A}B\bar{C} + \bar{B}C$$

While this equation prescribes five gates, complements of all input variables are also needed. The networks of Fig. 2.56(a) uses only six gates.

Direct Analysis

Actually redrawing NAND/NOR logic diagrams is an unnecessary chore. We can write AND-OR equations from logic diagrams of NAND and/or NOR gates only (NOTs are 1-input NANDs or NORs) with the path tracing method, using the following additional rules:

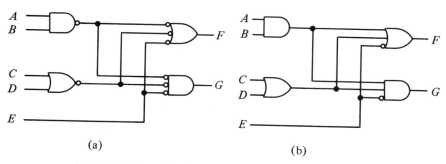

(a)   (b)

FIG. 2.55   Graphic Analysis of the Network of Fig. 2.53

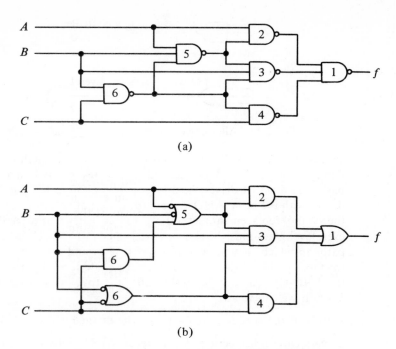

(a)

(b)

FIG. 2.56   Analysis of a NAND Network by
Symbolic Replacement

1. As each path is traced, treat NAND and NOR gates as AND and OR
   gates at each level according to Table 2.10.
2. Write the complements of input signals that enter gates at odd levels.

Some of the paths of the logic diagram of Fig. 2.56(a), and the way that
we must think of the gates are:

				Gate				
Path			1	2	3	4	5	6
f 1 2 A			OR	AND				
f 1 2 5 A			OR	AND			OR	
f 1 2 5 6 B			OR	AND			OR	AND
f 1 4 6 C			OR			AND		OR
f 1 4 C			OR			AND		

Gate 2 ANDs input variable A, while gate 5 ORs its complement ($\overline{A}$).
Variable C drives gate 6. Along some paths that gate ANDs C is true form;
along those paths where gate 6 is at an odd level, it ORs $\overline{C}$ with other
variables.

Several facts should now be apparent. If the output gate is a NAND, a
"sum-of..." expression is obtained. A 2-level NAND network [gates 1, 2,
3, and 4 in Fig. 2.56(a)] is described by a *sop* expression, and we do not

TABLE 2.10. NAND AND NOR REPLACEMENTS
FOR PURPOSES OF DIRECT ANALYSIS

Gate Type	Odd Levels	Even Levels
NAND	OR	AND
NOR	AND	OR

have to think about gate replacements to analyze such networks in the future. By duality, a 2-level NOR network analyzes to a *pos* expression. An input signal that must pass through an even number of inverting gates influences the output signal in true form. If the signal must pass through an odd number of inverters, it affects the output signal in complement form.

The preceding analysis rules apply to networks of inverting gates only. Networks of noninverting (AND and OR) gates as well as inverting gates can be analyzed with the basic path-tracing, algebraic manipulation method. Symbolic replacements may be helpful. The NAND/NOR rules can be applied if the noninverting gates can be merged with inverting gates. For example, an AND gate driving a NAND gate is equivalent to a high fan-in NAND gate.

$$(A \cdot B) \uparrow C = \overline{(A \cdot B) \cdot C} = A \uparrow B \uparrow C$$

An OR gate driving a NOR gate may be replaced with a single NOR gate of increased fan-in. Pieces of a network can be analyzed by inspection, and the pieces properly combined, of course.

Dot logic poses additional problems for the analyst. It is necessary to know what electronics are being used in order to determine what logic is accomplished by wiring output terminals together. If two open-collector DTL or TTL NAND gates are wired as in Fig. 2.57(a), the common-emitter amplifiers form a NOR gate. The wired NAND gates form an AOI gate as shown in Fig. 2.57(b), and the AOI structure may be substituted in the logic diagram for purposes of analysis. Using deMorgan's theorem on an AOI expression suggests the "wired-AND" gate shown in Fig. 2.57(c).

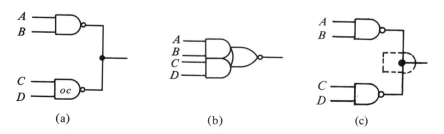

(a)   (b)   (c)

FIG. 2.57   Wiring Open-collector NAND Gates May
Create an AOI Structure

$$\overline{AB + CD} = \overline{AB} \cdot \overline{CD}$$

Wiring the emitter followers of ECL gates forms simple OR gates, and the OR symbol can be drawn over the dot on a logic diagram.

2.7 CANONICAL FORMS OF SWITCHING FUNCTIONS

Canonical expressions of switching functions are Boolean expressions written according to rules under which each function has only one expression. Two of these canonical forms were found in Section 2.5, where Th. 9 indicated that any switching function can be expressed as a sum of minterms or as a product of maxterms. These are the best known canonical forms. Others will be derived in this section.

Let $f(x_1, \ldots, x_n)$ be a single output function of n independent variables. If one of the variables is assigned a value $\alpha \in \{0, 1\}$, then f collapses to a new switching function of $n - 1$ variables. Let x_i be the assigned variable.

$$g(x_1, \ldots, x_{i-1}, x_{i+1}, \ldots, x_n) = f(x_1, \ldots, \alpha, \ldots, x_n) \qquad \textbf{(2.38)}$$

Specific examples make this notation clear.

$$F(A, B, C) = ABC + \overline{A}C$$
$$F(1, B, C) = 1 \cdot BC + \overline{1} \cdot C = BC = F_1(B, C)$$
$$F(0, B, C) = 0 \cdot BC + \overline{0} \cdot C = C = F_2(B, C)$$
$$F(A, 0, C) = A \cdot 0 \cdot C + \overline{A}C = \overline{A}C = F_3(A, C)$$

The original function can be expressed in terms of the new functions that were obtained by first setting a variable to 1 and then to 0. Doing so is referred to as expanding the function about the variable.

THEOREM 11. **Expansion**

$$f(x_1, \ldots, x_n) = x_i \cdot f(x_1, \ldots 1 \ldots, x_n) + \overline{x}_i \cdot f(x_1, \ldots 0 \ldots, x_n)$$
$$= [x_i + f(x_1, \ldots 0 \ldots, x_n)] \cdot [\overline{x}_i + f(x_1, \ldots 1 \ldots, x_n)]$$

Values may be assigned to more than one variable, of course. Using the the same example function as above:

$$F(1, 1, C) = 1 \cdot 1 \cdot C + \overline{1} \cdot C = C$$
$$F(0, B, 1) = 0 \cdot B \cdot 1 + \overline{0} \cdot 1 = 1$$

Also, a function may be expanded about more than one variable by expanding about the first and then expanding each subfunction about the next. If we assign values to all variables, the subfunctions reduce to constants. We used this result in Section 2.4 to evaluate Boolean expressions.

Let $d_1, d_2, \ldots, d_n$ be the bits of the binary expression of the decimal integer d. Assigning each x_i value d_i, and using f_d to abbreviate $f(d_1, \ldots, d_n)$ permits us to restate Th. 9 as

$$f(x_1, \ldots, x_n) = \sum_{d=0}^{2^n-1} f_d \cdot m_d \qquad (2.39)$$

$$= \prod_{d=0}^{2^n-1} (f_d + M_d) \qquad (2.40)$$

where m_d and M_d are minterms and maxterms as defined in Section 2.5. This theorem may be derived by completely expanding a generic function about all of its variables.

$$
\begin{aligned}
f(A, B) &= A \cdot f(1, B) + \bar{A} \cdot f(0, B) \\
&= A(B \cdot f(1, 1) + \bar{B} \cdot f(1, 0)) + \bar{A}(B \cdot f(0, 1) + \bar{B} \cdot f(0, 0)) \\
&= AB \cdot f(1, 1) + A\bar{B} \cdot f(1, 0) + \bar{A}B \cdot f(0, 1) + \bar{A}\bar{B} \cdot f(0, 0)
\end{aligned}
$$

Minterm and maxterm expressions are the best known and most often used canonical forms. They relate directly to truth tables and Karnaugh maps. Other canonical forms are easily derived from them by using deMorgan's theorem selectively. In the following expressions the range of d is always 0 through $2^n - 1$ and hence is omitted. From Eq. (2.39):

$$f(x_1, \ldots, x_n) = \overline{\overline{\sum f_d \cdot m_d}} \qquad \text{AND-OR}$$

$$= \overline{\prod \overline{f_d \cdot m_d}}$$

$$= \uparrow (f_d \uparrow m_d) \qquad \text{NAND-NAND} \qquad (2.41)$$

This is the NAND-NAND canonical form. Any switching function can be expressed as the NAND of the complement of the product of minterms and corresponding truth table entries f_d. Our example function may be so expressed.

$$
\begin{aligned}
F &= ABC + \bar{A}C \\
&= ABC + \bar{A}\bar{B}C + \bar{A}BC + 0 \cdot \bar{A}\bar{B}\bar{C} \qquad \text{AND-OR} \\
&= \overline{\overline{ABC} \cdot \overline{\bar{A}\bar{B}C} \cdot \overline{\bar{A}BC} \cdot \overline{0 \cdot \bar{A}\bar{B}\bar{C}}} \qquad \text{NAND-NAND}
\end{aligned}
$$

Applying deMorgan's theorem one more time converts Eq. (2.41) to

$$f(x_1, \ldots, x_n) = \uparrow (\bar{f}_d + M_d) \qquad \text{OR-NAND} \qquad (2.42)$$

The 1's in a truth table identify the maxterms that appear in an OR-NAND canonical expression. OR-gates-driving-NAND-gate networks are suggested by such expressions.

$$F = \overline{(\bar{A} + \bar{B} + \bar{C}) \cdot (A + \bar{B} + \bar{C}) \cdot (A + B + \bar{C})}$$

One more application of deMorgan's theorem produces the NOR-OR canonical form.

$$f(x_1, \ldots, x_n) = \sum (\overline{\bar{f}_d + M_d}) \qquad \text{NOR-OR} \qquad (2.43)$$

$$F = \overline{(\bar{A} + \bar{B} + \bar{C})} + \overline{(A + \bar{B} + \bar{C})} + \overline{(A + B + \bar{C})}$$

Figure 2.58 shows the form of logic networks prescribed by these four related canonical forms. It suggests how AND-OR networks can be directly transformed to equivalent networks that employ alternative technologies. Such transformations are of very practical, as well as theoretical, interest. For example, we simplified NAND network analysis by transforming NAND-NAND to AND-OR networks in the previous section. We will use the reverse transform in the next chapter to design NAND networks.

De Morgan's theorem may be used to transform sum-of-products as well as sum-of-minterms expressions of switching functions. Let π_1, π_2, ..., π_p be implicants of a switching function and sum terms σ_i be their complements, $\sigma_i = \bar{\pi}_i$. Proceeding as above

$$
\begin{aligned}
f(x_1, \ldots, x_n) &= \pi_1 + \pi_2 + \cdots + \pi_p && \text{AND-OR} \\
&= \overline{\bar{\pi}_1 \cdot \bar{\pi}_2 \cdot \cdots \cdot \bar{\pi}_p} && \text{NAND-NAND} \\
&= \overline{\sigma_1 \cdot \sigma_2 \cdot \cdots \cdot \sigma_p} && \text{OR-NAND} \\
&= \bar{\sigma}_1 + \bar{\sigma}_2 + \cdots + \bar{\sigma}_p && \text{NOR-OR}
\end{aligned}
\tag{2.44}
$$

The following example clarifies this notation and illustrates the "odd-level" rule encountered in the previous section. Notice also that each product term in the original expression of the function leads to a term in each of the other expressions. If the original expression is minimized, then the later ones are also.

$$
\begin{aligned}
f &= (A \cdot B) + (\bar{C} \cdot D) + E \\
&= \overline{\overline{A \cdot B} \cdot \overline{\bar{C} \cdot D} \cdot \overline{E}} \\
&= \overline{(\bar{A} + \bar{B}) \cdot (C + \bar{D}) \cdot \bar{E}} \\
&= \overline{(\bar{A} + \bar{B})} + \overline{(C + \bar{D})} + E
\end{aligned}
$$

From the 0's of a truth table, we learn which maxterms remain in the product-of-maxterms canonical form, Eq. (2.40). Using deMorgan's theorem repeatedly on that form gives the following canonical forms:

$$
\begin{aligned}
f(x_1, \ldots, x_n) &= \prod (f_d + M_d) && \text{OR-AND} \\
&= \downarrow \overline{(f_d + M_d)} && \text{NOR-NOR} \\
&= \downarrow (\bar{f}_d \cdot m_d) && \text{AND-NOR} \\
&= \prod \overline{(\bar{f}_d \cdot m_d)} && \text{NAND-AND}
\end{aligned}
\tag{2.45}
$$

For example,

$$
\begin{aligned}
G &= (A + B + C)(A + B + \bar{C}) \\
&= \overline{\overline{(A + B + C)} + \overline{(A + B + \bar{C})}} \\
&= \overline{\overline{A}\overline{B}\overline{C} + \overline{A}\overline{B}C} \\
&= \overline{\overline{A}\overline{B}\overline{C}} \cdot \overline{\overline{A}\overline{B}C}
\end{aligned}
$$

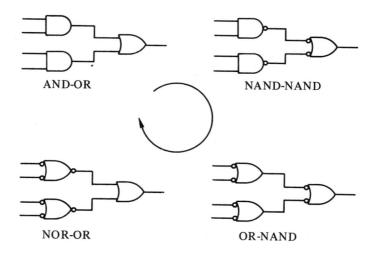

FIG. 2.58   Network Forms Directly Related to
Sum-of-Products Boolean Expressions

Each of these forms suggests a 2-level realization that uses a unique combination of gate types. Figure 2.59 suggests that product-of-sums expressions, as well as product-of-maxterms expressions, can be converted directly to these alternative technologies.

The forms that require input variables to be complemented are not necessarily undesirable. Three of the four independent variables appear complemented in the *sop* expression of the following function.

$$F = \bar{A}C + \bar{A}\bar{D} + \bar{B}C + \bar{B}\bar{D}$$

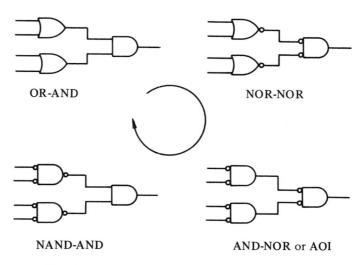

FIG. 2.59   Network Forms Directly Related to
Product-of-Sums Boolean Expressions

Therefore, three inverters must accompany an AND-OR realization, if variables are available in true form only. While OR-NAND and NOR-OR realizations require the same number of gates, only C need be explicitly inverted. Variables A, B, and D are brought directly to the second level gates of such realizations.

The Karnaugh map of this function shown in Fig. 2.60 reveals that fewer groups of 0's exist. Writing the function in *pos* form results in more economical realizations.

$$F = (\bar{A} + \bar{B}) \cdot (C + \bar{D})$$

Again three inverters are needed with an OR-AND realization. Only one inverter and three gates are needed if we elect to realize the function with AND-NOR (AOI) or NAND-AND networks.

This example illustrates another very important point. Using a Karnaugh map or other minimization techniques to reduce sum-of-products expressions also reduces NAND-NAND, OR-NAND, and NOR-OR expressions by the same amount. Reducing OR-AND expressions reduces NOR-NOR, AOI, and NAND-AND expressions. Thus, minimization techniques are more generally useful than we might have suspected.

Reed–Muller Expansions

If two Boolean expressions Be_1 and Be_2 never take the value 1 simultaneously, then their inclusive-OR and their exclusive-OR are equivalent.

$$\text{If } Be_1 \cdot Be_2 = 0 \quad \text{then} \quad Be_1 + Be_2 = Be_1 \oplus Be_2$$

The logic product of two different minterms is logic 0 (See Prob. 2.5-6), so the inclusive- and exclusive-sums of minterms are equivalent. An exclusive-sum of minterms canonical form may be proposed.

$$f(x_1, \ldots, x_n) = \oplus f_d \cdot m_d \qquad \textbf{(2.46)}$$

For $n = 2$

$$f(A, B) = f_0 \cdot \bar{A}\bar{B} \oplus f_1 \cdot \bar{A}B \oplus f_2 \cdot A\bar{B} \oplus f_3 \cdot AB$$

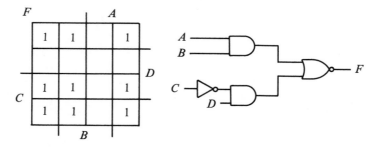

FIG. 2.60 Economy is achieved by writing the minimum *pos* expression and then converting to AOI technology to reduce the number of inverters required.

All complements $\bar{x}_i$ may be eliminated from such expressions by replacing them with $1 \oplus x_i$.

$$
\begin{aligned}
f(A, B) &= f_0 \cdot (1 \oplus A)(1 \oplus B) \oplus f_1 \cdot (1 \oplus A)B \oplus f_2 \cdot A(1 \oplus B) \oplus f_3 \cdot AB \\
&= f_0 \oplus (f_0 \oplus f_1)B \oplus (f_0 \oplus f_2)A \oplus (f_0 \oplus f_1 \oplus f_2 \oplus f_3)AB \\
&= g_0 \oplus g_1 \cdot B \oplus g_2 \cdot A \oplus g_3 \cdot AB
\end{aligned}
$$

where

$$
\begin{array}{ll}
g_0 = f_0 & g_2 = f_0 \oplus f_2 \\
g_1 = f_0 \oplus f_1 & g_3 = f_0 \oplus f_1 \oplus f_2 \oplus f_3
\end{array}
$$

In general, switching function may be expressed:

$$
f(x_1, \ldots, x_n) = g_0 \oplus g_1 x_n \oplus g_2 x_{n-1} \oplus g_3 x_{n-1}x_n \oplus \cdots \oplus g_{2^n-1}x_1 x_2 \ldots x_n \tag{2.47}
$$

where the 2^n coefficients g_i may be calculated from the truth table entries f_i. Even greater generality exists. For example, rather than eliminating all complements as was done above, we may elect to have each variable appear exclusively in true or exclusively in complement form. An expression like that of Eq. (2.47) may be developed for each of the 2^n possible sets of true and complemented variables. The coefficients will differ for each such set. These expressions are Generalized Reed-Muller (GRM) canonical forms for switching functions.[2]

We will concentrate on the GRM expression of Eq. (2.47) and determine how to calculate the coefficients. Let g_i be the coefficient of the product of only those variables that correspond to the 1's in the n-bit binary expression of i. Thus, if $n = 3$,

g_3 is the coefficient of $x_2 x_3 (3_{10} = 011_2)$
g_5 is the coefficient of $x_1 x_3 (5_{10} = 101_2)$
g_6 is the coefficient of $x_1 x_2$

Define integer m to be a *subnumber* of $n(m \subseteq n)$ if each bit m_i in the binary expression of m is less than the corresponding bit n_i of n under the natural order $0 \subseteq 0$, $0 \subseteq 1$, $1 \subseteq 1$. Under this definition $2 \subseteq 6$ $(010 \subseteq 110)$, but $2 \not\subseteq 4$ $(010 \not\subseteq 100)$.

Now the g_i coefficients are easily calculated from the truth table entries f_i for a switching function.

$$
g_i = \bigoplus_{j \subseteq i} f_j \tag{2.48}
$$

First we find all subnumbers of i. This is easily accomplished by replacing 1's in i with 0's in all possible ways.

[2]L. T. Fisher, "Unateness Properties of AND-EXCLUSIVE OR Logic Circuits," *IEEE Transactions on Computers*, Vol. C–23, Feb. 1974, pp. 166–172.

$$5 = 101$$
$$100 = 4$$
$$001 = 1$$
$$000 = 0$$

Since zero is a subnumber of all integers including itself, it appears in all sets of subnumbers. Second, we form the exclusive-sum of the corresponding truth table entries.

$$g_5 = f_0 \oplus f_1 \oplus f_4 \oplus f_5$$

f_0 will appear in all such sums.

While the Reed-Muller expansions describe standard circuits of AND and XOR gates that could be used to realize switching functions, their main application to data has been in theoretical studies.

2.8 AN INTRODUCTION TO DDL

Using "design languages" to express switching functions and to imply networks that realize them has been deferred to this point because such languages are substantial generalizations of the Boolean notation used earlier in this chapter. We now are prepared to accept new operators and formal rules for writing and interpreting statements. Languages consist of: (1) an alphabet of characters that may be used to write statements, (2) rules (syntax) under which valid statements are formed, and (3) interpretations (semantics) of those statements. The formality of syntax rules and statement interpretations ensures reliable communication, not only between people but also between people and computers. The suggested generalizations permit more concise descriptions. We use formal languages to gain the advantages of reliable communication and brevity, which can enable us to more easily understand the operation and organization of large systems.

We will apply one of a number of design languages currently used as communication tools to drive computer programs that simulate or synthesize digital systems. This section concentrates on those parts of Digital Design Language (DDL) that describe combinational logic networks. Later chapters will illustrate how the full language is used to describe sequential logic networks. The appendix to Chapter 5 provides the alphabet and formal syntax of the language. This section is neither as complete nor as rigorous as that appendix. DDL symbols and statements are set in upright rather than italic type so that we can distinguish formal descriptions from informal notation.

The alphabet of DDL includes a number of new symbols as well as the decimal digits, upper- and lower-case English letters, and familiar punctuation. New symbols will be introduced as they become important to us.

Constants, Identifiers, and Lists

Constants are expressed in DDL with syntax

where *n* specifies the value of the constant via one of three number systems, *R* specifies that number system, and *k*, a positive decimal integer, specifies the number of bits in the binary expression of the constant. Constants may be written in binary with $R = B$, octal with $R = O$, and decimal with $R = D$. Thus a 9-bit sequence 000101110 is described in DDL with any of the following notations: 101110B9, 56O9 and 46D9. When a constant is expressed in binary, $R = B$, the question mark (?) may be used to indicate that the value of a bit is of no concern or unknown. Tuples such as $0101x1$ when used to describe product terms exemplify where the question mark becomes important.

$$0101x1 \text{ is } 101\,?1\text{B6 in DDL}$$

Tuples of all 0's, all 1's, or all ?'s may be described with the general syntax above, of course. However, if the length of the tuple is one or implied by its usage in a statement, then only 0, 1, or ? need be written to specify these special binary sequences.

When constants appear in Boolean expressions, they indicate that wires bearing signals of constant amplitude are to exist in a system. Before building a logic network, we usually simplify Boolean expressions, and constants are eliminated in the process. Rather than connecting logic 0 to an AND gate, we eliminate the gate since its output signal will always have value 0. To think of wires bearing logic constants is a valuable way of interpreting constants; those wires may not exist in the actual digital system.

In Chapter 1, we referred to terminals and registers with names. Individual wires and flip-flops were denoted by subscripting those names. Similar notation is used in DDL. Positive decimal integers are used to express subscripts. Zero is considered to be such an integer. Then a subscript *s* is defined in a "language-defining" language as:

$$\begin{aligned}
pd \quad &:: = \text{positive decimal integer} \\
s \quad &:: = pdi \\
s \quad &:: = pdi : pdi \\
s \quad &:: = s, s
\end{aligned} \tag{2.49}$$

which says: a positive integer is a subscript, two integers separated by a colon is a subscript (that expresses a range of values), and two subscripts separated by a comma are a subscript (that identifies a member or members of a multidimensional array of items). A name or basic identifier *bid* may be as simple as one letter. Or, it may consist of a letter followed by letters or digits.

$$\begin{aligned}
bid &:: = letter \\
bid &:: = bid\ letter \\
bid &:: = bid\ digit
\end{aligned} \tag{2.50}$$

Two forms of identifiers *id* are then

$$\begin{aligned}
id &:: = bid \\
id &:: = bid[s]
\end{aligned}$$

Subscripts are enclosed in square brackets in DDL, parentheses being used for other purposes. The following are all valid identifiers under these definitions.

$$A, A1, ab[3], P2[0 : 4], M[16, 1 : 35], x[3, 4, 1 : 29]$$

We cannot know what items of hardware these *ids* reference until we know what system component has been named A, what other component has been named $A1$, etc.

We often have collections of identifiers, equations or operation statements to be treated in the same manner. A record of such items separated by commas is a *list*. We listed a number of identifiers above with this syntax; subscript lists appear in two of the identifiers. The last recorded item is not followed by a comma; commas separate the items. Usually the order of appearance of items in a list is of no significance. All of the items are to be treated in the same manner and at the same time. This point is particularly important when operations are listed. Then all of the operations are to be performed simultaneously, not in the sequence in which they are listed. Order does have importance in a subscript list in which the items refer to the different dimensions of a multidimensional collection of items. The first subscript in a subscript list references the first dimension; the last subscript references the last dimension.

⟨Terminal⟩ Declarations

All hardware items are explicitly declared or implied to exist with declaration statements in DDL. The simpler declaration statements have the form

$$\langle DT \rangle \; body.$$

where DT symbolize " declaration type" and the body of the declaration indicates what hardware exists of type DT. This is done by listing names and dimensions of hardware items or by indicating how declared items are to be interconnected or how they are to interact. It takes logic networks to interconnect other logic networks so these later declarations imply the existence of hardware, while clearly indicating their function. A few specific examples will make these very general statements clear.

DT consists of at least the first two letters of words that describe the hardware items specified in a declaration. Thus, when terminals are being specified, at least " TE" must appear within angle brackets; and " TERMINAL " can appear if we choose to present the full word. We will use $\langle DT \rangle$ to abbreviate " declaration of type DT." Thus in English sentences read ⟨TE⟩ as "terminal declaration."

Wires are declared to exist and assigned names with the ⟨TE⟩. Not all wires need be so declared. In fact, usually most wires are implied to exist. The body of the ⟨TE⟩ may consist of just a list of identifiers with or without subscripts. This declaration, like all others in DDL, terminates with a period (.) just as English sentences usually do. Therefore, the last item on the list is followed by a period. If identifiers appear without subscripts, those identifiers name single wires. The subscripts of subscripted identifiers give the dimensions of terminals and the order assigned to their members. If i and j are subscripts, then $id[i:j]$ indicates that terminal id is of single dimension with $|i-j|+1$ members; id_i is the left member; id_j is the right member. More generally, $id[i_1:j_1, i_2:j_2]$ specifies a two-dimensional array of wires with $|i_1-j_1|+1$ rows and $|i_2-j_2|+1$ columns; i_1 is the top row while j_1 is the bottom row, i_2 is the left column, and j_2 is the right column. One special case is provided in DDL. If a single subscript i is listed rather than a range, the range is $1:i$. Thus, $id[i, j]$ specifies a two-dimensional array with i rows and j columns; rows being numbered $1, 2, \ldots, i$, and columns being numbered $1, 2, \ldots, j$.

$$\langle TE \rangle \ A, \ B[5], \ C[0:4], \ D[1:6, 5:1]$$

declares one wire A, five B wires with B_1 on the left and B_5 on the right, five C wires (C_0, C_1, C_2, C_3, and C_4), and 30 D wires organized in 6 rows and 5 columns.

Additional syntax is permitted in the body of a $\langle TE \rangle$. Two or more names may be assigned to the same terminal by separating identifiers or concatenations of identifiers with an equal sign ($=$). Recall that we used a small centered circle to express concatenation. The same symbol is used in DDL. Then the declaration

$$\langle TE \rangle X[10] = Y[4:0] \circ Z[5].$$

indicates the existence of only ten wires. The leftmost of these has names X_1 and Y_4; the next one has names X_2 and Y_3. The rightmost wire is identified by both X_{10} and Z_5. While it is never necessary to give two or more names to a wire, it is very desirable to do so when we wish to refer to different subsets of a terminal as well as the entire terminal. It is rather obvious that the total dimension of the concatenated identifiers on the right must equal the dimension of the identifier listed on the left.

⟨REgister⟩ Declarations

Memory elements are stated to exist with the REgister and other declarations to be discussed in a later chapter. The syntax of the $\langle RE \rangle$ is identical to that of the $\langle TE \rangle$. Only their semantics differ; the $\langle RE \rangle$ declares the existence of flip-flops in a system and assigns them names. A single flip-flop F, a 16 by 12 collection of memory elements M, and a 16-bit register R with three subregisters are declared by

$$\langle RE \rangle \ F, \ M[16, 0:11], \ R[16] = OP[4] \circ IX[2] \circ ADR[10].$$

The left four flip-flops of register R form the OP subregister. The next two flip-flops have names $R_5 = IX_1$ and $R_6 = IX_2$. The total dimension of the three subregisters equals the dimension of R.

⟨TIme⟩ Declarations

Clocks are declared to exist, their signals named and their periods given with this type of declaration. In the next chapters we will continue to think of clock signals as being the ideal signals shown in Chapter 1. Only their names and periods are then important. Clock periods are enclosed in parenthesis between the basic identifier and bracketed subscript range, if any, and expressed with notation like the FORTRAN E form; thus

$$⟨TI⟩ \ P(1E - 8), Q(5E - 6)[2].$$

declares that P is a clock with 10 nanosecond period and two 5 microsecond Q clocks exist. Phase relationships between P, Q_1, and Q_2 cannot be specified.

Computer hardware description languages such as DDL attempt to clearly indicate the function and structure of digital systems, while implying or ignoring many fine details of their construction and operation. We are familiar with, and accustomed to, using ideal models of electronic gates. Languages are tools for preparing concise, abstract, and idealized models of digital systems. They therefore are most useful at the early stages of system design, long before hardware construction begins. While we must never forget the imperfections and limitations of electronic circuits, introducing such detail is a hinderance at the early stages of design where system organization and operation are the pressing problems. Introducing electronic detail into design language models eliminates their advantages. Thus, while it is possible to describe exactly what gates and memory elements are to be used and how they are to be interconnected, introduce propagation delay and specify clock pulse width and clock phase relationionship, etc., DDL does not make it easy to do so.

Boolean Expressions

We begin to see very powerful operators expressed with simple symbols as we look at Boolean expressions, *Be*. Boolean expressions appear as conditions that determine whether operations are performed or not, and as descriptions of information sources and sinks in Boolean equation-like and register transfer statements. In all of these cases they declare combinational logic to exist by implication. They specify switching functions and therefore imply logic network structures, but final network designs are usually derived by simplifying Boolean expressions with the limitations and costs of the available technology in mind.

Identifiers in Boolean expressions refer to terminals, registers, clocks, etc., declared to exist via appropriate declarations. If $A[1 : 10]$ appears in a ⟨TE⟩, then A in a *Be* refers to the entire terminal and $A[1]$ refers to the leftmost member of the set of ten wires, while $A[10]$ identifies the rightmost. $A[2 : 5]$ references a subset of 4 wires. Notice that "$A[10]$" means one thing in a ⟨TE⟩ and another in a *Be*. If $R[0 : 15, 12]$ is declared in a ⟨RE⟩, then $R[1, 11 : 12]$ refers to the right two members of the second row of memory elements. $R[1, 1 : 12]$ refers to the entire second row. This is a very usual type of reference. It is so common that $R[1]$ is considered to make the same reference in DDL. If the entire range(s) of the least (lesser) significant subscripts is (are) referenced, this range(s) may be omitted from the subscript list.

A Boolean expression can be thought of as a name, the name of the output terminals of the combinational logic network prescribed by the *Be*. The operators that may appear within Boolean expressions are listed in Table 2.11 in their hierarchical order. This means, unless parentheses appear to indicate otherwise, all HEAD operations are to be performed first, all TAIL operations second, ... and all OR combinations are to be made last.

To illustrate the syntax and semantics of these operators we will assume that the following declaration prevails.

$$\langle TE \rangle \; A[10], \; B[10].$$

The HEAD and TAIL operators provide a means of selecting the leftmost and rightmost, respectively, k members of the left operand, where k, a

TABLE 2.11 LOGIC OPERATORS OF DDL

Operator	Symbol	Typical Syntax	Result Assuming A and B Are of Dimension n, and k is an Integer
Head	$\vdash$	$A \vdash k$	k leftmost elements of A
Tail	$\dashv$	$A \dashv k$	k rightmost elements of A
Extension	$\times$	$A \times k$	k copies of A
Concatenation	$\circ$	$A \circ B$	a set consisting of all elements of A and B
Complement	$\neg$	$\neg A$	bit-by-bit complement of A
Selection	$\backslash$	$A \backslash kDn$	selective complement and omit elements of A
Reduction	$/$	ϕ/A	$A_1 \phi A_2 \phi \cdots \phi A_n$ where ϕ is any of the operators AND to OR below
Add	$+$	$A + B$	arithmetic sum of A and B
Subtract	$-$	$A - B$	arithmetic difference of A and B
Less than	$<, >$	$A < B$, $B > A$	true if A is the smaller number
Less than equal	$\leq, \geq$	$A \leq B$, $B \geq A$	true if A is less than or equal to B
Equal	$\doteq$	$A \doteq B$	true if A and B are the same number
Not equal	$\neq$	$A \neq B$	true if A and B are different numbers
AND	$\wedge, \cdot$	$A \wedge B$	bit-by-bit logic product
NAND	$\uparrow$	$A \uparrow B$	bit-by-bit complement of logic product
NOR	$\downarrow$	$A \downarrow B$	bit-by-bit complement of logic sum
Exclusive-NOR	$\odot, \equiv$	$A \odot B$	bit-by-bit equivalence
Exclusive-OR	$\oplus$	$A \oplus B$	bit-by-bit exclusive-or
OR	$\vee$	$A \vee B$	bit-by-bit logic sum

SECTION 2.8   **An Introduction to DDL**

decimal integer, is the right operand. Then $A \vdash 4$ identifies the same subset of wires as $A[1 : 4]$. $B \dashv 3$ references $B[8 : 10]$. These subsets could have been assigned names in the $\langle TE \rangle$ also, so clearly the head and tail operators are not essential to DDL. They are very concise ways to reference special subsets.

The EXTENSION operator ($\times$) creates k copies of the left operand. Thus $A[1] \times 5$ indicates that wires A_1 is to be divided to five wires all bearing the same signal. A fan-out is prescribed.

CONCATENATION ($\circ$) permits combining items to form a larger set without explicitly naming that set. Then $A[3 : 7] \circ B[2 : 6]$ describes a set of ten wires, the left five being members of terminal set A and the right five being members of B. No hardware other than a fan-out of wires is implied by a concatenation operation.

We have used the overbar to denote logic complement and will continue to do so. In a formal language where statements are formed by writing characters from a fixed alphabet next to one another, overbars violate this basic formation rule. On paper we do not hesitate to violate it in many ways, but if we expect to punch statements into cards or paper tape for submission to a computer then we must adhere to the rule. DDL therefore provides a prefix COMPLEMENT ($\neg$) operator, meaning the operator symbol is written before the variable or Boolean expression to be complemented. If that operand has dimension greater than one, then it is complemented on a bit-by-bit basis. We begin to see that the DDL logic operators are those of the Boolean algebra of tuples, which were discussed in Section 2.3 and defined in Eq. (2.29).

SELECTION ($\backslash$) is a generalization of complementation. The left operand is a variable; the right operand is a constant with length equal to the dimension of the variable. Members of the variable and bits of the constant are combined in a bit-by-bit fashion and use the rules to follow. (These should be compared with Table 2.6.)

$$A_i \backslash 1 = A_i$$
$$A_i \backslash 0 = \bar{A}_i \qquad\qquad \textbf{(2.51)}$$
$$A_i \backslash ? = \text{vacuous}$$

$A[1 : 5] \backslash 7D5$ $(A[1 : 5] \backslash 00111B5)$ describes the vector variable $(\bar{A}_1, \bar{A}_2, A_3, A_4, A_5)$. Complements of A_1 and A_2 are selected while A_3, A_4, and A_5 are selected in true form. In terms of hardware this selection example describes a terminal of five wires bearing signals $\bar{A}_1$, $\bar{A}_2$, A_3, A_4 and A_5. If the complement signals are not available, then this selection operation implies the existence of two inverters. $A \backslash 10???????10B10$ specifies a terminal of 4 wires bearing signals A_1, $\bar{A}_2$, A_9 and $\bar{A}_{10}$. The selection operator thus forms a terminal set by omitting or including in true or complement form members of its left operand.

Selection is often denoted in switching theory literature with superscript notation. If x_i is the ith independent variable and b_i is the ith bit of a tuple of elements 0, 1, and x, then

$$x_i^{b_i} = \begin{cases} x_i & \text{if } b_i = 1 \\ \bar{x}_i & \text{if } b_i = 0 \\ \text{vacuous if } b_i = x \end{cases} \tag{2.52}$$

While we have not used this notation previously, we have related products terms and tuples with the selection idea.

Selection is often used in conjunction with the REDUCTION (/) operator, which combines the members of its right operand according to the logic operator given as its left operand. The "AND reduction" of A is then a logic product

$$\cdot /A = A_1 \cdot A_2 \cdot \cdots \cdot A_{10}$$

A high fan-in AND gate is implied to exist. Other types of reduction including NAND and NOR also imply high fan-in gates:

$$\uparrow/A = \neg (\cdot/A) = \overline{A_1 \cdot A_2 \cdot \cdots \cdot A_{10}}$$
$$\downarrow/A = \neg (v/A) = \overline{A_1 \vee A_2 \vee \cdots \vee A_{10}} \tag{2.53}$$

The result of reduction is a variable with unit dimension, a single wire.

Since selection precedes reduction in Table 2.11, parentheses are not needed when these operators are used together to describe logic for detecting special cases. For example, $\cdot /A[1:5]\backslash 3D5$ selects $\bar{A}_1$, $\bar{A}_2$, $\bar{A}_3$, A_4, and A_5 and then reduces to form the logic product $\bar{A}_1 \cdot \bar{A}_2 \cdot \bar{A}_3 \cdot A_4 \cdot A_5$. This product has value 1 only when $A[1:5]$ bears the bits of 3D5. In a sense the AND gate "detects" the presence of the number 3 on the $A[1:5]$ lines. Similarly, if $X = \{x_1, x_2, \ldots x_n\}$ is a set of binary variables and d is a decimal integer less than 2^n, then $\cdot /X \backslash dDn$ describes minterm m_d.

$$m_d = \cdot /X \backslash dDn \tag{2.54}$$

If some variables are omitted by the selection operation, then product terms with absent variables are formed by AND-reduction. We now have operators for translating tuples to product terms.

Selection in conjunction with OR-reduction forms sum terms with maxterms being a special case. The differences between Tables 2.6 and 2.7 and the manner in which maxterms are named in Table 2.9 require the use of the complement operator.

$$M_d = v /\neg X \backslash dDn$$
$$= v /\neg (X \backslash dDn) \tag{2.55}$$

Since complement appears above selection in Table 2.11, the first line indicates that we are to select from the set of complemented variables. The parentheses in the second line direct us to select first and then complement all selected items. If the selection constant contains ?'s, then the selected set will have dimension less than that of X, and a sum term with absent variables is specified.

ARITHMETIC and RELATION operators are included in DDL because they are common operations in digital systems. However, since the rules for adding, subtracting, and comparing numbers depend upon the code used to express those numbers, and a variety of codes are commonly used (as we will see more fully in the next chapter), the symbols for these operators cannot imply the hardware required to realize them. We would prefer different arithmetic and relation symbols for each code that we are likely to use, but the gain would be small. Arithmetic and comparison networks are relatively complicated and their performance strongly influences system performance. As a result they are designed with great care. We use the arithmetic and relation operators of DDL when we are being very abstract, and then when we move toward implementation we use Boolean expressions to describe how we intend to realize those operators.

The LOGIC operators of DDL are generalizations of the operators of the Boolean algebra of tuples studied in Section 2.3. Generally they operate on variables of the same dimension in a bit-by-bit fashion. One exception is recognized in DDL. If one operand has unit dimension, its extension to the dimension of the other operand is implied, thus

$$A \cdot B = (A_1 \cdot B_1, \ldots, A_{10} \cdot B_{10})$$

whereas

$$A \cdot B[1] = (A_1 \cdot B_1, A_2 \cdot B_1, \ldots, A_{10} \cdot B_1)$$

Please notice that OR is denoted with $\vee$ in DDL, $+$ being used to denote addition. Logic operators imply the existence of corresponding logic circuits. Operand-operator symbol sequences may also be viewed as names of the networks they imply and the signals these networks produce.

Boolean expressions can now be defined much as they were in Section 2.2. We begin by indicating that a constant or identifier is a Be, and that parentheses may be used to specify the order in which operators are executed.

$$\begin{aligned} Be &::= kRn \\ Be &::= id \\ Be &::= (Be) \end{aligned} \tag{2.56}$$

The operators of Table 2.11 may appear in Boolean expressions.

$$\begin{aligned} Be &::= Be \vdash k, Be \dashv k, Be \times k \\ &::= Be_1 \circ Be_2 \\ &::= \neg Be, Be \backslash kRn \\ &::= \phi / Be \\ &::= Be_1 + Be_2, Be_1 - Be_2 \\ &::= Be_1 < Be_2, Be_1 \le Be_2, Be_1 > Be_2, Be_1 \ge Be_2 \\ &::= Be_1 \doteq Be_2, Be_1 \ne Be_2 \\ &::= Be_1 \wedge Be_2, Be_1 \cdot Be_2, Be_1 \uparrow Be_2, Be_1 \downarrow Be_2 \\ &::= Be_1 \odot Be_2, Be_1 \oplus Be_2, Be_1 \vee Be_2 \end{aligned} \tag{2.57}$$

If parentheses do not indicate otherwise, the operators that appear in a Boolean expression act on their operands;

1. In order of decreasing operator rank, and
2. In their order of appearance from left to right if they are of equal rank.

Thus:

$$A \cdot B \vee C \quad \text{means} \quad (A \cdot B) \vee C$$
$$A \oplus B {\downarrow} C \quad \text{means} \quad A \oplus (B \downarrow C)$$
$$A {\downarrow} B {\downarrow} C \quad \text{means} \quad (A \downarrow B) \downarrow C \quad \text{equal rank}$$

If A, B, and C are all of dimension 10, then each Boolean expression above implies a network of 20 gates. Each network consists of 10 identical 2-gate cells.

⟨Boolean⟩ Declarations

The body of a ⟨BO⟩ is a list of Boolean equations. Boolean equations have syntax

$$Be_1 = Be_2$$

with the equal sign ($=$) symbolizing the CONNECTION operator. It connects the terminals of Be_2, the *source* of information, to Be_1, the *sink* of that information. This hardware function interpretation of the equal sign does not violate our previous applications of that symbol. A Boolean equation may be interpreted to specify a connection; in less formal notations we tend to think of the connection as permanent. We will soon find that in DDL a connection operation may be conditional and the operation performed (connection established) only when the condition is satisfied.

It is essential that terminals be identified on the left of the connection operator. We do not connect to registers. Further, only the head, tail, and concatenation operators of Table 2.11 may be used to identify the sink of information, Be_1. Identifiers of terminals, registers, clocks, and so forth, may appear in the source expression Be_2. If register names appear in Be_2, their output terminals are being referenced.

The circuitry of system SELECT of Section 1.2 with the typical cell of Fig. 1.8 may now be described with the following declarations.

⟨TE⟩ A, B[10], C[10], D[10], E[10], F[10].
⟨BO⟩ D = A · B, E = ¬ A · C, F = D ∨ E.

The A signal is extended to meet the dimension of B; $\bar{A}$ is extended to meet the dimension of C. If it is not important to explicitly name the intermediate D and E terminals, the network may be described with

⟨TE⟩ A, B[10], C[10], F[10].
⟨BO⟩ F = A · B ∨ ¬ A · C.

The same network of 20 AND gates and 10 OR gates is implied.

Conditions

When discussing system SELECT in Chapter 1, we found the English "If ... then ... else ..." structure to be a very clear way of indicating the source to be connected to the sink. In system COUNT it also clearly indicated that a register is to be loaded only when a condition (clock pulse present) is satisfied. *If-then* statements are often a very natural way of indicating that operations are to be performed only when a condition is satisfied. Hence they appear in DDL where "If" and "then" are symbolized with vertical lines enclosing a Boolean expression of the condition to be satisfied. "Else" is expressed with the semicolon (;) in DDL. A period is used to indicate the end of the domain over which the *if-then* condition prevails.

Extending the definition of a Boolean expression illustrates the use of these symbols. Conditional Boolean expressions are *Be*s.

$$Be ::= |Be_1| \ Be_2.$$
$$::= |Be_1| \ Be_2; \ Be_3. \qquad\qquad (2.58)$$

System SELECT may be described using the second of these forms.

$$\langle TE \rangle \ A, \ B[10], \ C[10], \ F[10].$$
$$\langle BO \rangle \ F = |A|B; C..$$

F is to be connected to *If* $A = 1$ *then B else C*. One period is used to close the condition; the second period closes the $\langle BO \rangle$. The expression $|A| \ B; C.$ is more abstract than $A \cdot B \lor \neg A \cdot C$; the hardware that it implies is less obvious, although we see that it can be found by translating $|A|B; C.$ to $A \cdot B \lor \neg A \cdot C$ (which does suggest a realizing network).

If-then may condition connection and register transfer operations as well as Boolean expressions. When the condition is satisfied, the operation is performed. When the condition is not satisfied, the operation is not performed. Another way of thinking about conditional operations emerges. When the condition is satisfied, the operation is *active*; otherwise, it is inactive. When reading a DDL description, we can concentrate on the active operations, ignoring those that are inactive. Usually only a few operations are active at a time so that our attention is focused on a small part of the description. This will be even more true of other ways to express conditions described in Chapter 5.

A "compatible set of operations" (*csop*) is a list of connection and transfer operations such that all operations may be performed simultaneously without hardware conflicts. Operations that connect different sources to the same sink or load a register from different sources are not compatible since the results of performing them simultaneously are not necessarily what we desire or expect. Compatible operations may be conditioned with syntax

$$|Be|csop_1; \ csop_2.$$

If *Be* is true, evaluates to 1, then the operations of $csop_1$ are to be performed; "else", i.e., $Be = 0$, those of $csop_2$ are to be executed. The multiplexer of system SELECT can be described

$$|A|F = B; F = C.$$

within other declarations than the $\langle BO \rangle$. If $csop_2$ is an empty set, the syntax is simplified to

$$| Be | csop.$$

No operations are active when $Be = 0$.

The definition of $csop$ may be generalized to

$$csop :: = | Be | \; csop.$$
$$csop = | Be | \; csop_1 ; csop_2 . \qquad \textbf{(2.59)}$$

This generalization admits the nesting of conditions. Punctuation is very important in properly expressing nested conditions. With the terminal declaration

$$\langle TE \rangle \; A[2], \; B[10], \; C[10], \; D[10], \; F[10].$$

the operation

$$| A[1] | \; | A[2] | \; F = B; \; F = C.; \; F = D.$$

indicates that if $A_1 = 1$, then if $A_2 = 1$, F is connected to B else F is connected to C, else ($A_1 = 0$) F is connected to D. This may not be as clear in English as it is in DDL since we do not usually nest conditions in English. What is meant is: If $A_1 = A_2 = 1$ then $F = B$; if $A_1 = 1$ and $A_2 = 0$ then $F = C$; if $A_1 = 0$ then $F = D$.

This example illustrates several things about DDL. First, connections need not be permanent; they are made only when prevailing conditions are satisfied. F is never connected to more than one of B, C, and D at any time. Second, conditioned operations must be fully and properly nested. The left period in the example above closes the *if-then* condition to its immediate left. The right period terminates the A_1 condition. The leftmost semicolon expresses *else* for the immediately preceding *if-then* condition. No other interpretation of the punctuation is possible in DDL under Eq. (2.59).

When a Boolean expression to be used as a condition has dimension greater than one, *if-then* clauses become verbose and tend to hide the control that the Be is exerting. The IF-VALUE clause more clearly and briefly describes the "decoding" of dimensional conditions. Think of the n bits of Be as expressing a positive integer via the binary number system. Let a *value* (va) be a positive decimal integer less than 2^n, a list of values, or a range of values.

$$va \; :: = pdi$$
$$:: = va, va$$
$$:: = pdi : pdi \qquad \textbf{(2.60)}$$

The DDL syntax

$$\lceil Be \lfloor va \; csop.$$

then says: "*If Be* has value *va*, then do the operations of *csop*" where $\lceil$ expresses "if" and $\lfloor$ says "then do."

The 3-way multiplexer described with nested *if-then* clauses above is more clearly described

$$\lceil A \lfloor 0 : 1 \; F = D \lfloor 2 \; F = C \lfloor 3F = B.$$

The period closes the *if-value* condition. When the decimal equivalent of the binary integer expressed by the two bits of terminal A falls in the range 0 to 1, then F is connected to D. If A expresses value 2, then $F = C$ is the active operation. When the value of A is 3, $F = B$. A semicolon may be listed as *va* to express the *else* value—all values from 0 to $2^n - 1$ not included in any of the other *va* values of the *if-value* clause.

$$\lceil A \lfloor 2 \ F = C \lfloor 3 \ F = B; \ F = D.$$

indicates that if the value of A is other than 2 or 3, $F = D$ is the active operation.

Boolean expressions may be conditioned with if-value clauses. The final extension of the definition of *Be* is:

$$Be \ :: = \lceil Be_1 \ value \ list. \tag{2.61}$$

where *value list* consists of entries of the form $\lfloor va_i \ Be_i$ not separated by commas since $\lfloor$ separates them. Again, the 3-way multiplexer illustrates the syntax.

$$F = \lceil A \lfloor 2 \ C \lfloor 3 \ B; \ D.$$

While the semantics of *if-then* and *if-value* clauses is clear because we use them in English, the logic implied by these relatively abstract constructions is not as obvious. Formal rules for transforming conditioned Boolean expressions and connection operations to Boolean expressions and equations are beyond our interests at this point. Therefore, we will take a more intuitive approach here and in the next chapter. If paths between sources B, C, and D and sink F are via an AND to OR gate pyramid, then signals that tell if A has a specific value, or not, can "control" by making one and breaking the other possible source-to-sink paths. Generating such control signals is called "decoding." In general, AND gates that realize minterms decode the signals on a terminal. In our example the range 0 : 1 is specified; that is, the two minterms $\bar{A}_1 \cdot \bar{A}_2$ and $\bar{A}_1 \cdot A_2$ can be combined to the simpler product term $\bar{A}_1$, which indicates whether the value of A falls in the range or not. Figure 2.61 shows a typical cell of the 3-way multiplexer.

Register Transfer

Design languages are often called "register transfer languages" because the register transfer operation with syntax parallel to that of the connection operation,

$$Be_1 \leftarrow Be_2$$

is one of the most frequently written and descriptive operations. The information sink (Be_1) of the register transfer operator ($\leftarrow$) must be expressed in terms of identifiers declared in $\langle RE \rangle^3$ and the concatenation, head and tail operators. Input terminals of flip-flops are referenced by

---

[3] Also $\langle ME \rangle$ and $\langle LA \rangle$ as discussed in Chapter 5.

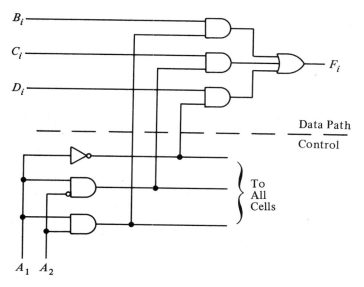

FIG. 2.61   The 3-Way Multiplexer Implied by
$F = \lceil A \lfloor 2C \lfloor 3B \; ; D.$

Be_1; if register identifiers appear in the source Boolean expression (Be_2) flip-flop output terminals are referenced.

A register transfer operation does not by itself imply all of the information required to load a register. It implies data flow logic. Control logic— the clocking signal—is implied by conditions on the register transfer statement. *If-then* and *if-value* conditions will be used now. We will encounter more elegant ways of expressing clock conditions later. The network of Fig. 1.9, system COUNT, may be described:

$$\langle TI \rangle \quad P(1E - 7).$$
$$\langle RE \rangle \quad A[4].$$
$$\langle TE \rangle \quad C[4] = C1 \circ CC[3], \text{ IN}.$$
$$C = A \cdot (CC \circ IN),$$
$$|P| \; A \leftarrow A \oplus (CC \circ IN).$$

Note that the members of C are given two names so that subscripted identifiers need not be used in the connection and transfer operations. The detailed description of the operation of this network previously given (in Section 1.2) applies to the DDL description above.

General Considerations

We are now familiar enough with DDL to appreciate a number of its general features. First, the blank is not a member of the DDL alphabet. We use blanks within and between statements when we write statements on separate lines, but such divisions do not alter the meaning of the statements. Blanks enhance human readability of a system description only;

their possible presence complicates computer programs prepared to read DDL descriptions.

Second, comments are as valuable in design language descriptions as they are in computer programs. Comments may be included in DDL descriptions as: (1) enclosed in parentheses following the period that terminates declarations beginning with $\langle DT \rangle$ and (2) in COmment declarations with syntax

$$\langle CO \rangle \ body.\langle$$

Comments of the first type may not include the right parenthesis ( )); it serves as the end delimiter of such comments. The body of a $\langle CO \rangle$ may not contain a left angle bracket ($\langle$); the period is not used as a delimiter here because it is useful within the body to punctuate English. The left angle is the first symbol of the next declaration, of course.

Third, DDL is a *block structured* language. It is usual to think of a digital system as composed of subsystems that are portions of the total hardware, which act in parallel with more or less independence. Subsystem hardwares may or may not be physically separated in many cabinets or portions of one cabinet. Each subsystem is then thought of as a collection of smaller units that act more or less independently. This sort of division is very valuable, and DDL permits dividing a description to reflect blocks within blocks and blocks in parallel. In fact, we saw the first examples of this block structure when discussing conditions. *If-then* statements can be written in parallel suggesting blocks of hardware that acts in parallel. Nesting of *if-then* statements emphasizes that smaller blocks must be entirely within larger ones. The boundaries of one block cannot cross those of another block.

The facilities (terminals, registers, etc.) declared in a block are *local* to it. They may be referenced within it without question even if facilities with the same names exist in parallel encircling blocks. These facilities are *global* to blocks within the block in which they are declared. In any block only local and global facilities may be referenced. Facilities declared in parallel blocks and their subblocks may not be referenced directly.

The block that encompasses all other blocks is declared with the SYstem declaration in DDL.

$$\langle SY \rangle \ head \ body.$$

Its heading part may take any of the following forms:

$$
\begin{aligned}
head \ &:: = id: \\
&:: = id:csop \\
&:: = id:Be: \\
&:: = id:Be:csop
\end{aligned}
\qquad \textbf{(2.62)}
$$

The system is named by *id*, which is usually free of subscripts. Compatible operations may be listed as part of the head. A condition to prevail over all register transfer operations (additional types will be presented in Chapter 5) may be listed as *Be* in the head. Connection operations are excluded since they are not subject to a clock signal, which is usually the significance of *Be* in the head of a ⟨SY⟩. In fact, it is possible to list all of the operations of a system in the head. The body of a ⟨SY⟩ consists of the ⟨TE⟩, ⟨RE⟩, etc., declarations that tell what facilities exist and how they are interconnected and process information.

Automata are relatively independent, disjoint portions of a digital system. An automaton is declared to exist with the AUtomaton declaration.

$$⟨AU⟩ \ head \ body.$$

The head may take any of the forms discussed above. Its components have the same meaning as in the ⟨SY⟩. We give the automaton a name that may be subscripted if identical parallel automata are to exist. We may state a condition on all register transfer operations. In Chapter 5, we will discuss several types of declarations that must appear within the body of AUtomaton declarations, if they are to appear. Since they and the ⟨AU⟩ are most useful for describing sequential switching circuits, we will do no more here than illustrate the block nature of DDL.

System $S1$ defined in Fig. 2.62 consists of three major blocks. The system itself is the largest block. Since its facilities such as P are global to the two automata blocks and all blocks within them, they may be referenced at any point and are therefore *public* facilities. Boolean expressions in the head of the ⟨SY⟩ must be expressed in terms of these facilities. The two parallel automata may use these facilities to communicate with each other. Facilities declared within an ⟨AU⟩ are local to that automaton.

⟨SY⟩ $S1$: $csop_1$
 ⟨TI⟩ $P(1E - 6)$, ...
 ⟨RE⟩ ... . ⟨TE⟩ ... .
 ⟨AU⟩ $A1$: P: $csop_2$
   ⟨TE⟩ ... . ⟨RE⟩ ... . . (end of A1)
 ⟨AU⟩ $A2$: P: $csop_3$
   ⟨TE⟩ ... . ⟨RE⟩ ... . . (end of A2)
. (end of S1)

  $S1$

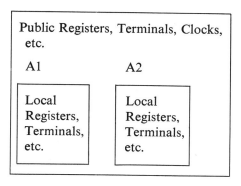

Fig. 2.62 Block Structure of DDL

⟨OPerator⟩ Declaration

At times it is desirable to identify a combinational logic network as a block. It is essential to do so if one logic network is to participate in different operations—included in different compatible sets of operations that are not active simultaneously. The OPerator declaration provides a means of defining a combinational logic block that is to combine one collection of signals at one time with the results going to one place, and to combine other input signals at other times with the results going to the same or different places. Such a network is *time-shared*.

The head of an ⟨OP⟩ is a list of identifiers that name and dimension the output signals of the OPerator network. These identifiers may have parenthetic arguments that appear as a list of identifiers, which are enclosed in parentheses between the basic identifier and bracketed dimensions. The body of the declaration consists of a
⟨TE⟩ that declares the parenthetic arguments to be local terminals, and a ⟨BO⟩ that defines the logic of the OPerator network. To illustrate and explain these definitions, we use the combinational logic of Fig. 1.9, which incremented or passed the number supplied to it.

$$⟨SY⟩\ S2:\ P:$$

|A| R ← COUNT(R, 1), |OVF| S ← 1D4..,
|B| S ← COUNT(S, 1).,
|C| S ← COUNT(R, 0).
⟨TI⟩ P(1E − 6).
⟨TE⟩ A, B, C.
⟨RE⟩ R[3], S[4].
⟨OP⟩ COUNT(X, IN) [4], OVF
⟨TE⟩ X[4], IN, C[4].
⟨BO⟩ C = X · (C[2:4]∘IN),
COUNT = X ⊕ (C[2:4]∘IN),
OVF = C[1] ... (end BO, OP, SY)

COUNT and *OVF* are the output terminals of the OPerator network within system *S2*. We might prefer to call this the *COUNT* block. Parenthetic arguments such as *X* and *IN* must be declared in a ⟨TE⟩ within the body of the ⟨OP⟩. They then are local to the OPerator network, as is *C* in the example, and may not be referenced outside of the ⟨OP⟩. Parenthetic arguments name some or all of the input terminals of the OPerator network. Boolean equations in the body of the block specify the combinational network in terms of parenthetic arguments and/or identifiers global to the ⟨OP⟩. The example indicates that four AND gates and four XOR gates are to be connected as in Fig. 1.9, but with the input terminals being *X* and *IN*, and the output terminals being *COUNT* rather than *D*, and *OVF* rather than C_1.

Connections are made to an OPerator block by using its output terminal identifier(s) in the source Boolean expression of a connection or register transfer operation. Arguments that are listed in parentheses in references indicate what variables are to be combined by the block—and must correspond in order and dimension to the parenthetic list in the head of the $\langle OP \rangle$. Thus in $S2$, if $A = 1$, then the output terminals of register R are to serve as X. A connection between R and the X input terminals of the combinational network is implied. Further logic 1 is to be supplied to the IN terminal. Output terminal OVF is also referenced, serving as a condition on a transfer to the S register. The $COUNT$ terminals must be connected to the R flip-flops so that register will be loaded with the P clock pulse.

When $B = 1$, $COUNT$ is used to increment the S register. Since $COUNT$ cannot increment two numbers at the same time, it is absolutely essential that A and B never have the value of 1 simultaneously. Finally, if $C = 1$, the contents of R are duplicated in S without modification since logic 0 is supplied to the IN terminal. $COUNT$ is being used as a data path between the registers to avoid additional wires.

Considerable logic is implied by the references to $COUNT$ in the operations of $S2$. When either A or C equal 1, the R flip-flops must be connected to the X terminals. When $B = 1$, the connection $X = S$ is needed. A multiplexer is suggested. It is explicitly described below.

$$X = |A \vee C| R. \vee |B| S.$$
$$= (A \vee C) \cdot R \vee B \cdot S$$

The multiplexer implied for the IN terminal can be simplified with Boolean algebra.

$$IN = |A \vee B| \ 1. \vee |C| 0.$$
$$= A \vee B$$

Registers R and S must be constructed to load from terminal $COUNT$ or to hold their content. Register S must also be able to load the constant 1D4. We will dwell on register design in a later chapter. The point here is that we are using one block of combinational logic to make counters out of two registers, thereby avoiding the cost of duplicating the combinational logic, yet incurring the expense of being unable to advance both registers simultaneously!

FOR, $\langle$IDentifier$\rangle$ and $\langle$ELement$\rangle$ Declarations

While the vector-like notation of DDL provides a very concise means of describing parallel information processing networks, there are times when a more explicit description is desired or necessary. When using subscripted variables, we often write an equation and indicate that it is valid "for", and then we give a list or range of values of an index variable used to express subscripts. The *for clause* permits us to use the same technique in DDL.

$\{index = value \ list\}$ *operations described in terms of index.*

The clause opens with braces ({ }) enclosing the index variable and its range or list of values. It closes with a period. We could have written

$$\{i = 1 : 3\}\ C[i] = C[i + 1] \cdot X[i],$$
$$COUNT[i] = C[i + 1] \oplus X[i].,$$
$$C[4] = X[4] \cdot IN,$$
$$COUNT[4] = X[4] \oplus IN$$

rather than

$$C = X \cdot (C[2{:}4]{\circ}IN),$$
$$COUNT = X \oplus (C[2{:}4]{\circ}IN)$$

in the description of *S2*. Perhaps you find the *for clause* description to be clearer, or does its length negate its advantages?

For clauses may also indicate that an automaton or operator block is to be replicated. For example, if a combinational logic network is to be replicated rather than time shared, the ⟨OP⟩ forms the body of a *for clause* with no list or range of index values. Each reference to the output terminals of the OPerator block implies the existence of another copy of that network.

The IDentifier declaration provides a means of giving a name to a compound facility or a compatible set of operations. Its body is a list of statements of either of the following forms:

$$id = \text{compound facility}$$
$$id = (csop)$$

Compound facil:.ies are described with the concatenation, head and tail operators only. Thus no hardware is implied by identifiers introduced in a ⟨ID⟩. They are simply short symbol sequences that may be used in place of larger ones.

We could have abbreviated previous DDL descriptions with the ⟨ID⟩, for example

$$⟨ID⟩\ a = C[2 : 4] \circ IN.$$
$$C = X \cdot a,$$
$$COUNT = X \oplus a$$

The Boolean equations become very concise and clear, but they are more abstract in that "*a*" is simply a symbol standing in place of "*C*[2 : 4]∘*IN*."

The ELement declaration provides a means of introducing a block into a system description without indicating how that block is constructed or what function it performs. Perhaps the block has not yet received any attention, and its function and organization are not yet well defined. Or, the block may be a standard network that is completely defined in other documents. The body of an ⟨EL⟩ is therefore a list of items of the form

$$id\ (\textit{output terminal list}: \textit{input terminal list})$$

While *id* names the block, it has no significance in the description. Connections may be made to the output and input terminals, if any, listed to the left and right of a colon, respectively.

We have examined a substantial fraction of DDL, a register transfer language for describing the organization and function of digital systems, and implying their logic. Examples were limited because many are available in the following chapters. Use this section as a reference as you encounter them. Syntax and semantics of the following types of declarations are available here.

constants, identifiers

$\langle TE \rangle$, $\langle RE \rangle$, $\langle TI \rangle$, $\langle BO \rangle$, $\langle CO \rangle$

Boolean expressions, logic operators

connection, register transfer

if-then, if-value, for

$\langle SY \rangle$, $\langle AU \rangle$, $\langle OP \rangle$, $\langle ID \rangle$, $\langle EL \rangle$

2.9   SUMMARY

The success of digital techniques in computer, communication, control, and instrumentation systems arises from the reliability, high speed, and low cost of binary amplifiers. The simplicity of binary operations permits simple circuits to perform reliability in the presence of noise and with component parameters that deviate from ideal values. While each of the logic circuit families (examined) have their own electrical characteristics such as propagation delay and power dissipation, all offer circuits that remember binary values and others that combine binary signals. When we ignore the detailed electrical characteristics and concentrate only on the logic behavior of such circuits, we form idealized models. Binary signals then become the mathematical variables; rules of combination become the mathematical operators of Boolean algebra. Such modeling is very valuable because the axioms and theorems of Boolean algebra clearly and concisely indicate what we can expect of combinational networks of logic circuits when they act in a purely binary, delay-free manner. In addition, since many Boolean algebras exist, we are free to use the one most appropriate to modelling a specific processing network.

Ideal combinational logic networks do to binary signals what Boolean functions do to binary variables, and vice versa, so the study of Boolean functions is a mathematical study of logic networks. We found that switching functions can be expressed with Boolean equations; truth tables, logic diagrams, n-space mappings, cubes, arrays, Karnaugh maps, and formal languages such as DDL. While truth tables offer an easy means of counting the different functions of n variables, or generating all of them, Boolean equations are most easily transformed to logic diagrams that indicate how a network is to be constructed. Transforming equations to truth tables or Karnaugh maps is slightly more difficult. The reverse process of finding equations from tables or maps requires minterms, maxterms, and simplification procedures. Analyzing logic diagrams by inspection is largely a matter of carefully tracing all paths in the diagram. When NAND and NOR gates appear, special gate type replacement rules can be used to arrive at AND-OR-NOT equations directly.

Register transfer languages such as DDL are a tool for preparing symbolic models of complete digital systems, not just combinational networks. Boolean algebra appears in a generalized form to express the combinational function and structure of systems. Other notation is used to express sequential (memory) function and structure. DDL permits the clear expression of functions at the expense of detailed logic network structure, which is given largely by implication because register transfer languages are most useful and necessary in the initial stages of design. (This is when processing algorithms and system organization must be established.)

This chapter is packed with concepts, notations, and techniques. Consider the range of topics treated: from the electronic details of one amplifier to algebraic, tabular, and topological mathematics or to abstract, symbolic models of systems that involve tens to tens of thousands of amplifiers. This is the breadth of knowledge used by the logic designer, not to speak of economics, sales, personal relations, and so forth. All following chapters are much more specific and aimed at the design process.

REFERENCES

Historical

1. G. BOOLE, *An Investigation of the Laws of Thought, on which are Founded the Mathematical Theories of Logic and Probability,* 1849, reprinted by Dover Publications, New York, 1954.
2. M. KARNAUGH, "The Map Method for Synthesis of Combinational Logic Circuits." *Communications and Electronics,* No. 9, Nov. 1953.
3. E. J. McCLUSKEY, Jr., "Minimization of Boolean Functions." *Bell System Tech. Jour.,* Vol. 35, No. 6, November 1956.
4. W. V. QUINE, "The Problem of Simplifying Truth Functions." *Am. Math. Monthly,* Fall, 1952.
5. J. P. ROTH, "Algebraic Topological Methods for the Synthesis of Switching Systems I." *Trans. of the Amer. Math. Soc.,* Vol. 88, July 1958.
6. C. E. SHANNON, "Symbolic Analysis of Relay and Switching Circuits." *Trans. of AIEE,* Vol. 57, 1938, pp. 713–723.

General

7. A. BARNA and D. I. PORAT, *Integrated Circuits in Digital Electronics.* New York: Wiley, 1973.
8. T. BARTEE, *Digital Computer Fundamentals, Ed. 3.* New York: McGraw-Hill, 1972.
9. T. L. BOOTH, *Digital Networks and Computer Systems.* New York: Wiley, 1971.
10. Y. CHU, *Digital Computer Design Fundamentals.* New York: McGraw-Hill Book Co., 1962.
11. D. L. DIETMEYER and J. R. DULEY, "Register Transfer Languages and Their Translation," Editor M. A. Breuer, *Digital System Design Automation*:

Languages, Simulation and Data Base. Woodland Hills, Calif.: Computer Science Press, 1975.

12. F. J. HILL and G. R. PETERSON, *Introduction to Switching Theory and Logical Design.* New York: Wiley, 1968, 1974.

13. G. A. MALEY and JOHN EARLE, *The Logic Design of Transistor Digital Computers.* Englewood Cliffs, N.J.: Prentice-Hall, 1963.

14. M. P. MARCUS, *Switching Circuits for Engineers.* Englewood Cliffs, New Jersey: Prentice-Hall, 1962; 1967; 1975.

15. E. J. McCLUSKEY, Jr., *Introduction to the Theory of Switching Circuits.* New York: McGraw-Hill, 1965.

16. E. MENDELSON, *Shaum's Outline of Theory and Problems of Boolean Algebra and Switching Circuits.* New York: McGraw-Hill, 1970.

17. R. L. MORRIS and J. R. MILLER, *Designing with TTL Integrated Circuits.* New York: McGraw-Hill, 1971.

18. J. B. PEATMAN, *The Design of Digital Systems.* New York: McGraw-Hill, 1972.

19. F. P. PREPARATA and R. T. YEH, *Introduction to Discrete Structures.* Reading, Mass.: Addison-Wesley, 1973.

20. V. T. RHYNE, *Fundamentals of Digital System Design.* Englewood Cliffs., N.J.: Prentice-Hall, 1973.

21. C. H. ROTH, Jr., *Fundamentals of Logic Design.* St. Paul, Minn.: West Publishing Co., 1975.

22. L. STRAUSS, *Wave Generation and Shaping, Ed. 2.* New York: McGraw-Hill, 1970.

PROBLEMS

2.1-1. Prepare a truth table and logic equation for each of the circuits below.

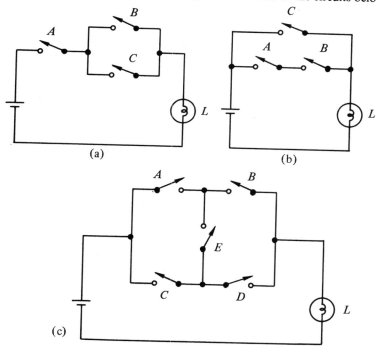

(a)

(b)

(c)

2.1-2. Motor M is controlled by relays A, B, and C. M is to run when A and B are both operated and also when B is not operated but both A and C are. Prepare a wiring diagram for this control. Use as many contacts as necessary on each relay, but try to be economical.

2.1-3. Three relays A, B, and C are to control two lamps, a red lamp denoted R and a green lamp, G, as follows: when all three relays are operated, the red lamp is to light; when A is not operated and either B or C is operated the green lamp is to light. For all other combinations, the red lamp is to be lit except when no relays are operated: then neither is to be lit. The lamps are not to be lit at the same time.

Prepare a wiring diagram for a circuit that will control R and G. Attempt to use as few contacts on each relay as possible.

2.1-4. Draw a relay contact network that realizes each of the following logic expressions, where A, B, and C represent contact pairs on three different relays.

(a) $A \cdot B + C$

(b) $\overline{A \cdot B + C}$

(c) $(A \cdot B) + (\overline{A \cdot C})$

2.1-5. In a logic system signal levels of $+10$ volts and 0 volts are used. Represent the $+10$ by 0 and the 0 volts by 1. Complete logic tables for the diode circuits of Figs. 2.7 and 2.8 with this encoding.

2.1-6. Three binary signal generators provide 0 and 10 volt levels via source resistance R. Diodes have a breakpoint value of 0.7 volts. Determine the output voltage of the circuit shown for each of the eight combinations of input signal levels. Identify worst case conditions, i.e., those that give the minimum high output level and the maximum low output level.

(a) $R = 0$

(b) $R = 100 \ \Omega$

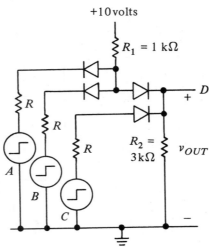

2.1-7. In the circuit of Prob. 2.1-6 above, assume $R_2 \gg R$.

(a) What ratio R_2/R_1 is required to ensure that the lowest high output voltage will equal 8 volts?

(b) What ratio R_1/R is required to ensure that the maximum low output voltage will equal 0.5 volts?

2.1-8. Common-emitter amplifiers of Fig. 2.10 are constructed with $V_{CC} = 10$ volts, $R_B = 2$ kΩ, $R_C = 1$ kΩ, and transistors with $V_{CE(SAT)} = 0.2$ V, $V_{BEO} = 0.6$ V and h_{fe} in the range 50 to 200.

(a) What ranges of gain A and $V_{IN(SAT)}$ are expected?

(b) One amplifier is to drive n others. What is the largest value of n such that all driven transistors will be saturated when the driving transistor is cut-off?

(c) As a safety factor each driven transistor is to receive a base current that is 3 times the minimum required to saturate it. Now what maximum fan-out is permitted?

(d) Repeat parts (a) and (b), assuming that the resistor values may differ from the nominal values listed by as much as 10%.

2.1-9. We wish to guarantee $V_{OH} = 3$ volts and $V_{OL} = 0.4$ volts for the amplifiers of Prob. 2.1-8(d).

(a) What fan-out limit should be published?

(b) Draw the transfer characteristics of an unloaded gate with minimum gain and a maximum gain gate with the fan-out of part (a). Estimate the noise margins from these worst case characteristics.

2.1-10. With no connection to the B input terminal, the DTL circuit of Fig. 2.20 acts as an inverting amplifier.

(a) Assuming its gain from terminal A to C is -100, prepare transfer characteristics assuming that 0, 1, 2, and 3 level shifting diodes are placed between the diode AND gate and the transistor base. What effect do these diodes have on the noise margins of the amplifiers?

(b) Prepare an input characteristic for the amplifier as shown in Fig. 2.20, assuming $R_1 = 2$ kΩ. Current into the amplifier is positive.

2.2-1. Use the axioms and theorems of Boolean algebra presented in Section 2.3 to prove the following theorems.

(a) $(A \cdot B) + (A \cdot \bar{B}) = A$

(b) $(A \cdot B) + (\bar{A} \cdot C) + (B \cdot C) = A \cdot B + \bar{A} \cdot C$

(c) $A\bar{B} + \bar{A}B = (A + B) \cdot (\bar{A} + \bar{B}) = (A + B) \cdot (\overline{AB})$

(d) $A + \bar{A} \cdot B \cdot C = A + B \cdot C$

2.2-2. Use the *simplification* theorem primarily to simplify the following Boolean expressions.

(a) $A + \bar{A}B + (\overline{A + B})C + (\overline{A + B + C})D$

(b) $A\bar{B} + AC + BCD + \bar{D}$

2.2-3. Use the *absorption* theorem primarily to simplify the following Boolean expressions.

(a) $\bar{A} + \bar{A}\bar{B} + BC\bar{D} + B\bar{D}$

(b) $A\bar{B}C + (\bar{B} + \bar{C})(\bar{B} + \bar{D}) + \overline{A + C + D}$

2.2-4. Two closed operators on the set {0, 1, 2} are defined below. Which axioms and theorems of Boolean algebra are satisfied by these operators? Which are not, if any?

$\cdot$	0	1	2
0	0	0	0
1	0	1	1
2	0	1	2

$+$	0	1	2
0	0	1	2
1	1	1	2
2	2	2	2

2.2-5. Write and run a FORTRAN program that will print the cases, if any, in which the operators of Prob. 2.2-4 above violate:
(a) the associative theorem,
(b) the distributive theorem.

2.2-6. Write the dual of each Boolean expression. (Insert parentheses so that the result is properly interpreted.)

(a) $A + B \cdot C$

(b) $A \cdot (B + \bar{C}) + \bar{A}(\overline{B \cdot C})$

(c) $A + B + (A + B)C + (\overline{A + B + C})D$

2.2-7. The AND and XOR operators of Fig. 1.6 form a mathematical system on $P_2 = \{0, 1\}$. But is that system a Boolean algebra? If not, what axioms and/or theorems are violated?

2.2-8. De Morgan's Theorem can be generalized in several ways.
(a) Prove that the following generalization is valid.

$$\overline{A \cdot B \cdot C} = \bar{A} + \bar{B} + \bar{C}$$

(b) Shannon [5] presented the following generalization.

$$\bar{f}(\{x_i\}, +, \cdot, 0, 1) = f(\{\bar{x}_i\}, \cdot, +, 1, 0)$$

Thus to obtain the complement of a function:
(1) Complement all literals;
(2) Replace $+$ with $\cdot$, and $\cdot$ with $+$;
(3) Replace 0 with 1, and 1 with 0.

It may be necessary to parenthesize the original expression to ensure proper interpretation of the result. Write the complement of each of the following switching functions.

(i) $f = A + \bar{B} \cdot C; \bar{f} =$

(ii) $g = A + \overline{\bar{B} \cdot C}; \bar{g} =$

(iii) $h = A \cdot \bar{B} + \bar{A} \cdot (\bar{C} + \overline{D + E}); \bar{h} =$

2.3-1. Let U be a set with n members. Let $\hat{U}$ be the set of all subsets of U.
(a) Show that $\hat{U}$ has the same number of members as P_2^n and that each member of $\hat{U}$ is related in a natural manner to one member of P_2^n, and vice versa.
(b) Show that the set *union* ($\cup$) and *intersection* ($\cap$) operators are Boolean operators on $\hat{U}$.
(c) Illustrate these results with $n = 3$ and $U = \{A, B, C\}$ by listing $\hat{U}$, the complement of each member of $\hat{U}$, and an example of each axiom and theorem of Boolean algebra.

2.3-2. Let the set P_4 of a Boolean algebra contain four elements denoted $0, 1, x, y$: $P_4 = \{0, 1, x, y\}$. Which of the four elements is the complement of x? Prove your answer.

2.3-3. Figure 2.35 defines Boolean operators as P_2^2 with $(0, 0)$ as the **0** element and $(1, 1)$ as the **1** element.
(a) Prepare similar tables for a Boolean algebra on P_2^2 in which $(0, 1)$ is the **0** element and $(1, 0)$ is the **1**.
(b) Repeat (a) with $(1, 0)$ as the **0** and $(0, 1)$ as the **1**.
(c) State in general how the $+$ and $\cdot$ operators are defined in terms of the components of the **0** element. Write these rules in terms of P_2^n rather than P_2^2.

2.3-4. Find the truth table for each Boolean expression by operating on defined columns to its left.

A	B	C	$\bar{A}$	$\bar{A}B$	$\bar{C}$	$\bar{A}B\bar{C}$	$A+B$	$(A+B)C$	$\bar{A}B\bar{C}+(A+B)C$
0	0	0							
0	0	1							
0	1	0							
0	1	1							
1	0	0							
1	0	1							
1	1	0							
1	1	1							

2.3-5. Table 1.4 indicates that each hexadecimal digit encodes a binary 4-tuple.
(a) From the operator definitions of Eq. (2.29), prepare tables of Boolean operators on the set H of all hexadecimal digits.
(b) Each truth table column of Prob. 2.3-4 can be expressed with two hexadecimal digits. Thus, A is expressed "0F"; B is encoded "33"; C = "55". Find the hexadecimal expressions of the remaining columns using the operators of H algebra.

2.4-1. For each of the following switching functions give (a) the circuit diagram representation, (b) the truth table representation, (c) the cube (topological) representation, and (d) the Karnaugh map representation.
(i) $F_1 = A + B + C$
(ii) $F_2 = \bar{A}\bar{B}\bar{C} + \bar{A}BC + A\bar{B}C + AB\bar{C}$
(iii) $F_3 = \bar{A} + C(\bar{B} + \bar{D})$

2.4-2. The topology of three switching functions is shown below. For each function prepare ON and OFF arrays, a truth table, and a Karnaugh map.

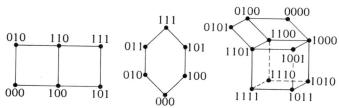

2.4-3. Karnaugh maps of three switching functions are shown below. Prepare ON and OFF arrays, the topology, and the truth table of each function.

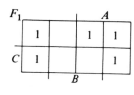

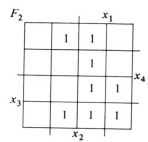

2.4-4. List the four Boolean functions of one variable.

2.4-5. (a) Show that the AND, OR, and NOT can be expressed in terms of the NAND operator ($\uparrow$) only.
(b) Draw networks of NAND gates only that act as AND, OR, and NOT gates.

2.4-6. Transform each Boolean function as written into a NAND operator expression (complements of variables may be expressed $\bar{A}$). Draw the logic net dictated by each NAND expression.
(a) $F_1 = A\bar{B} + ACD$
(b) $F_2 = \bar{A} + B\bar{C}D + ACD$
(c) $F_3 = (B + D) \cdot (\bar{A} + \bar{C}) + AD$

2.4-7. Repeat Prob. 2.4-5, using NOR in place of NAND.

2.4-8. Transform each Boolean function as written into a NOR operator expression (complemented variables may be expressed $\bar{A}$). Draw the logic net dictated by each expression.
(a) $F_1 = (A + \bar{B})(A + C + D)$
(b) $F_2 = \bar{A}(B + \bar{C} + D)(A + C + D)$
(c) $F_3 = (BD + \bar{A}\bar{C})(A + D)$

2.4-9. Prove the validity of the following NAND (NOR) counterparts of familiar theorems.

$$0 \uparrow A = 1 \qquad\qquad 0 \downarrow A = \bar{A}$$
$$1 \uparrow A = \bar{A} \qquad\qquad 1 \downarrow A = 0$$
$$A \uparrow A = \bar{A} \qquad\qquad A \downarrow A = \bar{A} \text{ (Idempotent?)}$$
$$(A \uparrow B) \uparrow (A \uparrow \bar{B}) = A \quad (A \downarrow B) \downarrow (A \downarrow \bar{B}) = A \quad \text{(Minimization)}$$

2.4-10. Prove that the EXCLUSIVE-OR and EQUIVALENCE operators are commutative and associative.

2.4-11. Which, if any, of the following distributive laws are valid?
(a) $A \oplus (B \cdot C) = (A \oplus B) \cdot (A \oplus C)$ $\qquad A \equiv (B + C) = (A \equiv B)$
$\qquad\qquad\qquad\qquad\qquad\qquad\qquad\qquad\qquad\qquad + (A \equiv C)$
(b) $A \cdot (B \oplus C) = (A \cdot B) \oplus (A \cdot C)$ $\qquad A + (B \equiv C) = (A + B)$
$\qquad\qquad\qquad\qquad\qquad\qquad\qquad\qquad\qquad\qquad \equiv (A + C)$
(c) $A \uparrow (B \cdot C) = (A \uparrow B) \cdot (A \uparrow C)$ $\qquad A \downarrow (B + C) = (A \downarrow B)$
$\qquad\qquad\qquad\qquad\qquad\qquad\qquad\qquad\qquad\qquad + (A \downarrow C)$
(d) $A \cdot (B \uparrow C) = (A \cdot B) \uparrow (A \cdot C)$ $\qquad A + (B \downarrow C) = (A + B)$
$\qquad\qquad\qquad\qquad\qquad\qquad\qquad\qquad\qquad\qquad \downarrow (A + C)$

2.4-12. (a) Show how AND, OR, and NOT can be expressed in terms of the INHIBIT operator only.
(b) Express the function $F = A + \bar{B} \cdot C$ in terms of the INHIBIT operator only.

2.4-13. Several forms of functions of three variables are given. How many functions of three variables does each form represent?

(a) $\ell_1 \ell_2 \ell_3 + \bar{\ell}_1 \bar{\ell}_2 \bar{\ell}_3$
(b) $\ell_1 (\ell_2 \ell_3 + \bar{\ell}_2 \bar{\ell}_3)$
(c) $\ell_1 \oplus (\ell_2 \ell_3)$

2.4-14. A 2-way multiplexer module realizes $ab + \bar{a}c$. Draw networks of those modules only that realize each of the following functions. The input variables are available in true form only.

(a) $F_1 = AC + B\bar{C}$
(b) $F_2 = \bar{A}\bar{C} + \bar{B}C$
(c) $F_3 = A\bar{B} + A\bar{C} + \bar{B}C$
(d) $F_4 = (\bar{A} + B)(A + \bar{C})$

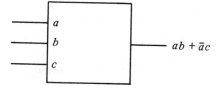

2.5-1. Find an ON-array and then the truth table of each function.
   (a) $F_1 = ABC + \bar{A}BC + \bar{B}\bar{C}$
   (b) $F_2 = A + B + \bar{C}$
   (c) $F_3 = AB + \bar{B}CD$
2.5-2. Prepare a Karnaugh map for each function of Prob. 2.5-1.
2.5-3. Write an OFF-array and then the truth table and Karnaugh map of each function.
   (a) $F_1 = (\bar{A} + \bar{B})(B + C)$
   (b) $F_2 = A(B + C)(\bar{B} + \bar{C})$
   (c) $F_3 = (A + B)(\bar{A} + C)(\bar{B} + C)$
2.5-4. (a) Expand each function of Prob. 2.5-1 to a sum of minterms algebraically.
   (b) Expand each function of Prob. 2.5-3 to a product of maxterms algebraically.
2.5-5. Express each function as (a) sum of minterms and (b) product of maxterms.

A	B	C	F_1	F_2	F_3	F_4
0	0	0	0	0	0	1
0	0	1	0	0	0	1
0	1	0	0	0	0	1
0	1	1	0	0	1	1
1	0	0	0	1	1	1
1	0	1	1	1	1	1
1	1	0	1	1	1	1
1	1	1	1	1	1	1

2.5-6. Prove:
   (a) $m_i \cdot m_j = 0$ for $i \neq j$.
   (b) $M_i + M_j = 1$ for $i \neq j$.
2.5-7. (a) Express each function as a product of maxterms.
   (i) $F_1 = \sum 0, 1, 2, 3, 7$
   (ii) $F_2 = \sum 1, 3, 5, 7$
   (b) Express the complement of each function as a sum of minterms.
   (c) Express the complement of each function as a product of maxterms.
2.5-8. (a) Write simplified *sop* equations from the Karnaugh maps of Prob. 2.4-3.
   (b) Write simplified *pos* equations.
2.5-9. Use the Karnaugh map to find simple, if not minimal, *sop*, *pos*, and AOI expressions for each function   How many gates and inverters are required to realize each expression?
   (a) $F_1 = \sum 0, 2, 4, 6, 7, 8, 10, 12, 13, 14$
   (b) $F = \bar{A}C\bar{D} + \bar{A}\bar{B}\bar{D} + \bar{A}B\bar{C} + B\bar{C}D + A\bar{B}D$
   (c) $F = A + D(\overline{BCD})$

2.6-1. Analyze each network writing a Boolean equation.

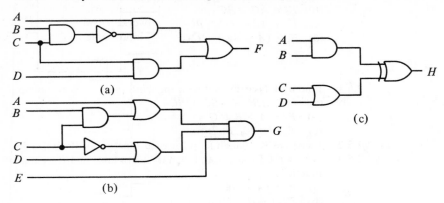

(a)

(b)

(c)

2.6-2. Simulate each network of Prob. 2.6-1 to find their truth tables.

2.6-3. If I and J are FORTRAN integer variables with assigned values from
{0, 1}, then logic operations may be simulated with FORTRAN computation in several ways, one of which is:

AND gate	MIN (I,J)
OR gate	MAX (I.J)
NOT	1-I
XOR gate	MOD (I + J,2)

Write and execute a FORTRAN program that will calculate the truth table of each network of Prob. 2.6-1.

2.6-4. Analyze each network by:
(a) Basic path tracing with algebraic manipulation,
(b) Symbolic replacement, and
(c) Gate type replacement and path tracing.

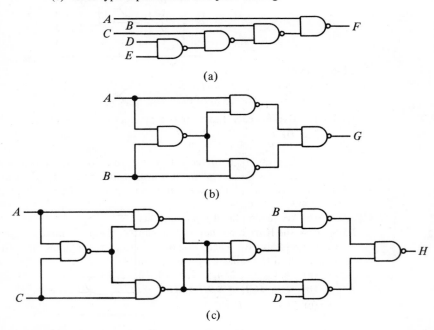

(a)

(b)

(c)

2.6-5. Repeat Prob. 2.6-4 with all gates being NOR gates.

2.6-6. Write Boolean equations for the output variables. From the equations prepare a truth table for the network. Do the *WXYZ* output tuples appear to be related to the *ABCD* input tuples?

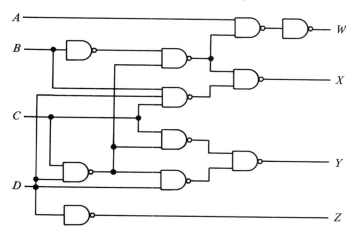

2.6-7. A portion of the logic of a medium scale integrated (MSI) circuit is shown. Write *sop* Boolean equations for the output variables.

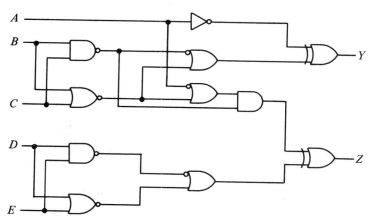

2.6-8. The network shown uses ECL technology. Write Boolean equations for the output variables.

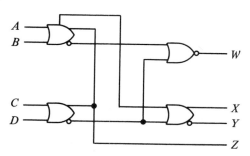

2.7-1. (a) Expand each of the Boolean functions about A, about B, and about C.
(b) Expand each about A; expand this expansion about B, etc.
    (i) $F_1 = ABC + \bar{A}BC + \bar{B}\bar{C}$
    (ii) $F_2 = A + B + \bar{C}$
    (iii) $F_3 = (A + B)(\bar{A} + C)(\bar{B} + C)$

2.7-2. Given $F(A, B, C) = A + \bar{B}\bar{C}$, write a simplified Boolean expression for:
(a) $F(1, B, C)$ (b) $F(A, 1, C)$
(c) $F(0, 0, C)$ (d) $F(1, 0, 1)$

2.7-3. Use Axiom 5, $A + \bar{A} = 1$, and the distributive laws to expand each function to canonical sum-of-products form.
(a) $F_1(A, B, C) = AB + \bar{B}C$;
(b) $F_2(A, B, C, D) = AB + \bar{B}CD$;
(c) $F_3(x_1, x_2, x_3, x_4) = x_1 + \bar{x}_2 x_4 + x_2 \bar{x}_3 \bar{x}_4$.

2.7-4. Use Axiom 5, $A \cdot \bar{A} = 0$, and the distributive laws to expand each function to canonical product-of-sums form.
(a) $F_1(A, B, C) = (\bar{A} + \bar{B})(B + C)$;
(b) $F_2(A, B, C, D) = \bar{A}(B + C + D)$;
(c) $F_3(x_1, x_2, x_3, x_4) = (\bar{x}_1 + \bar{x}_2)(\bar{x}_3 + \bar{x}_4)$.

2.7-5. Write out each minterm (maxterm), using the minimum number of literals in each case.
(a) m_4, m_{12}, m_{52};
(b) M_8, M_{15}, M_{66}.

2.7-6. Express each function in canonical form. Attempt to simplify the expression of each function.
(a) $F_1(A, B, C) = \sum 0, 1, 2, 3, 7$
(b) $F_2 = \sum 0, 2, 8, 10, 11, 14, 15$
(c) $F_3 = \sum 0, 1, 13, 14, 16, 17, 29, 31$

2.7-7. For each of the three functions of Prob. 2.7-6, express the complementary function as a sum of decimal integers.

2.7-8. Prove that any switching function may be expressed:

$$f(x_1, \ldots, x_n) = x_i \cdot g + \bar{x}_i \cdot h + k$$

where g, h, and k are functions of the $n-1$ other variables, and $g \cdot h = g \cdot k = h \cdot k = 0$.

2.7-9. (a) Prove that any switching function may be expressed:

$$f(x_1, \ldots, x_n) = x_i \cdot \ell \oplus m$$

where ℓ and m are functions of the $n-1$ other variables.
(b) Are ℓ and m disjoint, i.e., $\ell \cdot m \overset{?}{=} 0$?

2.7-10. $F = AB + BC$
(a) Express F in each of the eight canonical forms.
(b) Realize F with networks of each of the canonical forms of Figs. 2.58 and 2.59, starting with the minimum *sop* and minimum *pos* expressions of F.
(c) Input signals are available in true form only. What is the most economical realization of F if:
    (i) ECL OR-NOR gates are used, and
    (ii) TTL NAND and NOR gates are used?

2.7-11. Find the GRM coefficients for $F = AB + BC$ via Eq. (2.48). Draw a reduced network based on the GMR expansion of F.

2.7-12. (a) Derive the GRM expansion for $f(A, B, C)$ in which A, $\bar{B}$, and C appear.

(b) Express the coefficients g_i in terms of truth table entries for this GRM expansion.

(c) Which f_j appears in the expression of all g_i?

(d) Change the definition of "subnumber" to fit this expansion.

2.8-1. Write the binary string specified by each of the DDL constants.

(a) 17D6

(b) 23O6

(c) 1776O15

(d) ?10B5

2.8-2. Write the decimal and octal DDL constants for each of the following binary strings.

(a) 010110

(b) 010011

(c) 0001011010

2.8-3. Draw and label a line for each wire specified to exist in the following declarations.

(a) ⟨TE⟩ A[0:4], B, C[5].

(b) ⟨TE⟩ A[0:3, 2], B[6] = C[2]∘D[3:0].

2.8-4. Given two registers,

$$⟨RE⟩ \ A[10], B[10].$$

and that the A register holds the constant 113O10 while the B register holds 1726O10; what constant appears on each of the following terminal sets?

(a) B⊢4∘A⊣2

(b) B[1] × 10∘(¬A[1:5])

(c) ·/A\0 v · /A\1777O10

(d) (A[1:5] ⊕ B[6:10]) ∘ (A[6:10] ↑ B[1:5])

(e) A∘(⊕/A) · B[1]

2.8-5. Draw a diagram of the logic implied by each Boolean expression of Prob. 2.8-4.

2.8-6. Given that $C1 = 0$, $C2 = 1$ and

$$⟨TE⟩ \ C1, C2, A[10], B[10], C[10], F[10].$$

What is connected to F in each case?

(a) F = |C1| A ⊕ B · |C2|C; ¬C.; A v B ⊕ C.

(b) |C2|F = |C1| A · B; A · C.; |C1| F = B; F = A ⊕ C..

(c) F = |¬C1 · C2| A⊕ |C2|B; C.; B · |C1|A; C..

2.8-7. Use if-value clauses to express the connections of Prob. 2.8-6 more clearly.

2.8-8. ⟨OP⟩ X(IN, MODE)[16]

    ⟨TE⟩ IN[16], MODE, C[16] = CX∘CC[15].

    ⟨BO⟩ C = MODE · (IN v CC∘0),

        X = IN⊕C..

Evaluate X for the following input binary strings.

(a) IN = 2D16, MODE = 1

(b) IN = 10777O16, MODE = 0

(c) IN = 10770O16, MODE = 1

2.8-9. ⟨OP⟩ OUT (A, B)[2]

    ⟨TE⟩ A[10], B[10], C[10], D[10].

    ⟨ID⟩ X = 0∘C[1:9], Y = 0∘D[1:9].

    ⟨BO⟩ C = X v ¬A · B · ¬(X v Y),

        D = Y v A · ¬B · ¬(X v Y),

        OUT = C[10] ∘ D[10]..

(a) Compute *OUT* for the following supplied data.
  (i) $A = 13D10$, $B = 29D10$
  (ii) $A = 29D10$, $B = 13D10$
  (iii) $A = 29D10$, $B = 29D10$

(b) If the supplied A and B binary strings express integer magnitudes in the binary number system, what is the significance of $OUT = 00, 01, 10, 11$, if in fact these pairs can be generated?

2.8-10. A combinational logic block is required that receives a 24-bit word and a mode control signal. If mode is 0, the received data word is to appear on the output lines; if mode is 1, the output lines are to bear the complements of the input lines.

(a) Specify the required logic block in DDL.

(b) Draw a logic diagram of a typical cell of a network implied by your description.

2.8-11. ⟨TE⟩ A[10], B[10], C.

{i = 1:10} B[i] = |C| A[i]; A[11 − i]..

What binary sequence appears on B given:

(a) $A = 37O10$, $C = 1$.

(b) $A = 525O10$, $C = 0$.

ANSWERS

2.1-6.

A	B	C	$R = 0$ D	$R = 100$ D
0	0	0	0	.434
0	0	10	9.3	9.0
0	10	0	0	.82 ← worst

2.1-7. (a) $R_2/R_1 = 6.15$
        (b) $R_1/R = 17.6$

2.1-8. (a) $-100 \leq A \leq -25$
        $.692 \leq v_{IN(SAT)} \leq .992$ volts
        (b) 45
        (c) 14

2.1-9. (a) 4
        (b) $N_0 = 0.4 - 0.2 = 0.2$ V
        $N_1 = 3 - 1.1 = 1.9$ V

2.2-1. (a) $A \cdot B + A \cdot \bar{B} = A \cdot (B + \bar{B}) = A \cdot 1 = A$
        (d) $A + \bar{A}BC = A \cdot 1 + \bar{A}BC$
        $= A(1 + BC) + \bar{A}BC$
        $= A \cdot 1 + ABC + \bar{A}BC$
        $= A + (A + \bar{A})BC = A + 1 \cdot BC$
        $= A + BC$

2.2-3. (a) $\bar{A} + B\bar{D}$

2.2-5. Notice that the · operator gives the smaller, and the + operator gives the larger, of its operands. FORTRAN details differ from location to location; you may have to modify the following program that checks associativity to include local constraints.

$$\text{DO } 2 \text{ I} = 0,2$$
$$\text{DO } 2 \text{ J} = 0,2$$
$$\text{DO } 2 \text{ K} = 0,2$$
$$L = \text{MIN(I, MAX(J,K))}$$
$$M = \text{MAX (MIN(I,J), MIN(I,K))}$$
$$\text{IF(L.NE.M)WRITE(6, 1)I,J,K}$$
$$1 \text{ FORMAT(5X,3I4)}$$
$$2 \text{ CONTINUE}$$

2.2-6. (b) $(A + B \cdot \bar{C}) \cdot (\bar{A} + \overline{(B + C)})$

2.2-8. (b) $\bar{g} = \bar{A} \cdot (\overline{B + \bar{C}}) = \bar{A} \cdot \bar{B} \cdot C$

2.3-3. (c) Let $e = (e_1, \ldots, e_n)$ be the 0 element, and a and b be any two elements. Then $c = a \cdot b$ is calculated

$$c_i = \begin{cases} a_i \cdot b_i & \text{if } e_i = 0 \\ a_i + b_i & \text{if } c_i = 1 \end{cases}$$

where $\cdot$ and $+$ are the P_2 operators.

2.3-5. (b) $\bar{A}B = \text{"30"}$

2.4-2. The hexagon:

$$\text{ON} = \begin{cases} 0 & 0 & 0 \\ 0 & 1 & 0 \\ 0 & 1 & 1 \\ 1 & 0 & 0 \\ 1 & 0 & 1 \\ 1 & 1 & 1 \end{cases} \qquad \text{OFF} = \begin{cases} 0 & 0 & 1 \\ 1 & 1 & 0 \end{cases}$$

1	1	0	1
0	1	1	1

2.4-3. F_1

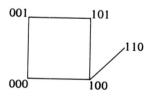

2.4-5. (a) $\bar{A} = A \uparrow A = A \uparrow 1$

$A \cdot B = (A \uparrow B) \uparrow 1$

2.4-6. $F_1 = (A \uparrow \bar{B}) \uparrow (A \uparrow C \uparrow D)$

$F_3 = [(\bar{B} \uparrow \bar{D}) \uparrow (A \uparrow C)] \uparrow (A \uparrow D)$

2.4-10. $A \oplus B = A\bar{B} + \bar{A}B$

$= B\bar{A} + \bar{B}A$

$= B \oplus A$

$(A \oplus B) \oplus C = (A \oplus B)\bar{C} + \overline{(A \oplus B)}\, C$

$= (A\bar{B} + \bar{A}B)\bar{C} + (\bar{A}\bar{B} + AB)C$

$= A\bar{B}\bar{C} + \bar{A}B\bar{C} + \bar{A}\bar{B}C + ABC^{10}$

$= A(\bar{B}\bar{C} + BC) + \bar{A}(B\bar{C} + \bar{B}C)$

$= A(\overline{B \oplus C}) + \bar{A}(B \oplus C)$

$$= A \oplus (B \oplus C)$$
$$= A \oplus B \oplus C$$

2.4-12. (a) $\quad \bar{A} = 1IA$
$$A \cdot B = AI(1IB)$$
(b) $\quad 1I[(1IA)\ I(CIB)]$

2.4-14. (d)

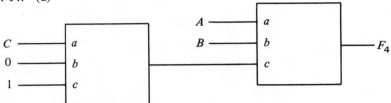

2.5-1.
$$\mathrm{ON}_1 = \begin{Bmatrix} 111 \\ 011 \\ x00 \end{Bmatrix} \qquad \mathrm{ON}_2 = \begin{Bmatrix} 1xx \\ x1x \\ xx0 \end{Bmatrix} \qquad \mathrm{ON}_3 = \begin{Bmatrix} 11xx \\ x011 \end{Bmatrix}$$

2.5-3.
$$\mathrm{OFF}_1 = \begin{Bmatrix} 11x \\ x00 \end{Bmatrix} \qquad \mathrm{OFF}_2 = \begin{Bmatrix} 0xx \\ x00 \\ x11 \end{Bmatrix} \qquad \mathrm{OFF}_3 = \begin{Bmatrix} 00x \\ 1x0 \\ x10 \end{Bmatrix}$$

2.5-5. $\quad F_1 = A\bar{B}C + AB\bar{C} + ABC$
$\quad F_4 = 1$ (sum of all minterms; product of none)

2.5-7. $\quad F_1 = \prod 4, 5, 6$
$\quad \bar{F}_1 = \sum 4, 5, 6$
$\quad \bar{F}_1 = \prod 0, 1, 2, 3, 7$

2.5-9. $\quad F_1 = \bar{D} + AB\bar{C} + \bar{A}BC$ $\qquad$ 3 gates $\quad$ 3 inverters
$\qquad = (B + \bar{D})(\bar{A} + \bar{C} + \bar{D})(A + C + \bar{D})$ $\quad$ 4 gates $\quad$ 3 inverters
$\qquad = \bar{B}D + ACD + \bar{A}\bar{C}D$ $\qquad\qquad$ 4 gates $\quad$ 4 inverters

2.6-1. (a) $\quad F = A(\overline{BC}) + CD$
$\qquad = A\bar{B} + A\bar{C} + CD$

2.6-3. Details of FORTRAN vary; it may be necessary to modify the following code.
(c) $\qquad$ INTEGER A, B, C, D, H
$\qquad$ DO 1 $\quad$ A = 0,1
$\qquad$ DO 1 $\quad$ B = 0,1
$\qquad$ DO 1 $\quad$ C = 0,1
$\qquad$ DO 1 $\quad$ D = 0,1
$\qquad$ H = MOD (MIN(A,B) + MAX(C,D), 2)
$\qquad$ 1 WRITE (6,2) A,B,C,D,H
$\qquad$ 2 FORMAT (5X, 4I3, I6)
$\qquad$ END

2.6-4. (a) $\quad F = \bar{A} + B(\bar{C} + DE)$

2.6-6. The integer expressed by $WXYZ$ is 3 less than that of $ABCD$ for many cases.

2.6-7. $\quad Y = \bar{A} \oplus (BC + \overline{B + C})$
$\qquad = \bar{A}(BC + \bar{B}\bar{C}) + A(BC + \bar{B}\bar{C}) = \bar{A} \oplus (\overline{B \oplus C}) = A \oplus B \oplus C$

$$= \bar{A}B\bar{C} + \bar{A}BC + A\bar{B}\bar{C} + ABC$$
$$= A \oplus B \oplus C \text{ (see Prob. 2.4-10)}$$

2.7-1. (a) Expanded about A only
$$F_1 = A(BC + \bar{B}\bar{C}) + \bar{A}(BC + \bar{B}\bar{C})$$
$$F_2 = A(1) + \bar{A}(B + \bar{C})$$
$$F_3 = A(C(\bar{B} + C)) + \bar{A}(B(\bar{B} + C))$$

2.7-3. (a) $F_1 = AB(C + \bar{C}) + (A + \bar{A})\bar{B}C$
$$= ABC + AB\bar{C} + A\bar{B}C + \bar{A}\bar{B}C$$
  (b) $F_2 = AB(C + \bar{C})(D + \bar{D}) + (A + \bar{A})\bar{B}CD$

2.7-4. (a) $F_1 = (\bar{A} + \bar{B} + C\bar{C})(A\bar{A} + B + C)$
$$= (\bar{A} + \bar{B} + C)(\bar{A} + \bar{B} + \bar{C})(A + B + C)(\bar{A} + B + C)$$

2.7-6. (a) $\sum 0,1,2,3,7 = \bar{A}\bar{B}\bar{C} + \bar{A}\bar{B}C + \bar{A}B\bar{C} + \bar{A}BC + ABC$
$$= \bar{A} + BC$$

2.7-10. (a) $F = \bar{A}BC + AB\bar{C} + ABC$
$$= (A + B + C)(A + B + \bar{C})(A + \bar{B} + C)(\bar{A} + B + C)(\bar{A} + B + \bar{C})$$
$$= B(A + C)$$

2.7-11. 00010011

2.7-12. (a) $B = 1 \oplus \bar{B}$
  (b) $g_0 = f_2$
$$g_1 = f_2 \oplus f_3$$

2.8-1. (a) 010001
  (b) 010011

2.8-2. (a) $22D6 = 26O6$

2.8-4. (a) 111111
  (b) 111111111111101

2.8-6. (a) $F = A \vee B \oplus C$
  (b) $F = A \cdot C$

2.8-7. (c) $F = [\text{C1} \circ \text{C2} [0\ \text{B} \cdot \text{C} [1\ \text{A} \oplus \text{B} ;\ \text{A} \cdot \text{B}.$

2.8-8. (a) 17774O16
  (b) 10777O16

2.8-9. (a) $A = 0000001101$
$$B = 0000011101$$
$$X = 0000001111$$
$$Y = 0000000000$$
$$C = 0000011111$$
$$D = 0000000000$$
  Form from left to right
$$OUT = 10 \qquad A < B$$

2.8-11. (a) 37O10

3

Synthesis of Combinational Logic Networks

To synthesis a digital system is to convert a *desire for* into the *desired* system by showing how primitive logic blocks are to be interconnected. In this chapter we wish to connect gates to form networks that process information. We begin with specifications of desired network performance that may be incomplete or ambiguous. We end with a logic diagram from which a network can be constructed. Here we will arrive at that end by informal means, using techniques of earlier chapters and taking advantage of prior efforts. We will use previously found results when they help solve new problems. Since we are familiar with arithmetic operations and arithmetic networks are very important parts of most digital systems, they are extensively used for purposes of example. To prevent confusion + will be used for arithmetic, and v will be used to denote the OR operation.

3.1 FIRST SYNTHESIS EXAMPLES

A design process must begin with the *specifications* of a desired product. Specifications may be written or verbal, or they may appear while working on an encompassing problem. If specifications of a required combinational switching network are presented as English statements, their conversion to one of the several representations of switching functions must be the first step of design. In order to arrive at the Boolean expression that describes the most desirable network, in terms of cost or other criteria, it may be necessary to transform the original representation. The intermediate manipulative steps of design must look ahead to the following step of drawing, from the Boolean expression, the logic block diagram that describes the circuitry to be assembled.

Other engineering and manufacturing steps must follow before a product can be supplied to satisfy the needs of a customer. The engineer will check his design through analysis, simulation either by hand or with the assistance of a digital computer, or through laboratory tests performed on a prototype. Drawings, wiring lists, parts lists, and other documentation, which place the designed and tested network within the overall switching system, must be prepared. Manufacturing steps may then be taken.

The connectives of an English statement often reveal required logic. Consider the following specifications.

> A counter provides the periodic signals shown in Fig. 3.1. A network is needed that generates a signal Z that is high ($Z = 1$) when either A and B are both high or neither B nor C are high.

Translating the words to Boolean notation gives the equation

$$Z = AB \lor \overline{B \lor C}$$

A network consisting of an AND and a NOR gate driving an OR gate is suggested immediately. If cost is measured by the number of gates in a network, this network is a low-cost realization of Z. If the suggested gates are not available for construction of the network, we may use Boolean algebra and/or Karnaugh maps to manipulate the equation to one that describes a network of available gates.

$$Z = AB \lor \overline{B}\overline{C}$$
$$= \overline{\overline{A}}\overline{B} \lor \overline{B}\overline{C}$$
$$= (A \lor \overline{B}) \cdot (B \lor \overline{C}) \quad \textit{and so forth}$$

In the process of deriving these equations, we might recognize that the functional form of Z is the same as that of the two-way multiplexer first seen in Fig. 1.8 (also see Prob. 2.4-14). Thus a two-way multiplexer (which

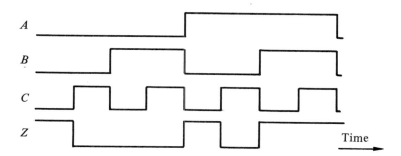

FIG. 3.1 We have signals A, B, and C. We need a network that will generate Z.

occupies one-fourth of a 14-pin integrated circuit) and an inverter for C may also be used to satisfy the requirements.

The next section will show how NAND and NOR networks may also be used. While the search for the most economical realization of a switching function is an important part of design, it will not be emphasized in this chapter.

English specifications are not usually so clear and easily transformed to networks. Often specifications are given in a piece-wise fashion with each piece expressed via a sentence. Designers must first assemble the pieces as the following example illustrates.

> Signals R and G, which depend on input signals A, B, and C, are never high simultaneously. Both are low simultaneously only when all input signals are low. When all input signals are high, $R = 1$. $G = 1$ when $A = 0$, and either or both $B = 1$ or $C = 1$. In all other cases $R = 1$.

To determine exactly what these specifications say, and if they are complete and consistent, we examine each sentence in turn. Findings will be recorded in a truth table. The "Both are low . . ." requirement tells the entries to be shared in the first row; the "When all are high . . ." sentence places an entry in the last row.

A	B	C	R	G
0	0	0	0	0
0	0	1		
0	1	0		
0	1	1		
1	0	0		
1	0	1		
1	1	0		
1	1	1	1	

The "$G = 1$" specification clearly supplies entries for the second, third, and fourth rows.

A	B	C	R	G
0	0	0	0	0
0	0	1		1
0	1	0		1
0	1	1		1
1	0	0		
1	0	1		
1	1	0		
1	1	1	1	

What does the "In all other cases . . ." statement mean? Could it mean that all remaining blanks in the R column should be filled? Probably not, since such entries contradict the "never high simultaneously" requirement; however, this is a point to review with the author of the specifications. We will take it to mean "for all input symbols not mentioned."

A	B	C	R	G
0	0	0	0	0
0	0	1		1
0	1	0		1
0	1	1		1
1	0	0	1	
1	0	1	1	
1	1	0	1	
1	1	1	1	

The "never high simultaneously" requirement now permits us to complete the truth table.

A	B	C	R	G
0	0	0	0	0
0	0	1	0	1
0	1	0	0	1
0	1	1	0	1
1	0	0	1	0
1	0	1	1	0
1	1	0	1	0
1	1	1	1	0

Equations may be written from this truth table with little difficulty.

$$R = A$$
$$G = \bar{A}(B \lor C)$$

And logic networks are easily derived from these.

If R is to be high whenever $A = 1$, why did the original specifications clearly not say so? The answer is that the specifications were not analyzed as just done, but were collected. Various special cases and requirements were recorded and the collection presented as specifications. This often used approach to generate specifications can lead to incomplete and/or inconsistent specifications.

"If, which, when, unless, else, but . . ." all can be translated into Boolean connectives. To see this, first suppose that the following is a complete specification.

If signal C is high, then output signal F is to have the same value as input signal A.

The truth table generated from these specifications is not complete.

C	A	F
0	0	
0	1	
1	0	0
1	1	1

When $C = 0$, no value of F is indicated. Since the specifications were said to be complete, we can conclude that when $C = 0$ the value of F is of no concern, and any value is acceptable to the author of the specifications. The input symbols for such cases are called *don't cares* as a result. A "d" is usually entered in truth tables and Karnaugh maps to indicate that the value of the dependent variable is optional.

C	A	F
0	0	d
0	1	d
1	0	0
1	1	1

A truth table or Karnaugh map containing ds records a *partial* switching function. If r ds appear, then any of 2^r fully specified switching functions is a *completion* of the partial function and satisfy the network specifications from which the partial function was derived. These completions are obtained by replacing the ds with 0's and 1's in all possible ways. Four functions complete our example partial function; an equation may be written for each.

C	A	F		C	A	F		C	A	F		C	A	F
0	0	0		0	0	0		0	0	1		0	0	1
0	1	0		0	1	1		0	1	0		0	1	1
1	0	0		1	0	0		1	0	0		1	0	0
1	1	1		1	1	1		1	1	1		1	1	1

$$F_1 = CA \qquad F_2 = A \qquad F_3 = \overline{C}\overline{A} \vee CA \qquad F_4 = A \vee \overline{C}$$

While the completions are all different switching functions, they all satisfy the given specifications. We as designers are therefore free to realize any of the completions.

One of the assignments of ds clearly gives a "best" network. F_1 indicates that an AND gate may be used; the AND gate passes signal A when $C = 1$ as required, and holds F at 0 otherwise. The OR gate of F_4 holds F at 1 when $C = 0$.

The following specifications each describe a two-way multiplexer.

(a) If $C = 1$, then $F = A$, else $F = B$.

(b) F is connected to A when $C = 1$ and B otherwise.

(c) If $C = 1$, then $F = A$. If $C = 0$, then $F = B$.

Let us examine the third form above; it consists of two pieces, the first of which we have already considered in detail. The second piece dictates the following partial function.

C	B	F
0	0	0
0	1	1
1	0	d
1	1	d

Putting the two pieces together, since they both must be satisfied, gives a complete function.

C	A	B	F
0	x	0	0
0	x	1	1
1	0	x	0
1	1	x	1

C	F
0	B
1	A

$$F = C\,A \vee \overline{C}B$$
$$= (\overline{C} \vee A) \cdot (C \vee B)$$

Each "if-then" statement may be realized with an AND gate; an OR gate is then used to combine the signals from the AND gates. If OR gates are used to realize "if-then" statements, then an AND gate is required to form the output signal. Perhaps you find it most natural to translate:

> "If condition, then connect source to sink."

to

> "connect source AND condition to sink."

The following specification is a useful variation on the two-way multiplexer:

> If $C = 1$, then $F = A$.
> If $D = 1$, then $F = B$.

Each of the statements gives a partial description of switching function F. Together they give an incomplete and inconsistent description. When both C and D equal 0, no value for F is dictated. Thus, don't cares exist that admit a variety of acceptable networks. The designer is free to pick the most desirable one.

When C and D both equal 1, then F is to take the value of both A and B, which is not possible if these values differ. Inconsistent specifications cannot be realized; they must be modified or clarified before synthesis can proceed. Suppose that we add to the above specifications: "C and D will never be high simultaneously," which is very possible since we do not know the origins of C and D. The inconsistency is replaced by additional don't care conditions. Since $C = D = 1$ will never happen, we care not what value a realization of the specifications provides should the impossible happen. A consistent truth table for F can be prepared; an abbreviated table and full Karnaugh map are shown in Fig. 3.2(a).

Two of the 16 possible completions of F are shown in the remainder of Fig. 3.2. These assignments of ds were so chosen as to obtain minimum cost *sop* and *pos* equations that are very much like those of the two-way multiplexer. NOTE: Since the maps are different, F_1 and F_2 are different Boolean functions that satisfy the specifications. Function F_1 gives $F = 0$ when both C and D equal 0. It would give $F = A \vee B$ if both C and D were to equal 1 at the same time. F_2 gives quite different results in these cases.

Specifications are sometimes most naturally stated in terms of the number held by a register. Decoders are very natural parts of networks that

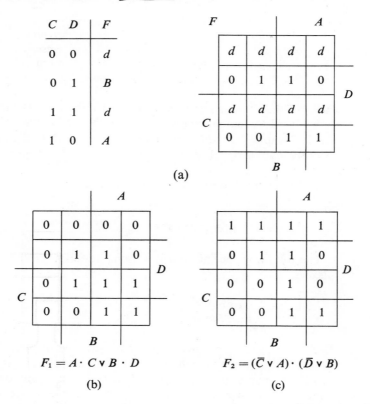

$$F_1 = A \cdot C \vee B \cdot D$$

(b)

$$F_2 = (\bar{C} \vee A) \cdot (\bar{D} \vee B)$$

(c)

FIG. 3.2   A Data Routing Network with Independent Control Signals

are based upon such specifications. The following specifications state the function of decoders.

✗ If the n input signals express in binary the decimal integer d, then a 1 is to be present on the dth output line. All other lines are to bear 0.

Let $n = 2$ for purposes of example. Two input variables can encode the integers 0, 1, 2, and 3. Let Z_0, Z_1, Z_2, and Z_3 be the output lines of the decoder. A table that expresses the above specifications reveals that each output variable may be described with a minterm.

$$Z_i = m_i \qquad\qquad (3.1)$$

and hence, realized with an AND gate driven by the appropriate set of true and complemented input variables. Fig. 3.3 presents this table and the logic network of a 2-to-4 wire decoder.

Counters may be used to control activity within a digital system. Each count value commands the performance of a different set of activities. The sequence of these sets accomplishes desired information processing, and the counter establishes that sequence. The following specifications illustrate that command signals (F and G) may depend on other conditions as well as the count.

The counter waveforms of Fig. 3.1 repeatedly present the integer sequence 0, 1, 2, ..., 7. Let K be the integer presented at a specific instant. Output variables F and G take nonzero values only as follows:

When K is less than 2, $F = D$.

When $K = 3$ or 7, $F = E$.

When $K = 4$ or 5, $F = 1$.

When $K = 2$, $G = E$.

When $K = 4$, $G = D$.

When $K = 6$, $G = D \vee E$.

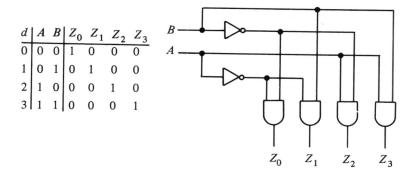

d	A	B	Z_0	Z_1	Z_2	Z_3
0	0	0	1	0	0	0
1	0	1	0	1	0	0
2	1	0	0	0	1	0
3	1	1	0	0	0	1

FIG. 3.3   A 2-to-4 Wire Decoder

The following table summarizes these specifications and reveals that they are consistent and complete.

K	F	G
0	D	0
1	D	0
2	0	E
3	E	0
4	1	D
5	1	0
6	0	$D \vee E$
7	E	0

With a decoder and the data routing networks of previous examples in mind, we may sketch the logic of Fig. 3.4 directly from this table. When the count is either 0 or 1, the top AND gate is enabled to pass D to output terminal F; all other AND gates are blocked. When the count is 4 or 5, F is set to 1 by the direct connections from the decoder to the F OR gate.

While the design of Fig. 3.4 was easily derived and is very understandable—and these characteristics have high practical significance in some situations, it is not a particularly low-cost design. Transforming the table derived from the specifications into the Karnaugh maps of Fig. 3.5(a), in which more general Boolean expressions than just 0 and 1 are entered, leads to the following equations for F and G.

$$F = (\bar{A}\bar{B})D \vee (BC)E \vee A\bar{B}$$
$$G = (B\bar{C})E \vee (A\bar{C})D$$

While D and E are independent variables, F and G infrequently depend upon them. Rather than prepare five variable maps for F and G, we can

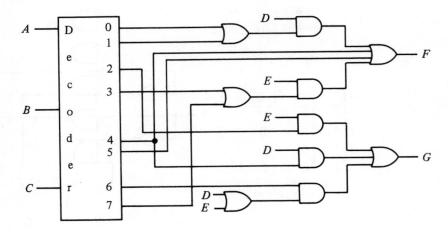

FIG. 3.4   A Decoder Realization

place Boolean expressions of infrequent, independent variables in the cells of a map and group adjacent cells containing the same entries to obtain reasonable, if not minimum, cost networks. The Boolean expressions within a group of cells must be ANDed with the labels of the group area. In the F map the cells containing Ds, those containing Es, and those containing 1's were combined to form the three product terms listed above. Comparing the first and third terms with the simplification theorem reveals a simplification of the first product term.

$$F = \bar{B}D \vee BCE \vee A\bar{B}$$

In the right cells of the F map, the 1's could be replaced with $1 \vee D$, which is logically equivalent. Then a group of four cells containing Ds appears to display graphically this simplification. If the Boolean expression within a cell(s) is an OR of product terms, these product terms may be included in different groups of cells. We see this in the map for G also. The network suggested by these equations are shown in Fig. 3.5(b). Decoding is specialized and combined with data routing, so operation of the network is not as easily understood, but the cost of meeting the specifications has been reduced.

Tabular Specifications

Tables most clearly express network specifications in some cases. If the dependent and independent variable entries in a table are encoded with binary tuples, a truth table results, and the remaining synthesis steps can be taken. The following examples provide very important results as well as illustrating synthesis from tabular specifications.

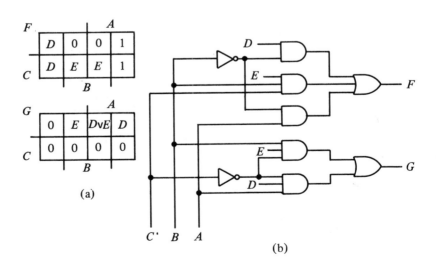

(a)

(b)

FIG. 3.5   A Realization Based Upon Karnaugh Maps with Infrequent Variables

SECTION 3.1   First Synthesis Examples

A network is required that generates sum and carry bits from two data bits in the manner of the binary addition table, Table 1.5.

The binary addition table is repeated in the form of a truth table in Fig. 3.6, with A and B being the data variables and S and C being the sum and carry variables, respectively. Equations may be written from this table by inspection.

$$C = A \cdot B$$

$$S = A \oplus B$$

(3.2)

Networks that realize these two equations are known as *half-adders* (HA).

The HA can serve only to add the least significant bits of two binary numbers. In all of the more significant bit positions, three bits must be considered—one from each number and a carry from the position on the right.

$$
\begin{array}{rccccc}
\text{Carry} = & 0 & 0 & 1 & 1 & \\
A = & 0 & 1 & 0 & 1 & 1 \\
B = & 0 & 0 & 0 & 1 & 1 \\
\hline
\text{sum} = & 0 & 1 & 1 & 1 & 0
\end{array}
$$

3 bits must be added

A full-adder network capable of adding three bits giving the correct sum and carry signals is most exactly and concisely specified by Fig. 3.7(a). Number the N bit positions of each number from left to right. Then A_i refers to the ith bit of the number A, and C_{i+1} denotes the carry bit to be added to A_i and B_i to obtain sum bit S_i and carry bit C_i.

The Karnaugh map for S_i (Fig. 3.8(a)) suggests that no combination of minterms to simplify the Boolean expression is possible, but manipulation of S_i leads to an interesting result.

$$
\begin{aligned}
S_i &= \bar{A}_i \bar{B}_i C_{i+1} \vee \bar{A}_i B_i \bar{C}_{i+1} \vee A_i \bar{B}_i \bar{C}_{i+1} \vee A_i B_i C_{i+1} \\
&= \bar{A}_i (\bar{B}_i C_{i+1} \vee B_i \bar{C}_{i+1}) \vee A_i (\bar{B}_i \bar{C}_{i+1} \vee B_i C_{i+1}) \\
&= \bar{A}_i (B_i \oplus C_{i+1}) \vee A_i \overline{(B_i \oplus C_{i+1})} \\
&= A_i \oplus (B_i \oplus C_{i+1})
\end{aligned}
$$

(3.3)

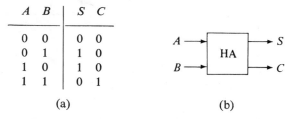

A	B	S	C
0	0	0	0
0	1	1	0
1	0	1	0
1	1	0	1

(a)

(b)

FIG. 3.6   Half-Adder Specifications and Symbolism

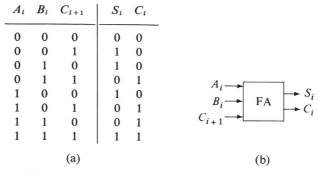

A_i	B_i	C_{i+1}	S_i	C_i
0	0	0	0	0
0	0	1	1	0
0	1	0	1	0
0	1	1	0	1
1	0	0	1	0
1	0	1	0	1
1	1	0	0	1
1	1	1	1	1

(a)   (b)

FIG. 3.7   Specifications of a Full-Adder

Thus the sum output S_i can be realized from its minterm expression with an AND-OR or NAND pyramid, or with two 2-input EX-OR gates—two half-adders.

Expression of the carry output, C_i, can be simplified by combining the adjacent minterms that appear in the map of Fig. 3.8(b).

$$C_i = A_i B_i \lor A_i C_{i+1} \lor B_i C_{i+1} \qquad (3.4)$$

C_i may also be expressed in an interesting non-sum-of-products form.

$$C_i = \bar{A}_i B_i C_{i+1} \lor A_i \bar{B}_i C_{i+1} \lor A_i B_i \bar{C}_{i+1} \lor A_i B_i C_{i+1}$$
$$= B_i C_{i+1} \lor A_i (B_i \oplus C_{i+1}) \qquad (3.5)$$

This expression of C_i is of interest if HA's are used to realize S_i. Half-adders provide the AND function as well as the EX-OR. Two HAs together with an OR gate then will generate both S_i and C_i if we interconnect them as in Fig. 3.9(a).

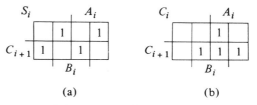

(a)   (b)

FIG. 3.8   Maps of the Sum and Carry Outputs of the Full-Adder

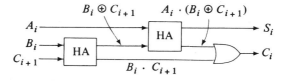

FIG. 3.9   Two Half-Adders Make a Full-Adder

A *half-subtractor* and *full-subtractor* can be developed in the same manner as we developed the HA and FA. Let D_i now denote the difference output bit and β_i the borrow. Then, to compute $D = A - B$:

$$D_i = A_i \oplus (B_i \oplus \beta_{i+1}) \tag{3.6}$$

$$\beta_i = B_i \cdot \beta_{i+1} \vee \bar{A}_i \cdot (B_i \oplus \beta_{i+1}) \tag{(3.7)}$$

Note that the expression for D_i is identical to that for S_i, Eq. (3.3); the borrow expression is of the same functional form as the carry C_{i+1}. In the β_i expression variable A_i appears complemented where it appears uncomplemented in the expression for C_i, Eq. (3.5).

Could the same equipment be used to both add and subtract according to the value of an add-subtract control signal *ASC*? The similarity of the equations suggests the possibility. It is necessary to use A_i when $ASC = 0$ (addition) and $\bar{A}_i$ when $ASC = 1$ (subtraction). An exclusive-OR gate can serve to provide the selective complementing of A_i. Figure 3.10 shows how the pieces must be put together to form an *Adder-Subtractor*: C_i denotes either carry or borrow according to the value of *ASC*; S_i denotes either the sum or difference (result).

$$S_i = A_i \oplus B_i \oplus C_{i+1} \tag{3.8}$$

$$C_i = (A_i \oplus ASC) \cdot (B_i \oplus C_{i+1}) \vee B_i C_{i+1} \tag{3.9}$$

This is certainly not a very formal design procedure, but it is one that is often used. We design small networks or select existing networks with which we are familiar, and then interconnect them to satisfy a specification. Complete familiarity with the characteristics of existing networks and means of designing others is mandatory. Their proper interconnection is more difficult and often a matter of experience. Very bad designs can easily result if we are not careful in making a decision whether to interconnect or design an entirely new network. It is essential to test a proposed design if we choose to interconnect, and may require an exorbitant amount of effort if "corrections" are made to a faulty design with less than careful consideration and analysis.

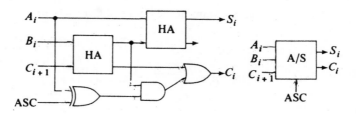

FIG. 3.10   Adder/Subtractor Network

Translation of Conditional DDL Connection Statements

Specifications of combinational networks stated in Digital Design Language (DDL) can be converted to Boolean equations by applying a number of identities. The process is so straightforward and algorithmic that a computer can be programmed to carry it out efficiently. Studying these identities can also help us convert English specifications to Boolean equations.

The following symbols are used to describe the transformations to be made:

$\Rightarrow$ "is replaced by"
C_i the ith of the Boolean expressions serving as conditions.
be_i the ith Boolean expression on the right of a connection operator.
co_i the ith connection operation.
l_i the left (sink) operand of co_i.
$coco_i$ the ith list of compatible connection operations.
v_i the ith value.

If-then clauses with Boolean expressions are reduced with the following transformations.

1. $|C| be . \Rightarrow C \cdot be$
2. $|C| be_1 ; be_0 . \Rightarrow C \cdot be_1 \vee \neg C \cdot be_0$

The full-adder-subtractor of Fig. 3.10 is described with the declarations:

$$\langle TE \rangle \ A, B, C, C_{out}, S.$$
$$S = A \oplus B \oplus C,$$
$$C_{out} = (|ASC| \ \neg A; A.) \cdot (B \oplus C) \vee BC$$

Transform (2) above converts the equation for C_{out} to:

$$C_{out} = (ASC \cdot \neg A \vee \neg ASC \cdot A) \cdot (B \oplus C) \vee BC$$

Conditional connection operations are reduced to Boolean equations with the following transforms, which are not necessarily applied in the order listed.

3. $|C| coco_1 ; coco_0 . \Rightarrow |C| coco_1 ., |\neg C| coco_0 .$
4. $|C| co_1, co_2, \ldots, co_m . \Rightarrow |C| co_1 ., |C| co_2 ., \ldots, |C| co_m .$
5. $|C_1| \ |C_2| \ \ldots |C_k| coco. \ldots . \Rightarrow |C_1 \cdot C_2 \cdot \ldots \cdot C_k| coco.$
6. $| C | \ell = be. \Rightarrow \ell = C \cdot be$

The following DDL statement will be used to illustrate these transforms.

$$|X| \ F = |Y| \ A; B., G = C; |Z| \ F = A \cdot C, G = B; F = B..$$

Applying transform (2) to the Y condition gives:

$$|X| \ F = Y \cdot A \vee \neg Y \cdot B, G = C; |Z| \ F = A \cdot C, G = B; F = B..$$

Applying transform (3) to the Z conditions yields:

$|X|$ F = Y $\cdot$ A $\vee \neg$ Y $\cdot$ B, G = C; $|Z|$ F = A $\cdot$ C, G = B., $|\neg Z|$ F = B..

The X condition is split with the same tranform.

$$|X| \ F = Y \cdot A \vee \neg Y \cdot B, G = C., |\neg X| |Z| \ F = A \cdot C,$$
$$G = B., |\neg Z| \ F = B..$$

The Z condition may be distributed with transform (4).

$$|X| \ F = Y^! A \vee \neg Y \cdot B, G = C., |\neg X| |Z| \ F = A \cdot C.,$$
$$|Z| \ G = B., |\neg Z| \ F = B..$$

Now the X and $\neg X$ conditions are distributed.

$$|X| \ F = Y \cdot A \vee \neg Y \cdot B., |X| \ G = C., |\neg X| |Z| \ F = A \cdot C..,$$
$$|\neg X| |Z| \ G = B.., |\neg X| |\neg Z| \ F = B..$$

Transform (5) reduces the multiple conditions.

$$|X| \ F = Y \cdot A \vee \neg Y \cdot B., |X| \ G = C., |\neg X \cdot Z| \ F = A \cdot C.,$$
$$|\neg X \cdot Z| \ G = B., |\neg X \neg Z| \ F = B.$$

Finally transform 6 eliminates all *if-then* conditions.

$$F = X \cdot (Y \cdot A \vee \neg Y \cdot B), G = X \cdot C, F = (\neg X \cdot Z) \cdot A \cdot C,$$
$$G = (\neg X \cdot Z) \cdot B, F = (\neg X \cdot \neg Z) \cdot B$$

While we have applied the transforms in turn to individual conditions, they can be applied simultaneously or in different order to obtain the same result, which is still a piece-wise description of a network.

The pieces are gathered with the following transform.

7. $\ell = C_1 \cdot be_1, \ell = C_2 \cdot be_2, \ldots, \ell = C_k \cdot be_k \Rightarrow \ell = C_1 \cdot be_1 \vee C_2 \cdot be_2$
$\vee \ldots \vee C_k \cdot be_k$

For compatibility, if $be_i \neq be_j$, then $C_i \cdot C_j = 0$ must be true. Applying this transform to the example above gives complete equations for F and G from which networks can be derived

$$F = X \cdot (Y \cdot A \vee \neg Y \cdot B) \vee (\neg X \cdot Z) \cdot A \cdot C \vee (\neg X \cdot \neg Z) \cdot B$$
$$G = X \cdot C \vee (\neg X \cdot Z) \cdot B$$

These transforms accomplish what we informally accomplished in the first examples of this section. Comparing the equations that resulted above with the original DDL statement reveals that they emphasize sum-of-products equations and assume that output signals are to have the value of 0 when no connection is specified. Early in this section we argued that "If $C = 1$ then $F = A$" was a partial specification leading to a partial switching function, don't cares, and a variety of networks. The transforms do not make the same assumption. Don't care conditions must be described using the question mark (?); terminals are assumed to bear logic 0 when no connection to them is specified.

Reducing *if-value* syntax proceeds in a similar fashion. Let C, the Boolean expression serving as the condition, be of dimension n. Then a value v_i is an integer from the range 0 through $2^n - 1$, a list of integers, or a subrange. Let j and k denote the first and last of the integers expressed by v_i. The *decoder function*, δ_i, on C is then

$$\delta_i = \cdot /C \backslash jDn \vee \ldots \vee \cdot /C \backslash kDn \tag{3.10}$$

Each of the AND-reductions describes a minterm; if v_i expresses more then one integer, then δ_i is the logic sum of more than one minterm.

Conditional clauses within Boolean expressions are reduced with:

8. $\lceil C \lfloor v_1 be_1 \lfloor v_2 be_2 \ldots \lfloor v_m be_m. \Rightarrow \delta_1 \cdot be_1 \vee \delta_2 \cdot be_2 \vee \ldots \vee \delta_m \cdot be_m$
9. $\lceil C \lfloor v_1 be_1 \ldots \lfloor v_m be_m; be_x. \Rightarrow \delta_1 \cdot be_1 \vee \ldots \vee \delta_m \cdot be_m \vee \delta_x \cdot be_x$
   where $x = \{0,1, \ldots, 2^n - 1\} - \{v_1, v_2, \ldots, v_m\}$

Transform 9 includes an else case; all permitted values not included in any va_i form value list x.

If-value conditioned compatible sets of connection operations are reduced to if-then conditioned operations with:

10. $\lceil C \lfloor v_1 coco_1 \ldots \lfloor v_m coco_m. \Rightarrow |\delta_1| coco_1., \ldots, |\delta_m| coco_m.$
11. $\lceil C \lfloor v_1 coco_1 \ldots \lfloor v_m coco_m; coco_x. \Rightarrow |\delta_1| coco_1., \ldots, |\delta_m| coco_m.,$
    $|\delta_x| coco_x.$
    where $x = \{0,1, \ldots, 2^n - 1\} - \{v_1, \ldots, v_m\}$

Previously defined transforms are used after these to further reduce the condition syntax and gather connections to each information sink.

The specification leading to the network of Fig. 3.4 may be stated in DDL as

$$\lceil A \circ B \circ C \lfloor 0: 1F = D \lfloor 2 G = E \lfloor 3 ,7 F = E \lfloor 4 F = 1,$$
$$G = D \lfloor 5 F = 1 \lfloor 6 G = D \vee E.$$

Transform 10 reduces this expression to:

$$|\delta_{0:1}| F = D., |\delta_2| G = E., |\delta_{3,7}| F = E., |\delta_4| F = 1, G = D.,$$
$$|\delta_5| F = 1., |\delta_6| G = D \vee E.$$

Since we are decoding $A \circ B \circ C$

$$\delta_{0:1} = \bar{A} \cdot \bar{B} \cdot \bar{C} \vee \bar{A} \cdot \bar{B} \cdot C$$
$$\delta_2 = \bar{A} \cdot B \cdot \bar{C}, \quad \text{etc.}$$

Specifications for combinational logic networks may take any of a variety of forms. Boolean functions must be derived from specifications by using tables, maps, arrays, or equations first, whichever seems to be most natural. After ambiguities and inconsistencies are clarified, the Boolean functions may be expressed in other forms to search for Boolean expressions that suggest optimum networks. In some situations first derived or more easily understood networks are optimum. In other cases costs must be minimized subject to technological constraints such as types of gates available, fan-in and fan-out limitations, and speed requirements. The next section and later chapters will treat these constraints and discuss cost minimization. We will also see that the searching can be accomplished in an algorithmic manner using formal transform rules, just as DDL condition syntax was reduced in this section.

3.2 NOR AND NAND NETWORK SYNTHESIS

Most of the basic electronic circuits reviewed in Section 2.1 realize either the NAND or the NOR operator. While AND and OR circuits can be formed by adding an inverting amplifier to NAND and NOR circuits, propagation delay, area, and power consumption are all increased by so doing. Realizing switching functions with networks of NAND and/or NOR gates is therefore very practical, and not significantly more difficult, as we shall see in this section.

Both the NAND and NOR operators possess the unusual property of being *functionally complete*—any Boolean function can be expressed entirely in terms of the NAND (NOR) operator. Recall the definition of the NAND (NOR) operator:

$$A \uparrow B = \overline{A \cdot B} \qquad A \downarrow B = \overline{A \vee B} \tag{3.11}$$

Then the NOT operator may be expressed in terms of either the NAND or NOR operator.

$$A \uparrow A = \overline{A \cdot A} = \overline{A} \qquad A \downarrow A = \overline{A} \tag{3.12}$$

The AND operator may be expressed in terms of the NAND (NOR) operator.

$$A \cdot B = \overline{\overline{A \cdot B}} = \overline{A \uparrow B} = (A \uparrow B) \uparrow (A \uparrow B)$$
$$A \cdot B = \overline{\overline{A} \vee \overline{B}} = (A \downarrow A) \downarrow (B \downarrow B) \tag{3.13}$$

And finally, the OR operator also may be so expressed.

$$A \vee B = \overline{\overline{A} \cdot \overline{B}}$$
$$= \overline{A} \uparrow \overline{B}$$
$$= (A \uparrow A) \uparrow (B \uparrow B) = (A \downarrow B) \downarrow (A \downarrow B) \tag{3.14}$$

Since we are able to express AND, OR, and NOT functions in terms of the NAND (NOR) operator, we are able to express any Boolean function in terms of the NAND (NOR) operator only. In terms of circuits, any switching function may be realized with a network of NAND (NOR) gates only. Any function can be realized with a network of AND and OR gates and INVERTERS. Each such logic block may be replaced with the logically equivalent NAND (NOR) network of Fig. 3.11.

An algebra involving the NAND (NOR) operator is commutative.

$$A \uparrow B = B \uparrow A \qquad A \downarrow B = B \downarrow A$$
$$\overline{A \cdot B} = \overline{B \cdot A} \qquad \overline{A \vee B} = \overline{B \vee A} \tag{3.15}$$

But such an algebra is *not* associative.

$$A \uparrow (B \uparrow C) = A \uparrow (\overline{B\,C}) \qquad A \downarrow (B \downarrow C) = \overline{A \vee \overline{(B \vee C)}}$$
$$= \overline{A \cdot (\overline{BC})} \qquad\qquad = \bar{A} \cdot (B \vee C)$$
$$= \bar{A} \vee BC$$

$$(A \uparrow B) \uparrow C = AB \vee \bar{C} \qquad (A \downarrow B) \downarrow C = (A \vee B) \cdot \bar{C}$$

We can construct three and more input NAND (NOR) gates that realize $\overline{A \cdot B \cdot C}$ $(\overline{A \vee B \vee C})$, which is denoted without parentheses by $A{\uparrow}B{\uparrow}C$ $(A{\downarrow}B{\downarrow}C)$. The function of this Boolean expression is not equal to that of either of the expressions above.

$$A \uparrow B \uparrow C \neq A \uparrow (B \uparrow C) \neq (A \uparrow B) \uparrow C \neq B \uparrow (A \uparrow C) \tag{3.16}$$

As a result, manipulating functions expressed in terms of the NAND (NOR) operator is very difficult, and we must conclude that this algebra is not a very useful tool for analyzing and synthesizing switching networks.

How then can we design networks of NAND (NOR) gates? A first suggestion might be to work in terms of the AND, OR, and NOT operators

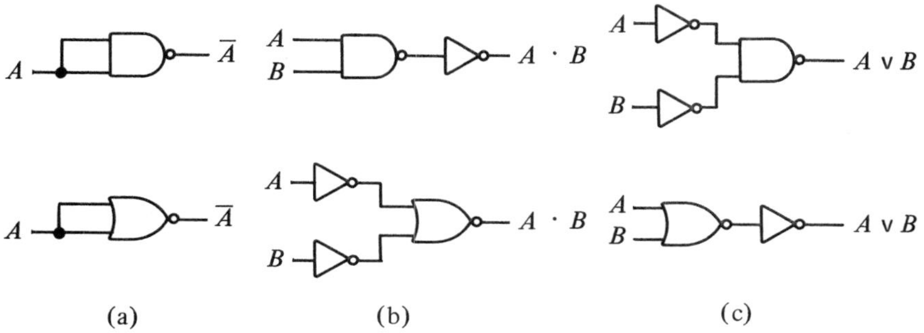

(a) (b) (c)

FIG. 3.11 NAND and NOR Realizations of (a) NOT, (b) AND, and (c) OR

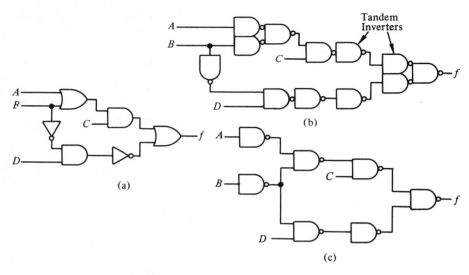

FIG. 3.12   Direct Network Transformation

until we place each Boolean function in a "desirable" form (for we have facility with AND, OR, and NOT) and then *transform* it to a NAND (NOR) expression using Eqs. (3.12) through (3.14): or transform the AND-OR-NOT network by replacing each block with the equivalent NAND (NOR) network of Fig. 3.11. The AND-OR-NOT network of Fig. 3.12(a) could be thus transformed by drawing the diagram of Fig. 3.12(b), which pictures a large number of gates. Some of these can be removed. If we look carefully, we see several places where one inverter follows another. Such a pair of gates serves no logic purpose and can be eliminated. We simplify the network to Fig. 3.12(c).

If we carefully clean up a network transformed in this direct manner, we may obtain a satisfactory result. But we can often obtain better results, and gain insight into how we should attempt to express Boolean functions prior to transformation, by studying another transform method that requires no subsequent cleaning-up, and less effort.

To find this better transform method, let us first treat a function expressed in *sop* form. Placing two complement bars over this expression does not alter it. Applying De Morgan's theorems once transforms the expression to one that involves the NAND (NOT of AND) operator only.

$$F = A \lor BC \lor DE$$

We place two bars over the function and apply De Morgan's theorems.

$$F = \overline{\overline{A} \cdot \overline{BC} \cdot \overline{DE}}$$

We recognize the complement of $B \cdot C$ and of $D \cdot E$; these are NAND terms and may be expressed with the NAND operator. But the complement of the product of these terms, $(\bar{A})$, $(B \uparrow C)$, and $(D \uparrow E)$, is also specified. Thus one can express F as the NAND of these 3 terms.

$$F = \bar{A} \uparrow (B \uparrow C) \uparrow (D \uparrow E)$$

Comparing the original and final expressions for F above, or the networks they suggest as shown in Fig. 3.13, reveals that sum-of-products expressions (AND-OR networks) can be converted to NAND expressions (networks) by:

1. Parenthesizing the product terms.
2. Converting all AND and OR operators (gates) to NAND operators (gates).
3. Complementing any single input variable product terms (primary input signals to the OR gates).

Inverting the analysis procedure of Section 2.6 gives the same results.

Generalizing, suppose that we have an expression in sum-of-products-of-sums-of-products form. It suggests a network of AND gates, feeding OR gates, feeding AND gates, ..., feeding an OR gate. The OR gates are at odd logic levels with respect to the output terminal; the AND gates are at even levels. The transform to NAND gates illustrated above can now be applied in turn to AND-to-OR gates subnetworks until the entire network is converted. As an example, consider the switching function

$$Z = A \vee B(C \vee D(E \vee F))$$

Applying deMorgan's theorem once converts the level 1 and 2 operators (gates) to NAND operators (gates).

$$Z = \overline{\overline{A \vee B(C \vee D(E \vee F))}}$$
$$= \overline{\bar{A} \cdot \overline{B(C \vee D(E \vee F))}}$$
$$= \bar{A} \uparrow [B \uparrow (C \vee D(E \vee F))]$$

A sum-of-products-of-sums expression remains to be converted. Two additional levels of operators (gates) are converted by repeating the above transform on the subexpression $C \vee D(E \vee F)$.

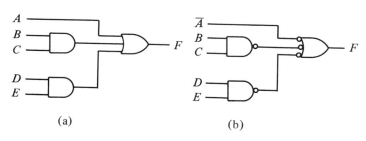

(a)   (b)

FIG. 3.13   AND-OR to NAND Network by Transformation Rules

$$Z = \bar{A} \uparrow [B \uparrow \overline{(C \vee D(E \vee F))}]$$
$$= \bar{A} \uparrow [B \uparrow \overline{(\bar{C} \cdot \overline{D(E \vee F)})}]$$
$$= \bar{A} \uparrow [B \uparrow (\bar{C} \uparrow \{D \uparrow (E \vee F)\})]$$

Repeating one more time converts the remaining sum term.

$$Z = \bar{A} \uparrow [B \uparrow (\bar{C} \uparrow \{D \uparrow (\bar{E} \uparrow \bar{F})\})]$$

The corresponding networks shown in Fig. 3.14 also suggest that to convert to a NAND network, start with AND gates at even levels and OR gates at odd levels, replace all gates with NAND gates, and complement primary input signals that drive odd level (OR) gates.

These examples indicate that if we wish to realize a Boolean function with a network of NAND gates then that function should be expressed in sum-of-product-of . . . form. Then the NAND network can be drawn directly from the expression, or an AND-OR network may be drawn as an intermediate step and converted to a NAND network by reversing the graphic substitution procedure of Section 2.6. (This procedure converts all gates to NAND gates and complements primary input signals to gates at odd levels.)

If the complements of network input signals are not available and we wish to minimize the number of inverters required in a NAND network, we should attempt to express functions so that complemented variables are combined with OR operators. Expressing function f in sum-of-products form as

$$f = \bar{A}\bar{C} \vee \bar{A}\bar{D} \vee \bar{B}\bar{C} \vee \bar{B}\bar{D}$$

suggests the need for five NAND gates and four inverters. In product-of-sums form

$$f = (\bar{A} \vee \bar{B}) \cdot (\bar{C} \vee \bar{D})$$

all complemented variables are combined by OR operators, but the transform does not apply. Adding logic 0 places the expression in the required sum-of-products-of . . . form.

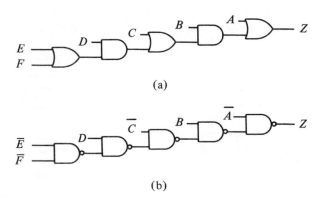

(a)

(b)

FIG. 3.14 Transforming a Multiple Level Network

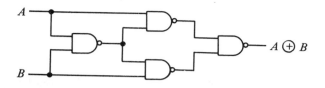

FIG. 3.15   NAND Realization of the Exclusive-OR Function

$$f = (\bar{A} \vee \bar{B}) \cdot (\bar{C} \vee \bar{D}) \vee 0$$

Only four 2-input NAND gates are called-for by this expression. Since the 0 is combined by an OR operator, its complement must be supplied to the output level gate, which therefore is acting as an inverter. We are realizing the function in AOI form.

$$f = \overline{\overline{AB} \vee \overline{CD}}$$

In some cases one NAND gate can serve in place of two or more inverters. The exclusive-OR function provides a well-known example of such economy.

$$f = A\bar{B} \vee \bar{A}B$$

Rather than using two inverters as suggested, we add redundant terms so as to induce a repetition of OR combinations of complemented variables.

$$f = A(\bar{A} \vee \bar{B}) \vee (\bar{A} \vee \bar{B})B$$

The NAND term, $\bar{A} \vee \bar{B} = A \uparrow B$, appears twice. If fan-out is permitted, this function can be realized with the 4 gate network of Fig. 3.15.

Duality permits us to conclude that the transformation and techniques of NAND network design apply in dual form for NOR networks. If we seek a NOR gate realization of a switching function, we should express that function in product-of-sums-of-products . . . form. The odd level rule continues to apply; now AND gates occupy the odd level positions. Then a NOR network may be drawn directly, or an OR-AND network may be drawn as an intermediate step and converted to a NOR network by symbolic replacement.

$$Z = A(B \vee C(D \vee EF))$$
$$= \overline{\overline{A(B \vee C(D \vee EF))}}$$
$$= \bar{A} \downarrow (B \downarrow \overline{\overline{[C(D \vee EF)]}})$$
$$= \bar{A} \downarrow (B \downarrow [\bar{C} \downarrow (D \downarrow \overline{\overline{EF}})])$$
$$= \bar{A} \downarrow (B \downarrow [\bar{C} \downarrow (D \downarrow \{\bar{E} \downarrow \bar{F}\})])$$

Mixed NAND/NOR Networks

The previous transformations produce networks of one gate type only. They require that the switching function to be realized be expressed within a constrained form. If both NAND and NOR gates are available for network construction, then arbitrary networks of AND, OR, and NOT gates can be converted to inverting logic. Again the transformation is made by gate substitution. Some modification of the original AND, OR, and NOT gate network may be required. Anticipating this transformation when preparing the original AND, OR, and NOT gate network can lead to superior results, as with the previous transforms.

To convert a network of AND, OR, and NOT gates to a NAND/NOR network:

1. Mark each output gate as at level 1 (odd).
2. Trace back from each output terminal, replacing gates according to Table 3.1. While inverters are deleted, their presence is noted by advancing the level count.
3. If an already converted gate is encountered, and the level at which it was converted is of the same parity (even or odd) as its position on the path currently being traced, no further conversion or tracing of that path is required. If the gate was converted at the "wrong" parity, insert an inverter at the output terminal of the converted gate in the path being traced; and cease tracing that path.
4. Insert inverters in input lines of odd level gates that are driven by primary input signals; delete inverter pairs via the Involution Theorem.

TABLE 3.1   GATE REPLACEMENTS FOR NOR/NAND SYNTHESIS

Gate Type	Odd Level Replacement	Even Level Replacement
AND	NOR	NAND
OR	NAND	NOR
NOT	delete	delete
Primary Input	Add Inverter	

The network of Fig. 3.16(a) illustrates most aspects of this transformation. Tracing the paths from the F terminal partially converts the network as shown in Fig. 3.16(b). Note that AND gate 5 was converted at an even level. Inverter 6 is at an odd level; it is deleted, but since it is driven by a primary input signal it is immediately reinstated. Gate 2 is converted at an odd level. Inverter 3 is then deleted, but it does contribute a level by placing gate 5 at odd level 3. Since that gate was converted at an even level, an inverter must be introduced in the path from gate 5 to gate 2. Thus inverter 3 is reinstated. An inverter is also introduced in the lower input line of gate 2, since a primary input signal drives that line. Figure 3.16(c) shows the completely converted network.

If we trace the paths from G first, the network is changed to that of Fig. 3.17(a). Inverter 3 is dropped and AND gate 5 is transformed at an odd level into a NOR gate. An inverter must be added in the C line of gate 5. Inverter 6 is at an even level and therefore is simply deleted. When

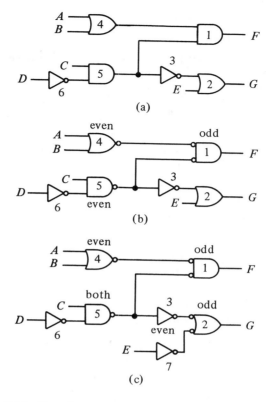

(a)

(b)

(c)

FIG. 3.16   Transforming by Tracing the Paths from F First

we transform the gates that realize F, gate 5 is found at an odd level so an inverter must be introduced between gates 5 and 1. Three NOR gates and one NAND gate appear in the final network of Fig. 3.17(b), whereas two NOR gates and two NAND gates are shown in Fig. 3.16(c). Clearly the

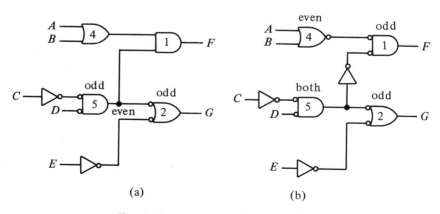

(a)   (b)

FIG. 3.17   Transforming from G First

SECTION 3.2   NOR and NAND Network Synthesis

transform techniques does not necessarily give a unique result for multiple output networks.

Table 3.1 should be compared with Table 2.10. We are inverting the direct NAND/NOR network analysis procedure here. On the other hand this transform method may be thought to place inverting circles at all input terminals of all odd level gates and at all output terminals of all even level gates. If an inverter is placed at one end of a line, another must be placed at the other end so as to preserve the Boolean function of the network. While this is accomplished for lines driven and terminated by gates, lines driven by primary input signals and gates at both even and odd levels require special actions.

3.3 MULTIPLEXER, ROM, AND PLA REALIZATIONS

Minterm and maxterm canonical expressions of switching functions suggest the use of a large number of high fan-in gates. If switching functions are to be realized with discrete gates, we minimize such expressions with Karnaugh maps or Boolean algebra to reduce the number of gates and interconnections needed. Two types of integrated circuits permit us to avoid such minimization by making the canonical realization of many switching functions economical. First we will examine the "multiplexer" as a function generator.

A *multiplexer* with n control terminals, $C_{n-1}, \ldots C_0$, and 2^n data terminals, D_i, connects the ith data signal to its output terminal f when the control signals present the number i. Multiplexers consist of 2^n $(n + 1)$-input AND gates that drive a 2^n input OR gate, or any logically equivalent network. Each AND gate receives one data signal and a unique combination of true and complemented control signals. Each AND gate forms the product of one minterm of the control variables and a data variable. A multiplexer is thus described by Th. 9(a), Eqs. (2.35) and (2.39), with the f_d's being data signals.

$$f = \bigvee_{i=0}^{2^n-1} D_i m_i \qquad (3.17)$$

If we use the truth table entries for a switching function as the data values and the independent variables as the control variables, the multiplexer realizes the switching function. In Fig. 3.18, the truth table expresses $F = A \oplus B \oplus C$. One 8-to-1 line multiplexer realizes this function when data terminals are tied to logic 0 or 1 as the truth table dictates.

An 8-input multiplexer can realize any 3 variable switching function; a 16-input multiplexer can realize any 4 variable switching function with the very direct design technique illustrated in Fig. 3.18. Larger multiplexers are not available as single ICs. Often functions can be realized with less than 2^n-input multiplexers by using an inverter or gate in addition to a smaller multiplexer. In the truth table of Fig. 3.18, notice that $F = C$ in the first two rows ($AB = 00$) and the last two rows ($AB = 11$). $F = \bar{C}$ in the middle four rows.

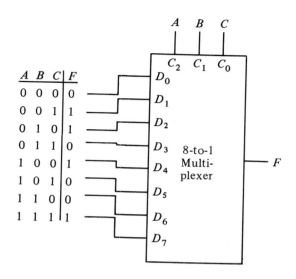

A	B	C	F
0	0	0	0
0	0	1	1
0	1	0	1
0	1	1	0
1	0	0	1
1	0	1	0
1	1	0	0
1	1	1	1

FIG. 3.18   An 8-Input Multiplexer Can Generate Any Switching Function of Three Variables

$$F = \bar{A}\bar{B}(C) \vee \bar{A}B(\bar{C}) \vee A\bar{B}(\bar{C}) \vee AB(C)$$
$$= \bar{A}\bar{B} \cdot F(0, 0, C) \vee \bar{A}B \cdot F(0, 1, C) \vee A\bar{B} \cdot F(1, 0, C) \vee AB \cdot F(1, 1, C)$$

A 4-input multiplexer and an inverter that provides $\bar{C}$ may be used in place of the 8-input multiplexer of Fig. 3.18. (A 4-input multiplexer is typically one-half of an available integrated circuit.) Figure 3.19 shows this realization of the exclusive-OR. The second 4-input multiplexer is used to realize the carry equation.

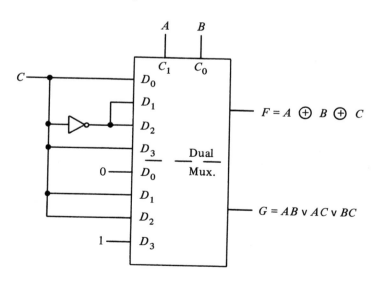

FIG. 3.19   A Dual 4-Input Multiplexer Realization of the Full-adder Equations

Equations developed in the previous example suggest how we may find realizations using smaller multiplexers. Suppose that an n-variable function is to be realized with a 2^m-input multiplexer. A subset of m of the input variables must be selected for use as control variables. Expanding the function about these m variables leads to an expression that includes 2^m functions of the remaining $n - m$ variables. These functions must be realized outside of the multiplexer. We desire simple subfunctions so that little additional circuitry is required.

The following functions were used in Section 3.1 to illustrate "infrequent" variables in Karnaugh maps.

$$F = \bar{A}\bar{B}D \lor BCE \lor A\bar{B}$$
$$G = B\bar{C}E \lor A\bar{C}D$$

Let us attempt to realize these functions with a dual 4-input multiplexer IC. We might select A and B as the control variables. Expanding then gives

$$F = \bar{A}\bar{B}(D) \lor \bar{A}B(CE) \lor A\bar{B}(1) \lor AB(CE)$$
$$G = \bar{A}\bar{B}(0) \lor \bar{A}B(\bar{C}E) \lor A\bar{B}(\bar{C}D) \lor AB(\bar{C}E \lor \bar{C}D)$$

The subfunctions D, CE, $\bar{C}E$, $\bar{C}D$, and $\bar{C}E \lor \bar{C}D$ can be observed in the Karnaugh maps of Fig. 3.5. Four gates are necessary to realize them, as shown in Fig. 3.20(a).

If we expand about variables B and C, we find

$$F = \bar{B}\bar{C}(\bar{A}D \lor A) \lor \bar{B}C(\bar{A}D \lor A) \lor B\bar{C}(0) \lor BC(E)$$
$$\quad = \bar{B}\bar{C}(A \lor D) \lor \bar{B}C(A \lor D) \lor B\bar{C}(0) \lor BC(E)$$
$$G = \bar{B}\bar{C}(AD) \lor \bar{B}C(0) \lor B\bar{C}(E) \lor BC(0)$$

Only two subfunctions, $A \lor D$ and AD, are required, as Fig. 3.20(b) emphasizes. While we are not sure that selection of some other subset such as A and C for use as control signals would not yield an even better realization, B appears most frequently, and D and E appear least frequently in the original equations. Therefore, B should probably be one of the control variables, and D and E should not be so used.

Many multiplexer ICs have an *enable* input terminal with the complement of the enable input signal being presented as an additional input signal to all AND gates. Thus, when enable = 0, the multiplexer performs as previously discussed. When enable = 1, however, the output signal has value 0 regardless of all other input signals. This terminal is very useful when the function to be realized can be written in the form

$$f(x_1, \ldots, x_n, x_{n+1}) = x_{n+1} \cdot g(x_1, \ldots, x_n)$$

Variable x_{n+1} (or its complement) is supplied to the enable terminal; the rest of the multiplexer is used to realize $g(x_1, \ldots, x_n)$.

ROM Realizations

A switching function may also be realized in canonical form by using an n-to-2^n-line decoder (see Section 3.1) and an OR gate. Selected output

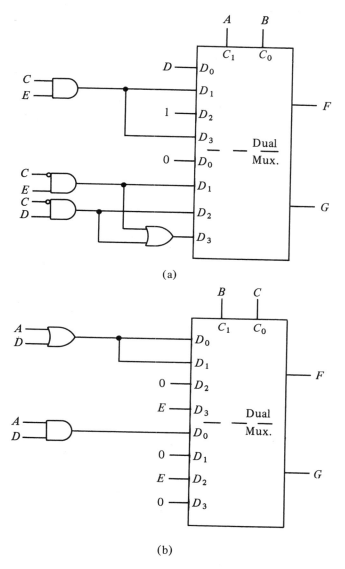

(a)

(b)

FIG. 3.20 Multiplexer Realizations of Functions with Infrequent Variables

terminals of the decoder are connected to the OR gate. While this approach requires little design effort and such networks are easily modified when a different function is required, it is not as economical of integrated circuits and requires many interconnections between the decoder and the OR gate.

A ROM consists of a complete decoder on n " address " lines and m OR gates that provide output signals. Any decoder output line may be connected to an OR gate. Thus a ROM has the structure discussed above, but AND to OR interconnections are made within the ROM integrated circuit rather than between separate ICs. Determining and making the interconnections that are to exist is called " programming " the ROM.

We supply n bits to a ROM; it responds by providing m bits after signals propagate through the 2-level network. This is the action of a memory unit. We think of the ROM as storing m-bit words in locations referenced by the n input signals. The m-bit words are written into the memory when the gate interconnections are made. This process is very slow compared with the time needed to retrieve these words. Hence the name *Read-Only Memory* is very descriptive.

ROMs realize switching functions in sum-of-minterms form, the least economical approach if discrete gates are used. ROMs are economically attractive because they consist of a large number of identical cells, both in the decoder and in the interconnect matrix that forms the OR gates. Once a "blank" ROM (all or no interconnections are made) is designed, it can be manufactured in large quantities and then specialized at an incremental cost. While the user may pay much more for a ROM than a gate, he does gain in other respects. The price of inspecting, assembling, testing, and wiring an IC in place is almost constant and independent of the logic provided by the IC. Thus, i a ROM can replace a large "random" network of gates, these and other costs are reduced. Further, should the logic be in error or the specifications of a system changed, the ROM can be replaced without modifying the printed circuit board mounting, whereas the random network (and its PC board) may require complete redesign.

ROMs are subject to the same limitations as other ICs. Heat must be dissipated. External connections are limited to between 14 and 24 pins usually. Fewer pins require less PC board area. In a 16-pin package we have several options for storage: 256 bits (32 words × 8 bits per word) when using 5 address pins and 8 output pins, 2048 bits (512 × 4 bits per word), or 4096 bits (4096 × 1). Several pins must be reserved for power supply and control signals.

ROMs may be programmed by the manufacturer. Only the mask used to form interconnections must be specialized. A few applications have been identified as being so common that IC manufacturers produce programmed ROMs to meet them. Alphanumeric code converters and character generators are examples. Suppose that we store the following truth table in a ROM.

Address	ROM Word
000	00000
001	11110
010	10001
011	10001
100	11110
101	10000
110	10000
111	10000

If the 5-bit words are read in turn and each is used to modulate the intensity of a CRT at 5 points (0-black, 1-light), the letter P will be displayed. A 5 × 7 dot matrix is used in video terminals and some printing equipment to form letters. Sixty-four characters are commonly available. Six address bits select the character; three more select a row or column of dots.

Mask programmed ROMs may be ordered from IC manufacturers. The truth table to be stored must then be specified, usually in the form of a deck of punched cards that may be read by a computer program that designs artwork for the interconnection network. While the initial charge with this approach may be high, cost per ROM decreases rapidly with volume.

ROMs that are programmed by the user at his location are referred to as PROMs—programmable ROMs. All interconnections are made in blank PROMs. The user may break interconnections by burning nichrome fuses with a piece of equipment that provides carefully controlled pulses on provided terminals. Figure 3.21 shows how fuses may be used in conjunction with diode and transistor matrices. Some technologies permit previous programs to be erased so that the PROM may be reprogrammed to correct errors or change the nature of a system.

The output amplifiers of ROMs are usually open-collector, as suggested in Fig. 3.21, or tri-state so that greater word capacity can be achieved with parallel ROM ICs. Most ROMs have one or more "enable" input terminals to facilitate this. When the enable input is high, all output transistors are off or tri-state amplifiers are in their high impedance state. Thus a 512 word, 4 bit per word ROM may be formed from 256×4-bit ROMs as shown in Fig. 3.22.

ROMs may appear to reduce logic design to the writing of correct truth tables. While their availability has greatly affected the design of digital systems, they are not universely applicable. Functions of a few variables or more than ten variables are often best realized with other types of logic blocks. The expense of ROMs may not be justified in one-of-a-kind or small quantity products.

Programmable Logic Arrays

While integrated circuits of a more general nature are also being called PLAs, here we define a *programmable logic array* (PLA) to be an integrated circuit with n input terminals that drive inverters, m output terminals driven by OR(NOR) gates, and p AND gates. Also included are interconnections between any or all input terminals (or their inverters) and the AND gates, and between any or all AND gates and each output gate. Each output gate may be either an OR or a NOR gate. Desired connections are established in the final metalization step of manufacture, or undesired connections are broken in a completely interconnected integrated circuit. If this final step can be performed after the IC is manufactured, the unit is a *field-programmable logic array*, or FPLA. As with ROMs, economy is attained by restricting the specialization of the integrated circuit to one final step of manufacture; and if extremely large quantities are to be produced then a specially designed integrated circuit is more economical than the PLA.

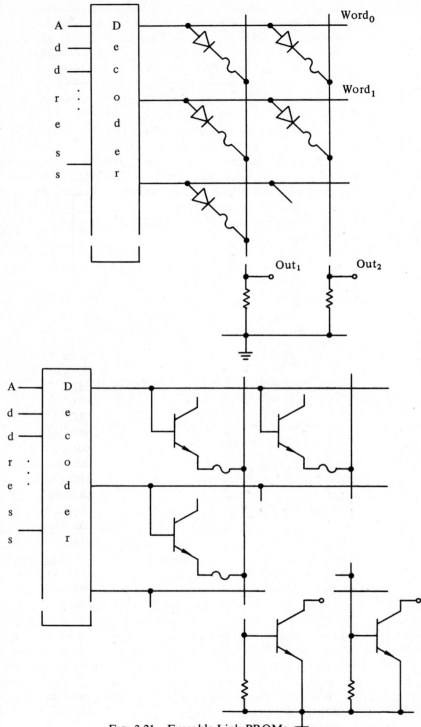

Fɪɢ. 3.21   Fuseable Link PROMs

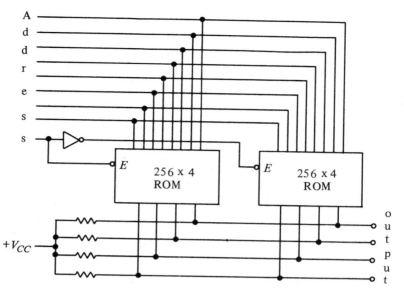

FIG. 3.22 ROM Expansion Using Open-Collector Output Amplifiers and Enable Input Terminals

The AND gates form product terms; the OR gates sum selected product terms to realize switching functions in either *sop* or AOI form (see Fig. 3.23). An n-input, m-output switching function that can be described with p or fewer product terms can be realized with a single PLA. Typical values of n, m, and p are: $n = 14$, $m = 8$, and $p = 40$ to 100.

PLAs are similar to ROMs; however, whereas ROMs completely decode the input variables to minterms, PLAs do not. The input variable set size is greater for PLAs while the number of gates is comparable with a 5 or 6 input ROM. Functions of n or fewer input variables that can be described with p or fewer product terms can be realized with one PLA rather than a network of ROMs or other combinational logic blocks.

Simplification or minimization (Chapter 10) of multiple-output switching functions is necessary in the programming of PLAs when obvious sum-of-products and AOI expressions of the output variables require more than p different product terms. It is not important that output variables be described in terms of prime implicants, only that the number of different implicants in use be less than the number of AND gates in the PLA.

In the tables below the left truth table specifies a 4-input, 4-output switching function; it is the programming one must prepare if the function is to be realized with a ROM. That ROM must have at least 16 AND gates within it. The table on the right is not a truth table. It is a *connection table* in that the entries in the output variable columns indicate:

0   no connection between the AND gate that realizes the input symbol and the OR gate that provides the output variable.
1   connection between AND and OR gate.

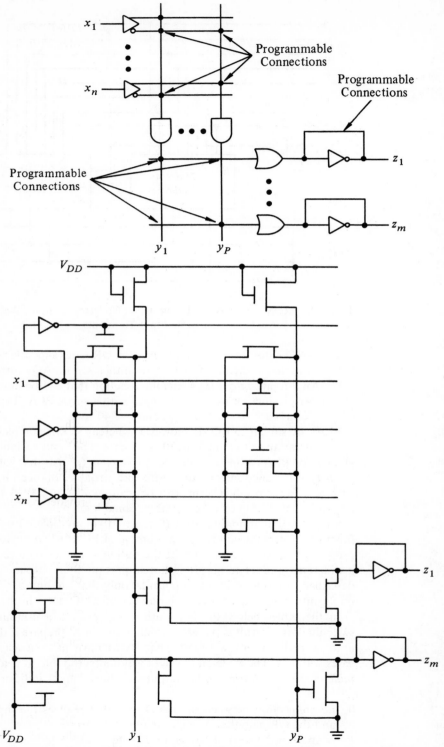

FIG. 3.23 The Logic and Technology of Programmable Logic Arrays. Either diodes or transistors can be used to form gates.

This table provides programming for a PLA realization of the same switching function. The PLA need only have 5 AND gates within it.

A	B	C	D	W	X	Y	Z
0	0	0	0	0	0	0	1
0	0	0	1	1	0	1	1
0	0	1	0	0	0	0	1
0	0	1	1	0	0	0	1
0	1	0	0	0	0	0	0
0	1	0	1	1	0	1	0
0	1	1	0	0	0	0	0
0	1	1	1	0	0	0	0
1	0	0	0	0	0	0	0
1	0	0	1	0	0	0	0
1	0	1	0	1	0	0	1
1	0	1	1	1	0	0	1
1	1	0	0	0	1	1	0
1	1	0	1	0	1	1	0
1	1	1	0	1	1	0	0
1	1	1	1	1	1	0	0

A	B	C	D	W	X	Y	Z
0	x	0	1	1	0	1	0
0	0	x	x	0	0	0	1
1	0	1	x	1	0	0	1
1	1	1	x	1	1	0	0
1	1	0	x	0	1	1	0

3.4 ITERATIVE NETWORKS

To iterate is to do again. An iterative network then is one that does the same task repeatedly. When repetitions occupy successive units of time, the iteration is through time. Spacial iteration is accomplished by many identical networks operating simultaneously. Time iteration is used in conjunction with the serial communication of data. Parallel communication is accompanied by the parallel processing provided by space iteration. Combinations of space and time iteration are possible, of course, and often are employed to obtain a desired processing-speed-to-processor-cost ratio.

An example of a time iterated (serial) processing system is provided by Fig. 3.24. Two shift registers provide the bits of two positive numbers, least significant bit first, to a network that consists of a full-adder and a D flip-flop. Sum bits provided by the full-adder are collected by a third shift register. Assuming that the carry flip-flop provides $C = 0$ when the A and B shift registers provide the least significant bits of numbers A and B, the bits of the sum of A and B are placed in the result register, R. The waveforms of Fig. 3.24 illustrate how the single full-adder repeatedly adds bits to form the sum of the 4-bit numbers $(2 + 3 = 5)$.

It is easy to understand how the serial adder works. We add binary numbers manually in a very serial fashion. The flip-flop delays the carry bit for one unit of time; it aligns it in time with the next more significant bits of the numbers being added. In terms of iteration we may view this as

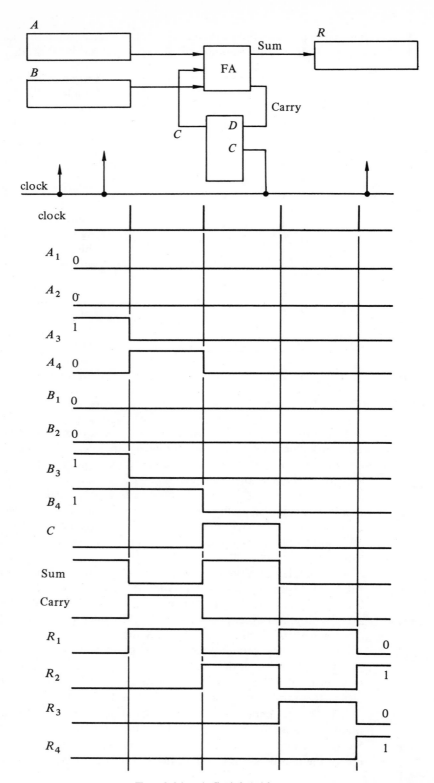

Fɪɢ. 3.24   A Serial Adder

providing the communication between repetitions that is required by the addition process.

If all bits of two numbers are available simultaneously, each on a separate wire, then a space iterated network of full-adders may be used to form their sum. Fig. 3.25 shows four full-adders operating simultaneously on two 4-bit numbers. Each passes its carry signal to its left neighbor. The carry out of the leftmost full-adder may be taken as a fifth bit of the sum or as an indication of overflow. The carry into the rightmost full-adder must be a zero, if the simple sum of A and B is desired.

Comparing these two adder systems reveals similarities and differences. Both realize the binary addition algorithm. Carries are propagated from iteration to iteration by the flip-flop in one case and intercell–cell communication lines in the other. Only one processing cell (full-adder) and n clock periods of time are required to add two n-bit numbers in the serial case. Parallel addition requires n full-adders, but very little time (gate delays). Thus cost can be exchanged for time by processing in a more serial fashion.

Iterative networks are of interest for a variety of reasons. These can be uncovered by further consideration of the problem of adding two 4-bit (n-bit in general) numbers. In principle we could write an 8-input, 9-output variable truth table that specifies addition of two 4-bit numbers. That table would suggest sum-of-product equations for the output variables. The number of gates suggested by these equations would be discouraging; very high fan-in requirements make the approach impractical. Minimizing the equations and forming multiple level networks that are more practical require great effort. On the other hand, designing time or space iterative adder networks requires modest design effort and very practical gate counts and fan-ins, regardless of n. It is easy to extend the network of Fig. 3.25 to add 18-bit numbers, but it is ridiculous to think of writing a complete 36-input truth table. Finally, iteration has advantages when digital systems are manufactured and maintained. Building and testing many identical networks can be far more efficient than dealing with that number of different networks.

Figure 3.26 provides a model of a one-dimensional iterative network; cells can be iterated in two or more dimensions also. In general, each cell

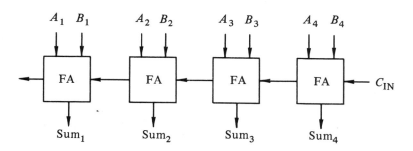

FIG. 3.25   A Parallel Adder of 4-Bit Numbers

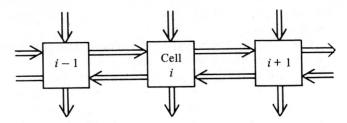

FIG. 3.26   A Space-Iterated Network

receives information from, and supplies results to, the environment of the network, and communicates with both adjacent cells. In specific situations some of the communication paths may be absent. Since all cells are identical, designing the network consists of a determination of what communication lines must exist and then the design of one cell. Often cells at the ends of the network can be simplified because they lack one neighbor.

From the specifications for a desired network and/or the processing algorithm that we envision to satisfy those specifications, we must determine the following: (1) the direction(s) of intercell communication, (2) the messages that a typical cell may receive from its neighbor(s), and (3) how it responds to those messages and the data provided by the environment. For purposes of example, we consider the following specifications:

> Two n-bit numbers, A and B, are to be compared and an indication
> of $A > B$, $A = B$, or $A < B$ given.

These specifications do not indicate how the comparison is to be made. Several algorithms suggest themselves. We may scan the two numbers from right (least significant) to left (most significant), or vice versa, to determine which is larger. Let us explore right-to-left scanning first. Messages are then passed to the left, and the answer must appear at the left end of an iterative network. In the ith bit position we compare A_i and B_i. If $A_i > B_i$ ($A_i = 1$, $B_i = 0$), then we must pass the finding "$A > B$ to the right of this point" to the left. If $A_i < B_i$, that finding must be passed. Finally, if $A_i = B_i$, then the message from the cell on the right must be passed to the cell on the left. The rightmost cell must be supplied with the message "all equal to the right." The message provided by the leftmost cell is the answer, of course; for example

| A | 0 | 1 | 1 | 0 | 1 | 0 | 1 | 0 | 1 | 1 | 0 |
| B | 0 | 0 | 1 | 1 | 0 | 1 | 1 | 0 | 0 | 1 | 0 |

∴ $A > B$

In summary, the messages that a cell may receive are:

(a) $A = B$ to the right.
(b) $A > B$ to the right.
(c) $A < B$ to the right.

The same messages are sent by the cells. A table may be used to express how a cell must respond to the message and data it receives.

Message Received	Message Sent $A_i B_i =$ 00	01	11	10
$A = B$	$A = B$	$A < B$	$A = B$	$A > B$
$A > B$	$A > B$	$A < B$	$A > B$	$A > B$
$A < B$	$A < B$	$A < B$	$A < B$	$A > B$

A coding problem exists. Three messages may be encoded with words of two or more bits in many ways. One rather natural code for the messages is used in the table below, which is organized in the form of a Karnaugh map so that the following equations for the cell may be derived with ease.

$$x_i = \bar{A}_i B_i \vee (\bar{A}_i \vee B_i)x_{i+1}$$
$$y_i = A_i \bar{B}_i \vee (A_i \vee \bar{B}_i)y_{i+1}$$

Message Received		Message Sent $A_i B_i =$ 00	01	11	10
$A = B$	00	00	10	00	01
$A > B$	01	01	10	01	01
	11	dd	dd	dd	dd
$A < B$	10	10	10	10	01

A NAND realization of these equations is shown in Fig. 3.27. The right-most cell may be simplified since it receives a fixed message $(x_{n+1}, y_{n+1}) = 00$. Using a full cell in the rightmost position may have manufacturing and maintenance advantages. The leftmost cell provides the desired result via the code of the above table.

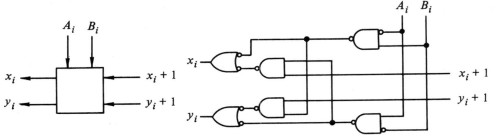

FIG. 3.27   Right-to-Left Number Comparison

If we compare numbers by scanning from the most to the least significant bit, the algorithm, and hence the logic, is quite different. The leftmost position in which the two numbers differ answers completely the question of which magnitude is larger. If the numbers are identical in all positions, then $A = B$. The messages that a typical cell may receive and send are therefore:

(a) all $A_j = B_j$ to the left
(b) $A > B$
(c) $A < B$

These messages are passed from cell to cell to the right end of the iterative network, where the final result appears.

A diagram may be used to summarize how a cell responds to the messages and data it receives. Such diagrams present exactly the same information as tables such as the one used in the previous example. Tables may be derived from diagrams, and vice versa. Thus, while diagrams are not absolutely necessary, they are valuable aids and more clearly reveal facts in some cases. To derive a suitable diagram, we draw a circle (square or whatever shape you prefer) for each message and place the message or its code word within the circle. Arrows labelled with input symbols connect the circles to indicate how the message and data received determine the message to be sent. The message at the tail of the arrow is the received message; the message sent is that at the tip of the arrow. In an example, if "all $A_j = B_j$ to the left" is received and $A_i = B_i$, we would draw

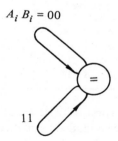

$A_i B_i = 00$

indicating that the transmitted message is the same as the received message when $A_i = B_i$. If "all $A_j = B_j$" is received, and $A_i > B_i$, then the "$A > B$" message must be sent. The diagram is extended to include this finding.

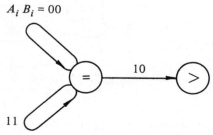

$A_i B_i = 00$

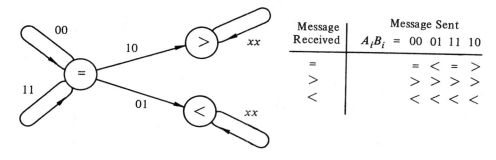

Message Received	Message Sent $A_i B_i =$ 00	01	11	10
=	=	<	=	>
>	>	>	>	>
<	<	<	<	<

FIG. 3.28   Message Transformation Diagram and Table for Left-to-Right Number Comparison.

A complete diagram for the left-to-right scan algorithm is given in Fig. 3.28, together with the table that is derived from it by preparing one row for each circle (message). Notice how the xx label is used to abbreviate the number of arrows that must be drawn.

Encoding this table with two or more bit code words permits us to derive Karnaugh maps, Boolean equations, a logic diagram, and a network for the typical cell. The leftmost cell may be specialized, or a standard cell used and the "equal" message permanently supplied to it. If the following straightforward code is used for the three messages, equations can be written by intuitive as well as formal methods.

Message	Code x_i y_i z_i
$A = B$	1 0 0
$A > B$	0 1 0
$A < B$	0 0 1

An equal message ($x_i = 1$) is to be sent only if it is received ($x_{i-1} = 1$) and $A_i = B_i$. An XOR gate can be used to tell whether or not $A_i = B_i$.

$$x_i = \cdot \ \overline{(A_i \oplus B_i)} \cdot x_{i-1}$$

The greater-than message is sent if it is received *or* if the equal message is received, and $A_i > B_i (A_i \overline{B}_i = 1)$. A similar statement may be made for the less-than message.

$$y_i = A_i \overline{B}_i x_{i-1} \vee y_{i-1}$$
$$z_i = \overline{A}_i B_i x_{i-1} \vee z_{i-1}$$

Inverters for A_i and B_i can be avoided since $A_i \oplus B_i$ is being formed.

$$A_i \cdot (A_i \oplus B_i) = A_i \overline{B}_i$$
$$B_i \cdot (A_i \oplus B_i) = \overline{A}_i B_i$$

Thus we may realize the typical cell from the equations.

$$x_i = \overline{(A_i \oplus B_i)} \cdot x_{i-1}$$
$$y_i = A_i(A_i \oplus B_i)x_{i-1} \lor y_{i-1} \qquad \textbf{(3.18)}$$
$$z_i = B_i(A_i \oplus B_i)x_{i-1} \lor z_{i-1}$$

The vector notation of DDL is ideal for describing complete iterative networks. If 16-bit numbers are to be compared, then terminal sets must be dimensioned to 16; propagation of signals to the right is expressed by displacing terminal sets with respect to each other in Boolean equations. Identifiers *XIN*, *YIN*, and *ZIN* are used to accomplish this in the DDL description of Fig. 3.29. Note how the "all equal to the left" message is presented to the left cell.

⟨OP⟩COMPARE(A,B) [3]

    ⟨TE⟩ A[16], B[16], W[16], X[16], Y[16], Z[16].

    ⟨ID⟩ XIN = 1 ∘ X[1 : 15], YIN = 0 ∘ Y[1 : 15], ZIN = 0 ∘ Z[1 : 15].

    ⟨BO⟩ $W = A \oplus B$,

          $X = \neg W \cdot XIN$,

          $Y = A \cdot W \cdot XIN \lor YIN$,

          $Z = B \cdot W \cdot XIN \lor ZIN$,

          COMPARE = X[16] ∘ Y[16] ∘ Z[16]. .

FIG. 3.29 A Network That Compares 16-Bit Numbers by Left to Right Scan

If desired information processing can be described in terms of the repeated performance of basis processing on portions of the data, the design of the processing network can be accomplished by designing a basic cell. If time permits, one cell may be used to iterate through time on the complete data—serial processing. When higher speed is desired, many interconnected cells may be used. The cell is designed by (1) determining the messages that it may receive and must send, and how it must respond to received messages and data, (2) encoding the messages, and (3) then designing the combinational logic of the cell. Since the cell has relatively few input and output terminals, designing it is substantially simpler than designing one large network.

3.5 MODULUS ARITHMETIC

Lists of integers, fractions, rationals, and so forth, are infinite in extent. Thus the binary number system provides us with a code for any such list only if code words of infinite length are used. This is not possible in a machine with a finite number of memory elements. In such machines the dimension of the registers is usually used as the length of code words and as a result is called the *word length* of the machine. A register of n memory elements can hold any of 2^n binary sequences of length n. If these sequences are thought to express magnitudes, integers in the finite range 0 through $2^n - 1$ can be expressed, or 2^n fractions in the range 0 through $1 - 2^{-n}$ can be expressed.

If we use digit sequences of fixed length n to express numbers of a finite range, and expect results of the same length (valid code words), then the familiar arithmetic operations must be modified. One way to modify the operations consists of simply accepting only the n least significant digits produced by ordinary arithmetic. This modified arithmetic as an example of *modulus arithmetic* produces results in some cases that differ from those of ordinary arithmetic and hence may appear to be strange initially. Suppose that we are using 5-bit code words to express integers; only magnitudes from 0 through 31 may be expressed. Adding such sequences and accepting only 5-bit results achieves modulo 32 arithmetic.

00110	6	11010	26	11111	31
10010	+18	10010	+18	00001	+ 1
11000	24	01100	12	00000	0

Modulus arithmetic may be performed in any R-based number system with respect to any integer *modulus m*. Expressed integers I and J are said to be *congruent* modulo m,

$$I \cong J \,(\text{mod } m) \tag{3.19}$$

if another integer K exists such that

$$I = J + mK \tag{3.20}$$

J is called a *residue* of I; the many values of J that satisfy Eq. (3.20) form the residue set. The use of the term "residue" usually refers to the least positive residue a, which must satisfy

$$0 \le a < m \tag{3.21}$$

The following examples use the decimal number system.

$$10 \cong 7 \cong 4 \cong 1 \cong -2 \,(\text{mod } 3)$$

$$8 \cong 6 \cong 4 \cong 2 \cong 0 \,(\text{mod } 2)$$

$$44 \cong 12 \,(\text{mod } 32)$$

$$32 \cong 0 \,(\text{mod } 32)$$

Given a set of N integers $I_i \cong a_i \,(\text{mod } m)$:

$$1. \quad \sum_{i=1}^{N} I_i \cong \sum_{i=1}^{N} a_i \,(\text{mod } m) \tag{3.22}$$

2. $\prod_{i=1}^{N} I_i \cong \prod_{i=1}^{N} a_i \pmod{m}$, and $\qquad (3.23)$

3. If $I_i \cong I_j$ and $I_j \cong I_k$, then $I_i \cong I_k$. $\qquad (3.24)$

Modulus m addition $\oplus$, subtraction $\ominus$, and multiplication $\otimes$ are arithmetic operations that give the least positive residue mod m of the normal sum, difference, and product, respectively. Using decimal integers, we obtain

$$6 \oplus 7 = 2 \oplus 3 = 1 \pmod{4}$$
$$6 \otimes 7 = 2 \otimes 3 = 2 \pmod{4}$$

We will often find it necessary to add binary digits modulo 2. Obtaining the least positive residue for bit addition is facilitated by Table 3.2. Note that this table is the same as the sum bit and difference bit entries in Table 1.5. Also note that the modulo 2 addition table and the truth table for the XOR logic operation are identical. Thus an XOR gate may be thought of as a unit that calculates the mod 2 sum of its input signals.

TABLE 3.2 MODULO 2 ADDITION

$\oplus$	0	1
0	0	1
1	1	0

$0 \oplus 0 = 1 \oplus 1 = 0$
$0 \oplus 1 = 1 \oplus 0 = 1$

Modulo 2^n Adders

If the registers of the serial system of Fig. 3.24 are of dimension n, then after n clock pulses register R holds the sum of A and B modulo 2^n.

$$R = A \oplus B \pmod{2^n} \qquad (3.25)$$

While the carry flip-flop holds the most significant bit of the true sum at that time, it is discarded in most systems in order to preserve word length; it may be used as an overflow indicator if simple magnitudes are being added. If the space iterated network of Fig. 3.25 is extended to n full-adders, that network provides

$$\text{Sum} = A \oplus B \oplus C_{in} \pmod{2^n} \qquad (3.26)$$

at the sum terminals. The carry-out of the most significant full-adder is again usually discarded to preserve word length. Figure 3.30 describes a 16-bit wide parallel adder. Equations (3.3) and (3.4) provide the Boolean equations of the full-adder.

The adders of Figs. 3.25 and 3.30 are called *ripple adders* because carry signals ripple from right to left through the full-adders. In the worst case, such as when we add

$$\langle OP \rangle \; \text{ADD(A,B,CIN)} \; [16]$$

$$\langle TE \rangle \; A[16], B[16], C[16], \text{CIN}.$$

$$\langle ID \rangle \; CC = C[2:16] \circ \text{CIN}.$$

$$\langle BO \rangle \quad C = A \cdot B \vee A \cdot CC \vee B \cdot CC,$$

$$\text{ADD} = A \oplus B \oplus CC..$$

FIG. 3.30   A 16-bit Wide Network of Full-Adders

$$0 = C_{in}$$

$$01111 \ldots 11$$

$$+ 00000 \ldots 01$$

$$\overline{10000 \quad 00}$$

the carry of 1 generated by the right cell propagates through all n full-adders to reach the leftmost sum terminal. If the carry portion of each full-adder is based upon Eq. (3.4), then the rightmost cell introduces 2 gate delays, $2 \Delta_g$, to generate its carry; and each of the other full-adders takes $2 \Delta_g$ units of time to pass the carry. In all, $2n \Delta_g$ units of time are required by a ripple adder in the worst case.

Asking how we can add numbers more rapidly provides an interesting example of intuitive synthesis. In principle, a 2 logic level network may be proposed; in practice we may not be able to obtain gates with sufficient fan-in. The problem resides with carry signals; to add faster we must provide correct carry values more rapidly to the more significant full-adders. Examine the carry signals for the rightmost two full-adders.

$$C_n = A_n B_n \vee A_n C_{in} \vee B_n C_{in}$$

$$= A_n B_n \vee (A_n \vee B_n) C_{in}$$

$$C_{n-1} = A_{n-1} B_{n-1} \vee (A_{n-1} \vee B_{n-1}) C_n$$

We could substitute for C_n in the equation for C_{n-1}, and use the distributive laws to obtain a sum-of-products expression for C_{n-1}. C_{n-1} could be generated with a 2-level network.

$$C_{n-1} = A_{n-1} B_{n-1} \vee A_{n-1} A_n B_n \vee A_{n-1} A_n C_{in} \vee A_{n-1} B_n C_{in} \vee B_{n-1} A_n B_n$$

$$\vee B_{n-1} A_n C_{in} \vee B_{n-1} B_n C_{in}$$

A fan-in of 7 is already encountered. If we continue in the same manner, we find that a fan-in 15 is required to realize C_{n-2} with a 2-level network. A fan-in of $2^{i+2} - 1$ is required to realize C_{n-i}; the approach is not practical even for modest values of i.

If we are willing to realize carry signals with more than 2 levels of logic, a very practical possibility exists. From the carry equation define

$$G_i = A_i B_i \quad \text{(generate carry)}$$

$$P_i = A_i \vee B_i \quad \text{(propagate carry)} \tag{3.27}$$

These signals are attractive because they are generated with one logic gate. Also, carry equations are easily written in terms of generate and propagate signals.

$$C_n = G_n \vee P_n C_{in}$$

$$C_{n-1} = G_{n-1} \vee P_{n-1} G_n \vee P_{n-1} P_n C_{in}$$

$$C_{n-2} = G_{n-2} \vee P_{n-2} G_{n-1} \vee P_{n-2} P_{n-1} G_n \vee P_{n-2} P_{n-1} P_n C_{in} \qquad (3.28)$$

$$C_{n-3} = G_{n-3} \vee P_{n-3} G_{n-2} \vee P_{n-3} P_{n-2} G_{n-1} \vee P_{n-3} P_{n-2} P_{n-1} G_n$$
$$\qquad \vee P_{n-3} P_{n-2} P_{n-1} P_n C_{in}$$

$$P_{Group} = P_{n-3} \cdot P_{n-2} \cdot P_{n-1} \cdot P_n$$

$$G_{Group} = G_{n-3} \vee G_{n-2} \cdot P_{n-3} \vee G_{n-1} \cdot P_{n-3} \cdot P_{n-2} \vee G_n \cdot P_{n-3} \cdot P_{n-2} \cdot P_{n-1}$$

A very practical, 2-level network can provide all of the carry signals of four adder positions, if the full-adders are modified to provide:

$$G_i = A_i B_i$$

$$P_i = A_i \vee B_i \qquad (3.29)$$

$$S_i = \bar{G}_i P_i \oplus C_{i+1}$$

Figure 3.31 shows how all of these ideas may be merged over 4-bit positions. The delay between the C_{i+1} carry input terminal and the C_{i-3} carry output terminal is $2\Delta_g$ rather than $8\Delta_g$ as found in the ripple adder. If the n full-adders are organized into p groups of q full-adders, each such that $pq = n$, and each group is bridged with a q-bit *carry-look-ahead* network, the worst case add time will be approximately $2p\Delta_g$ seconds. Substituting for p clarifies the advantage of carry-look-ahead.

$$\text{look-ahead delay} = 2(n/q)\Delta_g = \text{ripple delay}/q$$

If still greater speed is required, the 4-bit carry-look-ahead network is extended to provide the group generate and propagate signals of Eq.(3.28). A carry-

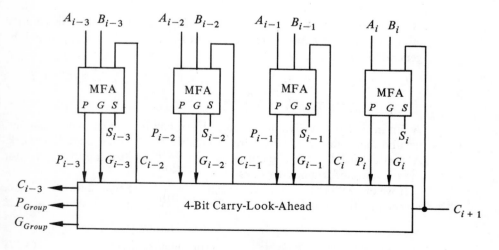

Fig. 3.31   Carry Look-Ahead Over 4-Bit Positions

look-ahead network is used to bridge four carry-look-ahead (16-bit position)-networks. Then carries are propagated over a group of adders with only a single gate delay. A group carry is generated with a three-gate delay from the A_i and B_i input terminals. One more level of carry-look-ahead propagates carries over 16-bit positions and gives a very fast 64-bit adder at an acceptable cost.

3.6 COMPLEMENT CODES FOR SIGNED NUMBERS

If quantities less than zero, *negative* numbers, are to be expressed, then new codes must be devised. One solution to expressing signed numbers consists of using a positional number system to express the magnitude of the number and a special symbol to express the *sign* of the number. This *sign-magnitude* code for expressing signed numbers can be used with binary as well as decimal expressions of magnitude. The two sign symbols (+ and −) must be encoded in terms of the electronic symbols (0 and 1); usually the plus is encoded 0 and the minus is encoded 1. Further, so that the *sign bit* can be located and treated in the special manner of " signs ", the sign bit is almost always the leftmost bit when binary sequences are communicated in parallel, and the first bit in time when serial communication is employed.

If fixed length sequences of n bits are used to express signed integers, then $n - 1$ bits express magnitude. 2^n sequences permit the expression of 2^n numbers; in the sign-magnitude code half of them are positive and half are negative. Two code words are assigned to zero. One may be called plus-zero, (0 0 ... 0) and the other minus-zero (1 0 ... 0). With $n = 4$, signed-integers in the range −7 through +7 may be expressed.

0111	expresses	+7
...		
0001	,,	+1
0000	,,	0
1000	,,	−0
1001	,,	−1
...		
1111	,,	−7

Each code for an alphabet of messages has its own characteristics and rules for processing those messages. If numbers are encoded, then addition, subtraction, and so forth, are accomplished with different algorithms for each code. Two codes with simpler arithmetic algorithms than the sign-magnitude code are considered in this section. The next section will then treat sign-magnitude arithmetic.

The *radix complement*, RC, of a digit of the radix R number system is obtained by subtracting that digit from R modulo R. Note that in any positional number system R, the radix of that system is expressed by the digit sequence 10 (one zero). Thus in decimal systems the "tens complement " of the digit 7 is 3; and in general

$$RC(d) = R \ominus d(\bmod R)$$
$$= 10_R \ominus d(\bmod R) \tag{3.30}$$

Notice that the radix complement of the digit 0 is 0 with this definition. The *diminished radix complement* DRC for a digit of the radix R number system is defined as

$$DRC(d) = R - 1 - d \tag{3.31}$$

Thus the "9's complement" of the decimal digit 7 is given by $9 - 7 = 2$. The "1's complement" of the binary digit 0 is $1 - 0 = 1$. The bit complement is thus the same as the diminished radix complement of a binary digit.

Radix complements and diminished radix complements of an n-digit sequence, S_n, are defined by generalizing the above definitions, with S_n *viewed as an integer*.

$$RC(S_n) = R^n \ominus S_n(\bmod R^n) \tag{3.32}$$

$$DRC(S_n) = R^n - 1 - S_n \tag{3.33}$$

Note first that R^n is expressed in the radix R number system as a 1 followed by n zeros. Thus in decimal form, with $n = 4$ for example, the " 10's complement" of 7428 is

$$10^4 - 7428 \quad \text{or} \quad \begin{array}{r} 10000 \\ -7428 \\ \hline \end{array}$$
$$RC(7428) = \quad 2572$$

The "9's complement" of the same sequence is

$$10^4 - 1 - 7428 \quad \text{or} \quad \begin{array}{r} 9999 \\ -7428 \\ \hline \end{array}$$
$$DRC(7428) = \quad 2571$$

In binary, we speak of the 2's and 1's complements. Again with $n = 4$, the 2's complement of 1010 is

$$2^4 - 1010 \quad \text{or} \quad \begin{array}{r} 10000 \\ -1010 \\ \hline \end{array}$$
$$2\text{'s comp}(1010) = 0110$$

The 1's complement of the same sequence is

$$2^4 - 1 - 1010 \quad \text{or} \quad \begin{array}{r} 1111 \\ -1010 \\ \hline \end{array}$$
$$1\text{'s comp}(1010) = 0101$$

If we examine these examples carefully, alternative short-cut procedures for obtaining the 2's and 1's complements emerge. First, $2^n - 1$ is always a sequence of n ones. Thus, extracting the 1's complement consists of subtracting from a sequence of all ones. Borrows can never develop in performing this subtraction; each bit is subtracted from 1, and we have seen this to be equivalent to adding 1 to the bit modulo 2. So the 1's complement of a sequence consists of a sequence of bit complements. We can write the 1's complement of a binary sequence without formally performing subtraction. We simply replace 0's with 1's and 1's with 0's.

$$DRC(101101110) = 010010001$$

Look back at the decimal examples and find that the 9's complement of a sequence consists of a sequence of the 9's complements of the original digits.

The RC is always one greater than the DRC of a digit sequence. So, with the DRC so easily computed, we may form the RC by first finding the DRC and then adding 1 to the DRC. Alternatively, note in the 2's complement example above that a borrow propagates from the rightmost 1 to the left end of the sequence. Thus we can obtain the 2's complement in the following way:

1. Copy the 0's to the right of the rightmost 1,
2. copy the rightmost 1, and
3. write the complement of all bits to the left.

These rules are simple enough so that the answer can be written directly with minimum effort and thought.

$$RC(10110101100) =$$
$$01001010100$$
complements   same

This same procedure, with suitable modification, can be applied to obtain the RC of a sequence of digits of any radix number system. Try it with decimal.

Each of these algorithms for finding the 1's and 2's complement of a binary sequence can be realized with a digital system that uses either serial or parallel communication. Each bit of a sequence must be inverted to form the 1's complement sequence; if flip-flops hold the orginal sequence the 1's complement is available at their complementary output terminals. The 2's complement can be formed from the 1's complement sequence via an adder and source of the number 1. While this may appear to be an expensive approach, we will find it to be very economical in a subsequent example. The final 2's complement algorithm above suggests the need for an iterative network. Whether the iteration is through time or space, the following messages must be passed between cells.

0   all less significant bits are 0's.
1   one or more less significant bits are 1's.

The following space-iterated network connects either the sequence given on input terminal A or the 2's complement of that sequence to output terminal B. Signal $COMP$ controls the mode of processing. When $COMP = 0$, all cell communicating signals $C_i = 0$, and hence $B_i = A_i$ for all i. When $COMP = 1$, the message $C_i = 1$ starts at the rightmost cell for which $A_i = 1$ and propagates to all cells to the left.

$$\langle SY \rangle \text{ COMP2}:$$
$$\langle TE \rangle \text{ A[16], B[16], C[16], COMP.}$$
$$\langle ID \rangle \text{ D} = \text{C}[2:16] \circ 0.$$
$$\text{C} = \text{COMP} \cdot (\text{A} \vee \text{D}),$$
$$B = \text{A} \oplus \text{D}.$$

Radix complement and DRC sequences are valuable as words of codes with simple addition and subtraction rules. We will illustrate this here for the $R = 2$ case and mention the decimal case in a later section.

2's Complement Convention

Let A be the signed integer in the range $-2^{n-1} \le A \le 2^{n-1} - 1$ held by register A. Let $\tilde{A}$ symbolize the magnitude of A and the $(n-1)$-bit expression of that magnitude in natural binary. In the 2's *complement code*:

$A \ge 0$ is encoded $0 \circ \tilde{A}$.
$A < 0$ is encoded $2^n - 0 \circ \tilde{A}$.

Thus positive numbers are encoded in natural binary with a leftmost bit of 0, just as in the sign-magnitude code. Negative numbers are encoded by the 2's complement of the sequence that encodes the positive number with the same magnitude. The leftmost bit is a 1 for negative numbers. Hence it can be viewed as a " sign " bit. With $n = 7$:

$+12$ is encoded 0001100	$+17$ is encoded 0010001
-12 is encoded 1110100	-17 is encoded 1101111

With numbers expressed in the 2's complement code:

> To change the sign of A, replace it with its 2's complement sequence. Check for overflow.

The range of integers that may be expressed in the 2's complement code is not symmetric; while -2^{n-1} is expressed with code word $10 \ldots 0$, $+2^{n-1}$ may not be expressed. Thus complementing this special code word is an attempt to generate a large positive number that fails because $10 \ldots 0$ is its own 2's complement. Overflow takes the form of a 0 carry into the sign bit position and a carry of 1 out of that position. Adding an AND gate to system COMP2 introduces overflow detection circuitry.

$$\text{OVF} = \text{C}[1] \cdot \neg \text{C}[2]$$

Advantages of the 2's complement code appear when we attempt to add and subtract.

> To add signed numbers A and B expressed with n-bit sequences via the 2's complement code, perform binary addtion on those sequences modulo 2^n; i.e., ignore carries into the $(n + 1)$ bit position. Check for overflow.

To add numbers, we perform binary arithmetic on the bits of the code words, including the sign bits. No checking, comparing, altering, and so forth, is required.

$$
\begin{array}{rl}
(+12) & 0001100 \\
+(+17) & 0010001 \\
\hline
+29 & 0011101
\end{array}
\qquad
\begin{array}{rl}
(+12) & 0001100 \\
+(-17) & 1101111 \\
\hline
-5 & 1111011
\end{array}
$$

$$
\begin{array}{rl}
(-12) & 1110100 \\
+(+17) & 0010001 \\
\hline
+5 & 0000101
\end{array}
\qquad
\begin{array}{rl}
(-12) & 1110100 \\
+(-17) & 1101111 \\
\hline
-29 & 1100011
\end{array}
$$

$$
\begin{array}{rl}
 & 01 \leftarrow \text{carries} \\
(+40) & 0101000 \\
+(+50) & 0110010 \\
\hline
-38 & 1011010 \\
 & \text{over-} \\
 & \text{flow}
\end{array}
\qquad
\begin{array}{rl}
 & 10 \leftarrow \text{carries} \\
(-40) & 1011000 \\
+(-50) & 1001110 \\
\hline
+38 & 0100110 \\
 & \text{over-} \\
 & \text{flow}
\end{array}
$$

Since we are dealing with finite code words, overflow is possible, but it does NOT take the form of a carry of 1 into the $n + 1$ position. Such carries are ignored, as in the examples above. Overflow can be detected in several ways; one way involves the carries into and out of the sign bit position. If those carries differ in value, then an overflow has occurred.

To see the validity of this addition algorithm, we break the problem $A + B$ into four special cases that correspond to the examples above.

Case 1. $(+\tilde{A}) + (+\tilde{B})$ calculates $0 \circ \tilde{A} + 0 \circ \tilde{B}$.

This is binary magnitude addition. The carry out of the sign position will be 0 since both sign bits are 0's; if a 1 is carried into the sign position the sum of magnitudes is too large to be expressed with $n - 1$ bits, which is the definition of overflow.

Case 2. $(+\tilde{A}) + (-\tilde{B})$ calculates

$$0 \circ \tilde{A} + (2^n - 0 \circ \tilde{B}) = 2^n + (0 \circ \tilde{A} - 0 \circ \tilde{B})$$
$$= 2^n - (0 \circ \tilde{B} - 0 \circ \tilde{A})$$

Magnitudes are subtracted so overflow is not possible. If $\tilde{A} > \tilde{B}$, then the result is shown above to be in excess by 2^n, which takes the form of a carry out of the sign bit position; this carry is ignored. Since one sign bit is a 0, the carry out of the sign position must equal the carry into that position. Thus the overflow detection rule is not violated. If $\tilde{A} < \tilde{B}$, then $2^n - (0 \circ \tilde{B} - 0 \circ \tilde{A})$ encodes a negative result. Since the sign bit of this result is 1, carry-in = carry-out = 0.

Case 3. $(-\tilde{A}) + (+\tilde{B})$ essentially the same as Case 2.

Case 4. $(-\tilde{A}) + (-\tilde{B})$ calculates

$$(2^n - 0 \circ \tilde{A}) + (2^n - 0 \circ \tilde{B}) = 2^n + [2^n - (0 \circ \tilde{A} + 0 \circ \tilde{B})]$$

Magnitudes are added. The result is too large by 2^n (ignored). The quantity in brackets encodes the negative result. Since both sign bits are 1, the sign of the result is 1 only if carry-in = carry-out = 1. Otherwise, an overflow has occurred.

To subtract $A - B$ when both signed numbers are expressed via the 2's complement code, binary subtract the sequence for B from that for A modulo 2^n; i.e., ignore borrows that propagate to the $(n + 1)$st position. Check for overflow.

$$
\begin{array}{rl}
(+12) & 0001100 \\
-(+17) & 0010001 \\
\hline
-5 & 1111011
\end{array}
\qquad
\begin{array}{rl}
(+12) & 0001100 \\
-(-17) & 1101111 \\
\hline
+29 & 0011101
\end{array}
$$

$$
\begin{array}{rl}
(-12) & 1110100 \\
-(+17) & 0010001 \\
\hline
-29 & 1100011
\end{array}
\qquad
\begin{array}{rl}
(-12) & 1110100 \\
-(-17) & 1101111 \\
\hline
+5 & 0000101
\end{array}
$$

Rather than proving this algorithm as we examined the addition algorithm, let us explore a number of interesting alternative algorithms.

To compute $A - B$ calculate $A + RC(B)$.
To compute $A + B$ calculate $A - RC(B)$.

Subtraction may be accomplished by a two-step process of complementing the subtrahend and adding the result to the minuend.

Case 1. $(+\tilde{A}) - (+\tilde{B}) = 0 \circ \tilde{A} - 0 \circ \tilde{B} = 0 \circ \tilde{A} \oplus 2^n - 0 \circ \tilde{B} \pmod{2^n}$
$$= 0 \circ \tilde{A} + RC(0 \circ \tilde{B})$$

When examining the other three cases, it is equally important to recall that n-bit sequences are being added and subtracted modulo 2^n, and since $2^n \cong 0 \pmod{2^n}$, 2^n may be freely introduced or removed from expressions.

Many of the previous ideas are incorporated in combinational logic block ADDSUB2C, which performs addition ($ASC = 0$) and subtraction ($ASC = 1$) according to the 2's complement rules. When subtracting, the X terminals bear the 1's complement of the B binary sequence, and a carry of 1 is supplied to the rightmost full-adder of the ripple-adder. The

2's complement is thus formed by 1's complementing and adding one; the "one" and A are added to the complement at the same time and by the same circuitry. Fig. 3.32(b) offers a block diagram of this system. While a ripple adder is described and shown in Fig. 3.32, parallel adders with carry-look-ahead can certainly be substituted to achieve faster arithmetic.

1's Complement Convention

Using n-bit words and integer A in the range $-(2^{n-1} - 1) \le A \le 2^{n-1} - 1$,

$$A \ge 0 \text{ is encoded } 0 \circ \tilde{A},$$
$$A \le 0 \text{ is encoded } 2^n - 1 - 0 \circ \tilde{A}$$

in the *1's complement code*. Again, positive numbers are encoded in natural binary, and the leftmost bit may be viewed as a sign bit. Since the acceptable range of A is symmetric, two words must be devoted to zero, as in the sign-magnitude code.

$$\langle \text{OP} \rangle \text{ ADDSUB2C}(A, B, ASC) [16], \text{ OVERFLOW}$$
$$\langle \text{TE} \rangle \text{ } A[16], B[16], ASC, C[16], X[16].$$
$$\langle \text{ID} \rangle \text{ } D = C[2 : 16] \circ ASC.$$
$$\langle \text{BO} \rangle \text{ } X = B \oplus ASC,$$
$$C = A \cdot X \vee (A \vee X) \cdot D,$$
$$\text{ADDSUB2C} = A \oplus X \oplus D$$
$$\text{OVERFLOW} = C[1] \oplus C[2]..$$

(a)

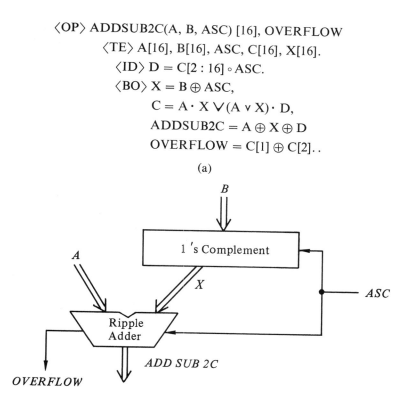

(b)

FIG. 3.32 2's Complement Arithmetic System ADDSUB2C

$$0 \ldots 0 \text{ encodes } +0$$
$$1 \ldots 1 \text{ encodes } -0$$

The code for a negative integer is the 1's complement of that for the positive integer with the same magnitude, and vice versa. Hence to change the sign of a number, one must replace its code word with the 1's complement of that code word.

Addition (subtraction) in the 1's complement convention uses binary arithmetic and the following rules.

> Add (subtract) the code words using the carry (borrow) out of the sign position as a carry (borrow) into the least significant bit position. Check for overflow.

The unusual carry rule is known as "end-around" carry (borrow). The following examples begin assuming the end-around carry (into the right bit position) is a 0.

$(+12)$ 0001100	$(+12)$ 0001100
$+(+17)$ 0010001	$+(-17)$ 1101110
$+29$ $\quad^0$0011101	-5 $\quad^0$1111010

(-12) 1110011	$-(12)$ 1110011
$+(+17)$ 0010001	$+(-17)$ 1101110
^{1}0000100	^{1}1100001
$\quad\quad\longrightarrow 1$	$\quad\quad\longrightarrow 1$
$+5$ $\quad$ 0000101	-29 $\quad$ 1100010

$(+12)$ 0001100	$(+12)$ 0001100
$-(+17)$ 0010001	$-(-17)$ 1101110
^{1}1111011	^{1}0011110
$\quad\quad\longrightarrow 1$	$\quad\quad\longrightarrow 1$
-5 $\quad$ 1111010	$+29$ $\quad$ 0011101

(-12) 1110011	(-12) 1110011
$-(+17)$ 0010001	$-(-17)$ 1101110
-29 $\quad^0$1100010	$+5$ $\quad^0$0000101

Four cases arise for addition; subtraction could be examined in similar fashion.

Case 1. $(+\tilde{A}) + (+\tilde{B})$ calculates $0 \circ \tilde{A} + 0 \circ \tilde{B}$.

Overflow exists if the sign bit of the result is 1; then carry into the sign position = 1; carry out = 0.

Case 2. $(+\tilde{A}) + (-\tilde{B})$ calculates

$$0 \circ \tilde{A} + (2^n - 1 - 0 \circ \tilde{B}) = 2^n - 1 - (0 \circ \tilde{B} - 0 \circ \tilde{A})$$

If $\tilde{A} < \tilde{B}$, the result encodes a negative number and carry-out = carry-in = 0. If $\tilde{A} > \tilde{B}$, $2^n - 1 + (0 \circ \tilde{A} - 0 \circ \tilde{B})$ describes a carry into the $(n + 1)$ position (2^n term), a deficiency of one (-1 term), and the difference between the magnitudes. The carry is used to compensate for the deficiency. In the sign position, carry-out = carry-in = 1.

Case 3. $(-\tilde{A}) + (+\tilde{B})$ is essentially the same as Case 2.

Case 4. $(-\tilde{A}) + (-\tilde{B})$ calculates

$$(2^n - 1 - 0 \circ \tilde{A}) + (2^n - 1 - 0 \circ \tilde{B}) = 2^n - 1 - [2^n - 1 - (0 \circ \tilde{A} + 0 \circ \tilde{B})]$$

The quantity in brackets is the desired result. The carry to the $(n + 1)$ position (2^n term) is used to compensate for the deficiency (-1 term). Overflow exists if the sign bit of the result is 0. Since both operand sign bits are 1, the carry out of the sign position must be 1. A result sign bit of 0 can appear only if the carry into the sign position is 0.

Thus overflow can be detected as it was in the 2's complement convention by comparing the carry out of and the carry into the sign bit position.

When adding $(+A) + (-A)$, a negative zero may be generated. When subtracting $(+A) - (+A)$, a positive zero may result.

$(+12)$	0001100		$(+12)$	0001100
$+(-12)$	1110011		$-(+12)$	0001100
-0	1111111		$+0$	0000000

These examples assume that all carry bits are 0 initially, an assumption that may not be valid in an electronic adder.

The end-around carry (borrow) suggests that the algorithm may not stabilize on an answer as the carries propagate forever. While unsuitable circuit delays can cause undesired effects, the algorithm itself does not fail. Suppose we begin to form the sum by assuming $C_1 = 0$, and begin adding at the least significant bit position. Assume that the carry into the ith position is 0 and that $A_i = B_i = 1$. Then a carry of 1 is *generated* in this position. It propagates to the left through all positions until position k, where $A_k = B_k = 0$. Here the carry *terminates*. If no such position k exists, the carry propagates out of the left end and into the right where it may propagate as far as the ith position. Since the carry out of the ith position is already a 1, no further change of the carry, and hence sum bits, is possible.

$$101110110$$
$$+\,010001101$$

| carry-in | $^{1}111111100 \leftarrow$ assume |
| sum | $\lfloor 000000011$ |

| carry-in | 111 |
| final sum | 000000100 |

Here the carry generated by the third from the right position is propagated by all other positions. Our assumed end-around carry of 0 was corrected ultimately. We could have assumed a carry into any bit position and started adding at that position.

Addition can be accomplished by complement-subtract in the 1's complement convention also.

$$0 \circ \tilde{A} + 0 \circ \tilde{B} = 0 \circ \tilde{A} - (2^n - 1 - 0 \circ \tilde{B})$$
$$= -2^n + 1 + (0 \circ \tilde{A} + 0 \circ \tilde{B})$$

The end-around borrow (-2^n) compensates for the excess $(+1)$. Also, subtraction can be performed by complement-add. Combinational logic block ADDSUB1C of Fig. 3.33 uses this technique. A block diagram of this system is very similar to that of Fig. 3.32.

We now have three codes for expressing signed integers and rules for performing addition and subtraction on words of the complement codes. Positive numbers are expressed with the same code words in all three. The words that represent a specific negative integer are different in the three codes. The range of representable integers is asymmetric for the 2's complement code, while as many positive as negative integers are encoded by the sign-magnitude and 1's complement codes. These codes and arithmetic algorithms apply to fractions as well as integers.

$\langle OP \rangle$ ADDSUB1C(A, B, ASC) [16], OVERFLOW

$\langle TE \rangle$ A[16], B[16], ASC, C[16], X[16].

$\langle ID \rangle$ D = C[2 : 16] $\circ$ C[1].

$\langle BO \rangle$ X = B $\oplus$ ASC,

$\qquad$ C = A $\cdot$ X $\vee$ (A $\vee$ X) $\cdot$ D,

ADDSUB1C = A $\oplus$ X $\oplus$ D,

OVERFLOW = C[1] $\oplus$ C[2]. .

Fig. 3.33 1's Complement Arithmetic Block ADDSUB1C

3.7 SIGN–MAGNITUDE ARITHMETIC

Section 1.4 showed how arithmetic may be performed manually on binary numbers. The familiar sign-magnitude code was employed, although the sign symbols were not encoded at that point. Table 1.6 summarized how we alter arithmetic problems when magnitudes must be subtracted so that the smaller is always taken from the larger. Section 3.1 provided us with the full-adder/subtractor in Fig. 3.10, a processing cell that may be used iteratively to add or subtract magnitudes. Therefore, we are well prepared to examine a digital system for adding or subtracting numbers expressed in the sign-magnitude code.

Algorithms for processing information by hand are not always well suited for machine implementation. Sign-magnitude arithmetic offers an excellent example. When subtracting magnitudes, we compare them and rearrange them if necessary to ensure that the smaller is taken from the larger. In Section 3.4, we developed iterative networks for comparing magnitudes; multiplexers may be used to interchange two numbers or not before presenting them to a subtractor. However, the comparing and switching circuitry is expensive and does take time to operate. What happens if we do not switch the operands before subtracting? Suppose that the magnitude of A is less than that of B and we compute $\tilde{A} - \tilde{B}$ modulo 2^{n-1} (the left bits express signs and must be treated separately in the sign-magnitude code).

$$\tilde{A} - \tilde{B} \,(\mathrm{mod}\ 2^{n-1}) = 2^{n-1} + \tilde{A} - \tilde{B} \,(\mathrm{mod}\ 2^{n-1})$$
$$= 2^{n-1} - (\tilde{B} - \tilde{A}) \qquad (3.34)$$

The correct difference is calculated, but it is expressed in the 2's complement code. To convert it to the sign-magnitude code, it must be complemented with a network such as $COMP2$ of Section 3.6, and the sign of the result adjusted properly.

Table 3.3 presents the information of Table 1.6 in encoded form. The signs of the numbers to be added, A and B, and the arithmetic command, ASC, to add or subract determine which arithmetic operation, AS, is to be actually performed on the magnitudes. From Table 3.3 we can see that

$$AS = \mathrm{Sign}_A \oplus \mathrm{Sign}_B \oplus ASC \qquad (3.35)$$

When magnitudes are added, the result always has the same sign as A.

TABLE 3.3. CONTROL OF A SIGN–MAGNITUDE PROCESSOR

ASC	Sign of A	Sign of B	AS	Sign of Result
0 (add)	0	0	0 (add)	0 (positive)
0	0	1	1 (sub)	?
0	1	0	1	?
0	1	1	0	1
1 (sub)	0	0	1	?
1	0	1	0	0
1	1	0	0	1
1	1	1	1	?

Overflow is possible only when adding magnitudes; the carry out of the leftmost adder may be used to detect overflow.

$$OVERFLOW = \overline{AS} \cdot C_1 \tag{3.36}$$

When magnitudes are subtracted, overflow is not possible. The borrow out of the leftmost subtractor cell may be used to indicate that the result

⟨OP⟩ ADDSUBSM(A,B,ASC) [16], OVERFLOW
    ⟨TE⟩ A[16],B[16],ASC,AS,OVERFLOW,C[15],COMP,E[15],
        SUMDIF [15].
    ⟨ID⟩ ASIGN = A[1], AMAG = A[2 : 16],
        BSIGN = B[1], BMAG = B[2 : 16],
        D = C[2 : 15] ∘ 0, F = E[2 : 15] ∘ 0.
    ⟨BO⟩ AS = ASIGN ⊕ BSIGN ⊕ ASC,
        C = (AMAG ⊕ AS)· (BMAG⊕ D) v BMAG· D,
        SUMDIF = AMAG ⊕ BMAG ⊕ D,
        OVERFLOW = ¬AS· C[1],
        COMP = AS· C[1],
        E = COMP· (SUMDIF v F),
        ADDSUBSM = (ASIGN ⊕ COMP) ∘ (SUMDIF ⊕F)..
(a)

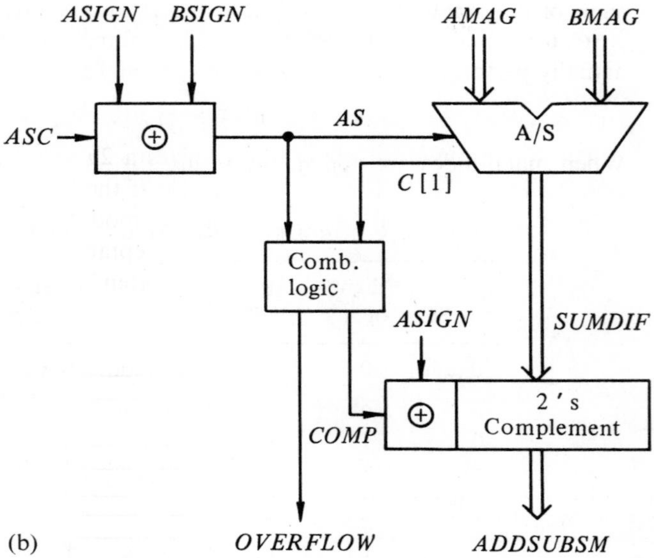

(b)

FIG. 3.34   Sign-Magnitude Arithmetic System ADDSUBSM

must be complemented and that the sign of the result is not that of A. Putting all of these " sign of result " arguments together gives

$$\text{Sign}_R = \overline{AS} \cdot \text{Sign}_A \lor AS \cdot (\text{Sign}_A \oplus C_1)$$
$$= \text{Sign}_A \oplus (AS \cdot C_1) \tag{3.37}$$

The arithmetic algorithm we are prepared to implement for sign-magnitude numbers is:

1. Determine the actual operation to be performed on the magnitudes.
2. Perform arithmetic on the magnitudes.
3. (a) If the magnitudes were added, check the result for overflow.
   (b) If the magnitudes were subtracted, check the result and correct, if necessary.
4. Affix the proper sign to the magnitude result.

Figure 3.34 provides a DDL description of a system based upon this algorithm with $n = 16$. Since the sign bits are handled with special circuitry, the adder/subtractor and 2's complementer are only 15 bits wide (carries in the complementing logic are propagated on terminal E). Comparing this sign-magnitude system with the 2's complement system of Fig. 3.32 reveal why the sign-magnitude code is seldom used any longer.

While the sign-magnitude, 2's complement, and 1's complement codes have all been used extensively in commercial digital computers, the 2's complement code is gaining in popularity at the expense of the other two. The complexity of sign-magnitude algorithms leads to more expensive, slower processing. Problems with the negative zero and complex multiplication algorithms discredit the 1's complement code.

3.8 MULTIPLICATION

Multiplying two n-digit sequences gives a $2n$ digit product in all positional number systems, including binary. If the binary sequences express integers, then the n-bit result produced by mod 2^n multiplication together with an overflow indication may be an acceptable substitute for true multiplication, since small integers are most often multiplied.

Integer   Fraction

Mod 2^n

Many computers generate $2n$-bit products, but make one or the other half-product sequences available as the answer. It may or may not be possible for the user of such machines to access the other half. We will concentrate on the problem of generating the full product in this section.

Multiplication illustrates additional facets of intuitive synthesis as well as being a very practical operation. It is a combinational operation: In principle we could derive a 2-level combinational multiplier network from a truth table that is nothing more than a binary encoded multiplication table. High fan-in requirements rule out this approach when word lengths are significantly long. However, multiplication can be accomplished by iterative combinational networks.

We will illustrate principles with 3-bit numbers.

$$
\begin{array}{rccccc}
 & & A_2 & A_1 & A_0 & \\
\times & & B_2 & B_1 & B_0 & \\
\hline
 & & A_2 B_0 & A_1 B_0 & A_0 B_0 & \\
 & A_2 B_1 & A_1 B_1 & A_0 B_1 & & \\
A_2 B_2 & A_1 B_2 & A_0 B_2 & & & \\
\hline
P_5 & P_4 & P_3 & P_2 & P_1 & P_0
\end{array}
$$

Basic bit products $A_i B_j$ can be formed with 2-input AND gates—n^2 (9 for $n = 3$) are required in general. We could use full-adders to sum bit products in much the same way that we add partial products by hand. Figure 3.35 shows a 2-dimensional iterative network of full-adders for our $n = 3$ example. In general, in each column bit products and carry bits from the right must be summed to form a product bit and carry bits. Approximately $n/2$ full-adders are required to add n bits. The total number of full-adders is therefore proportional to n^2.

The full-adder is a very small network to prepare as an IC cell in an iterative multiplier network. Using individual full-adders requires five signal connections per cell. The very large number of interconnections and printed circuit board spaces to mount the required full-adders makes this an unrecommended approach to multiplication. We need larger networks that complete a greater portion of the multiplication process with a modest number of interconnections to neighboring cells and data.

We iterate when we form partial product after partial product by hand. We might therefore think of a cell that forms a partial product and adds it to the sum of previous partial products. For our 3-bit example, we seek a cell that computes

$$
\begin{array}{rcccc}
 & K_2 & K_1 & K_0 & \\
+ & A_2 B_i & A_1 B_i & A_0 B_i & \\
\hline
S_3 & S_2 & S_1 & S_0
\end{array}
$$

A 9-input, 4-output terminal cell must be designed. We avoid design detail here by using three AND gates to form bit-products, and one half- and two

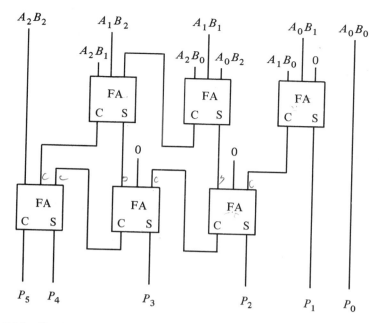

FIG. 3.35 Combinational Multiplication with Bit Multipliers and Full-Adders

full-adders as shown in Fig. 3.36 (a). Three such cells interconnected as in Fig. 3.36 (b) form $A \times B + C$ where the 3-bit constant C may, but neet not, be zero.

For larger value of n, supplying all multiplicand bits to each cell requires excessive connections. A cell that combines a few multiplicand bits and more than one multiplier bit gives a cell of suitable complexity for IC fabrication while keeping the pin requirements in bounds. A 4×2 multiplier cell spans 4 columns of bit products and hence requires that 4 constants be added to form a 6-bit result.

$$
\begin{array}{ccccccc}
 & & K_3 & K_2 & K_1 & K_0 & \\
 & & A_{i+3}B_j & A_{i+2}B_j & A_{i+1}B_j & A_iB_j & \\
 & A_{i+3}B_{j+1} & A_{i+2}B_{j+1} & A_{i+1}B_{j+1} & A_iB_{j+1} & & \\
\hline
S_5 & S_4 & & S_3 & S_2 & S_1 & S_0
\end{array}
$$

Such a cell requires 16 signal connections. Two such cells form the product of two 4-bit numbers. Figure 3.37 shows how the cells must be interconnected. Four cells can be interconnected in several ways to form an 8×4 bit multiplier. Eight cells are required to build an 8×8 bit multiplier.

Sign-Magnitude Multiplication

Time iteration becomes attractive when the cost of combinational multipliers is unbearable. Partial products can be formed, in turn, and added to

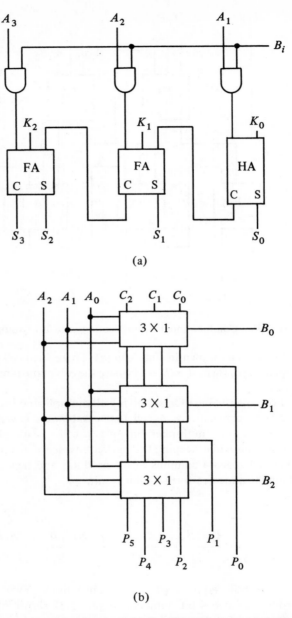

(a)

(b)

FIG. 3.36 Combinational Multiplication with 3×1 Multiplier Cells

the sum of previous partial products by using the existing adder of an arithmetic unit. Control of the sequences of more primitive operations is a new concern. In the following examples, observe how sequential algorithms may be translated into data processing and control logic networks. We begin with sign-magnitude multiplication.

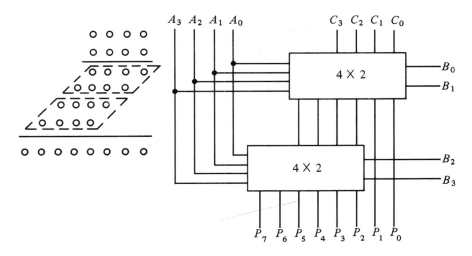

FIG. 3.37 Combinational Multiplication with 4×2 Multiplier Cells

Let A and B be two fractions expressed in the sign-magnitude code and registers that hold those fractions. We might number the flip-flops of these registers A_0, A_1, ..., A_{n-1} and B_0, B_1, ..., B_{n-1}, respectively, so that A_i holds the bit that is weighted 2^{-i}. A_0 and B_0 express the signs of the numbers, and $A_0 \oplus B_0$ expresses the sign of their product.

The algorithm for the manual binary multiplication shown in Section 1.4 requires writing $n - 1$ partial products shifted with respect to each other. These products may be symbolized

$$\tilde{A} \times B_{n-1} \times 2^{-(n-1)}$$
$$\tilde{A} \times B_{n-2} \times 2^{-(n-2)}$$
$$\vdots$$
$$\tilde{A} \times B_1 \times 2^{-1}$$

The 2^{-i} factors indicate what the shifting accomplishes. Then we add all of the partial products.

$$\sum_{i=1}^{n-1} \tilde{A} \times B_i \times 2^{-i} = \tilde{A} \times \sum_{i=1}^{n-1} B_i \times 2^{-i} = \tilde{A} \times \tilde{B} \qquad (3.38)$$

It is not essential that all partial products be formed and recorded before beginning to add them. The following algorithm accumulates partial products as each is formed.

1. Define $S_0 = 0$
2. Calculate $S_i = (\tilde{A} \times B_{n-i} + S_{i-1}) \times 2^{-1}$
   for $i = 1, 2, \ldots, n - 1$

The first partial product is formed by calculating S_1.

$$S_1 = (\tilde{A} \times B_{n-1} + 0) \times 2^{-1}$$

The second partial product is added to the first (shifted one place to the right), and the sum is shifted in the calculation of S_2.

$$S_2 = (\tilde{A} \times B_{n-2} + \tilde{A} \times B_{n-1} \times 2^{-1}) \times 2^{-1}$$
$$= \tilde{A} \times B_{n-2} \times 2^{-1} + \tilde{A} \times B_{n-1} \times 2^{-2}$$

S_3 is the sum of three shifted partial products; in general

$$S_j = \sum_{i=1}^{j} \tilde{A} \times B_{n-1-j+i} \times 2^{-i}$$

The last sum is then:

$$S_{n-1} = \sum_{i=1}^{n-1} \tilde{A} \times B_{n-1-(n-1)+i} \times 2^{-i}$$
$$= \sum_{i=1}^{n-1} \tilde{A} \times B_i \times 2^{-i} = \tilde{A} \times \tilde{B}$$

The calculation of S_i from S_{i-1}, A, and B_{n-1} involves three operations.

1. Multiply $\quad \tilde{A} \times B_{n-i}$
2. Add $\qquad \tilde{A} \times B_{n-i} + S_{i-1}$
3. Shift right $\quad (\tilde{A} \times B_{n-i} + S_{i-1}) \times 2^{-1}$

Multiplying a magnitude by a single bit is easy. If the bit is 0, the product is a binary sequence of 0's; if the bit is 1, the product is the magnitude sequence. We know how to add magnitudes and shift; we may combine these operations by placing the sum bits one place to the right in the receiving register. Counting to determine when each bit of the multiplier has been processed is a necessary control function. Shifting the multiplier with each step provides a low-cost way of examining the multiplier bits in turn.

The 16-bit system of Fig. 3.38 is based upon this algorithm. Some unshown operations place the multiplicand in the A register and the multiplier in the B register and then set terminal $GO = 1$ until automaton $MULT$ responds by setting $DONE = 1$. Register K and terminal KC form a 4-bit counter like that of Fig. 1.9. We assume that the count is zero when GO is set to 1. With a count of zero, $\text{v}/\text{K} = 0$ and the accumulator register ACC is cleared to hold S_0. Partial products are found on terminal X and added to the previous sum by the ripple-adder (terminals C and SUM). The sign of the product is formed at terminal SR. With each clock pulse during which the count is greater than zero, the counter advances, and the 16-bit accumulator and 15-bit $BMAG$ subregister are loaded as though they were one register. The 31 bits placed in these flip-flops are: (1) sign of product, (2) a 16-bit sum formed by adding two 15-bit magnitudes, one of which is either zero or the multiplicand magnitude, and (3) the left 14 bits held by the B register. The right-shift of the partial products and the multiplier are accomplished in this register transfer. With each shift, B holds one more partial product bit and one less multiplier bit.

⟨SY⟩ COMPUTE: ⟨TI⟩ P(1 E − 7)
   ⟨RE⟩ A[16] = ASIGN ∘ AMAG[15],
       B[16] = BSIGN ∘ BMAG[15],
       ACC[16] = ACCSIGN ∘ ACM[15].
   ⟨TE⟩ GO, DONE.
    ⋮

⟨AU⟩ MULT: P:
   ⟨RE⟩ K[4].
   ⟨TE⟩ X[15], C[15], KC[4], SUM[15], SR.
   ⟨ID⟩ D = C[2 : 15] ∘ 0.
   X = B[16] · AMAG,
   C = ACM · X ∨ ACM · D ∨ X · D,
   SUM = ACM ⊕ X ⊕ D,
   SR = ASIGN ⊕ BSIGN,
   KC = K · (KC[2 : 4] ∘ 1),
|GO| |∨/K| ACC ∘ BMAG ← SR ∘ C[1] ∘ SUM ∘ BMAG[1 : 14];
   ACC ← 0., K ← K ⊕ (KC[2 : 4] ∘ 1), | · /K| DONE = 1 ....

Fig. 3.38   A 16-Bit Sign-Magnitude Multiplier

SOURCE:	SR	C_1	SUM_1	...	SUM_{15}	B_2	...	B_{14}
SINK:		ACC_1	ACC_2	ACC_3		B_2	B_3	B_{15}

Figure 3.39 provides a block diagram of this digital system and an example of its operation using a 4-bit word length.

2's Complement Multiplication

The 2's complement code is a *weighted code*. In a weighted code a constant (weight) is associated with each bit position, and the encoded digit or number can be computed by evaluating a sum of products of bits and their respective weights, just as in a positional number system. If 2's complement code words represent signed integers, then the weights of the magnitude bits are powers of two; the weight of the sign bit is negative, -2^{n-1}. With fractions the sign bit is weighted -1 and the magnitude bits are weighted as in natural binary. Thus if A is a fraction expressed in the 2's complement code

$$A = -A_0 + \sum_{i=1}^{n-1} A_i \times 2^{-i} \tag{3.39}$$

Since magnitude bits do not display true magnitude in natural binary when $A < 0$, we would not expect the magnitude multiplication algorithm just examined to give correct results with 2's complement code words. We could convert negative numbers to sign-magnitude, multiply with a sign-magnitude system, and convert negative products back to 2's complement. An algorithm that takes the properties of the 2's complement code into account avoids the need for these expensive and time consuming conversions.

To derive the Booth algorithm [3] for multiplying 2's complement code words, extend the multiplier B by concatenating $B_n = 0$ on the right. Then the product of fractions is given by

$$
\begin{aligned}
A \times B &= A \times \left(-B_0 + \sum_{i=1}^{n-1} B_i \times 2^{-i}\right) \\
&= A \times (-B_0 + B_1 \times 2^{-1} + B_2 \times 2^{-2} + \cdots \\
&\quad + B_{n-1} \times 2^{-(n-1)} + 0 \times 2^{-n}) \\
&= A \times (-B_0 + (B_1 - B_1 \times 2^{-1}) + (B_2 \times 2^{-1} - B_2 \times 2^{-2}) \\
&\quad + \cdots) \\
&= A \times (B_1 - B_0) + (B_2 - B_1) \times 2^{-1} + \cdots \\
&\quad + (0 - B_{n-1}) \times 2^{-(n-1)}) \\
&= \left(A \times \sum_{i=1}^{n} (B_i - B_{i-1}) \times 2^{-i}\right) \times 2
\end{aligned}
\tag{3.40}
$$

This sum of partial products has the same form as Eq. (3.38) and hence can be iteratively accumulated as in the magnitude multiplier. The necessary iteration term when $S_0 = 0$ is:

$$
S_i = (A \times (B_{n-i+1} - B_{n-i}) + S_{i-1}) \times 2^{-1}
\tag{3.41}
$$

Again multiplication, addition, and shifting are necessary operations in all steps except the last, where shifting is not required [because of the final 2 in Eq. (3.40)]. The multiplication factor may now be any of $+1$, 0, and -1.

B_{n-i+1}	B_{n-i}	Multiplication Factor	Actual Operation
0	0	0	add 0
0	1	-1	subtract multiplicand
1	0	$+1$	add multiplicand
1	1	0	add 0

We are familiar with addition and subtraction in the 2's complement convention, but shifting has not been examined. To preserve the sign while dividing the magnitude by two, 2's complement code words must be shifted with the sign bit preserved as well as placed in the next right position. This is called an *extend shift*.

$$
\begin{array}{cccc}
11010 & 11101 & 00110 & 00011 \\
\Rightarrow & & \Rightarrow & \\
(-6) & (-3) & (+6) & (+3)
\end{array}
$$

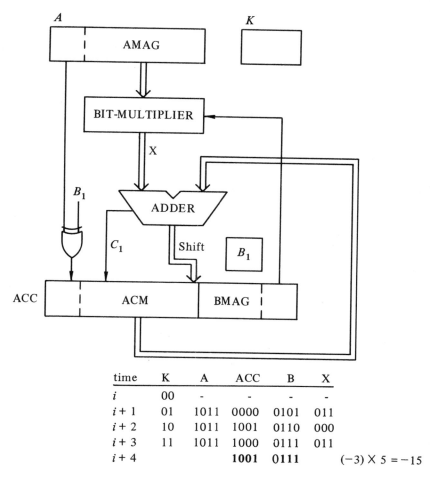

time	K	A	ACC	B	X
i	00	-	-	-	-
$i+1$	01	1011	0000	0101	011
$i+2$	10	1011	1001	0110	000
$i+3$	11	1011	1000	0111	011
$i+4$			**1001**	**0111**	

$(-3) \times 5 = -15$

FIG. 3.39   Sign-Magnitude Multiplication

Since a 0 is appended on the right of the multiplier, the first nontrivial arithmetic operation consists of subtracting the multiplicand from zero. Before we subtract again, the multiplicand must be added. If the bits of the multiplier alternate, the successive values of S_i are

i	$A(B_{n-i+1} - B_{n-i}) + S_{i-1}$	S_i
0		0
1	$-A$	$-\frac{1}{2}A$
2	$A - \frac{1}{2}A$	$\frac{1}{4}A$
3	$-A + \frac{1}{4}A$	$-\frac{3}{8}A$
$\vdots$	$\vdots$	$\vdots$

Overflow would not seem to be a problem, but it is in one special case. If the multiplicand is -1 (expressed $10\ldots0$), then $-A$ cannot be formed since it is not in the range of the 2's complement code. Overflow can be avoided by shifting before adding in Eq. (3.41), or by altering the shift when an overflow occurs so as to give the desired $S_i = -\frac{1}{2}A = -\frac{1}{2}(-1) = +\frac{1}{2}$. When the adder detects an overflow, a zero must be shifted into the sign position rather than extending the sign. If the multiplier is also -1, then the addition occurs in the final step that is not accompanied by a shift. In this case, the overflow is a multiplication overflow and must be reported together with the erroneous result.

Implementations of the Booth algorithm can take several forms. The shifting may be performed during separate clock cycles after each addition or subtraction. Or, a combinational shifter may follow the adder as suggested in Fig. 3.40 so that only n clock cycles are required to multiply

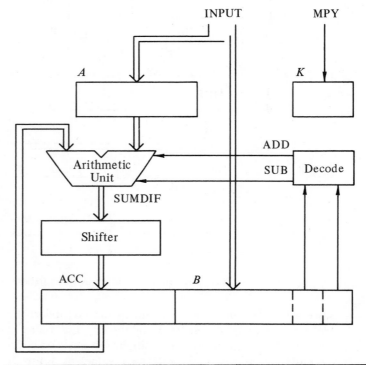

TIME	MPY	K	A	ACC	B	Arithmetic	SUMDIF
i	0	000	-	-	-		
$i+1$	1	000	-	-	-		
$i+2$	0	100	1011	0000	00110	sub	0101
$i+3$	0	011	1011	0010	10011	none	0010
$i+4$	0	010	1011	0001	01001	add	1100
$i+5$	0	001	1011	1110	00100	none	1110
$i+6$	0	000		1110	00100		

$$(-5/8) \times (3/8) = -15/64$$

FIG. 3.40 Booth Multiplier

n-bit numbers. The 2's complement adder/subtractor must be generalized so that it can add 0 or the multiplicand to the contents of the accumulator register. The 32-bit shifter must be able to:

1. Extend shift.
2. Left fill with 0 (overflow case).
3. Simply pass without shifting (final step).

This section introduces several new synthesis concepts. Typical cells may be iterated in two (or more) dimensions to achieve desired data processing. And, to find economical networks, a processing procedure must be studied in great detail looking for those variations that can be most economically realized and/or that give results most quickly. This section is not an exhaustive treatment of multiplication algorithms and networks, and says nothing about division. Consult the references for this, and other chapters for more specialized studies.

3.9 DECIMAL DIGIT CODES AND ARITHMETIC

Human users of information-processing equipment insist on using the decimal number system even though the binary system is more natural for the equipment. And we strongly desire that such machinery be able to manipulate alphabetic and special characters such as punctuation symbols in addition to the decimal digits. So information-processing systems have by one means or another been made to look as though they manipulate decimal or alphameric characters, even though their internal nature is binary: they merely utilize a binary sequence of several bits to represent each character.

We are concerned in this section with codes for the ten decimal digits. Codes can be tabulated, of course, but other representations exist. For instance, each code word of *n* bits can be thought of as specifying a point in *n*-dimensional space. Such a space is easily drawn and visualized for $n = 2$ and $n = 3$; a little imagination is needed for larger values. We could place a decimal digit or other symbol to be encoded next to each vertex to indicate that the binary sequence that locates that vertex is the code word to be used for the decimal digit.

An important concept appears from such a topological listing of a code —the concept of *distance*. Two code words are said to be *adjacent* if they are separated by unit distance when plotted in an *n*-dimensional space. Alternatively, realize that code words are adjacent if they differ in one bit position only. 011 and 111 are adjacent; 011 and 110 are separated by distance 2, not $\sqrt{2}$, as Fig. 3.41 suggests. Distance between code words measures the number of bit positions in which the code words differ. Only adjacencies and *all* adjacencies are shown as lines in the topological representation of a code. Thus if A and B are two *n*-bit code words, and $C = A \oplus B$, then the *Hamming* [8] *distance* between A and B is given by

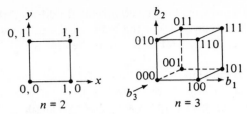

FIG. 3.41 N-space Representation of Codes

$$D = \sum_{i=1}^{n} c_i = \sum_{i=1}^{n} (a_i \oplus b_i) \qquad (3.42)$$

Rather than having a code word designate a point in space, we may let it determine an area on a map in which we place the symbol to be encoded. If we organize the map in the manner of Karnaugh maps, then the concept of distance remains. Cells that are physically adjacent represent adjacent code words. Cells with a common corner represent code words separated by a distance of 2.

The table, the cube, and the map all offer a means of associating a symbol with a binary sequence. In the table all symbols and all code words are explicitly stated. Code words are implied in the cube and map representations, which more clearly display distance relationships between code words.

Weighted Codes

Many useful 4-bit codes for the decimal digits may be generated by associating a weight with each bit position in the code word, much as a weight is associated with each digit position in a positional number system. In fact, the binary number system offers one of the most popular weighted codes—the 8421 code—in which the weights are successive powers of 2. Thus in the 8421 code, the code word 0111 encodes the decimal digit 7.

$$0 \times 8 + 1 \times 4 + 1 \times 2 + 1 \times 1 = 7_{10}$$

In general, if weights w_1, w_2, w_3 and w_4 are associated with bits b_1, b_2, b_3 and b_4, respectively, we may find the digit encoded by code word $b_1\, b_2\, b_3\, b_4$ by evaluating

$$d_{10} = \sum_{i=1}^{4} w_i \times b_i \qquad (3.43)$$

Thus in the 6321 weighted code, 0111 is one encoding of the digit 6.

$$0 \times 6 + 1 \times 3 + 1 \times 2 + 1 \times 1 = 6$$

Can you find another 6321 code word that encodes 6?

TABLE 3.4. THE 8421 WEIGHTED CODE

Decimal Digit Represented	$b_1 b_2\ b_3\ b_4$
	8421
0	0000
1	0001
2	0010
3	0011
4	0100
5	0101
6	0110
7	0111
8	1000
9	1001

		b_1	
0	4	x	8
1	5	x	9
3	7	x	x
2	6	x	x

x marks the six code words not employed

Some weighted codes for the decimal digits employ only positive weights. The sum of weights in such codes can not be less than 9 nor greater than 15. One of the weights must be 1, and another must be either 1 or 2. If negative weights are employed, it is not necessary that any of the weights be 1. The restriction of 9 as the minimum sum of weights is also relaxed. Convince yourself that these restrictions are valid.

The 8421 code is so widely used that it is commonly known as *BCD* for *binary coded decimal*, although all of the codes considered here encode decimal digits as binary sequences.

A number of weight sets provide multiple representations of a decimal digit. The 4311 code of Table 3.5 serves as an example; this set of weights

TABLE 3.5. A REFLECTED, WEIGHTED CODE

General		One Possible Code	
	4311		4311
0	0000	0	0000
1	0001 / 0010	1	0001
2	0011	2	0011
3	0100	3	0100
4	1000 / 0101 / 0110	4	1000
5	1001 / 1010 / 0111	5	0111
6	1011	6	1011
7	1100	7	1100
8	1110 / 1101	8	1110
9	1111	9	1111

generates 36 codes, only 18 of which can be considered as fundamentally different since the others are obtained by simply interchanging the last two columns.

Some codes, whose weights sum to 9 (necessary but not sufficient condition), make it possible to obtain the 9's complement of an encoded digit by extracting the 1's complement of the code word. The one specific 4311 code of Table 3.5 possesses this property: for example, the 9's complement of 2 is 7; the 1's complement of 0011 is 1100. Such a code is known as a *reflected* code; note the symmetry about its middle values.

Gray Codes

On occasion it is desirable to encode an ordered set of symbols so that adjacent symbols are represented by adjacent code words. In attempting to convert information in analog form to a digital form, problems arise that are solved by the use of such codes. *Gray* codes have the desired property; several are shown in Table 3.6. We have already used the 2-bit Gray code to label the rows and the columns of the Karnaugh map. The other Gray codes given are easily generated from the 2-bit code; examine these and note the appearance of the 2-bit code in both forward and backward order.

TABLE 3.6. GRAY CODES

2-bit Code	3-bit Code	4-bit Code	
00	000	0000	1100
01	001	0001	1101
11	011	0011	1111
10	010	0010	1110
	110	0110	1010
	111	0111	1011
	101	0101	1001
	100	0100	1000

Other Gray four-bit codes can be generated easily with the aid of a map. Fig. 3.42 (a) gives a map sketch of the four-bit code of Table 3.6, as well as other complete, four-bit Gray codes.

Incomplete Gray codes, which do not use all 16 code words, can also be obtained easily. The decimal digits might be encoded by one of the codes

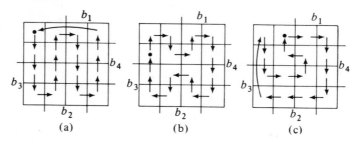

FIG. 3.42   Some Complete Gray Codes

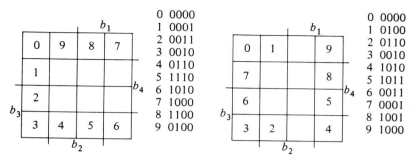

FIG. 3.43   Two Gray Decimal Codes

of Fig. 3.43. Note that only an even number of symbols can be encoded so that the code word for the last is adjacent to that for the first.

Other Codes

The *excess*-3 code listed in Table 3.7 gets its name from the relation of its code words to natural binary. Each code word expresses in natural binary an integer that is three greater than the encoded digit. The excess-3 code is reflected, which means that 9's complementation can be accomplished with ease. Other advantageous properties are illustrated later.

The *one-hot* code may be used to encode alphabets of any size. While it uses a large number of bits, its simplicity is often very useful and natural. A keyboard of ten push-button switches provides this code, for example. A number of digit display devices must be driven with words of this code. Counting is accomplished with a simple circulating shift register called a *ring counter*, in which one flip-flop is set and all others are reset at all times.

The *creeping code* is more economical of bits. Counting is also accomplished with a circulating shift register, but with the complement of the right bit being loaded to the left position. Figure 3.44 shows such a "twisted ring-counter" or "switch-tail counter." Any digit or range of digits may be detected with one 2-input gate, or less.

TABLE 3.7.   OTHER DECIMAL DIGIT CODES

Digit	Excess-3	One-Hot	Creeping
0	0011	1000000000	00000
1	0100	0100000000	10000
2	0101	0010000000	11000
3	0110	0001000000	11100
4	0111	0000100000	11110
5	1000	0000010000	11111
6	1001	0000001000	01111
7	1010	0000000100	00111
8	1011	0000000010	00011
9	1100	0000000001	00001

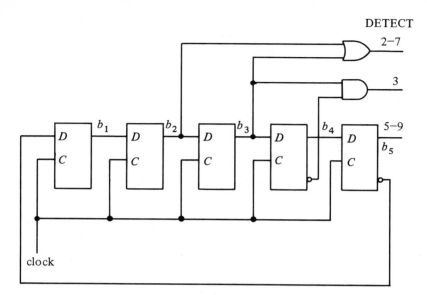

FIG. 3.44   Creeping-Code Decimal Counter

Decimal Arithmetic

With decimal digits encoded via four or more bit code words, decimal sequences are encoded via sequences of digit code words. Using the 8421 code for digits, the decimal integer 123 is expressed with the 12-bit sequence

$$0001 \circ 0010 \circ 0011$$

which represents the magnitude 291 if it is viewed via natural binary. We desire to process sequences of decimal digit code words in order to produce the sequence of code words that expresses the decimal "answer." Such processing can be broken into steps from which single or pairs of decimal digits are processed in turn. Addition is an example. We memorize the digit addition table and use it repeatedly to add digit pairs when summing decimal numbers. In principle the code word for each digit could be written in place of its digit in the addition table, which would provide a truth table that describes decimal digit adding logic. This would be a very large table and would require very extensive design effort. Therefore, in practice we design in a less formal manner by attempting to interconnect familiar networks so as to synthesize the required network.

Codes that permit the use of binary addition to accomplish decimal digit addition are said to be *additive*. The 8421 code is additive. Binary addition of 8421 code words gives the correct sum in natural binary, but does not give the correct encoded decimal sum when the magnitude exceeds nine.

3	0011	7	0111
+6	0110	+5	0101
9	1001	12	1100	← not a valid word

TABLE 3.8. 8421 CODE WORD SUMS AND DESIRED RESULTS

Decimal Sum	Natural Binary	Decimal Result	
0	0,0000	0,0000	no correction needed
⋮			
9	0,1001	0,1001	
10	0,1010	1,0000	
11	0,1011	1,0001	correction necessary
⋮			
18	1,0010	1,1000	
19	1,0011	1,1001	

When the sum exceeds nine, a correction must be made to derive the correct code word and a decimal carry bit. Fortunately this correction can be made using familiar concepts. Comparing the results of binary addition with the desired results shown in Table 3.8 reveals that the desired 5-bit results are greater than the results of binary addition by six. Thus adding six via binary arithmetic accomplishes the necessary correction and generates the necessary decimal carry bit. Figure 3.45 illustrates a system that uses two 4-bit binary adders to accomplish decimal digit addition. The presence of a carry flip-flop suggests that this system adds decimal numbers in a serial-by-decimal digit, parallel-by-code word fashion.

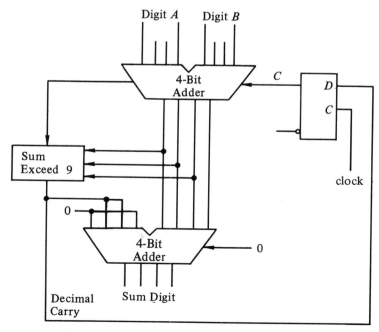

FIG. 3.45 An 8421 Decimal Digit Adder

A correction is always necessary when performing arithmetic on digits encoded in excess-3. If the sum of two digits is less than 10, the binary sum of excess-3 code words is in excess by six (6). Hence the correction of *subtracting* 3 (0011) must be made. If the sum of digits is greater than nine, the binary sum is again in excess by six. However, utilizing the binary overflow, which represents 16 units, as a decimal carry (representing 10 units) in effect throws six units away, making the sum digit representation in excess by zero. The correction called for in this case is to *add* 3 units or 0011 (the excess-3 encoding of 0). Thus a correction is always necessary and 0011 is always involved. And it is extremely easy to determine whether we are to add or subtract that sequence.

```
 +3        0110                          + 7       1010
 +6        1001                          + 5       1000
 ──        ────                          ───       ────
         0 1111  ← No overflow           +12      1 0010  ← Overflow
         −0011     correct by                   ╱ +0011     correct by
         ─────     subtracting         decimal ╱  ─────     adding
          1100                          carry      0101
```

```
          037      0011     0110     1010
         +065      0011     1001     1000      Add
         ────      ────     ────     ────
         +102        1╲       1╲
                 ┌─0 0111  ╲─1 0000  ╲─1 0010
                 │  −0011   ╲ +0011   ╲ +0011    Correct
                 ╰─ ─────     ─────     ─────
                    0100      0011      0101
```

Signed decimal numbers may be encoded using sign-magnitude, 9's complement, or 10's complement decimal codes. These codes have properties like their binary counterparts. The following example use the excess-3 code and 10's complement arithmetic techniques to compute 65–37 by the complement-add algorithm. Signs are encoded via a decimal digit, which is extravagant.

```
              0037      0011     0011     0110     1010 Bit
                        1100     1100     1001     0101 comple-
                                                    +1  ment
                                                   ──── 
  −37         9963      1100     1100     1001     0110
  +65         0065      0011     0011     1001     1000 Add
  ───        ─────      ────     ────     ────     ────
  +28       1 0028        1╲       1╲       0╲
              ↑       ┌─1 0000  ╲─1 0000  ╲─1 0010  ╲─0 1110
             sign     │  +0011   ╲ +0011   ╲ +0011   ╲ −0011  Correct
             digit    ╰─ ─────     ─────     ─────     ─────
                         0011      0011      0101      1011
                          ↑
                        +result
```

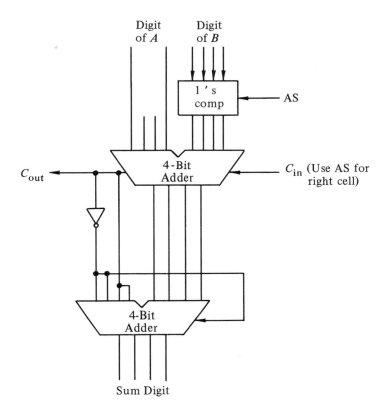

FIG. 3.46   An Excess-3, Tens-Complement Arithmetic Unit Cell

Figure 3.46 suggests the structure of a typical cell of the 10's complement arithmetic unit. Note that the subtractive correction is made by adding 1101, the 2's complement of 0011.

3.10   REDUNDANCY AND RELIABILITY

Every time information is encoded, transmitted, received, stored or manipulated in any way, it must be processed by fallible circuits, and there is a finite probability that, due to error, the information will be distorted. Many proposals for increasing the reliability of information manipulation have been made; they all involve either raising the signal-to-noise level in the electronic equipment, or adding redundancy in the equipment, in the processing, or in the information itself.

Most natural languages, including English, contain a surprisingly large amount of redundant information: they use a great many more symbols, sounds, and repetitions than appear to be necessary to convey the information, in order that under "noisy" conditions the intended information can be received even though much of the message itself is lost. Cross out the vowels, or every other letter in written English and see if you can't "get the message."

Table 3.4 gave a four-bit code for the decimal digits, which leaves six of the sixteen possible code words unused. The appearance of one of these unused bit patterns may then be taken as a signal that a mistake has occurred: a one has been changed to a zero or a zero to a one. Unfortunately, a mistake that transformed one valid word to another could not be caught by this system. And an inspection of all ten BCD codes reveals that of the forty possible ways a single mistake can occur, only ten will transfer a valid word into an unused one; the other thirty merely produce a bit sequence that is a valid code word.

To get a complete check against all possible one-bit errors (a zero becoming a one or vice versa), it is necessary to use five bits per digit. One scheme is to use a five-bit code in such a way that two of the five are always required to be ones and the rest zeros; such a two-out-of-five code is fairly easy to check, but not particularly easy to manipulate, and arithmetic is almost impossible to perform on the words of such a code.

As a consequence, the most common technique is to use a fifth, redundant bit to indicate the sum modulo two, or *parity*, of the other four. Such a code, designed to give "even parity," is shown in Table 3.9; the P or parity bit is always chosen so as to make the total number of ones in the character an even number—either zero, two, or four. With the addition of this one bit we will be able to detect when a single error, actually an odd number of errors, has been made in transmitting a code word. A circuit to generate this additional bit is cheap and easy to install; by transmitting all five bits and regenerating the P bit from the other four as received, we are always in a position to indicate immediately if some bit has been changed, and sound the appropriate alarm.

For example, suppose that the sequence 7 8 4 6 5 is to be sent from one place to another. The transmitter would emit the 5-bit code words shown

TABLE 3.9. BCD WITH EVEN PARITY

Decimal digit	Information bits b_1 b_2 b_3 b_4	Parity bit $b_5 = P$
0	0 0 0 0	0
1	0 0 0 1	1
2	0 0 1 0	1
3	0 0 1 1	0
4	0 1 0 0	1
5	0 1 0 1	0
6	0 1 1 0	0
7	0 1 1 1	1
8	1 0 0 0	1
9	1 0 0 1	0

TABLE 3.10. ERROR DETECTION WITH A SINGLE PARITY BIT

8 4 2 1 P	N	8 4 2 1 P	recalculated check bits
0 1 1 1 1	7	0 1 1 1 1	1
1 0 0 0 1	8	1 0 0 0 1	1
0 1 0 0 1	4	0 1 0 0 1	1
0 1 1 0 0	6	0 0 1 0 0	1
0 1 0 1 0	5	0 1 1 0 0	0
message sent		message received	
(a)		(b)	

in Table 3.10 (a). If a mistake is made during the transmission of the fourth digit, it might be received as shown in (b). It is a simple matter to add checking circuitry at the receiving station to recalculate the check bit at this time and compare with the check bit as received. If an odd number of errors has been introduced, even in the check bit itself, the received and recomputed parity bits will disagree, a situation that is easily identified.

Just what action should be taken when an error is detected can vary with the digital system and with the situation. Detection of an error can variously set an indicator, trigger an alarm, or automatically cause the transmission to be repeated. Some computers have been constructed so that they stop whenever an error in transmission is detected.

While considering the use of the parity check bit for error detection, it is appropriate to ask about the size of the group of bits to which such a check bit should be attached. Although this bit could be used in principle to check for single mistakes in a message of any size, as the length of the message goes up, so does the probability of double or higher multiplicity mistakes. So there is obviously a limit, and not a very high one at that, to the size of the message group that should be checked with a single bit. Commonly the smallest regularly occurring division of information within the computer—one digit, one character, or one word, ranging from four to fifty bits—is taken as the message group, the block of information to be treated as a unit for checking purposes.

Why does the parity scheme work and how might one use additional parity bits to check more thoroughly the validity of a received code word? We might view the effect of adding a parity bit as illustrated in Fig. 3.47, which presents only five of the ten valid code words of the BCD code. Without parity the valid code word 0000 is adjacent to valid code words 0001, 0010, 0100, 1000. Thus any single bit change in the transmitted code word 0000 produces another valid code word. With the parity bit included, the valid code word 00000 is at distance two or more from all other valid code words. Thus a single bit change in that code word produces not another valid code word, but rather a code word that can be recognized at the receiver as invalid. Reception of the code 00001 is a clear indication that an error has been made. Unfortunately, it is not possible to determine

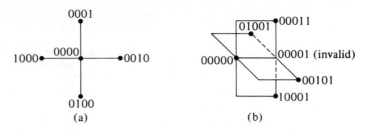

(a)   (b)

FIG. 3.47   BCD, in part, (a) without and (b) with an even parity bit appended. Emphasized vertices denote valid code words.

whether the transmitted code was 00000, or 00011, or one of the other valid codes adjacent to 00001.

If BCD-with-parity code word 00000 is altered in transmission to 00011, the receiver must accept this word and assume it was the code word transmitted. Double errors are not detected with a single parity bit. But we can now see the requirements on a code capable of detecting double errors. Suppose, as suggested by Fig. 3.48, that all valid code words are separated by distance 3, by adding two suitable redundant bits. Then if 000011 were received when 000000 had been sent, the receiver could determine that an error had been made. Either a double error had been made on the code word 000000 or a single error had been made on 000111 (referring to Fig. 3.48 only). Knowing that single errors occur far more frequently than double errors, might not the receiver now assume that a single error had occurred and that the transmitted code word was actually 000111? Such an assumption would be correct in a great majority of the cases if single errors were, in fact, far more common than double errors. The possibility of a code that allows the receiver to *correct* errors as well as detect errors can be seen. If valid code words are further separated, by additional redundant bits, as in Fig. 3.49, double errors can be detected and single errors corrected.

Can we construct codes in which valid words are sufficiently far apart? Suppose we wish to add two redundant bits to 4-bit message code words. Then six invalid code words must surround each valid word to achieve the idea of Fig. 3.48. If the ten decimal digits are to be encoded, we need ten

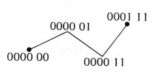

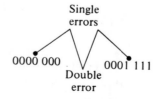

FIG. 3.48. Single errors can be corrected if valid codes are at distance 3.

FIG. 3.49. Single errors can be corrected and double errors detected if valid code words are at distance 4.

Digit Encoded	Hamming Code Words						
	7	6	5	4	3	2	1
	m	m	m	k_4	m	k_2	k_1
0	0	0	0	0	0	0	0
1	0	0	0	0	1	1	1
2	0	0	1	1	0	0	1
3	0	0	1	1	1	1	0
4	0	1	0	1	0	1	0
5	0	1	0	1	1	0	1
6	0	1	1	0	0	1	1
7	0	1	1	0	1	0	0
8	1	0	0	1	0	1	1
9	1	0	0	1	1	0	0

6-bit sequences for the valid code words and 60 sequences for the invalid words, or a total of 70 code words. But only 64 binary sequences of six bits exist. Clearly we must add at least three redundant bits to be able to achieve single-error detection and correction.

R. W. Hamming first showed the details of practical codes based on these ideas. Hence codes such as that in Table 3.11 are known as *Hamming codes*. Note in the sample code that the check bits (k) are interspersed with the message bits (m) of each 8421 code word. The check bits are in positions 1, 2, and 4 of each code word: this is not by accident. Each check bit is an even parity bit, but each checks a different set of the seven bits of a code word, according to the following table.

Check Bit	Positions Checked
K_1	1, 3, 5, 7
K_2	2, 3, 6, 7
K_4	4, 5, 6, 7

Each check bit checks itself and three message bits, but no other check bit. Again, the assignment of positions checked by each check bit is not accidental; they are chosen to make it easy to determine which bit (if any) of a code word has been changed in transmission. Convert the integers 1, 3, 5, and 7 to binary. What do these binary numerals have in common? Their least significant bit is a 1. What do all binary representations of the integers 2, 3, 6, and 7 have in common?

Figure 3.50 illustrates the correction of a single error made in transmission. The digit 6 is to be sent. In transmission a single error has been made in the sixth position, but this information can be obtained from the received code word only. Recompute the parity bits. The first, k_1, should be a 1; a 1 was received. The second, k_2, checks positions 2, 3, 6, and 7. It should be a 0, but a 1 was received: something is wrong! Finally, k_4 should be a 1, but a zero was received. If a 1 is written whenever the recomputed check bit disagrees with that received, and a 0 is written when agreement

Positions	7	6	5	4	3	2	1
Transmitted Code Word	0	1	1	0	0	1	1
Received Code Word	0	0	1	0	0	1	1
Recomputed Check Bits				1		0	1
Agree or Disagree				1		1	0 $= 6_{10}$
This message sent was	0	1	1	0	0	1	1

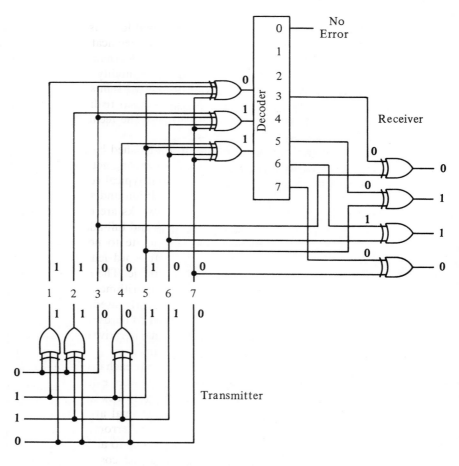

FIG. 3.50. Example of Error Correction with a Hamming Code

CHAPTER 3 **Synthesis of Combinational Logic Networks**

is noted, $110_2 = 6_{10}$ is formed, indicating that the sixth position of the received message is in error.

Many other schemes for adding redundant bits to code words so as to achieve error detecting and error correcting properties have been discovered, explored mathematically, and used in digital communication and processing systems. For that matter, many binary encoding schemes in addition to those presented in this chapter have been developed also. The reader is encouraged to expand in depth and breadth his knowledge and understanding of number systems and codes by referring to the extensive literature concerned with these topics. The references at the end of the chapter are some of the more easily read and understood elements of that literature; they constitute a good place to start such an expansion.

3.11 SUMMARY

This chapter emphasizes combinational logic aspects of network design via informal methods. Initial network specifications are translated into Boolean equations, truth tables, and/or Karnaugh maps. Techniques of the previous chapter are then used to modify Boolean equations that reduce network cost or optimize the network in some other way. If NOR and/or NAND technology is to be used to realize the specifications, an AND-OR-NOT network prepared with the technology in mind is transformed into a NAND-NOR network.

When data processing can be accomplished by performing a basic processing step repeatedly, design of a processing network takes the following two forms: (1) designing a suitable and typical cell and (2) using that cell iteratively through time or space. Addition and subtraction are examples of such processes, and arithmetic networks are used extensively for purposes of example in this chapter. Half- and full-adders are iterated to form ripple adders; modified full-adders iterate to networks that add faster. When data is limited to a fixed, finite word length, ordinary arithmetic must be replaced with modulus arithmetic. The code used for expressing signed numbers influences the design of arithmetic units; complement and sign-magnitude codes are most frequently used. Each code has unique arithmetic rules that can be translated to hardware. Multiplication introduces the need for either two-dimensional iterative combinational networks, or iteration through both space and time.

Decimal oriented digital systems encode each decimal digit with a four or more bit code word. While very many codes exist, only a few have been used. Some are well suited to arithmetic; rules for performing arithmetic on code words can be translated to decimal arithmetic networks. If redundant bits are appended to code words, errors can be detected and corrected. While it has been common practice to add a single parity bit, then single errors can be detected, the reduced cost of logic circuits has increased interest in the use of more redundant bits in codes that permit correction after the detection of errors.

While it may not be apparent, this chapter serves only as an introduction to a number of topics. For example, iterative networks, arithmetic algorithms and networks, and error detecting and correcting codes are explored elsewhere in much greater detail. The reader should consult the references of the previous chapter as well as those of this chapter.

REFERENCES

1. N. M. ABRAMSON, *Information Theory and Coding*. New York: McGraw-Hill Book Co., Inc., 1963.
2. A. M. ABD-ALLA and A. C. MELTZER, *Principles of Digital Computer Design*. Englewood Cliffs, N.J.: Prentice-Hall, 1976.
3. A. D. BOOTH and K. H. V. BOOTH, *Automatic Digital Calculators* 2d ed. New York: Academic Press, 1956.
4. W. J. CADDEN, "Binary Numbers, Codes, and Translators," *A Survey of Switching Circuit Theory*. New York: McGraw-Hill, 1962, pp. 15–30.
5. K. R. FIALKOWSKI, "The $\bar{2}$'s Complement and Other Semi-Systematic Binary Codes," *IEEE Trans. on El. Comp.*, Vol. EC-15, No. 4, Aug. 1966, pp. 604–605.
6. R. G. GALLAGER, *Information Theory and Reliable Communication*. New York: J. Wiley and Sons, 1968.
7. H. L. GARNER, "The Residue Number Systems," *IRE Trans. on Electronic Computers*, Vol. EC-8, No. 2, June 1959, pp. 140–147.
8. R. W. HAMMING, "Error Detecting and Error Correcting Codes," *Bell Sys. Tech. Jour.*, Vol. 29, No. 2, April 1950, pp. 147–160.
9. F. C. HENNIE, *Finite-State Models for Logical Machines*. New York: John Wiley and Sons, 1968.
10. G. K. KOSTOPOULOS, *Digital Engineering*. New York: John Wiley and Sons, 1975.
11. S. LIN, *An Introduction to Error-Correcting Codes*. Englewood Cliffs, N.J.: Prentice-Hall, 1970.
12. R. W. LUCKY, J. SALZ, and E. J. WELDON, JR., *Principles of Data Communication*. New York: McGraw-Hill, 1968.
13. W. W. PETERSON and E. J. WELDON, JR., *Error-Correcting Codes*, 2d ed. Cambridge: MIT Press, 1972.
14. J. R. PIERCE, *Symbols, Signals, and Noise: The Nature and Process of Communication*. New York: Harper & Brothers, 1961.
15. M. L. STEIN and W. D. MUNRO, *Introduction to Machine Arithmetic*. Reading, Mass.: Addison-Wesley, 1971.
16. N. S. SZABO and R. I. TANKA, *Residue Arithmetic and Its Applications to Computer Technology*. New York: McGraw-Gill, 1967.

PROBLEMS

3.1-1. A milling machine is to operate only if a master switch is open and both operator switches are closed, or if the master switch is closed and both operator switches are open. Write a Boolean equation which expresses when the machine is to operate. Draw a circuit diagram which indicates how the switches are to be interconnected.

3.1-2. None or only one of three generators is to be connected to a transmission line T at all times. Two control signals, A and B, determine which generator is connected, according to the following table.

A	B	Gen. Connected
0	0	None
0	1	$G1$
1	0	$G2$
1	1	$G3$

Write a Boolean expression for the required switching circuit (think of $G1$, $G2$, and $G3$ as binary variables). Draw a circuit diagram showing the required relay switching network.

3.1-3. If none or exactly two of the three input variables, A, B, and C, have value 1, then output variable $Z = 1$; otherwise, $Z = 0$. Prepare a truth table and Boolean equation for Z. Is the exclusive-OR operator useful in describing Z?

3.1-4. A five member committee requires a voting system. Each member sets his switch (1 for yes; 0 for no) and a master vote switch V is closed by the chairman. The pass light P or fail light F indicates the outcome. When V is open, both lights must be off. Prepare tables and equations that describe the required logic.

3.1-5. A counter derives A and B waveforms shown below from the *clock* signal. With these signals driving the decoder of Fig. 3.3, draw the waveforms of signals Z_0, Z_1, Z_2, and Z_3.

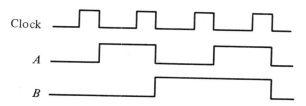

3.1-6. Integers in the range 0 through 15 are expressed with 4-bit code words. A network with four input lines is to generate $Z = 0$ when the received integer is 5 or less, and $Z = 1$ when the received integer is in the range 8 through 12. Prepare a truth table and Karnaugh map for Z. Write a minimum Boolean expression for Z.

3.1-7.
$$f(A, B, C, D) = \sum 11, 12, 13, 14, 15$$

(a) Use a 4-variable Karnaugh map to find a minimum sum-of-products expression for f.
(b) Use a 3-variable Karnaugh map with A treated as an infrequent variable to find the same minimum expression for f.
(c) Repeat part (b) using C as the infrequent variable.

3.1-8. A net with three input signals x_1, x_2, and x_3 has four output terminals z_1, z_2, z_3, and z_4. Output z_1 is to be at the 1 (high) level only when all inputs are at the 0 (low) level; z_2 is to be high only when exactly one input is high; z_3 is to be high when any two, but not all three inputs are high. Finally z_4 is to be high when all three inputs are high.

3.1-9. Design a logic network that will accept two 2-bit numbers, and present the 4-bit product on four output terminals.

3.1-10. Design networks to realize each of the following "addition" tables.

		A				A				A	
		0	1			0	1			0	1
B	0	01	10	B	0	11	10	B	0	00	01
	1	00	01		1	10	01		1	11	00

3.1-11. Translate each of the DDL statements of Prob. 2.8-6 to a sum-of-products expression for F.

3.2-1. Find a NAND network which requires a minimum number of gates for each function. Complements of variables must be generated if needed.
 (a) $F_1 = (B \lor D)(\bar{A} \lor \bar{C}) \lor AD$
 (b) $F_2 = \bar{A}BD \lor A\bar{B}C$ (4 gates)
 (c) $F_3 = \bar{A} \lor B\bar{C} \lor B\bar{D}E \lor B\bar{D}F$

3.2-2. Use only 3-input NAND gates to realize the following switching functions. Assume that unused input terminals are connected to 1.
 (a) $F_1 = A \cdot B \cdot C \cdot D$
 (b) $F_2 = A \lor B \lor C \lor D \lor E$
 (c) $F_3 = A \cdot B \cdot C \cdot D \lor E \cdot F \cdot G \cdot H$
 (d) $F_4 = \bar{A} \lor BC\bar{E} \lor BC\bar{F} \lor \bar{C}EG \lor E\bar{F}G$

3.2-3. Both NAND and NOR gates are available, but the complements of input signals are not. Derive minimum gate networks that realize the following functions.

 (a) $A \lor B \lor CD$
 (b) $\overline{A \oplus B}$
 (c) Half-adder with $\overline{\text{carry}}$ rather than carry.
 (d) Full adder.

3.2-4. Transform the decoder of Fig. 3.3 to a minimum NAND/NOR network.

3.2-5. A 3-input, 3-output terminal network is required that realizes

$$G = A \cdot B$$
$$P = A \lor B$$
$$S = C \oplus \bar{G}P$$

Design a NAND/NOR/NOT network.

3.3-1. (a) Draw the logic diagram of a dual 4-input multiplexer such as the one that appears in Fig. 3.19. Use NAND-NAND technology. Each input terminal must constitute a unit load.
 (b) Add enable input terminals for each multiplexer, subject to the same fan-out limit.

3.3-2. Using minimum external circuitry, show how to realize

$$F = A\bar{C}D \lor A\bar{B}D \lor \bar{B}C\bar{D} \lor \bar{A}B\bar{D} \lor \bar{A}B\bar{C}$$

using:
 (a) A 16-input multiplexer,
 (b) An 8-input multiplexer, and
 (c) A 4-input multiplexer.

If it costs $0.25 per pin to buy, mount, and power integrated circuits, which realization is the most economical? (Count only logic pins used.)

3.3-3. Repeat Prob. 3.3-2 for

$$F = A \oplus B \oplus C \oplus D$$

Use more than one 4-input multiplexer if you like.

3.3-4. Draw the logic diagram of a blank 8-word, 4-bit per word ROM using NAND technology. Each input terminal must present unit load. To simplify the diagram, use the following convention. Show a single line into each output gate. If a decoder gate is to be connected to an output gate, a dot is placed at the intersection of its output line and the input line of the output gate.

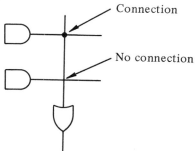

3.3-5. Program your ROM of Prob. 3.3-4 to realize the following switching functions:

$$F = A \oplus B \oplus C$$
$$G = AB \vee AC \vee BC$$
$$H = \bar{A}B \vee \bar{A}C \vee BC$$
$$I = \overline{C(A \vee B)}$$

Show how to use a 3-to-8 line decoder and four OR gates to realize these functions.

3.3-6. Show how four 256×4 ROMs with tri-state output amplifiers may be used to build a:

(a) 512×8 ROM, and a

(b) 1024×4 ROM.

3.3-7. An 8-word, 6-bits per word ROM is connected to the synchronized 3-bit A register as shown. The output signals of the A register provide the address to the ROM. The left three output signals serve as input signals to the A register and hence determine the next address of the system. The contents of the ROM are shown in the truth table.

(a) Assuming that all flip-flops of the A register are reset initially, show in tabular form how the address and output signals of the ROM change with successive clock pulses.

time	A	$A_1 A_2 A_3 B$ C D
0	000	
1		
2		
⋮		

(b) Repeat (a), assuming the A register holds 010 initially.

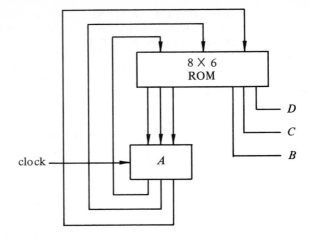

Address	A_1 A_2 A_3 B C D
0 0 0	1 1 0 0 0 0
0 0 1	0 0 0 0 0 0
0 1 0	1 0 1 1 1 1
0 1 1	0 1 0 1 1 0
1 0 0	0 0 1 1 1 0
1 0 1	0 1 1 1 0 1
1 1 0	1 1 1 0 0 1
1 1 1	1 0 0 0 1 1

3.3-8. Show the organization and ROM program for a system that counts in synchronism with a clock in the following fashion: 0,1, ..., 6,7, 6, ..., 1,0,1, ... Clearly indicate register and ROM dimensions, how they are interconnected, and on which wires the desired count is available.

3.3-9. A modified PLA has 2-bit "decoders", rather than inverters, between the input terminals and the AND gates. Maxterms of pairs of variables are generated. The number of lines presented to the AND gates is the same.

(a) Find the conventional PLA programming for the following switching function.

$$X = B \oplus D$$
$$Y = \bar{A}\bar{B}C \vee A\bar{B}C \vee \bar{A}C\bar{D} \vee A\bar{C}\bar{D} \vee \bar{A}B\bar{C}D \vee ABCD$$
$$Z = AC \vee ABD \vee BCD$$

(b) Indicate how a modified PLA must be programmed to realize this function by drawing a logic diagram of the programmed PLA. Compare results with those of part (a).

3.4-1. The serial adder of Fig. 3.24 is to be modified by deleting the R register and collecting the sum in the A register. Draw a block diagram for the modified system. Draw waveforms, like those of Fig. 3.24, for the modified system.

3.4-2. How could we detect overflow in the serial adder system of Fig. 3.24?

3.4-3. An iterative network is to accept n input signals and generate $n + 1$ output signals. A 1 must appear on the output line corresponding to the leftmost input line carrying a 1. All other output lines provide 0. If all input lines bear logic 0, then the $n + 1$ (rightmost) output line will be at 1. Design a typical cell for the network; show special connections or circuitry for the end cells.

3.4-4. Repeat Prob. 3.4-3 with the output signals identifying the second leftmost 1 on the input lines. Clearly indicate the cell messages and their codes. Prepare a message diagram as well as a table for the typical cell.

3.4-5. Output lines of an iterative network are to be at 1 if their corresponding input lines are at 1, and at least one input line to the left and at least one input line to the right are at 1. Design a typical cell; show special end cell connections.

3.4-6. A network has seven input lines. If the number of input lines at 1 is odd, the center of these lines is to be identified by a 1 on the corresponding output line. All other output lines are at 0, of course. Prepare a table for the typical cell that indicates how it responds to data and messages.

3.4-7. (a) Realize the typical compare cell of Eq. (3.18) with NAND and NOR gates.

(b) Find a more economical, faster comparator by dealing with a pair of cells. Modify the message codes of the left member of a pair so that the delays from x_{i-2} to x_i, y_{i-2} to y_i, and z_{i-2} to z_i are reduced.

3.5-1. What positive integers are congruent to: (a) 7 modulo 3; (b) -4 modulo 25?

3.5-2. (a) Find the least positive residue for each integer modulo, each m given.

modulus

	2	3	5
10			
14			
24			

(b) Recognizing that 24 is the sum of 10 and 14, do you find any relationship to exist between the least positive residues?

3.5-3. Form a modulo 2 binary multiplication table. Is this table like any other table you have encountered?

3.5-4. Unsigned integers are expressed in binary.

(a) Design a network which will add two such numbers and give the least positive residue of the sum modulo m.

length of numbers	m
2 bits	3
3 bits	5
3 bits	7

(b) Design a network which will perform modulo m multiplication on the same numbers.

(c) Convert your networks to NAND/NOR technology.

3.5-5. (a) Design a modified full-adder, Eq. (3.29), of NAND and NOR gates. To minimize delays, provide G_i and P_i in complement form, i.e., $\bar{G}_i$ and $\bar{P}_i$.

(b) Design a carry-look-ahead network, Eq. (3.28), of NAND and NOR gates. To be compatible with the adder of part (a) it must accept $\bar{G}_i$ and $\bar{P}_i$, and provide $\bar{G}_{group}$ and $\bar{P}_{group}$. You need not generate C_{i-3}.

3.5-6. (a) Show with a block diagram how 16 modified full-adders and five carry-look-ahead networks must be interconnected to form a fast 16-bit adder.

(b) How many gate delays exist between the $C_{n+1} = C_{in}$ and the C_1 (leftmost) carry out terminal?

(c) If only four carry-look-ahead networks were used, what would be the the answer to part (b)?

3.6-1. Let $R = -2$ be the radix and $\{0, 1\}$ the digits of a positional number system.
   (a) What integers can be expressed with 5-bit sequences? 6-bit sequences?
   (b) What are the rules for adding integers expressed in this number system? Demonstrate by computing $19 + (-5)$. Subtraction? Compute $19 - 5$.

3.6-2. Give three 6-bit (including sign) representations of each integer.

	sign-magnitude	1's complement	2's complement
-14	101110		
-30			
$+17$			

3.6-3. (a) Represent the numbers in 1's complement form, and perform the indicated arithmetic in binary. Check answers. Discuss any difficulties.

$$A = +10, \quad B = -22$$

   (i) $A + B =$
   (ii) $A - B =$
   (iii) $-A + B =$

   (b) Repeat the above, performing subtraction by complementing and adding.
   (c) Repeat all of the above using the 2's complement representation of signed numbers.

3.6-4. (a) Draw a block diagram of the 1's complement arithmetic unit of Fig. 3.33.
   (b) Draw a logic diagram of the typical cell of *ADDSUBIC*. Indicate how the end cells differ.

3.6-5. Register A holds 16-bit, 1's complement code words. Design an iterative network that detects whether A holds the number zero or not.

3.8-1. Show how AND gates, half-, and full-adders can be interconnected to form the product of two 4-bit numbers.

3.8-2. The 4×2 multiplier cells of Fig. 3.37 require adders that accept 5 bits and presents their sum in binary on three output lines.

   (a) Design a suitable adder using half- and full-adders.
   (b) Design a suitable adder from truth tables. Use NAND and NOR gates only; keep the number of logic levels low.
   (c) Show how the adders must be interconnected to form the 4×2 multiplier.

3.8-3. Simulate system *MULT* of Fig. 3.38, assuming the registers initially hold

$$A = 0\ 0\ 0\ 0\ 1\ 1\ 0\ 1\ 1\ 0\ 1\ 0\ 0\ 0\ 0\ 0$$
$$B = 1\ 0\ 0\ 1\ 0\ 1\ 0\ 0\ 0\ 0\ 1\ 0\ 0\ 0\ 0\ 1$$

3.8-4. Simulate Booth multiplication as illustrated in Fig. 3.40. Use 4-bit 2's complement code words for the signed integers and point out the correctness of your answers.

   (a) $+7$ (b) -7
        $\times -6$ $\times -1$

(c) -1 (d) -1
$$\underline{\times\ -6}$$ $$\underline{\times\ -1}$$

3.9-1. Find the distance between code words of each of the following pairs.

   (a) 001   (b) 1010   (c) 100110
      011   1110   101101

3.9-2. Encode the decimal digits with 4-bit weighted codes using the following sets of weights.

   (a) 5 4   2   1
   (b) 2 4   2   1
   (c) 8 4 -2 -1

3.9-3. Prove that 14 is the minimum number of 1's with which the decimal digits can be encoded using 4-bit code words. How many different 4-bit codes containing fourteen 1's exist?

3.9-4. Define a code to be *analytic* if its code words satisfy

$$N = \sum_{i=1}^{n} w_i \times b_i + \beta$$

where the w_i are weights, β is a bias constant, and N is the number encoded by binary sequence $b_1, b_2, \ldots, b_n$. Give the weights and the bias for those of the following codes that are analytic.

(a) Any weighted code.
(b) Excess-3 code (Table 3.7).
(c) One-hot code (Table 3.7).
(d) Creeping code (Table 3.7).
(e) 1's complement code.
(f) 2's complement code.

3.9-5. BCD and a Gray code for the decimal digits are shown below. Find a set of rules (algorithm) for generating each Gray code word from the corresponding BCD code work. *Hint:* $\oplus$.

	BCD	Gray
0	0 0 0 0	0 0 0 0
1	0 0 0 1	0 0 0 1
2	0 0 1 0	0 0 1 1
3	0 0 1 1	0 0 1 0
4	0 1 0 0	0 1 1 0
5	0 1 0 1	0 1 1 1
6	0 1 1 0	0 1 0 1
7	0 1 1 1	0 1 0 0
8	1 0 0 0	1 1 0 0
9	1 0 0 1	1 1 0 1

(b) Find the algorithm for generating the BCD code from the Gray code.

3.9-6. The counter of Fig. 3.44 can generate the code words of Table 3.7.
   (a) Starting with all flip-flops storing 0, draw a clock signal and corresponding waveforms for b_1, b_2, b_3, b_4, b_5, and the detector gate signals.
   (b) Repeat, assuming the flip-flops hold 1 0 1 0 1.

3.9-7. (a) Perform the addition suggested below by replacing each digit by its BCD code word, adding in binary, correcting as necessary, and checking results.

$$732 + 895 = 1627$$

(b) Convert 732 to binary. Compare with the BCD encoding of 732.

(c) Repeat part (a) using the excess-3 encoding. Your answer should be in the excess-3 code.

3.9-8. In Fig. 3.45, the least significant uncorrected sum bit does not enter the "sum exceeds 9" detector.

(a) Is this a mistake? Justify your answer with a simple argument.

(b) Prepare a truth table that describes the "sum exceeds 9" detector.

(c) State in English when the output signal of the "sum exceeds 9" detector is to be logic 1. Convert your English statement to a Boolean equation.

3.9-9. Every 4-bit binary sequence encodes a decimal digit under weights 2421 (like weights 4311, as shown in Table 3.5).

(a) Write the decimal digit encoded by each of the 16 code words of the 2421 code. Is the code reflected?

(b) This code is additive; we can perform decimal digit addition by binary adding code words and by correcting, if necessary. What corrections must be made and under what conditions? *Hint*: Carries into and out of the left-bit adder.

(c) Determine how the decimal carry must be formed. Summarize all findings by a logic design (like Figs. 3.45 and 3.46) of a digit adder/subtracter for use in a ten's-complement arithmetic unit. You need not show the logic of the 4-bit adders, but show all other logic at the gate level of detail.

3.10-1. Add a parity bit to each excess-3 code word so that the 5-bit code word has (a) *even* parity, (b) *odd* parity.

3.10-2. Do the following code words constitute a weighted code with parity? If so, find the weights of four columns and the position of the parity bit.

Digit	Code Word
0	0 0 0 0 0
1	1 0 0 1 0
2	1 0 0 0 1
3	1 1 1 1 0
4	1 1 1 0 1
5	0 0 1 0 1
6	1 0 1 1 1
7	0 1 0 0 1
8	1 1 0 1 1
9	0 0 0 1 1

3.10-3. Show how mod 2 addition is used to establish a parity bit to be associated with a code word, and how it is used to determine if a received code word is or is not error free.

3.10-4. A message of decimal digits encoded using the Hamming code of Table 3.11 is received as shown below. What was the decimal sequence?

```
0 1 0 1 0 1 0
0 0 0 0 1 1 1
0 0 0 0 0 0 1
0 0 1 0 0 0 0
0 1 0 1 1 0 1
0 1 1 0 0 1 1
0 1 0 1 0 0 1
0 1 1 1 1 0 1
```

3.10-5. A 7-bit code has been constructed by augmenting a certain 4-bit Gray code so that it can detect and correct any single error in the transmission of a decimal digit. The added bits indicate independent *even* parity checks. Note that this user likes to number positions from left to right. When the message 0, 1, 2, 3, 4, ..., 9 is sent, the message received is:

Message
Received

$k_1\ k_2\ k_3$

	1 2 3 4 5 6 7
0	1 1 0 1 0 0 1
1	1 0 0 0 1 1 1
2	0 1 1 0 0 0 1
3	1 0 1 1 1 1 0
4	0 1 0 1 0 1 0
5	1 1 0 0 1 1 0
6	1 0 1 1 1 0 0
7	0 0 1 0 0 0 0
8	1 0 1 1 0 0 0
9	0 0 1 1 0 0 1

Reconstruct the code.

3.10-6.   ⟨RE⟩ A[4 : 1] = AL[3] ∘ AR, B[7 : 1], C[7 : 1], D[4 : 1], E[3 : 1].

(a) Register *A* holds an 8421 encoded decimal digit. Register *B* is to hold the corresponding Hamming code word (see Table 3.11). Give the register transfer that specifies the circuitry needed to generate the *B* word from the *A* word; i.e.,

$$B \leftarrow AL \circ (\oplus/AL) \circ \ldots$$

(b) The Hamming code word is transferred from register *B* to register *C*. Sometimes errors are made in this transfer.

$$C \leftarrow B + noise$$

The position of the error, if any, is to be placed in the *E* register. Prepare the DDL register transfer that prescribes the required circuitry:

$$E \leftarrow (\oplus/C[7 : 4]) \circ \ldots$$

(c) Finally, the correct 4-bit code word is to be placed in the *D* register Complete $D \leftarrow$

3.1-1. $RUN = \bar{M} \cdot A \cdot B \vee M \cdot \bar{A} \cdot \bar{B}$

3.1-3. $Z = \bar{A}\bar{B}\bar{C} \vee AB\bar{C} \vee A\bar{B}C \vee \bar{A}BC$

$\qquad = \overline{A \vee B \vee C}$

3.1-6. $Z = A$

3.1-7. $f = AB \vee ACD$

3.1-8. $z_1 = \bar{x}_1\bar{x}_2\bar{x}_3$

$\qquad z_2 = x_1\bar{x}_2\bar{x}_3 \vee \bar{x}_1x_2\bar{x}_3 \vee \bar{x}_1\bar{x}_2x_3$

3.1-9.

I	J	$I \times J$	
$\overbrace{x_1 x_2}$ $\overbrace{x_3 x_4}$		$y_1 \; y_2 \; y_3 \; y_4$	
0 0	0 0	0 0 0 0	$0 \times 0 = 0$
0 0	0 1	0 0 0 0	$0 \times 1 = 0$
0 0	1 0	0 0 0 0	$0 \times 2 = 0$
$\vdots$		$\vdots$	
1 1	1 1	1 0 0 1	$3 \times 3 = 9$

3.2-1. $F_1 = AD \vee B(\bar{A} \vee \bar{C}) \vee D(\bar{A} \vee \bar{C})$ 5 NAND gates

$\qquad = D \vee B(\bar{A} \vee \bar{C})$ 4 NAND gates

3.2-2. $F_1 = \overline{(A \uparrow B \uparrow C) \uparrow D \uparrow 1}$ 4 gates

3.2-3. (a) $A \vee B \vee CD = (A \vee B) \vee CD = (A \downarrow B) \uparrow (C \uparrow D)$

3.3-2. (c) Minimize F with a Karnaugh map.

$\qquad 11 \times .25 = \$2.75$

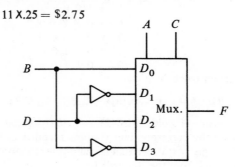

3.3-9. (a) 11 AND gates   (b) 7 AND gates

3.4-3.

Message	Code	
Only 0's to left	0	$x_i = x_{i-1} \vee A_i$
1's to left	1	$B_i = \overline{x_{i-1}} \cdot A_i$

3.4-5.

Messages Propagating Right	Code	Messages Propagating Left	Code
no 1's to left	0	no 1's to right	0
some 1's to left	1	some 1's to right	1

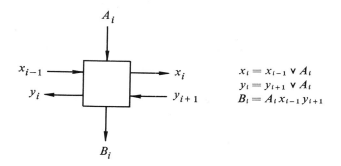

$$x_i = x_{i-1} \vee A_i$$
$$y_i = y_{i+1} \vee A_i$$
$$B_i = A_i x_{i-1} y_{i+1}$$

3.5-1. (a) $7 \cong \cdots \cong -5 \cong -2 \cong 1 \cong 4 \cong 7 \cong 10 \cong 13 \cong 16 \cong \cdots 7$
$\cong 3n + 1 \pmod 3$ $n = -\infty, \ldots, 0, \ldots, \infty$

3.5-2.

		2	3	5	
10		0	1	0	$0 \oplus 0 = 0 \bmod 2$
14		0	2	4	$1 \oplus 2 = 0 \bmod 3$
24		0	0	4	$0 \oplus 4 = 4 \bmod 5$

See Garner [7].

3.6-2.

	$S—M$	1's Comp.	2's Comp.
-14	101110	110001	110010
-30	111110	100001	100010
$+17$	010001	010001	010001

3.6-3. $+10 = 001010$, $+22 = 010110$, $-22 = 101001$

(a)
$$\begin{array}{ll} +10 & 001010 \\ +(-22) & +101001 \\ \hline & 110011 \rightarrow -12 \text{ in 1's comp. convention} \end{array}$$

(b) Adding by complementing and subtracting.
$$\begin{array}{ll} +10 & 001010 \\ -(+22) & -010110 \\ \hline & 110100 \\ & \underline{-1} \quad \text{end-around borrow} \\ & 110011 \rightarrow -12 \text{ in 1's comp. convention} \end{array}$$

(c)
$$\begin{array}{ll} +10 & 001010 \\ +(-22) & +101010 \\ \hline & 110100 \rightarrow -12 \text{ in 2's comp. convention} \end{array}$$

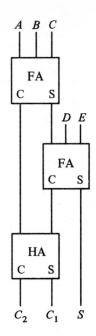

3.8-2.

3.9-1. 1, 1, 3

3.9-3. There is one code word with all 0's, 0000. There are four code words with one 1 each, 0001, 0010, ... The remaining five code words must be selected from the six code words containing two 1's each. Hence there is a minimum of

$$0 + 4 + 5(2) = 14 \text{ 1's in any 4-bit code.}$$

Pick any 5 of the six code words containing two 1's. For each choice, associate any of the ten code words with 0, any of the remaining nine with 1, etc. Thus $6 \times 10!$ codes exist; some will be related (permutation of columns, for example).

3.9-5. Let $b_1 b_2 b_3 b_4$ be a BCD code word and $g_1 g_2 g_3 g_4$ be the corresponding Gray code word. Test the following rules.

$$g_1 = b_1$$
$$g_2 = b_1 \oplus b_2$$
$$g_3 = b_2 \oplus b_3$$
$$g_4 = b_3 \oplus b_4$$

Note: The code is not truly cyclic because the encoding of 9 is not adjacent to that of 0.

3.9-7.

| 732 | → | 0111 | 0011 | 0010 |
| 895 | → | 1000 | 1001 | 0101 |

$732_{10} = 1011011100_2$, requires 2 fewer bits than BCD.

```
3.9-7.  732 → 0111    0011    0010     732₁₀ = 1011011100₂,
        895 → 1000    1001    0101     requires 2 fewer bits
       ────       ────    ────    ────     than BCD.
       1627       1111    1100    0111
                  0110    0110
                       1 ← 1 0010
          0001 ← 10110
             1       6       2       7
```

3.9-9 (b)

C_4	C_3	Correction
0	0	none, decimal carry $= 0$
0	1	add 0110
1	0	subtract 0110 or add 1010
1	1	

3.10-2. Code words 1, 2, 5, 7, 9 define the following equations expressed in terms of weights.

$$w_1 + w_4 = 1$$
$$w_1 + w_5 = 2$$
$$w_3 + w_5 = 5$$
$$w_2 + w_5 = 7$$
$$w_4 + w_5 = 9$$

A solution to these equations exists.

$w_1 = -3$, $w_2 = 2$, $w_3 = 0$, $w_4 = 4$, $w_5 = 5$

If all other code words satisfy these weights, then the code is a weighted code. They do.

3.10-5 For 0:

```
       1  1  0  1  0  0  1
       √  √     √
For 1: 1  0  0  0  1  1  1
       ×  √  ×  1
```

all OK: ∴ 0 is encoded 0001

bit 5 is suspect ∴ 1 is encoded 0011

4

Sequential Concepts and Components

A logic network is *combinational* if its static behavior can be described by a set of Boolean equations, with each output variable expressed as a function of only the input variables. In this chapter we begin to work with logic networks that do not satisfy this definition. We will find that their static behavior can be described in terms of a sequence of input variable values. As a result we call such networks *sequential*. Time enters as a variable in the equations and tables used to describe sequential logic networks.

This chapter first explores time and ways of modeling time and signal delay. Then elementary memory elements called latches and more complex memory elements called flip-flops are examined. The last sections emphasize the timing of activity in sequential networks and sources of delay in electronic circuits.

4.1 SEQUENTIAL ACTIVITY

We normally consider time to be a continuous variable. When analyzing electronic circuits, we solve equations to show how an output voltage varies as a function of an input voltage that varies with time. We conclude with an equation that shows how the output voltage varies with time.

$$v_{OUT}[v_{IN}(t)] \rightarrow v_{OUT}(t)$$

Thinking of time as continuous is practical for small electronic circuits, but logic networks often contain many electronic amplifiers. Analysis to

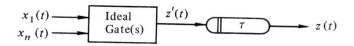

FIG. 4.1 A model of a Gate or Gate Network

obtain analytical solutions while taking all of the capacitance, resistance, inductance, transconductance, and so forth, into account is very difficult. Numerical analysis with the aid of a computer is practical only for the most modest logic networks. To analyze at lower cost, we must find and use models of digital amplifiers that summarize behavior. Results of such analysis are never any better than the models used, of course. If the models are crude, then the results will be of poor quality. But for many purposes such results are entirely adequate. We have used very crude models for logic gates in previous chapters, models that include only the logic behavior of gate circuits. Results using such models have been acceptable to this point.

It is now essential that we begin to include other characteristics of gates in our models. Gating circuits do not act instantly. They respond through time to input signals in complex ways. Wishing to avoid as much complexity as possible in modelling gates, yet reflecting their non-instantaneous action, we most often think of an actual gate as consisting of an ideal (logic behavior, instantaneous) gate followed by a delay block as shown in Fig. 4.1. The ideal gate instantaneously provides values of the fictitious, intermediate signal $z'(t)$. The delay block carries values of $z'(t)$ through time and may change them to derive values for the output signal $z(t)$.

The delay block is a fiction introduced to account for the propagation delay seen in gates and gate networks. We may think of the delay block as characterized by one parameter τ and as introducing transport delay; the input signal is transported through time without modification.

$$z(t + \tau) = z'(t)$$
$$z(t) = z'(t - \tau)$$

(4.1)

This model of delay is used most often. Tau is assigned the value of propagation delay known or assumed to exist. Or, such delay may be normalized, and all gates assumed to have the same delay. We rarely know the delay of a specific gate and do know that all gates have a delay near a specific value. It is quite reasonable to set $\tau = 1$ for all gates. This is the *unit delay* model of a network. Real networks do not always respond as an ideal model predicts. For example, the unit delay model of the network of Fig. 4.2 reveals that the output signal will briefly take the value of 1 when x_2 goes from 0 to 1 while x_1 and x_3 are held at 0.

Working with the unit delay model is relatively easy. The output value of each gate is a Boolean function of the value its input signals had one unit of time earlier. For Fig. 4.2, this is

$$u(t) = x_1(t - 1) \lor x_2(t - 1)$$
$$v(t) = \bar{x}_2(t - 1)$$

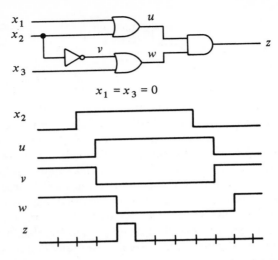

FIG. 4.2 The unit delay model predicts that z will take the value of 1 for a short time. The ideal model predicts $z = 0$.

$$w(t) = v(t-1) \vee x_3(t-1)$$
$$z(t) = u(t-1) \cdot w(t-1)$$

Or, since we are breaking time into unit intervals and do not expect (hence permit) signal values to change within an interval, we may speak of time steps and number them with integers.

$$u(i) = x_1(i-1) \vee x_2(i-1)$$
$$v(i) = \bar{x}_2(i-1)$$
$$w(i) = v(i-1) \vee x_3(i-1)$$
$$z(i) = u(i-1) \cdot w(i-1)$$

We evaluate signal values for one interval from the known values for the previous interval only. Having found all signal values, we advance our interest to the next interval, we increment i. Figure 4.2 shows the results of such a unit delay simulation.

In a network not all gates will have the same delay. In fact, it is likely that no two gates will have precisely the same delay. We can assign to gates different integer delay values, or assign maximum and minimum values of delay for each gate. Simulation then becomes a sizable bookkeeping problem, and computer assistance may be welcomed.

Transport delay fails to model some details of real circuit response. If the rise and fall time of a gate differ significantly, a two-parameter delay model may be used in which a 0-to-1 transition is delayed τ_r units of time, and a 1-to-0 transition is delayed τ_f.

The output pulse is shorter than the input pulse if $\tau_r > \tau_f$, and longer if $\tau_r < \tau_f$. A pulse of duration less than $\tau_r - \tau_f$ will not appear at the output with $\tau_r > \tau_f$.

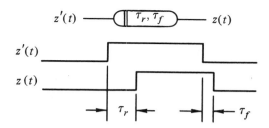

The mechanisms that introduce propagation delay in a gate circuit do not permit storing a long sequence of bits, as is possible with the mathematical transport delay. All real circuits have a cut-off frequency above which they fail to pass signals. The *inertial* delay attempts to model this characteristic. Such models may have one or more parameters. An inertial delay with parameters j and τ such that $j \le \tau$ passes input changes spaced greater than j units of time apart with transport delay τ. If the input signal changes two or more times within an interval of j units of time, the changes are ignored, and the output signal value is not altered until the input signal is stable for j or more units of time. Brief pulses or bursts of oscillation are eliminated, as they are in real circuits. Inertial delays are more easily simulated with a computer than are transport delays. Only a single memory location is used to store data. It is used as a counter. Each time the input signal changes value, the counter is reset to 0. If the input signal does not change during a unit interval of time, the counter is advanced. When the count reaches τ, the output signal value is reassigned.

A number of models are also used for gate networks. The *ideal* model includes no delays and has been used in previous chapters. In the *detailed* model a delay is associated with each gate. The detailed model requires considerable computation and is only used where the detail of the results it gives is required. A detailed model was used in Fig. 4.2. For most purposes where delay must not be ignored, the *fundamental* model in which a single delay is viewed as existing in each output line is adequate. See Fig. 4.1. Any of our gate delay models may be used, but since the single block represents the complex delay properties of the network it is difficult to justify the use of a complex delay model. The delay parameter may represent the longest delay accumulated along any path through the combinational network, or an average value. At times a circuit to provide a delay that is much larger than the gate delays is placed in the output lead. This masks uncertain gate delays. Only then does the value of τ have physical significance.

A Sequential Network

It is possible to interconnect gates so that the value of the output signal of the circuit at one instant is determined by previous values assumed by the input signals, as well as the present values of these signals. The interval between the appearance of these influential values to the present may be many times that of the internal circuit delays. The response of circuits of this type to time-varying input signals is determined by the "sequence" of these signals and not just their current values, as in combinational circuits. These circuits are thus known as *sequential switching circuits*. Sequential circuits present many new problems: they can not be represented with the same notations as combinational circuits, for example. But they also possess very desirable properties and we can not build very significant logic networks without them. But before we develop tractable representations of sequential circuits, and investigate their applications, let us analyze a sequential circuit to observe and emphasize the differences between it and a combinational circuit.

Figure 4.3(a) displays a logic network we might attempt to analyze with Boolean algebra. We can write:

$$z = x_2 \downarrow \alpha = x_2 \downarrow (x_1 \downarrow z)$$
$$= \bar{x}_2 \cdot (x_1 \vee z)$$

but we have not written an explicit expression for z. We might try to find an expression for z by examining all the functions of two variables or by further algebraic manipulation. But note that if $x_1 = x_2 = 0$, the above equation reduces to $z = z$, and either $z = 0$ or $z = 1$ is a solution. The value of z is not uniquely determined by x_1 and x_2 in this case.

We can obtain this result in another way. If $x_1 = x_2 = 0$, the circuit of Fig. 4.3(a) can be replaced by that of Fig. 4.3(b). This is an *autonomous* circuit (no input signals and hence self-governing) and x_1 and x_2 can not determine the value of z, for they do not appear. What value will z have? We can assume a solution to this problem and then verify it. If we assume $z = 0$, then we find $\alpha = 1$ and $z = 0$. The circuit preserves this value of z, if it exists. Similarly, if we assume $z = 1$, then $\alpha = 0$ and $z = 1$. This value is also preserved. Of course, z can not have both values simultaneously. What then determines the actual value of z at a specific instant?

In Fig. 4.3(c), gate delays are introduced. This gives a more realistic model from which we will be able to determine a specific value for z. Suppose that $x_1 = x_2 = 0$ and $z = 0$ in this network, and that these conditions have persisted for a long time so that $\alpha = 1$. Then the input and output signals of the delay elements have the same values, i.e., $z' = z$ and $\alpha' = \alpha$. We say that the circuit is in a *stable state*; no signals are tending to change their values.

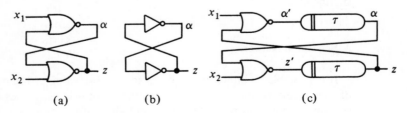

(a)   (b)   (c)

Fig. 4.3   A Sequential Circuit

Now we set $x_1 = 1$. Then $\alpha' = 0$ and τ seconds later $\alpha = 0$. During this interval $\alpha' \neq \alpha$. We know α will eventually change its value so we say that the circuit is "unstable" or in an *unstable state*. When $\alpha = 0$, $z' = 1$ is required; again the circuit is in an unstable state. In τ seconds, $z = 1$, which does not alter the value of α', and the circuit has reached another of its stable states. The circuit will remain in this state as long as we continue to hold $x_1 = 1$, $x_2 = 0$.

When we return x_1 to the value 0 ($x_1 = x_2 = 0$), neither z nor α changes its value. By setting $x_1 = 1$ for a brief time, we cause the circuit to switch from a stable state with $z = 0$ to another stable state with $z = 1$. Similar analysis will show that setting $x_2 = 1$ for a time greater than 2τ seconds causes the circuit to switch back to its $\alpha = 1$, $z = 0$ stable state, and remain in that state when x_2 is returned to 0. We see now what determines the value of z when $x_1 = x_2 = 0$.

A light switch on the wall of a room may be viewed as remembering whether its handle was last pushed up or down. The sequential circuit may be thought of in a similar manner: it "remembers" which input variable currently has or last had the value of 1. If x_1 had the value of 1 after x_2 last had that value, then $z = 1$. When $z = 0$, we know that either $x_2 = 1$, or x_2 had that value more recently than x_1 did. Internal variable $\alpha = \bar{z}$ in these cases, and we will soon show both α and z as output signals and think of them as complements. But one exceptional case must not be overlooked. If $x_1 = x_2 = 1$, then $\alpha = z = 0$ and the circuit is in a very unusual and generally undesirable stable state. We will tend to avoid this state by insisting that x_1 and x_2 never have the value of 1 simultaneously.

But why is the $\alpha = z = 0$ state so undesirable? First, if we use α as the complement of z in combinational logic driven by this sequential circuit, when α has the same value as z that combinational logic can not be expected to produce the correct signals. Second, suppose that after $x_1 = x_2 = 1$, x_1 and x_2 simultaneously take the value of 0. In this case we can not predict the stable state the circuit will ultimately reach. Small variations in the transistors and other components of the circuit will enter, and the circuit may always go to its $z = 0$, $\alpha = 1$ state, or always go to its $z = 1$, $\alpha = 0$ state. However, we can not know this in advance and circuits that appear to be identical may not go to the same stable state. This nondeterministic behavior must be avoided. We want to be able to predict what our circuits will do, and we wish them to do the same thing each time that they encounter the same situation.

We can and do show input- and output-signal sequences with a timing chart such as that of Fig. 4.4. But when we deal with networks involving many gates and signals, timing charts become extremely tedious to prepare and interpret. They allow little flexibility; if a certain situation is not already pictured, a new chart must be prepared. We can parallel the timing diagram with algebraic notation by expressing $z(t)$ as a function of $x_1(t)$, $x_1(t - \tau)$, $x_1(t - 2\tau)$, ..., $x_2(t)$, $x_2(t - \tau)$, ..., but this approach is even less satisfying than the diagram. Expressions become very lengthy and do not reveal input–output relations as easily as a diagram does.

257   **SECTION 4.1**   **Sequential Activity**

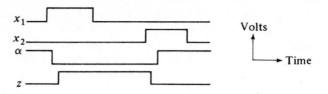

FIG. 4.4   Timing of Events in the Circuit of Fig. 4.3(c)

A Sequential Network Model

We can obtain a more concise notation and an extremely useful concept by examining the question, "What information in addition to the present values of the input variables will enable us to predict future values of the output variables?" In the sequential circuit we have examined, the signals at the output terminals of the delay elements, together with the input signals to the circuit at each instant determine the signals at the input terminals to the delay elements. Hence, the future values of the delay-element output signals are determined. This is more generally true, as we will see, when we consider larger sequential circuits. We formalize this concept by defining the *present state* (often we will say just *state*) of a sequential switching circuit modeled with p delay elements to be the *p-tuple of delay-element output signals*. The *next state* of the circuit is then the *p-tuple of delay-element input signals*.

We may generally view a sequential switching network as shown in Fig. 4.5. The delay elements are detailed, with output signals named $y_1, y_2, \ldots, y_p$ and referred to as *state variables*, and input signals denoted $y_1', y_2', \ldots, y_p'$. The present state of the network is then given by the p-tuple of state variables $(y_1, y_2, \ldots, y_p)$; the next state is $(y_1', y_2', \ldots, y_p')$. Some or all of the output signals, z_i's, may be equivalent to y_j's, but need not be. In this and the next chapters we will come to appreciate the very general nature of this model of a sequential switching circuit.

If we construct a sequential network, the delays introduced will vary from gate to gate. In Fig. 4.5, a different subscript is appended to each delay to emphasize this possibility. If we make no attempt to equalize these delays or to correlate the times at which the input n-tuple and state p-tuple change, the network is said to be operating in an asynchronous fashion. We refer to it as an *asynchronous* sequential switching circuit. Such circuits are the subject of Chapters 12 and 13.

There are advantages to operating a network in an asynchronous fashion, but in general, designing the network to achieve reliable performance is rather difficult. A glimpse at the type of problem encountered when designing asynchronous circuits is provided by Fig. 4.6 in which the combinational circuit of Fig. 4.2 is combined with the sequential circuit of Fig. 4.3. Suppose the input signals vary as they did in Fig. 4.2, and that initially $z = 0$. Then a pulse may appear on line α when x_2 changes its value. It may or may not be of sufficient width to make the sequential

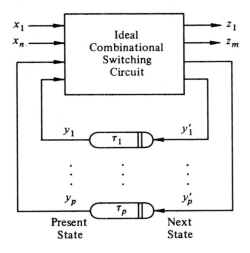

FIG. 4.5 A Very General Model of a
Sequential Network

circuit respond by changing its state to $z = 1$. We can not be certain how the overall network will perform, and, in general, this is an unacceptable type of design.

When we force all delays in a sequential network to be equal, i.e. $\tau_1 = \tau_2 = \ldots = \tau_p$ in Fig. 4.5, and force the input n-tuple and state p-tuple to change at the same time, we operate that network in the *synchronous* mode. As we find practical ways of achieving synchronism, we will find that the values of input and state signals are important only at certain selected instants. Thus pulses of uncertain duration generated by hazardous circuits such as that of Fig. 4.2 will be totally ignored if they occur between these instants. In general, synchronous sequential networks are much more easily designed and made operational than asynchronous networks. For this reason, we will deal almost entirely with the synchronous mode of operation in this and the next chapters.

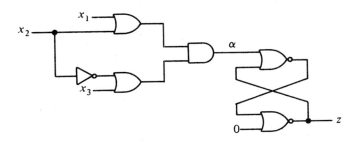

FIG. 4.6 An Asynchronous Network

If we construct a sequential circuit of p gates and associate a delay element with each gate, we view that circuit as having 2^p states. This number increases so rapidly with p that the model becomes unwieldy for circuits even of modest size. For many purposes a far more tractable model, with fewer delay elements than gates, will display the important characteristics of a circuit.

In this section we will be concerned with sequential circuits that resemble the circuit of Fig. 4.3 in that they consist of a number of logic gates that are primarily interconnected to form one loop. We can draw these circuits so that one signal appears as a *feedback* signal. The circuit of Fig. 4.3 is redrawn in Fig. 4.7(a) to illustrate this point. We see two delay elements in this loop. Might we replace the pair of delay elements that are connected in tandem, in a sense, with one delay element, as in Fig. 4.7(b)?

A bit of analysis is required to show that the model of Fig. 4.7(b) predicts activity very similar to that of the model of Fig. 4.7(a), as shown in Fig. 4.4. Comparison with the actions in a real circuit will show both to be very good models. We will, therefore, work with the simpler two-state model of this circuit, and view this and other sequential circuits with two predominant states in the manner of Fig. 4.7(c). Because such circuits possess two states and can be viewed as "remembering" something about the sequence of input signals they receive, we refer to them as *binary memory elements.*

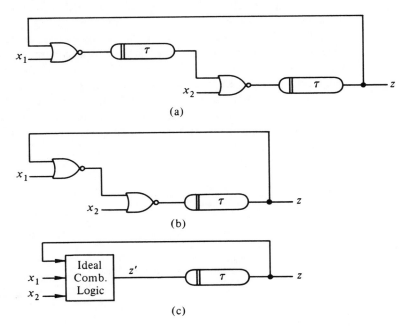

FIG. 4.7 Models of a Binary Memory Element

A comparison of Figs. 4.7(c) and 4.1 emphasizes that, unlike the gate, the output signal of a binary memory element at a time in the future is a function of the present value of that variable (present state of the circuit) as well as of the present values of the input signals.

Latches

A *latch* is a binary memory element whose behavior is combinational for some mixtures of input signal values. The NOR network is a first example. In Fig. 4.7, if $x_1 = 1$ and $x_2 = 0$, then $z = 1$. Input values alone determine the output value. With $x_1 = 0$ and $x_2 = 1$, $z = 0$. While gate delay is present, the present value of z does not influence its next value for these input symbols.

We use tables and equations to describe the logic of binary memory elements, but since time is a variable we do not refer to them as truth tables and Boolean equations. They show how the state of a network changes, and hence are called *state transition tables* and *equations*. Table 4.1 gives the logic of the NOR-latch. Present input and state variable values are listed on the left. The right entries indicate the future value of z. Note that the combinational behavior discussed above is recorded as well as the memory activity when $x_1 = x_2 = 0$ (then $z' = z$). While the $x_1 = x_2 = 1$ input symbol produces $z' = 0$, problems with this input symbol were discussed in the previous section.

We speak of a binary memory element as *set*, or in its *set state*, when its output terminal bears logic 1. It is *reset* when $z = 0$. The x_1 terminal of the NOR-latch is often labelled S for set, since a 1 on that terminal alone forces the latch to its set state. The x_2 terminal is then the R or reset terminal, and the latch is an *RS*-latch with the symbol of Fig. 4.8(a). The state transition table is usually abbreviated to emphasize the memory versus combinational input symbols. Notice in Fig. 4.8(b) that the output value is said to be undefined for the (1, 1) input combination. If the *RS*-latch is realized with NOR gates, we know that $z' = 0$ and also that the

TABLE 4.1. LOGIC OF THE NOR-LATCH

Present Inputs $x_1 x_2$	Present State z	Next State z'	
0 0	0	0	} memory
0 0	1	1	
0 1	0	0	
0 1	1	0	
1 0	0	1	} combinational
1 0	1	1	
1 1	0	0	} warning
1 1	1	0	

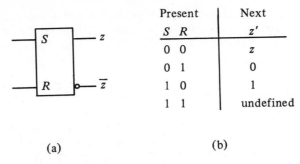

Present	Next
S R	z'
0 0	z
0 1	0
1 0	1
1 1	undefined

(a)   (b)

FIG. 4.8   The RS-Latch

two output signals shown in Fig. 4.8(a) are not complements as the logic symbol suggests. With different technology other results may be obtained in the (1, 1) case; so the "undefined" entry serves as a warning to avoid the input symbol, or be prepared for the consequences.

If two NAND gates are connected in a loop to form a *NAND-latch*, analysis parallel to that of the previous section will reveal the state transition logic of Fig. 4.9(b). Zeros on the input lines cause the setting and resetting action. The latch remembers with the (1, 1) input symbol, and the (0, 0) symbol is to be avoided. A suitable symbol for this latch is shown in Fig. 4.9(c) and it is sometimes called an RS-latch, but the differences from Fig. 4.8 would suggest that this is unwise.

One other type of latch logic is commonly encountered. The network of Fig. 4.10 provides this logic. When the C (control or clock) line bears logic 0, both of the left NAND gates provide 1's to the NAND-latch on the right, and it remembers. With $C = 1$, $\bar{D}$ and D are provided to the input terminals of the NAND-latch, and it sets if $D = 1$ and resets with $D = 0$.

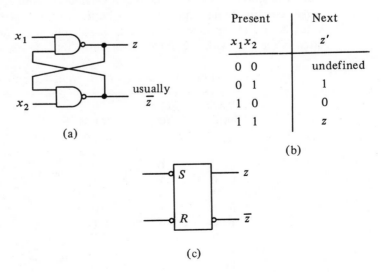

Present	Next
$x_1 x_2$	z'
0 0	undefined
0 1	1
1 0	0
1 1	z

(a)

(b)

(c)

FIG. 4.9   The NAND-Latch

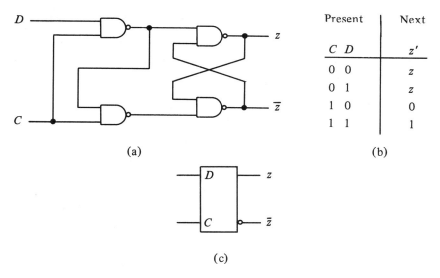

Present		Next
C	D	z'
0	0	z
0	1	z
1	0	0
1	1	1

(a) (b)

(c)

FIG. 4.10   The *D*-Latch (Gated-Latch)

Thus, the data bit on the *D* line is captured by the memory element under control of the *C* line. This is much like the ideal *D* flip-flops that were used in Chapter 1, and this latch is a *D-latch* or *gated-latch*.

We have said little about the magnitudes of the delays in our models except that they are not necessarily equal. To achieve synchronous network operation we must equate the delays of various binary memory elements in the same network. But we usually do this *in effect* and not *in fact*. Suppose we supplement the familiar NOR pair with two AND gates as shown in Fig. 4.11. For the moment we will think of the *clock* signal as a repetition of very short[1] pulses. During the brief duration that the clock signal is at the 1 level, input signals x_1 and x_2 are connected to the input terminals of the NOR-pair asynchronous memory element, and it will respond in the manner that we have already discussed. We thus "sample" the input signals at the "clock times." Most of the time, the clock signal has the value 0. The input signals are not passed to the pair of NOR gates during these intervals, and the NOR pair can not respond to x_1 and x_2 even though these signals have the value 1. Thus the times at which this circuit, this synchronized memory element, may change its state are controlled.

The values of x_1 and x_2 at one clock time determine the state that the synchronized memory element will have at the next clock time. Thus the effective delay of a synchronized memory element is the interval between clock pulses. Input signals and the state of the memory element change between the clock times, but only the values of these variables *at* the clock times are important. We refer to a network of more than one memory

[1] Ideally we can think of the clock pulses as being impulses of zero width. With our discussion of the circuit of Fig. 4.3 particularly in mind, we must have the clock-pulse width equal or exceed the time it takes the NOR-gate pair to change state as described in the previous section.

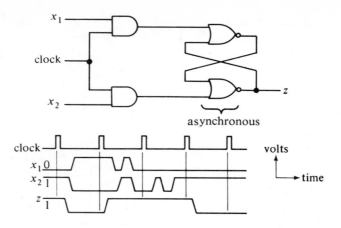

FIG. 4.11   A Synchronized Memory Element

element all of which are synchronized by the same clock signal as a *synchronous* network of synchronous memory elements. If input signals demand that one memory element in a synchronous network change its state at the ith time, other memory elements will not recognize that change until the $(i + 1)$th clock time. The effective delay of the memory element thus equals the clock interval and is controlled, even though its actual internal delays are much shorter, and of unknown (to an extent) magnitude.

Flip-Flops

A *flip-flop* is a binary memory element that exhibits only sequential activity. Even though it is synchronized by a clock signal, the memory element of Fig. 4.11 is not a flip-flop under this strict definition. When that clock $= 1$, the logic input signals influence the value of output signal z in a combinational fashion, and it is therefore a latch according to the definitions of this chapter. Not all authors and logic designers define "flip-flop" as carefully as we have, and the memory elements that we call latches are referred to as flip-flops in many publications.

To achieve the pure sequential activity demanded by the definition above, flip-flop networks examine input signal values at one point in time, store intermediate results internally, and then change output signal values at a later time. Gate networks for attaining this behavior will be examined in the next section. For now we accept that the intervals during which input signals are examined and output signal values are changed can be separated by one (or both) edge of a clock pulse. A 0-to-1 (1-to-0) transition of a clock signal, rather than the logic level of the signal, is the significant characteristic. This is denoted with a *wedge* within a logic symbol. This "dynamic indicator" symbol was first used in Prob. 1.2–9. We will find it at the clock input terminal of flip-flop symbols that are identical to latch symbols in other respects. Algebraic and tabular notation for flip-flops is also identical to that for latches, but where "present" and "next" are separated in time by gate delays for latches, they are separated by clock periods for flip-flops. In the synchronized case, time is made discrete by the clock signal.

RSFF

We have seen the *RS*-latch. The *RS flip-flop* has very similar logic. If at a given clock time $S = 1$ and $R = 0$, the flip-flop will go to its set state if it is not already there. A "reset input," 1 on the reset (lower) input terminal at a clock time, causes the flip-flop to assume its reset state, or "resets the flip-flop."

The logical behavior of this unit is summarized more completely in Table 4.2. The clock signal is not shown directly as an input signal in state transition tables. It provides the time variable and separates "present" from "next." Note that the response of the RSFF to simultaneous S and R inputs is presumably unknown or indeterminate. (See Prob. 4.2-5.) With additional components, "set-dominate" and "reset-dominate" flip-flops can be constructed that respond in a definite manner to simultaneous S and R signals.

From the state-transition table of Table 4.2 or the equivalent *state-transition map* of Fig. 4.12, we may write the state-transition equation for the RSFF,

$$z' = S \vee z\overline{R} \tag{4.2}$$

subject to the restriction that

$$R \cdot S = 0 \tag{4.3}$$

Alternatively, we can describe the behavior of this memory element by

$$\overline{z}' = \overline{S}(R \vee \overline{z}) \tag{4.4}$$

Equation (4.2) says that an RSFF can be placed in its set state by $S = 1$ at a clock time. It remains set, even if additional set signals ($S = 1$) follow the first, until $R = 1$ at a clock time. Equation (4.4) indicates that a reset RSFF remains reset as long as a set signal is not logic 1. Thus, as we have previously seen, this FF type indicates by its state whether an S or R input signal was most recently applied.

TABLE 4.2. LOGIC OF THE RSFF

Present Inputs S	R	Next State z'
0	0	z
0	1	0
1	0	1
1	1	undefined

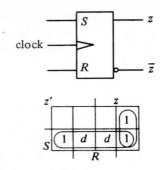

FIG. 4.12   Karnaugh Map Minimization of RSFF Logic

TFF

The *TRIGGER* flip-flop, TFF, is a single-input memory element with the state transitions of Table 4.3. This element reverses its state at each clock time that the trigger input signal has the value of 1.

$$z' = \overline{T} \cdot z \vee T \cdot \overline{z} \qquad (4.5)$$
$$= z \oplus T$$

Although special circuits may be built to accomplish the action of the TFF, such an element can be constructed of logic gates and an RSFF. In addition to Eqs. (4.2) and (4.5), the new requirements to be satisfied are

$$z' = S \vee z \cdot \overline{R} = z \oplus T \qquad (4.6)$$
$$R \cdot S = 0$$

Solution of these simultaneous equations for S and R is required. Table 4.4 is most helpful: with z and T as independent variables, list $z \oplus T$; Eq. (4.6) requires that this be equivalent to $S \vee z\overline{R}$. Values of S and R that satisfy this table must now be found.

In the first row $z' = 0$. Hence $S = 0$, and since $z = 0$, R may take either value, 0 or 1. Thus a don't care is entered into the table. In the second row S may take either value but R must equal 0 since we wish the flip-flop to remain set. If $S = 1$, then $R = 0$ to satisfy $R \cdot S = 0$. In the third row $z' = 1$. Since $z = 0$, the second term of $S \vee z\overline{R}$ cannot take the value 1. Hence, $S = 1$, and R must be assigned the value of 0 to satisfy $R \cdot S = 0$.

TABLE 4.3. STATE TRANSITIONS OF THE TFF

Present input T	Next state z'
0	z
1	$\overline{z}$

TABLE 4.4 REALIZING A TFF FROM AN RSFF

T	z	$z' = z \oplus T$ $= S \vee z\bar{R}$	S	R
0	0	0	0	d
0	1	1	d	0
1	0	1	1	0
1	1	0	0	1

From this truth table, we can write minimum expressions for S and R.

$$S = \bar{z}T \qquad R = zT$$

Thus a TFF may be constructed as shown in Fig. 4.13.

JKFF

The *JK flip-flop* is probably the most widely used type of flip-flop. Its logical behavior is similar to that of the RSFF and the TFF. The two input signals, J and K, are allowed to occur simultaneously, and when they do, the effect is the same as that of a T input; i.e., the FF changes state. When either J or K alone equals 1, the effect is that of an S or R input, respectively. Table 4.5 summarizes this activity.

$$z' = J\bar{z} \vee \bar{K}z \tag{4.7}$$

TABLE 4.5. STATE TRANSITIONS OF THE JKFF

J	K	z'
0	0	z
0	1	0
1	0	1
1	1	$\bar{z}$

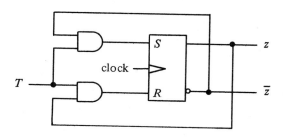

FIG. 4.13   Realization of a Synchronous TFF from an RSFF

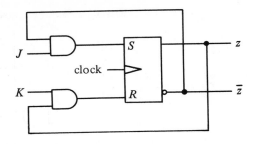

FIG. 4.14 A Realization of the Synchronous JKFF

When we construct a JKFF from a synchronous RSFF using the technique exhibited in the development of the TFF, Fig. 4.13, the circuit of Fig. 4.14 is produced. Note the similarity between this circuit and that of a TFF in Fig. 4.13. If we tie the J and K inputs together electrically, a TFF results. On the other hand, if we never allow $J = K = 1$, we satisfy Eq. (4.3) and the JKFF acts as an RSFF. The JKFF is thus very versatile: it can be used without modification in place of a TFF or RSFF, and is more flexible than either.

DFF

The *DELAY flip-flop*, DFF, is described by

$$z' = D \qquad\qquad \textbf{(4.8)}$$

where D is the single input signal. This type of flip-flop is also easily constructed from an RSFF or JKFF, as in Fig. 4.15.

GLFF

The *Gated Latch*, GLFF, assumes a state equal to the value of the L input signal when the gate input G is present. As long as $G = 0$, the FF does not change its state. Table 4.6 summarizes this action more precisely.

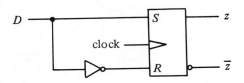

FIG. 4.15 A Realization of the DFF

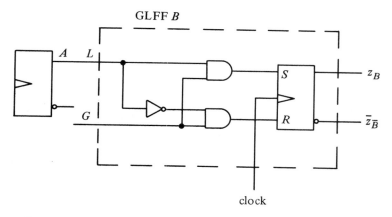

FIG. 4.16   Information Transfer Using the Gated Latch Flip-Flop

$$z' = \bar{G}z \vee GL \tag{4.9}$$

TABLE 4.6. STATE TRANSITIONS OF THE GATED LATCH

G	L	z'
0	x	z
1	0	0
1	1	1

This FF type is especially useful when information must be transferred from one memory element to another. In Fig. 4.16 the content of (0 or 1 stored in) flip-flop A is transferred to B each time the gate signal G is present (equal to 1). This transfer of a bit of information can be accomplished if other flip-flop types are used, but additional logic (as shown) that forms the GLFF from these other types is then required.

Asynchronous Input Terminals

Most flip-flops provided in integrated circuit form have additional input terminals that connect directly to those gates forming latches within flip-flops. These terminals then provide a means of asynchronously setting and resetting the flip-flop, and such action overrides commands provided via the synchronized input terminals (R, S, J, K, etc.). With TTL flip-flops, 0's on the *direct-set* or *preset* and *direct-reset* or *clear* terminals cause the suggested action. This suggests that NAND-latches are used within TTL flip-flops. The symbolism of Fig. 4.17 applies to such flip-flops. The small circles indicate that logic 0, rather than 1, causes the setting or clearing action, and not that an inverter exists within the flip-flop.

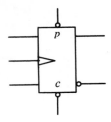

FIG. 4.17 TTL Flip-Flops Respond Asynchronously to 0's on the Preset and
Clear Terminals

4.3 SYNCHRONIZATION OF FLIP-FLOPS

The totally sequential behavior, which is demanded by the definition of
a flip-flop, can be achieved in a variety of ways; only two of which are
widely found in integrated circuit flip-flops. Some of the other techniques
are used when flip-flops are built of discrete components.

Master-Slave Flip-Flops

The *master–slave* variety of flip-flop consists of two synchronized latches
and a means of providing them with complementary clock signals. An
inverter is used to produce the complementary clock signal in Fig. 4.18(a).
In the flip-flop of Fig. 4.18(b), which includes the feedback paths of Fig.
4.14 as required to form a JKFF from an RSFF, the mechanism of presenting the complementary clock signal to the second latch is not as
obvious.

The left latch is called the *master*; its state is determined by the input
signal values [and the state of the second latch in the JKFF of Fig. 4.18(b)].
The right latch takes the state of the left latch; this then is the *slave* latch.
Thresholds of the gates and inverters are adjusted so that actions take place
at four points in a clock cycle, as shown in Fig. 4.19. With the clock at 0,
input signals are disconnected from the master latch, and the slave latch is
connected to the master and takes its state. As the clock signal begins to
rise, the slave is disconnected from the master. Later as the clock approaches
the logic 1 level, the input signals are connected to the master latch via its
sampling gates. The input terminals remain connected until the clock signal
begins to drop. When the clock falls to the lower threshold level, the slave
is again connected to the master. Note that when the master is responding
to the input signals, the slave is disconnected and holding the original state
of the master. When the slave is connected to the master, the master is disconnected from the input signals. At no time can the input signals simply
pass through gates to reach the output terminals. Combinational activity
is excluded.

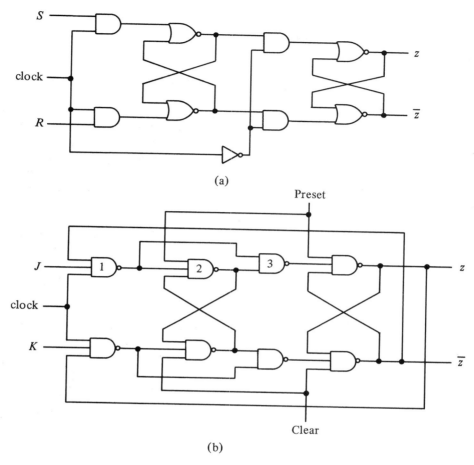

(a)

(b)

FIG. 4.18   Two Master–Slave, Flip-Flop Designs

Since two latches appear in the master–slave flip-flop, one might think of it as a four-state network. All four states are entered in normal operation. When the clock is low, the master and slave latches have the same state; the network is in either state (master, slave) = (0, 0) or (1, 1). When the clock is high, the master and slave may be in different states; the network may be in any of its four possible states. Since only the z and $\bar{z}$ terminals are available for observation in IC flip-flops, the state of the flip-flop is that of the slave latch, and we normally think of the master–slave flip-flop as a binary memory element.

The flip-flops of Fig. 4.18 change their output signal value after the clock falls to 0. Master–slave FF's are also constructed that change their state value after the rise (0 to 1 transition) of the clock. Figure 4.20 shows the symbolism used to distinguish these two synchronization methods.

Flip-flops do not behave per the ideal fashions of the previous section's tables unless restrictions are placed on their use. Two time intervals are

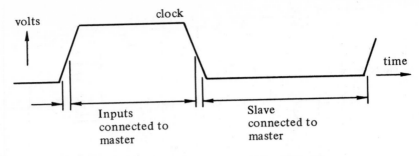

FIG. 4.19   Activity in the Master–Slave Flip-Flop

usually specified for flip-flops. The *set-up* time measures an interval in advance of the point in the clock waveform after which the output signal may change value. For proper operation input signals must not change value during the set-up interval. A *hold* time measures an interval after the critical point in the clock waveform during which the input signals must not change. Often the hold time is zero; for some flip-flops a negative hold time is specified.

IC manufacturers reveal their conservatism when they specify that the set-up time is equal to the clock pulse width for master–slave flip-flops. This is a very safe and recommended approach, but with the design of Fig. 4.18(a), we see that the master-latch will respond to changes in S and R that arrive after the clock is at 1 (but several gate delays before the clock returns to 0). The design of Fig. 4.18(b) yields more complex activity. Suppose the flip-flop is reset ($z = 0$) and $J = 1$ when the clock signal goes to 1. Gate 1 produces a 0, which sets the master-latch and goes beyond it to disconnect the slave from the master (gate 3). Now while clock = 1, if J drops to 0, then gate 1 produces a 1, which does not affect the master-latch but does cause gate 3 to produce a 0. This zero sets the slave latch. Output signals z and $\bar{z}$ change value, but not in synchronism with the clock. A drop in J with clock = 1 causes the flip-flop to set asynchronously. A drop in K will cause it to reset asynchronously. This is called *edge-triggering* the flip-flop, and while it is sometimes useful it does violate our definition of a flip-flop. To get the JK logic of the previous section, the J and K signals must not drop during the clock pulse.

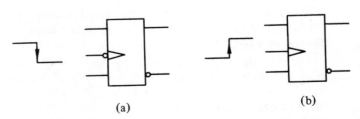

(a)   (b)

FIG. 4.20   The Master–Slave Flip-Flop Changes Its State Variable Value After the (a) Fall and (b) Rise of the Clock Signal

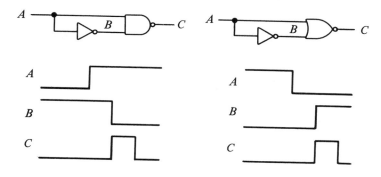

FIG. 4.21   Edge-Detection Networks

Waveform edges can be detected with the circuits of Fig. 4.21, where the inverter delay determines the width of the pulses produced. Differences in rise and fall delays will lengthen or shorten the pulses. If they are shortened sufficiently, inertia may wipe them out entirely. If circuit parameters are carefully controlled, such circuits may be used to synchronize latches such as the one shown in Fig. 4.11. However, this does not produce a reliable flip-flop for it does not truly disconnect the input lines while the state variable is changing value. In the master–slave flip-flop with *data-lockout*, a circuit of Fig. 4.21 provides a pulse to the master-latch on one edge of the clock waveform. The slave is connected to the master on the other clock transition. The input signals are sampled only at one edge rather than throughout the entire clock pulse. The set-up interval of the master–slave flip-flop with data-lockout is centered on the sampling edge and is much smaller than for a simple master–slave flip-flop.

Edge-Triggered Flip Flops

"Edge-triggering" is the second major mode of synchronizing IC flip-flops. As the clock rises (falls), the input signals are sampled, and the next state is computed and placed in storage. Then the input signals are released and the output variable value is made to correspond to that next state. Time intervals between these actions are measured in gate delays rather than clock pulse widths.

Modern edge-triggered flip-flops do not consist of an edge-detector followed by a latch. While we must wait until Chapter 12 for full analysis tools, we can now see a number of latches in the DFF of Fig. 4.22. These latches provide the temporary storage of the next state that is required. Unit delay simulation can give us some idea of how these latches enter into the edge sensitivity of this network. First, with (clock) $C = 0$, internal signals β and γ are fixed at logic 1, and the latch providing the state signal holds its previous state. Internal signals α and σ take complementary values and input signal D determines what those values are.

C	D	$\alpha\beta\gamma\sigma$
0	0	0111
0	1	1110

Suppose that D changes from 0 to 1 while $C = 0$. Then σ changes first, followed by α.

Time	C	D	$\alpha\beta\gamma\sigma$
i	0	0	0111
$i+1$	0	1	0111
$i+2$	0	1	0110
$i+3$	0	1	1110

Two gate delays elapse before the circuit reaches the other of the two stable states with $C = 0$. Such changes must take place in advance of a change in C from 0 to 1. The maximum time required to make these internal changes establishes the set-up time of the flip-flop (the time in advance of a 0 to 1 change of C during which D must not be changing).

Now let us simulate the response of a change in C, assuming that $z = 0$ (FF is reset) before the change.

Time	C	D	$\alpha\beta\gamma\sigma$	z	$\bar{z}$
$i+3$	0	1	1110	0	1
$i+4$	1	1	1110	0	1
$i+5$	1	1	1010	0	1
$i+6$	1	1	1010	1	1
$i+7$	1	1	1010	1	0

Because $D = 1$ before the change in C, $\alpha = 1$ at the change; β rather than γ drops to 0, which ultimately sets the output latch ($z = 1$). Three gate delays elapse before the circuit stabilizes. D can return to logic 0 as early as time $i + 5$ without interfering with the setting of the flip-flop. Hold time is therefore approximately one gate delay. With $CD = 10$, σ takes the value of 1, but the upper latch firmly remembers that D had the value of 1 when C changed to 1.

When C returns to 0, $\beta = \gamma = 1$, and the output latch is in effect disconnected from the rest of the circuit. Where C goes to 1 with $D = 0$, we can expect γ to change value.

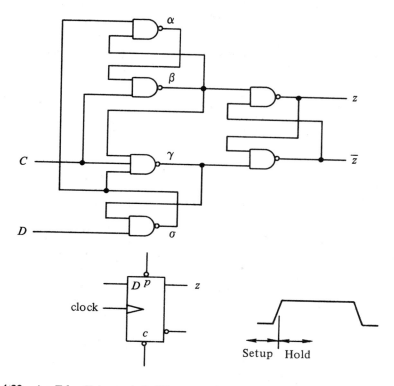

FIG. 4.22 An Edge-Triggered D Flip-Flop. Preset and clear connections are not shown in the logic diagram.

time	C	D	$\alpha\beta\gamma\sigma$	z	$\bar{z}$
$i+20$	0	0	0111	1	0
$i+21$	1	0	0111	1	0
$i+22$	1	0	0101	1	0
$i+23$	1	0	0101	1	1
$i+24$	1	0	0101	0	1

Signal γ must go to 0 and remain there for a sufficient length of time for the output latch to reset. D changing to 1 as early as $i+22$ does not force the lower input latch to forget that D equaled 0 when the clock changed.

We see that this edge-trigger design really samples the D input signal immediately before the sampling edge and promptly ignores D after that edge, as the output latch changes the state variable value. Set-up and hold restrictions must be satisfied. If they are not, then very erratic and unexpected behavior has been demonstrated for circuits like this one [4].

Figure 4.23 summarizes the set-up intervals of flip-flops and the edge, following which the state variable may change value. The wedge identifies

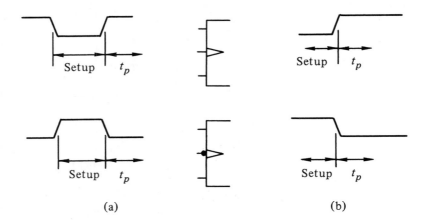

FIG. 4.23 The Setup Intervals of (a) Master–Slave and (b) Edge-Trigger Flip-Flops, and the Edge After Which the State Variable Changes

the clock terminal. The circle indicates that the state variable changes following the 1 to 0 transition of the clock. Propagation delay t_p appears here as it does in gates. In the ideal D flip-flops used in earlier chapters, all of these practical details were avoided by using clock impulses.

Figure 4.23 suggests a uniform way of thinking about flip-flop timing. Given the edge after which the state variable changes, both master–slave and edge-triggered flip-flops require that the input signals do not change during the preceding set-up interval. They both exhibit propagation delay after that edge. If we show the sensitive edge as an impulse, all of the waveforms of Fig. 4.23 are summarized as shown below.

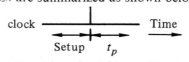

While the set-up time is fixed for edge-triggered flip-flops, only a minimum value is specified for master–slave flip-flops.

4.4 ACTIVITY IN SYNCHRONOUS SEQUENTIAL NETWORKS

Synchronous sequential networks of flip-flops and gates take the form of Fig. 4.24. A comparison of this figure and Fig. 4.5 shows that the flip-flops serve as delay elements. The combinational portion of the network is composed of actual gates that offer delay; it combines the output signals of the flip-flops with primary input signals from outside the network to form: (1) input signals to the flip-flops and (2) the signals that leave the network. We assume that the primary input signals are generated by another sequential network that is synchronized by the same clock, and that they therefore change value at the same time as the state of the network under examination. We also assume that all flip-flops are synchronized by the same clock and sensitive to the same edge of the clock waveform. To establish the

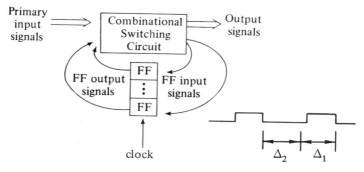

FIG. 4.24 Synchronous Sequential Networks

basis for mathematical models of such networks, we now examine in detail activity within a network of the form of Fig. 4.24.

Let us assume that all FFs are sensitive to the 1-to-0 transition of the clock. Clock period is given by $\Delta_1 + \Delta_2$. Δ_1 measures the longest set-up time of the flip-flops in the network. But now, how long must the Δ_2 interval be? During this interval the FFs must change state, and the new FF output signals and new values of the primary input signals must propagate through the combinational circuitry to the FF input terminals. For reliable operation we must wait until all transient variations on the FF input lines have disappeared before we have the FFs examine these signals. If Δ_f measures the propagation time of the FFs, and we assume that each gate in the combinational circuit introduces Δ_g seconds of delay and the combinational circuit has N or fewer gates between each of its input and output terminals, then for reliable operation we must select Δ_2 to satisfy

$$\Delta_f + N \cdot \Delta_g < \Delta_2 \tag{4.10}$$

Delays Δ_f and Δ_g are determined by the electrical characteristics of the logic circuits to be used. We may not have much control over these quantities. We can control N, however, and thereby influence Δ_2. If a short clock period (high clock frequency) is selected to achieve a high-speed network, then N may be greatly limited, and we must design combinational circuits with a low number of logic levels.

If FF input signals are to be invariant during the set-up interval, the primary input signals must change value when FF output signals are changing value. This requirement is satisfied when the primary input signals are provided by FFs of other sequential networks that are synchronized by the same clock signal as the network under consideration. Such a structure is so common that we will assume it to be the case in most examples.

In summary, we see that activity in a synchronous network is cyclic.

1. Clock goes to 1: (a) FF input signals are examined. (b) The next states of the FFs are stored internally. FF input signals must not change during this interval.

2. Clock goes to 0: (a) FFs assume new states. Primary input signals

change their values. (b) New input signals propagate through the combinational network, which computes new values of the output variables and FF input signals.

If we select Δ_2, or design combinational circuits, so that Eq. (4.10) is satisfied, then transients may arise on the output and FF input lines as a result of hazards in the combinational circuitry, but they will disappear before the FFs sample the signals on these lines. Given sufficient time actual combinational circuits provide the signal values that Boolean algebra predicts and ideal combinational logic circuits provide. We can treat the combinational logic portion of a sequential network as ideal. This then is the primary reason for the widespread usage of the synchronous mode of operating sequential logic.[2]

Let us look more closely at this cycle of activity. Fig. 4.25 shows a portion of a synchronous sequential network, and waveforms of activity that could occur in the circuitry. We assume that Eq. (4.10) is satisfied: all signals generated by combinational circuitry will reach the value predicted by the Boolean equations that model the circuitry before these signals are sampled by the flip-flops. Only the generation of signal S_1 is detailed.

$$S_1 = \bar{y}_2 \downarrow J = \bar{y}_2 \downarrow (x_1 \downarrow (x_2 \downarrow y_1))$$
$$= y_2(x_1 \vee \bar{x}_2 \bar{y}_1)$$

Suppose that we stand in the set-up interval before an imminent clock edge. We will refer to this as the "present" time. In Fig. 4.25 we see that the input signals x_1 and x_2 are both at the 1 level at the present time. We will refer to this 2-tuple of input signals $(x_1, x_2) = (1, 1)$ as the *present input symbol* of the network.

Only two flip-flops are shown in Fig. 4.25. While the complete combinational network is not shown, we will take these to be the only FFs of the network and refer to the collection of FF states at the present time as the *present state* of the network. In Fig. 4.25 $(y_1, y_2) = (1, 0)$ is illustrated. The value of this 2-tuple during the next set-up interval will be called the *next state* of the network. Of course, as time advances so does our interest, and that which we now call "next" will be "present" after one clock period has elapsed.

Two output signals, z_1 and z_2, are shown in Fig. 4.25. The 2-tuple $(z_1, z_2) = (1, 1)$ constitutes the *present output symbol* of the network. Note that the elements of the present output symbol may be expressed as Boolean functions of the input and flip-flop state variables.

$$z_1 = y_2 \downarrow J = y_2 \downarrow (x_1 \downarrow (x_2 \downarrow y_1))$$
$$= \bar{y}_2(x_1 \vee \bar{x}_2 \bar{y}_1)$$
$$z_2 = y_1$$

The present input symbol and present network state uniquely determine the values of: (1) the FF input signals and (2) the present output symbol.

---

[2] One might say that in a truly synchronous network the logic gates are clocked as well as the flip-flops. Then transients will be entirely avoided. Such systems have been used in the past and may be used in the future as new technologies appear.

Immediately before a clock edge all of these signals have settled to their final values if the clock signal satisfies Eq. (4.10).

During the set-up interval, the FF input signals are sampled. ($R_1 = J_2 = 1$ must be assumed in Fig. 4.25.) Next states are stored within the FFs. None of the signals shown in Fig. 4.25 change their values. Then the clock edge occurs. Now the FFs switch their state. They need not, and in an actual circuit will not, change state at precisely the same time and rate. But this is of no concern in a synchronous network if the clock period is sufficiently long to satisfy Eq. (4.10). Input signals x_1 and x_2 may be changing their values at this time also. These new signal levels begin to propagate through the combinational circuitry. Some reach the output terminals before others. Signal S_1 is shown to change its value three times before the ultimate value dictated by the Boolean equation is reached.

And the cycle repeats. As each clock time becomes the present time, the

(1) PRESENT input symbol and (2) PRESENT network state
uniquely determine the
(1) PRESENT output symbol and (2) NEXT network state.

A great deal of effort is required to generate detailed timing charts, such as the one shown in Fig. 4.25. And we now see that it is not necessary to

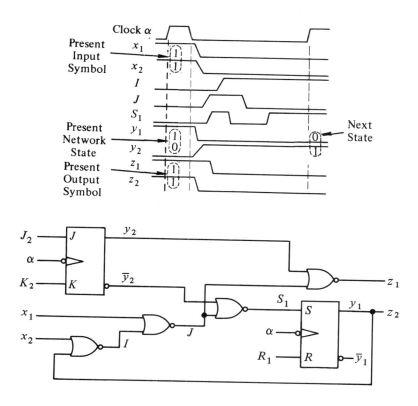

FIG. 4.25   A Portion of Synchronous Sequential Network

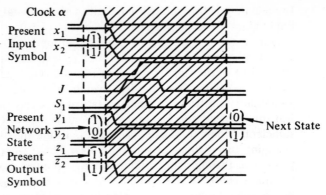

FIG. 4.26 Discrete View of Fig. 4.25

generate timing charts for synchronous sequential networks. If all transients generated by delays in the combinational circuits are ignored by the flip-flops, it is not necessary to determine whether or not these transients will even appear. We can treat the combinational circuitry as ideal and model it with Boolean equations. This is why we have been so free to do so in previous chapters. We can ignore the values of signals at all times except immediately before (during the set-up interval) the sensitive clock edge, as Fig. 4.26 attempts to emphasize. We can treat time as a discrete variable. State and output variables may reach their final values well before the discrete time at which we choose to examine them. We can not be sure that they will and usually do not attempt to determine if they actually do. We know that they will have their final values when they are sampled. In the next chapter we will formalize and generalize these definitions and ideas.

4.5 SOURCES AND APPLICATIONS OF DELAY

A variety of physical phenomena give rise to what we have called "delay" in electronic circuits. A corresponding variety of models and methods of device and circuit analysis are available in texts[3] on electronics and electromagnetic theory. Storage of energy or electrical charge is a part of these phenomena; usually the amount of energy or charge stored can not be changed instantaneously and as a result some time is required to increase or decrease the amount of stored energy or charge. We will consider only the capacitor as a storage device because it is almost universally the most important storage element in logic circuits.

Consider a source of electrical signals supplying energy to a receiving circuit or load. The output impedance of the source and the input impedance of the load often contain a capacitive component. In addition, the interconnecting wires introduce additional stray capacitance, and a discrete capacitor may be placed across such wires. We will summarize all

---

[3] The reader is referred to any of many excellent texts on electronics for a more detailed analysis of these circuits and a treatment of design techniques [5, 7].

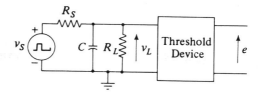

FIG 4.27 Delay Circuit

of these capacitive components with a single capacitor C in Fig. 4.27. The signal source may then be modeled as an ideal voltage generator in series with a resistance R_s, which models the resistive component of the output impedance of the source. The load is taken to be resistor R_L in Fig. 4.27. The box labeled "threshold device" is to be ignored for the moment.

Now suppose that the source voltage v_s is held at 0 volts for a long time so that capacitor C stores no energy, and then v_s is instantaneously increased to $+V$ volts. Voltage v_L will increase with time according to:

$$v_L(t) = \frac{R_L V}{R_s + R_L}\left(1 - \exp\left[-t/C\left(\frac{R_s R_L}{R_s + R_L}\right)\right]\right) = V_f[1 - \exp(-t/\tau)] \quad \textbf{(4.11)}$$

Thus v_L increases exponentially toward a final value of $V_f = VR_L/(R_s + R_L)$ with a *time constant* of $\tau = C(R_s R_L/(R_s + R_L))$. If τ is very small, v_L changes rapidly and the energy stored by the capacitor changes rapidly, but an infinite amount of time is still required for v_L to reach its final value. That final value is a fraction of V, the size of the source voltage increase. If we wait until $v_L = V_f$, then we experience an infinite delay.

The threshold device of Fig. 4.27 is assumed to have no energy storage devices and to produce an output signal $e = 0$ volts whenever $v_L > V_T$ and $e = E$ volts whenever $v_L < V_T$ where V_T is the *threshold level* of the device. Thus the threshold device is an idealized inverter. It produces a low output voltage in response to a high input voltage, and a high output-voltage level in response to a low input level.

Solving Eq. (4.11) for t gives:

$$t = \tau \ln\left(\frac{V_f}{V_f - v_L}\right) \quad \textbf{(4.12)}$$

Thus v_L increases exponentially from 0 volts and reaches a value of V_T in the range $0 < V_T < V_f$

$$t_T = \tau \ln\left(\frac{V_f}{V_f - V_T}\right) \quad \textbf{(4.13)}$$

seconds after v_s changes its value from 0 to V volts. Both v_s and e are binary signals like those of logic circuits. But e goes to its low value t_T seconds after v_s changes so we view this circuit as introducing t_T seconds of delay.

The magnitude of the delay introduced by this circuit is determined primarily by the magnitudes of R_s, R_L, C, and V_T. Often R_s, C, or V_T can be reduced, to reduce t_T. While reducing R_L also reduces the time constant τ of the circuit, it also reduces V_f and hence the range of V_T, and is not usually a desirable approach to decreasing the delay of the circuit. Alternatively, increasing R_s or C increases t_T and can be used to increase the delay introduced by the circuit. In those instances where delay must be increased, a discrete capacitor is often placed in parallel with existing stray capacitance, increasing greatly the value of C and hence the time constant of the circuit.

If v_s is held at V volts for a long time, say more than 5τ units of time, and is then instantaneously reduced to 0 volts, $v_L(t)$ decays exponentially from a value of V_f to 0 volts,

$$v_L(t) = V_f\, e^{-t/\tau} \qquad\qquad (4.14)$$

where V_f and τ are as previously defined, and the time required for v_L to reach the value of V_T is

$$t_{T}' = \tau\, \ln(V_f/V_T) \text{ seconds} \qquad\qquad (4.15)$$

The value of t_T' will differ from that of t_T unless

$$\frac{V_f}{V_f - V_T} = \frac{V_f}{V_T} \quad \text{or} \quad V_T = V_f/2 \qquad\qquad (4.16)$$

Thus if the circuit is to provide the same delay for 0 to V volt and V to 0 volt changes of v_s, the threshold voltage should lie near the middle of the range of v_L. If V_T satisfies Eq. (4.16), then $t_T = t_T' = 0.69\tau$.

Suppose now that $V_T = V_f/2$ and that v_s goes to V volts for a short time, say 0.5τ seconds, and then returns to 0 volts. Signal v_L will begin to increase toward V_f, but will only reach a value of $V_f(1 - e^{-.5}) = 0.312V_f$ before v_s returns to 0 volts, forcing v_L to decrease as shown in Fig. 4.28. Note that v_L never exceeds V_T, and as a result the threshold device will not switch. This then is an inertial delay element. The brief transient in v_s is not reproduced in signal e in Fig. 4.28.

In logic networks one gate serves as the signal source and another is the load and threshold device. The transistor is the central component; diodes

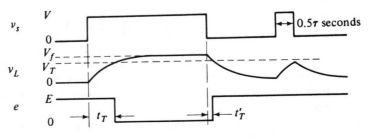

Fig. 4.28 Waveforms for the Circuit of Fig. 4.27

and resistors may be used to couple signals to the transistor while providing the gate with a more desirable threshold voltage than that provided by the lone transistor. All of the gate circuits of Section 2.1 may be reviewed from this viewpoint, but studying the common-emitter amplifier of Fig. 2.10 will show the points to be made.

First let us regard the CE amplifier of Fig. 4.29 as a load and threshold detector. As long as v_s is less than the base-to-emitter voltage, V_{BET}, at which the transistor begins to conduct significantly, the transistor is cut-off and presents a high resistance between its base and emitter. As a load, this amplifier presents a very high resistance. As a detector, it firmly decides that a low input voltage is being presented and it indicates this decision via a high output signal level. If v_s is increased above V_{BET}, the transistor presents a substantially lower resistance between its base and emitter; v_{BE} does not vary substantially. The difference between the threshold value and saturation value V_{BES} is usually less than 0.2 volts. Base current i_B varies significantly and is given by

$$i_B = \frac{v_s - V_{BET}}{R_s} \quad \text{for} \quad v_s > V_{BET} \tag{4.17}$$

The transistor saturates when the base current reaches the level

$$I_{BS} = \frac{I_{CS}}{\beta} = \frac{(V_{CC} - V_{CES})}{\beta R_C} \tag{4.18}$$

with source voltage

$$V_{sS} = I_{BS} R_S + V_{BET}$$

$$= \frac{R_S}{\beta R_C}(V_{CC} - V_{CES}) + V_{BET} \tag{4.19}$$

Still higher input voltages cause i_B to exceed I_{BS} and drive the transistor deep into saturation.

As a load with $v_s > V_{sS}$, the CE amplifier presents a resistance that is less than a few hundred ohms. As a detector it indicates the presence of a high input signal level via a low output level V_{CES}. For intermediate input levels, $V_{BET} < v_s < V_{sS}$, the input resistance is intermediate. The circuit amplifies with gain $\beta R_C/R_S$ and presents intermediate voltage levels at the output terminal; that is, $V_{CES} < v_L < V_{CC}$. In short, the detector cannot decide. We can reduce the region of indecision by making the gain large [look at Eq. (4.19) for this point of view], but we cannot reduce the range to a single point so as to achieve an ideal threshold detector.

As a source, the common-emitter amplifier has a higher resistance of R_C ohms when the transistor is cut-off, and a very low resistance when the transistor is saturated. To be more specific, let us review the response of a CE amplifier that drives a capacitive load, as suggested in Fig. 4.29. If the transistor has been saturated for some time so that $v_L = V_{CES}$, and then v_s drops to turn the transistor off, v_L rises exponentially toward V_{CC} with time constant $\tau = R_C C$. Equation (4.11) describes this rise and Fig. 4.29(b) illustrates it. When v_s increases so that base current $i_B > I_{BS}$,

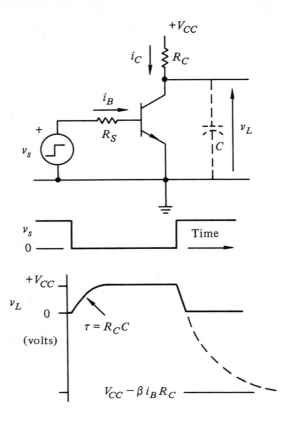

FIG. 4.29   The Common-Emitter Amplifier

the transistor turns on but does not saturate immediately. Capacitance C prohibits its collector-to-emitter voltage from dropping instantly. The transistor acts as a βi_B current generator between its collector and emitter. The capacitor is discharged toward $V_{CC} - \beta i_B R_C$ volts with the same time constant. After

$$t = \tau \ln\left(\frac{V_{CC} - \beta i_B R_C - V_{CC}}{V_{CC} - \beta i_B R_C}\right)$$

$$= \tau \ln\left(\frac{\beta i_B R_C}{\beta i_B R_C - V_{CC}}\right) \tag{4.20}$$

seconds, $v_L = 0$ and the transistor saturates, terminating the exponential decay. The rise and fall times of the waveform provided by the CE amplifier differ substantially. If equal delays in reaching threshold voltage V_T are to be observed, then

$$V_T = \frac{V_{CC}}{1 + \beta i_B R_C / V_{CC}} \tag{4.21}$$

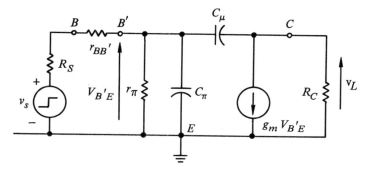

FIG. 4.30   A High Frequency Equivalent Circuit for the CE Amplifier

If we just bring the transistor to saturation, then $V_T = V_{CC}/2$. Usually we drive the transistor deep into saturation, and a low value of V_T must be used or the rise and fall delays will not be equal.

Load capacitance is not the only source of delay in the CE amplifier; so let us ignore it now. The active transistor is often modeled at high frequencies with the equivalent circuit of Fig. 4.30. Resistance $r_{BB'}$ represents the 50 to 100 ohms provided by the material of the base region. The capacitors represent junction capacitance, which does vary with the dc operating point of the transistor as does the base-to-emitter resistance r_π. Transconductance g_m is related to r_π and the relatively constant short circuit current gain β of the transistor by

$$\beta = g_m r_\pi \tag{4.22}$$

and to the total collector current by

$$g_m = 40 I_C \tag{4.23}$$

at room temperature. Resistance r_π and the junction capacitances cause β to drop with increasing frequency; at the beta cut-off frequency f_β, β is reduced from its low frequency value β_0 to $\beta_0/\sqrt{2}$.

$$f_\beta = \frac{1}{2\pi r_\pi (C_\pi + C_\mu)} \tag{4.24}$$

This model of the active transistor cannot be used to account for the properties of the saturated transistor. If the base current i_B exceeds the value needed to just bring the transistor to saturation, I_{BS}, then extra minority charge carriers are provided to the base region. While these minority carriers (electrons in an npn transistor) recombine with majority carriers at an exponential rate with life-time constant τ_B, the excess base current replenishes them. In the steady state, excess charge is *stored* in the base region. This excess may be removed with a reverse base current or, given sufficient time, by recombination. The storage time constant τ_S is used to reflect the rate at which excess charge (above that required to saturate the transistor) decays. While the saturated transistor can be considered as operating in both its forward and reverse active modes, and τ_S related to the small signal parameters of the forward and reverse modes, we will take τ_S to approximately equal τ_B, which is approximately related to f_β by

$$\tau_S \simeq \tau_B \simeq \frac{1}{2\pi f_\beta} \tag{4.25}$$

If we carefully measure the response of the CE amplifier to a pulse, waveforms like those of Fig. 4.31 are observed. When the input signal steps up from $V_2 < V_{BET}$ to $V_1 > V_{BES}$, the base current jumps to $I_{B1} = V_1/R_S$. Collector current does not change immediately. First the base voltage must be brought to V_{BET}. Then the first charge must move from the emitter through the base to the collector of the transistor. The time required to charge the junction capacitances to V_{BET} is given by

$$t_d = (R_S + r_{BB'})(C_\pi + C_\mu) \ln\left(\frac{V_1 - V_2}{V_1 - V_{BET}}\right) \tag{4.26}$$

Then with the transistor active, analysis of Fig. 4.30 reveals that collector current increases exponentially toward βI_{B1}, which is greater than $\beta I_{BS} = I_{CS}$, with a time constant

$$\tau_a = \frac{1}{2\pi f_\beta} + \beta R_C C_\mu$$

$$= \tau_B + \beta R_C C_\mu \simeq \tau_B \tag{4.27}$$

It reaches a threshold value $I_T = (V_{CC} - V_T)/R_C$ after a rise delay of

$$t_r = \tau_a \ln\left(\frac{\beta I_{B1}}{\beta I_{B1} - I_T}\right) \tag{4.28}$$

Collector current stops short of βI_{B1} at I_{CS}, where the transistor saturates (unless we have a capacitive load). The total time to turn on the transistor to $i_C \geq I_T$ is then

$$t_{on} = t_d + t_r \tag{4.29}$$

When the input signal steps down to V_2 volts, which may be zero or negative, the excess charge in the base region of the transistor holds the base voltage constant near V_{BES}. Base current reverses to

$$I_{B2} = (V_2 - V_{BES})/R_S \tag{4.30}$$

Collector current remains constant at I_{CS} during the *storage time*, while the reverse base current removes excess charge.

$$t_s = \tau_S \ln\left(\frac{I_{B2} - I_{B1}}{I_{B2} - I_{BS}}\right) \tag{4.31}$$

Then with the transistor active, collector current drops exponentially toward βI_{B2} with time constant τ_a. It reaches the threshold value after a fall delay of

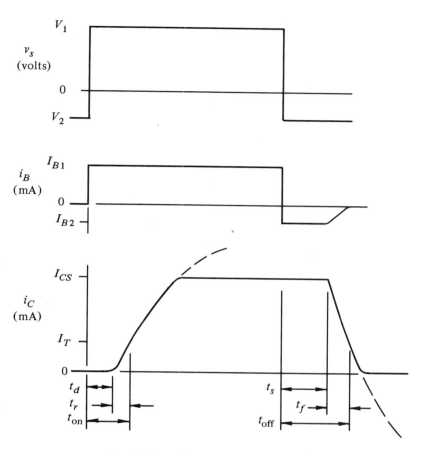

FIG. 4.31 Switching the CE Amplifier

$$t_f = \tau_a \ln\left(\frac{\beta I_{B2} - I_{CS}}{\beta I_{B2} - I_T}\right) \tag{4.32}$$

and, of course, stops dropping when the transistor cuts off and the base current returns to zero. The total time to reduce i_C to I_T is

$$t_{\text{off}} = t_s + t_f \tag{4.33}$$

A given transistor can be turned on rapidly by using small values of R_S (reduce t_d) and R_C (reduce τ_a and hence t_r) and a large value for V_1, and hence, I_{B1}. If we make $V_2 = V_{BET}$, then the delay time reduces to zero. We can reduce the turn-off time by making $I_{B1} = I_{BS}$; then the transistor just saturates and $t_s = 0$. Further, if we make V_2 large and negative, then the fall time is short. All of these desires cannot be satisfied simultaneously. For example, V_1, and hence I_{B1}, cannot be both large to reduce t_{on} and small to reduce t_{off}. We must compromise and learn to live with rise and fall delays in logic amplifiers.

The Capacitor as a Coupling Device

If a capacitor is used to couple a source to a load as shown in Fig. 4.32, v_L varies in response to changes in v_s in a substantially different manner. Again stray capacitance parallels the source and load, but usually is orders of magnitude smaller than the size of the coupling capacitor C and therefore will be neglected. If $v_s = 0$ volts for a substantial time, the capacitor will be completely discharged and $v_L = 0$ volts. A jump of v_s to $+V$ volts is then accompanied by a change in v_L described by

$$v_L(t) = \frac{VR_L}{R_s + R_L} \exp[-t/(R_s + R_L)C] = V_i e^{-t/\tau} \qquad (4.34)$$

That is, v_L jumps to an initial value $V_i = VR_L/(R_s + R_L)$, which has the same magnitude as V_f in previous equations, and then decays to 0 volts with a time constant $\tau = (R_s + R_L)C$, which differs substantially from the time constant of previous equations. A subsequent return of v_s to 0 volts is accompanied by

$$v_L(t) = -V_i e^{-t/\tau} \qquad (4.35)$$

A diode, perhaps the base-to-emitter junction of a transistor, placed in parallel with R_L has the effect of varying the magnitude of R_L. When the diode is forward biased, it serves as a very small resistance; in effect R_L has a very small value and V_i and τ also have small values. The diode shown in Fig. 4.32(a) has the effect of greatly attenuating positive excursions of v_L, shown by the broken line.

If the time constant τ of a capacitor-coupled circuit is much less than the duration of the positive excursion of v_s, then the waveform of v_L is a good

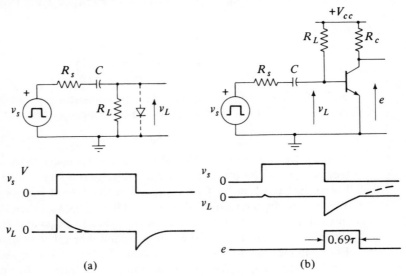

(a)   (b)

FIG. 4.32 Capacitor-Coupled Circuits

approximation to the derivative of v_s. The leading and trailing edges of v_s are converted into very brief positive and negative pulses, respectively. Such pulses, which mark the time at which the source signal changes, are useful. Older types of flip-flops differentiate the set and reset signals they receive and use these very brief pulses to initiate state switching, for example. Diodes may be included to suppress either the positive or negative pulses.

In Fig. 4.32(b) a transistor is included as a threshold device. R_L is returned to the positive supply so that the transistor is normally turned on and its base-to-emitter diode is normally forward biased. Positive excursions of v_s thus have little effect on v_L, but drops of v_s from V to 0 volts turn the transistor off and hence are not suppressed. Then as the capacitor discharges, v_L increases according to

$$v_L(t) = V_{cc} - (V_{cc} + V)e^{-t/\tau} \tag{4.36}$$

and reaches the cut-in voltage of the transistor V_{BET} at time

$$t_T = \tau \ln\left(\frac{V_{cc} + V}{V_{cc} - V_{BET}}\right) \tag{4.37}$$

To simplify, let us assume that $V = V_{cc}$, $V_{BET} = 0$, and $R_s \ll R_L$.

$$t_T = \tau \ln(2) = 0.69\tau \tag{4.38}$$

Note that the step size V and the time constant τ of the circuit, and hence the values of R_s, R_L, and C, determine the interval over which the transistor is turned off. With a fixed step size, this circuit then emits a pulse of a fixed duration. Such fixed duration pulses are often used for timing purposes in switching systems. The edge-detection circuits of Fig. 4.21 provide the same results if inverter delay can be increased sufficiently by adding load capacitance.

Monostable Multivibrators

If pulse duration must be nearly invariant, then it is desirable to place an amplifier that provides voltage changes of constant magnitude between the source and the RC timing components of Fig. 4.32(b). Also note that the transistor of Fig. 4.32(b) will not be abruptly switched from its cut-off to saturated condition because v_L does not rise abruptly from 0.6 volts to V_{BES}, and thus signal e that the transistor provides will not change abruptly from a high to a low voltage.

The circuit of Fig. 4.33 overcomes these problems. An amplifier precedes the RC timing circuit, and the two transistor amplifier stages are coupled in a loop so that once transistor $T2$ begins to turn on, transistor $T1$ is turned off, which rapidly turns on $T2$. Capacitor $C1$ is a small capacitor used to differentiate input signal x.

The circuit normally resides in its stable state with transistor $T1$ turned off and $T2$ turned on. If input signal x changes value from logic 0 to 1, the

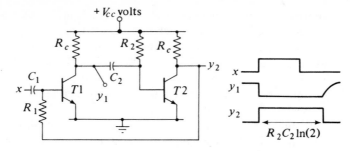

FIG. 4.33 The Monostable Multivibrator

positive pulse that results from differentiation turns $T1$ on. The resulting sudden drop in y_1 turns $T2$ off, and it remains off until its base potential rises exponentially to its cut-in level. Then it begins to turn on; y_2 decreases, which begins to turn $T1$ off. The resulting increase in y_1 is coupled to the base of $T2$ and turns it on more fully. This regenerative action very rapidly returns the circuit to its stable state. Thus state transitions are performed rapidly by the circuit, but it remains in its unstable state for a precisely controlled length of time.

Latches provide regenerative switching and may be used to form monostable multivibrators as shown in Fig. 4.34(a). When x goes to 1, z goes to 1. Some time later x_d goes to 1 and z then returns to 0, regardless of whether x has returned to 0 or not. If it is desirable to have a separate gate drive the delay element to ensure that a source voltage step of constant amplitude is presented to the RC circuit, then signal α may be used in place of $\bar{x}$ to drive an inverter as in Fig. 4.34(b).

Most integrated circuit logic families include monostable multivibrators that give a pulse of very fixed width in response to an edge of a triggering signal or a logic combination of signals. External resistance and capacitance determines their pulse width according to equations and tables supplied by the manufacturer. Pulses from 50 nanoseconds to many milliseconds may be generated. The monostable multivibrator of Fig. 4.35 generates a pulse each time $A \cdot \bar{B}$ goes to logic 1, either because A goes to

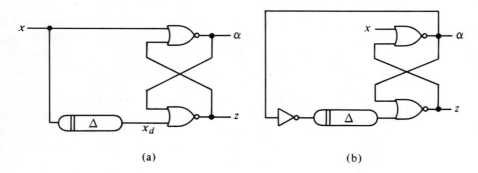

(a)   (b)

FIG. 4.34 Gate Monostable Multivibrators

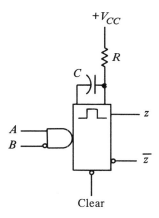

FIG. 4.35 The IC Monostable Multivibrator is Edge-Triggered, Retriggerable, and May Be Asynchronously Cleared

1 with $B = 0$ or B drops to 0 with $A = 1$. Some of these IC multivibrators have an additional advantage over the circuits of Figs. 4.33 and 4.34. If a second trigger edge occurs before the first pulse is completed, these multivibrators begin the timing anew. The output signal does not drop until the fixed time after the second (last) triggering edge. They are "retriggerable."

Astable Multivibrators

An *astable* multivibrator that provides a periodic waveform may be constructed by connecting two monostable multivibrators together so that the output signal of each serves as the input signal for the other. The two monostable multivibrators need not provide pulses of equal duration; if they do not, then the astable multivibrator will provide a rectangular rather than a square waveform. Figure 4.36(a) shows a traditional astable multivibrator circuit, which consists of two of the circuits of Fig. 4.32(b) connected so that each serves as the source for the other. If both transistors are turned on simultaneously, the circuit remains in a stable state. Since there is no way to get the circuit out of this state, it is important that it never be allowed to attain it.

But astable multivibrators may also be constructed in other ways. Suppose that three or any larger odd number of inverters are connected in a loop as in Fig. 4.36(b). Also assume for that figure that $y_1 = 1$. Then y_2 must equal 0 and $y_3 = 1$, which requires that $y_1 = 0$. This result contradicts our original assumption. The circuit does not tend to maintain y_1 at a value, but constantly tends to change it.

Note that this activity repeats, and that each state variable takes the value 1 for half of the states. This circuit might thus be employed as a source of periodic waveforms. If the delays of the gates vary substantially from one

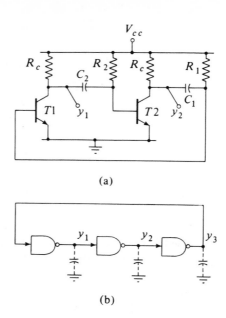

(a)

(b)

FIG. 4.36   Astable Multivibrators

another, then the time during which a variable is at 1 can differ substantially from the time during which it equals 0. The gate delays and hence the period of the waveforms may be controlled by adding appropriate capacitance, of course. It is a simple matter to replace one of the inverters with a NAND or NOR gate so that a stop-start control signal may be introduced.

If frequency stability is important, then a superior astable multivibrator may be constructed using two IC monostable multivibrators so connected that each triggers the other by the completion of its pulse. A crystal oscillator built with gates to provide amplification gives even greater frequency stability.

Schmitt Trigger Circuit

The final circuit to be considered provides a much more nearly ideal threshold device than a single transistor amplifier. Figure 4.37(a) shows the schematic diagram of a *Schmitt trigger* circuit; Fig. 4.37(b) suggests typical terminal characteristics of this circuit. If the input signal v_i increases from a value of 0 volts, transistor $T1$ remains cut-off and $T2$ remains fully turned on until v_i attains a magnitude of V_2 volts. At that point $T1$ turns on and $T2$ is turned off very rapidly with the regenerative switching action which we have seen in previous circuits. As long as v_i remains greater than V_2, $T2$ remains turned off and the output signal v_o of the circuit has a high value. When v_i is decreased, $T2$ remains turned off until $v_i = V_1 < V_2$ is reached, at which point $T1$ is turned off and $T2$

is turned on again. Since $V_1 \neq V_2$, this circuit's output versus input characteristic exhibits hysteresis.

The threshold voltages of this circuit, V_1 and V_2, are primarily determined by the values of resistors R_1, R_2, R_3, and R_e. Parameters of the transistors influence the threshold values also. If the circuit is to be used as a pure threshold device, then it is desirable to design it so as to minimize $V_2 - V_1$. But in many applications the hysteresis of this circuit is an advantage. For example, the signal provided by a mechanical switch may include considerable noise as a result of contact bounce. The RC inertial delay element discussed earlier can be used to attenuate noise pulses of brief duration. A Schmitt trigger circuit with appropriate hysteresis can be used to "remove" noise pulses of limited amplitude.

REFERENCES

1. A. BARNA and D. I. PORAT, *Integrated Circuits in Digital Electronics.* New York: John Wiley and Sons, 1973.
2. J. C. BOYCE, *Digital Logic and Switching Circuits.* Englewood Cliffs, N.J.: Prentice-Hall, 1975.
3. D. CASASENT, *Digital Electronics.* New York: Quantum Publishers, 1974.

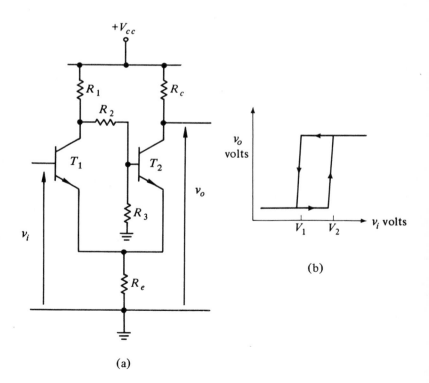

(a)

(b)

FIG. 4.37 The Schmitt Trigger Circuit

4. T. J. Chaney and C. E. Molnar, "Anomalous Behavior of Synchronizer and Arbiter Circuits," *IEEE Transactions in Computers*, Vol. C-22, April, 1973, pp. 421–422.

5. J. Millman and H. Taub, *Pulse, Digital, and Switching Waveforms*. New York: McGraw-Hill, 1965.

6. R. L. Morris and J. R. Miller, *Designing with TTL Integrated Circuits*. New York: McGraw-Hill, 1971.

7. L. Strauss, *Wave Generation and Shaping* (Second Edition). New York: McGraw-Hill, 1970.

PROBLEMS

4.1-1. To demonstrate that amplification is necessary in a sequential network, show that (a) output signal z of the logic shown below will not depend on the sequence of inputs if a diode gate is employed, or more generally (b) that a memory element can have two stable states only if its open loop gain is greater than one for some value of input potential as shown.

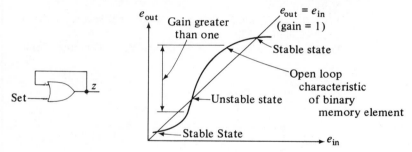

4.1-2. Assume input signal x in the circuit below changes its value every 10 units of time.
(a) While feedback, delay and amplification are present in this circuit, do you consider it to be a combinational or a sequential logic circuit?
(b) Model each gate per Fig. 4.1; assume each gate introduces 1 unit of delay. Express z at the nth time, $z(n)$, in terms of previous values of x and z, i.e. $x(n-1)$, $x(n-2)$, ..., $z(n-1)$, ... (c) Assume a waveform for x and draw the corresponding waveform for z and intermediate signals. (d) Model the entire circuit per Fig. 4.1 assuming it introduces 2 units of delay. Again express $z(n)$ in terms of previous values of x and z. Draw waveforms of x and z and carefully compare with those of (c).

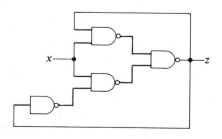

4.1-3. Express the state of the circuit as the 4-tuple $(\alpha, \beta, \gamma, z)$. Each gate introduces unit delay.

(a) For each of the possible pairs $(x_1, x_2) = 00, 01, 10,$ and 11 determine the states of the circuit in which no variable is tending to change.

(b) Which of these states are such that a change in x_1 will cause an eventual change in z? Which of these states are such that a change in x_2 will cause an ultimate change in z?

(c) Do pairs of states exist such that for the same (x_1, x_2) z has different values? How can we cause the circuit to change from one of these states to another? How long will it take for the circuit to change from one such state to another?

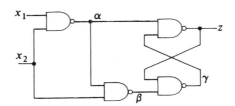

4.1-4. Model the circuit of prob. 4.1-3 with one delay (2 time units) per Fig. 4.1. Let the value of z measure the "state" of the circuit. Repeat problem 4.1-3 with this model. Carefully compare all results, and also the effort required by the two approaches. Draw conclusions based upon these comparisons.

4.1-5. Signal x changes very slowly compared with the unit delay introduced by each gate. Find the stable states of the network. If x is a periodic square waveform, draw the corresponding z waveform.

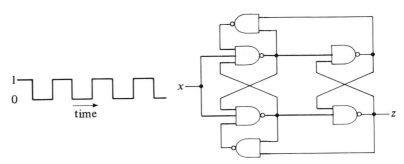

4.1-6. (a) Assume $\tau = 2$ units of time in the network below. Input signal x is the same as in prob. 4.1-5. Draw the waveform for z and compare with the results of prob. 4.1-5.

(b) Find the value of τ which causes the response of this network to be as close as possible to that of prob. 4.1-5.

(c) We construct the network of prob. 4.1-5 of NAND gates whose delay can not be precisely known. Typical delays are in the range 10 to 20 nanoseconds. Does the network of this problem constitute a suitable engineering model of that of prob. 4.1-5, and if so, what value of τ is appropriate?

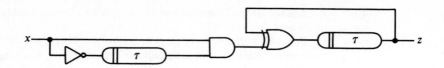

4.1-7. Assume each gate introduces unit delay and that $x_1 = x_2 = z = 0$ has existed for many units of time. Now x_1 goes to 1.

(a) Show that $SET = 1$ for only a very short duration, and the circuit in effect responds to the 0 to 1 transition.

(b) With $x_1 = x_2 = z = 0$, what happens when both x_1 and x_2 simultaneously assume the value of 1?

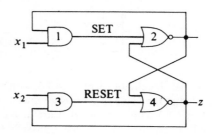

4.1-8. The circuit of prob. 4.1-7 is modified as shown. Again assume each gate introduces unit delay, and that $x = z = 0$.

(a) Draw a timing diagram showing how z varies if x is a square waveform with a period of 20 units of time.

(b) How would the diagram of (a) change if the inverter introduced 5 or 10 units of delay?

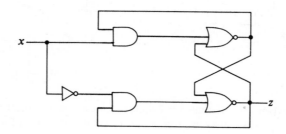

4.1-9. All gates introduce unit delay. Signal x is a square waveform with a period of 20 units of time.

(a) Draw the waveform of signal z assuming $z = 0$ initially.

(b) How is z related to x?

(c) Two of these circuits are connected in tandem. Draw waveforms for z_1 and z_2. How are they related to x?

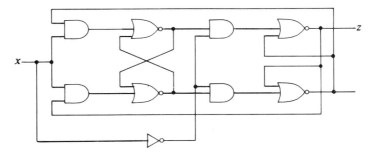

4.1-10. Relays introduce delay which is usually measured in milliseconds. Thus if we close switch S below ($y' = 1$) we may have to wait for 1 ms before the contacts close ($y = 1$, $\bar{y} = 0$).

(a) Write the excitation equations (A', B', T') for relays A, B, and T for the circuit of Fig. P4.1-10, in terms of variables A, B, and T and their complements, and x.

(b) Assume relay T introduces no delay and relays A and B have a 1 ms response time. No relays are energized. Then we close switch x for 10 ms, open it for 10 ms, close it again for 10 ms, and then open it. Draw a timing diagram which indicates when coils are energized and when contacts are closed and open. Identify the states of the network. Which of these would you refer to as "stable" states? Relate z to x.

(c) Repeat (b) assuming relay T introduces 2 ms of delay. Compare the results.

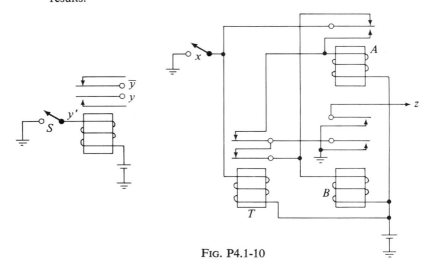

Fɪɢ. P4.1-10

4.2-1. Each of the NOR gates in Figs. 4.7(a) and 4.11 introduce unit delay. The AND gates of Fig. 4.11 introduce no delay. The duration of the clock pulses shown below is 3 units; 7 units of time separate pulses. Signals x_1 and x_2 shown below are applied to the asynchronous circuit of Fig. 4.7(a) and the synchronous circuit of Fig. 4.11. Carefully draw the waveforms of the output signals of both circuits and compare them. Assume $z = 0$ initially.

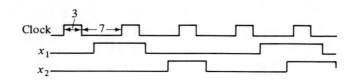

4.2-2. Find the state transition table for the latches (?) shown. If normal latch entries cannot be placed in a table, describe what the network will do.

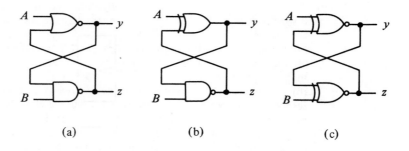

(a)   (b)   (c)

4.2-3. If each gate in Fig. 4.10(a) introduces unit delay, what is the minimum time that $C = 1$ must be maintained for reliable latch action? (Assume that D does not change when $C = 1$).

4.2-4. If all the gates in Fig. 4.10 were NOR gates, would the network be a latch? If so, show its state transition table and a suitable logic symbol.

4.2-5. All gates in Fig. 4.11 introduce unit delay. Input signal values remain fixed as the clock goes from 0 to 1 (5 units of time) and back to 0. Perform unit delay simulation to find the final state of the latch, assuming $z = 0$ initially.

(a) $x_1 = 0$, $x_2 = 0$
(b) $x_1 = 1$, $x_2 = 0$
(c) $x_1 = 1$, $x_2 = 1$

4.2-6. The following sequences of inputs are presented to a synchronous RSFF. Indicate the state of the FF at each time.

TIME	1	2	3	4	5	6	7	8	9	10
SET	0	0	0	1	1	0	0	1	1	0
RESET	0	0	1	0	0	0	1	1	0	0
STATE	0									

4.2-7. An FF is characterized by the following state transition equation.

$$z' = z \oplus x_1 \oplus x_2$$

(a) Construct this FF from an RSFF.
(b) Construct this FF from a JKFF.

298   **CHAPTER 4**   **Sequential Concepts and Components**

4.2-8. Signals A and B have the values indicated at each clock time. Show the resulting FF states for each of the networks.

$$\text{time} \quad \rightarrow$$

A	0	0	1	1	0	1	0
B	0	0	0	0	1	1	1
y_1	0						
y_2	0						

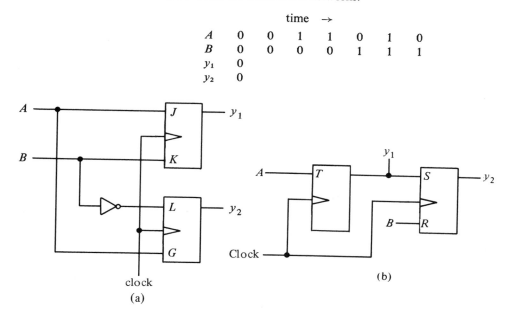

(a)

(b)

4.3-1. The clock signal applied to the master–slave RSFF of Fig. 4.18(a) is a square wave with a period of 20 units of time. All gate delays are less than one unit of time in length. The following set and reset signals are applied. Show how z, the state of FF, and the state of the master latch vary.

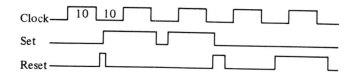

4.3-2. Extend the logic of Fig. 4.18(b) as necessary to form a synchronous GLFF. There is no limit on gate fan-in.

4.3-3. Show via unit delay simulation that the JKFF of Fig. 4.18(b) will toggle if J and K drop simultaneously while clock = 1.

4.3-4. (a) Does the flip-flop of Fig. 4.18(a) respond in any way to changes in S or R when clock = 1? (b) Add the feedback of Fig. 4.18(b) to form a JKFF, and repeat part (a). (c) Why might data-lockout be desirable in master–slave flip-flops?

4.3-5. Add *preset* and *clear* input terminals and necessary interconnections to the logic of Fig. 4.22(a).

4.3-6. In the unit delay simulation of the edge-triggered DFF, what is the earliest time in the range $i + 4$ to $i + 7$ at which D can return to 0 without interfering with the setting of the flip-flop? What then is the hold time of this flip-flop?

4.4-1. A certain manufacturer indicates that his NAND gates and master–slave flip-flops introduce propagation delays of from 7 to 18 ns., with the average delay being 10 ns. If a maximum of 10 levels of NAND gating is to be allowed between flip-flops and the minimum clock pulse width is 20 ns., what clock period would you suggest for reliable synchronous operation?

4.4-2. Two 4-bit registers supply operands to a 4-bit wide parallel adder. The sum is captured by a third register. All gates and flip-flops have an average propagation delay of 5 ns. The flip-flops have an 8 ns. set-up time. What is the maximum clock frequency that may be used if the adder is: (a) a ripple adder of Fig. 3.25? (b) A look-ahead adder of Fig. 3.31?

4.4-3. Master–slave flip-flops and gates, all with 30 ns. propagation delay, are used to realize the counter of Fig. 1.9. What is the maximum permissible frequency of a symmetric clock signal? Repeat assuming that the XOR gates are constructed of NAND gates, as shown in Fig. 3.15.

4.5-1. The circuit of Fig. 4.27 has been constructed, and t_T of Eq. (4.13) found to be 1 microsecond. The values of capacitance and resistance are not precisely known, but when a 100 picofarad (10^{-10} farads) capacitor is placed in parallel with C of Fig. 4.27, the circuit introduces 2.2 microseconds of delay. What is the value of the original capacitor C?

4.5-2. A transistor amplifier that acts as the signal source for the circuit of Fig. 4.27 is modeled as shown below.
(a) Find the time constant of the circuit with the transistor cut off and with the transistor turned on. Your answers need to be accurate only to within five percent.
(b) Find the value of V_f for the transistor turned off.
(c) Assume $V_T = 1$ volt. What delay does this circuit introduce when the source transistor turns on? turns off? Accurately draw waveforms like those of Fig. 4.28.

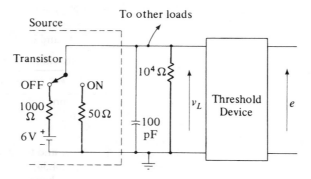

(d) What value of V_T will cause $t_T = t_T'$? See Eqs. (4.13) and (4.15).

4.5-3. The source of Prob. 4.5-2 drives n loads, each of which adds 50 pF of capacitance and 10^4 ohms of resistance in parallel with the load elements shown. Assume $V_T = 1.5$ volts.
(a) For $n = 1, 3, 5,$ and 7, determine the delay when the transistor turns off. Plot these values. Repeat for the transistor turning on.
(b) Find the largest value of n for which $V_f \geq V_T$.
(c) Discuss your findings of parts (a) and (b) in terms of logic design with gates with the characteristics of the source amplifier.

4.5-4. A transistor amplifier source drives a DTL gate, that may be modeled as shown below. Assume $V_T = 1.5$ volts, and that the diode is ideal.
(a) Show with equations and plots of waveforms how v_L varies with time following the source transistor being turned off and following its being turned on. Indicate the delay introduced by the circuit in both cases.
(b) The source transistor is turned on. What is the value of v_L? Assume $n = 1, 3, 5, 7$ identical loads are connected to the source. Show how the value of v_L varies with n, and discuss the significance of this result.

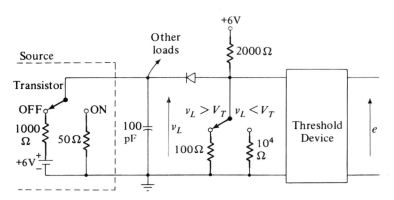

4.5-5. Ignore the transistor in the circuit of Fig. 4.32(b). Assume $V_{cc} = +10$ volts, $R_L = 2000$ ohms, $R_S = 100$ ohms, and $C = 100$ pF. At widely spaced intervals v_s jumps between 0 and 5 volts.
(a) What is the time constant of the circuit?
(b) Show how v_L varies with time, following changes in v_s that are 10 microseconds apart.
(c) With the transistor in place and $V_{BET} = 0.5$ volts, what is the duration of the pulse on line e following a drop in v_s?
(d) Repeat part (c) for $V_{BET} = 0$ volts and compute the sensitivity:

$$\frac{\text{change in pulse width}}{\text{change in } V_{BET}}$$

(e) Repeat part (c) for v_s changing from 6 to 0 volts and compute the sensitivity: $\dfrac{\text{change in pulse width}}{\text{change in } v_s}$.

4.5-6. Assume each gate introduces unit delay and the delay elements introduce 10 units of delay in the circuits of Fig. 4.34.
Show how α and z vary in response to changes in x for the two circuits.

4.5-7. In the circuit of Fig. 4.36(a) $C_1 = 100$ pF, $C_2 = 500$ pF, $R_1 = 1000$ ohms, and $R_2 = 5000$ ohms. Draw waveforms for y_1 and y_2, paying particular attention to the time scale.

4.5-8. DTL or TTL gates with $V_T = 1.4$ volts are used to construct a delay box as shown. The open collector transistor is driven on with $\beta i_B = 80$ mA. Ignoring gate delays find:
(a) the 0 to 1 and 1 to 0 delays τ_r and τ_f.
(b) the 0 to 1 to 0 inertia and 1 to 0 to 1 inertia.

4.5-9. A CE amplifier has a silicon transistor with $V_{BET} = 0.5$, $\beta = 50$, $f_\beta = 1$ MHz, and a collector resistor $R_C = 1$ kΩ. The source has a resistance of 1 kΩ when it presents 5 volts, and 50 Ω when it presents 0.1 volts $= V_{CES}$. Find the turn on and off times, t_{on} and t_{off}.

ANSWERS

4.1-2. $z = xz \vee x\bar{z} = x$

(a) Since z can be expressed as a Boolean function of the input variable alone, this circuit is thought of as a combinational circuit.

(b) $z(n) = x(n-2) \cdot [z(n-2) \vee \bar{z}(n-3)]$
$\qquad = x(n-2) \cdot$

(c) The following waveform for z is based upon the expression for $z(n)$.

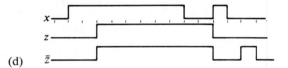

(d)

4.1-3. (a) With $(x_1, x_2) = 00$, $(1, 1, 0, 1)$ and $(1, 1, 1, 0)$ are stable state.

(b) With $x_2 = 0$, a change in x_1 will have no effect on z. With $x_1 = 0$ and the circuit in state $(1, 1, 0, 1)$ when x_2 changes to 1, the state of the circuit will switch to $(1, 1, 1, 0)$.

4.1-6. (a)

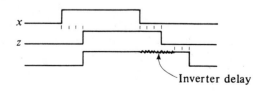

(b) We might use an average value of 2.5. While z in Prob. 4.1-5 changes its value 2 or 3 time units after a 0 to 1 transition of x, the circuit does not reach a stable state that rapidly.

4.1-8.

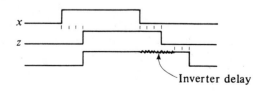

Inverter delay

4.1-10. $A' = TA \vee \bar{T}B$; $B' = T\bar{A} \vee \bar{T}B$; $T' = T \vee \bar{T}\bar{A}B$.

4.2-2. (a) Remembers with $AB = 01$.
(b) Oscillates.

4.2-4. D-latch that loads when $C = 0$.

4.2-5. (c) Oscillates.

4.2-6.
Time:	1	2	3	4	5	6	7	8	9	10
State:	0	0	0	0	1	1	1	0	?	1

4.2-8. (a)
A	0	0	1	1	0	1	0		
B	0	0	0	0	1	1	1		
y_1	0	0	0	1	1	0	1	0	
y_2	0	0	0	1	1	1	0	0	

4.3-2. One inverter must be introduced, and the fan-in of the left NAND gates increased.

4.3-4. (a) A brief $S = 1$ sets the master latch and ultimately the flip-flop.
(c) Data lockout prevents flip-flop response to brief, perhaps unexpected and undesired, values on the input signal lines.

4.3-6. $i + 4$, hold time $= 0$.

4.4-1. Using maximum delays is safe: $20 + 18 + 18 \cdot 10 = 218$ ns. period, or 4.58 MHz. If we could be sure that all gates gave minimum delay, the clock frequency could be 10.3 MHz. Average delays suggests a 7.7 MHz frequency. Most systems should work at this frequency.

4.4-3. Signal A_4 passes through four gates to reach D_1, $4 \times 30 + 30 = 150$ ns. With a symmetric clock waveform period $= 300$ ns.; $f = 3.3$ MHz.

4.5-1. 90.9 pF

4.5-2. (a) $\tau_{off} = 90$ ns.; $\tau_{on} = 5$ ns.

(b) $V_f = \dfrac{10\,k}{10\,k + 1\,k} \times 6 = 5.45$ V.

(c) Turning on delay $= 8.5$ ns.; turning off delay $= 18$ ns.

4.5-5. (a) 210 ns.
(c) 92 ns.

4.5-8. (a) $\tau_r = 46$ ns., $\tau_f = 328$ ns. (ignoring gate delays).
(b) x drops to 1.4 volts in 46 ns., so a 45 ns. ⊓ pulse would not be detected by the second gate.

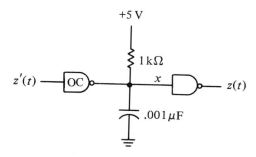

5

Synchronous Sequential Networks

Section 4.4 introduced a model for synchronous sequential networks while emphasizing the time relations and restrictions on activity in such networks. This chapter deals with the other side of the coin. The state model of synchronous sequential networks will be formally developed and used to analyze them. Timing will be ignored, but not forgotten. Then the analysis process will be reversed and a synthesis algorithm explored. Finally, the chapter presents syntax of the DDL language not presented in Section 2.8. This syntax is very dependent on state models of sequential networks.

5.1 ANALYSIS OF SYNCHRONOUS SEQUENTIAL NETWORKS

We found that the response of an FF depends upon the history or sequence of input signals that it receives. We said an RSFF was much like a wall switch. Suppose such a switch has been in use for one year and flipped several times a day. To record its actual history of received input signals would be a great task. To express *all* of the possible input sequences that it might possibly have received would be almost impossible. Certainly the wall switch is not able to record its entire actual history nor can it be prepared to respond to possible histories in a very sophisticated manner. The designer of the switch did not consider all possible histories and prepare his product to respond to each in a different manner. He viewed all of its possible sequences in a very simple fashion: all sequences that end with the handle being pushed up were treated as equivalent. All input sequences that end with the handle being pushed down form another set of equivalent histories. With each set of equivalent histories we associate a "state."

The combination lock is a more sophisticated mechanical switch. Its response depends on a sequence of input symbols, so we can think of it as a sequential network. Again, many possible input sequences exist, each of which is in effect placed in one of a small number of sets by the lock. All histories that end with the dial being spun are equivalent. All sequences that end with the dial set on the first integer of the combination after being spun are equivalent. All sequences that end with two specific integers form a third set, etc. We find that the lock views input sequences as being members of one of four sets (three-integer combination assumed). As a result, we may think of the lock as a four-state mechanical sequential network.

As we turn the dial of a combination lock, its state may change, particularly if we know the combination to open the lock. A diagram often helps to visualize possible state transitions. In Fig. 5.1 there is a box for each state. Possible state changes are indicated by arrows accompanied by the input signal that promotes the transition. If we know the combination we can cause the lock to go from state to state by dialing the integers of the combination in sequence until the lock reaches state 3, where it responds as we desire.

What physical form, if any, do "states" take within a combination lock? We might well find two binary mechanical switches within the device. Initially (state 0) both switches prevent opening the lock; both might be said to be reset. Then when the first correct integer is dialed, one of the two withdraws (is set). The lock is then in state "one switch reset, the other set," or we might say, more briefly, state "01", or in decimal (with Fig. 5.1 in mind) state 1. With the second correct integer, the states of the

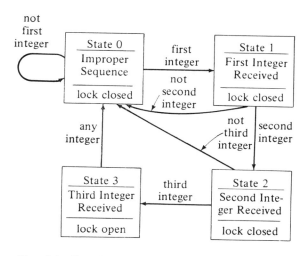

FIG. 5.1 State Transitions of the Combination Lock

switches are altered to 10; the lock is now in state 2. Finally, with a correct sequence of three integers both switches withdraw so that the lock may be opened, and we say the machine is in state 3.

Electronic sequential networks operate in a very similar fashion. Flip-flops act as the mechanical switches; each has two states. The state of the network is then taken as the collection of the states of the flip-flops. But to see this better, let us model synchronous sequential networks.

The State Model

By a sequential *machine* we will mean a mathematical model of a sequential network. A synchronous sequential machine with n binary input variables $x_1, x_2, \ldots, x_n$ and m binary output variables $z_1, z_2 \ldots, z_m$ may be considered to consist of:
(1) ideal combinational logic, and
(2) p synchronous memory elements
interconnected as in Fig. 5.2. The n variables from outside the machine and the p variables from the memory elements serve as input variables to the combinational logic, which in turn forms the m output variables and p next-state variables. Unless DFFs are utilized as the memory elements, the combinational logic must actually form more than p flip-flop input signals (S, R, J, K, G, L, etc.), which are related to the next-state variables of the machine by the state equations of the FFs. We are ignoring the variations between FF types at this point in order to keep the model as general and as simple as possible.

Now we can define the *state* of a sequential machine as the p-tuple of memory element states. We will find that this definition is consistent with the idea of sets of equivalent input sequences. A machine thus has 2^p states, although it may not be possible for the machine to enter some of them. Let i be a decimal integer in the range $0 \leq i \leq 2^p - 1$. A very natural name to give each state is then the p-tuple, binary equivalent of i, y^i.

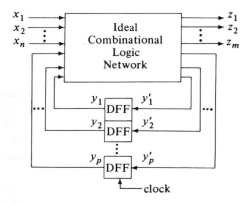

FIG. 5.2 Model of Synchronous Sequential Machines

$$y^0 = (0, 0, \ldots, 0) \qquad \text{all flip-flops reset}$$
$$y^1 = (0, 0, \ldots, 1) \qquad \text{first flip-flop set}$$
$$y^2 = (0, 0, \ldots, 1, 0) \qquad \text{second FF set}$$
$$\vdots$$
$$y^{2^p - 1} = (1, \ldots, 1) \qquad \text{all flip-flops set}$$

Other more descriptive names will also be used to identify states, for the name we attach to a state is not of fundamental importance. The set of all machine states, Y, is known as the *state alphabet*.

$$Y = \{y^0, y^1, \ldots\}$$

In a similar fashion, and as in Section 2.4, let the *input symbol*, x^i, be an n-tuple of valued input variables x_i. Some of the 2^n possible input symbols of a machine may never be presented to that machine, and fewer than 2^n symbols may appear in the *input alphabet*, X.

$$X = \{x^{i_1}, x^{i_2}, \ldots\} \subseteq \{x^0, x^1, \ldots x^{2^n - 1}\}$$

An *output symbol*, z^i, of a machine is then an m-tuple of valued output variables, z_i, and the *output alphabet* of a machine is the collection of all symbols that machine is able to generate.

$$Z = \{z^{i_1}, z^{i_2}, \ldots\} \subseteq \{z^0, z^1, \ldots, z^{2^m - 1}\}$$

Our model of Fig. 5.2 suggests that combinational logic generates a unique output symbol $z^i \in Z$ and a next state $y' \in Y$ for each given input symbol $x^j \in X$ and present state $y^k \in Y$. Thus the combinational logic of our model defines two Boolean functions.

$$z^i = \zeta(x^j, y^k).$$
$$\zeta: X \times Y \to Z \qquad (5.1)$$

$$y' = \psi(x^j, y^k)$$
$$\psi: X \times Y \to Y \qquad (5.2)$$

Mapping ψ determines the next machine state given the present input symbol and present state of the machine. Or, in terms of logic networks, ψ is an $n + p$-input, p-output switching function. The states of the FFs and the values of the input variables just before the ith clock time are combined by ψ into signals that determine what the states of the FFs will be before the $(i + 1)$st clock time.

Function ζ maps input symbol, machine-state pairs onto the output alphabet of the machine. Our network model considers ζ to be an $n + p$-input, m-output switching function. This is known as a *Mealy* model; the output symbol given by ζ is a function of both the input symbol and machine state. Other models can be proposed. The *Moore* model, for example, treats ζ as a p-input, m-output switching function.

$$z^i = \zeta'(y^k)$$
$$\zeta' : Y \to Z \tag{5.3}$$

Here the output symbol is determined entirely by the machine state. Our combination lock (while not synchronous) was modeled in this fashion in Fig. 5.1. The lock provided the output symbol "open" when in state 3. In all other states the output symbol "closed" was generated. The integers dialed effect state transitions, but do not directly influence the output symbol of the mechanical sequential network.

While we can build a network to do a specific job from either a Mealy or a Moore model, Eq. (5.2) in both cases and either Eq. (5.1) or Eq. (5.3) respectively, networks based upon the Mealy model can require fewer memory elements and we will most often use it.

A *sequential machine* is then the 5-tuple (X, Y, Z, ψ, ζ). Alphabets are conveniently expressed with tables. Boolean functions ψ and ζ can also be expressed in tabular form, or equations, maps, and directed graphs may be used. We will usually refer to just a presentation of ψ and ζ as a machine, for we are considering completely specified[1] machines, and all input symbols, states, and output symbols thus appear.

Analysis Examples

Let us now analyze several synchronous sequential networks and attempt to determine and express in a compact, descriptive fashion exactly how they respond to sequences of input symbols. We then can "operate" the models to see more clearly how the actual network will perform. Network $\mathcal{N}1$ shown in Fig. 5.3 is very simple: it has a single input signal, a single output signal, and a single memory element. Thus the input, output, and state alphabets of $\mathcal{N}1$ are as in Table 5.1.

TABLE 5.1. ALPHABETS OF NETWORK $\mathcal{N}1$

Input Alphabet		Output Alphabet		State Alphabet	
Symbol	x	Symbol	z	State	y
x^0	0	z^0	0	y^0	0
x^1	1	z^1	1	y^1	1

[1] In incomplete machines ψ, ζ, or both may not be defined for some present-state–input-symbol pairs. Synthesis of such machines parallels but is significantly more complicated than the procedure we will investigate.

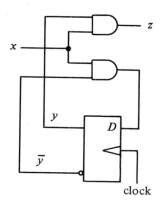

FIG. 5.3   Synchronous Sequential Network $\mathcal{N}1$

The output and next-state mappings are also easily expressed for network $\mathcal{N}1$.

$$z = x \cdot y \qquad\qquad y' = x \cdot \bar{y}$$
$$\zeta(x^0, y^0) = z^0 \qquad\qquad \psi(x^0, y^0) = y^0$$
$$\zeta(x^0, y^1) = z^0 \qquad\qquad \psi(x^0, y^1) = y^0$$
$$\cdot\,\zeta(x^1, y^0) = z^0 \qquad\qquad \psi(x^1, y^0) = y^1$$
$$\zeta(x^1, y^1) = z^1 \qquad\qquad \psi(x^1, y^1) = y^0$$

This is a very formal presentation, which illustrates the definitions given earlier. The tables of Table 5.2 express ζ and ψ in a much clearer and more useful fashion. Several tables are shown, only to illustrate the various formats commonly used in the literature to present this information. We will refer to a tabular representation of ψ as a *state-transition table*. A tabular presentation of ζ is nothing more than a conventional truth table.

TABLE 5.2.   VARIOUS TABULAR FORMS OF ψ AND ζ

ψ			ζ					
x	y	y'	x	y	z	x	y	y' z
0	0	0	0	0	0	0	0	0 0
0	1	0	0	1	0	0	1	0 0
1	0	1	1	0	0	1	0	1 0
1	1	0	1	1	1	1	1	0 1

x	y	y'/z
0	0	0/0
0	1	0/0
1	0	1/0
1	1	0/1

		y'/z	
y		$x = 0$	$x = 1$
0		0/0	1/0
1		0/0	0/1

Now let us see how this network responds to input symbol sequences. Let us assume that when we begin our examination the FF is reset and $x = 0$. Then from either the network of Fig. 5.3 or the state-transition tables of Table 5.2 we see that the next state of $\mathcal{N}1$ is y^0, i.e. the FF remains reset. Now we advance to the next clock time. If we assume $x = 1$ with the FF reset, the network and the tables tell us that the next state of the machine will be y^1, i.e. the FF will be set. If we hold $x = 1$ with $\mathcal{N}1$ in state y^1, we find that $z = 1$ and the next state is y^0. Thus the combination which "opens the lock" ($z = 1$) consists of two successive x^1 input symbols. One state transition remains to be examined. With the FF set and $x = 0$, $y' = 0$, so that the next state of the machine is y^0. Figure 5.4 illustrates the electrical signals that might be observed in network $\mathcal{N}1$ and the corresponding notation we might use to express more briefly the synchronous behavior of $\mathcal{N}1$. This last information can be more easily obtained from Table 5.2.

A graphic representation of ψ and ζ is also very helpful, as we have seen in Fig. 5.1. A *state-transition diagram* is a connected graph with nodes corresponding to machine states and connecting arrows indicating allowed transitions. We may display ψ and ζ for $\mathcal{N}1$ as in Fig. 5.5(a). An input-symbol/output-symbol notation is associated with each transition arrow to indicate the input symbol that causes the transition and the corresponding machine response. Fig. 5.5(b) illustrates the same diagram with the abbreviated notation that we will more commonly employ.

If we think of $\mathcal{N}1$ as a physical entity, then Table 5.1 and either Table 5.2 or Fig. 5.5 constitute machine $\mathcal{M}1$, a mathematical model of $\mathcal{N}1$. If $\mathcal{N}1$ is constructed of reliable components and operated with a clock waveform that satisfies Eq. (4.10), $\mathcal{M}1$ provides an excellent representation of $\mathcal{N}1$ as far as synchronous terminal characteristics are concerned. Hence, in the future we will seldom distinguish between "machine" and "network."

Figure 5.6 provides us with the logic of synchronous machine $\mathcal{M}2$. JKFFs are used as memory elements of $\mathcal{M}2$ so we can not identify next state variables y_i', but must deal with FF input signals. Since $\mathcal{M}2$ is a single input machine, we can easily identify the input alphabet as

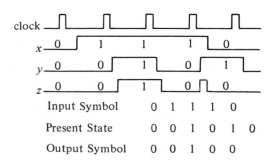

FIG. 5.4 Behavior of $\mathcal{N}1$

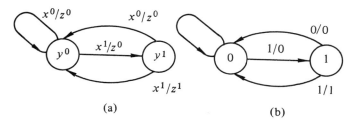

(a)   (b)

FIG. 5.5   State Diagram of $\mathcal{N}1$

$X = \{x^0, x^1\}$. $\mathcal{M}2$ must have four states because it is constructed of two binary memory elements. With both FFs reset, $y_1 = y_2 = 0$, we say that $\mathcal{M}2$ is in state $y^0 = (00)$. The other states are named in parallel fashion and listed in Table 5.3.

To find the output alphabet of $\mathcal{M}2$, we write Boolean equations for the output variables of $\mathcal{M}2$.

$$\zeta: z_1 = \bar{y}_1 \cdot \bar{y}_2$$
$$z_2 = x \cdot y_1 \cdot \bar{y}_2$$
$$z_3 = y_1 \cdot y_2$$

With $\mathcal{M}2$ in state y^0, we see that $z_1 = 1$. Thus y^0 is a Moore model type of state, since machine response is independent of the input symbol. With $\mathcal{M}2$ in state y^2, z_2 has the value of x, and y^2 is a Mealy model type of state. We thus model the entire machine in the Mealy form. Note in the above equations that no two z_i may simultaneously have the value of 1. Thus the output alphabet of this machine is brief, consisting of only four rather than $2^m = 8$ symbols. This alphabet is also indicated in Table 5.3.

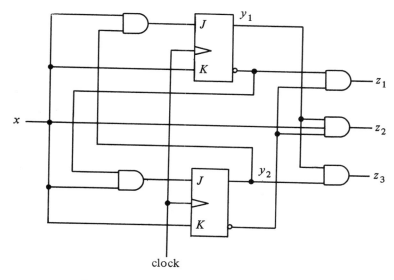

FIG. 5.6   Synchronous Sequential Machine $\mathcal{M}2$

TABLE 5.3 ALPHABETS OF $\mathcal{M}2$
NOTE THAT Z IS INCOMPLETE

X	x	Y	y_1	y_2	Z	z_1	z_2	z_3
x^0	0	y^0	0	0	z^0	0	0	0
x^1	1	y^1	0	1	z^1	0	0	1
		y^2	1	0	z^2	0	1	0
		y^3	1	1	z^4	1	0	0

To facilitate finding ψ for $\mathcal{M}2$ we write the FF input equations.

$$J_1 = x \cdot y_2 \qquad J_2 = x \cdot \bar{y}_1$$
$$K_1 = x \qquad K_2 = x$$

These expressions may be substituted into Eq. (4.7), the state-transition equation of the JKFF.

$$\psi : y_1' = (xy_2)\bar{y}_1 \vee (\bar{x})y_1$$
$$y_2' = (x\bar{y}_1)\bar{y}_2 \vee (\bar{x})y_2$$

These equations may be converted to tabular form or we may find the entries of such a table by repeated analysis of $\mathcal{M}2$. We assume the machine to be in one of its states, and for each input symbol of X find the next state of $\mathcal{M}2$ by determining the next states of the two flip-flops. For example, with $\mathcal{M}2$ in state y^0 (both FFs reset) and $x = 0$, the output symbol will be z^4 (only $z_1 = 1$), and $J_1 = K_1 = J_2 = K_2 = 0$. Both FFs remain reset so the next state is y^0, the same as the assumed present state. This finding, and others found similarly, are recorded in Table 5.4 and the state diagram of Fig. 5.7.

We see from Table 5.4 or Fig. 5.7 that state y^3 is unusual in that no transitions from other states lead to it. Thus $\mathcal{M}2$ can not get into state y^3 by ordinary means. When we turn on the power, $\mathcal{M}2$ may be in state y^3, or we may be able to force $\mathcal{M}2$ into state y^3 manually by introducing artificial input signals, but under normal operating conditions $\mathcal{M}2$ will be in one of three states. Thus we may think of $\mathcal{M}2$ as being a three-state machine for many purposes.

TABLE 5.4 STATE-TRANSITION TABLE FOR $\mathcal{M}2$

Present State	Next State $x = 0$	1	Output 0	1
y^0	y^0	y^1	z^4	z^4
y^1	y^1	y^2	z^0	z^0
y^2	y^2	y^0	z^0	z^2
y^3	y^3	y^0	z^1	z^1

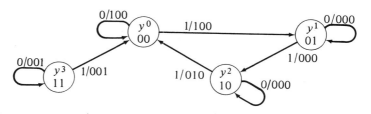

FIG. 5.7   State-Transition Diagram of $\mathcal{M}2$

As long as $x = 1$, $\mathcal{M}2$ progresses from state y^0 to y^1 to y^2 to y^0 to .... $\mathcal{M}2$ is a *modulo-3 counter*: it counts mod-3 the number of clock times that $x = 1$. When the count is 0, $z_1 = 1$. Advancing the count from 2 to 0 is indicated to the world outside $\mathcal{M}2$ by $z_2 = 1$. If $x = 0$, $\mathcal{M}2$ remains in a state, remembering the count to that point. (The FF states give the count in binary, i.e. 00, 01, 10, 00, ...)

Let us "operate" $\mathcal{M}2$ to see this action. We must assume an initial state for $\mathcal{M}2$ and a sequence of input symbols. We can then determine the corresponding sequence of states and output symbols the machine will generate. Assume $\mathcal{M}2$ is initially in state y^3 because that is the only way that it can get there—unless a component fails or a wiring error was made in the construction of $\mathcal{M}2$.

Time $\rightarrow$

x 0 0 1 1 1 1 0 1 0 0 . . .

State $y^3 y^3 y^3 y^0 y^1 y^2 y^0 y^0 y^1 y^1 y^1$ . . .

Output $z^1 z^1 z^1 z^4 z^0 z^2 z^4 z^4 z^0 z^0$ . . .

Either the ST table or diagram may be used to determine the sequence of states and output symbols of $\mathcal{M}2$. Using either, the present input symbol and present state determine the NEXT machine state, and PRESENT output symbol. Apply other input sequences to $\mathcal{M}2$ using the circuit diagram, ST table, and ST diagram as representations of the machine.

Finally, let us examine an abstract machine $\mathcal{M}3$, for which we do not have a network diagram. The next section will be concerned with the problem of obtaining a network that corresponds to a given machine. Let the ST table or diagram of Fig. 5.8 define $\mathcal{M}3$.

Many theoretical and practical questions concerning machines such as $\mathcal{M}3$ can be asked and answered. Perhaps an obvious question is: "How can we open the lock?" To answer, we must be given a starting point. For example, if $\mathcal{M}3$ starts in state y^0, then the input sequence 1111 will provide an output $z = 1$. Other sequences will also work; for example, the first input symbol of the four-symbol sequence is not important: 0111 also opens the lock, if we start the machine in state y^0.

We may ask "How do we get from state y^i to state y^j?" Answers are of practical importance when we build and test a network, to ensure that it

Present State	Next State $x=0$	1	Output z $x=0$	1
y^0	y^2	y^2	0	0
y^1	y^0	y^0	0	1
y^2	y^5	y^3	0	0
y^3	y^1	y^1	0	0
y^4	y^0	y^0	0	0
y^5	y^4	y^1	0	0
y^6	y^4	y^3	0	0
y^7	y^5	y^5	0	0

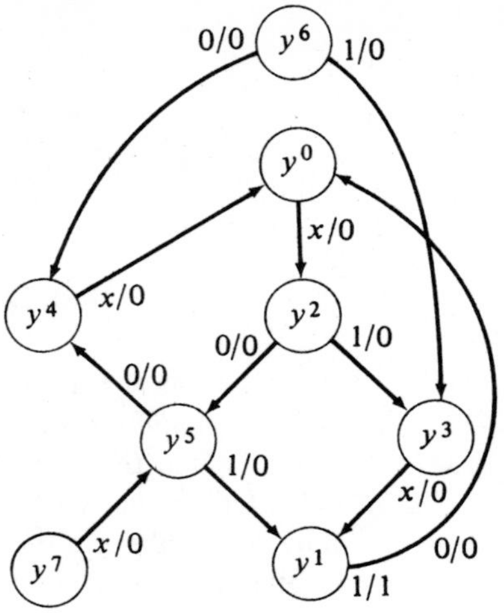

FIG. 5.8   Machine $\mathcal{M}3$

corresponds to desired action as expressed in a model. We may find our network in an unexpected state and ask "How did we get here?" If we find a network based upon $\mathcal{M}3$ in either state y^6 or y^7, we know our design is faulty or components have failed, for no transitions lead to these states.

With this deeper understanding of the state model of synchronous sequential networks, and various formats for expressing such models, let us proceed to the synthesis problem of designing a network from a state model.

5.2 SYNTHESIS OF SYNCHRONOUS SEQUENTIAL NETWORKS

We now wish to investigate the design problem of obtaining a synchronous sequential network from specifications of desired sequential behavior. A step-by-step procedure will be presented. If we were to explore this procedure in depth, we would find that chapters could be (and have been) written on each step. So this section is clearly an introduction. In summary, the design algorithm proceeds as follows.
1. Model provided specifications
2. Minimize the number of states in the model
3. Assign a binary p-tuple to each state
4. Select the type of flip-flop to be used and write flip-flop input equations
5. Design the combinational circuitry.
Steps in the design of synchronous sequential machines will be illustrated with the following example.

A single-input, single-output synchronous sequential machine $\mathcal{M}4$ is required, which determines whether the input bits at four

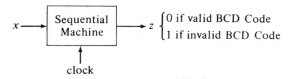

$$x \longrightarrow \boxed{\begin{array}{c} \text{Sequential} \\ \text{Machine} \end{array}} \longrightarrow z \begin{cases} 0 \text{ if valid BCD Code} \\ 1 \text{ if invalid BCD Code} \end{cases}$$

clock

FIG. 5.9 The Design Example Machine $\mathcal{M}4$

successive times constitute a valid BCD decimal digit representation. When the output is zero ($z = 0$) at the fourth-bit time, it indicates that the previous three input bits and present bit constitute a valid BCD decimal digit representation. When it is one ($z = 1$) it indicates that an invalid digit representation has been encountered. The output is to be zero at all other times. Least significant bits of the code words appear first in time.

Step 1. Model the Specifications

As our first step we attempt to find a machine, i.e. the 5-tuple (X, Y, Z, ψ, ζ), with the input-sequence–output-sequence behavior of the specifications. Many machines may satisfy this requirement and one may be better than the others. But at this point we will be satisfied if we can express any machine that satisfies the given specifications. In doing so we may find that the specifications are vague or ambiguous, and may wish to ask the writer of those specifications, "What do you want the response of the machine to be in this case?"

Finite machines can be classified to a limited extent. For example, many machines have a state or states through which they pass periodically, regardless of the input-symbol sequence. Often it is natural to begin our deliberation by assuming the machine to be in this state, and hence it is known as an "initial" state. Our combination lock is an example: we spin the dial before dialing the combination to ensure that the lock is in the proper starting state. Our BCD verifier is another example. We want our network to be in a specific state when we present the first of the four bits of a code word, and to return to that state after it has received four input symbols, in preparation for the next code word.

The BCD verifier also exhibits another property of a first category of machines. Besides the initial state, it must deal with input subsequences of fixed length. We will not expect the verifier to respond to 3-bit or 5-bit code words. It is to respond to 4-bit input subsequences only. If, after receiving three of the four bits, it is able to decide on the validity of the code word, it still must accept the fourth bit before presenting its decision.

While the concept of "equivalent history" sets can be used to decide upon the states of a machine of this first type, we will demonstrate a brute force technique for finding such a machine, which is based upon the fact that only fixed-length subsequences are to be recognized. If λ measures the length of these subsequences, then only 2^λ sequences need be considered, and we can form a state sequence (a path in an ST diagram) for each. We will illustrate this procedure in terms of the ST diagram and then prepare

315   **SECTION 5.2**   **Synthesis of Synchronous Sequential Networks**

an ST table. But in general the problem to be solved and the preferences of the individual will dictate which format is to be used.

We recognize the need for an initial state in $\mathcal{M}4$. Let us give it a name: "INITIAL", "START", "1", etc. could be used. We choose to name it "a" for brevity. To what states will $\mathcal{M}4$ go following receipt of the first bit of a code word? If we were to think, we could come up with a clever answer. But the brute force approach suggests that we simply draw a state transition for each possible input symbol! Let the machine go to state "b" if $x = 0$ and "c" if $x = 1$. We as designers are free to propose as many states as we feel are required to perform the specified task, but with each new state we increase the bookkeeping effort we must expend and the complexity of the network to be synthesized. Hence we should always be looking for the possibility of going to an already defined state.

It is natural to refer to states b and c as *following* state a. Now we ask what states follow states b and c. Again we propose new states: d, e, f, and g as shown in Fig. 5.10. Consideration of all the 16 possible code words of 4 bits and the output that is to accompany receipt of each leads to the remainder of that figure. Note that many states have state a as their next state: these are the possible states of the machine just before it receives the fourth bit of a code word. Receipt of that bit causes the machine to return to its initial state in preparation for the next code word. We must not postulate an infinite number of states and hence should constantly be looking for the possibility of utilizing already postulated states as next states.

Generation of the state diagram or equivalent state table (Table 5.5) requires that the designer consider *all* of the conditions or situations that the machine must recognize and respond to. Each such condition or situation leads to a state of the machine.

The second category of machines differs from the first in that recognized subsequences need not be of fixed length. The combination lock falls into this classification. If we initialize it, dial the first digit, and then dial an erroneous second digit, the lock returns to its initial state without con-

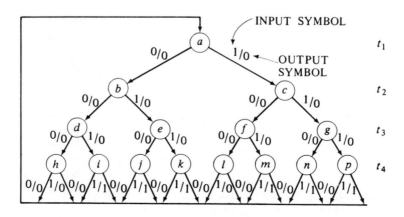

FIG. 5.10 State Diagram Representation of $\mathcal{M}4$

TABLE 5.5 STATE TABLE OF $\mathscr{M}$4

Present	Next state		Output	
state	$x = 0$	$x = 1$	$x = 0$	$x = 1$
a	b	c	0	0
b	d	e	0	0
c	f	g	0	0
d	h	i	0	0
e	j	k	0	0
f	l	m	0	0
g	n	p	0	0
h	a	a	0	0
i	a	a	0	1
j	a	a	0	1
k	a	a	0	1
l	a	a	0	0
m	a	a	0	1
n	a	a	0	1
p	a	a	0	1

sidering the third digit dialed. See Fig. 5.1 again. The brute force method can be used to form an ST diagram or table for such machines, but we must recognize that subsequences shorter than those of maximum length may cause a return to an already established state.

The final category of finite state machines consists of those for which no state exists through which the machine *must* pass. We may still refer to one state as "initial" in that we think of the machine as starting in that state, but it need never return to that state. As an example of a machine of this type let us consider a variation of the specifications of $\mathscr{M}$4.

$\mathscr{M}$5: At each clock time the output signal is to reflect the validity of the three immediately preceding bits and present input bit as an 8421 code word.

Now rather than examining four bits, making a decision, disregarding those bits, and resetting in preparation for the next sequence, as $\mathscr{M}$4 does, $\mathscr{M}$5 is to be monitoring and making decisions continually. The present input bit serves as the most significant bit of the presently considered code word, and this bit will serve as the second most significant bit in the code word to be considered at the next clock time, etc.

Identification of equivalent-history sets usually suggests the states of a machine of this third type. Here all input sequences of three or more bits that end in the same triple may be treated as equivalent. At most, $\mathscr{M}$5 must retain information about the three preceding input bits; it must store these bits either directly or in some coded fashion. But how does a machine "store" or "remember" bits? A machine stores a particular input symbol sequence by being in a particular state.

Thus eight states are suggested for $\mathscr{M}$5, although careful thought may reveal that some of these 3-bit sequences can be treated as equivalent. The procedure of the next step will reveal this, if we can not now recognize that some sequences are equivalent. We might number these equivalent history sets 0 through 7. With $\mathcal{M}5$ in state 5 we consider it to be remembering the past input sequence 101. We might begin to draw the state diagram with state 0, not because it is an initial state to which the machine returns, but because it is probably the state we prefer the machine to be in when we first turn on the power, i.e., it is the "power–on" state. (We may include explicit "reset" circuitry not called for by the ST diagram to ensure that state 0 is the power–on state.)

When the first input symbol is received, we discard the least significant bit of the stored 3-bit sequence and add the present input symbol as the most significant bit to obtain the next state of the machine. Fig. 5.11 shows the resulting ST diagram and table. No state can be recognized as an initial state. Closed paths of length 2, 3, 4 and higher are easily identified in the ST diagrams, but $\mathcal{M}5$ does not necessarily pass through any state at regular intervals. Note the uniformity of the ST table, and symmetry of the ST diagram.

While procedures have been suggested for obtaining a model from specifications, this first step in our design procedure is still more art than science. Experience will enable us to do a better job of arriving at a model, with less effort.

Step 2. State Minimization

Unless we possess outstanding insight, our first state description of a machine will contain *equivalent* states. Two states of a machine are said to be equivalent if it is impossible to distinguish between them by submitting input sequences and observing the output sequences generated by the machine. Suppose we are told that $\mathcal{M}5$ is at present in either state 0 or state 1, and will be returned to that unknown state as often as we wish. Can we determine which state $\mathcal{M}5$ is in without observing signals internal to $\mathcal{M}5$? We apply $x = 0$ and observe $z = 0$. It tells us nothing, and from Fig. 5.11 we see that any longer input sequence beginning with 0 will give us no additional information, for after the first 0 we are in state 0 in either case. So we ask that the machine again be placed in the unknown state and apply $x = 1$ observing $z = 0$. Again we have the same response and both states 0 and 1 are followed by state 4, so longer input sequences beginning with 1 are also futile. We say states 0 and 1 are equivalent states because $\mathcal{M}5$ treats all input sequences ending with 000 as being equivalent to those ending with 001. But it is supposed to do so, for 0000, 1000, 0001, and 1001 are all valid code words.

The equivalence of states 0 and 1 is very easy to see in the ST table of Fig. 5.11. The next state and output entries for present states 0 and 1 are identical. We also see that states 2 and 3 are equivalent, as are states 4 and 5, and states 6 and 7. Detecting state equivalence is not usually such a simple matter.

It is often desirable, from economic and other viewpoints, to eliminate the duplication that equivalent states present. One of each pair of identical rows may be deleted. All references to the deleted row must be changed

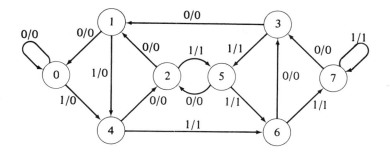

Present	Next state		Output	
state	$x = 0$	1	$x = 0$	1
0	0	4	0	0
1	0	4	0	0
2	1	5	0	1
3	1	5	0	1
4	2	6	0	1
5	2	6	0	1
6	3	7	0	1
7	3	7	0	1

FIG. 5.11   Machine $\mathcal{M}5$

to refer to the retained row. New names may be given to the retained states if it is helpful to do so. Thus $\mathcal{M}5$ can be constructed as the four-state machine of Fig. 5.12, possibly at lower cost. As far as terminal characteristics are concerned, we will never be able to tell a four-state version of $\mathcal{M}5$ from an eight-state version based upon Fig. 5.11. Analyze the state diagram of Fig. 5.12 to convince yourself of this, and that the least significant bit of a code word does not affect its validity and hence need not be stored. This is the fact upon which the four-state model of Fig. 5.12 is based.

It is possible to find equivalent states in the model of $\mathcal{M}4$ given in Fig. 5.10. We might try to locate them by examination of random pairs of states. But an algorithm that will guarantee that all equivalent states have

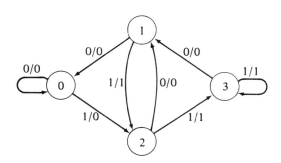

FIG. 5.12   Four State Model of $\mathcal{M}5$

been found is far more efficacious. One such algorithm proceeds as follows.

(1) Partition the set of all states into sets such that all members of a set have identical output rows in the ST table. For $\mathcal{M}4$ this step takes the form:

$$S1 = \{a, b, c, d, e, f, g, h, l\}, \qquad S2 = \{i, j, k, m, n, p\}$$

This step suggests that by giving an input $x = 1$ and observing the output signal it is possible to distinguish whether $\mathcal{M}4$ is presently in one of the states of $S1$ or $S2$. If $z = 0$ is observed, the machine was in one of the states of set $S1$; if a 1 is observed the state was a member of $S2$. See Fig. 5.10.

(2) Under each state, for each input symbol record the number of the set of which the following state is a member.

$$\{ a, \quad b, \quad c, \quad d, \quad e, \quad f, \quad g, \quad h, \quad l \ \}$$
$$1, 1 \quad 1, 1 \quad 1, 1 \quad 1, 2 \quad 2, 2 \quad 1, 2 \quad 2, 2 \quad 1, 1 \quad 1, 1$$
$$\{ i, \quad j, \quad k, \quad m, \quad n, \quad p \ \}$$
$$1, 1 \quad 1, 1 \quad 1, 1 \quad 1, 1 \quad 1, 1 \quad 1, 1$$

State a is followed by states b and c; both are members of $S1$, so we write 1, 1.

(3) Divide existing state sets so that all members of a new state set possess the same subscripts. When no new sets are formed the algorithm terminates; if new sets are formed we repeat step 2.

$$S1 = \{a, b, c, h, l\}, \quad S2 = \{d, f\}, \quad S3 = \{e, g\}, \quad S4 = \{i, j, k, m, n, p\}$$

This second group of sets says that it is possible to tell whether the machine was initially in a state of $S1$ or of another set with a two-symbol input sequence. The input sequence 1, 1 will distinguish state "d" from state "a", for example, and hence states "a" and "d" are placed in different state sets.

Repeating steps 2 and 3 ultimately gives the following state sets.

$$\{a\} \quad \{h, l\} \quad \{b, c\} \quad \{d, f\} \quad \{e, g\} \quad \{i, j, k, m, n, p\}$$
$$A \qquad B \qquad C \qquad D \qquad E \qquad F$$

We can not tell state "h" from "l"; must these be different states of $\mathcal{M}4$? No, each set of equivalent states is represented by a single state in the *minimum state* model of a machine. We thus can model $\mathcal{M}4$ with 6 states $A, B, \ldots, F$ as shown in Fig. 5.13. State A corresponds directly to state "a" and is hence our initial state. Any input sequence that we apply to the minimum machine starting in state A gives the same output sequence as the 15-state model of $\mathcal{M}4$ starting in state "a". This is more generally true: if we start in state B we get the same response as starting in either state h or l, and a corresponding statement can be made for each of the 6 states in Fig. 5.13. We thus say that Fig. 5.13 models a network with synchronous

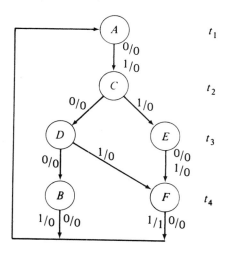

Present	Next state		Output	
state	$x = 0$	$x = 1$	$x = 0$	$x = 1$
A	C	C	0	0
B	A	A	0	0
C	D	E	0	0
D	B	F	0	0
E	F	F	0	0
F	A	A	0	1

FIG. 5.13 Minimum State Representations of $\mathscr{M}4$

terminal behavior, which is equivalent to that of the network modeled by the machine of Fig. 5.10.

Given the sets of equivalent states, we can obtain the minimum ST diagram or table with ease. Suppose $S = \{s_1, s_2, \ldots\}$ is a set of equivalent states. We are to replace all of these states with a single state: let us call it s. Then all state transitions to $s_1, s_2, \ldots$ must now be drawn to state s, and all transitions from $s_1, s_2, \ldots$ to other states must now emanate from s. We must do this for each set of equivalent states. Duplicate transitions need be shown only once, of course.

Step 3. State Assignment

A machine with α states can only be realized with β or more binary memory elements where α and β satisfy:

$$\alpha \le 2^{\beta} \tag{5.4}$$

Thus at least 3 flip-flops will be required to realize $\mathscr{M}4$. The importance of finding the reduced machine representation is now clear.[2] Four flip-flops

[2] Reduction of the number of states, to minimize the number of memory elements required, may not actually minimize the cost of the machine. In some cases the use of more than the minimum number of memory elements can reduce the required combinational logic. (See Prob. 5.2-6 and then Probs. 5.3-1(c) and 5.3-2).

would be required to realize the 15 states of $\mathcal{M}4$'s original description; only 3 are actually required to distinguish the unique states of machine $\mathcal{M}4$. There is no advantage at this point in reducing the number of states to less than 8, however.

States of the three flip-flops to be used in the machine realization must now be assigned to each of the states A, B, ..., F. The code chosen is important since it affects the required combinational logic, but no simple method is available for determining which assignment will lead to minimum cost combinational logic. Although complex procedures are available, we will look at state-assignment techniques that are more easily applied.

A state assignment may be suggested by the problem or state diagram. Machine $\mathcal{M}5$ offers an example. We have four states, each of which corresponds to the machine remembering two previous input symbols. At least two flip-flops must be used in any realization of $\mathcal{M}5$. It is rather natural to assign the flip-flop states to correspond directly to the input bits we wish to store. Thus the first FF state may indicate the previous input bit, while the second FF records the second previous input bit. Table 5.6 details this state assignment.

This very natural state assignment for $\mathcal{M}5$ may actually be the best possible assignment, not because of circuit costs, but because the direct correspondence between FF states and previous input symbols facilitates communication and understanding of how $\mathcal{M}5$ operates. Preparation of manuals for the machine and explanation of the operation of $\mathcal{M}5$ and significance of the FF states to maintenance technicians will be simplified. They will probably perform their tasks more effectively than if we had selected a state assignment that led to simpler circuitry, but in which FF states were not associated with equivalent history sets in such a direct manner. Overall we may save a great deal more money by reducing maintenance costs than we could have saved by hardware simplification.

Actually the state assignment for $\mathcal{M}5$ that we have selected leads to reasonably low circuit and design costs. The required state transitions are realized by interconnecting two FFs in a "shift register" configuration, a rather regular and common structure, which we will encounter in a later

TABLE 5.6

A NATURAL STATE
ASSIGNMENT FOR $\mathcal{M}5$

Present machine state	Present FF states	
	y_1	y_2
0	0	0
1	0	1
2	1	0
3	1	1

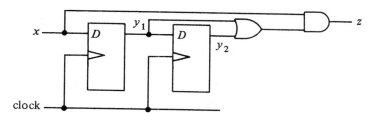

FIG. 5.14   A Realization of $\mathcal{M}5$

section. The combinational circuitry required to generate the output symbol of $\mathcal{M}5$ is easily designed. Figure 5.14 illustrates the simplicity that results from this state assignment.

If the state assignment is not suggested by the problem, we may make an arbitrary state assignment. The following rules of thumb may enable us to arrive at reduced cost networks.

1. Use the minimum number of states.
2. Assign adjacent code words to a state and the state that follows it.
3. If two present states have the same next states, assign those present states adjacent code words.

These rules may be contradictory and impossible to satisfy in all cases. To the extent that they can be satisfied, they tend to facilitate minimization of the state-transition equations of the machine.

For example, if we apply the first two rules to the minimum state representation of $\mathcal{M}4$, we might obtain the assignment shown in Fig. 4.35. Most state transitions are then accomplished by only one FF state change. The transition from state C to state D is an exception.

If we now replace each state name in Fig. 5.13 with the 3-tuple assigned to it, we obtain Table 5.7 from which state-transition equations may be written. For $\mathcal{M}4$ with the state assignments of Fig. 5.15, prepare a Karnaugh map for each next state variable and one for output variable z. From these maps, we write

$$y_1' = \bar{x}y_1 y_3 \vee \bar{x}y_2 \bar{y}_3$$
$$y_2' = \bar{y}_1 \bar{y}_2 \bar{y}_3 \vee x\bar{y}_1 \bar{y}_3$$
$$y_3' = y_2 \vee xy_1 y_3$$
$$z = x\bar{y}_1 \bar{y}_2 y_3$$

The unassigned state triples are used as don't cares to simplify these state-transition equations. If DFFs are to be used to realize $\mathcal{M}4$, we now have the Boolean equations required to design the combinational logic of our model of Fig. 5.2. If other FF types are to be used, an additional design step is required.

State	y_1	y_2	y_3
A	0	0	0
B	1	0	0
C	0	1	0
D	1	0	1
E	0	1	1
F	0	0	1
not used	1	1	0
not used	1	1	1

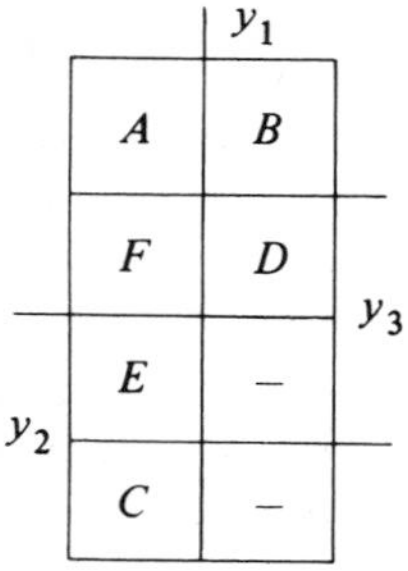

Map (with y_1 across top, y_3 at right, y_2 at left):

	y_1	
A	B	
F	D	y_3
E	—	
C	—	

FIG. 5.15 A State Assignment for $\mathcal{M}$4 based upon Two Rules of Thumb

Would application of rule 3 lead to lower cost state-transition equations? Apply it, obtain an alternative state assignment, and see.

Finally, several "standard" state assignments exist and possess merit. Some of these are more advantageous when designing asynchronous machines and hence are treated in a later chapter. But let us consider one such assignment, which uses as many memory elements as there are states. State minimization is obviously desirable if this assignment is to be used. With a *one-hot* code the ith FF is set when the machine is in the ith state, all other FFs being reset. Thus the states of $\mathcal{M}$5 shown in Fig. 5.12 might be encoded:

State	y_1	y_2	y_3	y_4
0	1	0	0	0
1	0	1	0	0
2	0	0	1	0
3	0	0	0	1

This assignment is desirable from a maintenance point of view. We can identify the state of the machine by identifying the flip-flop which is set. But another advantage also exists. Suppose the network of Fig. 5.16 is available at low cost, perhaps on an integrated circuit chip. With this

TABLE 5.7 FLIP-FLOP STATE TABLES FOR $\mathcal{M}$4

Present state y_1, y_2, y_3	Next states y_1', y_2', y_3'		Output z	
	$x = 0$	$x = 1$	$x = 0$	$x = 1$
A 000	010	010	0	0
B 100	000	000	0	0
C 010	101	011	0	0
D 101	100	001	0	0
E 011	001	001	0	0
F 001	000	000	0	1

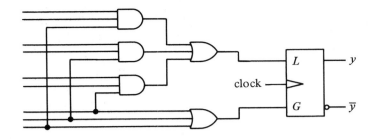

FIG. 5.16 Module for Sequential Machine Synthesis

module we can synthesize directly from the state-transition diagram. One module is required for each state. For each arrow into a state we connect the input symbol signals to terminals of an AND gate. The present-state signal is also connected to one of the AND gate terminals and to one of the OR-gate terminals. Figure 5.17 shows in part the network corresponding to Fig. 5.12. Output signals must be realized independently.

Step 4. Flip-Flop Input Equations

If other than DFFs or special modules are used, we must transform state transition equations into R, S or J, K, or T flip-flop input equations. This transformation is relatively complicated and so details are deferred to the next section. Before we can perform the transformation we must know what type of FF is to be used in the realization of a sequential machine. In some cases we may be told "Our company uses JKFFs only". If we insist on using some other type of FF, we must construct them from JKFFs. In other cases a variety of types of FFs may be available. In any case, the decision of what type of FF will be used is very significant. It

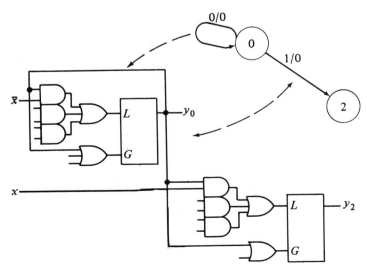

FIG. 5.17 Module Synthesis from the State Diagram

greatly influences the amount of combinational circuitry required to realize the machine. To complete the design of $\mathscr{M}4$, a JK, an RS, and a TFF were selected to illustrate the transformation for each of these FF types. The results provided by the transformation are

$$J_1 = \bar{x} y_2 \bar{y}_3 \qquad S_2 = \bar{y}_1 \bar{y}_2 \bar{y}_3$$
$$K_1 = x \vee \bar{y}_3 \qquad R_2 = \bar{x} y_2 \vee y_3 \qquad \text{(obtained in the next section)}$$
$$T_3 = y_2 \bar{y}_3 \vee \bar{x} \bar{y}_2 y_3 \vee \bar{y}_1 \bar{y}_2 y_3$$

There is no reason at this point to believe that this selection of FF types leads to minimum cost combinational logic.

Step 5. Implement the Network.

Given the above equations, we can implement $\mathscr{M}4$ using all of the design techniques of previous chapters. The types of gates we are to use, the delay they introduce, their fan-in and fan-out limits, etc., must all be taken into account. A straightforward implementation is shown in Fig. 5.18. Figure 5.19 shows typical waveforms for this circuit. Check the circuit and waveforms to verify that the original specifications of $\mathscr{M}4$ are satisfied.

What happened to the two unused code words of Fig. 5.15? Don't cares were assigned in the process of arriving at the flip-flop input equations. We can now investigate the circuit to see what it actually will do if it

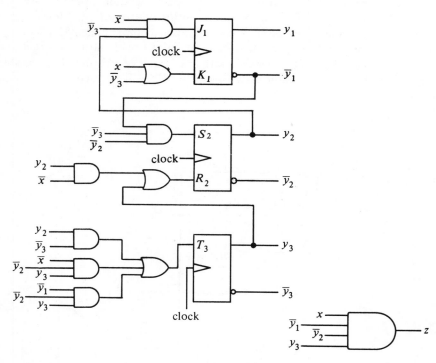

FIG. 5.18   Implementation of $\mathscr{M}4$
Not all interconnections are shown

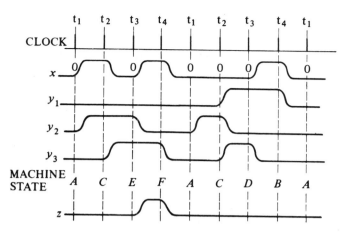

FIG. 5.19   Typical Waveforms of $\mathcal{M}4$

somehow arrives in one of the two unused states. Assume all FFs are set and $x = 0$, for example. Such analysis should reveal the state diagram of $\mathcal{M}3$ shown in Fig. 5.8.

While this five-step algorithm gives a realization of given specifications, we have no assurance that it is of minimum cost. The state diagram, number of FFs, state assignment, and types of flip-flops that we elect to use each influence the cost of the final network. And these factors interact. While each step attempts to optimize, we can not yet take the interactions into account in general.

5.3   DERIVATION OF FLIP-FLOP INPUT EQUATIONS

In this section we wish to obtain the FF input equations for $\mathcal{M}4$ shown in the previous section, but more generally, we wish to establish Marcus'[11] procedure for deriving FF input equations from a ST Table. We will find that this procedure is not complex, but it does require that we understand (1) exactly what a state-transition table is and (2) the logic of each type of flip-flop. We will use $\mathcal{M}4$ for purposes of illustration; its ST table is repeated below.

TABLE 5.8   STATE TRANSITIONS OF $\mathcal{M}4$

Present state			Next state $y_1'\, y_2'\, y_3'$		
y_1	y_2	y_3	$x = 0$	$x = 1$	
0	0	0	010	010	Must set FF2
1	0	0	000	000	Must reset FF1
0	1	0	101	011	
1	0	1	100	001	Reset FF3, Reset FF1
0	1	1	001	001	
0	0	1	000	000	

Let us examine this table row by row and note what FF state changes are specified by each row. The first row tells us that if all FFs are reset at present, then regardless of the value of input signal x appropriate signals are to be presented to FF2 *now*, so that it will be in its set state at the next clock time. The input signals supplied to FF1 and FF3 must be such that neither will change its state.

Exactly what FF input signals are "appropriate" depends on the type of FF we expect to use. If FF2 is to be an RSFF, for example, then the input signals required to satisfy the first row of the ST table are clear. We want FF2 to change to the set state so we must supply a set signal ($S_2 = 1$) and not supply a reset signal ($R_2 = 0$). If FF2 is to be a JKFF the situation is slightly different. To set a JKFF we must supply a J_2 input signal ($J_2 = 1$). We may but need not have $K_2 = 1$. Thus a don't care situation with respect to K_2 arises. If we can simplify logic by having $K_2 = 1$ when

$$y_1 = y_2 = y_3 = 0,$$

we are allowed to do so with the JKFF. This freedom did not exist with the RSFF.

What input signals must we supply to FF1 and FF3 so that they will not change state? Again, the FF type must be known before we can answer. If FF1 is to be an RSFF, then we may not have $S_1 = 1$. We may but are not required to have $R_1 = 1$. Thus a don't care situation with respect to R_1 arises. If FF1 is to be a TFF, we have no choice. The logic of the TFF requires that $T_1 = 0$ if its state is not to change. These findings are summarized in Table 5.9, which shows the values that FF input variables must take to accomplish state transitions.

Machine $\mathcal{M}4$ is described by the entire table and not just the first row. We must examine each row and analyze it in the same manner. We might first examine each row and locate only the situations that call for a FF to change state. These are the situations in which we clearly *must* supply input signals. In row 2 of Table 5.8 we see that when $y_1 = 1$, $y_2 = y_3 = 0$, and we must supply a signal to cause FF1 to reset.

TABLE 5.9   INPUT VARIABLE VALUES TO ACHIEVE
FF STATE CHANGES

Desired Transition $y \rightarrow y'$	RSFF S R	TFF T	JKFF J K	GLFF G L
0 0	0 d	0	0 d	0 d
				d 0
0 1	1 0	1	1 d	1 1
1 0	0 1	1	d 1	1 0
1 1	d 0	0	d 0	0 d
				d 1

In row 3 we see that several FFs are required to change state, and that the value of input signal x must enter our deliberations. If $x = 0$, then all three FFs must change state. With $x = 1$, only FF3 changes state.

After examination of all rows and determination of what input signals are required, we can gather together, as a Boolean sum-of-products expression, all of the situations that require a specific value of a specific input variable. To illustrate this point, let us arbitrarily choose FF1 to be a JKFF. In Table 5.10 the $y_1{}'$ columns of Table 5.8 are accompanied by notation of the required action.

We really have ill-formed truth tables for J_1 and K_1. Given these truth tables, we can employ all of the design and minimization techniques of previous chapters to arrive at the combinational logic which supplies the input signals to FF1. Fig. 5.20 shows these truth tables and the Karnaugh maps we use to arrive at the minimum input equations for FF1 given in the previous section.

No entries for J_1 and K_1 appear in four rows of the truth table. These rows correspond to the unused code words in Fig. 5.15. ($\mathcal{M}4$ is a six-state machine.) We never expect these conditions to arise, so it is reasonable to utilize them as don't care conditions. Doing so in the Karnaugh maps of Fig. 5.20 (denoted – to distinguish the origin of the don't cares) permits greater minimization.

Some designers prefer to locate on an ST table only situations where FF states must be changed, and develop input equations from only these conditions. Doing so assigns all don't cares the value of 0, and minimum Boolean expressions cannot be expected. If we ignore don't cares for FF1 we must write

$$J_1 = \bar{x}\bar{y}_1 y_2 \bar{y}_3$$
$$K_1 = y_1 \bar{y}_2 \bar{y}_3 \lor x y_1 \bar{y}_2$$

The cost of circuitry to generate K_1 is greatly increased.

Transforming ST Maps

We used the Karnaugh map in a last step to minimize equations. Could we not also perform our examination on the Karnaugh map? Yes, and it is valuable to do so if we are free to use various types of FFs to construct a machine. To see how we can use maps, let us draw the state-transition

TABLE 5.10 STATE TRANSITIONS OF FF1

Present state	Next State $y_1{}'$		Required action	
y_1 y_2 y_3	$x = 0$	$x = 1$	$x = 0$	$x = 1$
0 0 0	0	0	$J = 0 : K = d$ (don't care)	$J = 0, K = d$
1 0 0	0	0	Reset: $K = 1, J = d$	Reset : $K = 1, \ J = d$
0 1 0	1	0	Set : $J = 1, K = d$	$J = 0, K = d$
1 0 1	1	0	$J = d, K = 0$	Reset : $K = 1, \ J = d$
0 1 1	0	0	$J = 0, K = d$	$J = 0, K = d$
0 0 1	0	0	$J = 0, K = d$	$J = 0, K = d$

x	y_1	y_2	y_3	J_1	K_1
0	0	0	0	0	d
0	0	0	1	0	d
0	0	1	0	1	d
0	0	1	1	0	d
0	1	0	0	d	1
0	1	0	1	d	0
0	1	1	0		
0	1	1	1		
1	0	0	0	0	d
1	0	0	1	0	d
1	0	1	0	0	d
1	0	1	1	0	d
1	1	0	0	d	1
1	1	0	1	d	1
1	1	1	0		
1	1	1	1		

J_1 map (rows y_2, y_3; columns y_1, x):

0	d	d	0
0	d	d	0
0	$-$	$-$	0
1	$-$	$-$	0

$J_1 = \bar{x}\,y_2\,\bar{y}_3$

K_1 map (rows y_2, y_3; columns y_1, x):

d	1	1	d
d	0	1	d
d	$-$	$-$	d
d	$-$	$-$	d

$K_1 = \bar{y}_3 \lor x$

FIG. 5.20   Signals J_1 and K_1

map for y_2' of $\mathcal{M}4$ as in Fig. 5.21. We simply transfer the entries for y_2' in Table 5.8 to the map format.

Now let us consider cells just as we considered the rows of the original ST table. In the upper left corner cell we find a 1. This cell is in the $\bar{y}_2$ half of the map. The y_2 boundary has been darkened to emphasize this point. This cell and all others in the $\bar{y}_2$ half of the map correspond to situations in which FF2 is reset at present. The 1's in the upper corner cells indicate that FF2 is to be set at the next clock time following that at which the conditions that identify these cells (present state and input symbol) are encountered. We must take appropriate action now, and to record and emphasize this let us replace those 1's with S's as in Fig. 5.21(b).

S means " must take setting action." All other entries in the $\bar{y}_2$ half of the map are 0's indicating that FF2 is to remain reset when any of these six conditions are encountered.

In the y_2 (lower) half of the y_2' map of Fig. 5.21(a) we find three cells which contain 0's. All cells in this half of the map correspond to situations

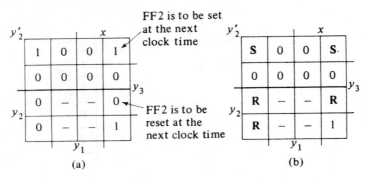

FIG. 5.21   ST Maps for y_2' of $\mathcal{M}4$

in which FF2 is set at present. The 0's indicate that FF2 is to be reset and again we must supply input signals to accomplish this state change. An **R** in the map of Fig. 5.21(b) reminds us that we must supply FF input signals that cause it to reset.

R means "must take resetting action." The 1 appearing in the lower half of the map indicates that FF2, set at present, is to remain set. And again we see the don't cares that arise from unassigned code words.

We can write FF input equations from the transformed map of Fig. 5.21 (b). But we must interpret this map in a different fashion for each type of FF. So let us consider the FF types in turn, and draw intermediate maps to emphasize the difference in interpretation of our transformed ST map.

RSFF

Assume FF2 is an RSFF. Then those cells which contain **S**'s dictate the conditions under which S_2 must have the value of 1. **S** cells constitute the ON-array for S_2 and OFF-array for R_2 for $S_2 = 1$, $R_2 = 0$ is the only way to set an RSFF which is reset at present. In a dual fashion, **R** cells make up the ON-array for R_2 and the OFF-array for S_2.

Now what can we say about the cells of the transformed map that contain 1's? These cells denote situations in which FF2 is set at present and is to remain set. May or must we have either S_2 or R_2 equal to 1? We *may* have $S_2 = 1$, but this is a don't care condition with respect to S_2. We *must* have $R_2 = 0$. Hence 1 cells are part of the DC-array of S_2 and the OFF-array of R_2. Cells containing 0's become don't care conditions with respect to R_2 and must be included in the OFF-array of S_2.

Don't-care cells of the transformed map are also don't-care cells for both S_2 and R_2. Let us draw maps for S_2 and R_2 with all of these considerations in mind. From these we can write minimum expressions for S_2 and R_2.

$$S_2 = \bar{y}_1 \bar{y}_2 \bar{y}_3$$
$$R_2 = y_3 \vee \bar{x} y_2$$

With experience we will be able to write these equations directly from the transformed ST map.

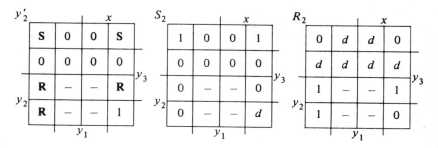

FIG. 5.22   Transformed *ST* Map for y_2', and S_2 and R_2

In summary, cells containing S's make up the ON-array of the set input equations for an RSFF. Cells containing don't cares and 1's make up its DC-array. Let us symbolize this

$$\text{Set} = \bigvee \mathbf{S} \vee \bigvee_d 1, d \qquad \textbf{(5.5)}$$

The reset input equation is determined by $\mathbf{R}$ cells with don't care and 0-cells used as don't cares.

$$\text{Reset} = \bigvee \mathbf{R} \vee \bigvee_d 0, d \qquad \textbf{(5.6)}$$

JKFF

The JKFF differs from the RSFF in that simultaneous $J = K = 1$ is allowed, and can often be used to advantage. But as a result of this difference, we must interpret the transformed ST map for y_2' in a different manner. In particular, $\mathbf{S}$ cells denote situations where we must take setting action by $J = 1$. We may have $K = 1$ in these cases if it is advantageous to do so. Hence, $\mathbf{S}$ cells become don't care cells in a map for K. Similarly $\mathbf{R}$ cells are don't care cells with respect to J-input equations.

$$J = \bigvee \mathbf{S} \vee \bigvee_d \mathbf{R}, 1, d \qquad \textbf{(5.7)}$$

$$K = \bigvee \mathbf{R} \vee \bigvee_d \mathbf{S}, 0, d \qquad \textbf{(5.8)}$$

These equations are used in drawing the J_2 and K_2 maps of Fig. 5.23 from the transformed y_2' map. But must we actually draw J and K maps to be able to write minimum equations? Try to write these equations directly from the transformed y_2' map.

$$J_2 = \bar{y}_1 \bar{y}_3$$
$$K_2 = \bar{x} \vee y_3$$

TFF

To change the state of a TFF, we must supply $T = 1$. Both $\mathbf{S}$ and $\mathbf{R}$ cells denote situations in which FF state is to be changed. Thus both $\mathbf{S}$ and $\mathbf{R}$ cells make up the ON-array of T. Cells containing 0's and 1's denote situations in which the FF state is not to change. These cells thus make up the OFF-array of T.

$$T = \bigvee \mathbf{S}, \mathbf{R} \vee \bigvee_d d \qquad \textbf{(5.9)}$$

We will not draw an explicit map for T_2, but write the minimum equation directly from the transformed ST map.

$$T_2 = y_2 y_3 \vee \bar{x} y_2 \vee \bar{y}_1 \bar{y}_2 \bar{y}_3$$

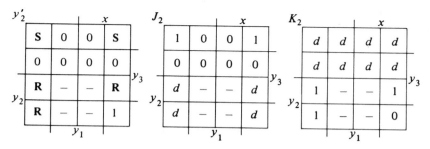

FIG. 5.23 J and K Maps for y_2'

GLFF

The transformed ST map is most difficult to interpret for the GLFF. Some aspects are straightforward. To set a presently reset GLFF (S cells), we must have $G = L = 1$. Thus S cells are a part of the ON-arrays of both G and L. To reset a set GLFF, we must have $G = 1, L = 0$. Thus **R** cells are also a part of the ON-array of G, but are a part of the OFF-array of L. And don't care cells are straightforward.

But now let us look at the 1 cells. The FF is to remain set. This can be accomplished in several ways. If $G = 0$, L may have any value. If $G = 1$, L must have the value of 1. Thus 1-cells may be thought of as don't-care cells with respect to both G and L, but a dependency exists. If in minimizing G we assign one of these don't care cells the value of 1, we must do the same for L, which does not always result in a minimum expression of L. The experienced designer can view maps for G and L simultaneously, and weigh the advantages of assigning a 1-cell to the G ON-array.

To make the matter worse, the 0 cells must be viewed in a different manner. Here the FF is to remain reset: we can have either $G = 0$ with any value for L, or $G = 1, L = 0$. Thus 0 cells are don't care cells with respect to both G and L, but if we assign one of these don't cares to the ON-array of G, it must be assigned to the OFF-array of L.

$$G = \bigvee S, R \vee \bigvee_d d, 1^*, 0^{**} \qquad (5.10)$$

$$L = \bigvee S \vee \bigvee_d d, 1^*, 0^{**} \qquad (5.11)$$

G and L maps for FF2 of $\mathscr{M}4$ are shown in Fig. 5.24. We might write the following sets of equations from these maps, but feel uneasy about calling them absolute minimum cost equations because of the don't care dependencies. (Section 10.5 offers a means of obtaining minimal equations.)

$$G_2 = \bar{y}_1 \qquad\qquad G_2 = \bar{x} \vee \bar{y}_2 \vee y_3$$

or

$$L_2 = \bar{y}_3(x \vee \bar{y}_2) \qquad L_2 = \bar{y}_1 \bar{y}_2 \bar{y}_3$$

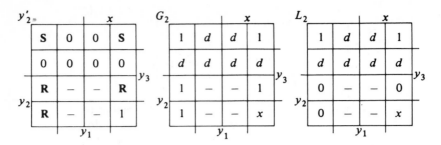

FIG. 5.24 G and L Maps for y_2'

Three types of don't cares appear: d's in the G_2 map are related to those in the L_2 map; the x in the G_2 map is related to the x in the L_2 map in a different manner.

We have interpreted the transformed ST map for y_2' in four different ways, corresponding to the four most popular FF types. If we really have these four types of FFs available when constructing $\mathscr{M}4$, we can compare the input equations to determine the FF type requiring minimum cost combinational circuitry. We see in our example that a JKFF requires less circuitry than an RSFF, as we should expect, since the JKFF can act as an RSFF but is more versatile. The GLFF requires circuitry of very nearly the same cost as the JKFF and we might consider using a GLFF.

In Fig. 5.18 we show FF3 to be a TFF. Is that really the best type of FF to use? We now have a means of evaluating the merit of each type of FF. But it is a great deal of work to draw S, R, J, K, etc. Karnaugh maps, and really not necessary. Draw the transformed map for y_3' and write minimum cost S, R, J, K, etc., FF-input equations directly from it. With practice this can be done with relative ease.

Direct, Heuristic Approaches

Back in Section 4.2, we designed a TFF from an RSFF. This was synthesis and the flip-flop input equation was solved in Table 4.4 by examining all possible cases, much as we have done in this section. To solve that problem, we equated the characteristic equation of the flip-flop to be used and the Boolean expression of desired next state behavior.

Table 4.4 was used as a means of solving the equation for two unknowns. While solving Boolean equations is difficult, we can often use this approach informally to obtain flip-flop input equations more easily. For example, ST equations for machine $\mathscr{M}4$ were derived in the previous section. Suppose that we decide to use a JKFF to realize y_1. Then equating its characteristic equation and the expression for y_1' gives

$$y_1' = J\bar{y}_1 \lor \bar{K}y_1 = \bar{x}y_1y_3 \lor \bar{x}y_2\,\bar{y}_3$$

Using the Expansion theorem on the right side brings these expressions to a form where answers can be written by inspection.

$$J\bar{y}_1 \vee \bar{K}y_1 = y_1(\bar{x}y_3 \vee \bar{x}y_2\,\bar{y}_3) \vee \bar{y}_1(\bar{x}y_2\,\bar{y}_3)$$
$$J_1 = \bar{x}y_2\,\bar{y}_3$$
$$\bar{K}_1 = \bar{x}y_3 \vee \bar{x}y_2\,\bar{y}_3$$
$$= \bar{x}(y_3 \vee y_2)$$
$$K_1 = x \vee \bar{y}_2\,\bar{y}_3$$

These equations are more expensive to realize than those derived from Fig. 5.20 because we did not take advantage of don't cares. But, note how easily they were obtained. The effort of introducing don't cares as specified by Eqs. (5.7) and (5.8) eliminates this advantage.

Many practical sequential networks have numerous input and output variables. Tables and maps become ponderous. The many minimization techniques of Chapter 9 may be used, or equations of reasonable cost may be written directly from the state diagram. The state diagram of Fig. 5.25 provides an example with different, but descriptive notation. Rather than complete input symbols, Boolean expressions are used to label the transitions. Output symbols are replaced by a list of the output variables that are to take the value of 1 under the conditions of each label. Thus if machine $\mathcal{M}6$ leaves state α under input condition $A \cdot B = 1$, then $U = 1$.

A reasonable state assignment for $\mathcal{M}6$ is provided by Fig. 5.25. We will call the state variables, y_1 and y_2, and realize y_1 with a DFF and y_2 with a JKFF to illustrate how the flip-flop type enters into the considerations. For y_1, we must identify when it is to be set and when it is to remain set. For all other cases, it is to be reset. Then y_1 must be set when $\mathcal{M}6$ is in state α and $AB = 1$. It remains set when $\mathcal{M}6$ is in state β and $C = 0$.

$$D_1 = \alpha AB \vee \beta\bar{C}$$
$$= \bar{y}_1\bar{y}_2\,AB \vee y_1\bar{y}_2\,\bar{C}$$

For a JKFF we must identify when the flip-flop is to be set and when it is to be reset. With extra effort don't cares can be identified and used to

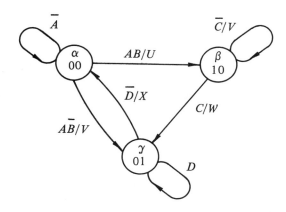

FIG. 5.25   Machine $\mathcal{M}6$

simplify equations. Flip-flop y_2 is to be set when state γ is entered. Two transitions lead to state γ.

$$J_2 = \alpha A \bar{B} \vee \beta C$$

Flip-flop y_2 is to reset when $\mathcal{M}6$ exits state γ.

$$K_2 = \gamma \bar{D}$$

The following don't cares can also be identified from Fig. 5.25.

Don't cares for J_2	Don't cares for K_2
γD, remaining in γ	$\alpha \vee \beta$, y_2 already reset
$\gamma \bar{D}$, leaving γ	or about to be set
y_2, already set	$\bar{y}_2$, already reset

When the input labels involve different input variables, minimization is seldom possible and the don't cares are of little value, with the exception of the final entries. The characteristic equation $z' = J\bar{z} \vee \bar{K}z$ guarantees that z need never appear in either true or complement form in the expressions for J and K. Minimization to the extent of removing the state variable from expressions for J and K is always possible.

$$
\begin{aligned}
J_2 &= \alpha A \bar{B} \vee \beta C \vee (y_2)_d \\
&= \bar{y}_1 \bar{y}_2 A \bar{B} \vee y_1 \bar{y}_2 C \vee (y_2)_d \\
&= \bar{y}_1 A \bar{B} \vee y_1 C \\
K_2 &= \gamma \bar{D} \vee (\bar{y}_2)_d \\
&= \bar{y}_1 y_2 \bar{D} \vee (\bar{y}_2)_d \\
&= \bar{y}_1 \bar{D}
\end{aligned}
$$

Equations for output variables can also be written directly from the state diagram. Figure 5.25 indicates that $U = 1$ only when $\mathcal{M}6$ is in state α and $AB = 1$; thus

$$
\begin{aligned}
U &= \alpha AB = \bar{y}_1 \bar{y}_2 AB \\
V &= \beta \bar{C} \vee \alpha A \bar{B} = y_1 \bar{y}_2 \bar{C} \vee \bar{y}_1 \bar{y}_2 A \bar{B} \\
W &= \beta C = y_1 \bar{y}_2 C \\
X &= \gamma \bar{D} = \bar{y}_1 y_2 \bar{D}
\end{aligned}
$$

If we trust that state $y_1 = y_2 = 1$ will never be entered, then the product $y_1 y_2$ may be used as a don't care term. The expressions for V, W, and X may all be simplified quickly through the use of this don't care. For example

$$X = \bar{y}_1 y_2 \bar{D} \vee (y_1 y_2)_d = y_2 \bar{D}$$

This technique is even easier to carry out when a one-hot code word is assigned to each state. $\mathcal{M}6$ of Fig. 5.25 could have been encoded using three flip-flops. We may call their output variables, α, β, and γ, with the agreement that $\alpha = 1$ only when $\mathcal{M}6$ is in state α, and so forth. Then the α flip-flop must be set whenever state α is entered. Using JKFFs

$$J_\alpha = \gamma \bar{D}$$
$$K_\alpha = AB \vee A\bar{B} = A$$
$$J_\beta = \alpha AB$$
$$K_\beta = C$$
$$J_\gamma = \alpha A\bar{B} \vee \beta C$$
$$K_\gamma = \bar{D}$$

by inspection.

5.4 SEQUENTIAL SYNTAX OF DDL

Section 2.8 concentrated on the description of combinational logic with design language DDL. Registers, the register transfer operator, automata, and systems were introduced so that processing systems with registers could be described. Having an expanded knowledge of memory elements and the state model of sequential networks available, we are now able to expand on these declarations and consider the remaining syntax of DDL, which is closely related to the state model.

Let us begin with a review. Registers are considered to consist of synchronized flip-flops in DDL. We declare registers to exist with the ⟨REgister⟩ declaration. Name(s) and a dimension are specified for each register. The dimension specification establishes the subscripts to be used to reference individual flip-flops or portions of a register. Either ascending or descending subscript ranges may be specified. Information is placed in a register with the register transfer operator ←. The left operand of the transfer operator identifies the destination of information—the register(s) to be loaded. Register transfers must be conditioned operations. The condition expression must include a synchronizing (clock) signal and may include other system variables.

Now we know a great deal about synchronizing techniques and flip-flop timing restrictions. DDL does not require or encourage the recording of such information. It is not of great concern to the designer when he is working at the level of abstraction where DDL is most valuable. We simply keep in mind that timing restrictions must be satisfied in networks derived from DDL descriptions. We write descriptions for which timing restrictions can be satisfied. We may indicate how they are to be satisfied in several ways. For example, since transfers are to be synchronized, we may identify the synchronizing variable by declaring it in a special way to be seen later. For now we will declare that signal to be a clock with the ⟨TIme⟩ declaration.

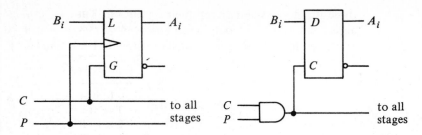

FIG. 5.26 Semantics of $|P \cdot C|\ A \leftarrow B.$ for (a) Registers and (b) Latches

⟨TI⟩ P(1E-6).
⟨TE⟩ C.
⟨RE⟩ A[10], B[10].
C = Boolean expression in other system variables,
$|P \cdot C|\ A \leftarrow B.$

The condition on the register transfer is the logic product of two Boolean expressions, one of which is obviously the synchronizing signal. It is important to identify that signal because flip-flops provide a special terminal to which it must be connected. The other Boolean expression indicates whether the flip-flops are to respond to incoming data or not. The design of registers is the subject of the next chapters; we will apply this condition in a straightforward manner here by recalling Fig. 4.16. The GL flip-flop is especially well suited for constructing registers. Condition C serves as the G signal, while data source B provides the L signal. P synchronizes the flip-flops. Figure 5.26(a) illustrates the semantics of the conditional register transfer statement. While we may actually realize the A register with some other type of flip-flop, that final design must be logically equivalent to Fig. 5.26(a).

Shift and Count Operators

Two types of register transfers are encountered so often in digital systems that they justify the inclusion of special operator symbols in a description language. In a shifting register bits of information are synchronously moved one or more cells to the left or right. DDL provides descriptive symbols for shift operations.

$$x\underset{\rightarrow}{\big\downarrow} \qquad \underset{\leftarrow}{\big\downarrow}x$$
shift right   shift left

The left operand, if any, is a constant enclosed in parentheses[3], which indicates the number of positions of shift. A default shift of one bit position is indicated by no left operand. The right operand is the register(s) to

---

[3] Necessary to prevent ambiguity.

⟨RE⟩ R[1 : n]. ⟨TE⟩ T[1 : k] and $k < n$

Shift Type	Shift Operation	Register Transfer Equivalent
save	(k) ⊥ R	R[k + 1 : n] ← R[1 : n − k]
serial load	(k) ⊥ T∘R	R ← T∘R[1 : n − k]
0-fill	(k) 0⊥ R	R ← 0 × k∘R[1 : n − k]
1-fill	(k) 1⊥ R	R ← 1 × k∘R[1 : n − k]
circulate	(k) c⊥ R	R ← R[n − k + 1 : n]∘R[1 : n − k]
extend	(k) e⊥ R	R ← R[1] × k∘R[1 : n − k]

be shifted. The symbol qualifier x specifies the nature of the shift. Table 5.11 indicates the permitted qualifiers and the register transfer specified by right shift operations. A very similar table can be developed for left shift operations.

Counting is also a register transfer operation with one register serving as both the source and sink of information. A constant is added to, or subtracted from, the content of the register to derive the new content. While Section 2.8 indicated that addition, subtraction, and relational operators are a part of DDL, they are particularly abstract operators in that their semantics depend upon the number system that the author has in mind. The same thing can be said of counting. Are we counting in BCD, binary, or 2's complement? Thus while these operators may be used, they cannot be translated to hardware without more information. An ⟨OPerator⟩ declaration is usually used to specify hardware, as we saw in Chapter 3.

The DDL count operators are

$$⊄ \qquad ⊄$$

Count up   Count down

A left operand may appear. As with shift operators it consists of an integer enclosed in parentheses. The right operand identifies the register to be incremented or decremented by 1 in the default case and the magnitude of the left operand in general.

⟨LAtch⟩ and ⟨DElay⟩ Declarations

A set of latches is declared to exist with the ⟨LAtch⟩ declaration, which has the same syntax as the ⟨TErminal⟩ and ⟨REgister⟩ declarations. The register transfer operator is also used to describe the loading of latch registers. The semantics of |P · C| A ← B. differ slightly if A has been declared to be a latch rather than a register. Since latches seldom have a special synchronizing terminal, the full conditional expression must be used as the control signal for a D-latch (gate signal for a gated-latch), as shown in Fig. 5.26(b).

Flip-flops are used as memory elements in synchronous sequential networks. We will find later that latches may be so used in asynchronous sequential networks. The very general model of Fig. 4.5 indicated that delay elements serve in general. The ⟨DElay⟩ declaration provides a means of introducing delay blocks into a network description. In addition to dimensional information provided with the same syntax as for terminals, registers, and so forth, delay magnitude is indicated with the syntax of the ⟨TIme⟩ delcaration; thus

$$\langle DE \rangle \ D1(1E\text{-}6) \ [10], \ D2(27E\text{-}9) \ [5].$$

declares ten 1-microsecond delays, D1, and a bank of five 27-nanosecond delay elements, D2. The mechanism for achieving these delays is not indicated, and no detailed parameters such as rise and fall time or inertia may be given.

⟨STate⟩ Declarations and State Statements

Sequential networks are often used to control activity in a digital system or subsystem. The ⟨AUtomaton⟩ declaration has a *head* and a *body* part. The head can take any of the following forms:

$$head \ :: = id:$$
$$:: = id: csop$$
$$:: = id: Be:$$
$$:: = id: Be: csop$$

The automaton is named by identifier *id*. Compatible operations to be performed may be listed as *csop* in the head, or after the body of the declaration, but most operations are usually expressed within the body of ⟨AU⟩ declarations. If a Boolean expression, *Be* enclosed with colons, is included in the heading, it serves as a condition on all register transfer type operations (←, ⌐, and others to be discussed) within the ⟨AU⟩ declaration. The Boolean expression is most often a system clock variable for use in synchronizing sequential activity in the entire subsystem. The clock need not be repeatedly written as a condition for every register transfer statement, if it is recorded once in the head part of an ⟨AU⟩ declaration (where it is easily observed).

The body of an ⟨AU⟩ declaration may contain declarations of facilities local to the automaton. It may also contain a ⟨STate⟩ declaration, which has very simple syntax.

$$\langle STate \rangle \ state \ statement \ list.$$

Note that the period is again used to terminate the declaration. State statements are of either of the following forms:

$$state \ statement \ :: = id: csop.$$
$$:: = id: Be: csop.$$

A state is named by identifier *id*, which may be subscripted and may include the state assignment enclosed in parentheses. The Boolean expression *Be* provides a condition over all operations in the set *csop*. Connection as well as transfer operations are conditioned by *Be*, if it is included. Since each state statement is terminated with a period, commas are not used to separate state statements in the body of the ⟨STate⟩ declaration.

An operator is required so that state transitions may be expressed. For each state, we describe its next state(s) with the "go to" operation

$$\rightarrow id$$

where the only operand *id* names a state of the automaton.

Machine $\mathcal{M}6$ is described with the following DDL statements.

⟨TE⟩ A, B, C, D, U, V, W, X.

⟨TI⟩ P1 (1E–6).

⟨AU⟩ $\mathcal{M}6$: P:

⟨ST⟩ ALPHA: A: |B| U = 1, → BETA; V = 1, → GAMMA..

BETA: |C| W = 1, → GAMMA; V = 1, → BETA..

GAMMA: ¬D: X = 1, → ALPHA...

If $\mathcal{M}6$ is in state *ALPHA*, it is to stay there and issue no output signals until $A = 1$ (at a clock time, of course, since $\mathcal{M}6$ is synchronized by P). With $\mathcal{M}6$ in state *ALPHA* and $A = B = 1$, $U = 1$ and the three-state machine advances to state *BETA*. These operations are not performed in sequence, but at the same time. If $A = 1$ and $B = 0$, then $V = 1$ and *GAMMA* is the next state of $\mathcal{M}6$. Each state of the state diagram is described in a natural way by one state statement.

Before the logic of a digital system can be detailed, the algorithms that subsystems follow to process information must be developed. The ⟨STate⟩ declaration is valuable for documenting such algorithms. We might call it the ⟨STep⟩ declaration because algorithms are usually described as a sequence of steps.

The ⟨STate⟩ declaration implies that a sequential network exists and controls the activity of an automaton. While the state transitions are clearly specified, the sequential network cannot be designed until a state assignment is made and the flip-flops of the network are declared as a "state sequencing register." The state sequencing register is identified by preceding its name in a ⟨REgister⟩ declaration with the symbol #. For example, the declaration

⟨RE⟩ A[5], #B[2], C[10].

indicates that the *B* register provides the flip-flops for the sequential network that controls activity in the automaton. While DDL provides several ways for specifying state codes, only the one previously mentioned will be illustrated here. The state code is listed as a parenthetical part of the state identifier. The 2-bit state assignment of Fig. 5.25 is indicated:

⟨AU⟩ 𝓜6: P:

    ⟨RE⟩ #Y[2].

    ⟨ST⟩ ALPHA (00B2): A: |B| → BETA; → GAMMA..

           BETA (10B2): |C| → GAMMA: → BETA..

           GAMMA (01B2): ¬D: → ALPHA...

Only the state transitions are listed as operations above so that we can consider the design of sequential networks from ⟨STate⟩ declarations more easily. State transition operations are special forms of register transfer operations. The operation → *id* indicates that the code for the identified state is to be placed in the state sequencing register. Let "#*id*" denote the code for the state *id* and *ssr* be the state sequencing register; then

$$→ id \quad \text{is equivalent to} \quad ssr ← \#id$$

The state transitions of 𝓜6 could be described as explicit transfers, but such a description is more difficult to read.

⟨AU⟩ 𝓜6: P:

    ⟨RE⟩ #Y[2].

    ⟨ST⟩ ALPHA (00B2): A: |B| Y ← 10B2; Y ← 01B2..

           BETA (10B2): |C| Y ← 01B2; Y ← 10B2..

           GAMMA (01B2): ¬D: Y ← 0...

Section 3.1 summarized the translation of DDL conditional syntax. We may now extend that translation to cover register transfer operations. First, the global condition expressed in the heading of an ⟨AU⟩ declaration applies to all transfer operations; it is an *if-then* condition that applies to all transfer operations. Second, the operations in a state statement are performed only when the controlling network is in the state that identifies the statement. Another *if-then* structure is suggested with the condition being a signal that takes the value of 1 only when the control network is in the state. A decoder may be used to form such signals from the state variables. Third, if the state statement includes a Boolean expression, another condition applies to the operations of that statement. These three levels of control are illustrated for the state transition of 𝓜6 below.

⟨AU⟩ 𝓜6:

    ⟨RE⟩ #Y[2].

    |P| | ·/Y\00B2| |A| |B| Y ← 10B2; Y ← 01B2...,

        | ·/Y\10B2| |C| Y ← 01B2; Y ← 10B2..,

        | ·/Y\01B2| |¬D| Y ← 0....

Connection operations are not shown and cannot be shown within the *if* P *then* clause because the automaton condition does not apply to them. While this description is not easily read and perhaps should never be written by hand, it does illustrate the meaning of and emphasize the clarity of the more sophisticated forms of expressing conditions provided by the state statement. And, it does transform the earlier descriptions of $\mathcal{M}6$ to the point where the transforms of Section 3.1 may be applied. Since we are now dealing with register transfer rather than connection operations, a control, as well as the data expression, must be formed. As a first step we distribute the most local conditions.

$\langle$AU$\rangle$ $\mathcal{M}6$:

$\quad\langle$RE$\rangle$ $\#$Y[2].

$\quad\quad$|P| | $\cdot$/Y\00B2| |A $\cdot$ B| Y $\leftarrow$ A $\cdot$ B $\cdot$ 10B2.,

$\quad\quad\quad\quad\quad\quad\quad\quad\quad$ |A $\cdot$ $\neg$B| Y $\leftarrow$ A $\cdot$ $\neg$B $\cdot$ 01B2.. ,

$\quad\quad\quad$ | $\cdot$/Y\10B2| |C| Y $\leftarrow$ C $\cdot$ 01B2.,

$\quad\quad\quad\quad\quad\quad\quad\quad$ |$\neg$C| Y $\leftarrow$ $\neg$C $\cdot$ 10B2.. ,

$\quad\quad\quad$ | $\cdot$/Y\01B2| |$\neg$D| Y $\leftarrow$ $\neg$D $\cdot$ 0....

Distributing the remaining conditions but treating the synchronizing signal in special fashion, since it will go to a special terminal on the flip-flop, gives:

$\langle$AU$\rangle$ $\mathcal{M}6$:

$\quad\langle$RE$\rangle$ $\#$Y[2].

$\quad\quad$|P $\cdot$ ($\cdot$/Y\00B2 $\cdot$ A $\cdot$ B v $\cdot$/Y\00B2 $\cdot$ A $\cdot$ $\neg$B v

$\quad\quad\quad$ $\cdot$/Y\10B2 $\cdot$ C v $\cdot$/Y\10B2 $\cdot$ $\neg$C v

$\quad\quad\quad$ $\cdot$/Y\01B2 $\cdot$ $\neg$D)|

$\quad$ Y $\leftarrow$ $\cdot$/Y\00B2 $\cdot$ A $\cdot$ B $\cdot$ 10B2 v $\cdot$/Y\00B2 $\cdot$ A $\cdot$ $\neg$B $\cdot$ 01B2

$\quad\quad\quad$ v $\cdot$/Y\10B2 $\cdot$ C $\cdot$ 01B2 v $\cdot$/Y\10B2 $\cdot$ $\neg$C $\cdot$ 10B2

$\quad\quad\quad$ v $\cdot$/Y\01B2 $\cdot$ $\neg$D $\cdot$ 0..

The operation of the control network has been reduced to one expression of the form:

$$|clock \cdot Be_1|ssr \leftarrow Be_2.$$

This development was not presented to frighten the reader but to show the semantics of the $\langle$STate$\rangle$ declaration. It also indicates that all of the transfers to a register can be collected to a single transfer in an algorithmic manner. A computer can be programmed to perform this transformation. We would not necessarily follow this procedure when designing by hand. While we did follow this procedure to a great extent for individual flip-flops in the previous section, we introduced the logic of the flip-flops to be used initially and then used more conventional notation. For example, $\bar{y}_1\bar{y}_2$ was used to express $\cdot$/Y\00B2.

$\langle$SEgment$\rangle$ Declarations

Often the states of an automaton can be divided into subsets in a very natural way. Some of the states are concerned with one phase of data processing, others with a second phase, still others with phase three, etc.

Natural divisions of a state set often exist, and it may be very desirable actually to partition a state set so that one designer might work on one phase, while another concentrates on others, for example.

A SEgment declaration describes a portion of an automaton. Usually it declares a subset of the state set of the automaton. It may also include declarations of facilities that are involved only in the operations of the states of the segment. Those facilities and states declared within a segment declaration are "private" to that segment and the names of such facilities and states are not recognized outside of the segment declaration. This allows the different designers working on different segments to use the same names for states or facilities of their segments, but requires that the expression of state transitions between states of different segments be more complicated.

The syntax of a segment declaration,

$$\langle SE \rangle \ head \ body.$$

is very much like that of the automaton declaration. The "head" section of the declaration may take any of several forms, but all begin with a name for the segment followed by arguments enclosed in parentheses, if any, followed by a colon. The name identifies the segment; no two segments of an automaton may have the same name. The parenthetical arguments are dummy facility names used in the description of the segment. Actual facility names must be given when references to a segment with arguments are made. Thus in Fig. 5.27 the segment named SEG2 has an argument X. The actual register to be used when performing the arithmetic operation of state R1 is not determined by the declaration of segment SEG2. References to segment SEG2 made in segment SEG1 indicate that actually register B is to be used in some cases and register C in other cases.

The first colon of the head of a segment declaration may terminate the head section, or it may be followed by:

(1) A Boolean expression followed by a colon; or
(2) A compatible set of operations, *csop*, or
(3) Boolean expression—colon—*csop*.

In segment SEG1 we see an example of form 1—Boolean expression:—. The Boolean expression is very simple: just the variable GO. As with the automaton declaration, no operation declared within the segment declaration (except a connection operation) is to be performed unless this global condition is satisfied. Thus no information-transfer operation of SEG1 is to be performed unless $GO = 1$ (and $P = 1$ from the encompassing automaton declaration).

In the declaration of SEG2 a compatible set of operations consisting of just the count operation $\notin K$ appears in the head. This operation is to be performed each time any operation of the segment is performed. Thus this operation could be written in every compatible set of operations declared in the segment, but it is clearly desirable to express it only in the segment heading.

The body of a segment declaration consists of facility and state declarations. Only the segment transition operations that appear in Fig. 5.27 are new and require explanation. The double arrow, $\Rightarrow$, is used to denote a transition to a state of another segment. In state R1 of segment SEG2 we see the notation "$\Rightarrow$SEG1," which may be read "go to segment SEG1." But which state of SEG1 is to follow state R1? It is assumed that the first state listed in the state declaration is desired. If some other state is to follow R1, then that state must be specified following "$\rightarrow$" as an argument. Thus in state S1 of SEG2, "$\Rightarrow$SEG1($\rightarrow$ S2)" may be read "go to segment SEG1, state S2."

In state R2 of SEG2 we find just the double arrow. This brief notation, which may be read "return," indicates that the next state of the automaton is to be declared in the segment in which the transition to the segment containing the return was specified. The next state may be specified, following $\Rightarrow$, as an argument of the original segment transition operation. If it is not so specified it is assumed to be the next state declared after the state containing the original transition operation. Thus in SEG1, state S1, $\Rightarrow$ SEG2(B, $\Rightarrow$S2) calls for a transition to the first declared state of SEG2 using register B in place of the dummy argument X. If a return operation is executed, the return is to be to state S2. State S2 of SEG1 is followed by state R2 of SEG2, with register C used in place of dummy argument X, and unspecified returns are to be to state S1 of SEG1.

That two states of automaton AU1 are identified by the same name introduces no confusion or ambiguity. Intrasegment state transitions are specified with a single arrow. Thus "$\rightarrow$S1" means "go to state S1 of this segment." Intersegment transitions are specified with the double arrow.

Automaton AU1 must actually have more states than the five declared in Fig. 5.27. SEG2 is declared with an argument. Thus the finite-state control machine must include a memory element that remembers whether register B or register C is actually to be used in the operations of the

$\vdots$

$\langle$AU$\rangle$ AU1 : P :
$\langle$ME$\rangle$ A[10], B[10], C[10], K[5], FF, GO.

$\langle$SE$\rangle$ SEG1 : GO :
    $\langle$ST$\rangle$ S1 : $csop_1$, |FF| $\rightarrow$ S2; $\Rightarrow$ SEG2(B, $\Rightarrow$ S2)..
           S2 : $csop_2$, $\Rightarrow$ SEG2(C, $\rightarrow$ R2, $\Rightarrow$ S1)...

$\langle$SE$\rangle$ SEG2(X) : $\notin$K $\langle$RE$\rangle$ X[10].
    $\langle$ST$\rangle$ R1 : A $\leftarrow$ A + X, | $\cdot$ /K| $\rightarrow$ R2; $\Rightarrow$ SEG1..
           R2 : $csop_3$, |X[1]| $\rightarrow$ S1; $\Rightarrow$..
           S1 : $csop_4$, $\Rightarrow$ SEG1($\rightarrow$ S2)....

FIG. 5.27   Segment State Transitions

$\vdots$

$\langle$AU$\rangle$ AU1 : P :

$\langle$ME$\rangle$ A[10], B[10], C[10], K[5], FF, GO.

$\langle$ST$\rangle$ SEG1S1 : GO : $csop_1$, |FF| → SEG1S2; →SEG2R11..

SEG1S2 : GO : $csop_2$, → SEG2R22.

SEG2R11 : ⊄ K, A ← A + B, | · /K| → SEG2R21; → SEG1S1..

SEG2R21 : ⊄ K, $csop_3$, |B[1]| → SEG2S11; → SEG1S2..

SEG2S11 : ⊄ K, $csop_4$ → SEG1S2.

SEG2R12 : ⊄ K, A ← A + C, | · /K| → SEG2R22; →SEG1S1..

SEG2R22 : ⊄ K, $csop_3$, |C[1]| → SEG2S12; → SEG1S1..

SEG2S12 : ⊄ K, $csop_4$, → SEG1S2...

Fig. 5.28   AU1 Declared without Segments

segment. Hence, that finite-state machine must have more states. If AU1 were declared without use of the segment declaration, SEG2 would have to be duplicated in effect, as suggested by Fig. 5.28, although equivalent states might be found and deleted. A suitably programmed computer can do this sort of "duplicating" as well as a logic designer.

As this example suggests, segment declarations are best used only when transitions between the states of a segment are common and transitions between states of different segments are rare. When these conditions are not satisfied, the "go to segment" notation is awkward, and little is gained by partitioning the state set.

5.5 SUMMARY

If clock periods are sufficiently long, false values on signal lines, resulting from propagation delays, are ignored by the flip-flops of synchronous sequential networks. They operate "properly." As a result it is relatively easy to design reliable synchronous sequential networks, and they are often used.

We record the "proper" operation of a synchronous sequential network with a state model that consists of input, output and state alphabets, and next state and output functions. Either a state transition table, diagram, or equations may be used to record the functions. The alphabets are usually so obvious that we fail to formally record them.

To analyze a given network, we break down the combinational logic. Flip-flop input equations are then substituted into the characteristic equations to introduce the flip-flop logic. The resulting equations are easily transferred to an ST table, just as Boolean equations are transferred to truth tables. The state diagram is easily drawn from the ST table since both present the same information.

To design a synchronous sequential network, we reverse the process. From given specifications we find an ST table or diagram that models the specifications. We may simplify by minimizing the number of states that appear in the model. Given the number of states to be realized and the number of flip-flops to be used, a binary code word for each state may be established. Many codes are available, and the code used effects the cost of the ultimate combinational logic. Several codes may be explored or other considerations used to select the final code. State transition equations, transformed ST maps, or heuristic procedures may be used to obtain flip-flop input equations from the coded state table or diagram. The set of flip-flop input equations and output variable equations are realized by using the combinational design techniques of previous chapters.

While these notational schemes and design techniques are adequate for small networks, describing and designing a full digital system is another matter. The vector notation of DDL introduced in Chapter 2 is augmented in this chapter so that sequential activity can be concisely described with sufficient precision for network design directly from it. Extensive design examples are not included because we will make much more use of DDL in subsequent chapters. There we will consider larger portions and finally full digital systems. Not all details of DDL have been considered, for it is a very extensive language. Please consult the appendices of this chapter for the full syntax and a partial dictionary of the semantics of DDL.

REFERENCES

Historical

1. D. A. HUFFMAN, "The Synthesis of Sequential Switching Circuits," *Journal of Franklin Institute*, Vol. 257, No. 3 and 4, March and April 1954, pp. 161-190 and 275–303.
2. G. H. MEALY, "A Method for Synthesizing Sequential Circuits," *Bell System Tech. Jour.*, Vol. 34, Sept. 1955, pp. 1045–1079.
3. E. F. MOORE, "Gedanken-Experiments on Sequential Machines," *Automata Studies, Annals of Math. Studies*, No. 34, Princeton University Press, New Jersey, 1956, pp. 129–153.

General

4. A. BARNA and D. I. PORAT, *Integrated Circuits in Digital Electronics*. New York: John Wiley and Sons, 1973.
5. T. L. BOOTH, *Digital Networks and Computer Systems*. New York: John Wiley and Sons, 1971.
6. M. A. BREUER, editor, *Digital System Design Automation: Languages, Simulation and Data Base*, Ch. 2. "Register Transfer Languages and Their Translation." Woodland Hills, California: Computer Science Press, 1975.
7. C. R. CLARE, *Designing Logic Systems Using State Machines*. New York: McGraw-Hill, 1973.
8. Y. CHU, *Digital Computer Design Fundamentals*. New York: McGraw-Hill, 1962.
9. F. J. HILL and G. R. PETERSEN, *Introduction to Switching Theory and Logical Design*, 2nd Ed. New York: John Wiley and Sons, 1974.

10. M. Krieger, *Basic Switching Circuit Theory.* New York: MacMillan, 1967.

11. M. P. Marcus, *Switching Circuits for Engineers.* Englewood Cliffs, New Jersey: Prentice-Hall, 1962; 1967.

12. E. J. McCluskey, Jr., *Introduction to the Theory of Switching Circuits.* New York: McGraw-Hill, 1965.

13. C. H. Roth, Jr., *Fundamentals of Logic Design.* St. Paul: West Publishing, 1975.

Advanced

14. T. L. Booth, *Sequential Machines and Automata Theory.* New York: John Wiley, 1967.

15. A. Gill, *Introduction to the Theory of Finite-State Machines.* New York: McGraw-Hill, 1962.

16. S. Ginsburg, *An Introduction to Mathematical Machine Theory.* Reading, Mass.: Addison-Wesley, 1962.

17. M. A. Harrison, *Introduction to Switching and Automata Theory.* New York: McGraw-Hill, 1965.

18. J. Hartmanis and R. E. Stearns, *Algebraic Structure Theory of Sequential Machines.* Englewood Cliffs, N. J.: Prentice-Hall, 1966.

19. F. C. Hennie, *Finite-State Models for Logical Machines.* New York: John Wiley, 1968.

20. R. E. Miller, *Switching Theory, Volume II: Sequential Circuits and Machines.* New York: John Wiley, 1965.

21. R. E. Prather, *Introduction to Switching Theory: A Mathematical Approach.* Boston, Mass.: Allyn and Bacon, 1967.

PROBLEMS

5.1-1. For each of the networks of Prob. 4.2-8 determine the input alphabet, state alphabet, and next state mapping, ψ.

5.1-2. The network shown has no input signals and hence is autonomous.

(a) Express its state alphabet and state mapping ψ via a state-transition table and state diagram.

(b) This network is initially in state $y_1 y_2 = 00$. Find its state sequence.

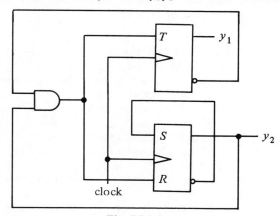

Fig. P5.1-2

5.1-3. Redraw the sequential machine of Fig. P5.1-3 to conform with the model of Fig. 5.2.

(a) Determine the set of states of the network.
(b) Determine the input and output alphabets.
(c) Determine the state table.
(d) Draw a state-transition diagram of the network.

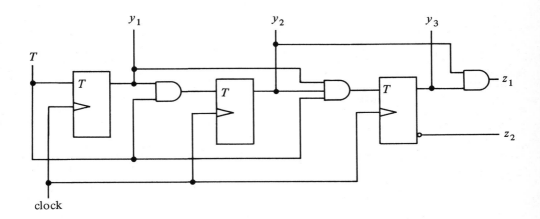

Fig. P5.1-3

5.1-4. Draw the state diagram for the network of Fig. P5.1-4. Is this network related to $\mathcal{M}2$?

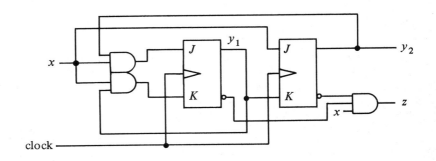

Fig. P5.1-4

5.1-5. The sequential machine described by the state diagram of Fig. P5.1-5 is presently in state α. What will its next states and outputs be for the given input sequence? Propose other initial states and input sequences, and repeat your analysis.

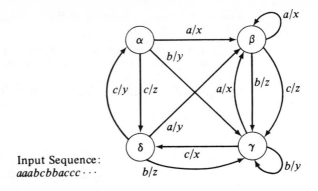

Input Sequence:
aaabcbbaccc · · ·

Fig. P5.1-5

5.1-6. The machine of Fig. P5.1-5 is in an unknown state. An input sequence is presented and the output of the machine observed in an attempt to determine that unknown state. For the input-output sequences given, try to determine the initial state and final state of the machine.
(a) *c/x, c/y*
(b) *c/z, b/z, a/x*
(c) *a/x, c/z, b/y, c/x*

5.1-7. Assuming that the machine of Fig. P5.1-5 is in one of two states, what input sequence would you give to determine the initial state of the machine for the following pairs of states?
(a) α, β (b) α, γ

5.1-8. It is necessary to force the machine of Fig. P5.1-5 from its presently unknown state to state β. Can you find an input sequence which will ensure that the machine ends up in state β if (a) the ouput sequence is not observed while the input sequence is being presented, and (b) the output sequence is observed and the input sequence altered as a result?

5.1-9. Machine $\mathcal{M}3$, Fig. 5.8, is in an unknown state. How would you get this machine into state y^0?

5.1-10. Machine $\mathcal{M}3$ is in state y^0. What input sequence or sequences cause $z = 1$ with their last input symbol? What is the final state of $\mathcal{M}3$ when each of these sequences is applied?

5.2-1. Which of machines $\mathcal{M}1$, $\mathcal{M}2$, and $\mathcal{M}3$ possess a state such that the machine must enter that state if any input symbol sequence is applied which is equal or greater in length than the number of states of the machine? Place each machine into one of the three classes discussed in the text.

5.2-2. A machine is said to be "strongly connected" if it is possible to get from any state to any other state, perhaps by passing through other states, by applying the proper input sequence. Which of $\mathcal{M}1$, $\mathcal{M}2$, and $\mathcal{M}3$ are strongly connected?

5.2-3. Draw state diagrams for the single-input, single-output machines specified.

$\mathcal{MA}$: An output $z = 1$ is to accompany the input symbol 1 if that 1 follows exactly two 0's, i.e., ...1001.... .

$\mathcal{MB}$: An output $z = 1$ is to accompany the input symbol 1 if that 1 follows two or more 0's.

$\mathcal{MC}$: An output $z = 1$ is to accompany an input symbol if the previous input symbol differs from that symbol.

$\mathcal{MD}$: An output $z = 1$ is to accompany an input symbol if the preceding two or more symbols are the same but differ from that symbol.

5.2-4. A machine with an input alphabet of 48 characters, including 26 letters, 10 numbers, the blank, and other punctuation, is required to signal the end of an English sentence by ringing a bell when it has detected the input sequence "period," "blank," "blank." Draw a state diagram for this machine. What is the output alphabet?

5.2-5. (a) Eliminate any redundant states in your answers to prob. 5.2-3.
(b) What is the minimum number of memory elements required to realize each machine?
(c) Make a state assignment for each machine, and write state equations. Show the logic for each machine using DFFs.
(d) Repeat (b) and (c) using a different state assignment. Attempt to achieve designs which require a minimum number of logic blocks.

5.2-6. Is the four-state model of $\mathcal{M}5$ given in Fig. 5.12 the minimum-state model? If not, draw a minimum-state diagram.

5.2-7. Eliminate equivalent states if they exist in the following machines.

Present state	Next state $x = 0$	1	Output $x = 0$	1
a	b	h	0	0
b	h	a	0	1
c	i	h	0	1
d	h	e	0	1
e	d	c	0	0
f	g	h	0	0
g	c	f	0	1
h	d	h	0	1
i	c	a	0	1

(a)

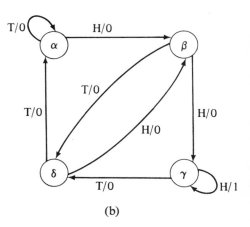

(b)

If in machine (b), T represents a tossed coin landing with tails showing and H represents heads showing, what does $z = 1$ signify?

5.2-8. A sequential machine is required to indicate by an output of 1 ($z = 1$) the time at which its single input changes from a value of 0 to a value of 1. The output is to be 0 at all other times.

(a) Prepare a state diagram for the required machine.

(b) Eliminate redundant states, if any, in your state diagram.

(c) Determine the number of memory elements required.

(d) Assign FF states to the machine states.

5.3-1. (a) Does the state diagram below describe a machine with which you are familiar?

(b) Write state-transition equations using the natural state assignment. Show a DFF network that realizes this machine.

(c) Redesign the network using JKFFs. Compare costs.

(d) The two networks that you have designed each have four or more states. Draw the complete state diagrams for your networks.

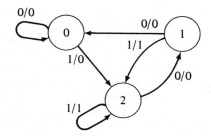

5.3-2. Attempt to find a four-state machine that does exactly what the machine of Prob. 5.3-1 does, but has a lower cost DFF realization. *Hint*: Prob. 5.2-6.

5.3-3. Design minimum cost JKFF networks for the machines of Prob. 5.2-5. Compare your networks with the original specifications. Do the networks perform as required?

5.3-4. Complete the design of the machine of Prob. 5.2-8 using
(a) RSFF, (b) JKFF. Compare costs.

5.3-5. If the cost of combinational logic is directly proportional to the number of gates required, which FF type would you use for each of the following state-transition maps?

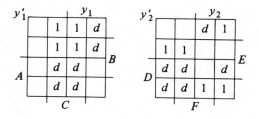

5.3-6. Design an autonomous decade counter that uses the 8421 weighted code. Four flip-flops are required. Determine the FF type that requires minimal combinational logic in each of the four positions.

5.3-7. Two synchronous sequential networks described by state diagrams below interact.

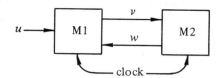

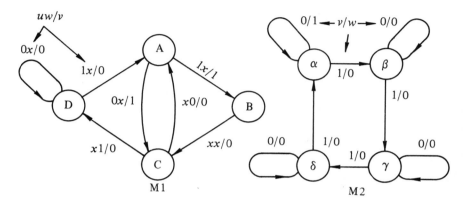

(a) Assume $M1$ is in state D, $M2$ is in state α at time 0, and input sequence $u = 0011100001\ldots$ is applied. Determine how the system responds (states of $M1$ and $M2$ and values of v and w at successive clock times).

(b) Design networks of NAND gates and JKFFs that realize $M1$ and $M2$. Demonstrate the validity of your designs by applying the input sequence $u = 0011100001$ and by showing the values of v and w, also the flip-flop states, at successive clock times.

5.4-1. Draw the state diagram of the machine described by these statements. Are all state transitions specified?

⟨ST⟩

S1 : |A| → S2; → S3 ..
S2 : AX1 : |A| → S4, AC2 = 1; → S5 ..
S3 : SX1 : → S4, SC2 = 1.
S4 : QX2 = 1, → S5.
S5 : DONE = 1, → S1..

5.4-2. Design the network of Prob. 5.4-1 using heuristic techniques and DFFs.
(a) Use a one-hot code.
(b) Use a 3-bit code.

5.1-1. (a)

	A	B		y_1	y_2		ψ	x^0	x^1	x^2	x^3
x^0	0	0	y^0	0	0		y^0	y^0		y^3	
x^1	0	1	y^1	0	1		y^1				
x^2	1	0	y^2	1	0		y^2		y^0		
x^3	1	1	y^3	1	1		y^3		y^1	y^3	

From Prob. 4.2-8(a).

5.1-2. $T_1 = \bar{y}_1\,y_2$
$S_2 = \bar{y}_2$
$R_2 = \bar{y}_1\,y_2$

y_1	y_2	y_1'	y_2'
0	0	0	1
0	1	1	0
1	0	1	1
1	1	1	1

$00 \rightarrow 01 \rightarrow 10 \rightarrow 11$

5.1-4. $J_1 = xy_2$
$K_1 = xy_1$
$J_2 = x$
$K_2 = y_1$
$z = x\bar{y}_1\bar{y}_2$

Present state	Next $x = 0$	1	Output $x = 0$	1
y^0	y^0	y^1	0	1
y^1	y^1	y^3	0	0
y^2	y^2	y^1	0	0
y^3	y^2	y^0	0	0

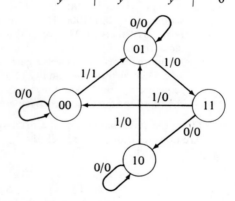

5.1-5.
a	a	a	b	c	b	b	a	c	c	c
α	β	β	β	γ	δ	γ	γ	β	γ	δ	α
x	x	x	z	x	z	y	x	z	x	y

5.1-7. Give input symbol b and observe output symbol.

5.1-10. $\left.\begin{array}{l} x1x1 \\ x011 \end{array}\right\}$ Machine is returned to y^0 by these 4-bit input sequences.

5.2-3.

MA: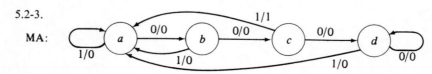

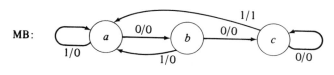

MB:

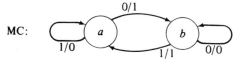

MC:

5.2-7. (a) $\{a, e, f\}$ $\{b, c, d, g, h, i\}$
$\{a, e, f\}$ $\{b, d, g, i\}$ $\{c, h\}$
    A   B   C

	$x = 0$	1	$x = 0$	1
A	B	C	0	0
B	C	A	0	1
C	B	C	0	1

5.3-3. The following assignment might be used for $\mathcal{MA}$. Observation of the cases in which each FF must change its state allows us to write the FF input equations below. Reflection on the don't care conditions suggests the simplification.

	y_1 y_2	y_1' $x = 0$	1	y_2' 0	1
a	0 0	0 1	0 0	0 0	
b	0 1	1 1	0 0	0 0	
c	1 1	1 0	0 0	0 1	
d	1 0	1 0	0 0	0 0	

$J_1 = y_2 \bar{x}$
$K_1 = x$
$J_2 = \bar{y}_1 \bar{x}$
$K_2 = y_1 \vee x = J_2$
$z = x y_1 y_2$

5.3-5. (a) DFF: $D = C$ Lowest cost
       RSFF: $S = C, R = A$ Same cost, 2 connections
       JKFF: $J = C, K = A$ Same cost, 2 connections
       TFF: $T = \bar{y}_1 C \vee y_1 \bar{C}$ Greatest cost
       GLFF: $G = 1, L = C$

5.3-6. FF4 should be a TFF.

5.4-1.

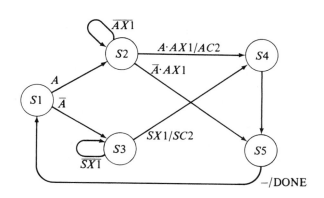

SYNTAX OF DIGITAL DESIGN LANGUAGE (DDL)

[Ref.6]

A language provides a set of symbols, rules for concatenating symbols to form allowed statements, accepted meaning of statements, and methods for generating statements that express information an author wishes to convey. The *syntax* of a language consists of the symbol alphabet and rules within which statements may be prepared. It is not concerned with the meaning of statements; that is the domain of *semantics*. The *pragmatics* of a language is concerned with the formation of statements.

This appendix presents only the syntax of DDL. The DDL alphabet includes 119 distinct symbols, none of which can be formed by concatenation of others. The "blank" or "space" is not a member of the alphabet, so separation of symbols has no significance, but may be used to make a statement more legible. It is assumed that statements are read from left to right and top to bottom of a page. The rightmost symbol of one line immediately precedes the leftmost symbol of the line immediately below.

Since the DDL alphabet includes the English alphabet, decimal digits, and many common punctuation, logic, and arithmetic symbols, it would be particularly awkward and confusing if the grammar of DDL were described without recourse to the symbols of the DDL alphabet. Thus English and Greek letters and decimal digits will be used to form names for *metalinguistic variables*, each of which describes a set of allowed DDL symbol sequences. Most *ml*-variables are defined as concatenation of other *ml*-variables and many take the form of any one of several such concatenations. The symbol ' is used as the concatenation operator. The concatenation of an *ml*-variable with a Greek name is written without the ' symbol, the concatenation being clearly implied. When alternative structures of an *ml*-variable exist, the underscore (_) is used to express "or." Thus ' and __ are *metalinguistic operators* of the language used to express the syntax of DDL.

A third *ml*-operator will be symbolized by a vertical line. The *ml*-variable named to the left of the line has the form described by the *ml*-expression that appears to the right of the line, much as an algebraic variable appears to the left of an equal sign and an expression for that variable appears to the right. Each line of the table then describes a syntatic rule of DDL.

English titles partition the table of syntatic rules, and line numbers and descriptive English names accompany each line to facilitate understanding and future reference. On the far right, line numbers offer limited cross-references.

Most *ml*-variable definitions are recursive, i.e., the *ml*-variable is defined in terms of itself and other *ml*-variables. Thus in line 3.1 a basic-bit string, *bbs*, is said to be either a bit $\beta \in \{0, 1\}$, or a basic-bit string followed by a bit, $bbs\beta$. Thus any sequence of 0's and 1's, regardless of length, is a *bbs* since it can be formed by concatenating in turn bits to the right starting just with the leftmost bit. Concatenating a bit on the left is not prescribed in line 3.1 but this limitation does not restrict the basic-bit strings that can be written. In all cases definitions must provide a non-recursive form to which other symbols may be attached via recursive definitions. It may be difficult to find this "starting point", but it does exist. Thus the definition

$$bbs \quad | \quad bbs\beta$$

would not be acceptable since it does not provide a way to form the first *bbs* to which bits may be appended.

0. *Alphabet*

A B C D E F G H I J K L M N O P Q R S T U V W X Y Z

a b c d e f g h i j k l m n o p q r s t u v w x y z

0 1 2 3 4 5 6 7 8 9 ?

⊢ ⊣ × ∘ ∧ · ↓ ↑ ◯ ≡ ⊕ ∀ ∨ ¬ /\ ≤ < ≥ > + −

$\downarrow$ $_0\downarrow$ $_1\downarrow$ $_e\downarrow$ $_c\downarrow$ $\downarrow$ $_\leftarrow0$ $_\leftarrow1$ $_\leftarrow e$ $_\leftarrow c$ ⊄ ⊅

( ) [ ] { } ⟨ ⟩

→ ⇒ ⇨ ← # * =

| ⌈ ⌊ : ; , .

1. *Characters and Operators*

1.1	letter	λ	A__B__C__ ⋯ __Z__a__b__ ⋯ __z
1.2	bit	β	0__1
1.3	octal character	oc	β__2__3__ ⋯ __7
1.4	digit	δ	oc__8__9
1.5	additive operator	ao	+__−
1.6	multiplicative operator	mo	*__/
1.7	logic operator	lo	∧__·__↑__↓__◯__≡__⊕__∀__∨
1.8	relation operator	ro	ao__≤__<__≥__>
1.9	count, shift operator	co	$\downarrow$ __ $_0\downarrow$ __ $_1\downarrow$ __ $_e\downarrow$ __ $_c\downarrow$ __ $\downarrow$ __ $_\leftarrow 0$ __ $_\leftarrow 1$ __ $_\leftarrow e$ __ $_\leftarrow c$ __ ⊄ __ ⊅

2. *Period*

2.1	comment character 1	com1	any member of the alphabet except right parenthesis).
2.2	comment string 1	cms1	com1__cms1'com1
2.3	period	π	. __ . (cms1)

3. *Numbers and Arithmetic Expressions*

3.1	basic bit string	bbs	β__bbsβ
3.2	octal string	os	oc__os'oc
3.3	unsigned decimal integer	udi	δ__udiδ
3.4	signed decimal integer	sdi	udi__ao'udi
3.5	basic floating point number	bfn	.udi__ao. udi__sdi__sdi . __sdi . udi
3.6	floating point number	fn	bfn__bfn'E'sdi
3.7	proper integer	pin	bbs'B__os'O__udi'D
3.8	binary string	bs	β__?__pin'udi__ # npid

4.3

3.9	basic value	*bva*	*udi__pin__udi : udi__pin : pin*	
3.10	value	*va*	*bva__va, bva*	
3.11	primary	*pri*	*udi__(ae)__ao'pri*	3.14
3.12	factor	*fac*	*pri__fac↑pri*	
3.13	term	*term*	*fac__term'mo'fac*	1.6
3.14	arithmetic expression	*ae*	*term__ae'ao'term*	1.5
3.15	field	*fld*	*ae : ae*	

4. *Identifiers*

4.1	basic identifier	*bid*	*λ__bid λ __bid δ*	1.1
4.2	unbracketed identifier	*uid*	*bid__bid(carg)*	5.1
4.3	nonparenthetical identifier	*npid*	*bid__bid[barg]*	5.4
4.4	identifier	*id*	*uid__uid[barg]*	
4.5	basic time identifier	*btid*	*bid(fn)*	3.6
4.6	basic delay identifier	*bdid*	*btid__bid(fn * bid)__bid(bid * f n)*	
4.7	delay identifier	*did*	*bdid__bdid[mda]*	5.9
4.8	basic operator identifier	*boid*	*uid__uid[mda]*	5.9

5. *Arguments*

5.1	call argument	*carg*	*Be__carg, Be*	7.15
5.2	declaration argument	*da*	*ae__fld*	3.15
5.3	basic bracket argument	*bba*	*da__id*	4.4
5.4	bracket argument	*barg*	*bba__barg, bba*	
5.5	auxiliary argument	*aarg*	*⇒__⇒npid*	
5.6	argument 1	*arg*1	*→npid__carg,→npid__arg*1, *carg*	
5.7	argument 2	*arg*2	*aarg, carg__carg, aarg__arg*2, *carg*	
5.8	argument	*arg*	*carg__arg*1__*arg*2 __*arg*1, *arg*2__*arg*2, *arg*1	
5.9	memory declaration argument	*mda*	*da__mda, da*	

6. *Clauses, Destinations, and Value Lists*

6.1	basic value clause	*bvc*	$\lfloor va^4$	3.10
6.2	if clause	*ifc*	|*Be*|	7.15
6.3	for clause 1	*fc*1	{*bid*}	4.1
6.4	for clause 2	*fc*2	{*bid = va*}	3.10
6.5	destination 1	*dst*1	?__*npid__ifc'dst*1; *dst*1π ___⌈*Be'vl*1π__⌈*Be'vl*1; *dst*1π	
6.6	value list 1	*vl*1	*bvc'dst*1'*bvc'dst*1__*vl*1'*bvc'dst*1	
6.7	destination 2	*dst*2	?__*bid__bid(arg)__ifc'dst*2; *dst*2π ___⌈*Be'vl*2π__⌈*Be'vl*2; *dst*2π	5.8
6.8	value list 2	*vl*2	*bvc'dst*2'*bvc'dst*2__*vl*2'*bvc'dst*2	
6.9	destination 3	*dst*3	*bid__bid(csop)__ifc'dst*3; *dst*3π ___⌈*Be'vl*3π__⌈*Be'vl*3; *dst*3π	8.8 7.15
6.10	value list 3	*vl*3	*bvc'dst*3'*bvc'dst*3__*vl*3'*bvc'dst*3	

[4] It is desirable to underline the entire value $\lfloor va$ when writing statements by hand. The underline delimits the value clause clearly. But this form is not a simple concatenation of symbols, and hence can not be submitted to a computer and is not a part of the formal syntax.

7. Boolean Expression

7.1	Boolean identifier	*Bid*	$id\_bs\_rhs\_(Be)$	8.4
7.2	head operand	*ho*	$Bid\_Bid \vdash udi$	
7.3	tail operand	*to*	$ho\_ho \dashv udi$	
7.4	extension operand	*ext*	$Bid\_Bid \times udi$	3.3
7.5	concatenation operand	*cop*	$ext\_ext \circ cop$	
7.6	Boolean primary 1	*Bp1*	$cop\_\neg cop^5$	
7.7	Boolean primary 2	*Bp2*	$Bp1\_Bp1 \backslash bs$	3.8
7.8	Boolean primary 3	*Bp3*	$Bp2\_lo/Bp2$	1.7
7.9	Boolean primary 4	*Bp4*	$Bp3\_Bp3'ro'Bp3$	1.8
7.10	Boolean secondary 1	*Bs1*	$Bp4\_Bs1 \wedge Bp4\_Bs1 \cdot Bp4$	
7.11	Boolean secondary 2	*Bs2*	$Bs1\_Bs2{\uparrow}Bs1$	
7.12	Boolean secondary 3	*Bs3*	$Bs2\_Bs3{\downarrow}Bs2$	
7.13	Boolean secondary 4	*Bs4*	$Bs3\_Bs4 \odot Bs3\_Bs4 \equiv Bs3$	
7.14	Boolean secondary 5	*Bs5*	$Bs4\_Bs5 \oplus Bs4\_Bs5 \veebar Bs4$	
7.15	Boolean expression	*Be*	$Bs5\_Be \vee Bs5$	
7.16	restricted Boolean expression	*rBe*	*Be* in which the leftmost character is not a δ.	

8. Operations

8.1	common concatenation	*cco*	$id\_cco \circ id$	
8.2	left of connection	*lc*	$cco = \_lc'cco =$	
8.3	left of transfer	*ltc*	$lc\_cco \leftarrow \_ltc'cco \leftarrow \_ltc'lc$	
8.4	right hand side	*rhs*	$ifc'Be; Be\pi\_\lceil Be'vdl\pi$	6.2
			$\_\lceil Be'vdl; Be\pi$	7.15
8.5	value delimiter list	*vdl*	$bvc'rBe'bvc'rBe\_vdl'bvc'rBe$	6.1, 7.16
8.6	basic operation	*bop*	$co'cco\_(udi)'co'cco\_{\to}dst\,1$	1.9
			$\_.{\Rightarrow}dst\,2\_ \Rightarrow dst\,3\_\Rightarrow$	
			$\_uid\_ltc'Be\_fc2'csop\pi$	6.2
			$\_ifc'csop\pi\_ifc'csop; csop\pi$	
			$\_\lceil Be'vos\pi\_\lceil Be'vos; csop\pi$	
8.7	operation	*op*	bop_*bop	
8.8	compatible set of operations	*csop*	op_csop, op	
8.9	value operation string	*vos*	$bvc'csop\_vos'bvc'csop$	6.1

9. Memory Declaration

9.1	memory tag	*mtg*	$ME\_RE\_mtg\ \lambda$	
9.2	basic memory declaration	*bmd*	$bid\_bid[mda]\_bmd = mrp$	5.9
9.3	memory right part	*mrp*	$bmd\_mrp \circ bmd\_(bmd)\_bmd \circ mrp$	
9.4	memory declaration body	*mdb*	$bmd\_mdb, bmd\_\#bmd\_mdb, \#bmd$	
9.5	memory declaration	*md*	$\langle mtg \rangle mdb\pi$	

---

[5] The overscore used throughout the text to denote logic NOT violates the basic DDL statement structure: it is not a simple concatenation of symbols. Hence the symbol $\neg$ is introduced to denote NOT in the formal syntax. An expression to be complemented must be enclosed in parentheses preceded by this symbol.

10. *Terminal Declaration*

10.1	terminal tag	*ttg*	TE__*ttg* λ
10.2	terminal declaration body	*tdb*	*bmd*__*tdb*, *bmd*	9.2
10.3	terminal declaration	*td*	$\langle ttg \rangle tdb\pi$

11. *Time Declaration*

11.1	time tag	*titg*	TI__*titg* λ
11.2	time list	*tl*	*btid*__*tl*, *btid*	4.5
11.3	time declaration	*tid*	$\langle titg \rangle tl\pi$

12. *Delay Declaration*

12.1	delay tag	*dtg*	DE__*dtg* λ
12.2	delay list	*dyl*	*did*__*dyl*, *did*	4.7
12.3	delay declaration	*dd*	$\langle dtg \rangle dyl\pi$

13. *Operator Declaration*

13.1	operator tag	*otg*	OP__*otg* λ
13.2	auxiliary declaration	*aux*	*Bd*__*Bd'td*__*td'Bd*	10.3, 16.3
13.3	operator declaration list	*odl*	*boid*__*odl*, *boid*	4.8
13.4	operator declaration	*od*	$\langle otg \rangle odl' \, aux\pi$
13.5	for operator argument	*foa*	*od*__*foa'od*

14. *Element Declaration*

14.1	element tag	*etg*	EL__*etg* λ
14.2	input-output list	*iol*	*tdb* :__*tdb* : *tdb*__ : *tdb*	10.2
14.3	element list	*el*	*bid(iol)*__*el*, *bid(iol)*	4.1
14.4	element declaration	*ed*	$\langle etg \rangle el\pi$

15. *Identifier Declaration*

15.1	identifier tag	*itg*	ID__*itg* λ
15.2	identifier right part	*irp*	*cco*__(*csop*)	8.1, 8.8
15.3	identifier list	*idl*	*id* = *irp*__*idl*, *id* =*irp*	4.4
15.4	identifier declaration	*idd*	$\langle itg \rangle idl\pi$

16. *Boolean Declaration*

16.1	Boolean tag	*Btg*	BO__*Btg* λ
16.2	Boolean equation	*Beq*	*lc'Be*__*Beq*, *lc'Be*__*fc2'Beq*π	6.4, 8.2
16.3	Boolean declaration	*Bd*	$\langle Btg \rangle Beq\pi$
16.4	for Boolean argument	*fBa*	*Bd*__*fBa'Bd*

17. *State Declaration*

17.1	state tag	*stg*	ST__*stg* λ
17.2	state operation	*so*	: *csop*π__ : *Be* : *csop*π	8.8
17.3	state head	*sh*	*npid*__*bid(bs)*__*bid(bs)*[*barg*]
17.4	state	*st*	*sh'so*__*st'so*
17.5	state string	*sts*	*st*__*sts'st*__*st'sts*__*fc2'st*π	6.4
17.6	state declaration	*sd*	$\langle stg \rangle sts\pi$

18. *Segment Declaration*

18.1	segment tag	*setg*	SE__*setg* λ
18.2	heading	*hd*	:__: *Be* :__: *Be* : *csop*__: *csop*
18.3	facility declaration	*fd*	*md__td__tid__dd__od__ed__idd*
			__*Bd__fc*1'*foa*π__*fc*2'*foa*π 16.4
			__*fc*2'*fBa*π
18.4	basic declaration	*bd*	*fd__sd__csop* 8.8, 17.6
18.5	declaration list	*dl*	*bd__dl'bd*
18.6	segment body	*seb*	*id'hd__id'hd'dl* 4.4
18.7	segment declaration	*sed*	⟨*setg*⟩*seb*π
18.8	for automaton argument	*faa*	*sed__faa'sed*

19. *Comment Declaration*

19.1	comment tag	*ctg*	CO__*ctg* λ
19.2	comment character 2	*com2*	any member of the alphabet except⟨ ·
19.3	comment string 2	*cms2*	*com2__cms2'com2*
19.4	comment declaration	*cd*	⟨*ctg*⟩ *cms2*π

20. *Automaton Declaration*

20.1	automaton tag	*autg*	AU__*autg* λ
20.2	basic automaton declaration	*bad*	*bd__sed__fc*2'*faa*π 18.4–8
20.3	automaton declaration list	*audl*	*bad__audl'bad*
20.4	automaton body	*aub*	*npid'hd__npid'hd'audl* 4.3, 18.2
20.5	automaton declaration	*ad*	⟨*autg*⟩*aub*π
20.6	for system argument	*fsa*	*ad__fsa'ad*

21. *System Declaration*

21.1	system tab	*sytg*	SY__*sytg* λ
21.2	basic system declaration	*bsyd*	*fd__ad__csop__fc*2'*fsa*π 18.3
21.3	system declaration list	*sydl*	*bsyd__sydl'bsyd*
21.4	system body	*syb*	*npid'hd__npid'hd'sydl* 4.3, 18.2
21.5	system declaration	*syd*	⟨*sytg*⟩*syb*π
21.6	system	*sy*	*syd__fc*2'*syd*π 6.4

APPENDIX 5.2
SUMMARY OF THE SEMANTICS OF DDL

The semantics of all of the possible statements that fall within the syntax of a complex language can not be listed *a priori* because of the large number of such statements and because the meaning of a statement may depend upon its context. Only the meanings of English words are listed in dictionaries. Often several definitions are offered, and in some cases the list may not be complete or revealing. The meaning of an English sentence must be developed from the dictionary listings and the surrounding sentences. DDL is also a complex language in which the meaning of a "word" is not always independent of the "words" which surround it, i.e., DDL is not a *context-free* language. Thus the meaning only of some "words" can be indicated here. This brief "dictionary" can not substitute for the semantics offered throughout this text. Paragraph numbers below refer to corresponding line numbers in the syntax tables of Appendix 5.1. Study of that table will reveal the richness of DDL and the brevity of this appendix.

2. Period: The period (.) is required terminal punctuation for all declarations. But any such use of the period may be followed by a comment enclosed in parentheses. Hence *ml*-variable π appears frequently in Appendix 5.1.

3.6. Floating point number: A floating number, as opposed to an integer, contains a decimal point. The E exponent notation of FORTRAN may be used. Examples: .71, $-.89$, $+17.$, -4.29, $-18.3 - E17$

3.8. Binary string: A bit, or the don't care symbol ?, or a sequence of the form *nRl* where *R* is one of the letters B, O, D, and *n* is a binary, octal, or decimal integer, respectively, that expresses the binary sequence in the corresponding number system, and *l* is a decimal integer that expresses the length of the binary string. Thus 1011B5, 13O5, and 11D5 all express 01011. The special form *#npid* denotes the binary code assigned to the state identified after the # symbol.

3.10. Value: A binary or octal integer (identified by B or O), a decimal integer (D is optional), two integers separated by a colon (:), or a list of integers or pairs separated by commas. Examples: 10B, 17, 12 : 17 or 10B, 17, 12 : 17

3.14. Arithmetic expression: An integer expressed as the sum, difference, product, quotient, or exponentiation of unsigned decimal integers or other arithmetic expressions, enclosed in parentheses. Examples: 14, 14-4, 2*3, (14-4) $\uparrow$ (2*3).

3.15. Field: Two arithmetic expressions separated by a colon.

4.1. Basic identifier: A letter or letters followed by other letters or digits. Examples: A, P2Q, accumulator, START.

4.4. Identifier: A basic identifier, basic identifier followed by Boolean expressions enclosed in parentheses, or such structures followed by arithmetic expressions or fields enclosed in brackets. The basic identifiers name facilities; the bracketed information dimensions the named facility; the parenthetical information conveys operator or segment arguments. Examples:

A[1 : 17]—references 17 members of the memory element or terminal set named *A*

OP(X, Y)[13]—names (OP) and dimensions (13) an operator with input terminal sets named *X* and *Y*

5.9. Memory declaration argument: An arithmetic expression, field, or list of these items separated by commas. Such lists enclosed in brackets provide dimensional information for memory, terminal, operator, element, etc. names.

A[17]—a 17-item set *A*, with A[1] assumed to be the leftmost.
A[0 : 16]—a 17-item set *A*, with A[0] assumed to be the leftmost.
A[17 : 1]—a 17-item set *A*, with A[17] assumed to be the leftmost.
A[1 : 3, 1 : 17]—a 3×17 array *A* of items, with 3 rows and 17 columns.
A[2] references the entire second row. A[1, 2 : 5] references the 2nd through 5th items in the first row.

7.15 Boolean expression: An identifier, an extension of an identifier, a concatenation of identifiers, or Boolean expression in parentheses, a right hand side, or a binary string by itself or on the left and another on the right of a logic or relation operator. Parentheses are required when a hierarchy other than the assumed hierarchy of DDL, $\times$, $\circ$, $\neg$, $\backslash$, $/$, relation operator, $\cdot$, $\uparrow$, $\downarrow$, $\equiv$, $\oplus$, $\vee$, is to be specified. All identifiers combined by operators $\cdot$ through $\vee$ must be of the same dimension as others or have a dimension of 1.

Examples:

$A[1] \times 3$—extend $A[1]$ to a set of 3 terminals all bearing the same signal

$(A \vee B) \cdot C$—A, B, C must have the same dimension

$(\neg A \uparrow B) \cdot \downarrow/C$—Since $\downarrow/C$ describes a single terminal it is compatible with $(\neg A \uparrow B)$ by fan-out (implied extension)

$A \circ B \times 3 \uparrow C \oplus D$—The dimension of A must be 3 less than that of C and D. The implied hierarchy demands that we view this expression as $((A \circ (B \times 3)) \uparrow C) \oplus D$.

8.6. Basic operation—An expression of information transfer or connection. Examples:

(4) $0|_{\rightarrow} A$—shift register A to the right 4 places

$\rightarrow S1$—automaton state transition to state $S1$

$\Rightarrow SEG(\rightarrow PI)$—state transition to state PI of segment SEG

$\Rightarrow AUT(RR = 1)$—set public terminal RR to 1; automaton AUT is expected to respond

$\Rightarrow$—return to next state of the segment which preceded this segment

ID—perform operations identified by declared identifier ID

$A = B = C \vee D$—connect $C \vee D$ to terminal sets A and B

$A = B \leftarrow C \vee D$—transfer $C \vee D$ to register B, connect to terminal set A

$\{A = 1 : 7\}OP.$—for $A = 1, 2, \ldots, 7$ duplicate operator OP

$|A|CSOP.$—if $A = 1$ perform operator of compatible set $CSOP$;
 if $A = 0$ do nothing

$|A| CSOP_1; CSOP_0.$—if $A = 1$ perform $CSOP_1$ else perform $CSOP_0$

$[Be \ 1 \ \lfloor CSOP_1 \ \lfloor 2 \ CSOP_2 \cdots$ if Be expresses 1 decimal perform $CSOP_1$,
 if Be expresses 2 decimal perform $CSOP_2$,
 etc.

$CSOP_1$ and $CSOP_2$ can not be expressed with the leftmost character being a digit.

8.6. Operation: A basic operation or one preceded by a *. The star indicates that the following operation is to be performed whether the state, segment, automaton, or system heading condition is satisfied or not.

8.7. Compatible set of operations: a list of operations separated by commas which can be performed simultaneously without hardware conflict or ambiguity.

Example:

$$A \leftarrow B \vee C, \downarrow B, E = C \oplus D, \rightarrow PI$$

9. Memory declaration: statement with syntax $\langle ME \rangle$ body. or $\langle RE \rangle$ body. which declares the existence of memory elements in a segment, automaton, or system, names the elements, and gives them dimension.

Examples:

APPENDIX 5.2   Summary of the Semantics of DDL

⟨RE⟩ A, B[17], C[17 : 0] = D[7] ∘ E[10].

declares a single flip-flop named *A*, a register of 17 flip-flops named *B* with the leftmost flip-flop referenced by B[1] and the rightmost by B[17], and a register *C* of 17 flip-flops, each of which is given two names. Thus the leftmost flip-flop is referenced by C[17] or D[1]; the rightmost by C[0] or E[10].

⟨ME⟩ A, B[17], C[1 : 10, 12], D[10, 12, 20].

declares a single flip-flop *A*, a register *B*, a 10-row, 12-column matrix of flip-flops *C*, and a $10 \times 12 \times 20$ 3-dimensional array *D*. C[3] references the entire third register of *C*. C[3, 4 : 8] references the 4th through 8th flip-flops of the third register of *C*. D[2, 4] references the entire 4th register of the second bank of flip-flops of *D*. D[2] refers to that entire bank. D[2, 4, 8] references the 8th bit of D[2, 4].

⟨ME⟩ # SSR[10].

declares and dimensions the state sequencing register of an automaton. The state assignment may be declared with Boolean declarations such as ⟨BO⟩S1 = ·/SSR\1D10.

10. Terminal declaration: statement with syntax ⟨TE⟩ body. which declares the existence of terminal sets in an operator, segment, automaton, or system. Dimensional information has the same syntax and semantics as in memory declarations.

11. Time declaration: statement with syntax ⟨TI⟩ body. which declares the existence of one or more clocks and indicates their periods.
Example:

⟨TI⟩ P(10. E-8), PP(1.E-6).

declares clock *P* with a 100 nanosecond period and clock *PP* with a 1 μs period. No synchronism between these clocks is implied.

12. Delay declaration: statement with syntax ⟨DE⟩ body. which declares the existence of one or more delay elements and the magnitude of each delay.
Example:

⟨DE⟩ D1(P), D2(2.3 E-6).

in conjunction with the time declaration example above declares 100 ns delay element D1 and 2.3 μs delay D2.

13. Operator declaration: statement with syntax ⟨OP⟩ ID aux. which declares the existence of and describes the structure of one or more blocks of combinational logic circuitry. ID names and dimensions the output terminals of the block and may name dummy input terminals. The combinational logic of the blocks is described by TE and BO declarations, which make up the aux portion of the body.
Example:

⟨OP⟩ COMB(X, Y)[10], Q[4]
  ⟨TE⟩ X[10], Y[10].
  ⟨BO⟩ COMB = X ∨ Y, Q = R ⊕ S..

declares the existence of 10 OR-gates with output terminals named COMB[1], COMB[2], etc., and four EX-OR gates identified by *Q*. Input lines to the OR-gates are named *X* and *Y*, but since these names are declared within the OP declaration, they have no significance outside of the block. A reference such as COMB(A, B) indicates that terminal sets *A* and *B* are to be connected to the OR-gates, perhaps through implied switching matrices. (*A* and *B* must be dimensioned 10.) Since *R* and *S* are not declared in the OP declaration, they must be declared in the block that contains the OP declaration, and no other signals may be connected to the input terminals of the EX-OR gates.

14. Element declaration: statement declaring the existence of a block of equipment of unspecified design and function with syntax ⟨EL⟩ body. . The element and its output and input terminals are named in the declaration, but its internal structure is not documented. (It may be summarized in a comment.)
    Example:

    ⟨EL⟩ A(OUT[10], Q : IN[5], GO), B(: PQ).

    declares that element *A* has 11 output terminals and 6 input terminals, and that element *B* has a single input terminal only. Element *B* might be simply a console lamp, or an extensive communication network, the design of which is not to be detailed in a DDL document, but to which a connection is to be made.

15. Identifier declaration: statement with syntax ⟨ID⟩ body., which names a collection of identifiers, Boolean expressions, or compatible sets of operations so that the collection may be referenced repeatedly without repeatedly detailing the collection.
    Example:

    ⟨ID⟩ A = (B = C ∨ D, E ← F, b = 1), X(Z) = (Y ⊕ Z),
    ODDH =(H[1], H[3], H[5]).

    names as *A* a collection of transfer and connection operations, and as ODDH the odd members of a terminal or memory element set *H*. Any reference to *A* in the document in which this ID declaration appears is an abbreviation for replacing that reference with the three operations. *Z* is an argument of identifier *X* which is a name for the EX-OR of terminals *Y* and the terminals named by the argument. Thus reference X(Q) expresses $Y \oplus Q$, and X(X(Q)) expresses $Y \oplus (Y \oplus Q)$.

16. Boolean declaration: statement which lists Boolean equations separated by commas.
    Example:

    ⟨BO⟩ A = (B ⊕ C) · ↓/D, X = Y · Z.

    declares that the exclusive-or of *B* and *C* is to be connected to *A* when ↓/D has the value of 1, and the logical product of *Y* and *Z* is to be connected to *X*. Terminal sets *A*, *B*, and *C* must be of the same dimension; if that dimension is *n* and *i* is an index, the first Boolean equation of this Boolean declaration expresses the *n* Boolean equations: A[i] = (B[i] ⊕ C[i]) · ↓/D for *i* = 1, 2, ..., *n*. ↓/D has unit dimension and is therefore compatible with any terminal set by fanout.

17.2. State: statement with either of the following forms:

SID : *CSOP.*
SID : Be : *CSOP.*

that names (SID) a state of an automaton, gives the condition if any (Be) that must be satisfied before the specified operations (*CSOP*) are performed with the automaton in state SID.

Examples:

S0 : A ← B, $|C|\underset{\longrightarrow}{\rule{0pt}{1ex}\hspace{1em}}$ D, →S1; →S2..
S1 : GO · $\overline{\text{RESET}}$: A ← ADD1(R, Q), P2 = 1, ⇒SEQ(→X).

describe two states S0 and S1 and the operations to be performed with the automaton in each. When in S0, the content of register B is to be transferred to register A; if $C = 1$ then register D is shifted right one place, and the automaton is to transfer to state S1. When in state S1 and GO · $\overline{\text{RESET}} = 1$, load register A from combinational network ADD1 with the input terminals of ADD1 connected to registers R and Q, set terminal P2 to logic 1, and transfer to state X of segment SEQ of the automaton. If GO · $\overline{\text{RESET}} = 0$ remain in state S1 performing no operations.

17.6. State declaration: statement with syntax ⟨ST⟩ SL., which lists the states of an automaton or a segment where SL is a list of states. Since each state statement is terminated with a period, no additional punctuation need separate the states of SL.

18.2. Heading: a colon, a Boolean expression enclosed in colons, a Boolean expression enclosed in colons and followed by a compatible set of operations, or a colon followed by a compatible set of operations.

Example:

:, : P · GO :, : P: ∉ KOUNT, A = B ⊕ C, : ∉ KOUNT, A = $\overline{\text{B}}$

The colon by itself separates the identifier of a segment, automaton, or system from the body of a segment, automaton or system declaration, respectively. A Boolean expression such as P · GO above enclosed in colons is a universal condition on all the operations except connection operations in a segment, automaton, or system declaration. Thus state sequencing and information transfers are allowed only when P · GO = 1. P typically is a clock signal in a synchronous system. The compatible set of operations following a colon in a heading is to be performed each time the condition enclosed in colons, if any, is satisfied. Thus these operations are performed regardless of the state of an automaton; they could be repeated as a part of each state.

18.3. Facility declaration: a terminal, memory, time, delay, operator, identifier, element, or Boolean declaration, or an operator or Boolean declaration preceded by a "for" clause and followed by a period.

Example:

⟨TE⟩ A[10].
⟨ME⟩ B[10].
{i} ⟨OP⟩ MAJ(A, B, C) ⟨BO⟩ MAJ = A · B ∨ A · C ∨ B · C...

$$\{i = 1, 5\}\ \langle\mathrm{BO}\rangle S[i] = A[i] \oplus B[i] \oplus C[i-1],$$
$$C[i] = \mathrm{MAJ}(A[i],\ B[i],\ C[i-1]) \ldots$$

MAJ is an operator available without limit. Each reference to $\mathrm{MAJ}(A_1, A_2, A_3)$ specifies the existence of a unique majority gate with signals A_1, A_2, A_3 connected to its input terminals. Thus $C[1] = \mathrm{MAJ}(A[1],\ B[1],\ C[0])$ calls for one majority gate with its output terminal connected to terminal $C[1]$. $C[5] = \mathrm{MAJ}(A[5],\ B[5],\ C[4])$ specifies the fifth majority gate in the last example above.

18.7. Segment declaration: statement with syntax $\langle\mathrm{SE}\rangle$ ID Head Body., which usually specifies a portion of the facilities and states of an automaton. ID names the segment, Head lists the local conditions and *CSOP*, if any, and Body consists of facility declarations, state declarations or a compatible set of operations. Segments may also be declared with an argument list following the identifiers. The arguments serve to control state transitions to and from the segment, and prescribe operands for operations specified in the segment. If

$$\langle\mathrm{SE}\rangle\ \mathrm{SE1}(A, B)\ :\ \langle\mathrm{ST}\rangle\ \mathrm{S1}\ :\ C \leftarrow A,\ \rightarrow\mathrm{S2}.\ \mathrm{S2}\ :\underset{\rightarrow}{\underline{}}\,B,\ \Rightarrow\ldots$$

is declared, register C will be loaded from register D and register E will be shifted following the transition $\Rightarrow\mathrm{SE1}(D, E)$.

A transition from another segment directly to state S2 of segment SE1 may be made via the transition operation $\Rightarrow\mathrm{SE1}(D, E, \rightarrow\mathrm{S2})$. The return operator $\Rightarrow$ in state S2 of segment SE1 does not explicitly indicate the next state of the automaton. The next state will be the one in the state list after the state that prescribes the transition to SE1, unless the transition to SE1 indicates another return state by means of an argument. Example:

$$\langle\mathrm{SE}\rangle\ \mathrm{SE1}\ :\ \langle\mathrm{ST}\rangle\ \mathrm{S1}\ :\ B \leftarrow A,\ \rightarrow\mathrm{S2}.\ \mathrm{S2}\ :\ C \leftarrow D,\ \Rightarrow\mathrm{SE2}(\rightarrow\mathrm{P2},\ \Rightarrow\mathrm{S1}).$$
$$\mathrm{S3}\ :\ D \leftarrow B,\ \rightarrow\mathrm{S1}\ldots$$

$$\langle\mathrm{SE}\rangle\ \mathrm{SE2}\ :\ \langle\mathrm{ST}\rangle\ \mathrm{P1}\ :\ A \leftarrow 0,\ \rightarrow\mathrm{P2}.\ \mathrm{P2}\ :\ D \leftarrow B,\ A \leftarrow C,\ \rightarrow\mathrm{P3}.$$
$$\mathrm{P3}\ :\ |X|\ B \leftarrow A,\ \Rightarrow\ ;\ \Rightarrow\mathrm{SE1}(\rightarrow\mathrm{S3}).\ldots$$

The state sequence will be S1, S2, P2, P3, S1, etc. if $X = 1$ when the automaton reaches P3. If $X = 0$ at that point, the sequence will be S1, S2, P2, P3, S3, S1, etc.

20. Automaton declaration: statement with syntax $\langle\mathrm{AU}\rangle$ ID Head Body., which declares the private facilities, operations, and finite-state machine that controls the operation performed on the private facilities of the automaton and public facilities of the system. ID identifies the automaton and may not include arguments in parentheses. Again Head may specify conditions that must be met before any nonconnection operation is performed, and compatible operations that must be performed regardless of the state of the automaton. Body consists of facility and state declarations, and may include segment declarations.

21. System declaration: statement with syntax $\langle\mathrm{SY}\rangle$ ID Head Body., which declares the facilities, operation, and finite-state machines of the digital system. Facilities declared outside of all automaton declarations are public facilities subject to the control of all automata. Heading conditions prevail over all automata of the system.

6

Sequential Subsystems

Some sequential networks are ubiquitous and the logic designer must be familiar with their design. We are familiar with functions such as counting, shifting, selecting, transfering, etc., and their applications; this chapter concentrates on the variety of ways to realize such functions. While synchronous realizations of functions will be emphasized (such networks are easier to design and are reliable), asynchronous realizations are more economical in some cases. This chapter concludes with a summary of the organization and operation of memories—digital systems that are able to store large numbers of words of information.

6.1 COUNTERS

Counters are finite state, sequential machines that satisfy, to some extent. the following restrictions.

1. The output mapping is trivial: $\zeta : z_i = y_i$.
2. The state of the machine is interpreted as an integer modulo m.
3. The state sequence is highly independent of input variables.
4. The state sequence can be interpreted as an ascending or descending sequence of integers modulo m.

This definition is not intended to exclude some finite-state machines from the set of counting circuits, but rather to summarize characteristics of machines commonly referred to as counters.

Counters may be classified in a number of ways:

1. Synchronous, asynchronous, and mixed.
2. Binary and nonbinary.
3. Serial gating, parallel gating, and mixed.
4. Up-counter and down-counter.

We are familiar with the synchronous network; and the synchronous counter is just an example. In asynchronous and mixed synchronization counters, not all flip-flops will be clocked by the system clock signal, as we shall see. A "binary" counter has a state sequence that repeats with a period of 2^n. Binary counters count modulo 2^n, whereas nonbinary counters do not. Serial versus parallel gating is a matter of combinational logic design. Let us explain this point first for it influences the maximum frequency at which a network will count, and it applies to both synchronous and asynchronous counters.

Figure 1.9 presented our first counter example. Looking back at it, we see that T flip-flops were made from DFFs by associating an XOR gate with each flip-flop. With four flip-flops, the counting sequence had a period of 16; therefore, this is a binary synchronous counter. Now look at the AND gates; they are connected together in a serial fashion. The IN signal and new values of A_4 must pass through three AND gates and one XOR gate to reach the D terminal of the left flip-flop. If this network is to operate reliably in the worst case, the clock period must exceed the sum of the flip-flop set-up time, the flip-flop propagation delay, and the propagation delay of four gates.

$$\Delta_c \geq \Delta_{su} + \Delta_f + 4\Delta_g \qquad (6.1)$$

The maximum frequency at which this counter can operate is therefore

$$f_{\max} = \frac{1}{\min (\Delta_c)} \qquad (6.2)$$

This calculation does not provide for any time to propagate the state variables to other networks that respond to the count.

It is not necessary to connect the AND gates in the serial fashion of Fig. 1.9. We could write

$$C_4 = IN \cdot A_4$$
$$C_3 = C_4 \cdot A_3 = IN \cdot A_4 \cdot A_3$$
$$C_2 = C_3 \cdot A_2 = IN \cdot A_4 \cdot A_3 \cdot A_2$$
$$C_1 = C_2 \cdot A_1 = IN \cdot A_4 \cdot A_3 \cdot A_2 \cdot A_1$$

These equations describe a four gate network in which the gates act in parallel to calculate carry values. With this parallel gating, a mod 16 counter using TFFs would operate with a clock period

$$\Delta_c \geq \Delta_{su} + \Delta_f + \Delta_g \qquad (6.3)$$

If Δ_{su}, Δ_f, and Δ_g are all approximately equal, the maximum counting frequency is increased by a factor of two at a cost of AND gates with a higher fan-in. With four stages fan-in is no problem, of course, but fan-in increases linearly with the number of stages and in large counters may prohibit total parallel gating.

Previous chapters have made us very aware of how synchronous counters may be designed. Such counters may have controlling input signals, such as signal *IN* of Fig. 1.9, which starts and stops counting. They may have output signals other than the state variables that indicate some useful condition or other. Thus, in a decade (mod 10) counter we usually find a carry signal that takes the value of 1 when the count is 9, and the control input signal calls for advancing the count. Such input and output signals permit cascading mod 10 counters to form the mod 10^m counters found in digital voltmeters, counters, and other instruments.

In the synthesis procedure of the previous chapter, forming a state diagram or table from counter specifications is usually a trivial job. Further, state minimization is not possible unless the specifications are absurd. State assignment is another matter. Often a specific code is required, but if it is not then the use of unusual codes can lead to counters that require no gates. Many such counters have been designed and tables of such counters have been published. Often such counters are found within integrated circuits.

Let us take a brief example by designing a mod 3 synchronous counter that uses the count sequence 00, 01, 10, 00, ..., which is not at all unnatural. The state transition table is particularly simple.

$y_1 y_2$	$y_1' y_2'$
0 0	0 1
0 1	1 0
1 0	0 0
1 1	– –

Transforming this table, as shown in Section 5.3, is also easy.

$y_1 y_2$	$y_1' y_2'$
0 0	0 S
0 1	S R
1 1	– –
1 0	R 0

The rows have been reordered so that we can view this table as a map and write JKFF input equations.

$$J_1 = y_2 \qquad J_2 = \bar{y}_1$$
$$K_1 = 1 \qquad K_2 = 1$$

This design is shown in Fig. 6.1, along with other gate-free counters to be discussed.

The creeping-code counter of Fig. 3.44 is free of gates, although detector gates are shown in Fig. 3.44. With n flip-flops this type of network counts mod $2n$, unless modifications are made. For example, three flip-flops connected to form a shift register with the complement of the right stage fed to the left stage give the mod 6 gate-free counter of Fig. 6.1(c). What if we desire a mod 5 counter? We may alter the mod 6 counter code by deleting one entry. We might reset two flip-flops simultaneously to delete the count 001, for example.

$$
\begin{array}{ccc}
000 & 000 \\
100 & 100 \\
110 & \Rightarrow 110 \\
111 & 111 \\
011 & 011 \\
001
\end{array}
$$

This is easily accomplished if JKFFs are used as shown in Fig. 6.1(b).

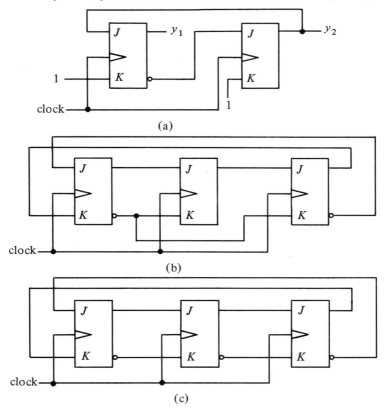

(a)

(b)

(c)

FIG. 6.1   Synchronous (a) Mod 3, (b) Mod 5, and (c) Mod 6 Gate-Free Counters

Hopefully this example will suggest how to design gate free mod x counters with ease. Please be careful when using the creeping code and one-hot codes for counters. With $x/2$ or x flip-flops and only x desired states, $2^{x/2} - x$ or $2^x - x$ states exist that are not to be entered. If the counter enters one of these undesired states, perhaps when turning on the power, it may behave in a totally unexpected manner. If the counter of Fig. 3.44 attains state 01010, for example, it counts mod 2 ($01010 \rightarrow 10101 \rightarrow 01010 \rightarrow \ldots$)! Gates to prohibit such action may be very worthwhile.

Asynchronous Binary Counters ($ABCs$)

A *ripple* counter consists of n toggle flip-flops with the output signal of the ith flip-flop serving as the clock signal for the $i + 1$ stage. The signal X, which provides transitions to be counted, serves as the clock signal for the first stage.

$$c_1 = X$$

$$c_i = y_{i-1} \quad i = 2, 3, \ldots, n$$

While the flip-flops may be simpler than master-slave or edge-triggered JKFFs, Fig. 6.2(a) shows JKFFs with all $J = K = 1$ that change their output signal value on the 1-to-0 transition of their clock signal. Then, with each transition of X from 1 to 0, the value of y_1 is changed; with each transition of y_1 from 1 to 0, the value of y_2 is changed; and so forth. Two changes of X from 1 to 0 cause y_1 to return to its original value: If X is periodic, y_1 has a period that is twice that of X; y_2 has a period that is twice that of y_1 and four times that of X. The nth flip-flop toggles with a period that is 2^n times that of X, or a frequency that is 2^{-n} times that of X. The ripple counter of Fig. 6.2(a) counts 1-to-0 transitions of X mod 2^n, with y_1 being the least significant bit of the count and y_n being the most significant bit of the count. The count, of course, is expressed in the binary number system.

Flip-flop state changes do not occur simultaneously in the ripple counter. The first stage, y_1, toggles Δ_f seconds after X changes; y_2 changes Δ_f seconds after y_1, and hence $2\Delta_f$ seconds after X. The nth stage toggles $n\Delta_f$ seconds after X. Activity propagates or ripples down the line of flip-flops. In the worst case with all flip-flops set (count $= 2^n - 1$) and following a 1-to-0 change of X, the first flip-flop resets, which causes the second to reset, $\ldots$, and which causes the nth flip-flop to reset. For an interval of $n\Delta_f$ seconds following the change of X, the instantaneous state of the counter does not measure true count mod 2^n.

The worst case interval during which true count is not measured by network state is called the *turn-over time*, t_o, of the counter. Other circuits of a system must have an opportunity to sample true count. Let t_s be the *sampling time* required by these other circuits. The interval T between successive 1 to 0 changes of X must exceed the sum $T = t_o + t_s$; the maximum counting frequency of the ripple counter is then $1/T$.

$$f < \frac{1}{t_o + t_s} = \frac{1}{n\Delta_f + t_s}$$

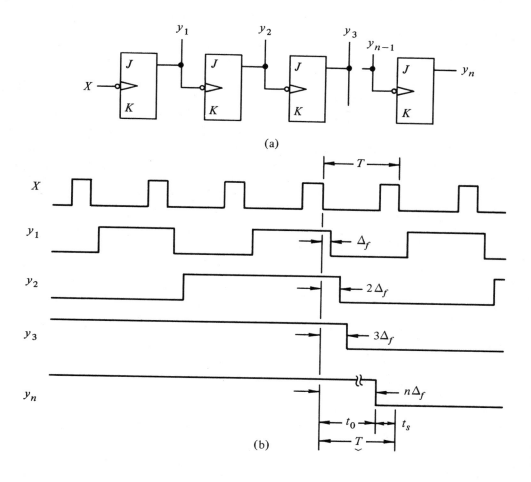

FIG. 6.2 Ripple Counter and Turn-over Waveforms

Table 6.1 summarizes characteristics of the ripple counter and other counters to be discussed.

The turn-over time of an ABC can be reduced by either serial or parallel gating, or a combination of them. In the extreme, parallel gating associates an AND gate with each flip-flop, except the first. Each AND gate forms the logic product of X and all prior state variables; thus

$$c_1 = X$$
$$c_i = X \cdot y_1 \cdot y_2 \cdots y_{i-1} \quad \text{for } i = 2, 3, \ldots, n$$

The first flip-flop turns over Δ_f seconds after X; all other stages turn over $\Delta_f + \Delta_g$ seconds after X. Thus in the worst case $t_o = \Delta_f + \Delta_g$, $T \geq \Delta_f + \Delta_g + t_s$. Turn-over time is reduced substantially, but at great cost. Gates with fan-in of n must be available to construct this counter. High fan-out is also required.

TABLE 6.1 SUMMARY OF COUNTER CHARACTERISTICS

Counter Type (n Stages)	Turn-over Time t_o	Minimum Clock Period T	Gates	Clock Fan-out	Maximum Gate or Flip-Flop	
					Fan-in	Fan-out
Synchronized Binary Counters						
Parallel gating every stage	Δ_f	$\Delta_f + \Delta_g + \Delta_{su}$	$n-2$	n	$n-1$	$n-1$
Serial gating every stage	Δ_f	$\Delta_f + (n-2)\Delta_g + \Delta_{su}$	$n-2$	n	2	3
Asynchronous Binary Counters						
Ripple	$n\Delta_f$	$n\Delta_f + t_s$	0	1	$-$	1
Parallel gating every stage	$\Delta_f + \Delta_g$	$\Delta_f + \Delta_g + t_s$	$n-1$	n	n	$n-1$
Parallel gating every jth stage ($n = jk$)	$j\Delta_f + \Delta_g$	$j\Delta_f + \Delta_g + t_s$	$k-1$	k	$n-j+1$	$k-1$
Serial gating every stage	$\Delta_f + (n-1)\Delta_g$	$\Delta_f + (n-1)\Delta_g + t_s$	$n-1$	2	2	2
Serial gating every jth stage	$j\Delta_f + (k-1)\Delta_g$	$j\Delta_f + (k-1)\Delta_g + t_s$	$k-1$	2	$j+1$	2
Mixed Binary Counters						
Parallel gating every jth stage ($j \ge 2$)	$j\Delta_f$	$j\Delta_f + \Delta_g + \Delta_{su}$	$k-1$	k	$n-j$	$k-1$
Serial gating every jth stage ($j \ge 2$)	$j\Delta_f$	$j\Delta_f + (k-1)\Delta_g + \Delta_{su}$	$k-1$	k	j	3

Intermediate turn-over times can be achieved at moderate costs with parallel gating. Let j be a factor of n so that $n/j = k$, where j and k are integers. Then k j-stage ripple subcounters can be used to form an n-stage counter with AND gates used to drive each subcounter, as shown in Fig. 6.3.

$$\left.\begin{aligned} c_1 &= X \\ c_{ij+1} &= X \cdot y_1 \cdot y_2 \cdot \ldots \cdot y_{ij} \\ c_{ij+2} &= y_{ij+1} \\ &\vdots \\ c_{ij+j} &= y_{ij+j-1} \end{aligned}\right\} \text{for } i = 0, 1, 2, \ldots, k-1$$

The parallel gating philosophy does not inherently require that all ripple subcounters be of the same length, but turn-over time of the entire counter must be taken as the maximum turn-over time of all the subcounters. Suppose a 10-stage ABC is formed of 4-, 3-, 2-, and 1-stage subcounters. The turn-over times of the subcounters are then, respectively, $4\Delta_f$, $\Delta_g + 3\Delta_f$, $\Delta_g + 2\Delta_f$, and $\Delta_g + \Delta_f$. The final stage may well turn over before the fourth stage or the seventh stage. We must locate the flip-flop that toggles last when calculating turn-over time.

The high fan-in and fan-out requirements of parallel gated counters can be avoided by *serial gating*. In serial gating AND gates in tandem, and hence in sequence, form flip-flop clock signals. Serial gating every stage of an ABC leads to a turn-over time of $t_o = (n-1)\Delta_g + \Delta_f$ and the costs listed in Table 6.1. The 1-to-0 change of X must propagate through $n-1$ AND gates to reach the clock terminal of the final flip-flop. While only 2-input gates are required, turn-over time is reduced with respect to the ripple counter by $(n-1)(\Delta_f - \Delta_g)$, which may be small or negative.

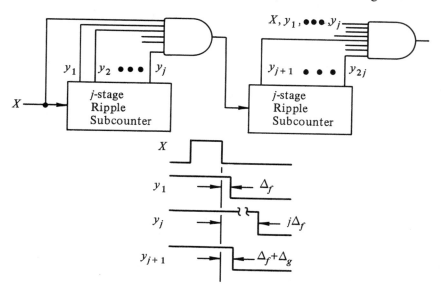

FIG. 6.3   Parallel Gating Every Jth Stage

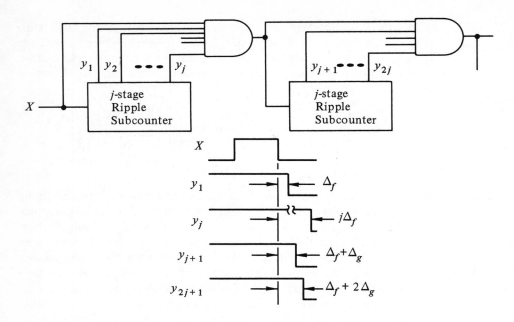

FIG. 6.4 Serial Gating Every Jth Stage

With $n = jk$ (both integers), k j-stage subcounters may be serially gated with a turn-over time of $t_o = (k - 1)\Delta_g + j\Delta_f$, as shown in Fig. 6.4. The change in X must propagate through $k - 1$ AND gates and j flip-flops before y_n changes. Fan-in of $j + 1$ is required.

If the ripple subcounters are not of the same length, turn-over time for the entire counter is the maximum of all subcounter turn-over times. Assume a 10-stage counter is formed of 4-, 3-, 2-, and 1-stage serially gated subcounters. These have turn-over times of $4\Delta_f$, $\Delta_g + 3\Delta_f$, $2\Delta_g + 2\Delta_f$, and $3\Delta_g + \Delta_f$. If $\Delta_g = \Delta_f / 2$, the fourth flip-flop will be the last to turn over; if $\Delta_g = \Delta_f$, then the 4th, 7th, 9th, and 10th stages turn over simultaneously.

Serial and parallel gating techniques can be combined to obtain a counter with acceptable turn-over time, gates of available fan-in and fan-out, and low cost.

Mixed Synchronization Counters

Asynchronous binary counters do not utilize the full capabilities of JKFFs, and can be constructed of simpler, perhaps cheaper, flip-flops. If JKFFs are to be used to construct counters, turn-over time can be reduced by using all of their capabilities. If input signal X is supplied to the clock terminal of every flip-flop, then a synchronous binary counter with minimum turn-over time of Δ_f is formed. The J_i and K_i signals can be formed by serial or parallel gating; the technique used affects the maximum frequency of operation, as we have seen.

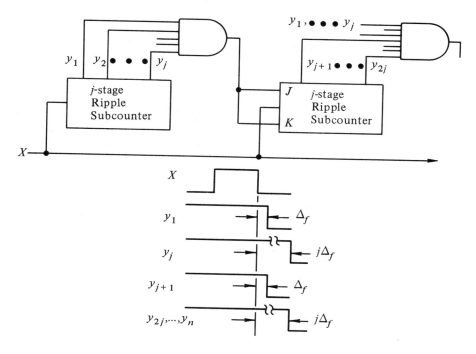

FIG. 6.5 Parallel Gating and Synchronizing Every Jth Stage

Counters with different speed and cost characteristics can be obtained by synchronizing some, but not all, flip-flops with the count signal X. If $n = jk$ (both integers), then k j-stage ripple subcounters can be used with the first stages of the subcounters being synchronized.

$$c_{ij+1} = X \qquad \text{for } i = 0, 1, \ldots, k-1$$
$$c_i = y_{i-1} \qquad \text{for all other } i$$
$$J_{ij+1} = K_{ij+1} = y_1 \cdot y_2 \cdots y_{ij} \qquad \text{for } i = 0, 1, \ldots, k-1$$
$$J_i = K_i \quad = 1 \qquad \text{for all other } i$$

The subcounters have a turn-over time of $j\Delta_f$ seconds; Δ_g seconds are required for the state variables to propagate to J and K terminals. Thus the minimum period of X is $\underline{T} = j\Delta_f + \Delta_g + \Delta_{su}$, assuming sampling is done after the ripple subcounters have settled and while new values of J_i and K_i are being found. Table 6.1 again is the reference for costs. Figure 6.5 illustrates the network and its timing. Serial gating of j-stage synchronized ripple subcounters is also possible, of course. Finally, if variable SS is included in the product terms for $J_{ij} + 1$ and $K_{ij} + 1$, then counting may be controlled. It is stopped by $SS = 0$.

These methods of gating and synchronizing subcounters are used extensively to create large counters from smaller ones. To examine some of the possible connections, suppose that we desire a mod p counter. If p is a prime number, then we must design a single, preferably synchronous, counter. If p is not prime, however, then it can be expressed as the product of two or more smaller integers. For purpose of example, assume $p = q \cdot r$. If q and r are small, we may already have suitable counters in IC form, or we may have designs available. This section has already illustrated mod 2, 3, 4, 5, 6, and 2^n counters for example. If mod q and mod r counters are connected so that the r counter advances by 1 each time the q counter goes through one complete count cycle, a mod p counter results.

Proper interconnection of cascaded counters may take several forms. Suppose the r counter has only a clock input terminal. It may be synchronous or asynchronous internally. We must provide a single clock edge for each q count cycle to be used as the clock input signal of the r counter. For example, the y_2 signal of the mod 3 counter of Fig. 6.1(a) could be used as the clock signal for the mod 5 counter of Fig. 6.1(b) to achieve a mod 15 counter with an unusual 5-bit code.

count	code
0	00 000
1	01 000
2	10 100
3	00 100
4	01 100
5	10 110
⋮	

If a suitable state variable is not available, then a signal must be formed. An AND gate might be used to detect the maximum count of the q counter, for example. A carry signal out of a decade counter, as previously discussed, is very suitable for cascading decade counters in an asynchronous fashion.

If the r counter has both a stop/start and clock terminal, then the r counter can be synchronized to the q counter. The q counter must provide a carry signal that takes the value of 1 for one clock period of each count cycle. If the q counter also has a stop/start input terminal, then the entire counter may be controlled. Figure 6.6 summarizes these two interconnection techniques.

It is as easy to count down as up, of course. Synchronous counters may be designed to count either up or down according to the value of a control signal. Also, synchronous counters may be designed to load a count value from input terminals each time a specific count is reached. This technique is used to create variable modulo counters. A binary down counter is modified so that when the count of zero is reached the next count value is taken from input lines, perhaps driven by switches. Figure 6.7 shows a mod 16 synchronous down-counter with input lines I_1, I_2, I_3, and I_4. The count of $I_1 I_2 I_3 I_4$ follows the count of zero. Stop/start and carry circuitry can be added so that this counter could be cascaded with others.

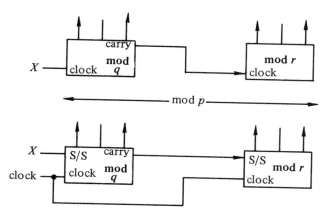

FIG. 6.6   Cascading Counters

Finally, if the flip-flops have preset and clear terminals, other designs are possible. Counters are commonly made asynchronously resettable by using the clear terminals. The count sequence of ripple counters may be interrupted by detecting a specific count and resetting all flip-flops. The clear pulse must have sufficient length, and a multivibrator or latch must be used to provide it reliably.

The great variety of counter considerations and design techniques may appear to be overwhelming, and perhaps unwarranted. Counting is a very fundamental operation and the variety of counter applications is very large. Counters are ubiquitous.

6.2   MOVING INFORMATION

While the state models and design procedure for sequential networks developed in the previous sections may always be applied, such formality is not always required. Many if not most of the memory elements in an information processing system are used in a less sophisticated manner simply to provide temporary *storage* for bits of information. The state of a memory element used to achieve storage is usually in direct correspondence with the bit being stored. We will begin to speak of the "content" of an FF or collection of FFs, which implies that FFs are able to hold or contain something. We view them as holding binary encoded information with the state of the collection of FFs being a direct expression of the stored binary sequence.

Suppose that a circuit generates a signal, x, the value of which changes from one clock time to the next, and that we will be required at some time in the future to know the present value of x. It is not convenient, economical, or perhaps possible for the circuit that generates x to retain the current value until the time at which we require it. We clearly must introduce circuitry capable of retaining or storing the present value of x

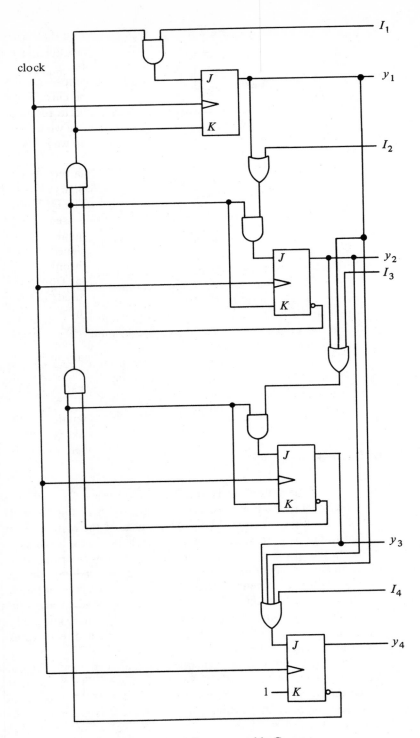

FIG. 6.7   A Programmable Counter

until we no longer require it, and perhaps require that a new value of x then be retained. We know that memory elements must be a part of such circuitry, for our problem is sequential in nature.

Formal design is not really required. We simply want a flip-flop to assume the state that directly corresponds to the current value of x, and then ignore subsequent values of x until it is told to respond again to the then-current value of x. If $x = 0$ at present, then we wish the FF to be reset before the next clock time. If $x = 1$ currently then we require the FF to be set.

GL flip-flops have a control as well as a data terminal and hence are natural solutions to the storage problem. Figure 4.16 indicated how RS and JK flip-flops may be gated to achieve GL logic. The DFF retains a bit only until it receives another clock pulse. It can be used to retain a bit for several system clock periods by gating its clock signal. System synchronization will be lost if this gating is not carefully designed. Figure 6.8 shows a scheme that (attempts to) maintains system synchronization by generating the system clock and flip-flop gated clock signal from the same master source. Gates with (nearly) identical delays are used. In the remainder of this chapter we will assume all flip-flops are synchronized to the system clock, and in most cases we will not show that clock signal.

Two signals are shown in Fig. 6.8. We refer to signal x as a *data* signal: it has significance (perhaps it is one bit of a binary number) or we would not be interested in retaining it. *LOAD* is a *control* signal: it determines at each clock time whether the current value of x is stored or ignored. When $LOAD = 1$ the value of x is stored.

A set of flip-flops and closely associated combinational logic that simultaneously perform very similar if not identical tasks is known as a *register*. A register might be used to store a binary number, with each FF storing

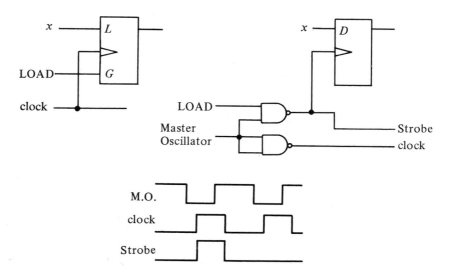

FIG. 6.8   The Value of x May Be Retained for Many Clock Periods

one bit, including the sign bit if there is one. The counters of the previous section may be thought of as fairly complex registers. If we wish to load a register from terminals or another register of the same dimension, we may simply iterate the cell of Fig. 6.8 to design the destination register.

A more typical situation is described by:

$$\langle RE \rangle \ A[15], \quad B[15], \quad C[15].$$

$$|ATOC| \ C \leftarrow A.$$

$$|BTOC| \ C \leftarrow B.$$

Here we are to transfer the content of register A to register C when control signal $ATOC = 1$. The contents of register B are to be transferred to C when control $BTOC = 1$. When $ATOC = BTOC = 0$, the contents of register C are not to change. It is very usual for the circuitry that generates the control signals to prohibit $ATOC = BTOC = 1$, because we cannot simultaneously store two 15-bit words of information in register C. The logic of the typical cell must now be more complicated and we might use formal design procedures, but it is almost ridiculous to do so. We desire that the typical cell routes either A_i or B_i to FF C_i, the sort of task that a multiplexer performs. Let us express this task in another way, which clearly suggests the required logic.

$$|ATOC \ \lor \ BTOC| \ C \leftarrow ATOC \cdot A \ \lor \ BTOC \cdot B.$$

Figure 6.9 illustrates the multiplexer as well as the control circuitry common to all 15 cells of register C. The Boolean condition within the *if ... then* bars must be supplied to the G input terminal, while the AND-OR logic routes data to the L terminal.

Generalization to provide for transfers from any registers A, B, C, or D to register E is relatively easy. The data-routing switch must be enlarged.

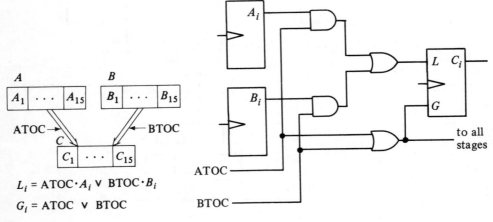

$$L_i = ATOC \cdot A_i \ \lor \ BTOC \cdot B_i$$

$$G_i = ATOC \ \lor \ BTOC$$

FIG. 6.9 Selective Register to Register Transfer

Simultaneous transfers into and out of the same register require that we use flip-flops rather than latches as the memory elements.

$$\langle RE \rangle\ A[15],\ B[15],\ C[15].$$

$$|CIRCULATE|\ B \leftarrow A,\ C \leftarrow B,\ A \leftarrow C.$$

At the clock time when $CIRCULATE = 1$, the contents of A are transferred to B, the contents of B are passed on to C, and the contents of C are passed to A. Figure 6.10(a) illustrates the typical cell required to accomplish this simultaneous transfer. Further, the transfer $A \leftarrow \bar{A}$ may be accomplished if flip-flops are employed.

$$|COMP|\ A \leftarrow \neg A.$$

The binary sequence stored by register A is complemented, and returned to A. Figure 6.10(b) shows the typical cell required to accomplish this transfer. Note that in all of these cases we did not design using a formal procedure. When we understand what is to be accomplished, we can usually draw the required typical cell directly.

The bundles of wires between registers are called *data paths*. If each register is fed by separate data paths, simultaneous information transfers may take place, and information processing may take place at a maximum rate. But this speed is not gained without cost. If the wires are of any length, transmission line (driving and receiving) amplifiers may be required. The size of the required data-routing switch grows with the number of data paths that serve a register. These costs encourage a system designer to seek information processing algorithms that require a minimum number of different data paths.

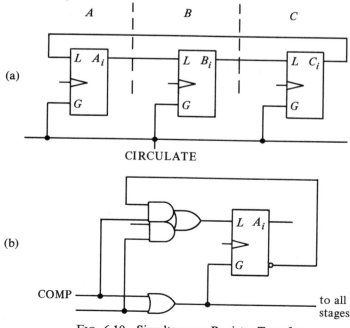

FIG. 6.10   Simultaneous Register Transfers

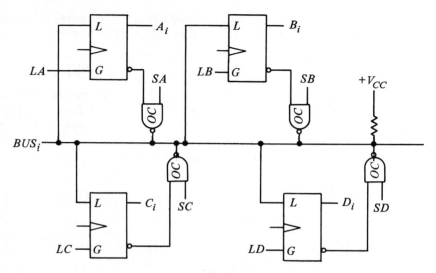

FIG. 6.11   A Single Bus System

A speed–cost trade is possible. Suppose we wish the freedom to transfer information from any of our registers to any other, but are content to perform only one transfer at a time. Then the data paths and selection switches can be replaced with a *bus*, a bundle of wires that may be driven by any of the registers. Figure 6.11 shows a typical cell of a 4-register, single bus system. The bus line is connected to the *L* terminal of all GLFFs, of all registers. Clock signals are not shown. Each register has a load control signal; *LA*, *LB*, *LC*, and *LD* are shown. Thus any one or more of the registers may simultaneously load from the bus. Each of the registers has a send control signal, *SA*, *SB*, *SC*, and *SD*. As long as the send control for a register is low, the open-collector NAND gates of that register exert no influence on the voltage level of the bus lines. When a single send control is high, the information held by the register of that send control is placed on the bus lines. Wired logic is used to form an "OR" gate from physically separated NAND gates.

Note the simplicity, similarity, and hence low cost of the register cells in Fig. 6.11. Some flip-flops include open collector or tri-state output amplifiers so that separate bus driver gates are not required. Other variations are also possible. *D* flip-flops may be used if the load control is combined with the clock signal to form a strobe signal (see Figs. 5.26 and 6.8). Since we would not expect a register to transmit and receive simultaneously in this single bus system, latches may be used in place of flip-flops.

Shifting

Information transfers that involve a single register are not uncommon. In fact, cell-to-cell information transfers within a register are so common that registers capable of performing such transfers are given the name *shift register*.

We are familiar with the concept of shifting and networks to accomplish shifting. Right shifting a 5-bit register under control of signal *SHR* is described with the DDL "for" clause by:

$$\{i = 1 : 4\} \, |SHR| \, B[i + 1] \leftarrow B[i]..$$

Note that this description offers no indication of the state of $B[1]$ after the shift operation is complete. An information bit from several sources is commonly entered into this cell; let us explore some of the possibilities. First we might transfer a 0 (or 1) to this cell.

$$\{i = 1 : 4\} \, |SHR| \, B[1] \leftarrow 0, \; B[i + 1] \leftarrow B[i]..$$

This action is common enough to warrant the DDL notation

$$|SHR| \quad 0 \!\!\downarrow_{\!\!\rightarrow} \quad B.$$

which suggests the direction of the shift as well as what is to be placed in the leftmost cell.

Alternatively, we may desire that the state of FF $B[1]$ not change with the shifting operation.

$|SHR| \; e\!\!\downarrow_{\!\!\rightarrow} \; B.$ denotes $\{i = 1 : 4\} \, |SHR| \, B[1] \leftarrow B[1], \; B[i + 1] \leftarrow B[i]..$

This is particularly appropriate if we view the binary sequence stored by register B as expressing a signed number in 1's complement representation. Retaining the most significant bit (the sign bit) while shifting then gives the 1's complement representation of one-half the original number. Shifting one bit-position to the right divides by 2, but we must retain and "extend" the sign bit to satisfy the 1's complement convention.

Third, the contents of $B[5]$ (the rightmost cell, in general) may be transferred to $B[1]$ (the leftmost cell). This type of shifting is referred to as *circulating* a binary sequence, and is specified by the notation $c\!\!\downarrow_{\!\!\rightarrow} \; B$.

Finally, a network independent of the shift register may provide on a terminal, say A, the bit to be placed in $B[1]$. This we denote $\downarrow_{\!\!\rightarrow} \; A \circ B$. Terminal A may be part of a logic structure outside of the register that contains no memory elements. But if A names another register, then both registers are to be shifted right together and some indication of what is to be placed in the left cell of A must be given. With this and several of the other types of shifts described above, the logic network of the leftmost cell will differ slightly from that of the typical cell.

Figure 6.12 shows the contents of shift register B at successive clock times for several of these types of shift operations. Unless we circulate B or make other provisions, the rightmost bit is lost with each shift.

We can construct a left-shifting register also. Here we must achieve $B_i \leftarrow B_{i+1}$ and be concerned with what information is transferred to the rightmost cell. Again we can transfer a constant or variable to that cell or circulate. And we can put the circuitry of left- and right-shifting cells together to form a register that can shift in either direction. Usually two

time	$0 \downharpoonleft B$	$1 \downharpoonleft B$	$c \downharpoonleft B$	A	$\downharpoonleft A \circ B$ B
n	10111	10111	10111	0	10111
$n+1$	01011	11011	11011	1	01011
$n+2$	00101	11101	11101.	1	10101
$n+3$	00010	11110	11110	0	11010
	00001	11111	01111		01101

FIG. 6.12   Types of Right Shifts

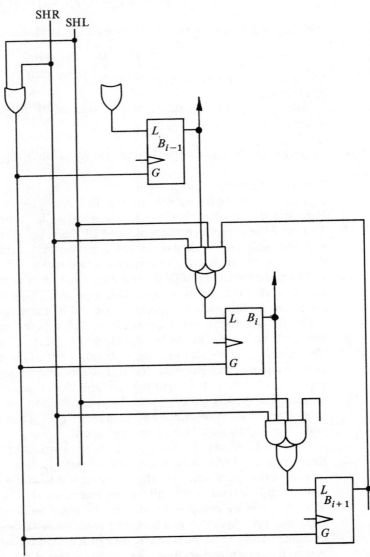

FIG. 6.13   Typical Cell of a Bilateral Shift Register

control signals, *SHR* and *SHL*, are used to control the shifting. When $SHR = SHL = 0$ no shifting takes place. Should $SHR = SHL = 1$ the results may be meaningless and we guard against this case. Figure 6.13 shows a typical cell.

If we combine the abilities to load and shift, very useful tasks can be performed. Suppose we construct circuitry able to comply with the following declarations.

$$\langle RE \rangle\; A[4],\; B[4].$$

$$|LOAD|\; B \leftarrow A.$$

$$|SHR|\; \underline{0|}_{\rightarrow}\; B.$$

Figure 6.14 shows the required typical cell for register *B*. We will not be concerned at the moment with how information such as a BCD code word is placed in the *A* register.

Now suppose that the *A* and *B* registers contain the 4-bit sequences shown in Fig. 6.15 and the control signals are as indicated. We first transfer a 4-bit word to the *B* register and then shift right three times. Signal B_4 then is a serial representation of the 4-bit word we originally transferred in parallel. We have used a shift register to perform *parallel-to-serial* conversion of information.

Conversion of serially presented binary sequences to a parallel form is as easily accomplished with a shift register. But rather than detailing this application, let us turn to another variation of shift register. The registers we have been considering shift a binary sequence one place to the right or left each clock time that a shift control signal is presented. In very high

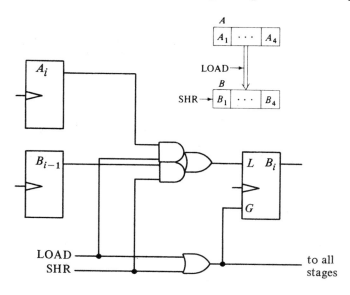

Fig. 6.14   A Load and Shift Register

LOAD	SHR	A	B	
0	0	0111	1000	
1	0	0111	1000	↓ time
0	1		0111⌉	
0	1		0011	
0	1		0001	
1	0	0010	0000⌋	
0	1		0010	

FIG. 6.15 Parallel-to-Serial Conversion

speed information-processing systems that deal with long binary sequences, it may take many clock periods to accomplish a long shift of many positions. This is intolerable in truly high speed processors, and we require a register capable of shifting a sequence an arbitrary number of positions in one clock period. To develop such a register, let us first consider the simpler problem of designing a register able to shift a sequence either one or two positions to the right under the control of appropriate signals. Figure 6.16 illustrates the typical cell of such a register. We may find this structure familiar, for it is very similar to the other register cells that we have examined. When control $SR1 = 1$ we route signal B_{i-1} to the input terminal of the ith stage. When $SR2 = 1$ we route B_{i-2} to that point. A selection switching matrix is used to accomplish this routing. Extension of this design to achieve shifts of a greater number of positions is easily accomplished with a larger selection switch. This is an expensive design if we have many positions in register B and we may choose to implement a more restricted shift register and take more time to accomplish a longer shift.

If shifting is desired in a bus oriented digital system, then at least two buses are required. One bus conveys information from a source register to a shifting network; the second bus carries the shifted information to a destination register or registers. Examples of such systems will be seen in the next section, which treats structures to realize information manipulating processes of which shifting is an example.

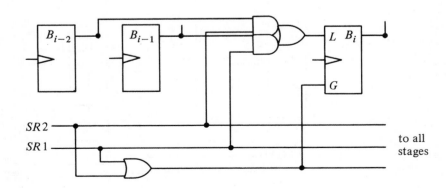

FIG. 6.16 Multiple Cell Shifting

Asynchronous Transfers

If total system synchronization is not essential and the flip-flops of registers possess preset and clear terminals (or latches are to be used), then several asynchronous techniques may be used to load registers. Figure 6.17(a) shows the cell design for a two-step process for loading the flip-flops. First the *CLEAR* line is dropped to 0 and all flip-flops are reset. Then *CLEAR* is returned to 1 and *LOAD* is raised to 1. Information on the *X* terminals sets those flip-flops driven by *X* lines bearing 1. If the *CLEAR* and *LOAD* signals are derived from the system clock signal, two clock periods are required to complete the transfer, but circuitry cost is low and the synchronized input terminals are available to realize other desired register actions. While JKFFs are shown in Fig. 6.17(a), NAND-pair latches may also be loaded with this technique.

In Fig. 6.17(b), two NAND gates are associated with each flip-flop. The NAND gates are interconnected, as in Fig. 4.10 for the gated latch. As long as $LOAD = 0$, the flip-flops act according to their synchronous input signals. When $LOAD = 1$, data on the *X* lines appears complemented at the preset terminals and in true form at the clear terminals of the flip-flops. If $X_i = 0$, then flip-flop A_i is reset asynchronously, regardless of the clock and synchronous input signal values. If $X_i = 1$, then A_i is set. This method of loading flip-flops requires only one step and is called "jam transfer". Information is jammed into the flip-flops.

The synchronous terminals in Fig. 6.17(b) are shown connected to accomplish right shifting when control $SHR = 1$. The clock signal presented to the flip-flops is not truely synchronized to the system clock signal. The resulting clock "skew" may lead to system timing problems. When it does not, the simplicity of this register design is very appealing.

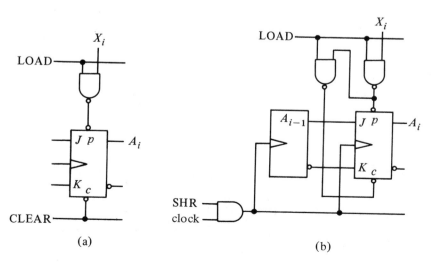

(a)

(b)

FIG. 6.17   Two Asynchronous Techniques for Loading
Flip-Flop or Latch Registers

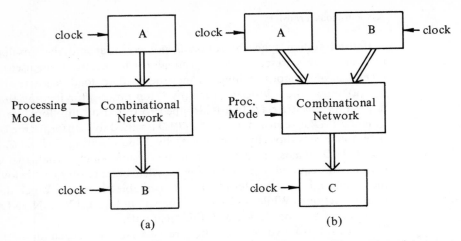

FIG. 6.18 (a) Unary and (b) Binary Processing Systems. The result register may be one of the operand registers if flip-flops are used.

6.3 PROCESSING INFORMATION

Data processing consists in the main of transferring information from one register to another or back to the original register through combinational logic that combines or alters it in some way. In a large system, many such transfers may be taking place simultaneously and some may influence the nature of others. Shifting and counting are simple examples in which information held by a set of flip-flops is altered as it is passed through relatively simple combinational logic and then returned to the same set of storage elements. Figure 6.18(a) presents a model of such *unary processing systems*. While two registers are shown, A and B may be the same register if flip-flops are used.

Some of the more important manipulations combine two binary sequences to form a third new sequence, which is passed to a register for temporary storage. Figure 6.18(b) models *binary processing systems*. As a design example, we develop a register able to form and then store the logical sum or product, on a bit-by-bit basis, of its contents and the contents of another register of equal length.

$$\langle RE \rangle\ A[15],\ B[15].$$
$$|AND|\ B \leftarrow A \cdot B.$$
$$|OR|\ \ B \leftarrow A \vee B.$$

Logic to perform these manipulations is easily visualized as a result of our work with registers in the previous section.

$$\{i = 1 : 15\}\ LB[i] = A[i] \cdot B[i] \vee OR \cdot (A[i] \vee B[i]),$$
$$GB[i] = AND \vee OR.$$

The first term of each equation provides for the logic-product manipulation, while the second terms obviously apply to the OR combination of the binary sequences. Why doesn't the control signal AND appear in the first term for $LB[i]$?

The design procedure of Sections 5.2 and 5.3 can also be used, for we are designing a sequential machine. This procedure may give better results than simply drawing the obvious circuitry. Figure 6.19 shows the truth table and state map for the typical cell of the desired register. Alternative designs, which are not as obvious are suggested. For example, (since all cells are identical we need not detail subscripts):

$$LB = OR,$$
$$GB = A \cdot \overline{B} \vee \overline{A} \cdot AND.$$

Here the L input equations are very simple: we are to place a 1 on the L input terminal only when performing the OR operation. Is this reasonable; do we ever set an FF when performing the AND operation? No. If an FF of the B register is not already set, then it will never be set during formation of the logical product. If it is set and the corresponding A FF is reset, then we must reset the B FF by placing a 0 on its L terminal and a 1 on its G input terminal. Hence we find the term $\overline{A} \cdot AND$ in the GB equation above.

If the B register is to be capable of performing other manipulations in addition to AND and OR, this design possesses an inherent disadvantage in the term $A \cdot \overline{B}$, which can have the value of 1 when neither control signal is presented to the register. An erroneous manipulation could result, so we return to Fig. 6.19 for a third design.

The third design requires the same amount of combinational circuitry, but can be easily coupled with partial designs that achieve other manipulations.

$$LB = A,$$
$$GB = \overline{B} \cdot OR \vee \overline{A} \cdot AND.$$

These equations not only describe more desirable logic circuitry, but are also instructive. Consider the OR operation. The state of a memory

AND	OR	A_i	B_i	B_i'
0	0	x	x	B_i
0	1	0	0	0
0	1	0	1	1
0	1	1	0	1
0	1	1	1	1
1	0	0	0	0
1	0	0	1	0
1	0	1	0	0
1	0	1	1	1
1	1	x	x	d

State map (B_i'):

B_i' (rows) vs $\dot{B}_i$ / A_i (columns), with OR and AND labels:

		$\dot{B}_i$		
0	0	1	1	
0	S	1	1	OR
d	d	d	d	
0	0	1	R	AND

A_i

FIG. 6.19 Specifications of a Register to Form the Logical Product or Sum of Its Contents and Another Binary Sequence

element of the B register need only be changed if that element is reset, and the corresponding memory element of the A register is set. If memory element B_i already stores a 1, the content of A_i need not even be considered. When performing the AND operation, B_i must be reset if $A_i = 0$; otherwise its state can remain at its present value. Perhaps the LOAD and SHIFT designs of the previous section should be reviewed with these arguments in mind.

To summarize and indicate the complexity of the designs we have realized, let us detail the typical cell of a register B that can perform the following manipulations.

$$\langle RE \rangle\ A[15],\ B[15].$$
$$|LOAD|\ B \leftarrow A.$$
$$|SHR|\quad \underline{0|}_{\rightarrow}\ B.$$
$$|SHL|\quad \underline{0|}\ B.$$
$$|AND|\quad B \leftarrow A \cdot B.$$
$$|OR|\quad B \leftarrow A \vee B.$$

Assuming that no two control signals will be simultaneously presented, designs in this and the previous section allow us to write, for the typical cell,

$$LB_i = B_{i-1} \cdot SHR \vee B_{i+1} \cdot SHL \vee A_i \cdot (LOAD \vee OR)$$
$$GB_i = SHR \vee SHL \vee LOAD \vee \overline{B}_i \cdot OR \vee \overline{A}_i \cdot AND$$

Figure 6.20 then shows the NAND pyramids described by these equations. Note that some combinations of control signals may be formed once (7

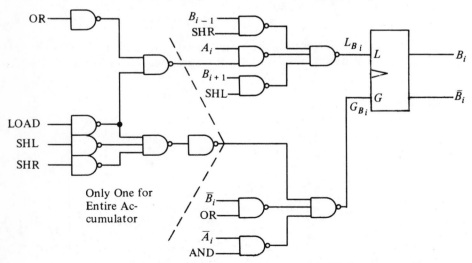

FIG. 6.20 **NAND Implementation of the Typical Cell of a LOAD, SHIFT, AND, and OR Register**

gates) and supplied to all cells of the *B* register. The 15 cells of the register are constructed of $7 \times 15 = 105$ NAND gates, and 15 GLFFs.

Bus Oriented Processing Systems

The arithmetic networks of Chapter 3 are very natural candidates for inclusion in binary processing systems. Review those networks with Fig. 6.18(b) in mind. Some of the figures of Chapter 3 show source and destination registers. It is easy to add registers to the other arithmetic processing blocks.

Two or more buses are usually used in bus oriented processing systems. At least one bus carries an operand to the processing block, while another carries the result from the processing block to a destination register. Registers may be: (1) tied bidirectionally to one bus, (2) tied bidirectionally to more than one bus at greater cost, or (3) tied unidirectionally to two buses. Figure 6.21(a) illustrates these possible register connections in a two-bus system. The *A* register is bilaterally connected to the *B*1 bus with the cells of Fig. 6.11. The *ACC* register is similarly connected to the *B*2 bus, but in addition it provides one of the operands to the arithmetic logic processing unit (*ALU*). The *BUFF* (buffer) register loads from *B*2 and sources information to the *B*1 bus. The cell of Fig. 6.11 can be used with obvious modification. Finally, the *B* register is bilaterally connected to both buses. A suitable cell for the *B* register is shown in Fig. 6.21(b). The cost of this cell makes this very general interconnection less attractive.

Processing is accomplished in Fig. 6.21 with the ALU and shifter combinational blocks connected in tandem to show the possibility of performing two operations during the same clock period. In Chapter 3 we saw the need to follow an arithmetic step with a shifting operation. If all processing were done by one block, then separate clock periods would be required to add and shift. To be generally useful, the ALU might have the following modes of operation, plus others.

Pass either operand to the output terminals.
Add 1 to either operand.
Subtract 1 from either operand.
Add operands.
Subtract operands.
AND, OR, XOR operands.

Flip-flop *X* is connected to the shifter. If it is assumed that only single position shifts are performed, this single bit register captures the bit that would otherwise be lost. One mode of shifting would fill the vacated bit position from this flip-flop. To clarify the function of flip-flop *X*, let us assume the following shift modes and mode codes.

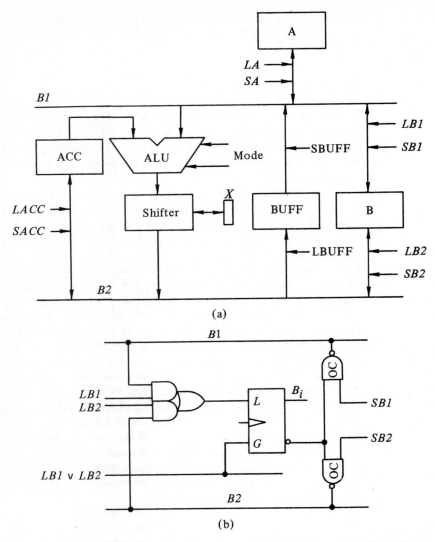

FIG. 6.21 (a) A Two-Bus System Permits Processing as well as Transferring Information and (b) a Typical Cell of the Bilaterally Tied B Register

code $b_1 b_2 b_3$	mode
0 x x	pass
1 x x	shift right 1 position
1 0 0	0-fill
1 0 1	1-fill
1 1 0	extend
1 1 1	fill from X

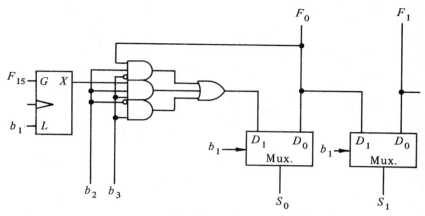

FIG. 6.22   The Shift-Save Flip-Flop and the Left Cells of the Shifter

The typical cell of the shifter is then a 2-input multiplexer, which is controlled by the left mode control bit, b_1. Figure 6.22 shows circuitry to drive the unusual leftmost cell. When $b_1 = 0$, shifter output bit S_0 equals the leftmost ALU bit F_0. Otherwise, bits b_2 and b_3 are decoded and either a 0, 1, X, or F_0 is presented to the multiplexer for connection to output terminal S_0.

The X flip-flop permits shifting bits from one register to another. Suppose we first pass the contents of the ACC register through the ALU, shift right, which extends the sign bit, and return the results to the ACC. Then we pass the B register contents through the ALU, shift right, filling from X, and return the result to the B register. The overall effect of this two step process is to shift $ACC \circ B$ to the right, using one 16-bit shifter and two clock periods. When discussing multiplication in Chapter 3, we saw the need for such extended shifts, but did not concentrate on the economics of the shift network.

Memory elements might well be attached to the ALU also. Such flip-flops would capture status information such as overflow, sign of result, zero result, and so forth. These are facts we often wish to know. Capturing information at this central point of the system eliminates the need to attach special detection circuitry to a register or registers. Duplicating detection circuitry is expensive, and the attachment of detection networks to different registers restricts the way in which those registers may be employed to realize processing algorithms. This is a point of systems design. If we know exactly what our system will be expected to do tomorrow, and how it will do it, then we optimize our designs based upon that knowledge. However, it is more common to know that we do not know what and how we will desire our system to process tomorrow. To prepare, we strive to organize systems so that their performance may be more easily altered tomorrow. The next chapters will have more to say on this point.

6.4 CONTROL UNITS

Now we turn to the question, "Where do all of the control signals that we have assumed to exist originate?" While the answer, "a sequential network", is quickly given, decisions about the structure of control networks are not so easily made. A great variety of system organization and logic design options exist. To illustrate some of the possibilities, we will develop various control units for the following data collection system.

> A keyboard presents ASCII code words, $b_8 b_7 b_6 \cdots b_1$, and signal $KD = 1$ for the many milliseconds that a single key is depressed. KD is synchronized to the 1 MHz system clock. If an octal digit key is depressed, the bits of that digit $b_3 b_2 b_1$ are to be placed on the right of a 24-bit D register, all previously placed bits being shifted left 3 places. If the special $CLEAR$ key is depressed, the D register is to be cleared. If the special $XMIT$ key is depressed, signal AVailable $= 1$ until a receiving unit indicates that it has captured the word from the D register by setting $GOT =$ (less than $50\mu sec$ wait). GOT will be returned to 0 soon after AV is returned to 0. All other key depressions are to be ignored.

The data processing circuitry is minimal in this system so that we can concentrate on control.

$\langle RE \rangle$ D[24].

$\langle TE \rangle$ KD, OCT, XMIT, CLEAR, LS, AV, OUT, RESET, b[8:1].

$\quad$ OCT = KD $\cdots$ /b[8:4]\6O5,

$\quad$ XMIT = KD $\cdot$ b[8] $\cdot$ b[7],

$\quad$ CLEAR = KD $\cdot$ b[8] $\cdot$ b[6],

$\quad$ |LS| (3) $\downarrow$ D $\circ$ b[3:1].,

$\quad$ |RESET| D $\leftarrow$ 0.

We begin designing a control unit by recording the processing algorithm in detail. Our example does not possess the long sequences of steps often found in processing algorithms, but short sequences are necessary.

Step 1. If no key is depressed do nothing.
$\quad$ a. If an octal key is depressed, load-shift register D and go to step 2.
$\quad$ b. If the $CLEAR$ key is depressed, reset D.
$\quad$ c. If the $XMIT$ key is depressed, issue $AV = 1$ and go to step 3.
Step 2. Wait until the key is released; then go to step 1.
Step 3. Issue $AV = 1$ until $GOT = 1$; then go to step 2.

Step 2 is necessary to ensure that the actual digit is loaded only once, regardless of how long the key is depressed.

Chapter 5 indicated that the first step of synthesis consists of finding a suitable state diagram. The processing algorithm dictates the state diagram

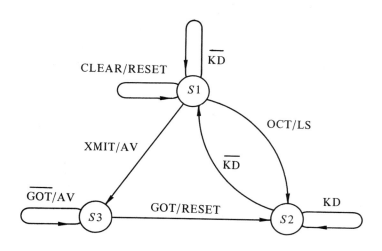

$$\langle ST \rangle \ S1: KD: |OCT| \ LS=1, \ \rightarrow S2.,$$
$$|XMIT| \ AV=1, \ \rightarrow S3.,$$
$$|CLEAR|RESET=1, \ \rightarrow S1..$$
$$S2: \neg KD: \ \rightarrow S1.$$
$$S3: |GOT|RESET=1, \ \rightarrow S2; \ AV=1, \ \rightarrow S3...$$

FIG. 6.23   State Models of the Control Algorithm

of a control unit. The steps above are easily translated into the state diagram and DDL description of Fig. 6.23.

State minimization is clearly not possible. Each state corresponds to a very unique situation. State assignment is an important consideration. In many processing algorithms a sequence of steps is performed repeatedly, and a counter may be used as the control network. In our example a simple counter is not suitable. We may use two flip-flops; then the following code has the desirable property that the network can be placed in its idle state $S1$ with an asynchronous clear operation.

state	code $y_1 y_2$
$S1$	0 0
$S2$	0 1
$S3$	1 0

We may also design the network so that it returns to the idle state when no key is depressed, and it is in some other state such as the $y_1 y_2 = 11$ state.

With the code above, JKFF input equations and output variable equations can be written directly from the state diagram of Fig. 6.23.

$$J_1 = XMIT \qquad\qquad\qquad J_2 = OCT \vee y_1 \cdot GOT$$
$$K_1 = GOT \qquad\qquad\qquad K_2 = \overline{KD}$$

$$S1 = \bar{y}_1 \cdot \bar{y}_2$$
$$LS = S1 \cdot OCT$$
$$AV = S1 \cdot XMIT \vee y_1 \cdot \overline{GOT}$$
$$RESET = y_1 \cdot GOT \vee S1 \cdot CLEAR$$

This is a design of very reasonable cost. The ease with which it was obtained is not entirely typical. While equations may be written directly from the state diagram, as we saw in Section 5.3, usually more terms must be collected for each equation. Using another code would make the equations for the example control unit more complex.

A control unit based upon a one-hot code is called a *sequencer*. More flip-flops are required, but designing sequencers is easy; state decoders are not needed. Modifications to the control algorithm result in corresponding modification of the sequencer rather than complete redesign of the control unit. If $S1$, $S2$, and $S3$ are flip-flops, as well as states, and the following code is used,

state	code $S1$	$S2$	$S3$
$S1$	1	0	0
$S2$	0	1	0
$S3$	0	0	1

a sequencer using D flip-flops is easily designed from Fig. 6.23.

$$D_1 = S1' = S1 \cdot \overline{KD} \vee S1 \cdot CLEAR \vee S2 \cdot \overline{KD}$$
$$D_2 = S2' = S1 \cdot OCT \vee S2 \cdot KD \vee S3 \cdot GOT$$
$$D_3 = S3' = S1 \cdot XMIT \vee S3 \cdot \overline{GOT}$$
$$LS = S1 \cdot OCT$$
$$AV = S1 \cdot XMIT \vee S3 \cdot \overline{GOT}$$
$$RESET = S1 \cdot CLEAR \vee S3 \cdot GOT$$

This design is not self-starting. If several of the flip-flops assume the set state when power is first applied, the network may not reach the idle ($S1$) state. If all flip-flops power up to the reset state, then the system will not respond to the keyboard. Either additional circuitry is required to manually/automatically place the control unit in state $S1$ when power is first applied, or the design must be modified to ensure that $S1$ is reached soon after power is applied. Figure 6.24 shows how an RC network may be used to initialize the control register in state $S1$ when power is applied. Note that the logic of Fig. 6.24 may be drawn directly from the state

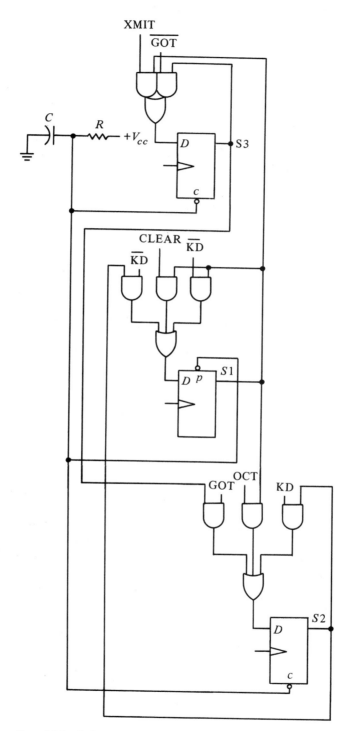

FIG. 6.24 A Sequencer with Automatic Initialization Circuitry

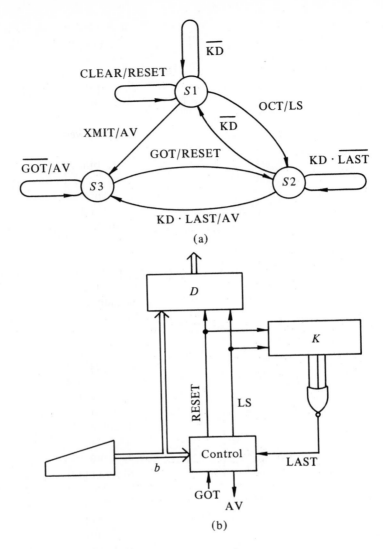

FIG. 6.25 A Counter Is Used to Expand the Set of Control States

diagram, as the equations above were. If a transition in the state diagram were changed, the wiring of an AND gate would have to be rerouted in a parallel manner.

An idea! Why not have the control unit transmit data automatically after 8 octal digits have been entered and the D register filled[1]? This is an example of the sort of modification that can be suggested after a design is completed. (Keep in mind that we are dwelling on the first steps of design only. After the logic design is finalized, it must be converted to available

---

[1] Error correction on the last digit is not possible with automatic transmission.

technology, parts must be placed on a printed circuit board, wire routes need to be determined on that board, and so forth.) It also gives an opportunity to consider a frequently encountered control situation. Clearly we must count digits loaded to the D register and respond to the eighth digit in a unique manner. Is this counting a part of controlling the system or not? In our example counting within the control unit would require that states $S1$ and $S2$ appear eight times each. We would have a 17-state machine to design. If the range of counting were smaller, this approach might be more attractive.

If we think of using a separate 3-stage binary counter, K, the control unit must be modified to direct it. The counter must be cleared when the D register is ($RESET = 1$). It must be advanced when the D register is loaded ($LS = 1$). We have suitable control signals already available. Now we must change the processing algorithm and state diagram to go from state $S2$ to $S3$ when the count is 0! With the first digit loaded, the counter is advanced from 0 to 1; with the second digit it is advanced from 1 to 2. With the eighth digit, it is advanced from 7 to 0. We chose to advance the counter before testing it. Figure 6.25 shows the modifications that must be made to the state diagram and a block diagram of the overall system.

ROM Control

Numerous are the reasons to use a ROM realization of the combinational logic in a control unit. "Engineering changes" are a fact of life. If a control unit is fabricated by using individual gates and a modification or correction is required, all of the original hardware, artwork, and documentations will require modifications, or perhaps total replacement. Modifying and replacing entire circuit boards may be very expensive, compared with replacing one integrated circuit—the ROM. Costs must be carefully compared in each situation, of course. A ROM centered control unit may require less space, power, and cooling. It may be slower than a conventional design. The hardware design effort is replaced with a microprogramming effort, which is not necessarily less extensive or cheaper.

ROM control units generally have the form of Fig. 6.26. A register provides the ROM with the present state of the system. In memory terminology the present state of the system is referred to as the "address." The ROM replies by providing a number of bits of information. Some are used to determine the next state (address) of the control unit. Others direct the data registers and counters to be controlled. The address modification logic is a conventional combinational logic network that forms the actual next address of the control unit from system input and status signals as well as next address information from the ROM. To gain the advantages of ROMs, this block should be kept as simple and low-cost as possible.

While the data collector system may be too small to economically justify a ROM control unit, it will be used to illustrate the problems of ROM centered system design. Again we begin with the processing algorithm

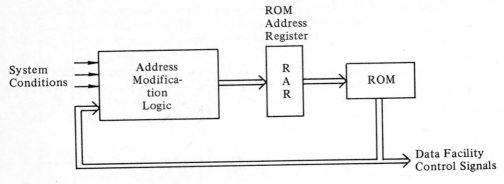

ROM
Address
Register

System
Conditions

Address
Modifica-
tion
Logic

R
A
R

ROM

Data Facility
Control Signals

FIG. 6.26   A ROM-Centered Control Unit

or state diagram, but we may find it desirable to modify them. In Fig. 6.25, we see only three states. A two or more way branch is made at each state. If we thought of using 3 addresses specified by two bits, the address modification logic would have to generate both bits—the full address—to accomplish a 3-way branch. The AML does nearly the full job of determining next state rather than just modify what the ROM provides.

Address modification usually consists of modifying a small number of bits of a single address provided by the ROM, or selecting one of several next addresses provided by the ROM. To illustrate address modification, let us assume that the ROM provides two next addresses and that the AML selects one as the actual next address. Each address can then be followed by either of two addresses, one or both of which may be the present address. The state diagram of Fig. 6.25 must be altered to satisfy this restriction. Only two-way branches must be specified. States $S1$ and $S2$ present problems. Rather than testing for key depression and the type of key depressed all at one point, we must make successive tests. This is not really a problem since a key will be depressed for milliseconds, and our system clock frequency might well exceed a megahertz. Figure 6.27 details this state diagram change and the expansion of $S2$ to test variables $LAST$ and KD in turn.

With an algorithm or state diagram that satisfies address modification limits, we turn to the state assignment problem. Our decision to select one of two next addresses makes this an easy problem to solve. To reduce the size of the address register and ROM, we use the minimum number of bits necessary to encode the states. Any assignment within that limit will work with our AML.

How does the AML know which input signal to use in selecting the next address? It must be provided that information, and the ROM is called upon to provide the selection control bits. Since the diagram of Fig. 6.27 branches on five different signals, KD, OCT, $XMIT$, $CLEAR$, GOT and $LAST$, at least three bits will be required from the ROM.

Output signal values provided by the ROM of Fig. 6.26 are determined by the address (present state). Figure 6.27 is a Mealy model of a sequential

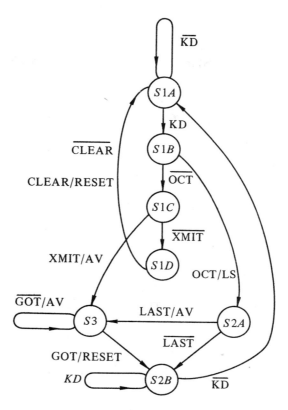

FIG. 6.27 Restricted Branching Forces Expansion of the State Set

network. Before we can complete the ROM design, we must convert it to a
Moore model in which output signal values are determined entirely by the
present state. (Modifying control signals as we modify next addresses is
generally undesirable as it would slow system performance.) While states
may be added to the diagram of Fig. 6.27 to document a Mealy-to-Moore
transformation, the diagram is no longer an attractive format. Table 6.2
records the change in a manner that is very close to the ROM contents.
Note that signal AV is not issued until state (address) $S3$ is reached.

Now we see that ten addresses are to be used. Four address bits are
required. The ROM must have a word length of $3 + 4 + 4 + 3 = 14$ bits,
as shown in Table 6.2. The codes used for address modification and output
signals are rather obvious. Figure 6.28 shows how the AML may be constructed from two multiplexers. Only four ICs are required to build this
control unit—two multiplexer ICs and two 16 word, 8-bit per word
ROMs.

Each line in Table 6.2 is a *microinstruction*, since it instructs the entire
system for the smallest unit of time—the clock period. Preparing Table 6.2
is *microprogramming*, since it is very much like preparing digital computer
programs. Please do not conclude from the previous development that

TABLE 6.2. ROM CONTENT IN MNEMONIC AND BINARY FORM

Present Address	Test Signal	Next Address 0	1	Output Signal	IN	OUT			
S1A	KD	S1A	S1B		0000	001	0000	0001	000
S1B	OCT	S1C	S4		0001	010	0010	0111	000
S1C	XMIT	S1D	S3		0010	011	0011	0110	000
S1D	CLEAR	S1A	S5		0011	100	0000	1000	000
S2A	LAST	S2B	S3		0100	101	0101	0110	000
S2B	KD	S1A	S2B		0101	001	0000	0101	000
S3	GOT	S3	S6	AV	0110	110	0110	1001	100
S4	none	S2A	S2A	LS	0111	000	0100	0100	010
S5	none	S1A	S1A	RESET	1000	000	0000	0000	001
S6	none	S2B	S2B	RESET	1001	000	0101	0101	001

microprogramming is an impossible task. We started with a diagram best suited for synthesizing a hard-wired control unit and modified it as necessary to emphasize the restrictions placed by the structure of Fig. 6.26 and our decision to use a two address instruction format. If we had started with the instruction format

Present Address	Test Signal	Next	Addresses	Output

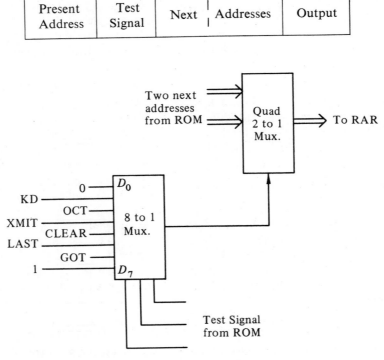

FIG. 6.28 The Address Modification Logic

and the significance of an entry in each field, we could have written the microprogram much more directly. Section 8.3 expands upon ROM-centered control units.

6.5 SEQUENTIAL PLA'S

Section 3.3 defined a PLA to be an integrated, programmable AND-to-OR gate array. ICs with the organization of Fig. 6.29 and programmable AND-to-OR logic are also being called PLAs. Such integrated circuits must be generally useful since Fig. 6.29 is almost identical with the general model of a synchronous sequential network, in Fig. 4.24. Parameter values for a commercial PLA are usually of the order of $n = 14$, $m = 8$, $p = 8$, and $q = 100$.

The sequential PLA is a generalized IC that is specialized to a user's "program" in the final step of manufacture. If production volume is moderate-to-high, it is a very economical and highly reliable alternative to a random logic design. When very high production volume is expected, a specially designed integrated circuit is more economical. The sequential PLA does not introduce any new concepts, or require theory or design techniques that we have not seen, although its optimum utilization in a system may challenge the system designer.

To illustrate the problem of programming a PLA, let us return to the data collector of the previous section and first program an 8-bit data register that can clear and shift left three places. The feedback register will be used as the data register. To make this data available, the output register will be slaved to the data register. Elements of both registers will be numbered 1 through 8 from left to right. LS, $RESET$, and I_6, I_7, I_8 will be input signals. The right three stages are described by the following state equations.

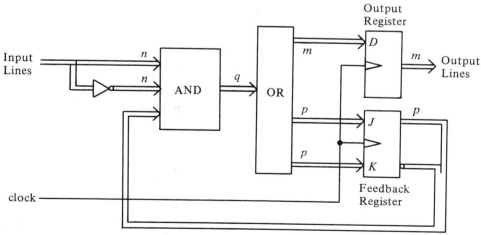

FIG. 6.29 The Sequential PLA. The AND and OR gates may be programmed. All dimensions and other connections are fixed.

$$y_i' = LS \cdot I_i \vee \overline{RESET} \cdot \overline{LS} \cdot y_i$$
$$= LS \cdot I_i \cdot \bar{y}_i \vee (LS \cdot I_i \vee \overline{RESET} \cdot \overline{LS})y_i$$

J and K input equations may now be written.

$$J_i = LS \cdot I_i$$
$$\overline{K}_i = LS \cdot I_i \vee \overline{RESET} \cdot \overline{LS}$$
$$K_i = LS \cdot \bar{I}_i \vee \overline{LS} \cdot RESET$$

For stages 1 through 5, I_i must be replaced with y_{i+3} in the above equations. Equations for all stages may be recorded in a table that is a "connection array." Each row in Table 6.3 expresses a product term and the output variables that require that term. Ones and zeroes in an input variable column indicate that the variable is to appear in true and complemented form, respectively. Blanks here indicate that the variable does not appear in the product term. The output variable columns specify AND-to-OR gate programming. A one dictates a connection; a blank indicates that no connection is to be made. First the J equations are specified. The K logic is then given. Finally the next state equations are completed for the output register. Twenty-five AND gates are utilized. Three of these PLAs may be interconnected to form the 24-bit data register.

Sequential PLAs are ideal for realizing control units, but note that the output signals are buffered by synchronized D flip-flops. Output signals are therefore delayed by one clock period with respect to the state models of previous chapters. It may be necessary to modify the state diagram, taking this delay into account, in order to obtain desired control signals. The data collector keys are depressed infrequently and held for many clock periods. Thus we will program a PLA directly from Fig. 6.25 by using the 2-bit state assignment that was first illustrated in the previous section.

$$S1 = \bar{y}_1 \bar{y}_2$$
$$S2 = \bar{y}_1 y_2$$
$$S3 = y_1 \bar{y}_2$$

Five bits from the keyboard, KD, GOT, and $LAST$[2], constitute the set of input signals. Recall that

$$OCT = KD \cdot \bar{b}_8 \bar{b}_7 b_6 b_5 \bar{b}_4$$
$$XMIT = KD \cdot b_8 b_7$$
$$CLEAR = KD \cdot b_8 b_6$$

From Fig. 6.25, we write J and K input equations for the two control flip-flops.

$$J_1 = S1 \cdot XMIT \vee S2 \cdot KD \cdot LAST$$
$$K_1 = GOT$$

---

[2] Later we will generate $LAST$ within the PLA.

TABLE 6.3 PLA PROGRAM FOR A CLEAR, LEFT-SHIFT RESISTER (BLANK ENTRIES INDICATE NO CONNECTION)

LS	RESET	I6	I7	I8	Y1	Y2	Y3	Y4	Y5	Y6	Y7	Y8	J1	K1	J2	K2	J3	K3	J4	K4	J5	K5	J6	K6	J7	K7	J8	K8	D1	D2	D3	D4	D5	D6	D7	D8
1								1					1																1							
1									1						1															1						
1										1							1														1					
1											1								1													1				
1												1									1												1			
1		1																					1											1		
1			1																						1										1	
1				1																							1									1
1								0						1																						
1									0							1																				
1										0								1																		
1											0									1																
1												0										1														
1		0																						1												
1			0																							1										
1				0																								1								
0	1													1		1		1		1		1		1		1		1								
0	0				1																								1							
0	0					1																								1						
0	0						1																								1					
0	0							1																								1				
0	0								1																								1			
0	0									1																								1		
0	0										1																								1	
0	0											1																								1

AND →

OR ↓

$$J_2 = S1 \cdot OCT \vee S3 \cdot GOT$$

$$K_2 = \overline{KD} \vee S2 \cdot KD \cdot LAST$$

Output equations now drive the D flip-flops of the output register.

$$D_1 = LS = S1 \cdot OCT$$

$$D_2 = AV = S1 \cdot XMIT \vee S2 \cdot KD \cdot LAST \vee S3 \cdot \overline{GOT}$$

$$D_3 = RESET = S1 \cdot CLEAR \vee S3 \cdot GOT$$

Since these equations are all in sum-of-products form, it is an easy but tedious task to form the PLA program of Table 6.4.

The only part of the data collector that remains to be realized in PLA form is the mod-8 counter. Three of the feedback flip-flops of the control PLA will be used. A suitable clearable counter is described as follows:

TABLE 6.4 PLA PROGRAM FOR THE CONTROL OF FIG. 6.25

KD	b8	b7	b6	b5	b4	GOT	LAST	Y1	Y2	J1	K1	J2	K2	D1	D2	D3
1	1	1						0	0	1				1		
1							1	0	1	1		1		1		
						1					1					
1	0	0	1	1	0			0	0			1	1			
							1	1	0			1				1
0													1			
						0		1	0						1	
1	1		1					0	0							1

$$J_3 = LS$$
$$K_3 = LS \lor RESET$$
$$J_4 = LS \cdot y_3$$
$$K_4 = LS \cdot y_3 \lor RESET$$
$$J_5 = LS \cdot y_3 \cdot y_4$$
$$K_5 = LS \cdot y_3 \cdot y_4 \lor RESET$$
$$LAST = \bar{y}_3 \cdot \bar{y}_4 \cdot \bar{y}_5$$

LS and RESET must be replaced by their sum-of-products expressions, and LAST in the equation for J_1 must be replaced by its product terms rather than bringing it out of the PLA and then back in again. Table 6.5 provides the programming for the total control unit.

Building the data collector from four ICs, three of which have identical programming, is very attractive in terms of space, power consumption, and reliability. Design requires that a system be broken into modules suitable

TABLE 6.5 TOTAL CONTROL PROGRAM

KD	b8	b7	b6	b5	b4	GOT	Y1	Y2	Y3	Y4	Y5	J1	K1	J2	K2	J3	K3	J4	K4	J5	K5	D1	D2	D3
1	1	1					0	0				1										1		
1							0	1	0	0	0	1	1									1		
						1						1												
1	0	0	1	1	0		0	0								1		1	1				1	
							1	1	0							1			1	1	1			1
0															1									
							0	1	0														1	
1	1	1					0	0								1		1	1					1
1	0	0	1	1	0		0	0	1									1	1					
1	0	0	1	1	0		0	0	1	1										1	1			

for PLA realization. This is not a new problem, just a different one. Freedom from worry about fan-in limitations is countered by the need to reduce all logic to sum-of-products form. The sequential PLA is a very powerful logic building block and may well become ubiquitous.

6.6 MEMORIES

A flip-flop stores a bit of information. A register holds a word of information. If we desire to store a number of words and are content to change or copy these words one at a time, we may use a *register file* with the organization of Fig. 6.30. Each register is identified by a number. That number is the *address* of the register and its contents. If the address is n bits long, then the file may contain up to 2^n registers. The address bits drive a decoder that together with *WRITE* and *READ* command signals cause the selected register to load from the *DATAIN* bus or connects the selected register to the *DATAOUT* bus. While a demultiplexer and multiplexer may be used to route data to and from the selected register, economy suggests the use of bus techniques. While the file may be synchronized by a clock, usually the registers are sensitive to the edge of the *WRITE, READ,* or other control signals.

Files of 8 or 16, 4-bit registers are available in integrated form. Several are concatenated to build a file with the desired word length. Such files are

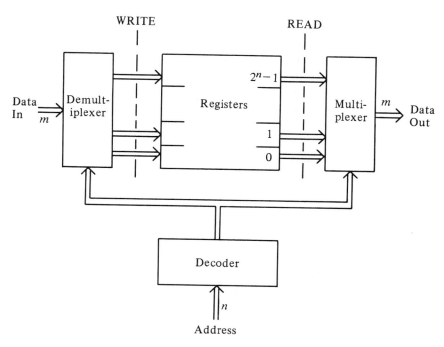

FIG. 6.30 Register File Organization

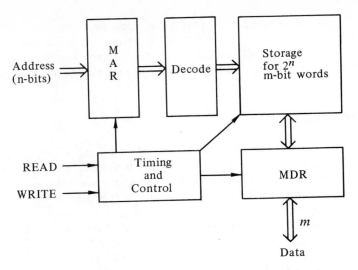

FIG. 6.31 The Organization of Read–Write Memories

used where we have previously shown a single register. For example, a file rather than a single accumulator is often associated with the ALU of a computer.

A file is declared in DDL with the ⟨MEmory⟩ declaration. Subscript ranges are separated by commas with the most significant subscript range being given on the left. If the following register and file are declared

$$\langle RE \rangle \ FAR[3], \ A[16].$$
$$\langle ME \rangle \ FILE[8,1:16].$$

then the transfer statement

$$A \leftarrow FILE[FAR]$$

specifies loading A from the file register specified by the File Address Register. The decoding and multiplexing is implied by the statement. The file may be the destination of a word of information.

$$FILE[FAR] \leftarrow A$$

In both cases the correct address must be placed in the FAR during a previous clock period.

When we wish to store larger numbers of words, we must search for the most economical and reliable technology. Magnetic materials have been, and continue to be, used very extensively. Semiconductors are in use also. But before we look at the details, let us model a memory and define a number of its parameters. Figure 6.31 offers a model for read–write memories. Two registers are usually considered to be a part of a memory. The memory address register (MAR) holds the n-bit address of the word of information to be replaced or copied. The maximum possible memory size is then 2^n words. Each of these words consists of m bits. These m-bit words

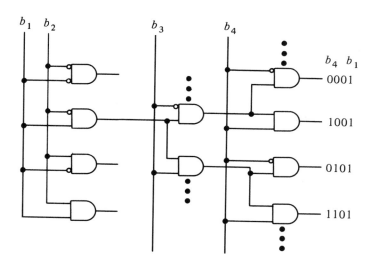

Fig. 6.32 The Structure of a Tree Decoding Network

are passed to and from the memory via the memory data register (MDR).

The decoder may take any of a great variety of forms, which depends upon the storage technology and organization. In a register file we could think of using an AND gate with n input terminals for each register. Now, since n may be substantially larger, we must review that decision. With $n = 15$, 32,768 15-input AND gates would be required. Nearly one-half million diodes ($n \times 2^n$) are required if the straightforward approach is taken. Figure 6.32 suggests how a tree switching matrix is organized. Two variables are combined at the first level; one additional variable is introduced at each additional level. A total of $2^2 + 2^3 + \cdots + 2^n$ 2-input AND gates are required. While the diode (transistor) saving can be substantial, the numbers of logic levels, and hence delay, is large. Fewer diodes and logic levels are needed when variables are combined in groups of two and three, and then the combinations are combined. Figure 6.33 shows examples of this approach.

This last approach is especially interesting for memories in which words are stored at different points in space, also where the technology permits the memory cells to act as the final level AND gates. Then the address is broken into two parts and each part is decoded. Memory cells are arranged in a matrix with one decoder selecting a row of cells and the other decoder selecting a column. The cell at the intersection of the selected row and column is the selected cell. This is *coincidence selection*. The two decoders required are substantially smaller than the single complete decoder required for *linear selection*.

If words are stored at different points in time rather than space, then the decoder involves counters and comparators to determine when the address to be referenced is accessible. Magnetic disks, drums, and tapes are well-known examples of *sequential access memories*. Addresses are best refer-

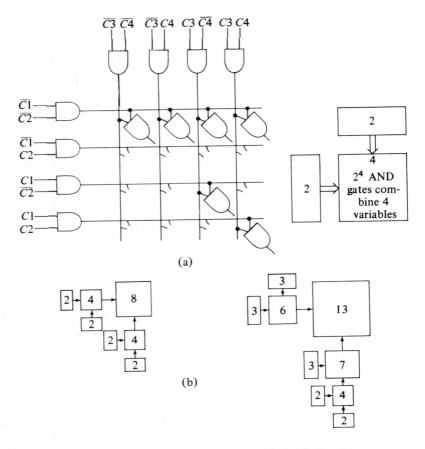

FIG. 6.33 Decoders with Intermediate Logic Levels
and Minimum Diode Requirements

enced in order. If they are not, a wait time from zero to some maximum is experienced. Average access time is measured in milliseconds or seconds for such systems. This is quite different from storage-in-space memories. There all addresses are accessed in the same time and that time is measured in nanoseconds or microseconds. Any sequence of addresses to be accessed is optimum in such memories, as their name *random access memory* (RAM) is meant to imply.

Semiconductor Memories

MOS amplifiers avoid resistors and use transistors that require very little area and few manufacturing steps. These conditions permit the placing of a great number of transistors on a silicon chip of economical size. The registers, decoders, control logic, and memory element matrix can all be formed as one integrated circuit. Figure 6.34 shows a six transis-

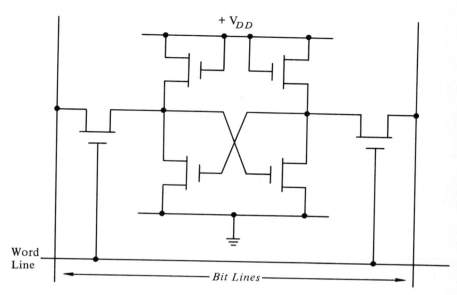

FIG. 6.34   A Six-Transistor, Static MOS Memory Cell

tor MOS memory cell that is suitable for use in a linear select memory. *m*
such cells are driven by the same word line; each is connected to a unique
pair of bit lines. Two amplifiers form a latch. The side transistors connect
or disconnect the latch to the bit lines. When the word line is low, the
latch is isolated from the bit lines by the high resistance of these transistors.
When the word line is high, the selected row of cells are connected to the
bit lines. They drive these lines in a read operation. They are driven by
those lines to write information into the selected address. A single sense
and drive amplifier is required for each pair of bit lines.

The MOS cell of Fig. 6.34 will retain information as long as power is
supplied. It is a *static* memory cell in that the information does not deter-

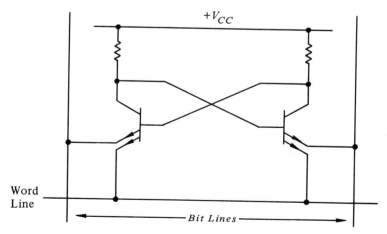

FIG. 6.35   A TTL Memory Cell

iorate with time and require periodic renewal. *Dynamic* memory cells use a capacitor as the storage device. While leakage currents can be made very small, the capacitor will discharge in time, thus it must be recharged periodically to prevent loss of information. When the power is interrupted, semiconductor memories lose stored information. They are, therefore, *volatile* memories.

While bipolar technology requires more area, and power, it is faster than MOS technology. Figure 6.35 shows a TTL memory cell that is suitable for use in a linear selection memory. If a third emitter were placed on each transistor, the added emitters could be tied to a second select line, and coincidence selection used. As long as the select line is at a lower voltage than the bit lines, the cell is disconnected from the bit lines by reverse biased base-to-emitter junctions. If the select line is raised, emitter current is transferred to the bit lines; here it is sensed for a read operation and switched for a write operation.

Semiconductor memory cell design is a search for simplicity. Transistors and interconnections require area, which is undesirable. We would like to place an infinite number of cells on a tiny chip and have the memory system run at fantastic speeds with no heat generation. Manufacturers have not yet reached these goals.

Magnetic Core Memories

The magnetic core of Fig. 6.36(a) is made of a material that provides the rather rectangular hysteresis loop of Fig. 6.36(b). If a current of $+I$ amperes exists in the winding on that core for a sufficient time and then is removed, the core will retain a residual flux density $+B_r$. Following the existence of a current $-I$, the core will retain a flux density $-B_r$. The core will remain in either of two states of magnetization as long as no current greater than I_{min} amperes exists in the winding on the core. The core is capable of acting as a memory element.

An emf will be induced in a second winding on a magnetic core as the operating point of the core is forced to move along its hysteresis loop. Assume a core stores $-B_r$ before current $+I$ exists in the drive winding of Fig. 6.37. The flux density in the core changes from $-B_r$ to something greater than $+B_r$ when the current exists, and falls back to $+B_r$ when the

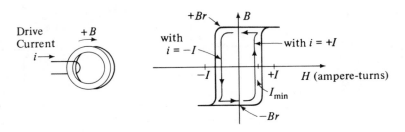

FIG. 6.36   Basic Magnetic Core Operation

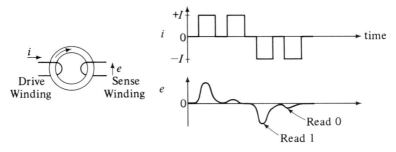

Fig. 6.37   Induced emf

current is reduced to zero. The relatively large change in flux density induces a relatively large emf in the sense winding. A subsequent pulse of $+I$ amperes will cause only a minor change in flux, and a relatively small emf will be induced. But a subsequent current $-I$ causes a large change from $+B_r$ to $-B_r$ in flux density and again a large emf will be induced into the sense winding. Subsequent current pulses of $-I$ amperes cause minor induced emf's.

These basic facts can be used in a variety of ways to achieve a large capacity memory, but only one organization of magnetic core memories will be detailed here. Agree to the convention of interrogating a core by applying a drive current of $-I$ amperes. If the core stored $+B_r$, the large induced emf can be sensed, amplified, and shaped into the voltage level used to represent a 1 throughout a computing system. If the core stored $-B_r$, the small induced emf can be converted to the voltage representing 0. In either case the core ends up storing a 0. The read-out is *destructive*; the 1 must be written again with a $+I$ current if the memory is to appear to be nondestructive.

One core stores one bit of information and is not particularly valuable by itself. But many cores can be used to store many bits in any of several ways. Consider the following possibility. Rather than having a single drive winding, place drive windings on each of many cores arranged in a matrix fashion as in Fig. 6.38. Let one X-drive line and one Y-drive line each carry $I/2 < I_{min}$ amperes of current. Then all but one of the cores looped by the selected X-winding and all but one of the cores looped by the Y-line will *not* receive sufficient magnetomotive force to change their residual flux density. Only the core at the intersection of the selected X- and Y-lines will be switched to storing a 1. If n X-lines and m Y-lines exist, such an arrangement requires $n + m$ drive amplifiers to select one of $n \times m$ cores; the cores in addition to providing memory participate in the selection process. A magnetic core memory with this organization is referred to as a *coincident current* memory.

Further economy exists in the coincident current memory. Two small switching matrices are required to decode an address; one matrix selects an X-line while the other selects a Y-drive line. The two smaller decoding matrices require fewer components than one large matrix able to select one

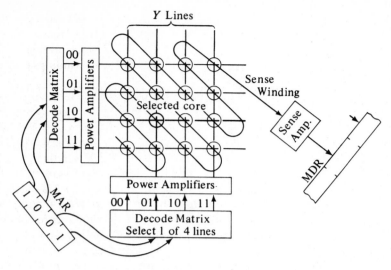

FIG. 6.38 Coincident Current Magnetic Core Plane

of $n \times m$ lines. Also, one sense winding usually links all cores of a core plane so only one sense amplifier is required for each core plane.

By applying $-I/2$ amperes to an X- and a Y-drive winding and looking at the emf induced in the sense winding a selected core can be interrogated. With $+I/2$ amperes on the appropriate X- and Y-lines, that core can be switched to store a 1. Reading and writing are possible.

If p-bit words are to be stored, p core planes are stacked and common drive windings used, as shown in Fig. 6.39. Each plane has its own sense winding and sense amplifier, but no additional drive amplifiers or address

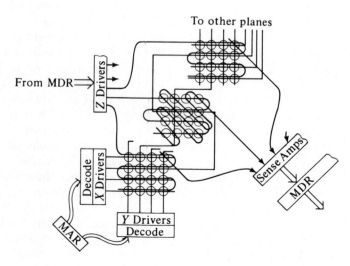

FIG. 6.39   Coincident Current, Magnetic Core Stack

decoding circuits are required. Placing half-select currents on one X- and one Y-drive line, then selects one core in each plane. During reading, all selected cores are switched to store 0; each sense winding provides the bit of the word stored in its plane. Together the sense windings provide the bits of the word in parallel.

But a problem exists when attempting to write a word into a memory location. Placing $+I/2$ amperes in each of a selected X- and Y-drive winding will cause all fully selected cores to switch to store a 1. Thus only the word $11\ldots1$ can be written into a location. This is not tolerable. One often used solution calls for a common fourth winding, the Z winding, to be placed on all cores of a plane. The addressed word is read resetting all of the selected cores to store 0. Then a current $-I/2$ is made to exist in the Z windings of the planes whose cores are to store 0 at the same time as the X and Y lines are energized. The selected cores add three mmfs obtaining a resulting mmf of

$$ +\frac{I}{2} \quad +\frac{I}{2} \quad -\frac{I}{2} = +\frac{I}{2} $$
$$ X \qquad Y \qquad Z $$

which is not sufficient to switch them. A 1 is written into selected cores of planes with no Z current. Obviously the sense of the X-, Y-, and Z-windings must be such that this inhibiting effect takes place and such that the Z-drive and X-drive (Z- and Y-drives) do not add in any nonselected core and cause it to be switched to storing a 1.

Shift Register Memories

The number of access terminals and the amount of selection logic found in a register file can be drastically reduced by using serial communications. Any number of shift cells may appear between one data input terminal and one data output terminal. The cells are identical and well-suited to integration. Thus we find one long shift register or several shorter shift registers available as single integrated circuits. How do we use them?

Suppose we have one N-bit SR. It can be thought to store N 1-bit words or p q-bit words, where $p \times q = N$. A 1024-bit SR may be thought of as storing 128 8-bit words, 64 16-bit words, and so forth. To read a specific word, it must be shifted to the output terminal. Recirculating the preceding words prevents their loss. The read word may also be recirculated, or the loop may be broken and a new word inserted at the selected address. With the SR constantly shifting, the physical position of a stored word is constantly changing, but at each instant of time one can speak of the address that is available at the input and output terminals. We select through time rather than space. A q-stage counter indicates which bit of a word is currently at the output terminal; a p-stage counter indicates which word is currently available. Figure 6.40 shows a 1024 bit SR that is thought to hold 128 8-bit words. The 7-bit word count is compared with an address held by the *MAR*. When the two numbers are the same, the desired address is at the output terminal. For a read operation the selected word is routed to an 8-bit recirculating memory data register, where it is held until needed. For a write operation the recirculating loop is broken, and the word held by the data register is substituted for the original word.

In the system of Fig. 6.40, a selected address appears at the output terminal for 8 clock periods. We call this the word time of the system. On the average we must wait $128 / 2 = 64$ word times, or 512 clock periods, for a desired word to be available. Sequential memories are well suited to systems that can tolerate slow response.

Moving Magnetic Surface Memories

In tape, drum, and disk memory systems a surface of ferromagnetic material is moved past reading and writing "heads." Since mechanical motion is involved, such memories are much slower than the memories considered above, and sequential rather than random access is dictated. But the tape, drum and disk constitute very economical means of achieving memories of very great capacity.

Consider the drum first. A steel cylinder coated with a ferromagnetic material (thick film) is rotated, commonly at 3600 rpm. Spots of the material are magnetized in one of two directions by a field concentrated

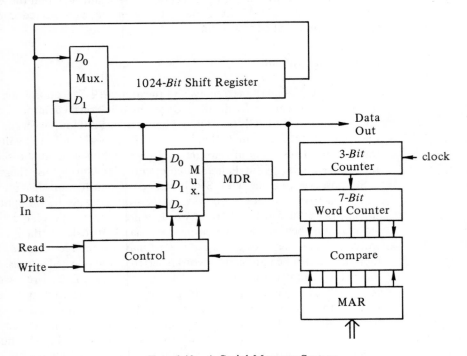

Fig. 6.40 A Serial Memory System

at the moving surface by a magnetic circuit of a soft magnetic material with an air gap very close to the surface. The air gap and spacing between the head and moving magnetic surface are of the order of .001 inches.

If a "write-1" current is established in a write winding on a head, the magnetic material moving under the head while the current exists will be magnetized in what will be called the *1 direction*. Material passing under the head while a "write-0" current exists is magnetized in the 0 direction. Sequences of 1's and 0's can be placed in a "track" around the circumference of the drum by the appropriate sequence of write-1 and write-0 pulses of current. A sequence of 1's may be written by a continuous write-1 current, in which case a continuous strip of material will be magnetized in the 1 direction, or a sequences of write-1 pulses may be applied that magnetize spots in the 1 direction on a surface nominally magnetized in the 0 direction. Other possibilities exist in fact; only the "spot magnetization" approach will be considered further here.

If information has been recorded on a track, that information may be recovered by sensing the emf induced in a read winding on a head by the motion of the magnetized surface under the head. Figure 6.42 shows the flux pattern for the recorded sequence 1011: note that abrupt change in the flux pattern cannot be achieved because of the motion of the surface as a write current is established. The induced emf is proportional to the time rate of change of flux as shown. Electronic differentiation of the induced emf facilitates wave-shaping to recover the recorded sequence.

To achieve greater memory capacity, many heads are placed along the axis of the drum. Each head provides a means of reading and writing on one track. Track densities of 100 tracks per inch and bit densities of 1000 bits per inch are common and higher densities will probably appear in the future.

The bits of a word may all be placed on one track, together with many other words. To read or write in a particular word location, a track must be selected (space selection as in the random access memories considered), and then a word on that track must be selected by recognizing the sensed bits only at the proper time (time selection). Access time is therefore not a constant. (Fifteen milliseconds maximum is typical.) To determine the time at which a specific word will pass under a read head, one or more tracks on the drum may permanently record timing information. Bits of the word are then read serially in time.

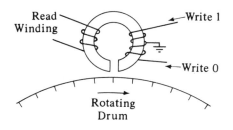

FIG. 6.41   A Head on a Drum

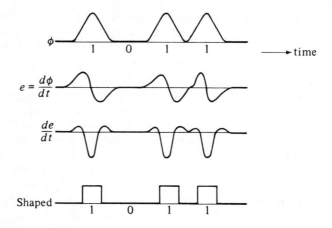

FIG. 6.42 Reading from a Drum

Disk memories consist of many rotating disks coated on both sides with a ferromagnetic material. Movable heads select a track (concentric circles on the surface of the disk) by their position over the disks. Thus far fewer heads are required than on a drum, but these heads must be very accurately and quickly positioned. One hundred millisecond access times are common. Again bits of a word can be located on separate recording surfaces, and written and read in parallel.

Some disk memories allow manual removal of the "disk pack," thereby effectively extending the memory capacity almost without limit, and providing a means of protecting information stored upon a disk pack.

A reel of magnetic tape may also be removed from a tape drive, thereby providing almost unlimited, slow access storage. Usually seven or nine bits, one byte of information and one parity bit, are placed in a row across a 1/2 inch tape. Successive bytes of a word are placed in successive rows at a density of 200 to over 1000 rows per inch. Usually blocks of rows are separated by gaps in which no information is recorded. The tape drive accelerates and stops the tape within these gaps. Access time can be very long if the desired information is at the opposite end of the tape from that point currently under the read-write heads. It is therefore important to place information on tape in the sequence in which it will be needed.

Tapes, drums and disks are most commonly utilized as secondary stores in conjunction with high-speed, random access, main stores. Bits of a word are placed and read in parallel with each bit on a different track. Blocks of words rather than individual words may be read or written at very high speed from the main memory.

REFERENCES

1. A. M. ABD-ALLA and A. C. MELTZER, *Principles of Digital Computer Design*, Volume 1. Englewood Cliffs, N.J.: Prentice-Hall, 1976.

2. A. Barna and D. I. Porat, *Integrated Circuits in Digital Electronics.* New York: John Wiley and Sons, 1973.

3. J. C. Boyce, *Digital Logic and Switching Circuits: Operation and Analysis.* Englewood Cliffs, N.J.: Prentice-Hall, 1975.

4. H. Fleicher and L. I. Maissel, "An Introduction to Array Logic," *IBM Jour. of Research and Development,* Vol. 19, March, 1975, pp. 98–109.

5. S. J. Hong, R. G. Cain, and D. L. Ostapko, "MINI: A Heuristic Approach to Logic Minimizations," *IBM Jour. of Research and Development,* Vol. 18, Sept. 1974, pp. 443–458.

6. J. W. Jones, "Array Logic Macros," *IBM Jour. of Research and Development,* Vol. 19, March, 1975, pp. 120–126.

7. G. K. Kostopoulos, *Digital Engineering.* New York: Wiley, 1975.

8. J. C. Logue, et al., "Hardware Implementation of a Small System in Programmable Logic Arrays," *IBM Jour. of Research and Development,* Vol. 19, March, 1975, pp. 110–119.

9. R. L. Morris and J. R. Miller, *Designing with TTL Integrated Circuits.* New York: McGraw-Hill, 1971.

10. D. L. Ostapko and S. J. Hong, "Generating Test Examples for Heuristic Boolean Minimization," *IBM Jour. of Research and Development,* Vol. 18, Sept., 1974, pp. 459–464.

11. J. B. Peatman, *The Design of Digital Systems.* New York: McGraw-Hill, 1972.

12. V. T. Rhyne, *Fundamentals of Digital Systems Design.* Englewood Cliffs, N.J.: Prentice-Hall, 1973.

PROBLEMS

6.1-1. A 10-stage ripple counter is constructed of FFs that require 50 nanoseconds to change state. What is the maximum time during which the count indicated by this counter is in error? What is the maximum rate at which the input signal may vary such that the counter always reaches the correct count before it is again required to advance?

6.1-2. Show the logic of a three-stage, ripple down-counter, i.e., one which counts 0, 7, 6, ..., 1, 0, 7, ...

6.1-3. A binary counter must always attain a new state in 30 nanoseconds or less. FFs introduce 5 nanoseconds of delay each; logic gates introduce 3 nanoseconds delay. Design the most economical 10-stage counter.

6.1-4. (a) A synchronous decade counter in which FF states express decimal digits in the 2421 weighted code is required. Design using JKFF. Would some other FF type be better suited to this application?

(b) The counter is required to indicate when it reaches a count of 9 so that decades can be cascaded to achieve decimal counting over a wider range. Append the necessary circuitry.

6.1-5. A decimal up-down counter using the 8421 weighted code is to provide the carry signal when it reaches a count of 9 while counting up ($UD = 0$), and when it reaches a count of 0 while counting down ($UD = 1$). Design the counter using any FF type, and show that your design may be cascaded with itself to achieve decimal counting over any number of decades.

6.2-1. Three n-bit registers A, B, and C are constructed of synchronous RSFFs. They are to be interconnected to respond to control signals LFA and LFB (load from B) in the following manner.

$$|\,LFA\,| \qquad C \leftarrow A.$$
$$|\,LFB\,| \qquad C \leftarrow B.$$

Design the required interconnecting circuitry giving both the Boolean equations and a block diagram for a typical cell of register C.

6.2-2. $\langle RE \rangle$ $A[5]$, $B[5]$, $C[5]$.

$$|\,Q_1\,| \qquad A \leftarrow B, \; C \leftarrow \bar{C}.$$
$$|\,Q_2\,| \qquad C \leftarrow A.$$
$$|\,Q_3\,| \qquad C \leftarrow A, \; B \leftarrow C.$$

(a) Show typical cells of registers A, B, and C.
(b) At a certain clock time register A contains 11011, B holds 01101, and C holds 00001. Control signals are supplied at successive clock times in the sequence $Q1$, $Q2$, $Q1$, $Q3$. What does each register hold after this sequence of controls has been applied?

6.2-3. A push-down "stack" is described by:

$$\langle RE \rangle \qquad A[15], \; B[15], \; C[15], \; D[15].$$
$$\langle TE \rangle \qquad X[15].$$
$$|\,PUSH\,| \qquad A \leftarrow X, \; B \leftarrow A, \; C \leftarrow B, \; D \leftarrow C.$$
$$|\,POP\,| \qquad A \leftarrow B, \; B \leftarrow C, \; C \leftarrow D.$$

(a) Show typical cells of registers A and B.
(b) If we push more than four times, data will be lost from the D register. Show how an up-down counter might be used to monitor the number of words currently in the stack and prevent further entries if the stack is full.

6.2-4. (a) Register A of synchronous RSFFs is to shift right one position when control $SR1$ is presented. Design a typical cell and special end cells. Zeros are to fill the left cells.

$$|\,SR1\,| \qquad \underline{0\,|}\!\!\rightarrow A.$$

(b) The same register is to shift its contents *two* places to the right each bit time that control $SR2$ is present. Redesign register A.

$$|\,SR1\,| \quad (1) \; \underline{0\,|}\!\!\rightarrow A.$$
$$|\,SR2\,| \quad (2) \; \underline{0\,|}\!\!\rightarrow A.$$

(c) Register A stores the bit pattern 10101010 when both control signals are presented to it. What does your register store one bit time later?

6.2-5. Four-bit registers are connected to a bus, as in Fig. 6.11. The registers initially hold: $A = 1011$, $B = 0100$, $C = 1111$, $D = 0011$. The following control signals equal 1 at successive clock times.

Time	Controls
1	LA, SD
2	LC, SB
3	LA, LD, SC

What are the register contents after each command set has been executed?

6.2-6. (a) Design a 4-bit synchronous right shifting register A, which can be loaded from register B and emits a signal *STOP* when all flip-flops of the register store 1's. The unusual shift of the register is specified below.

$$|\text{LOAD}| \qquad A \leftarrow B.$$
$$|\text{SHIFT}| \qquad A_1 \leftarrow A_3 \oplus A_4, \ \{i = 1 : 3\} \ A_{i+1} \leftarrow A_i.$$

(b) If A is loaded with 0111 and *SHIFT* continuously applied, how many bit times will elapse before $STOP = 1$?

(c) Suppose the *STOP* signal is used to form the *LOAD* and *SHIFT* signals in the following manner:

$$\text{LOAD} = \text{STOP}$$
$$\text{SHIFT} = \overline{\text{STOP}}$$

What must register B contain if the *STOP* signal is to equal 1 every (i) 7th bit time, (ii) 13th bit time, (iii) never?

6.2-7. The register of Prob. 6.2-6 is altered slightly.

$$|\text{STOP}| \qquad A \leftarrow B.$$
$$|\overline{\text{STOP}}| \qquad A_1 \leftarrow A_2 \oplus A_4, \ \{i = 1 : 3\} \ A_{i+1} \leftarrow A_i.$$

What is the maximum possible period of *STOP*? What must B contain to achieve the maximum period?

6.2-8. Shift registers can often be used to advantage in synchronous sequential machines. Design the machine specified below in each of the two ways indicated, and compare the resulting designs.

Input Alphabet		Output Alphabet				
	x		z_1	z_2	z_3	z_4
S — Stop	0	z^1	1	0	0	0
C — Count	1	z^2	0	1	0	0
		z^3	0	0	1	0
		z^4	0	0	0	1

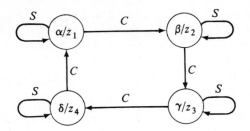

(a) Design using two flip-flops and the design procedure of Section 5.2.

(b) Design using four flip-flops and a 1-out-of-4 code for state assignment. Does a shift register structure appear? This structure is known as a *ring counter*.

6.2-9. A machine is to give the output sequence of length 8 shown, in response to an input command.

> Input   000 10000000 00 . . .
>
> Output   000 01011011 00 . . .

(a) Draw a state diagram for the required machine.

(b) Design the machine using a shift register to advantage if possible.

6.3-1. (a) Register A contains a signed number in sign-magnitude representation. The sign bit is stored by flip-flop A_1. Upon receipt of a *CNVT* control signal, the 2's complement representation of the number is to be formed and placed in register A. Design register A and document your design with the design language.

(b) If a number in 2's complement representation is stored in your register A when the *CNVT* control is given, what will your register contain one bit time later?

(c) Repeat for 1's complement representation and compare the designs.

6.3-2. Only synchronous JKFFs and NOR gates are to be used in the design of two eight bit registers, R and ACC, interconnected to perform the following operations:

|LOAD|	$ACC \leftarrow R.$	
|ADD|	$ACC \leftarrow R + ACC.$	(+ means 1's complement addition)
|SUB|	$ACC \leftarrow \bar{R} + ACC.$	($\bar{R}$ means 1's complement of number stored in R register)

6.3-3. A 15-bit register must form and store the 1's complement sum when given the command ADD, and the bit-by-bit exclusive–or when given the command $EXOR$. Design a typical cell of the register attempting to minimize the required logic. Draw a block diagram of your cell.

6.3-4. Two 4-bit registers, A and B, each hold Excess-3 encoded decimal digits. When the command ADD is received the correct sum digit (Excess-3 encoded) is to be formed and stored in B. A decimal carry-in is to be accepted; a decimal carry-out signal is to be generated. Design the B register.

6.3-5. Register A, constructed of 32 synchronous GLFFs, is capable of holding 8 encoded decimal digits. Each bit time that control $CNVT$ (convert) is present, the four bits in the rightmost positions are to be converted from a BCD to an Excess-3 code word, all stored bits shifted to the right four positions, and the Excess-3 code word stored in the left four positions of the register. If the right four bits do not constitute a valid BCD code word, an error signal $ERER$ is to be set to 1 and 0's placed in the four left positions of the register. Design the register.

6.4-1. Two 24-bit registers, A and B, are to be designed so that upon receiving an $SRADD$ command, A circulates its contents 3 positions to the right and B shifts its contents 3 bits to the right with the 3-bit sum of the rightmost 3 bits of the A and B registers shifted into the 3 leftmost bit positions of B. A carry flip-flop stores the carry from one 3-bit sum to the next.

(a) Design a typical cell of A. Show the complete structure of the 3 rightmost cells of B.

(b) Design a counter to work in conjunction with these registers and indicate when 8 shifts and adds have been performed.

(c) Design a control unit that will accept the command ADD, issue signals to the registers and counters as necessary to achieve 2's complement addition, and then generate a $DONE$ signal. Show a state diagram. Include provisions for placing the carry flip-flop in the proper initial state. How may overflow be detected?

6.4-2. Two numbers reside in the A and B registers of Fig. 6.21(a). What sequence of control signals must be issued to place their sum in the B register?

6.4-3. Replace the final, conditioned statement of Fig. 3.38 with a state declaration of a control unit so that the GO signal need only be high for one clock period and the counter be properly initialized at the beginning of multiplication. Draw a state diagram. Design a suitable control unit.

6.4-4. The Booth multiplier sketched in Fig. 3.40 requires a control unit. Design a control unit and counter; assume the data registers hold 24-bit numbers.

6.4-5. The data collector is to be ROM controlled, but no address modification logic is to be used. System condition signals LS, OCT, $XMIT$, and so forth, are to be fed directly to the ROM address register together with next address bits from the ROM.

(a) Determine the number of state variables and the state assignment to be used.

(b) Determine the size of the RAR and the ROM.

(c) Write the ROM program. List only those words of the ROM that will actually be addressed. Keep in mind that two keys cannot be depressed simultaneously, and the electronics is far faster than fingers.

6.5-1. Program a sequential PLA with 4-bit registers to add a 4-bit magnitude on input lines to the magnitude held in the feedback register and a carry-in bit. When input signal $A = 0$, then the first output line must give the overflow bit from the adder; otherwise, the sum is to appear on these lines. The sum is always placed in the feedback register.

6.5-2. Show how the programmed PLAs of Prob. 6.5-1 may be cascaded to add 16-bit magnitudes. Describe the control problem. Program a PLA to act as a control unit so that a 17-bit sum ultimately appears on output lines.

6.5-3. Program a sequential PLA with 5-bit registers to act as a decimal up–down counter. The counter is to have a stop/start control and is to provide a carry that equals 1—when the count is 9 and the unit is counting up and when the count is 0 and the unit is counting down. These PLAs should be cascadable.

6.6-1. Four registers are to be used to provide storage in a memory with a four word capacity and five bits per word. An MAR and MDR are also to be a part of that memory. All registers are synchronous. Show the logic design of a random access, nondestructive memory capable of responding to $READ$ and $WRITE$ commands.

6.6-2. A semiconductor memory has a capacity of 32,768 48-bit words. How many memory cells are required?

6.6-3. (a) A coincident select, 4K memory with coincident select memory cells requires four decoding matrices, each capable of selecting one of eight lines, and two matrices that are able to select one of 64 lines (from two sets of 8 lines). If these matrices are constructed of diodes, how many diodes will be required?

(b) A coincident select, 4K memory uses two single logic-level diode decoding matrices, each of which selects one of 64 drive lines. How many diodes are required?

(c) A single diode matrix is required to decode a 12-bit address in a linear select 4K memory. How many diodes are required?

ANSWERS

6.1-1. $10 \cdot 50 = 500$ ns, $f = 1/500 \times 10^{-9} = 2$ mHz.

6.1-4. Many 2421 codes exist; each will require a slightly different design. Two codes and designs are given below; the maps are for the second design. Signal x is the input signal.

	z_4	z_3	z_2	z_1	
0	0	0	0	0	$J_1 = K_1 = x$
1	0	0	0	1	$J_2 = xz_1$
2	0	0	1	0	$K_2 = xz_1(\bar{z}_3 \vee z_4)$
3	0	0	1	1	$J_3 = xz_1z_2$
4	0	1	0	0	$K_3 = xz_1z_4$
5	0	1	0	1	$J_4 = xz_1z_2z_3$
6	0	1	1	0	$K_4 = xz_1$
7	0	1	1	1	
8	1	1	1	0	
9	1	1	1	1	

	z_4	z_3	z_2	z_1	
0	0	0	0	0	$J_1 = K_1 = x$
1	0	0	0	1	$J_2 = xz_1z_4$
2	0	0	1	0	$K_2 = xz_4(z_1 \vee \bar{z}_3)$
3	0	0	1	1	$J_3 = xz_4$
4	1	0	1	0	$K_3 = xz_1z_2$
5	0	1	0	1	$J_4 = x(z_1z_2 \vee z_3)$
6	1	1	0	0	$K_4 = xz_1z_2$
7	1	1	0	1	
8	1	1	1	0	
9	1	1	1	1	

z_2' z_4

d	0	d
0	S	d
d	R	d
d	1	R

(with 0,0,1,1 column at left and z_1, z_2, z_3 labels)

z_3'

0	d	1	d
0	1	1	d
0	d	R	S
0	d	1	d

z_3

z_4' z_4

0	d	1	d
0	S	1	d
S	d	R	R
0	d	1	d

z_3

6.2-2. (a)

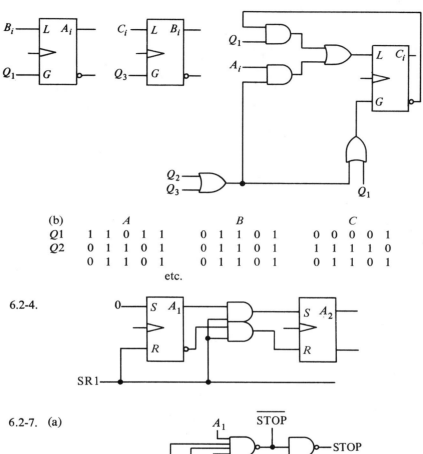

(b)

			A						B						C		
Q1	1	1	0	1	1		0	1	1	0	1		0	0	0	0	1
Q2	0	1	1	0	1		0	1	1	0	1		1	1	1	1	0
	0	1	1	0	1		0	1	1	0	1		0	1	1	0	1

etc.

6.2-4.

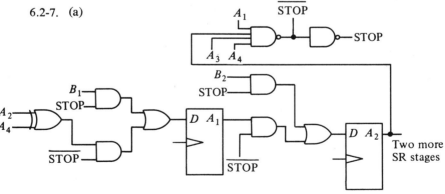

6.2-7. (a)

Suppose B holds 1100. One bit time is required to load A. Then shifting begins.

1100
1110   Period $= 4$ in this case
1111

6.3-1. Let A_1 serve as the control signal for the $COMP2$ network developed in Section 3.6. The contents of A_1 are not to be changed so the $COMP2$ network will have one less cell than the A register. $CNVT$ must be used to control loading A from $COMP2$.

6.3-3. Alter the full-adder so that carry $= 0$ when $EXOR = 1$.

6.3-5. Place a BCD-to-excess-3 converter between the rightmost and leftmost cells of the register.

6.4-1. This is a serial, octal adder. A 3-cell, parallel adder accepts bits from the right cells of A and B; sum bits are routed to the left 3 cells of B.

6.6-2. $32,768 \times 48 \simeq 1.57$ million

6.6-3. (a) 4 matrices $\times$ 3 bits $\times$ 8 lines $+ 2 \times 2 \times 64 = 352$ diodes

(b) 2 matrices $\times$ 6 bits $\times$ 64 lines $= 768$ diodes

(c) 1 matrix $\times$ 12 bits $\times$ 4096 lines $= 49,152$

7

An Elementary Digital Computer

Electronic information processing systems are large assemblages of (1) combinational logic, (2) sequential logic including bulk storage memories, (3) transducers that enable the electronics to interact with its environment, and (4) other essential items such as power supplies, cabinets, oscillators, cables, etc. While we will continue to be concerned primarily with the logic of systems, these last two items are no less important, for our logic cannot operate without power or a clock signal, and a data processing system that is not capable of accepting data to process and presenting results to us or other machines is of no value.

As an example of digital system design, this chapter defines, designs, and programs a very elementary digital computer (EDC). A digital system is defined by its *architecture*—that of the system seen by the user. To design the system, a configuration of registers and combinational processing blocks and processing algorithms capable of realizing the architecture must be found before detailed logic design begins. Those items constitute the *organization* of the system. A *program* is a body of knowledge that initializes the system for operation in a desired manner. We have previously spoke of "programming" ROMs and PLAs. Disregarding cost, the quality of the architecture and organization are directly proportional to the ease with which programs are prepared and the speed with which they are executed. Architecture and organization should, therefore, be based upon the programs we expect to write—the work we expect the machine to do.

The architecture and organization of EDC are simple so that we concentrate on principles without being overwhelmed by details. Preparing and examining a program for EDC will reveal its low quality (as defined above) and will provide motivation for the more advanced concepts presented in the next chapter.

7.1 ARCHITECTURE AND ORGANIZATION OF EDC

Almost all of the digital computers constructed to date are "von Neumann machines." We are all more or less familiar with the calculator. A keyboard provides a means of entering "operands," i.e., numbers. One or two registers capable of holding other operands may be a part of the calculator. We inform the calculator how to manipulate these numbers by means of "instruction" keys. Thus to use such a machine we might enter the instruction "clear the accumulator register" by pressing one key, then enter a number in the keyboard, and then give the instruction "add." This sequence of instructions and operands loads the accumulator with a specific number. Then if we enter a second number in the keyboard and again command "add," we obtain the sum of the numbers.

Calculators operate more rapidly than their human operators. It takes us longer to write down the sum than it took the machine to form it. We cannot perform substantially more computation on a calculator that adds faster for it still takes us seconds to enter operands and instructions and write down intermediate results. While the calculator may be able to perform one million additions per second, we cannot supply more than one or two operands and instructions per second.

Addition of more registers or a bulk memory to the calculator provides only limited improvement. We may not be required to write down intermediate computational results, and if we can place operands in this memory before beginning to compute, these numbers can be retrieved at high speed by the machine. But we must still push instruction buttons, so the machine is still constrained to operate at human speeds.

John von Neumann[1] is credited with proposing that operands and instructions be essentially indistinguishable and placed in a common high-speed memory so that both can be retrieved at electronic speeds. Both instructions and operands are then binary sequences. We arrange that some sequences are viewed by the machine as instructions, while others are treated as operands. But if there is no fundamental difference between the two, then one sequence may be viewed as an operand at one point in time and as an instruction at another, and it becomes possible for the machine to alter its instructions. We can only appreciate this very powerful concept after we are familiar with von Neumann-type general purpose digital computers.

Digital computers have traditionally consisted of four main parts, as shown in Fig. 7.1. The calculator consists of a keyboard and some sort of display; those items allow man–machine communication and con-

[1] See [4] in Chapter 8.

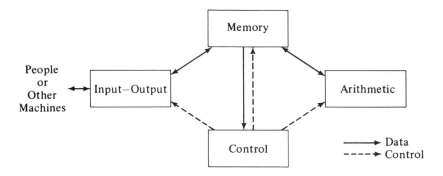

FIG. 7.1   Traditional Computer Organization

situte elementary examples of input-output equipment. The calculator also has some sort of arithmetic section. In addition to input-output equipment and an arithmetic section, the digital computer possesses a bulk memory for storing instructions and operands, and a control section that interprets instructions and issues appropriate commands to all of the other blocks. The human operator with pencil and paper serves as the control and bulk memory for the calculator. It should be noted that an actual digital computer may be packaged in one or many physical cabinets and the four sections are not usually fabricated as separate entities. The control and arithmetic sections together form the *central processing unit*, or CPU.

To see more clearly the details of the organization, operation, and use of a von Neumann-type digital computer, we will examine an elementary but realistic machine. A number of assumptions will be made to keep this machine as simple as possible. These can be challenged as unrealistic, and we will remove them in the next chapter after we have gained experience and confidence. For example, we will initially make no provisions for input and output equipment; we will assume that console push buttons somehow set certain flip-flops.

The specifications of this machine will appear to be rather arbitrary. But again, we can not properly judge the merit of these specifications until we see their effect on the organization and detailed logic design, and have some knowledge as to what sort of tasks the machine will be asked to perform. Then we will be in a position to envision the change in organization, cost, and usefulness that a proposed revision of specifications will cause.

Let us name the computer we are about to examine EDC, for "elementary digital computer." EDC has a 16-bit word length. Thus operands will be 16-bit binary sequences. The arithmetic circuits of EDC will perform 2's-complement addition and subtraction on these sequences, with the overflow value being placed in the OVF flip-flop.

A 16-bit word may expresss non-numeric information in a code known only to the programmer. EDC has limited "logical" word manipulating features. Two 8-bit ASCII character codes may be packed into one 16-bit word. Alphanumeric data are stored and processed using this "byte" format.

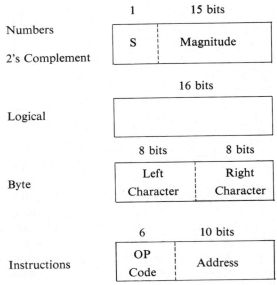

Numbers
2's Complement

1	15 bits
S	Magnitude

Logical

16 bits

Byte

8 bits	8 bits
Left Character	Right Character

Instructions

6	10 bits
OP Code	Address

Instructions will also be 16-bit sequences. EDC interprets the leftmost six bits of a sequence as expressing the type of operation to be performed. The remaining 10 bits of a 16-bit sequence that EDC views as an instruction will be interpreted in a variety of ways. But in many cases these ten bits will be interpreted as an address of the 1024 word memory of EDC. EDC is thus a *single address* machine, since only one field of bits in an instruction specifies a memory address.

All information transfers and manipulations are parallel in EDC. A clock with period of 1 microsecond synchronizes all flip-flop activity.

The activity within EDC, like that in most digital computers, is cyclic in nature. Many internal operations are performed periodically. A single cycle of this internal operation can be divided into two phases. During the first phase a word is obtained from the memory and placed in a special register of the control portion of EDC, where it is interpreted as being an instruction. This phase is thus known as the *instruction-fetch* phase. During the second phase of a cycle of operation, commands are issued to the other blocks of the computer in accordance with the instruction just fetched from the memory. Internal operations that fulfill instructions such as "load a number in the accumulator" or "add" are performed during this phase and it is thus referred to as the *execution phase*.

Let us first examine the instruction-fetch phase in greater detail. A word is to be retrieved from the memory and placed in a special register, which we will call the instruction register (IR). But we can only obtain a specific word from memory if we know its address. We must have an answer to the question, "What is the address of the next instruction to be executed?" The answer is provided in single-address digital computers by a counting

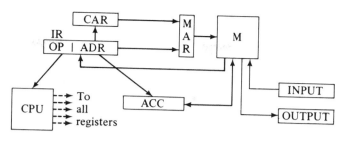

FIG. 7.2 Facilities and Data Paths of EDC

register, the current-address register (CAR). In single-address digital computers such as EDC, instructions are placed in memory in the sequence in which we wish them to be performed. (We will require means of breaking this rule.) With the first instruction in location 0, the second in location 1, etc. we begin operation with the CAR reset to all 0's. Once the first instruction has been fetched, CAR counts once in preparation for fetching the next instruction, from the next memory location.

Figure 7.2 shows the registers and main data paths of EDC. Fig. 7.3 offers full detail. All registers are public in this elementary machine. The instruction-fetch phase is represented by the states named IF1, IF2, and IF3.

Assume three console pushbuttons START, STOP, and RESET exist and are connected so that depressing START sets flip-flop RUN; depressing STOP resets RUN; and depressing RESET sets flip-flop CLEAR. Circuitry to achieve this is not specified in Fig. 7.3. The states of FF's RUN and CLEAR are examined when EDC is in the first state of the instruction-fetch phase. If RUN is set, the contents of CAR are transferred to the memory address register, MAR, in preparation for fetching a word. The count held by CAR is increased at the same time. If CLEAR is set (the RESET button must have been depressed), RUN and CLEAR are reset, as well as all flip-flops of CAR and IR. Then the machine jumps to state IF3, where the bits held by the OP subregister are examined.

If the RESET button has not been depressed, the instruction-fetch phase continues in state IF2. Here the word stored in memory M at the address specified by the contents of MAR is transferred to the instruction register. In state IF3 the address field of the instruction just fetched is placed in the MAR, and the leftmost four bits of that instruction are examined, if EDC is running. Execution of the instruction is about to begin.

Now let us turn to the execution phase of EDC's cycle of operation. But first, what instructions is EDC capable of executing? Table 7.1 offers the *instruction set* of EDC. While this instruction set is not as large as that of most commercial computers it does include instructions typical of those found in larger machines. Working with this smaller set reduces the detail we must master initially, and does not require us to be concerned with more complicated machine organization and design. Multiplication,

⟨SYSTEM⟩ EDC: ⟨TIME⟩ P(1E-6).
   ⟨RE⟩ IR[16] = OP[0 : 3] ∘ IX[2] ∘ ADR[10], CAR[10],
     ACC[16], MAR[10], INPUT[16], OUTPUT[16], OVF,
     RUN, CLEAR, IN, OUT.
    ⟨ME⟩ M[0 : 1023,16].
    ⟨AU⟩ CPU: P: ⟨STATE⟩
      IF1: |CLEAR| RUN ← 0, CLEAR ← 0, CAR ← 0,
       IR ← 0, →IF3;
       |RUN|MAR ← CAR, ¢ CAR, →IF2; →IF1 ...
      IF2: IR ← M[MAR], → IF3.
      IF3: RUN: MAR ← ADR, ⌊OP ⌊0 →EXINP
       ⌊1 →EXOUT
       ⌊2 ADR ← 1D10, ACC ← 0D6 ∘ ADR, →EXBIN
       ⌊3 →EXACC ⌊4 →EXLDA
       ⌊5 →EXADD ⌊6 →EXSUB ⌊7 →EXSTO
       ⌊8 →EXAND
       ⌊9 →EXTRA ⌊10 →EXIMM ⌊11 →EXSHF;
       → IF1 ..
     EXINP: IN: M[MAR] ← INPUT, IN ←0, →IF1.
     EXOUT: ¬OUT: OUTPUT ← M[MAR], OUT ← 1,
      →IF1 .
     EXBIN: |v/ACC| MAR ← ADR, ACC ← ACC − 1D16,
      →EXBIX; →IF1..
     EXBIX: IN: M[MAR] ← INPUT, ¢ ADR, IN ←0,
      →EXBIN.
     EXACC: →IF1, ⌊IX ⌊0 ACC ← 0 ⌊1 ACC ← ACC + 1
      ⌊2 ACC ← ACC + ACC ⌊3 ACC ← 0 − ACC .
     EXLDA: ACC ← M[MAR], →IF1.
     EXADD: ACC ← ACC + M[MAR], →IF1.
     EXSUB: ACC ← ACC − M[MAR], →IF1.
     EXSTO: M[MAR] ← ACC, →IF1.
     EXAND: ACC ← ACC · M[MAR], →IF1.
     EXTRA: →IF1, ⌊IX ⌊0 CAR ← ADR
      ⌊1 |↓/ACC| CAR ← ADR.
      ⌊2 |ACC[1]| CAR ← ADR.
      ⌊3 |OVF| CAR ← ADR ...
     EXIMM: →IF1, ⌊IX ⌊0 ACC ← ADR[1] × 6∘ADR
      ⌊1 ACC ← ACC + ADR[1] × 6∘ADR ..
     EXSHF: |v/ADR| →EXSHF, ¢ ADR,
      ⌊IX ⌊0 e| ACC ⌊1 0| ACC ⌊2 c↴ ACC
      ⌊3 (8) c↱ ACC; →IF1.
  ... (end of ST, AU and SY)

FIG. 7.3 An Elementry Digital Computer

division, and floating-point arithmetic instructions are desirable but would require much greater complexity.

Note that a unique 4-bit *operation code* (opcode) is assigned to individual instructions and groups of similar instructions in Table 7.1. The IX bits are both zeros in many instructions; the next chapter is concerned with the utilization of IX bits. The four transfer instructions all have the same operation code and are distinguished by different IX bits.

The last step of the instruction-fetch phase loads the MAR with the address portion of the instruction in preparation for the execution of many instructions. See Fig. 7.3. Simultaneously, the OP bits are "decoded," causing CPU to transfer to a state where execution of a single type of instruction takes place. Let us examine the execution of each instruction type in turn to see exactly what can be accomplished with each type of instruction and how it is accomplished.

INP (0000)

The input instruction offers a means of getting one word (16 bits) from a set of terminals named INPUT to the memory of the computer. The address field of the instruction indicates the location into which the data word is placed. Note that when we depress the RESET button, we "clear" the IR: set zeros in the OP, IX, and ADR subregisters. Thus after the RESET button is depressed, EDC begins to operate by performing an INP instruction. This then is how we begin to place instructions and data in the memory of EDC.

We assume that some unit of input equipment places a word in the INPUT register. Usually such equipment operates much more slowly than the CPU—central processing unit. A problem could arise if the CPU were asked to execute an INP instruction repeatedly more rapidly than the input equipment can supply words of information. The input equipment must "tell" CPU when a new word is available in the INPUT register, and CPU must tell the input equipment when it has accepted a word and can place the next word in the INPUT register. Flip-flop IN provides this communication. EDC assumes that when a new word is available in INPUT, the input equipment sets FF IN. We see in Fig. 7.3 that execution of the INP instruction in state EXINP includes transfer of the word in the INPUT register and reset of FF IN. This last action notifies the input equipment that it is now appropriate to place the next word in the INPUT register.

OUT (0001)

This instruction is in many ways the dual of the INP instruction. A public register, OUTPUT, is declared and CPU places a word taken from a specific memory location in this register. Again we assume without offering details that a unit of output equipment accepts this word and converts it to a form that can be read by a human. The word may be

TABLE 7.1   INSTRUCTION SET FOR EDC

Instruction	Mnemonic	Encoding		Action to be taken
		OP	IX	
Input	INP	0000	00	When a word becomes available in the INPUT register, transfer it to the memory location specified by the address field of the instruction.
Output	OUT	0001	00	Transfer the word MLSAFI[1] to the OUTPUT register. Notify the output equipment.
Block Input	BIN	0010	00	Transfer a block of n words[2] from the INPUT register to memory locations 1 through n. Contents of the accumulator will be destroyed.
Zero	ZRO	0011	00	Set the ACC to zero.
Advance	ADV	0011	01	Add 1 to the ACC.
Double	DBL	0011	10	Double the ACC.
Change Sign	CSN	0011	11	2's complement the ACC.
Load Acc.	LDA	0100	00	Load ACC with the word MLSAFI.
Add	ADD	0101	00	Add the word MLSAFI to the word held by ACC. Place the sum in ACC.
Subtract	SUB	0110	00	Subtract the word MLSAFI from the word held by ACC. Place the difference in ACC.
Store	STO	0111	00	Store MLSAFI the word held by ACC.
Mask	AND	1000	00	AND the word MLSAFI and the word held by ACC. Place product in ACC.
Transfer	TRA	1001	00	Take the next instruction from the word MLSAFI.
Transfer-zero	TRZ	1001	01	If the word held by the ACC is zero, take the next instruction from the word MLSAFI. Otherwise continue.
Transfer-Negative	TRN	1001	10	If the word held by the ACC is negative, take the next instruction from the word MLSAFI. Otherwise continue.
Transfer-Overflow	TRO	1001	11	If the overflow flip-flop is set take the next instruction from the word MLSAFI. Otherwise, continue.
Enter ACC	ENA	1010	00	Load ACC with the address field extending the sign bit.
Increment	INC	1010	01	Increment the contents of the ACC by n.
Shift Right: Arithmetic	SRA	1011	00	Shift ACC right n places; extend sign.
Logical	SRL	1011	01	Shift ACC right n places; zero fill.
Circulate	CIR	1011	10	Circulate ACC right n places.
Byte Exchange	BEX	1011	11	Exchange the left and right bytes.

[1] MLSAFI abbreviates, "In the memory location specified by the address field of the instruction."

[2] n is the decimal equivalent of the address field.

printed, for example. But again, the output equipment is usually somewhat mechanical in nature and hence not capable of operating at the electronic speed of CPU. Thus flip-flop OUT is provided. The output equipment resets this flip-flop when it has completed its handling of a word and is ready to accept the next word. Thus we see that CPU waits in state EXOUT until FF OUT is reset before placing another word in OUTPUT.

BIN (0010)

While the INP instruction provides a means of placing one word into any location of the memory of EDC, we must be able to get many instruction and data words into the memory before we can ask EDC to perform profitable computation. The Block INput instruction, BIN, offers a means of loading n words into memory with one instruction. These words are placed in memory locations $1, 2, \ldots, n$.

Execution of the BIN instruction is relatively complicated. In essence, the INP instruction must be executed repeatedly; the address field of the instruction tells how many times. Thus counting is essential. With each new word, a new address must be generated and placed in MAR. Note that when a BIN instruction is encountered, the 10-bit representation of 1 is placed in the ADR, and the word count is placed in the ACC. Then in state EXBIN the word count is examined. If it is not zero, the ADR is transferred to MAR as an address and ACC is reduced. ACC thus is acting as a down-counter, and we have completed execution when the word count has been reduced to zero. In state EXBIX the INP instruction is executed, in effect, and the ADR is incremented, thus generating the address for the next word of the block. We repeatedly enter states EXBIN and EXBIX until n words have been transferred from the input equipment to the memory. Before each transfer CPU must wait in state EXBIX until the input equipment supplies a new word, as indicated by IN = 1.

The operation ACC ← ACC − 1D16 in the state EXBIN specification indicates that subtraction is to be performed, but does not detail exactly how it is to be accomplished. We will find similar operations in other state specifications. Before we can actually build EDC, we must specify the subtraction algorithm. This will be done when we design the logic of EDC.

Let us review now. Assume that we walk up to EDC, depress the RESET button to stop the machine and prepare it to serve us, and supply the

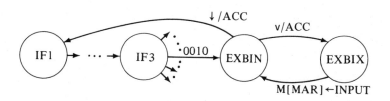

Fig. 7.4   Control of the BIN Instruction

SECTION 7.1   **Architecture and Organization of EDC**

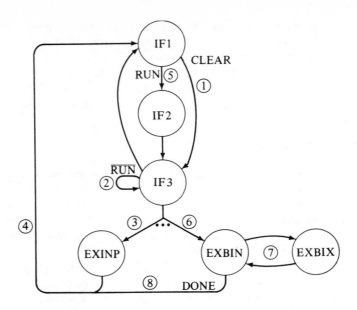

FIG. 7.5   Starting Sequence in EDC

word 0010000000000101 to the input equipment. Eventually EDC reaches state IF1 and detects that CLEAR = 1. It clears the CAR and IR, and the RUN and CLEAR flip-flops are reset. Then it goes to state IF3. Since RUN = 0, it remains in that state performing no operations until we depress the START button. Then RUN = 1 and the op code bits are examined.

Since the IR was just cleared (state IF1),the operation code is that of an INP instruction. In state EXINP our word is transferred from the input equipment to memory location 0 (the CAR was also cleared). Then CPU returns to state IF1 and proceeds through a normal instruction fetch. But note that the word from memory location 0 is fetched, and that is the one word we supplied. It is decoded as a BIN instruction. CPU will proceed to state EXBIN and attempt to load five words into memory. If we do not supply five more words to the input equipment this can not be accomplished and CPU will wait in state EXBIX until we supply a word and IN = 1.[2] This then is how we get a 5-instruction program loaded into memory. The first instruction is placed in memory location 1 and will be fetched and executed immediately after the BIN instruction has been completed. Fig. 7.5 shows the state diagram pertinent to this loading sequence.

The next four instructions of Table 7.1 all have the same op-code; the IX bits distinguish them. They are examples of instructions that process data already held by the registers of a system. Since only the ACC holds

---

[2] This may be an undesirable design in that we can not get CPU out of state EXBIX by depressing console switches. But it is a simple design and commercial machines that must complete an input-output operation before they respond to console commands have been marketed.

data in EDC, the ACC is the implied source of operands and sink for results.

ZRO(0011) (00)

The accumulator is cleared.

ADV(0011) (01)

The content of the ACC is treated as a 2's complement number to which one is added and the result returned to the ACC.

DBL(0011) (10)

The number held by the accumulator is doubled by adding it to itself.

CSN(0011) (11)

The binary sequence held by the ACC is replaced with its 2's complement sequence. If the original sequence is viewed as expressing a number, the sign of that number is changed by this instruction.

LDA (0100)

In computation, we often need to place a specific word in the accumulator, perhaps in preparation for adding it to another word. The LDA instruction accomplishes this. In state IF3 the address portion of the instruction was transferred to the MAR. Then in state EXLDA all that remains to be accomplished is to transfer the word provided by the memory to the ACC.

ADD (0101), SUB (0110)

These instructions call for adding (subtracting) the numbers held by the accumulator and a second number located in memory at the address specified by the address field of the instruction. Thus this second word must be fetched from memory before any arithmetic can be performed. In states EXADD and EXSUB the arithmetic is specified in a very functional way. Later we will consider how addition and subtraction are to be accomplished and how the OVF flip-flop is controlled.

STO (0111)

When an intermediate result to be used again in later computation has been formed in ACC, we must place it in the memory, from which it can be recalled at any later time. With a calculator we are forced to write such intermediate results on paper. Execution of the STO instruction thus consists of transferring the address field of the instruction to the memory address register, MAR ← ADR in state IF3, and then transferring the word held by the accumulator to the memory, which will put it away in the specified location, M[MAR] ← ACC in state EXSTO.

AND (1000)

The need for logic instructions, of which this is an example, may not be obvious. As long as we are concerned only with adding and subtracting numbers, such instructions are not needed. But computers are asked to do a great deal more than arithmetic, and data words may encode other than just numbers. As an example suppose that a 16-bit word contains two ANSICII encoded alphameric characters and we wish to examine one of those characters. While we can proceed in other ways, one way of "removing" the unwanted character is to AND the two-character word with a mask word established for the purpose. Using mask word 1111111100000000 retains the left character in place and replaces the right character code with zeros. We may desire this action to ensure that the original right character will not interfere with subsequent testing and processing of the left character.

Execution of the AND instruction is given in state EXAND, Fig. 7.3. We have already examined (Section 6.3) accumulator circuitry to accomplish the operator ACC ← ACC · M[MAR].

TRA (1001) (00), TRZ (1001) (01), TRN (1001) (10), TRO (1001) (11)

Suppose that we fill the 1024 word memory with instructions of the type we have already considered, and press the START button. Each instruction will be fetched and executed in turn. EDC with its 1 microsecond clock period will require 4 or 5 microseconds to handle each instruction. In about 5 *milliseconds* EDC will complete the given task and be out of work. To keep a modern computer busy doing useful work we often interrupt the fetching of instructions from sequential memory locations and have the computer perform the "same" instruction sequence again and again.

Herein really lies the great ability and value of a digital computer. Suppose we wish to know the value of y

$$y = x^4 + 4.7x^3 - 3.2147x^2 + 19x - 793.2$$

for $x = 7.2$. We can probably compute the required value by hand with a calculator or slide rule in less time and effort than is required to instruct a computer as to exactly what we wish it to do. It is unwise even to think of using a computer to do this job (unless our motive is instruction in how to use a computer).

But if we change the problem slightly and ask for the values of y for $x = 0.1, 0.2, \ldots, 24.9, 25$ the situation is entirely different. Now we wish to perform essentially the same computation 250 times. Only the value of x changes. This is now a very big job to approach by hand. We might start, but would very rapidly tire and begin to make mistakes. Once errors begin to enter our computation, our effort doubles or we accept incorrect results. A computer does not tire as we do, and seldom makes mistakes. (It is essential that we know when it does, of course.) And if we can instruct it to perform one computation, we can have it perform 250 or more almost identical computations with very little additional effort.

Execution of the *transfer* instructions is detailed in state EXTRA of Fig. 7.3. Let us look first at the *unconditional transfer* instruction, TRA. Its execution consists of only one information transfer: CAR ← ADR. The address portion of the instruction is placed in the current address register. Then CPU goes back to state IF1 and fetches the instruction from the location specified by CAR. That location is, of course, the address field of the TRA instruction just executed. Thus the TRA instruction allows us to interrupt the sequence of memory locations from which instructions are fetched. For example, having changed the value of x we might have EDC return to the instruction sequence for calculating y.

The TRZ instruction is a *conditional transfer* instruction. As we see in Fig. 7.3, the execution of a TRZ instruction consists of looking at the ACC: if any FF of the ACC is set, nothing is done except a return to state IF1. EDC makes a decision and this is about the most complex decision which any digital computer is able to make: if the ACC is cleared, the next instruction executed is located at the address specified by the address field of the TRZ instruction, otherwise the instruction which follows the TRZ is executed next.

The TRN conditional transfer instruction bases the decision on the sign bit of the number held by the accumulator. The TRO instruction bases it on the state of the overflow flip-flop. Transfer instructions may be based upon the state of any flip-flop or logic function of flip-flops.

ENA(1010)(00)INC(1010)(01)

These two instructions are examples of *immediate* type instructions. The address fields of these instructions does not indicate an address of data, but rather the data itself. Thus operands are immediately available in the instruction without the need for going to the memory again. A problem exists, however: the accumulator and memory of EDC deal with 16-bit data words, but an address field consists of only 10 bits. Thus the range of numbers that can be expressed in the address field is restricted, but in many circumstances this will not limit the usefulness of these instructions. To form a 16-bit data word from the given 10 bits, we copy the leftmost bit of the address field six times with the *extend operator* $\times$. In Fig. 7.3 ADR[1] $\times$ 6 ∘ ADR indicates that six copies of the ADR[1] terminal are to be concatenated to the left of the ten terminals of set ADR; this forms a set of 16 terminals, seven of which bear the same signal.

We can load the accumulator with a data word from memory with the LDA instruction. The LDA instruction occupies one memory location; the word to be loaded occupies another. In many cases, as we will see, we desire to load the ACC with an integer that can be expressed in less than 16 bits. The ENA instruction lets us do so without using an additional memory location to hold that constant. The address field with its leftmost bit extended is placed directly in the ACC. (See Fig. 7.3, state EXIMM.)

Extension of the leftmost bit of the address field preserves the 2's complement representation of the integer we wish to place in the ACC.

The INCrement instruction parallels the ADD instruction. The address field with leftmost bit extended is added to the contents of the accumulator.

SRA(1011) (00), SRL(1011) (01), CIR(1011) (10)

Logical and arithmetic shift and circulate instructions are typically found in digital computers. Their execution, detailed in state EXSHF, is somewhat involved. The ACC must be shifted to the right the number of places specified by the address field of the instruction. (Thus we see another example of the "address field" not really specifying an address.) EDC does not use complicated shifting circuitry, but rather repeated single position shifting of the ACC. The ADR subregister counts the number of shifts performed, counting down to zero.

BEX (1011) (11)

If computers are expected to process sequences of characters, then instructions that facilitate character processing should be included in the instruction set. This instruction is an example. It exchanges the left and right bytes held by the ACC. Since such an exchange is an eight position circulate, this instruction is not really necessary, but it is desirable in that it operates eight times faster than the equivalent CIR instruction. The address field should be set to 1 for this instruction, because of the control statements of state EXSHF in Fig. 7.3.

In summary, Fig. 7.6 offers a skeleton state diagram of automaton CPU. The normal state sequence and information transfers of the instruction fetch phase are as follows.

$$\text{IF1} \quad \text{MAR} \leftarrow \text{CAR}, \ \text{\textcent} \ \text{CAR}$$
$$\text{IF2} \quad \text{IR} \leftarrow \text{M[MAR]}$$
$$\text{IF3} \quad \text{MAR} \leftarrow \text{ADR}, \ \text{Decode}$$

The execution phase differs for each instruction type. The transfer to MAR in state IF3 is a part of the execution of many instructions.

7.2 LOGIC DESIGN OF EDC REGISTERS

While one reason for carefully defining and using a design language is that the logic design task then can be made algorithmic and performed by a digital computer, we will take a less formal approach and do the job by hand to see the steps that must be taken and to exercise design techniques of previous chapters.

Let us turn first to the registers of EDC. Assume for the moment that the controlling finite-state machine, which we have not yet detailed, is constructed so that only a single terminal of a set of sixteen terminals, one for each state of CPU, bears a 1 when CPU is in the state with the same

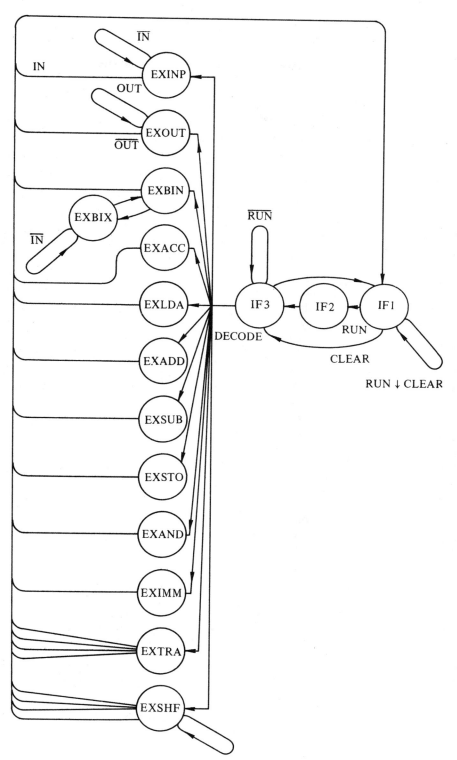

FIG. 7.6 State Diagram of CPU

name as the terminal. Hence, terminal IF1 will bear a 1 only when CPU is in state IF1. This dual use of names, to simplify and clarify our work, does not really pose a contradiction. We are not giving two units of hardware the same name, since a state is not a circuit component. Only the terminal appears in the logic diagram of the machine.

EDC is completely synchronous so P, the clock signal, should be expressed in each Boolean condition that governs whether or not a register transfer takes place. But it will not be so expressed because it is to be ubiquitous.

From Fig. 7.3 we compile a list of all transfers of information to a specific register and the condition under which each transfer is to take place. Fig. 7.7 shows this list. While it appears to be rather long and complicated, it can be generated from Fig. 7.3 with surprisingly little effort. Interpretation of this list is also straightforward.

Let us turn first to the MAR: we see that it must accept information from two sources. This is not particularly complicated, and the conditions under which each loading is to take place are not complicated either. We can easily visualize the typical cell of Fig. 7.8 from the conditional register-transfer statements for the MAR given in Fig. 7.7.

Note in Fig. 7.8 that several internal control signals have been given names Q1, Q2, etc. Looking ahead to the design of other registers and the control unit, we will find other occasions where these signals are needed. We propose to generate them once and use them in many places to reduce the logic of EDC. This is not to say that we are minimizing the combinational logic of EDC; no formal attempt will be made to minimize, for that is not the topic of prime concern here. But we will attempt to be efficient by simplifying Boolean equations and using already defined variables when it is obvious that we can do so.

$$Q1 = IF1 \cdot RUN \cdot \neg CLEAR$$
$$Q2 = IF3 \cdot RUN$$
$$Q3 = v/ACC$$
$$Q4 = EXBIN \cdot Q3$$
$$Q5 = Q2 \vee Q4$$

Since Q1 and Q5 cannot take the value of 1 simultaneously, we may use either to control a 2-input multiplexer and take advantage of MSI.

Problems arise when we begin to draw large assemblages of logic, which can be reduced by an introduction of new symbolism. Rather than show each and every input line to AND and OR gates we will use the following single line representation. Note that the dots and slashes where lines cross are very significant.

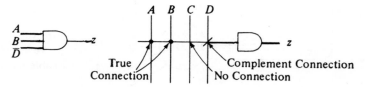

MAR
  |IF1·RUN·¬CLEAR| MAR ← CAR
  |IF3·RUN| MAR ← ADR
  |EXBIN· v /ACC| MAR ← ADR
CAR
  |IF1· CLEAR| CAR ← 0
  |IF1· RUN·¬CLEAR| ¢ CAR
  |EXTRA· (¬IX [1]·¬IX [2] v ¬IX[1]· IX [2]· ↓/ACC
    v IX[1]·¬IX[2]· ACC[1] v IX[1]· IX [2]· OVF) | CAR ← ADR

M
  |IN· (EXINP v EXBIX)| M[MAR] ← INPUT
  |EXSTO| M[MAR] ← ACC
OUTPUT
  |EXOUT·¬OUT| OUTPUT ← M [MAR]
IR
  |IF1· CLEAR| IR ← 0
  |IF2| IR ← M [MAR]
ADR (Subregister of IR)
  |IF3· RUN· ·/OP\2D4| ADR ← 1D10
  |EXBIX· IN| ¢ ADR
  |EXSHF· v / ADR| ¢ ADR
ACC
  |IF3· RUN· ·/OP\2D4| ACC ← 0D6∘ADR
  |EXBIN· v /ACC| ACC ← ACC − 1D16
  |EXACC·¬IX[1]·¬IX[2]| ACC ← 0
  |EXACC·¬IX[1]· IX[2]| ACC ← ACC + 1D16
  |EXACC· IX[1]·¬IX[2]| ACC ← ACC + ACC
  |EXACC· IX[1]· IX[2]| ACC ← 0 − ACC
  |EXLDA| ACC ← M[MAR]
  |EXADD| ACC ← ACC + M[MAR]
  |EXSUB| ACC ← ACC − M[MAR]
  |EXAND| ACC ← ACC· M[MAR]
  |EXIMM·¬IX[1]·¬IX[2]| ACC ← ADR[1]×6∘ADR
  |EXIMM·¬IX[1]· IX[2]| ACC ← ACC + ADR[1]×6∘ADR
  |EXSHF· v /ADR·¬IX[1]·¬IX[2]| e↓ ACC
  |EXSHF· v /ADR·¬IX[1]· IX[2]| 0| ACC
  |EXSHF· v /ADR· IX[1]·¬IX[2]| c| ACC
  |EXSHF· v /ADR· IX[1]· IX[2]| (8) c↓ ACC

Fig. 7.7 Compiled List of Information Transfers to Each Register of EDC

The CAR is only a bit more complicated than the MAR. We must be
able to load CAR from ADR, clear CAR, and have it count. Loading from
ADR and loading with zeros present no new problems, and we know how
to design counters.

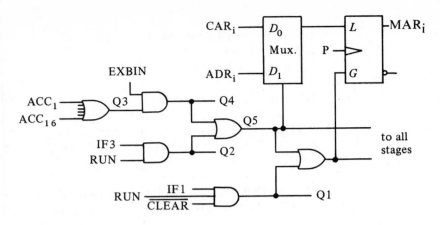

FIG. 7.8   Control and Typical Cell of the MAR

Synchronous, cascadable, 4-bit, and load-clear MSI counters are available for construction of the CAR. Three such counter ICs provide more stages than we need. Figure 7.9 shows one such IC and the necessary control circuitry, assuming that the counter IC requires separate COUNT, RESET, and LOAD signals in true form. The following control signals are provided by the gate network and decoder on the IX subregister.

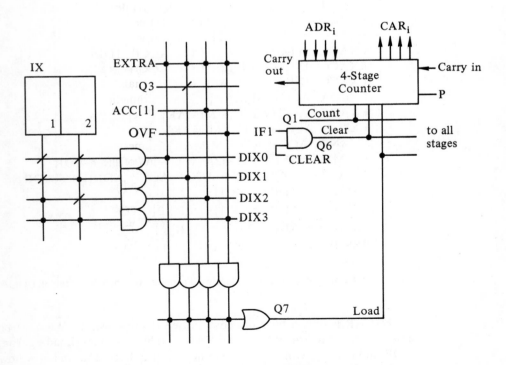

FIG. 7.9   Control and an MSI Realization of the CAR

$Q6 = IF1 \cdot CLEAR$

$Q7 = EXTRA \cdot (DIX0 \lor DIX1 \cdot \neg Q3 \lor DIX2 \cdot ACC[1] \lor DIX3 \cdot OVF)$

$$DIX0 = \overline{IX}[1] \cdot \overline{IX}[2]$$

$$DIX1 = \overline{IX}[1] \cdot IX[2]$$

$$DIX2 = IX[1] \cdot \overline{IX}[2]$$

$$DIX3 = IX[1] \cdot IX[2]$$

Q7 can also be realized by a 4-input multiplexer with an enable terminal.

The IR is similar to the CAR in that it must be able to load and clear. Counting involves only a portion of the register so not all cells of the IR will be identical. Up-down, loadable, and clearable counters are required to construct the ADR. The use of available MSI counters alleviates the design problem.

The accumulator is the most complicated and hence most expensive register. From Fig. 7.7 we see that the accumulator must be loaded from several sources including an adder-subtractor network, AND words, and shift right. Thus the ACC is not really complicated and no new design problems arise, but the amount of detail to be considered is large.

ALU Design

Rather than design a separate combination network to perform each of the processes associated with the ACC and rather than use extensive multiplexers to route operands to, and results from, these processors, economy dictates that we design one or a few blocks, each able to perform a variety of processing steps under control of mode signals. Arithmetic and logic processing is typically performed by an Arithmetic-Logic-Unit (ALU). Shifting may be performed by the ALU or a separate network. We will design separate ALU and SHIFTER networks.

Figure 7.7 indicates that the following arithmetic and logic operations must be performed with the ACC as destination of the result.

subtract 1, add 1, double, subtract from 0, add, subtract, AND

While ALU integrated circuits are available that perform many additional operations, they do not perform some of these. If we were going to actually construct EDC, we would be economically motivated to use existing ICs by adding and deleting entries from the instruction set. To see how ALUs are designed, we will insist that the operations above are required and develop a suitable ALU.

Addition is a fundamental part of the list of operations above. Any full-adder may be used as a starting point for ALU design, but one of the easiest ones to generalize is the modified full-adder of Fig. 3.31. The carry-look-ahead network provides all carry signals, and also ensures speed. To add two numbers A and B, each cell of a modified full-adder must generate three signals.

$$G_i = A_i \cdot B_i$$
$$P_i = A_i \vee B_i$$
$$S_i = \bar{G}_i P_i \oplus C_{i+1}$$

We will use these full-adder cells with the 2's complement algorithm suggested by Fig. 3.32. The carry into the least significant cell is C_{17}. When adding $A + B$, $C_{17} = 0$.

To add 1 to A, the cells must be modified so that the B bits are replaced by 0's. The 1 to be added may then be introduced by $C_{17} = 1$. The signals to be generated are then:

$$G_i = A_i \cdot 0 = 0$$
$$P_i = A_i \vee 0 = A_i$$
$$S_i = \bar{G}_i P_i \oplus C_{i+1}$$

In the following, S_i will not change, and hence, will not be repeated.

To subtract $A - B$, the B bits must be complemented and $C_{17} = 1$.

$$G_i = A_i \cdot \bar{B}_i$$
$$P_i = A_i \vee \bar{B}_i$$

Table 7.2 summarizes the above results and those for the other operations that we require.

The last two entries of Table 7.2 require additional discussion. We will soon see the advantage of passing the B operand to the output terminals of the ALU. The AND operation can be performed with G and P functions found earlier in Table 7.2. With $G_i = A_i \cdot B_i$ and $P_i = 1$

$$C_i = G_i \vee P_i \cdot C_{i+1}$$
$$= G_i \vee C_{i+1}$$

If $C_{17} = 1$, then all other carries will also equal 1. This is important when we examine the sum equation.

TABLE 7.2 GENERATE AND PROPAGATE SIGNALS FOR
DESIRED ALU OPERATIONS

Operation	G_i	P_i	C_{17}
$A + B$	$A_i \cdot B_i$	$A_i \vee B_i$	0
$A + 1$	$A_i \cdot 0 = 0$	$A_i \vee 0 = A_i$	1
$A + A$	$A_i \cdot A_i = A_i$	$A_i \vee A_i = A_i$	0
$A - B$	$A_i \cdot \bar{B}_i$	$A_i \vee \bar{B}_i$	1
$A - 1$	$A_i \cdot \bar{0} = A_i$	$A_i \vee \bar{0} = 1$	0
$0 - A$	$0 \cdot \bar{A}_i = 0$	$0 \vee \bar{A}_i = \bar{A}_i$	1
A	$A_i \cdot 0 = 0$	$A_i \vee 0 = A_i$	0
B	$B_i \cdot 0 = 0$	$B_i \vee 0 = B_i$	0
$A \cdot B$	$A_i \cdot B_i$	1	1

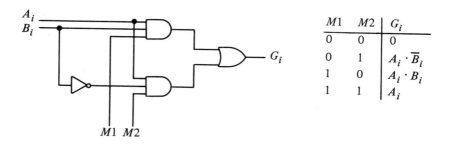

FIG. 7.10   The G Network of the ALU

$$S_i = \overline{G}_i P_i \oplus C_{i+1}$$
$$= \overline{A_i \cdot B_i \cdot 1} \oplus 1 = A_i \cdot B_i$$

Table 7.2 indicates that the following G functions must be available to perform different operations: 0, AB, $A\overline{B}$, A. Two operations are minterms; the other two are the sum of none and both of these minterms. Two control signals will be necessary to select which G function is to be used to perform an ALU operation. Figure 7.10 shows a small network that introduces mode control signals, M1 and M2, in a very obvious manner.

While the P network of commercial ALU ICs is as simple as the G network of Fig. 7.10, our ALU is more involved. The following P functions are listed in Table 7.2: $A_i \vee B_i$, A_i, $A_i \vee \overline{B}_i$, 1, $\overline{A}_i$, and B_i. The first four functions are easily generated, but six functions require three control signals and a more complicated network. Figure 7.11 shows one suitable network and the entire ALU cell. If a sixth mode control signal were introduced so that the carry into the XOR gate could be set to 1, then a great variety of logic functions could be provided. In ALU design we attempt to get a great variety of functions with a minimum of cell complexity and control signals. Table 7.3 indicates the control signal values required to

TABLE 7.3
ALU MODE CONTROL VALUES

Operation	$M_1 M_2 M_3 M_4 M_5$	C_{17}
0	0 0 0 0 0	0
$A + B$	1 0 1 0 1	0
$A + 1$	0 0 1 0 0	1
$A + A$	1 1 1 0 0	0
$A - B$	0 1 1 1 0	1
$A - 1$	1 1 1 1 1	0
$0 - A$	0 0 0 1 1	1
A	0 0 1 0 0	0
B	0 0 0 0 1	0
$A \cdot B$	1 0 1 1 1	1

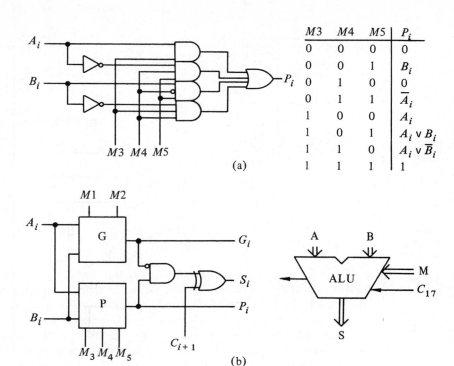

M3	M4	M5	P_i
0	0	0	0
0	0	1	B_i
0	1	0	0
0	1	1	$\overline{A}_i$
1	0	0	A_i
1	0	1	$A_i \vee B_i$
1	1	0	$A_i \vee \overline{B}_i$
1	1	1	1

(a)

(b)

FIG. 7.11   The P and ALU Networks

obtain the operations of Table 7.2. Figure 3.32 indicated that only one XOR gate is necessary to drive the OVF flip-flop.

$$|P| \quad OVF \leftarrow C[1] \oplus C[2].$$

Now we return to Fig. 7.7 to see what this ALU will do for us. Note that the contents of the ACC must be added to two words: the word from memory when executing the ADD instruction and the extended word from the ADR when executing the INC instruction. A multiplexer at the B input terminal of the ALU will be required. Since the ACC is loaded from these two sources and we have provided for passing the B operand through the ALU, all loading can be performed with the same circuitry. One detail requires design. Set-up for the BIN instruction requires that 0's be placed on the left of the ADR rather than its left bit being extended. The left six cells of the multiplexer must therefore pass 0's, ADR[1], or M[MAR]$_i$ to the ALU. This three-way branch is easily accomplished with the cell of Fig. 7.12(a). Multiplexing could have been considered to be an additional task for the ALU and more complex G and P networks designed. Greater speed, and possibly lower cost, would then be gained, but design effort would be increased.

The shift operations remain to be realized. While a shift network can be operated in parallel with the ALU, the possibility of performing both an

M6	M7	B Operand
0	0	0 D6 • ADR
0	1	M[MAR]
1	0	ADR[1] × 6 • ADR
1	1	not used

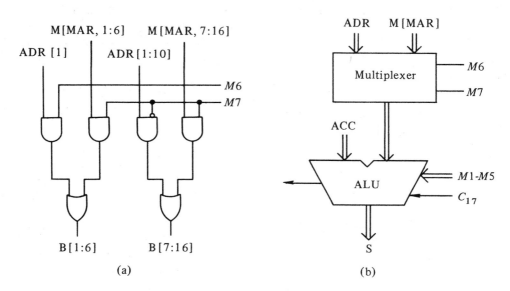

FIG. 7.12   The Modified Multiplexer and Its Connection to the ALU to Accomplish ACC Loading as well as Arithmetic

ALU and a SHIFTER operation in the same clock cycle is realized by placing the shifter between the ALU and the ACC. Then the shifter must be able to pass as well as shift, of course. Because of the 8-place byte shift, 3-input multiplexers with two control signals must be used. We see the cost of the byte shift instruction here; 2-input multiplexers would suffice if that entry were not in the instruction set. And since we have four shift modes as well as the pass mode, a total of three control signals will be required. Figure 7.13 provides a suitable design by using 4-input multiplexers to advantage. Only the leftmost cell requires the AND gate to distinguish arithmetic from logical shift.

Now we have a multiplexer-ALU-shifter that is able to perform all of the ACC operations of Fig. 7.7, and many more. Looking back, we should be tempted to start over again. We might revise the instruction set so as to take advantage of the capabilities and perhaps delete instructions that complicate the hardware, and hence increase the cost of EDC. Computer design is not a single-pass operation, and while revision so as to optimize EDC is entirely in order, we will continue with the accumulator design for it is not yet complete. All control signals remain to be defined and suitable control networks designed.

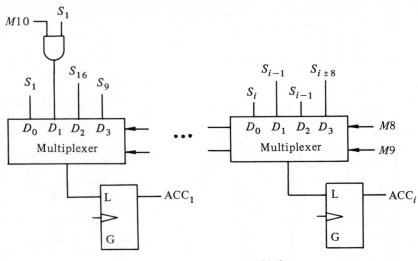

$M8$	$M9$	$M10$	Shift Mode
0	0	x	Pass
0	1	0	0-Fill
0	1	1	Extend
1	0	x	Circulate
1	1	x	Byte

FIG. 7.13   Shift Network

Table 7.4 repeats the accumulator operations of Fig. 7.7 and indicates the care and don't care values of mode control signals for each. The following equations are written from the 1's in Table 7.4; the don't care entries might be used to simplify some of these equations.

$M1 = Q4 \lor EXACC \cdot DIX2 \lor EXADD \lor EXAND \lor EXIMM \cdot DIX1$

$M2 = Q4 \lor EXACC \cdot DIX2 \lor EXSUB$

$M3 = Q4 \lor EXACC \cdot (DIX1 \lor DIX2) \lor EXADD \lor EXSUB \lor EXAND \lor$
$\qquad EXIMM \cdot DIX1 \lor EXSHF \cdot Q3$

$M4 = Q4 \lor EXACC \cdot DIX3 \lor EXSUB \lor EXAND$

$M5 = Q2 \cdot \cdot /OP\backslash 2D4 \lor Q4 \lor EXACC \cdot DIX3 \lor EXLDA \lor EXADD$
$\qquad \lor EXAND \lor EXIMM \cdot \neg IX[1]$

$M6 = EXIMM \cdot \neg IX[1]$

$M7 = EXLDA \lor EXADD \lor EXSUB \lor EXAND$

$M8 = EXSHF \cdot Q3 \cdot IX[1]$

$M9 = EXSHF \cdot Q3 \cdot (\neg IX[1] \lor IX[2])$

$M10 = EXSHF \cdot Q3 \cdot DIX0$

$C_{17} = EXACC \cdot DIX1 \lor EXACC \cdot DIX3 \lor EXSUB \lor EXAND$

TABLE 7.4 ACC Operations and Mode Control

Condition	Operation	M1	M2	M3	M4	M5	M6	M7	M8	M9	M10	C_{17}
Q2··/OP\2D4	ACC←0D6∘ADR	0	0	0	0	1	0	0	0	0	x	0
Q4	ACC←ACC−1D16	1	1	1	1	1	x	x	0	0	x	0
EXACC·DIX0	ACC←0	0	0	0	0	0	x	x	0	0	x	0
EXACC·DIX1	ACC←ACC+1D16	0	0	1	0	0	x	x	0	0	x	1
EXACC·DIX2	ACC←ACC+ACC	1	1	1	0	0	x	x	0	0	x	0
EXACC·DIX3	ACC←0−ACC	0	0	0	1	1	x	x	0	0	x	1
EXLDA	ACC←M[MAR]	0	0	0	0	1	0	1	0	0	x	0
EXADD	ACC←ACC+M[MAR]	1	0	1	0	1	0	1	0	0	x	0
EXSUB	ACC←ACC−M[MAR]	0	1	1	1	0	0	1	0	0	x	1
EXAND	ACC←ACC·M[MAR]	1	0	0	1	1	0	1	0	0	x	1
EXIMM·DIX0	ACC←ADR[1]×6∘ADR	0	0	0	0	1	1	0	0	0	x	0
EXIMM·DIX1	ACC←ACC+ADR[1]×6∘ADR	1	0	1	0	1	1	0	0	0	x	0
EXSHF·Q3·DIX0	e|↑ACC	0	0	1	1	0	x	x	0	1	1	0
EXSHF·Q3·DIX1	0|↑ACC	0	0	1	0	0	x	x	1	1	1	0
EXSHF·Q3·DIX2	c|↑ACC	0	0	1	0	0	x	x	1	0	0	0
EXSHF·Q3·DIX3	(8) c|↑ACC	0	0	1	0	0	x	x	x	1	x	0

Other Registers of EDC

The remaining registers of EDC are very straightforward. Figure 7.7 reveals that a 2-input multiplexer must precede the memory data input terminals. Control signal EXSTO may be used to control the multiplexer. The memory must perform a write operation under the following condition

$$Q8 = IN \cdot (EXINP \vee EXBIX)$$

$$Q9 = Q8 \vee EXSTO$$

In all other cases it performs read operations. The OUTPUT register is loaded from the memory under condition

$$Q10 = EXOUT \cdot \neg OUT$$

The INPUT register is never loaded by the CPU, but CPU does partially control the IN and OUT communication flip-flops. The IN flip-flop must be reset under condition Q8. OUT is set by Q10. If JK flip-flops are used, then simpler input signals could be expressed, but Q8 and Q10 are needed for register control purposes and therefore must be generated anyway. Figure 7.14 summarizes these last designs.

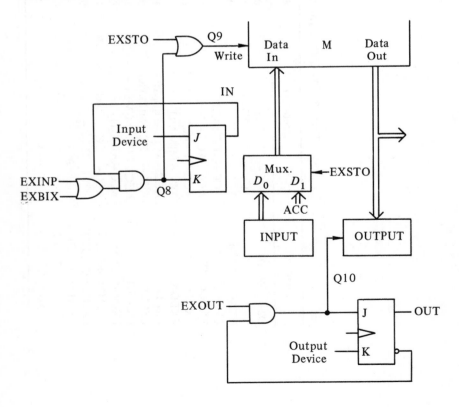

FIG. 7.14 Input/Output Data Paths and Control

The registers of EDC and their control networks have been designed with the assumption that the controlling state machine provides state signals. Twenty-five control signals were named; some intermediate signals went without names. Little attention was paid to logic network cost or speed, with the exception of the accumulator where an ALU, multiplexer, and shifter were designed so as to keep costs and propagation delays low if not minimal. A compromise was made between network cost–speed and design complexity. We could have designed a minimum 2-level network to perform all ACC operations, but the design effort would have been very great (and gate counts might be very high). Breaking this processing block into three blocks based upon function enables manual design. It also reveals many other operations that could be performed at very little additional expense and those initially desired operations that greatly increase the expense of the processing system. To obtain a computer with a greater performance-to-cost ratio, one should return to the register configuration, instruction set, and processing algorithm descriptions and modify them based upon our new familiarity with ALU and register design.

7.3   CONTROL OF EDC

Control circuitry for EDC may take any of many forms and hence is not as easily developed. We have a 16-state finite-state machine to synthesize. Many input signals are involved; many output signals must be generated, although these have in effect been realized in the control signals for the registers. But that development was based upon the existence of 16 state-related terminals. Our formal design procedure for finite-state machines using Karnaugh maps is not very effective when we face large problems and we must resort to more straightforward, but less effective (in terms of circuit cost) techniques.

Now how do we implement this 16-state machine? A minimum of four flip-flops is required. At the other extreme, we might use 16 flip-flops, one for each state, and design a sequencer. This state assignment greatly simplifies design effort; we need only find all transitions to a state and draw an AND-OR pyramid before a DFF for each state. (Initialization is a problem.) And with this approach one and only one flip-flop is set at all times. Each FF thus acts as the source for the state terminals IF1, IF2, etc., which we have assumed to exist.

A second approach might be to use 4 FFs and formally design the 16-state machine. This approach might well lead to a most desirable control unit. But the size of our current problem prohibits the use of Karnaugh maps and hand techniques. We can effectively design in this way only if appropriate computer assistance is available.

Still another approach recognizes that the OP subregister holds a different binary sequence for each instruction type and thus can be used to distinguish most of the EXecution states of CPU. We might build a 5-state machine with states IF1, IF2, IF3, and a new state EX, as shown in

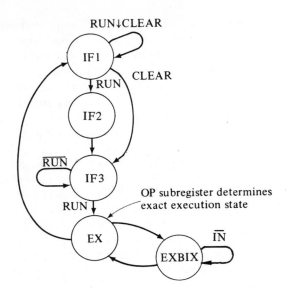

RUN↓CLEAR

IF1

↓RUN   CLEAR

IF2

$\overline{RUN}$

IF3

RUN↓

EX

OP subregister determines
exact execution state

$\overline{IN}$

EXBIX

FIG. 7.15   Skeleton of a 5-State Control

Fig. 7.15. When this 5-state machine is in state EX, the actual state of CPU is to be determined by the state of the OP subregister. State EXBIX is the fifth state unless it can be eliminated by redefining the BIN algorithm.

Figure 7.16 redefines EDC, using combinational decoding of the op-code rather than the sequential decoding shown in the previous section. One general EXecution state is shown; the OP subregister contents determine what operations are performed when CPU is in state EX. State EXBIX is not eliminated because doing so requires that the MAR and much control logic be redesigned. While some statements of Fig. 7.16 use different DDL syntax than corresponding statements of Fig. 7.3, no register design change is specified.

All of the conditions upon transitions to each state of Fig 7.16 are shown in Fig. 7.17. State signals assumed to exist in the previous section are used to keep condition expressions brief and clear. We will decode the OP register to generate these signals later. A sequencer can be designed directly from this information (with the exception of initialization circuitry). The IF1 flip-flop should receive additional design effort. Gating for the other flip-flops is rather simple. Let us see if a three flip-flop control unit is any more economical or difficult to design.

We might foresee a potential malfunction of a 3-FF control unit. EDC will react to the RESET button as previously described, only if it attains state IF1. When power is first supplied to a finite-state machine, flip-flops will set and reset almost at random. We must thus ensure that control circuitry can not attain a state from which it cannot get to state IF1. One way to "use up" extra code words is to give multiple assignments to some states. Table 7.5 below shows such an assignment for the state sequencing

⟨SY⟩ EDC: ⟨TI⟩ P(1E-6). ⟨ME⟩ M[0 : 1023,16].
    ⟨RE⟩ IR[16] = OP[0 : 3]∘IX[2]∘ADR[10], CAR[10],
        ACC[16], MAR[10], INPUT[16], OUTPUT[16], OVF,
        RUN, CLEAR, IN, OUT.
    ⟨AU⟩ CPU: P: ⟨ID⟩ Z = 0D16.
    ⟨ST⟩ IF1: |CLEAR| RUN←0, CLEAR←0, CAR←0, IR←0,
        →IF3;
        |RUN| MAR←CAR, ¢ CAR, →IF2; →IF1 ...
      IF2: IR←M[MAR], →IF3.
      IF3: RUN: MAR←ADR, ⌈OP ⌊2 ADR←1D10, ACC←
        0D6∘ADR, →EX ⌊12 : 15 → IF1; →EX ..
      EX: ⌈OP ⌊0 |IN| M[MAR] ← INPUT, IN ← 0, →IF1;
        →EX.
        ⌊1 |¬OUT| OUTPUT ← M[MAR], OUT ← 1,
        →IF1; →EX.
        ⌊2 |v/ACC| MAR ← ADR, ¢ ACC, →EXBIX;
        →IF1.
        ⌊3 →IF1, ACC ←⌈IX ⌊0 Z ⌊1 ACC+1D16
            ⌊2 ACC+ACC ⌊3 Z−ACC.
        ⌊4 ACC ←M[MAR], →IF1
        ⌊5 ACC←ACC+M[MAR], →IF1
        ⌊6 ACC ← ACC−M[MAR], →IF1
        ⌊7 M[MAR] ← ACC, →IF1
        ⌊8 ACC ← ACC· M[MAR], →IF1
        ⌊9 →IF1, ⌈IX ⌊0 CAR ← ADR
            ⌊1 |↓/ACC| CAR ← ADR.
            ⌊2 |ACC[1]| CAR ← ADR.
            ⌊3 |OVF| CAR ← ADR. .
        ⌊10 →IF1, ⌈IX ⌊0 ACC ← ADR[1] × 6∘ADR
           ⌊1 ACC ← ACC+ADR[1] × 6∘ADR.
        ⌊11 |v/ADR| → EX, ¢ ADR, ⌈IX ⌊0 e↓ ACC
          ⌊1 0↓ ACC ⌊2 c↓ ACC ⌊3 (8)c↓ ACC; →IF1....
     EXBIX: IN: M[MAR]← INPUT, ¢ ADR⃗, IN←0, →EX.
... (end of ST, AU and SY)

FIG. 7.16   Revised Definition of EDC That Uses Combinational Decoding

TABLE 7.5.   STATE ASSIGNMENT FOR SS REGISTER

State	SS[1]	SS[2]	SS[3]
IF1	0	0	0
IF2	0	0	1
IF3	0	1	x
EX	1	0	x
EXBIX	1	1	x

|IF1 · $\overline{(\text{CLEAR} \cdot \overline{\text{RUN}})}$ v IF3 · OP[0] · OP[1] v EXINP · IN v
EXOUT · $\overline{\text{OUT}}$ v EXBIN · ↓/ACC v EXACC v EXLDA v EXADD v
EXSUB v EXSTO v EXAND v EXTRA v EXIMM v EXSHF · ↓/ADR|
    →IF1
|IF1 · $\overline{\text{CLEAR}}$ · RUN| →IF2
|IF1 · CLEAR v IF2| →IF3
|IF3 · RUN · $\overline{\text{OP[0]}}$ · $\overline{\text{OP[1]}}$ v EXINP · $\overline{\text{IN}}$ v EXOUT · OUT v
    EXSHF · v/ADR v EXBIX · IN| →EX
|EXBIN · v/ACC| →EXBIX

FIG. 7.17 Transitions to Each State of EDC as Given in Fig. 7.16

SS register. This assignment provides a means to detect easily the phase of operation of EDC. When FF SS[1] is set EDC is in the execution phase, otherwise it is fetching an instruction.

Now we are finally in a position to detail the decoders on the SS and OP registers, which provide the state signals we have assumed to exist.

$$IF1 = \overline{SS}[1] \cdot \overline{SS}[2] \cdot \overline{SS}[3]$$
$$IF2 = \overline{SS}[1] \cdot \overline{SS}[2] \cdot SS[3]$$
$$IF3 = \overline{SS}[1] \cdot SS[2]$$
$$EX = SS[1] \cdot \overline{SS}[2]$$
$$EXBIX = SS[1] \cdot SS[2]$$
$$EXINP = EX \cdot \overline{OP}[0] \cdot \overline{OP}[1] \cdot \overline{OP}[2] \cdot \overline{OP}[3]$$
$$EXOUT = EX \cdot \overline{OP}[0] \cdot \overline{OP}[1] \cdot \overline{OP}[2] \cdot OP[3]$$
$$EXBIN = EX \cdot \overline{OP}[0] \cdot \overline{OP}[1] \cdot OP[2] \cdot \overline{OP}[3]$$
$$EXACC = EX \cdot \overline{OP}[0] \cdot \overline{OP}[1] \cdot OP[2] \cdot OP[3]$$
$$EXLDA = EX \cdot \overline{OP}[0] \cdot OP[1] \cdot \overline{OP}[2] \cdot \overline{OP}[3]$$
$$EXADD = EX \cdot \overline{OP}[0] \cdot OP[1] \cdot \overline{OP}[2] \cdot OP[3]$$
$$EXSUB = EX \cdot \overline{OP}[0] \cdot OP[1] \cdot OP[2] \cdot \overline{OP}[3]$$
$$EXSTO = EX \cdot \overline{OP}[0] \cdot OP[1] \cdot OP[2] \cdot OP[3]$$
$$EXAND = EX \cdot OP[0] \cdot \overline{OP}[1] \cdot \overline{OP}[2] \cdot \overline{OP}[3]$$
$$EXTRA = EX \cdot OP[0] \cdot \overline{OP}[1] \cdot \overline{OP}[2] \cdot OP[3]$$
$$EXIMM = EX \cdot OP[0] \cdot \overline{OP}[1] \cdot OP[2] \cdot \overline{OP}[3]$$
$$EXSHF = EX \cdot OP[0] \cdot \overline{OP}[1] \cdot OP[2] \cdot OP[3]$$

The state diagram of Fig. 7.15 and the state assignment of Table 7.5 provide the information necessary to design the SS register. We take each flip-flop in turn and require the conditions under which it must be set and reset. While JKFFs will be used for purposes of example, we will not take advantage of the toggling logic of such flip-flops to simplify the equations. The first flip-flop, SS[1], must be set when we enter the EX state from IF3.

$$J_1 = IF3 \cdot RUN \cdot \overline{OP[0]} \cdot \overline{OP[1]}$$

The first flip-flop must be reset when we enter IF1 from EX. (It is already reset for all other entries to IF1, IF2, and IF3.) The conditions under which this transition takes place are obtained from Fig. 7.17.

$$K_1 = EXINP \cdot IN \vee EXOUT \cdot \overline{OUT} \vee EXBIN \cdot \downarrow/ACC$$
$$\vee EXACC \vee EXLDA \vee EXADD \vee EXSUB \vee EXSTO$$
$$\vee EXAND \vee EXTRA \vee EXIMM \vee EXSHF \cdot \downarrow/ADR$$

This expression demands a very large fan-in OR gate. We can reduce that fan-in requirement by going back to the OP register and examining the instruction codes for the ACCumulator through the IMMediate instructions; then

$$Q11 = EXACC \vee \ldots \vee EXIMM$$
$$= EX \cdot (OP[0] \oplus (OP[1] \vee OP[2] \cdot OP[3]))$$

At this point we see the importance of assigning op-codes to instructions. If the ACCumulator instructions had been assigned code 1011 rather than 0011, this expression would have been much simpler. Now:

$$K_1 = EXINP \cdot IN \vee Q10 \vee EXBIN \cdot \overline{Q3} \vee Q11 \vee EXSHF \cdot \downarrow/ADR$$

Flip-flop SS[2] must be set when we enter IF3 and EXBIX.

$$J_2 = IF1 \cdot CLEAR \vee IF2 \vee EXBIN \cdot v/ACC$$
$$= Q6 \vee IF2 \vee Q4$$

It must be reset when we enter EX and IF1 from IF3, and EX from EXBIX.

$$K_2 = IF3 \cdot RUN \cdot \vee EXBIX \cdot IN$$
$$= Q_2 \vee EXBIX \cdot IN$$

The third flip-flop is set when CPU enters state IF2.

$$J_3 = IF1 \cdot \overline{CLEAR} \cdot RUN = Q1$$

This flip-flop must be reset when CPU returns to IF1, but the don't care entries of Table 7.5 provide a variety of points to reset SS[3]. The simplest resetting equation is

$$K_3 = 1$$

Thus we have all flip-flop input equations in short order. The state diagram expressed via the conditioned transitions to each state and the state assignment Table 7.5 make the writing of FF input equations rather simple. Fig. 7.18 shows a great deal of the control circuitry we have just developed.

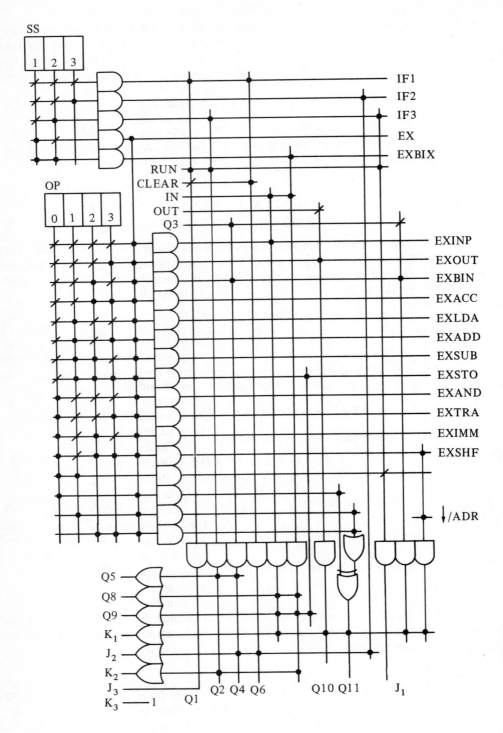

FIG. 7.18   Control Logic of EDC

We still have not completed the logic design of EDC, of course. We have not converted to the NAND or NOR technology that will actually be used. We have not emphasized fan-in, fan-out, and logic-level limitations. We have not expressed the input equations for each stage of every register. And we have not done many other things that must be done before we can begin to fabricate EDC. In particular, we have not checked, rechecked, and triple-checked every aspect of our design by looking for outright errors. We have not found ways in which we can improve EDC by extending or altering its organization, and refining its logic design.

7.4 PROGRAMMING EDC

Digital computers are designed, built, and sold to assist people in solving their problems. Their merit must be based upon their usefulness as computational tools and not upon the sophistication of their detailed logic design. As system architect and logic designer then, we must be familiar with how our products will be used.

We now have a computer named EDC. What must we do in order to get it to perform computation for us? First, we must have a problem appropriate to EDC. If the problem is too small, we will find that more effort is required to get EDC to assist us than to solve the problem with other tools. If the problem is too large we may find that the limited memory capacity of EDC prevents us from using it, or if many millions of computations are involved, that EDC can not perform them in an acceptable length of time.

Given a suitable problem, we must then find a step-by-step procedure for solving that problem. Not only must we have *an* algorithm, but that algorithm must be expressable in the instruction set of EDC and it must be possible for EDC to complete all steps of the algorithm in less than its mean time between failures. This first point is particularly important when establishing the instruction set for a digital computer. Instructions must be included that make it possible to perform the computations that many customers desire to perform. And if we wish our machine to be outstanding, we will provide instructions that make it very convenient to express algorithms. The instruction set of a digital computer is a very formal and rigid language: the grammar of that language must not be violated or the computer will not perform the desired computation. To be useful, that language must be as rich in vocabulary and flexible in grammar as possible.

The second point recognizes that machines do fail. Algorithms can be proposed that solve problems but require extremely large amounts of computation. If the machine fails before it completes the computation, and if it will always do so because of the extremely large amount of computation required, we do not really have an algorithm for solving the problem. For example, EDC takes four microseconds to fetch and execute most instructions. Suppose we present an algorithm that requires the execution of one billion instructions. Our machine must then function properly for over 66 minutes. A vacuum tube machine of a few years ago might well not be capable of doing this. Solid state machines today will usually perform for longer periods. But they too begin to fail when we require continuous operation without failure for months or years, as in space probe missions.

Expressing an algorithm in the instruction set of a machine is known as *coding* or *programming*. The list of instructions that expresses an algorithm is known as a *program*. Let us now attempt to write a program for the following (almost trivial) problem.

Add N integers, where N and the integers will be supplied later.

If N is small we can do this task by hand. But if N is 50 to 100 or larger we may prefer machine assistance. And if we are going to have many lists of numbers to be added, it is worthwhile to invest the effort to prepare a program.

A number of algorithms may be proposed. We might place all of the numbers into memory before beginning to add. Or we might add each number as it is placed in memory to the sum of the the numbers previously read from the input equipment. This approach will be taken, for if $N > 1024$ we clearly can not read all numbers into memory. And because the input equipment will operate more slowly than CPU, we might just as well perform computations between reading in the integers. The CPU and input equipment will then be working simultaneously and less time will be required to complete the job.

The basic computational steps that we wish to perform may thus be summarized:

(1) Read another integer into memory.
(2) Add it to the sum of previous integers.
(3) Repeat.

We can also picture this with a state-diagram–like notation referred to as a *flowchart*.

Unfortunately we can not be this brief or casual when preparing an algorithm for execution on a digital computer. For example, we really want to repeat only N times, not indefinitely. Yet nothing of this sort was mentioned above or shown in Fig. 7.19. How can we have EDC repeat only a certain number of times? We might have it count the number of additions it performs, and stop only when the count is equal to N. This is how we would accomplish the same thing if we were doing the task by hand, although often counting is such a natural sideline computation that we do not make a point of it or may not even be aware that we do it.

We do not wish to perform the counting manually, but to have EDC do it as a part of the algorithm for solving our problem. Thus we must add two blocks to our flowchart. One block calls for counting. The second shows that the count must be compared with N and a decision made. If the count is less than N, repetition is in order. When the count equals N we can

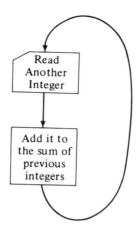

Fig. 7.19 Flowchart
of the Basic Algorithm

have EDC print the resulting sum, for the job is completed. This last step is also shown as a block in the flowchart of Fig. 7.20.

But wait! We still do not have a complete algorithm. No detail must be overlooked when programming a digital computer. How does EDC know the value of N, the initial value of the sum to which the first integer is to be added, and the initial value of the count? Our program must make explicit provisions for establishing these quantities, and we thus add still more blocks to the flowchart of our algorithm. Further, the flowchart of Fig. 7.21 shows the algorithm returning to its starting point after printing the sum. This step is simply looking forward to the time when we will have many lists of integers to be added. It would be very annoying to reload the

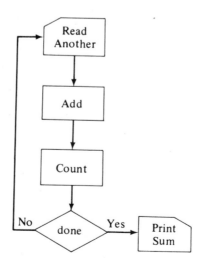

Fig. 7.20 A More Complete Flowchart

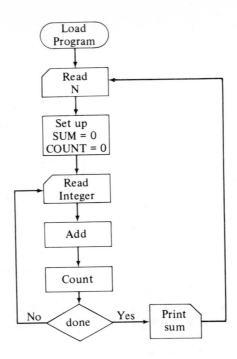

FIG. 7.21 The Complete Flowchart

entire program with each data set, and it is not necessary to do so if we prepare the program so that it will return to the beginning of its own accord.

Now we must express our algorithm using only the words that appear in the instruction set of EDC, Table 7.1. The first block in our flowchart calls for loading the program into the memory of EDC. This we accomplish with a BIN instruction. To write a BIN instruction we must know its OP and IX bits, and the 10-bit expression of the number of words of our program to be loaded. The first two items are provided by the instruction set—001000. But we cannot provide the address field of the instruction for we do not yet know how many instructions will make up our program. We must come back later and fill in this information. We know though that we want a BIN instruction to be placed in memory location 0_{10} or 0000000000_2.

Memory location	Word	Remarks
0	001000??????????	BIN to load program
1	0000001111111111	INP N

The second block of our flowchart calls for reading N from the input media to the memory of EDC. An INP instruction can do this job. Again we must supply an address field that gives the memory location in which we wish N to be placed. We can choose almost any of the 1024 available memory locations. We do not wish the word expressing N to be interpreted as an instruction, so let us place N in location 1111111111_2, for we do not expect our program to occupy all of memory. This location should be sufficiently out of the way. So we must write the instruction

$$0000001111111111_2$$

It will be stored in memory location 1_{10} by the BIN instruction in memory location 0.

Now we must initialize in memory two data words we will call SUM and COUNT. Both are to be given initial values of zero. Let us establish first that the word in location 1111111110 will be referred to as SUM and the word in 1111111101 as COUNT. We could use the INP instruction to place zeros into these two locations, but this would require that we have two data words consisting entirely of zero's in each and every data set. This can be a nuisance and an opportunity to get meaningless results should we forget these two words. A better approach consists of placing zero in the ACC with the ZRO instruction and then using STO instructions to give SUM and COUNT their desired initial values.

Memory location	Word	Remarks	
2	0011 00 ??????????	ZRO	Clear ACC
3	0111 00 1111111110	STO	Set SUM = 0.
4	0111 00 1111111101	STO	Set COUNT = 0.

Now we transfer the first of the integers to be added to memory with an INP instruction. This integer will be placed in location 1111111100 and referred to as NEW. Then we add SUM and NEW and return the sum to memory location 1111111110. Thus variable SUM is given a new value: its old value is destroyed, but we have no need for that value. The sequence of instructions to accomplish this is rather classic:

$$\begin{array}{ll} \text{LDA} & \text{SUM} \\ \text{ADD} & \text{NEW} \\ \text{STO} & \text{SUM} \end{array}$$

In the language that EDC recognizes, we express these statements as below.

Memory location	Word	Remarks	
5	0000 00 1111111100	INP	NEW
6	0100 00 1111111110	LDA	SUM
7	0101 00 1111111100	ADD	NEW
8	0111 00 1111111110	STO	SUM

SUM ← SUM + NEW

The next block of our flowchart calls for counting. This bookkeeping operation requires that we use the ACC to add 1 to variable COUNT. This is the reason it was necessary to store the new value of SUM; if we did not need ACC for purposes other than the accumulation of the sum of integers, we would not have to return SUM to storage.

Memory location	Word	Remarks	
9	0100 00 1111111101	LDA COUNT	COUNT ←
10	1010 01 0000000001	INC by 1	COUNT + 1
11	0111 00 1111111101	STO COUNT	

The new value of COUNT must be stored because we expect to use the ACC to add another integer to SUM.

Now we test to see if the correct number of integers has been accumulated. To compare N and COUNT we subtract, and since the value of COUNT is already in ACC (as well as storage) we form COUNT − N. If the difference is negative we must repeat; when the difference is zero, the correct number of integers has been accumulated. Thus the TRN instruction performs an appropriate test of the difference. Thus we may complete our program with the following words.

Memory location	Word	Remarks
12	0110 00 1111111111	SUB N forming COUNT − N
13	1001 10 0000000101	TRN to location 5
14	0001 00 1111111110	OUT SUM
15	1001 00 0000000001	TRA to location 1

Now we can go back and fill in the address field of the BIN instruction as 0000001111.

One nuisance we must have observed is that the language EDC accepts is not a desirable one from our point of view. We do not work well in binary! While 16-bit sequences are not very big in the digital computing world, they require excessive human effort and give excessive opportunity for making mistakes. Most people have found octal or hexadecimal representation to be much more satisfactory. Or even better, could we not use mnemonics such as LDA, ADD, etc. rather than giving all numbers? The answer is "yes," if we prepare *assembler* programs, which translate alphameric character sequences such as "LDA" to the correct binary sequence. In summary, the entire program is expressed in octal and mnemonic language in Table 7.6. Memory addresses are given in octal also. When we use six octal digits to express a 16-bit binary sequence, the leftmost octal digit can be either 0 or 1, but can not be a higher digit.

While expressing and reading a program is much easier using a mnemonic assembly language, we still must master the instruction set of EDC and pay a great deal of attention to detail. *Compiler* programs can be prepared for computers with more memory than EDC. These convert statements that satisfy the syntax of higher level programming languages such as FORTRAN or ALGOL into mnemonic statements, which in turn are assembled into machine language instructions. Higher level programming languages permit a programmer to express more easily the computation he wishes the computer to perform. Thus the following FORTRAN statements are more easily written and understood, and more descriptive of the computation we have been attempting to describe to EDC in this section.

```
1   READ, N
    SUM = 0
    DO 2 KOUNT = 1, N
    READ, XNEW
2   SUM = SUM + XNEW
    PRINT, SUM
    GO TO 1
    END
```

TABLE 7.6 PROGRAM TO ADD N INTEGERS

Memory Location	Word-Octal		Word-Mnemonic
0_8	02	0017	BIN 15 word program
1	00	1777	INP N
2	03	0000	ZRO
3	07	1776	STO SUM
4	07	1775	STO COUNT
5	00	1774	INP NEW
6	04	1776	LDA SUM
7	05	1774	ADD NEW
10_8	07	1776	STO SUM
11	04	1775	LDA COUNT
12	12	2001	INC 1
13	07	1775	STO COUNT
14	06	1777	SUB N
15	11	4005	TRN 5
16	01	1776	OUT SUM
17	11	0001	TRA 1
$\vdots$			
1774	Variables		NEW
1775			COUNT
1776			SUM
1777			N

Such character sequences mean nothing to EDC. They can be interpreted only by a compiler program, or a person who has learned the syntax of the FORTRAN language. Before EDC can perform the desired computation these statements must be compiled into the instruction sequence of Table 7.6 or its equivalent.

After we have written a program, we must put it in a form that the input equipment assumed to be associated with EDC can recognize. This may mean that we have to keypunch each word in a punched card, or express each word as a sequence of holes in paper tape. Ample opportunity for making errors exists in either case, and such mistakes will generally cause our program to fail. EDC will do something in response to an erroneous program, so we must check the results it gives to make certain that they are correct. Then we can approach EDC, submit our program and data set to the input equipment, press RESET, then START, and wait for answers. It is very rare to obtain the correct answers on the first attempt, for everyone makes mistakes in writing programs and placing the program in form that can be read by the machine.

We now have some appreciation of how EDC might be used to solve one rather straightforward problem. We must be very familiar with the instruction set of EDC and to some extent how it operates before we can prepare a program. A fair amount of detail must be specified when we speak to EDC in its language. In fact, programming a digital computer is not unlike designing one. We must be equally concerned with parallel details in both cases. Signals in binary circuitry are similar to variables such as SUM or COUNT. We use flip-flops to retain the value of signals, and memory locations to store numerical variables. Registers are paralleled by sets of memory locations storing numerical matrices, and the notation we use to refer to a register cell parallels matrix notation.

Now that we have some insight into how EDC will be employed, we might ask how we could alter EDC so that it would be easier to use. This is really unfair for we should write many tens or hundreds of programs concerned with a variety of algorithms before asking this question. But from this one program we can see one programming nuisance that will appear in almost all programs. Bookkeeping is almost always necessary. It would be very desirable to be able to perform counting and testing as a side operation not requiring the accumulator.

REFERENCES

See Chapter 8 for References.

PROBLEMS

7.1-1. An accumulator is to be a part of system SY1, in which numbers are represented with 16 bits in the 1's complement convention. R is to be a public register; ACC and a ripple adder are private facilities of synchronous automaton ARITH. Terminals LOAD, ADD, SUB, STO, and AVA are public terminals.

ARITH is to place the contents of R in the ACC register in response to a 1 on the LOAD terminal. The sum of the contents of the ACC and R registers is to be placed in ACC in response to ADD = 1. SUB = 1 is to cause the difference of the contents of ACC and R to be placed in the ACC. (Two clock periods are required to perfom addition and subtraction.) STO = 1 is to cause the contents of ACC to be transformed to the R register. When ARITH is performing one of these tasks, it is to hold AVA = 0 and ignore further commands. When ARITH is available to perform a task, it is to so indicate by causing AVA = 1.

(a) Use the design language to specify as much of system SY1 as possible from these specifications.

(b) Draw the logic diagrams of typical cells of the R and ACC registers. Show how the cells are interconnected.

(c) Carefully estimate the number of flip-flops and gates that will be required to implement the portion of SY1 documented.

7.1-2. System SY2 is to satisfy specifications similar to those for SY1 of Prob. 7.1-1, but is to operate in the serial mode so that the adder network can be replaced with a single full adder-subtractor and a carry flip-flop.

(a) Draw a block diagram of serial system SY2. Document your SY2. Attempt to reduce network cost whenever possible. Do not overlook the timing control problem and generation of signal AVA.

(b) How long will your system take to perform the LOAD, ADD, SUB, and STO tasks? For each of these tasks suggest how time could be exchanged for hardware.

(c) Compute the hardware cost (gate and flip-flop counts) of SY2 as documented in part (a). Compare with the result of Prob. 7.1-1(c).

7.1-3. Assume EDC exists as described in Fig. 7.3 and RUN =1, CLEAR =0. Show the successive states of EDC and the contents of all registers at 5 successive clock times following each of the initial conditions given below. All numbers and addresses are given in octal below. The initial contents of selected memory locations follow: indicate when the content of any memory location is altered.

Mem. Loc.	Contents		Initial State	CAR	MAR	IR	ACC
⋮							
0002	040020	(a)	IF1	0002	1777	001777	177776
0003	110003	(b)	IF3	0017	0016	122005	000073
⋮		(c)	IF3	0017	0016	110003	177776
0017	123123	(d)	IF1	0017	0017	116003	177776
0020	124007	(e)	IF3	0021	0020	124007	126261
⋮							

7.1-4. How many microseconds does EDC require to fetch and execute each of its instruction types? If the execution time of a particular instruction type is not fixed, give a formula so that it may be calculated easily.

7.1-5. Alter the description of EDC as given in Fig. 7.3 to correspond to a 24-bit word length. Discuss how the instruction format might be altered if the word length were increased to 24 bits, and the ramifications of such alterations.

7.1-6. EDC is to recognize and execute the following instruction types.

OP		IX	
LDA	0100	00	Load ACC with the word **MLSAFI** (Table 7.1)
LCM	0100	01	Load ACC with the complement of the word **MLSAFI**
LIN	0100	10	Load ACC from the **INPUT** register
LAD	0100	11	Load ACC from the **CAR** register

(a) What alterations must be made to Fig. Fig. 7.3?
(b) What alterations must be made to Fig. Fig. 7.2?
(c) Write a description of these new instruction types for potential users of EDC.

7.1-7. A $1 + 1$ *address instruction format* includes two address fields: the first gives the address of an operand while the second gives the address of the next instruction to be fetched and executed.
Assume a 24-bit word length and 10-bit address fields.
(a) Write an instruction set for a $1 + 1$ address machine that includes input, output, arithmetic, logic and shift, and transfer type instructions. (For purposes of comparison it is desirable to produce a set like that of Table 7.1, but it may not be possible or desirable to precisely match that table.) Briefly describe each instruction type.
(b) Draw a block diagram and specify with the design language a $1 + 1$ address machine that will execute this instruction set.
(c) Discuss the similarities and differences between your $1 + 1$ address machine and EDC. (Number of registers, their interconnection, speed of execution, etc.)

7.1-8. The two address fields of a *2-address instruction* format specify the memory locations of two operands. Thus a single instruction may say " add the numbers in location X and Y together."
The result must be routed to an assumed storage unit, which may be either a register such as the ACC of EDC, or a memory location such as X, the location of the first operand. Assume that this latter scheme is to be used with a 24-bit word length and 10-bit address fields.
(a) For a 2-address machine write an instruction set that parallels that of EDC to the extent of including an input, output, arithmetic, logic, shift, and transfer instruction type.
(b) Draw a block diagram and specify with the design language a 2-address machine that will execute this instruction set.
(c) Discuss the similarities and differences between your 2-address machine and EDC (and the $1 + 1$-address design of Prob. 7.1-7).

7.1-9. A computer is to have an *n*-bit word length. When selecting the value of *n*, the following items, as well as others, must be given consideration.

(1) The instruction format (see previous problems), instruction set size, memory size, modifier fields (IX bits of EDC).

(2) Numerical accuracy. If the n-bit word records fractions with an error less than 2^{-n}, the sum and product of such fractions may be in error by 2^{-n+1}. Repeated arithmetic performed on such results leads to even greater error.

(3) Cost. In a parallel machine, hardware costs increase directly with n. As a simple approximation, assume that costs increase linearly with n and are given by $\$1000 \times n$.

(a) For $n = 18$, 36, and 64 in turn:

(i) Propose an instruction format such that an instruction set of 50 to 100 instruction types may be implemented and a main memory of up to 65,536 words may be utilized.

(ii) Determine the accuracy of the sum of 100 fractions (assume overflow does not occur).

(b) Find a value of n and an instruction format that satisfy the following requirements to as great an extent as possible.

(i) Minimum cost.

(ii) At least 32 instruction types

(iii) At least 8192 words of memory

(iv) The sum of 10 fractions is to be accurate to at least 5 decimal digits.

7.2-1. Design the IR of EDC. Show typical cells, unusual cells, control circuitry, etc.

7.2-2. Problem 7.1-6 suggested an extension of EDC. Identify all modifications to the registers that must be made to include this extension. Determine the incremental cost (numbers of gates) of this extension.

7.2-3. All flip-flops of EDC are to be of a master-slave variety. The clock signal is not of $1\mu s$ duration as specified, but is to be at logic 1 for 50 nanoseconds and at logic 0 for x nanoseconds. If each gate and flip-flop introduces 50 nanoseconds of delay and EX–OR gates are constructed with 3 logic levels, what is the smallest value x may assume so that EDC will operate reliably? Give justification for your answer.

7.2-4. Design a minimum gate P network for an ALU that provides the following four functions and requires only two control signals: $A_i \vee B_i$, A_i, $A_i \vee \bar{B}_i$, and 1. Using this P network, and the G network of the text, what arithmetic and logic operations can be performed?

7.2-5. The ALU cell below has mode control signals so that any of the minterms of A_i and B_i may be combined, and the carry lines may be interrupted. If cells of this design are ripple-interconnected, what arithmetic and logic functions are available?

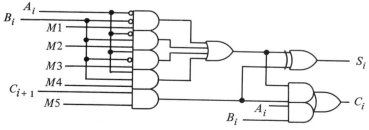

7.2-6. If the ALU of Fig. 7.11 can form $0 - A$, can it also form $B - A$? If so, what mode control signal values must be given? If not, why not?

7.2-7. Can the ALU of Fig. 7.11 form $A \oplus B$? If so, what mode control signal values must be given? If not, how would you change the cell design to provide this function (and possibly many more)?

7.3-1. Implementation of the 16-state machine described by Fig. 7.6 is to be investigated.

(a) Assuming a 4 flip-flop machine is to be designed, make a state assignment; justify your assignment.

(b) Design a decoding network based upon your assignment which will generate signals IF1, IF2, ..., EXBIN, ... EXSHF. Determine the gate cost of your decoder, and compare with the cost of the gates of Fig.7.18 which generate the same set of signals.

(c) Design the circuitry for at least one of the 4 flip-flops, again using your state assignment.

7.3-2. Alter Fig. 7.16 to eliminate state EXBIX. Draw a state diagram for the resulting 4-state control unit.

7.3-3. Design a 5-state sequencer control for EDC from Figs. 7.16 and 7.17.

7.3-4. Show how a 3-to-8 or a 2-to-4 line MSI decoder could be used to provide signals IF1 through EXBIX in Fig. 7.18.

7.4-1. The program of Table 7.6 can be shortened. Remove at least one word from the 20_8 word program. Be sure that corrections are made to the remaining instructions so that the program will still perform properly.

7.4-2. The sum of N numbers is to be calculated by reading all numbers into memory and then adding them. Flow chart and program this algorithm. Recognize and record any limitations of your program.

7.4-3. (a) Two positive numbers with magnitude $\leq 2^7$ have been placed in memory locations 1776 and 1777. Prepare a program that will form the product of these two numbers and place it in location 1775.

(b) Two signed numbers (2's complement convention) with magnitude $\leq 2^7$ reside in locations 1776 and 1777. Prepare a program which will form the product and place it in location 1775.

(c) What instruction or instructions would have made the above programs simpler and more efficient if they existed in the instruction set of EDC? Estimate how much additional hardware each instruction you propose would introduce.

(d) Your program of part (b) could be used as part of a larger program if it would work properly when stored in successive memory locations beginning with location 100_8. Change your program so that it will perform properly. (This process is known as *relocating* a program.)

7.4-4. A program to evaluate a polynominal such as $y = 7x^3 + 2x^2 - 4x + 3$ for a specific value of x must perform many multiplications. Rather than insert the program of Prob. 7.4-3 for each multiplication, it is desirable to place only one copy of such a *subprogram* in memory and prepare the main program and subprogram so that the main program may transfer to the subprogram many times and the subprogram will in turn transfer back to the appropriate point in the main program each time.

(a) What information must the main program transmit to the subprogram so that it can accomplish its task and return control to the proper point in the main program?

(b) Show the portion(s) of a main program that will transmit the necessary information to a "multiply" subroutine, and the portions of the subroutine that will effect a return of control to the main program. We expect to use many subroutines and these portions of programs should be amenable to standardization and use in all cases.

7.4-5. $N \leq 1000_8$ integers have been placed in memory locations 1000_8 through $(777 + N)_8$. The value of N has been placed in location 777_8. Write an EDC program that will sort these integers, placing the smallest in location 1000_8 and the largest in $(777 + N)_8$. Use a mnemonic language to describe your program.

8

Advanced Computer Concepts

In many respects EDC of the previous chapter falls on the ill-defined boundary between microcomputers and minicomputers. Few microcomputers have a word length as great as 16 bits. Minicomputers have word lengths of 16 bits and greater. Neither "minis" nor "micros" have instruction sets as small as EDC, but that of EDC was deliberately kept minuscule. The CPUs of minicomputers operate four or more times faster than their main memories; microcomputers are CPU-bound, as is EDC (again, the memory was artificial). Microcomputers execute instructions in few microseconds, as does EDC. Minicomputers complete two or more instructions per microsecond. While microprocessors in dedicated applications may utilize only 1024 words of main memory, minicomputers usually have 8 K to 64 K words (1 K = 1024 words) of main memory and secondary memory in the form of disks or tapes. Many micros are able to address up to 64 K of main memory. Microcomputers have relatively few peripheral devices, while the bulk of the many thousand dollars cost of a mini system is toward extensive input, output, and storage devices. Microcomputers cost a few hundred dollars and often occupy one printed circuit board.

Clearly the distinction between minicomputers and microcomputers is vague. It is also subject to change for both types of computers are continuously becoming more sophisticated and less expensive. If we compare the architecture and organization of both computers with those of EDC, we must conclude that EDC is neither, for many features found in both are absent in EDC. This chapter is concerned with those missing features. Emphasis is placed on architecture and organization rather than logic design. The concepts apply to both microcomputers and minicomputers.

8.1 ADVANCED REGISTER APPLICATIONS

Before we discuss register configurations that enhance data processing, let us consider a more fundamental problem—the instruction format. With a 16-bit word length and 10-bit address field, we cannot address a memory of significant size. And four op-code bits do not permit an instruction set of adequate size for a modern machine. Both fields must be expanded. Increasing the word length is an obvious solution; however, logic circuitry is not yet free and going to 24- or 32-bit word lengths will place the machine in a higher price category. To illustrate solutions that have been found for short word length machines, we will continue to use a 16-bit word length.

To address a memory of up to 65 K words requires an address of up to 16-bits—a full word. While several ways of obtaining 16-bit addresses will be shown later in this section, we can have a 16-bit address field only by using two words to express those instructions that require a true memory address. While techniques for reducing the number of such instructions will be shown, some instructions always remain that require an address field. The full first word of the pair of words expressing such instructions can be used to express the op-code. We will continue to call the left four bits the "op-code" and use the remaining 12 bits (4 octal digits) in a variety of ways. We will illustrate concepts without dwelling on deriving a consistent, complete instruction set.

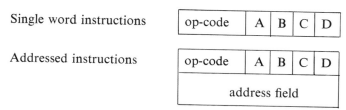

Using double-word instructions impacts the register configuration and the instruction fetch phase. A full 16-bit address register must be introduced (or the address cleverly placed in some other register). The CAR and MAR must be expanded to 16-bits. EDC of Fig. 7.16 can proceed to state EX from IF3 only if the op-code describes a single word instruction. Addressed instructions require a second reference to memory to obtain the second word. CAR provides the address of this word. After the address word is obtained from memory, the CAR must be incremented again so that it points to the following instruction. If *ai* summarizes the op-codes of double word instructions, the following declarations sketch the necessary instruction fetch phase.

$$\langle RE \rangle \ IR[16] = OP[0:3] \circ A[3] \circ B[3] \circ C[3] \circ D[3],$$
$$ADR[16], CAR[16], MAR[16].$$
$$IF1 : MAR \leftarrow CAR, \text{\textcent} CAR, \rightarrow IF2.$$

IF2 : IR ← M[MAR], →IF3.

IF3 : [OP[*ai* MAR ← CAR, ¢CAR, →IF4; →EX. .

IF4 : ADR ← M[MAR], →IF5.

IF5: MAR ← ADR, →EX.

State IF5 could be avoided by placing the address directly in the MAR in state IF4, but we will soon be speaking of processing this address before placing it in the MAR.

One way to drastically reduce the number of instructions that require addresses consists of introducing a temporary register, T, and including only "load T" and "store T" instructions in the set. Single word instructions then load ACC from T, form ACC + T, and so forth. Two instructions must be executed to transfer a word from memory to the accumulator. We will not elaborate on this scheme but rather seek other ways of reducing the use of double-word instructions.

The Register File

One way to reduce the constant loading and storing of the accumulator shown in the program of Section 7.4 is to have more than one accumulator. We could add numbers in one accumulator and count in another. Some memory references are required to obtain and store data, but the number is greatly reduced. Section 6.6 describes a *file* as a collection of registers with attached address decoder and multiplexers so that a word may be written to or read from any register. Actually, a variety of file designs are available for our selection. It is not necessary to construct the storage registers of master-slave flip-flops. Latch registers may be used for storage if an output register of latches is controlled to act as the slave. This economical design permits a variety of file organizations. If one address is supplied, the output register must be connected to the selected storage register. One memory location is simultaneously written and read with each clock pulse. With two addresses and their decoders, the slave need not be connected to the storage cell being written. Figure 8.1 shows the logic of a file that permits reading data from one location, processing it, and writing the result in a different location with each clock cycle. The write address decoder in Fig. 8.1 distributes clock pulses to addressed registers only when the write-enable permits this activity. Reading, without writing, is also possible. Further, if two read addresses, decoders, and output registers are provided, then a file may simultaneously read from two locations.

As a first example of a file oriented CPU, we simply replace the ACC of EDC with an 8-word, 2-address file, as shown in Fig. 8.2. Accumulator-type instructions may specify one file address or two. Two addresses permit different source and destination registers for ADV, DBL, etc., instructions. A file-register to file-register transfer instruction is certainly possible. LDA, ADD, ..., MASK instructions require an address word. Any register may be loaded, stored, or participate in an arithmetic operation,

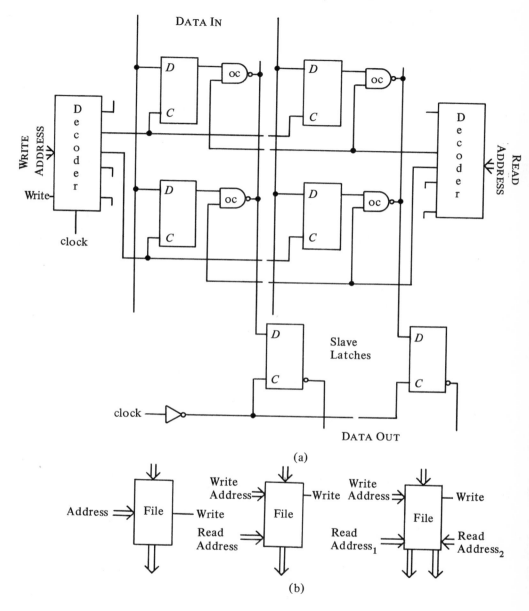

DATA IN

DATA OUT

Slave
Latches

(a)

(b)

FIG. 8.1   File Logic and Variations

with the result being returned to any file-register, if both source and destination file addresses are included in the instruction. Register-register arithmetic is not possible with the organization of Fig. 8.2.

While the transfer instructions did not involve the ACC, they may be made single-word instructions that involve the file in several ways. First, one of the registers of the file may be reserved for use as the CAR. Then, a data path from this register, or from the file in general, to the MAR must

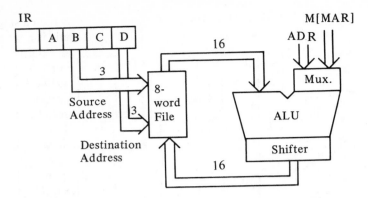

FIG. 8.2 Simply Replacing the Accumulator with a File
Provides Limited Improvement

be provided. Transferring then clearly impacts the file. Second, a file-address may be included in a single word transfer instruction. The addressed register then contains the address to be stored in the CAR, if the transfer is performed. Immediate instructions may be double word or, if a single file address is specified, the 8 or more remaining bits of the instruction may be extended to form an operand as in EDC. While the file or-

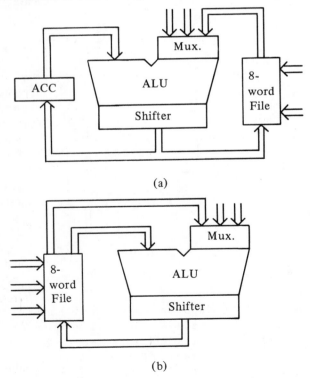

(a)

(b)

FIG. 8.3 Organizations That Permit Register-Register Processing

ganization of Fig. 8.2 is certainly more desirable than a single ACC design, its inability to perform register-register arithmetic is a severe limitation.

Register-register processing requires that an external register provide one operand or the file supply both operands to the ALU. Figure 8.3 shows two organizations. The ACC of Fig. 8.3(a) may be loaded from the file first, or it may already hold an operand from a previous calculation. Then the two operands may be combined, with the result returned to the ACC and possibly the file. Two instructions may be required to add two numbers. Using a double-output register file removes this necessity. Any register can provide the left and any other the right operand. The number of file-addresses required may be reduced by making the destination address the same as one or the other of the operand addresses.

Both of the organizations of Fig. 8.3 permit register-register arithmetic, as well as the other types of instructions previously discussed. With two file addresses and a memory address supplied in a double-word instruction, register-memory arithmetic may be performed, with the result stored in an addressed file-register. If a data path between the file and the MAR exists, then the CAR may reside in the file, and a file-register may be taken to contain either an operand or the address of an operand. Single-word register-memory, and even memory-memory, arithmetic instructions become possible, as do a variety of other applications for the file-registers. As a result of these varied applications, file-registers are usually called "general purpose" registers rather than accumulators.

Indexed Addressing

In computers with clock periods less than memory cycle times, register operations are executed more rapidly than instructions that reference memory. While we may attempt to minimize the number of memory reference instructions that are used in a program to increase the rate of processing data, memory references cannot be avoided entirely because the bulk of the instructions and data resides in memory. Several architectural features can further reduce memory references and/or permit referencing a large memory with small or modest address fields. *Indexed* addressing is one such feature. Suppose we have a memory reference instruction with a full or abbreviated address field, and a register IX that can address the full memory. In *indexed addressing*, the address referenced is specified by the sum of the address field and the index register content.

$$\text{effective address} = \text{instruction address} + \text{index}$$

A memory reference must thus be preceded by the register transfer

$$\text{MAR} \leftarrow \text{ADR} + \text{IX}$$

An adder is required to compute the effective address.

In searching a table of data, one instruction may be used repeatedly to reference successive table entries without changing its address, if the contents of the index register can be changed with a register operation. Often we go even further and couple the incrementing or decrementing of an index register with a transfer instruction that tests that register. Loops in programs are often controlled by placing a count in an index register and by test-decrementing that register as the last instruction of the loop. As long as the count exceeds zero, the program loops. When zero is reached, looping terminates. The index of the loop (contents of the index register) is also data in many programs. (Examine the DO loops in several of your FORTRAN programs, and observe that you use the DO index as data.)

In very high-speed systems, index registers are discrete or form a separate file. A separate arithmetic unit is dedicated to address formation. Economy is gained at the expense of time by using some or any of the general registers as index registers. This very flexible possibility is illustrated by the organization of Fig. 8.4. Instruction fields S and D address source and destination locations for operands in a variety of ways. Addressing mode fields MS and MD indicate the manner in which S and D fields specify locations. The modes to be examined now are:

Mode Field	Mode
0	Register
1	Register Indirect
2	Direct (address word required)
3	Indexed (address word required)

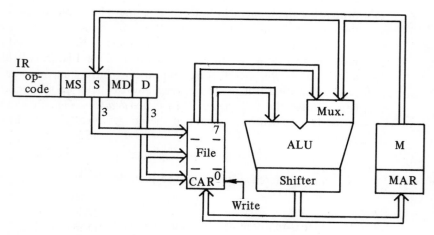

FIG. 8.4   Organization That Supports at Least Three
Modes of Addressing Memory

Table 8.1. ADD Instructions with Two Modifier Fields

MS	MD	Effect of Execution	Memory Cycles Required
0	0	$F[D] \leftarrow F[S] + F[D]$	1
0	1	$M[F[D]] \leftarrow F[S] + M[F[D]]$	3
0	2	$M[M[X^1+1]] \leftarrow F[S] + M[M[X+1]]$	4
0	3	$M[M[X+1] + F[D]] \leftarrow F[S] + M[M[X+1]+F[D]]$	4
1	0	$F[D] \leftarrow M[F[S]] + F[D]$	2
2	0	$F[D] \leftarrow M[X+1] + F[D]$	2
3	0	$F[D] \leftarrow M[M[X+1] + F[S]] + F[D]$	3
3	3	$M[M[X+2] + F[D]] \leftarrow M[M[X+1] + F[S]]$ $+ M[M[X+2] + F[D]]$	6

[1] X is the address of the ADD instruction.

Register addressing uses the file-registers as sources of operands and destinations of results. *Register indirect addressing* treats the file-register contents as memory addresses for operands and results. Address word(s) must be appended to the instruction for *direct addressing*, which treats these words as addresses of operands and results. Finally, *indexed addressing* uses addressed general-register contents as indices on address words that are appended to the instruction. Four addressing modes in each of two file addresses permit 16 variations on each instruction of this format. Selected examples of these variations are shown for the ADD instruction in Table 8.1. The file of Fig. 8.4 is named F. F[D] is the file register specified by the *D* field of the instruction held by the IR. F[S] is the specified source register.

Fetching and executing an ADD instruction is obviously complex. Assuming MD = 0 will simplify yet reveal how the system of Fig. 8.4 must act to fetch and execute 2-address instructions.

$$IF1 : MAR \leftarrow F[0]. \ (F[0] \text{ is CAR})$$
$$IF2 : IR \leftarrow M[MAR], F[0] \leftarrow F[0] + 1.$$
$$IF3 : \lceil OP \rfloor \ add \ \lceil MS \ \lfloor 0 \ F[D] \leftarrow F[S] + F[D], \ \rightarrow IF1$$
$$\lfloor 1 \ MAR \leftarrow F[S], \ \rightarrow EX$$
$$\lfloor 2 \ MAR \leftarrow F[0], \ \rightarrow IF5$$
$$\lfloor 3 \ MAR \leftarrow F[0], \ \rightarrow IF4...$$
$$IF4 : MAR \leftarrow F[S] + M[MAR], \ \rightarrow IF5.$$
$$IF5 : F[0] \leftarrow F[0] + 1, \ \rightarrow EX.$$
$$EX : F[D] \leftarrow F[D] + M[MAR], \ \rightarrow IF1.$$

The first steps of instruction fetch are slightly new to us. The organization of Fig. 8.4 does not permit loading the MAR and incrementing the CAR at the same time. All of OP, MS, and MD may be decoded in state IF3. With MS = 0, addition is performed immediately since the operands reside in registers. MS = 1 requires that the contents of file-register F[S] be passed to memory as the address of data. One additional memory cycle is required. With MS = 2, an address word must be obtained from memory before addition can be performed. The CAR must be incremented in state IF5. Finally, MS = 3 specifies indexing. The address word is obtained and added to the index register F[S] in state IF4; the CAR is incremented in IF5 in preparation for the next instruction fetch cycle.

Indirect Addressing

In the *indirect* mode of addressing, the address field of an instruction provides the address of the address of the data, rather than the address of the data. While indirect addressing requires an extra memory cycle to get the address of the data, it can simplify some types of programs, and it may be used to generate a full address from an abbreviated address field. Calling a subroutine illustrates the use of indirect addressing. A call with arguments may be translated into (1) an unconditional transfer to the first instruction of the subroutine, followed by (2) a list of addresses of the call arguments.

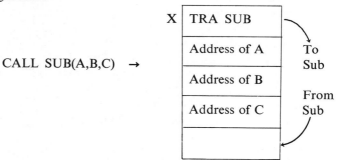

If the subroutine knows the address X of the calling transfer statement, then the arguments may be read or replaced with instructions that indirectly address locations X + 1, X + 2, and so forth. These very same instructions reference different arguments when the subroutine is called from a different point in memory (with a different argument list). It is not necessary for the subroutine to fetch argument values or fetch addresses, nor to distribute them to all of its instructions that process arguments before beginning to perform the computations the subroutine is intended to perform.

If short address fields are a part of instructions, then indirect addressing is a means to full memory addressing. Full addresses must be stored in the low address memory locations, which are addressable by the abbreviated address fields. The short address field directs the CPU to a low numbered memory location, where the full address of data resides.

When one bit of the instruction is used to indicate indirect-versus-direct addressing, that bit may be examined in the addressed word. If it indicates indirect addressing, another address is searched in memory.

Multiple level indirection is available on some machines, but its availability is not as important as that of single level indirection or indirection in combination with indexing. Either indexing the initial address, preindexing, or indexing the final address of the indirection search, postindexing, are possible modes of compound addressing. Usually only one is found in a computer.

Relative Addressing

In short word length machines, with instruction formats that permit only abbreviated address fields, full addresses may be found in ways other than indirect addressing and index registers. *Relative* addressing forms a full address from an abbreviated address and the contents of the CAR. First, the address field may be added to the CAR to form a full length address that follows the location of the instruction in memory. With an 8-bit address field any of the 64 locations that follow an instruction may be addressed by that instruction. If the address field is subtracted from the CAR, then addresses that precede the instruction in memory can be referenced. Many loops in programs consist of less than 64 instructions, and as a result relative addressing is often used in transfer instructions. If the left bit of the address field is extended before adding to the CAR, then locations on both sides of an instruction can be referenced by that instruction.

Second, a full address may be formed by concatenating the more significant bits of the CAR and the address field of instructions. An address adder is not required. Any location in the current "page" of memory may be addressed. With an 8-bit address field, a page consists of 64 memory locations. If an entire program and its data fits into 64 words, then no programming difficulties are encountered, but such programs are too small to warrant consideration in selecting the architecture of a computer. With large programs page boundaries must be crossed. Other addressing schemes must accompany this type of relative addressing.

A third approach to relative addressing utilizes a program alterable register, which supplies the necessary information to form an execution address by adding or concatenating the address field. Instructions that allow a program to alter the contents of the *address extension register* must be included in the instruction set.

Indirect and relative addressing will be illustrated with the original EDC instruction format and the CPU organization of Fig. 8.5. While the ACC, address extension register B, and CAR are shown as a small register-file, they may also be three separate registers with a suitable multiplexer at the input port of the ALU. The two IX bits permit four addressing modes. We assume that the modes of Table 8.2 are available for at least the ADD instruction.

Fetching and executing an ADD instruction is rather straightforward. The ALU is again used to increment the CAR. It is also used to form the

TABLE 8.2. ADDRESSING MODES OF THE
MACHINE OF FIG. 8.5

IX	Addressing Mode	Effect of Executing an ADD Instruction
00	Immediate	$ACC \leftarrow ACC + XADR$
01	Direct	$ACC \leftarrow ACC + M[ADR]$
10	Relative	$ACC \leftarrow ACC + M[B + ADR]$
11	Indirect	$ACC \leftarrow ACC + M[M[ADR]]$

effective address in the relative addressing mode. Indirect addressing requires an additional memory cycle.

$\langle ID \rangle$ XADR = ADR[1] × 6∘ADR, ZADR = 0D6∘ADR.

IF1 : MAR ← CAR, →IF2.

IF2 : IR ← M[MAR], CAR ← CAR + 1, →IF3.

IF3 : ⌈OP⌊ *add* ⌈IX ⌊0 ACC ← ACC + XADR, →IF1

⌊1 MAR ← ZADR, →EX

⌊2 MAR ← ZADR + B, →EX

⌊3 MAR ← ZADR, →IF4...

IF4 : MAR ← M[MAR], → EX.

EX : ACC ← ACC + M[MAR], →IF1.

Multiple level indirection may be provided with a slight alteration of this instruction fetch sequence. If state IF4 is changed to

IF4 : IX∘ADR ← M[MAR, 5 : 16], →IF3.

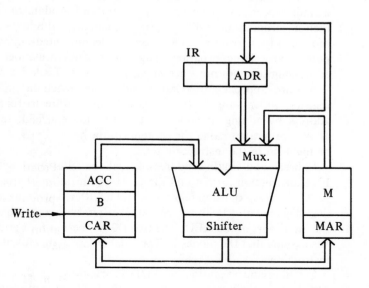

FIG. 8.5 An Organization to Illustrate Relative and Indirect Addressing

the word retrieved from memory is not taken as a full address, rather as an expression of new IX and 10-bit ADR fields for the instruction. These new fields are examined in state IF3. If the IX bits again call for indirection, another word is pulled from memory, placed in part in the IR, and examined. Ultimately (we hope), the IX field will specify one of the other three addressing modes. Thus indirection is followed by immediate, direct, or relative addressing with this instruction fetch algorithm. Indirect addressing is no longer a means of forming a full address from an abbreviated address.

Restricting the formation of full addresses is not undesirable in larger computing systems. The direct addressing of Table 8.2 permits references to page zero of memory. We may wish to eliminate this addressing mode or restrict its use also.

With these restrictions, B contains the lowest or *base* address of the 1024 memory locations that can be referenced without changing the contents of B. Thus, B is often called the *base register*. A " load B " instruction must be provided so that the base address can be changed. Then, when a portion of a program has been completed, the contents of B may be advanced so that other portions of memory can be accessed. Alternatively, if users are prevented from freely using the load B instruction, they can be prevented from disturbing the contents of other than their page of memory.

Relocation

To this point we have loaded programs by means of a BIN instruction in memory location 0. The first actual instruction of the program was placed in memory location 1, and other instructions were placed in the immediately following memory locations. This mode of programming and operating a computer is rare today. Usually the *object code*—actual machine language binary instructions to be executed—is not prepared by the programmer. He expresses his *source code* in a mnemonic assembly language or in a compiler language such as FORTRAN. Then he does not need to be concerned with the actual address of each instruction. In fact, he may have little or no control over the actual addresses of the instructions of the object code that is generated from his source code by the assembler or compiler program.

A program that is to be executed only once is of limited value and may not justify the expense of its preparation. Programs that are to be used over and over can not always be placed in exactly the same memory locations. In a time sharing environment where programs of many computer users may appear simultaneously in memory and from time to time may be set aside in drum or disk memory, the location in the main memory of a computer of a specific user's program may vary from minute to minute.

Address fields of instructions of a program must be altered if the program's location in memory, or that of data, is changed.

Assembler and compiler programs usually do not prepare true object code, but prepare code in which *relocation* information is interspersed, based upon the assumption that the first word of the program will be placed in memory location 0. The relocation information indicates whether or not an address may be altered. A *loader* program then takes this relocatable object code and transforms it into true object code with addresses modified in accordance with the memory location in which it actually places the first instruction. After a program has been relocated and loaded, a transfer to the first instruction of the program may be performed, to initiate execution of the program. Thus a source program must be assembled or compiled, and then relocated and loaded before its execution commences. If it has previously been assembled or compiled, its execution requires only that it be relocated and loaded.

The expense of relocating a program via software may be avoided by introducing a base register—also now called a *relocation* register. Suppose that an assembler or compiler prepared a program as though the first instruction were to be placed in memory location 0, but then actually loaded that instruction into memory location 100_8. If the B register is also loaded with this address, then all pertinent address fields will be automatically relocated when $B + 0D6 \circ ADR$ is formed and used as the execution address. Immediate and shift instructions will not be relocated because the unaltered contents of ADR are used in their executions. Usually a programmer would not be allowed to alter the contents of the B register. A master control program will ensure that the contents of B are altered appropriately when it is necessary.

Thus the inclusion of a base register in a computer is very worthwhile even when it is not necessary to resort to relative addressing to be able to reach any memory location. Again, as throughout this section, we find that software may be exchanged for hardware.

Stacks and Pointers

Hardware stacks are as valuable to the computer architect as they are to the computer programmer (see Section 9.5). A FILO (first-in-last-out) or "push-down" stack is a memory unit that appears to have a single address. Words of information may be pushed onto the top of the stack and popped from the top of the stack. Words stored in other than the top location are not available until the words above them are removed. Addressing is thus not a stack property. Real stacks have finite dimensions, of course, and a word may not be popped from an empty stack or pushed onto a full stack.

A stack may be constructed of bidirectional shift registers or from a register file and a pointer register. The pointer register supplies an address to the file, and this address determines which register holds the top entry. Figure 8.6 shows the organization of a small register stack that can respond to two commands. If the stack is not full, a push command results in

$$\text{POINT} \leftarrow \text{POINT} + 1$$

and then

$$S[\text{POINT}] \leftarrow \text{DATAIN}$$

A pop command can be executed only when the stack is not empty. It consists of

$$\text{DATAOUT} = S[\text{POINT}]$$

followed by

$$\text{POINT} \leftarrow \text{POINT} - 1$$

The POINT register is nothing more than an up-down counter. A control unit to sequence the steps of pushing and popping, and commanding POINT and an empty-full flip-flop, are easily designed.

Main memory, or a portion of it, may be used as the storage medium of a stack. The pointer may be a dedicated register or a general register of the CPU. Push and pop instructions are carried out as indicated above with memory references replacing stack S references. An additional clock period may be required to load the MAR from the pointer. Some speed and economy may be gained by having the top entries in the stack reside in registers and the lower entries in main memory. Two pointers must be maintained. Transferring words from (to) the register to main memory portions of the stack when the register portion is filled (empty) requires control algorithms; these must be implemented as part of the control unit.

One application of stacks is suggested by the previous discussion of calling subroutines. It is necessary for the subroutine to know the location of the instruction that transferred control to it. A great deal is accomplished if that transfer instruction is altered to place the contents of the CAR on a stack before placing the address of the subroutine in the CAR. Then a subroutine may find its arguments and ultimately return control to the calling program by referring to the top stack entry. Subroutines may call subroutines, which may call subroutines, and so forth. Each operates according to the same straightforward rules for finding arguments and returning. The return transfer must be accompanied by a pop stack operation.

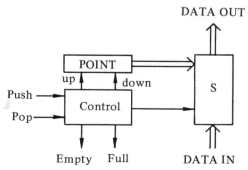

FIG. 8.6 The Organization of a Small Hardware Stack

A stack is very desirable if a computer is to respond to interrupt signals by stopping what it is doing, processing in response to the interrupt, and then resuming its original processing. When an interrupt is sensed, all of the CPU registers, including the CAR, are pushed onto the stack. Then they are reloaded to process the interrupt. When that processing is completed, the registers are reloaded from the stack and the original program continues. If the CPU is interrupted while processing an interrupt, the register contents are again placed on the stack. The second interrupt may then be processed, the first interrupt process reinstated and completed, and finally the original program is returned to active status by filling the CPU registers from the stack.

The organization and operation of several computers and calculators are centered on a data stack located largely in main memory. Very few instructions include an address field, and if the word length is short, these instructions may rely on relative and/or indirect addressing. "Push stack from memory" and "store from stack into memory" are examples of addressed instructions in such machines. Processing instructions like ADD and AND pop operands from the top of the stack (TOS) and push the result onto the stack. TOS is very much like an accumulator. Ordering data and instructions in Polish strings (with the algorithms seen in Section 9.5) is very important for efficient processing. For example, $A \times (B + C)$ is computed by following an add with a multiply instruction.

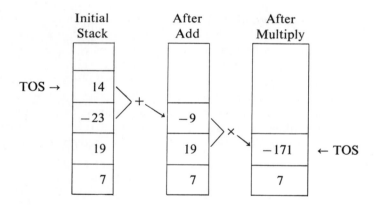

This section has examined (in isolation) a number of architectural and organizational concepts that are found in CPUs from calculators through to super computers. In the design of a specific machine, cost and speed restrictions must be satisfied by a compatible register configuration and instruction set. Some combination of these concepts will likely be included. If the cost of high speed logic circuits continues to drop, the emphasis will shift to the needs of the user, where it belongs. Other organizational and architectural concepts that facilitate the preparation of reliable programs are likely to emerge.

8.2 BUS ORGANIZATIONS

The designs discussed in previous sections are unrealistic in that main memories seldom operate with a cycle time identical to the clock period of processing logic. Bipolar gate delays are usually one or more orders of magnitude smaller than memory cycle times. Thus the clock frequency of bipolar processors is usually five or more times greater than the maximum memory frequency. If the processor and memory utilize the same technology, then gate delays and memory cycle times are comparable. The memory operates faster than the processor. The designs are also unusual in that they include all data paths that seem to be necessary. Data paths are expensive. This section will consider the problem of tying a memory and a CPU that operate at different rates and bus organizations of the CPU.

Autonomous Memory

To illustrate the problems and possibilities of an autonomous memory, one not synchronized to the CPU and serving other users in addition to the CPU, we will assume a static semiconductor RAM with attached address and data registers and control. The RAM itself has the following properties:

Read: 400 nsec after the address is provided, data is available for transfer to the MDR. The address must be held for at least 500 nsec. Control signal RW is held at 1 throughout the 500 nsec cycle.

Write: After the address and data are loaded in the MAR and MDR, RW is dropped to 0 for at least 300 nsec. Data must be held for 400 nsec; the address must be held for 500 nsec.

Figure 8.7 suggests the organization of the memory, as well as the buses of the first system to be explored. DC is a data channel that controls peripheral memory and input/output devices.

Since DC operates in conjunction with slow devices, and therefore is expected to rarely request the services of MEM, it will be given priority over CPU. The control of MEM will resolve simultaneous requests for service in favor of DC. A unit requests service by placing a 1 on its individual request line, D2R for CPU and D1R for DC. If MEM is available for service, indicated by MSTATUS = 0D4, then it responds to a request by changing the first two bits of MSTATUS to indicate which unit it is serving (01 for DC and 10 for CPU). The third bit of MSTATUS will be set to 1 when MEM is prepared to receive information over data bus MEMBUS. Three pieces of information are needed in the worst case. First, MEM must know if a read or write operation is being requested. Both require that an address be supplied to MEM. Finally, for a write operation a data word must be supplied via MEMBUS. The control bus is used by the requesting unit to indicate when and which piece of information is being transmitted on MEMBUS.

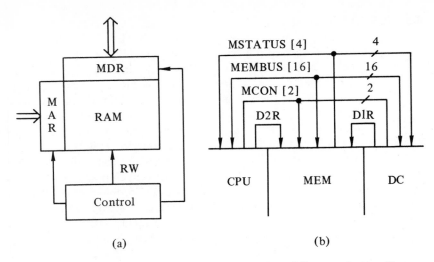

(a)   (b)

FIG. 8.7   The Memory System and a First Set of Communication Buses

Since the three units of this system may not be synchronized, and may be operating with clocks of different frequencies, it is important that an information receiving unit indicate it is prepared to receive, a transmitting unit acknowledge and transmit continuously, a receiving unit indicate completion of reception, and so forth. Raising and lowering request and acknowledge signals is called "hand shaking." In Fig. 8.7, MEM uses MSTATUS[3]; the CPU and DC signal via the MCON bus. Table 8.3 summarizes control bus codes.

Figure 8.8 provides the instruction fetch for CPU and an autonomous memory. Let us examine the memory subsystem first. It is idle until service is requested. Device 1 (DC) line is examined first. If it does not request service, then device 2 (CPU) is considered. Priority is expressed by nested *if-then* statements. If service is requested, hand shaking commences by setting STATUS[3]. Then MEM waits in state A for the op-code on MEMBUS[16]; 0, dictates a read operation is desired and 1 requests write. The op-code is captured in flip-flop OP. Op-code reception is signaled by setting STATUS[3] to 0 and waiting in state B for an acknowledgement. The address is received with MEM in state C. Write

TABLE 8.3. CONTROL BUS CODES FOR THE SYSTEM OF FIG. 8.7

MSTATUS		MCON	
00xx	MEM idle	—	device not transmitting
01xx	MEM responding to device 1	00	acknowledge
10xx	MEM responding to device 2	01	operation type on MEMBUS
0	MEM not ready for data	10	address on MEMBUS
1	MEM ready for data	11	data on MEMBUS
0	Read data not available		
1	Read data on MEMBUS		

⟨SY⟩ EDC2: ⟨TI⟩ P(75E-9), Q(1E-7).
　　　⟨TE⟩ MSTATUS[4], MEMBUS[16],
　　　　　MX[4] = D2R∘D1R∘ MCON[2].
⟨AU⟩ CPU: P: ⟨RE⟩ CAR[16], IR[16], ...
　　　⟨ST⟩ IF1 : ↓/MSTATUS[1 : 2]: D2R = 1, →IF2.
　　　　　IF2 : | · /MSTATUS[1 : 2]\2D2| MCON = 0, →IF3;
　　　　　　　D2R = 1, →IF2..
　　　　　IF3 : | · /MSTATUS[1 : 3]\3D3| MCON = 1D2,
　　　　　　　MEMBUS = 0, →IF4; MCON = 0, →IF3..
　　　　　IF4 : | · /MSTATUS[1 : 3]\2D3| MCON = 0, →IF5;
　　　　　　　MCON = 1D2, MEMBUS = 0, →IF4..
　　　　　IF5 : | · /MSTATUS[1 : 3]\3D3| MCON = 2D2,
　　　　　　　MEMBUS = CAR, →IF6; MCON = 0, →IF5..
　　　　　IF6 : | · /MSTATUS[1 : 3]\2D3| MCON = 0, →IF7;
　　　　　　　MCON = 2D2, MEMBUS = CAR, →IF6..
　　　　　IF7 : |MSTATUS[4]| IR←MBUS, ⊄CAR, MCON = 3D2,
　　　　　　　→IF8; MCON = 0, →IF7..
　　　　⋮ *Execution not detailed here.*
⟨AU⟩MEM: Q: ⟨RE⟩ MAR[16], MDR[16], STATUS[4], OP, RWFF.
　　　⟨EL⟩ RAM(OUT[16]: IN[16], RW).
　　　⟨BO⟩ IN = MAR, MSTATUS = STATUS, RW = RWFF.
　　　⟨ST⟩ IDLE: |D1R| STATUS←6D4, →A;
　　　　　　　|D2R| STATUS←10D4, →A;
　　　　　　　STATUS←0, RWFF←1, →IDLE...
　　　　　A : · /MCON\1D2: OP←MEMBUS[16],
　　　　　　　STATUS[3]←0, →B.
　　　　　B : ↓/MCON: STATUS[3]←1, →C.
　　　　　C : ·/MCON\2D2: MAR←MEMBUS,
　　　　　　　STATUS[3]←0, |OP|→D; →T1..
　　　　　D : ↓/MCON: STATUS[3]←1, →E.
　　　　　E : · /MCON\3D3: MDR←MEMBUS,
　　　　　　　STATUS[3]←0, →T1.
　　　　　T1 : |OP|RWFF←0, →T2.
　　　　　T2 : →T3.
　　　　　T3 : ↓/MCON: →T4.
　　　　　T4 : |OP|RWFF←1, STATUS←0, →IDLE;
　　　　　　　STATUS[4]←1, MDR←OUT, →T5..
　　　　　T5 : MEMBUS = MDR, | · /MCON|STATUS←0,
　　　　　　　→IDLE; →T5....
　　⟨AU⟩ DC: *Data channel not detailed here.*

FIG. 8.8　An Instruction Fetch and Memory Control Design
for the Bus System of Fig. 8.7.

operations use states D and E to pass the data to MEM. States T1 through
T4 time the RAM activity. A final hand shake is given in states T4 and
T5 for read operations. This ensures that the retrieved word is received by
the requesting unit before MEM returns to the idle, available status. The

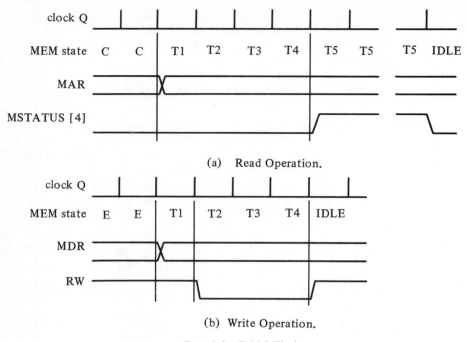

(a) Read Operation.

(b) Write Operation.

Fig. 8.9 RAM Timing

timing diagrams of Fig. 8.9 may be helpful in determining that the RAM is properly timed. We cannot be sure how long MEM resides in each of the preliminary states.

Instruction fetch in Fig. 8.8 is much slower than we are accustomed to find. CPU waits in IF1 until the memory becomes available; then it places a request. It then waits in IF2 until MEM recognizes that request. A DC request for service would lengthen this wait. Then the hand shaking, along with transmission of operation type and address, begins. Finally CPU waits in IF7 for the word it requested. It acknowledges receiving that word with the 11 code, which permits MEM to release MEMBUS and return to its idle state. A similar sequence of states will be needed to fetch operands and indirect addresses. An even longer sequence will be needed to execute the STORE instruction.

Speed may be obtained in the following ways: (1) by making the bus wider so that the op-code, address, and data may be passed to MEM simultaneously, (2) by performing the hand shaking asynchronously, or (3) by coupling the units more tightly with their synchronization. All of these approaches will enhance reliability also; nothing in the previous design prevents flip-flop input signals from changing during their set-up interval. The first approach is obviously more expensive. We will examine the last two approaches with the system of Fig. 8.10(a). While extendible bus control principles will be demonstrated, the system is kept simple enough as not to encounter some of the very difficult problems of bus management.

The three synchronized units of Fig. 8.10(a) must be able to send to and receive 16-bit words from each of the other units via one information interchange (II) bus. The bilateral II bus carries 16 bits of data, 2 bits of address, and 2 bits of instruction information. Information as to which units wish to transmit on the bus and which units are prepared to receive information is carried on the 6-bit arbiter bus (ARB). Descisions as to who "owns" the II bus for one clock period are simple enough in this system so that each unit will include decision circuitry rather than have a bus-arbiter unit. This decision logic must be designed to operate promptly, as we will see that a good deal of communication and many logic levels are involved. Table 8.4 gives codes and assignments for the II and ARB buses.

To obtain permission to transmit over the II bus, a unit must bid by promptly raising its bid line after a clock edge. Simultaneous bids are resolved according to the priorities of Table 8.4. MEM always gets the bus, if it wants it. CPU obtains permission only if the other two units do not bid. Bidding is unwise if the destination of the message is not receptive. Thus, if CPU wishes to transmit to MEM, it should check MEM's status (ARB[4]) and the bids of MEM and DC (ARB[1] and ARB[2]). If neither is bidding and MEM is receptive, transmission is permitted and will be completed.

$$CPU_{xmit} = MEM_{status} \cdot \overline{MEM}_{bid} \cdot \overline{DC}_{bid}$$
$$= ARB[4] \cdot \downarrow/ARB[1:2]$$

If it is necessary to know, a unit can determine who is transmitting to it by examining the bid lines.

The destination of a transmission is indicated with the IIA (address) wires of the II bus. Operations sent on the IIO wires vary with the destina-

TABLE 8.4. ASSIGNMENTS FOR THE BUSES OF FIG. 8.10

IIA code	To Address	IIO code	To MEM	To CPU, DC
00	noвe	00	Read for DC	Accept data from MEM
01	MEM	01	Read for CPU	Accept data from other
10	DC	10	Write for DC	Send data to MEM
11	CPU	11	Write for CPU	Send data to other

ARB Wire Assignments			ARB Codes		Bid Priority	
1	MEM	Bid	Bid Wire		1	MEM
2	DC	Bid	0	no bid	2	DC
3	CPU	Bid	1	bid	3	CPU
4	MEM	Status	Status Wire			
5	DC	Status	0	busy		
6	CPU	Status	1	receptive		

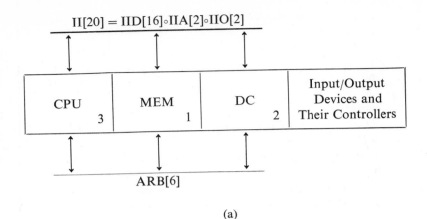

$$II[20] = IID[16] \circ IIA[2] \circ IIO[2]$$

| CPU 3 | MEM 1 | DC 2 | Input/Output Devices and Their Controllers |

ARB[6]

(a)

⟨SY⟩ EDC3: ⟨TI⟩ P(1E-7).
    ⟨TE⟩ II[20] = IID[16]∘IIA[2]∘IIO[2], ARB[6].
    ⟨AU⟩ CPU: P: ⟨RE⟩ CAR[16], IR[16], ..., STAT.
        ⟨BO⟩ ARB[6] = STAT.
        ⟨ST⟩ IF1 : ARB[4]· ↓/ARB[1 : 2]: ARB[3] = 1,
                 II = CAR∘5D4, STAT←1, ¢CAR, →IF2.
            IF2 : ↓/IIO· · /IIA: IR←IID, STAT←0, →IF3.
            IF3 : *decode, execute ...*
    ⟨AU⟩ MEM: P: ⟨EL⟩ RAM(OUT[16]: IN[16], RW).
        ⟨RE⟩ MAR[16], MDR[16], STAT, OP[2], RWFF.
        ⟨BO⟩ IN = MAR, RW = RWFF, ARB[4] = STAT.
        ⟨ST⟩ IDLE: · /IIA\1D2: MAR←IID, OP←IIO,
                 |IIO[1]| →A; STAT←0, →T1..
            A : ARB[1] = 1, IIA = 1∘OP[2], IIO = 2D2,
                MDR←IID, STAT←0, →T1.
            T1 : |OP[1]| RWFF←0., →T2.
            T2 : →T3.
            T3 : →T4.
            T4 : |OP[1]| RWFF ← 1, STAT←1, →IDLE;
                MDR←OUT, →T5..
            T5 : ARB[1] = 1, STAT←1, II = MDR∘1∘OP[2]∘
                0D2, →IDLE..

    ⟨AU⟩ DC: P: *not documented here..*

(b)

F<small>IG</small>. 8.10 Synchronized System

tion. MEM may be commanded to read or write. The 2-bit IIO code indicates the source of the instruction as well as the operation type. Thus MEM views IIO[1] as the operation type and IIO[2] as the source of the command. The source address is formed by 1∘IIO[2]. CPU and DC can be requested to send data; since the IIA and IIO wires are already in use, they respond by placing information on the IID (data) lines only.

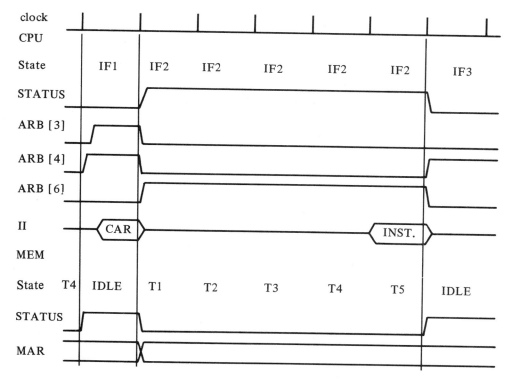

FIG. 8.11   Instruction Fetch Timing in the Synchronized System

The CPU instruction fetch and a complete specification of MEM are provided by Fig. 8.10(b). When the bus is available to it, CPU transmits the CAR and a read instruction to MEM. CPU proceeds to state IF2 where it waits for the command to receive data. When MEM provides a word on the II bus, CPU places it in the IR. Decoding and execution then follow, while MEM and DC do other things or wait for further instructions from CPU.

MEM is IDLE until an instruction is addressed to it. It requires that an address accompany that instruction. The instruction is decoded directly from the IIO bus so that state A may be skipped for read operations. If a write operation is requested, MEM enters state A, issues a send data command asynchronously, and captures that data in MDR. MEM commands the II bus without decision, and hence delay. The five timing states are as before. In T5, MEM places the word read from the RAM on the II bus, along with its proper address and command to accept that data. CPU and DC must remain alert after issuing a read request to MEM, and they must respond to these MEM specifications.

The timing of an instruction fetch and read cycle of MEM are shown in Fig. 8.11. DC is assumed to be inactive. The arbiter communication and decisions, ARB[4] to 1, ARB[3][1] to 1, must take place promptly so that

[1] None of the units look at ARB[3] because it has least priority.

the address may be propagated to MEM. Then the RAM requires 400 nsec to read the word. Since the II bus is in use during the T5 period, another instruction cannot be sent to memory, and further increase in speed cannot be gained without a substantial increase in the complexity of MEM.

Bus Oriented CPU

If the CPU must frequently wait for the memory unit to provide information, then it is not important to perform simultaneous register transfers. They may be performed in sequence during the waiting periods. Expensive data paths and multiplexers may be replaced with simple register cells and buses (see Section 6.2). To illustrate the use of buses within the CPU, we will retain the architecture of EDC from Chapter 7, but change the organization to include the synchronized MEMory and II bus and exclude the INPUT and OUTPUT registers, since they will be a part of the data channel. Since arithmetic must be performed on all registers of EDC, all are shown on the A-port side of the ALU in Fig. 8.12. All may load from the data or D bus; the CAR and ACC may source information to the A bus. The ADR subregister may be loaded from the D bus for performing shift operations, and either ZADR (0D6∘ADR) or EADR (ADR[1]×6∘ADR) may be sourced to the A bus. Each load and source operation is accomplished by setting the corresponding load (L) or source (S) control signal to logic 1.

One new register is introduced. BUF handles all communication with MEM via the II bus and temporarily holds B operands for ALU. BUF is loaded from either the D bus or the IID bus as signal LBUF commands: LBUF = 1 for BUF ← MEM. The BUF multiplexer replaces the ALU multiplexer of Chapter 7. The ALU and shifter from that chapter are used, but their mode control is more clearly described by signals ADD, SUB, INC, DECrement, DBL, CS, AND, SRExtended, SR0, CIR, CIR8, and operand transfers XA and XB, than by using the highly coded M signals of Chapter 7.

Instruction fetch and execution of the EDC instruction set by this CPU are detailed in Fig. 8.13. The memory interface requirements and procedures of Fig. 8.10 are merged with the EDC description of Fig. 7.16. First the terminals just discussed are declared. Then a list of eight connection and transfer operations is listed prior to the ⟨ST⟩ declaration. These operations are in the heading of the ⟨AU⟩, and hence the transfers are only subject to the global clock condition P. The connection operations are performed continuously. The first one describes the multiplexing performed by the A bus. We expect only one S control signal to have value 1 at any instant. The D bus connection description really describes the ALU and shifter. We see the significance of the ADD, SUB, ..., XB control signals and the functions that the ALU is expected to perform. The last six conditioned transfer and connection operations reveal the significance of the L control signals. If an L signal equals 1 at a clock time, its register is loaded. Note

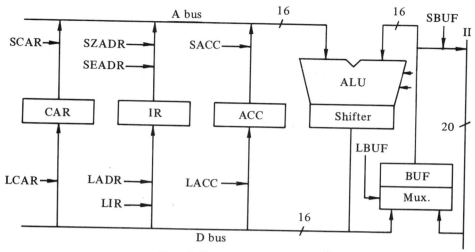

FIG. 8.12   Bus Oriented CPU for EDC

that BUF is loaded each clock cycle. Only the source of information is controlled.

The state statements indicate which control signals are to have value 1, for the most part. The large number of such signals may make the ⟨ST⟩ declaration difficult to read initially. Look for the following action in each state:

IF1 :	Get the current address to BUF.
IF2 :	If MEM and II are available to CPU, pass the read command and address to memory; otherwise, hold BUF.
IF3 :	Increment CAR while waiting for MEM.
IF4 :	Capture word from MEM when it becomes available.
IF5 :	Pass it to IR.
IF6 :	Decode. Some operations can be executed immediately; others require operand fetch. The ADR is placed in BUF for some instructions.
OF1 :	Since the address is in BUF, proceed as in IF2.
OF2 :	Same as IF4.
STO1 :	The address is in BUF. When MEM is available, pass it and a write command; get the data word to BUF. Otherwise, save the address in BUF.
STO2 :	When MEM demands, pass the data word.
EX :	If two operands are needed, the second one is now in BUF. The ACC is sourced and loaded for all arithmetic operations.
	Transfer operations route the ADR to the CAR. Counting for the shift instructions requires an additional state.
COUNT :	Decrement the ADR for shift instructions.

What has the introduction of buses accomplished? First, the registers are all very simple—FFs with open-collector or tri-state output amplifiers.

⟨SY⟩ BUSEDC: ⟨TI⟩ P(IE-7).
   ⟨TE⟩ II[20] = IID[16] ∘ IIA[2] ∘ IIO[2], ARB[6].
⟨AU⟩ CPU: P:
   ⟨RE⟩ CAR[16], IR[16] = OP[4] ∘ IX[2] ∘ ADR[10], ACC[16],
        BUF[16], STAT.
   ⟨TE⟩ A[16], LCAR, SCAR, LIR, LADR, SZADR, SEADR,
        LACC, SACC, LBUF, SBUF, ADD, SUB, INC, DEC,
        DBL, CS, AND, SRE, SR0, CIR, CIR8, XA, XB.
   ⟨ID⟩ ZADR = 0D6 ∘ ADR, EADR = ADR[1] × 6 ∘ ADR.
        A = SCAR · CAR ∨ SZADR · ZADR ∨ SEADR · EADR ∨
            SACC · ACC,
        D = ADD · (A + BUF) ∨ SUB · (A − BUF) ∨ INC · (A + 1) ∨
            DEC · (A − 1) ∨ DBL · (A + A) ∨ CS · (0 − A) ∨ AND ·
            (A · BUF) ∨ SRE · (A[1] ∘ A[1 : 15]) ∨ SR0 ·
            (0 ∘ A[1 : 15]) ∨ CIR · (A[16] ∘ A[1 : 15]) ∨ CIR8 ·
            (A[9 : 16] ∘ A[1 : 8]) ∨ XA · A ∨ XB · B,
ARB[6] = STAT,
        |LCAR| CAR←D., |LIR| IR←D., |LADR| ADR←D
        [7 : 16].,
        |LACC| ACC←D., |SBUF| IID = BUF., BUF←|LBUF| IID; D.
        ⟨ST⟩ IF1:   SCAR = XA = 1, →IF2.
             IF2: |ARB[4] · ↓/ARB[1 : 2]| ARB[3] = 1, II [17 : 20]
                         = 5D4, SBUF = 1, →IF3; XB = 1, →IF2. .
             IF3:   LCAR = SCAR = INC = 1, STAT←1, →IF4.
             IF4:   ↓/IIO· ·/IIA: LBUF = 1, STAT←0, →IF5.
             IF5:   LIR = XB = 1, →IF6.
             IF6:   [OP[0 : 2→EXIO⌊3, 9 : 11 → EX, SEADR = XA = 1
                    ⌊4 : 6, 8→OF1, SZADR = XA = 1⌊7→ STO1,
                    SZADR = XA = 1⌊12 : 15→IF1. .

In Chapter 7, the CAR and ADR were counting registers. Now the ALU performs all necessary arithmetic. Besides cost reduction, buses have also induced generality. The CAR and ACC are identical; they differ only in the way we use them. With different control statements to satisfy a different instruction set, we might find this to be very valuable. Buses have not increased the speed with which instructions are fetched and executed, nor do they slow it down very much. States IF1, IF5, and COUNT are marks against the bus structure in this respect. The biggest delays are introduced by MEM, not the buses.

8.3 MICROPROGRAM CONTROL

   Data register structure was greatly simplified and regularized in the previous section by introduction of the bus concept. Control circuitry

OF1: |ARB[4] · ↓/ARB[1 : 2]|ARB[3] = 1,
  II[17 : 20] = 5D4, SBUF = 1, STAT←1, →OF2;
  XB = 1, →OF1. .

OF2: ↓/IIO· ·/IIA : LBUF = 1, STAT←0, →EX.

STO1: |ARB[4]· ↓/ARB[1 : 2]|ARB[3] = 1,
  II[17 : 20] = 7D4, SBUF = SACC = XA = 1,
  STAT←1, →STO2; XB = 1, →STO1. .

STO2: | ·/IIA · · /IIO\2D2|SBUF = 1, STAT←0, →IF1;
  XB = 1, →STO2. .

EXIO: *issue I/O commands to data channel.*

EX: ⌈OP⌊ 3→IF1, LACC = SACC = 1, ⌈IX⌊ 1INC = 1
  ⌊2DBL = 1⌊ 3CS = 1.
  ⌊4→IF1, LACC = XB = 1
  ⌊5→IF1, LACC = SACC = ADD = 1
  ⌊6→IF1, LACC = SACC = SUB = 1
  ⌊8→IF1, LACC = SACC = AND = 1
  ⌊9→IF1, ⌈IX⌊ 0LCAR = SZADR = XA = 1
    ⌊1LCAR = SZADR = XA = ↓/ACC
    ⌊2LCAR = SZADR = XA = ACC[1]
    ⌊3LCAR = SZADR = XA = OVF.
  ⌊10→IF1, ⌈IX⌊ 0LACC = SEADR = XA = 1
    ⌊1LACC = SACC = ADD = 1
  ⌊11|↓/ADR|→IF1; →COUNT,
    ⌈IX⌊ 0LACC = SACC = SRE = 1
    ⌊1LACC = SACC = SR 0 = 1
    ⌊2LACC = SACC = CIR = 1
    ⌊3LACC = SACC = CIR8 = 1 . . .

COUNT: →EX, LADR = SZADR = DEC = 1 . . .
⟨AU⟩ MEM: *See Fig. 8.10.* ⟨AU⟩DC: *See Fig. 8.24.* .

FIG. 8.13 Bus Oriented CPU

structure is still highly irregular, however. Figure 8.13 indicates that a 13-state control machine is needed to provide 23 control signals, which dictate which switches are to be set and which registers are to be loaded at each clock time. Designing this finite-state control machine is not a trivial task, and it will be expensive to build because of its irregular structure. Furthermore, redesigning this unit to correct errors, alter the manner in which an instruction is executed, or extend the instruction set of the computer is equally expensive.

In this section we will investigate a means of making the control unit structure highly regular, and much more easily altered when a change or extension of the instruction set is required. While the highly regular form of a control unit to be studied is not economical when the instruction set is as small as we have been considering, it is more economical than an equivalent finite-state machine when the instruction set is as large as that of commercial computers. We will also introduce new concepts of digital computer operation and usage.

READ-ONLY Memory Control

The binary *n*-tuple of all control signals at any clock time is known as the *control word*. Rather than generate control words as they are needed, could we not determine all the control words that will be required, place them in a table, and look them up in a prescribed sequence? A memory is required to hold the table, but since we do not require the ability to write in this memory often or at high speed, ROMs may be used. Control words are written into the ROM at the time it is constructed. If we find it necessary to alter control words regularly, we may store the table in RAMs and arrange to load the RAMs from a nonvolatile memory, before expecting the CPU to function.

A finite-state machine might be used to activate the address lines in the proper sequence, but relatively little is accomplished unless the finite-state machine can have a particularly simple or regular structure, as we saw in Chapter 6. If the ROM word length is increased, the state table of the control-sequence generator can also be stored. For example, the address of the next control word may be stored with each control word. Some means of achieving conditional branching to one of two or more next addresses must be provided, but this can be relatively simple.

Figure 8.14 presents a CPU organization based upon these ideas. While many of the blocks shown have been examined in previous sections, some have not. The three main buses of the CPU are closely related with the ALU. The A and B buses carry operands to the ALU; the D bus carries ALU results to the other blocks of the CPU. The instruction register (IR) may be loaded from the D bus and sourced to the A bus. In addition, bit positions 8-to-10 and 14-to-16 of the IR are provided to multiplexers on the address ports of the file. These connections facilitate addressing registers of the file for some instruction formats. They bias the machine in favor of those formats.

The file has two output ports, which may be sourced to the A and B buses. The write address is the same as the B output port address. An independent write control permits reading from the file without writing into the addressed register. Fourteen registers may be addressed in the file, as we will see when the ROM word format is considered.

The ALU and shifter are simpler than those considered in previous sections. The ALU can add (with or without carry-in), subtract (2's complement, and with or without borrow-in), AND, OR, XOR, and XNOR the operands that are provided via the A and B buses. The ALU provides four signals that indicate, respectively: (1) the carry out of the most significant position C_x, (2) overflow OV, (3) if the result is zero or not Z, and (4) if the result is positive or not P. The shifter can pass ALU results or shift them left or right one position. The shift modes are as follows: (1) zero-fill, (2) extend-fill, and (3) circulate through a "shift save" flip-flop in the STATUS register. In addition, the shifter can circulate right one place. The shifter provides signal S_x that indicates the bit removed in the shift operation.

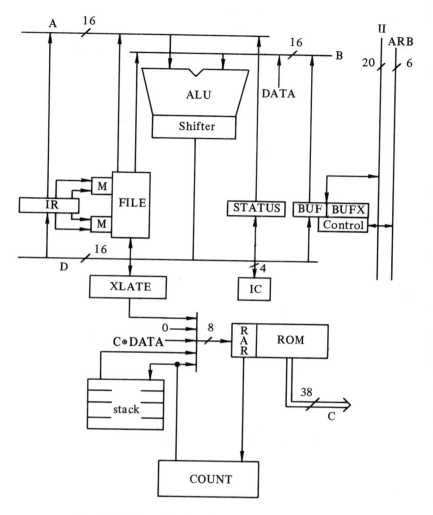

FIG. 8.14   Organization of a ROM Controlled CPU

Each flip-flop of the STATUS register has unique significance. Many have special wiring so that they may be set when certain conditions exist, as well as when STATUS is loaded from the D bus. Others are simply used in special ways by the microprograms that establish the instruction set of the computer. STATUS[1] is used in this way to determine the mode, either user or privileged, as discussed later. STATUS [3:7] is loaded under microinstruction control with signals C_x, S_x, OV, Z, and P from the ALU and shifter. STATUS[4] is the "shift save" flip-flop that is involved in some shift operations. STATUS[10], the DC flip-flop, is set when the data channel notifies CPU that it has completed an input/output task. The other interrupt flip-flops may be wired to other pieces of equipment that interact with CPU. The mask flip-flops provide a means of sensitizing CPU to interrupts, or not. The state of STATUS[16], the I flip-flop summarizes the states of the sensitized interrupt flip-flops. Flip-flop I is set if

$$\vee/ (STATUS [8:11] \cdot STATUS[12:15]) = 1$$

STATUS		Interrupt	Mask	
M	C_x S_x OV Z P	DC		I
1	3	8	12	16

Register BUF links the CPU to memory via the II bus, as previously discussed. While BUF may be assigned to source the B bus, a small sequential network centered on extension register BUFX controls the loading of BUF and its connection to the II bus. II bus address and operation codes may be placed in the BUFX register; when they are, the sequential network controls BUF to complete the transfer of information to or from MEM, or DC. The sequential network communicates with the main control of CPU via two signals. When the sequential network is idle and available to process another request to MEM or DC, signal IOI = 1. Signal DA is reset to 0 when a new request is entered into BUFX. Following a MEM read request, DA goes to 1 when the data word is available in BUF. A write request must be followed by a data word; DA goes to 1 when that data word has been loaded into BUF.

The 4-bit iteration counter IC may be loaded from the right of the D bus. It is used in conjunction with the ROM centered control unit. Let us turn our attention to it. The 256 word, 38-bit per word ROM is supplied an address by the ROM address register, RAR. Figure 8.15 shows the fields of each ROM word. Most fields have a single, fixed significance. This then is a *horizontally* microprogrammed ROM. A few fields have multiple meanings. In a *vertically* microprogrammed ROM, nearly all fields have a variety of meanings that are determined by the contents of other fields.

Fields N, C, and S are concerned with the next address of the ROM to be read. Table 8.5 indicates the entries that may be placed in each field and the significance of each possible entry. In all ROM fields the width of the field determines the base of the digits used to describe that field. An N field entry is described with an octal digit; a C field entry is specified with a hexadecimal digit. The RAR is loaded from either of the following: (1) the COUNT logic, (2) the C and *DATA* fields, (3) the STACK, (4) zero, or (5) the translation box XLATE, according to the N field entry. The COUNT box consists of the half adders of Fig. 1.9, for the most part. In the least significant position the half-adder is replaced with a full-adder so that the count may be increased by 0, 1, or 2. Table 8.5 reveals that the increment may depend on any of a number of CPU signals. The immediate C_x, S_x, OV, Z, and P values from the ALU-shifter may be used to step the count by either 1 or 2 (a 2-way branch). In mode $C = 6$, the count is not advanced until the iteration counter holds 0001; one microinstruction may be performed several times before proceeding to the next. The DA and IOI signals are tested with $C = 7$ and 8, respectively. The ROM address is held constant until the tested signal takes the value of 1. The remaining C modes test flip-flops of the STATUS register. Note: while the C field determines count mode when $N = 0$, it provides data when $N = 2$. This is a first example of vertical microprogramming.

The S field controls the four register STACK. When the stack is pushed, the current address held by RAR, plus 1, is placed on the top of this push-down stack to facilitate jumps to and from subroutines. The triple (N, C, S) provides a great variety of ways to form the next ROM address, and hence gives the microprogrammer a great deal of freedom. It is easier to describe the restrictions on next address formation than to list the variety of available address sequencing modes.

1. With $N = 0$ (*count* mode), the next address is restricted to be (a) the the present ROM address, (b) that address plus one, or (c) the second next address.
2. The *go to* mode ($N = 1$) may be used only when the *DATA* field is not required for other purposes. STACK is often pushed in conjunction with such a *go to* transfer.
3. STACK is usually popped in conjunction with a *return* mode ($N = 2$) transfer.
4. Translation ($4 \leq N \leq 7$) is very specialized.

The XLATE block of Fig. 8.14 is designed to ease the problem of decoding a specific CPU instruction set. (CPU instructions are called *macroinstructions* to distinguish them from microinstructions.) It has the task of converting the 16 bits presented on the D bus and 2 bits of the N field into "desired" 8-bit ROM addresses. We cannot begin to design this block until we adopt an instruction set for CPU. We cannot begin to say what "desirable" ROM addresses are until we write microcode to realize that instruction set.

The next three ROM fields of Fig. 8.15 select the sources of information for the A and B buses and the destination(s) of information from the D bus. Both the A and B buses may convey 0D16 to their ALU ports. The IR, file, and STATUS registers may be sourced to the A bus. 0D12∘DATA may be placed on the B bus as may the contents of the BUF register, or the file, of course. None, any, or all of IR, file, and STATUS may be loaded simultaneously from the D bus. BUF is also loaded from the D bus, but not under direct control of the ROM word.

File addresses are specified in the AA and BA fields. Entries from 0 through D directly address a file register. Entries of E and F address the register from the IR. Since only 3 bits are supplied by the IR to each file address multiplexer, only registers 0 through 7 may be addressed in this mode. [The address multiplexers select 3 bits from the IR or 4 bits from the C (control) bus carrying the ROM word.]

FIG. 8.15 Fields of the ROM word

Next Address		Bus Address			Bus Address		File Address				ALU Mode			DATA	BUFX
N	C	S	A	D	1	2	3	B	A	AA	BA	AL	SH	IC		

TABLE 8.5. ROM FIELD CODES

N	Mode
0	COUNT
1	C ∘ DATA
2	TOS
3	0
4-7	Xlate

S	Stack Mode
0	Hold
1	Push
2	Pop

C	Count Mode
0	$+1$
1	$+1+IC_x$
2	$+1+IS_x$
3	$+1+IOV$
4	$+1+IZ$
5	$+1+IP$
6	$+\cdot/IC\backslash ID4$
7	$+DA$
8	$+IOI$
9	$+1+C_x$
A	$+1+S_x$
B	$+1+OV$
C	$+1+Z$
D	$+1+P$
E	$+IRPT$
F	$+DCI$

A	A Bus Source
0	0
1	IR
2	File
3	STATUS

B	B Bus Source
0	0
1	$0 \times 12 \circ$ DATA
2	File
3	BUF

D	Register Mode
0	Hold
1	Load

D_1 — IR
D_2 — File
D_3 — STATUS

AA,BA	File Address
0-D	AA,BA
E	$0 \circ IR[8:10]$
F	$0 \circ IR[14:16]$

AL	ALU Mode
0,8	$A+B$
1,9	$A+\bar{B}+1$
2,A	AND
3,B	OR
4,C	$A+B+C$
5,D	$A+\bar{B}+C_x$
6,E	XOR
7,F	XNOR

8-F: Store C_x, S_x, OV,Z,P in STATUS

IC	Mode
0	Hold
1	Load
2	-1

SH	Shifter Mode
0	Pass
1	SRZ
2	SRE
3	SRS_x
4	CR
5	SLZ
6	SLE
7	SLS_x

BUFX	MODE
0	Hold
1	Load

The next group of ROM fields controls the ALU, shifter, and iteration counter. A fourth bit in the *AL* field determines whether the immediate carry, shift, overflow, etc., values are stored in STATUS, or not. Otherwise, the modes as listed in Table 8.5 are self-explanatory. The last group of fields provides *DATA* and a command to the sequential network that controls BUF. *DATA* has been mentioned twice previously; it may be used to express a next ROM address with $N = 1$ or it may be placed on the B bus with bus source field $B = 1$. In addition, if field $BUFX = 1$, *DATA* is placed in the BUFX extension register to command communication with MEM or DC. The sequential network will load this data to BUFX only if it is idle ($IOI = 1$), so a $BUFX = 1$ field entry is usually accompanied by $N = 0$, $C = 8$. Then the ROM control unit remains in the same state (ROM address does not change) until the load BUFX command has been accepted.

To avoid the complexity of timing problems, we assume all registers are synchronized by the same clock. Then the following sequence of activities must be completed between the end of one clock pulse and the beginning of the next:

1. When all flip-flops change state, a new address is presented to the ROM.
2. The ROM decodes the address and presents a new word on the control bus C.
3. Control signals propagate to the registers, file, ALU, counters, etc.
4. Data words are placed on the A and B buses.
5. A result is generated by the ALU and shifter, and passed via the D bus to the input terminals of the registers.

With the next clock pulse those registers load that are commanded to do so, and the cycle begins again. A great many signals pass simultaneously through a large number of logic levels and along many lines. The logic design must ensure that all signals stabilize before the next clock pulse; otherwise, the clock period must be increased, or the timing mode or organization must be altered to ensure reliable operations.

This microprogrammed computer (MC) is of the complexity of microcomputers and minicomputers. A great many details must be mastered before the system can be understood or programmed for use. It is possible that all of the ways of doing useful tasks most efficiently will never be discovered. Before we look at examples, let us review the hardware organizations from an engineering point of view. The registers, buses, file, ALU, and shifter of Fig. 8.14 are all rather conventional. Their design and fabrication using MSI or LSI are straightforward, if we are given obviously necessary control signals. However, the ROM does not always provide those signals. While the *D* field provides uncoded load commands to the IR, file, and STATUS registers, most other fields are coded, and these codes will not necessarily simplify logic design. Consider the ALU commands for purposes of example. In the last chapter we found that 8 M-signals were needed to control the ALU and shifter designed there. While we have a simpler ALU here, the 3 bits provided by the *AL* field may require decoding and recoding to match the ALU design. Code conversion can be avoided by placing the required ALU control words in the ROM word. This approach usually lengthens the ROM word and aggravates the preparation of microinstructions. File addressing provides a similar example. Detecting the special codes *E* and *F* can be avoided by placing extra bits in the ROM word. Desires to keep ROM words short and the logic simple are contradictory. Encoded ROM fields shorten the ROM word, but usually make decoders necessary at the processing blocks and registers. To find an optimum design, the cost and speed advantages of a longer ROM word must be balanced against the cost of decoders and a narrower C bus.

An Instruction Set

The hardware of Fig. 8.14 is not able to execute instructions until the ROM has been programmed. One of the major advantages of micro-programmed control appears. Given the fixed processing hardware, the substitution of ROMs programmed in different ways results in a CPU with different characteristics. Rather different instruction sets may be realized by altering microprograms rather than redesigning the hardware.

Only one instruction set can be illustrated here. It includes the four forms of instruction words shown in Fig. 8.16. The *S* and *D* fields specify source and destination registers, respectively. The *SM* and *DM* mode fields specify addressing modes. Table 8.6 indicates in octal the field entries to specify an addressing mode, the location of the source or destination operand, where the result is placed, and accompanying operations. With one exception the result replaces the destination operand. Immediate addressing

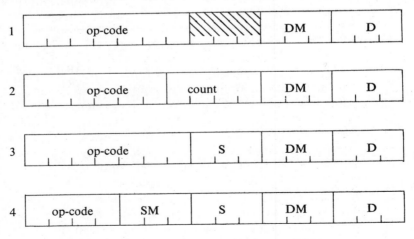

Fig. 8.16   Forms of Instruction Words. All may require one or two following address words.

places the result in the destination file register, rather than the memory location that holds the D operand. Addressing modes 1, 2, 3, 6, and 7 require an address word following the instruction word in memory. If both source and destination addressing require an address word, the source word must be the first of the two address words following the instruction word. Thus a complete instruction may occupy from one to three words of memory. Register-to-register, mixed, and memory-to-memory instructions are all possible with these eight addressing modes.

Table 8.7 lists the instructions of form 1. A word with the most significant bits being 0's is not taken as a valid instruction since data usually has many left 0's. CPU response to such words and the IO instruction will be discussed later. Instructions IO through SS in Table 8.7 are *privileged* instructions. They will not be executed unless the mode bit M of the STATUS register is a 0 or the base (relocation) register holds zero. General users of computers are seldom given complete access to the system. A program known as an *operating system* resides in memory along with users' programs. The operating system provides services to users' programs. Hardware and software restrictions are placed on users so that they cannot destroy the operating system; or more importantly, they cannot interfere with, or even observe, the programs of other users that may be in memory at the same time. The mode flip-flop STATUS[1] is set to 1 when execution of a user's program begins. The user is then prevented from altering some registers of CPU or performing input and output operations without the assistance of the operating system.

TABLE 8.6. ADDRESSING MODES

Mode	Name	Source of Operand	Destination of result	Accompanying Operations
0	Register	F[·][1]	F[D]	
1	Immediate	M[CAR]	F[D]	¢CAR
2	Direct	M[M[CAR]]	M[M[CAR]]	¢CAR
3	Indexed	M[M[CAR] + F[·]]	M[M[CAR] + F[D]]	¢CAR
4	Register-Indirect	M[F[·]]	M[F[D]]	
5	Pointer	M[F[·]]	M[F[D]]	¢F[·] for operand ¢F[D] for results
6	Indirect	M[M[M[CAR]]]	M[M[M[CAR]]]	¢CAR
7	Preindexed-Indirect	M[M[M[CAR] + F[·]]]	M[M[M[CAR] + F[D]]]	¢CAR

[1] F[·] stands in place of F[S] or F[D].

507   **SECTION 8.3**   **Microprogram Control**

TABLE 8.7. FORM 1 INSTRUCTIONS

Mnemonic	Op-Code	Description
INVALID	0000000	Invalid code
PUI	0000001	Pause until unmasked interrupt
PUDC	0000010	Pause until DC interrupt
IO,DM,D[1]	0000100	Input/Output
XEQ,DM,D[1]	0001000	Load IR and execute
LB,DM,D[1]	0001001	Load BASE
LU,DM,D[1]	0001010	Load UPPER
LS,DM,D[1]	0001011	Load STATUS
SIR,DM,D[1]	0001100	Store IR
SB,DM,D[1]	0001101	Store BASE
SU,DM,D[1]	0001110	Store UPPER
SS,DM,D[1]	0001111	Store STATUS
ZERO,DM,D[2]	0010000	Zero
COM1,DM,D[2]	0010001	1's Complement
COM2,DM,D[2]	0010010	2's Complement
DBL,DM,D[2]	0010011	Double
INC,DM,D[2]	0010100	Increment
DEC,DM,D[2]	0010101	Decrement
ADDC,DM,D[2]	0010110	Add C_x
SUBC,DM,D[2]	0010111	Subtract C_x

[1] Privileged Instruction.
[2] Load C_x, S_x, OV, Z, P.

The ZERO through SUBC instructions of Table 8.7 are user mode instructions; they may be used to alter the destination file or memory location specified by their DM and D fields. All memory addresses in these and all other instructions are relocated with the BASE register, and the relocated address is compared with the UPPER register before a memory reference is allowed. The relocated address must fall within the range BASE to UPPER. If it does not, the operating system must take over and terminate the user's program; this action protects other user programs that may be in memory at the same time.

Table 8.8 lists the instructions of the second form. The 4-bit count field determines the number of positions that the destination operand is to be shifted before being placed back into its original location. The TSB instruction places the destination operand in the CAR if the bit of STATUS designated by the count field is a 1. Thus transfers on overflow, positive, zero, and so forth, are all provided by this one instruction.

Table 8.9 presents the remaining instructions. Transfer to the location specified by the destination operand may be conditional on the sign and content of a file register. The subroutine transfer is unconditional. The contents of the CAR are placed in the file register specified by S before being replaced with the destination operand. The subroutine itself may easily find its arguments using that register, and when it has completed its

TABLE 8.8. FORM 2 INSTRUCTIONS

Mnemonic	Op-Code	Description[1]
SRZ,C,DM,D	001100	Shift Right, 0-fill
SRE,C,DM,D	001101	Shift Right, extend
SRS,C,DM,D	001110	Shift Right, S_x-fill
SRC,C,DM,D	001111	Circulate Right
SLZ,C,DM,D	010000	Shift Left, 0-fill
SLE,C,DM,D	010001	Shift Left, extend
SLS,C,DM,D	010010	Shift Left, S_x-fill
SLC,C,DM,D	010011	Circulate Left
TSB,C,DM,D	011000	Transfer on STATUS Bit Set

[1] Shift instructions load C_x, S_x, OV, Z, P.

task the subroutine return SRN instruction simply loads the CAR from this file register. The DTR and ITR instructions alter the register designated by S and test its new value simultaneously. If the new content of the file register will not be zero, the transfer is executed by loading the CAR with the destination operand. These are then ideal last instructions for program loops. The MOV through XOR instructions replace the destination operand with a combination of the source and destination operands in general. Immediate addressing of the destination operand is always an exception. The compare (CMP) instruction is another exception in that the result of subtraction is not stored but the C_x, S_x, OV, Z and P bits are set in the STATUS register. A CMP instruction is usually followed by a TSB instruction.

TABLE 8.9. FORM 3 AND 4 INSTRUCTIONS

Mnemonic	Op-Code	Description[1]
TRU,DM,D	0101000	Transfer Unconditional
TRP,S,DM,D	0101001	Transfer on S Positive
TRN,S,DM,D	0101010	Transfer on S Negative
TRZ,S,DM,D	0101011	Transfer on S Zero
STR,S,DM,D	0101100	Subroutine Transfer
SRN,S	0101101	Subroutine Return
DTR,S,DM,D	0101110	Decrement and Transfer
ITR,S,DM,D	0101111	Increment and Transfer
MOV,SM,S,DM,D	1000	$D \leftarrow S$
ADD,SM,S,DM,D	1001	$D \leftarrow D + S$
SUB,SM,S,DM,D	1010	$D \leftarrow D - S$
CMP,SM,S,DM,D	1011	Compare
AND,SM,S,DM,D	1100	$D \leftarrow D \cdot S$
OR,SM,S,DM,D	1101	$D \leftarrow D \vee S$
XOR,SM,S,DM,D	1110	$D \leftarrow D \oplus S$

[1] All form 4 instructions load C_x, S_x, OV, Z, P.

As a very limited example of the use of this instruction set, suppose that N numbers to be added are stored beginning in location 100_8 with the value of N being stored in 77_8. The following sequence of macroinstructions will then form the sum in general register 3. General register 0 is assumed to be the CAR.

$$
\begin{aligned}
&\text{MOV 7, 0, 0, 1} \\
&\text{000003} \\
&\text{ZERO 0, 3} \\
&\text{MOV 0, 0, 0, 2} \\
&\text{ADD 3, 1, 0, 3} \\
&\text{000077} \\
&\text{DTR 1, 0, 2} \\
&\qquad \vdots
\end{aligned}
$$

The first MOV instruction loads general register 1 from the location given by the contents of CAR + 3. Since the CAR is pointing to the ZERO instruction when this MOV instruction is being executed, 000077_8 (location of N) is the address of the operand and N is loaded into register 1. Register 3 is cleared to zero. Then the contents of CAR is moved to register 2; the CAR is pointing to the ADD instruction at this point. The ADD and DTR instructions form a two-instruction loop that accumulates the sum. New numbers are provided as source operands by means of indexed addressing in the ADD instruction. Then register 1 is decremented and tested, and a transfer is made to the address stored in register 2, the address of the ADD instruction. Only one word of this program need be altered if the value of N (and list of numbers) is located at other than 77_8.

Microprogramming

Microinstructions that provide the preceding instruction set will be described in terms of the following file register assignments: Register 0 is the CAR, register 8 is the base or *lower* bound register, and register 9 is the *upper* limit register. Registers A though D are not available to the macroprogrammer, but the microprogrammer may use them to advantage. We will adopt the following conventions. Source operands, if any, will be stored in register D. Destination operands and results reside in register C, while destination addresses, if any, are held in B. Finally, addresses will be formed in register A.

To process information as rapidly as possible, communication with MEM must be initiated as early as possible in a program. The microprograms discussed here do so to the extent that the code is not made unduly complex. Since it is very easy to specify a next ROM address of 00_{16} —all ROM addresses will be given in hexadecimal, the INSTRUCTION FETCH microprogram will begin in that location. Figure 8.17 indicates that INSTRUCTION FETCH begins by relocating the CAR by placing the sum of registers 0 and 8 on the D bus. This sum is held on the D bus until the

File

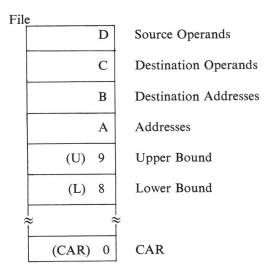

	D	Source Operands
	C	Destination Operands
	B	Destination Addresses
	A	Addresses
(U)	9	Upper Bound
(L)	8	Lower Bound
≈	≈	
(CAR)	0	CAR

sequential network loads it into BUF, and the *DATA* field into BUFX. $DATA = 5_{16} = 0101_2$ commands a read request to MEM. While the DDL description

$$IF1 : IOI : BUF \leftarrow CAR + L, BUFX \leftarrow 5.$$

does describe the effect of this first instruction, it does not convey the fact that CAR + L may be placed on the D bus for many clock periods before IOI takes the value of 1. We shall continue to take such liberties in DDL descriptions of microinstructions.

With the memory cycle initiated, the relocated address is placed in the *A* register, the next location is placed on the stack, and the next address is set at $C \circ DATA = 07_{16}$. This "call" to the BOUNDS CHECK subroutine is denoted ⇒ BC0. The segment transfer symbol of DDL is used as a reminder that we expect the program to return to the next instruction. The BOUNDS CHECKING subroutine increments the CAR (address 07), compares the address previously supplied to MEM with the upper limit (address 08), and compares it with the lower bound (address 0A). If all checks are passed, then the BOUNDS CHECKING subroutine "returns" (address 0C) to the address provided by STACK. If a bound is violated by the relocated address sent to MEM, the TRAP subroutine (discussed later) is executed.

CPU waits in address 02 until data is available from MEM (COUNT mode 7 is specified). When the word from MEM is available, it is routed from BUF via B bus through ALU via D bus to IR. A first examination of the word is made in address 03. At this point we do not know what next address the XLATE block will provide for each of the translation modes and instruction types. For now note that the IR is placed on the D bus in addresses 03 through 06. Further, the next address is placed on the top of the stack so that we can write programs to return to the next translation microinstruction. When an instruction is completely executed, the microprogram must go to address 00_{16} (IF1).

		Next		Bus			File		ALU			DATA	BUFX	
	ROM ADR	N	C	S	A	B	D	A	A	L	H	C			Comments
INSTRUCTION FETCH	00		8		2	2			8					5	IF1 : IOI : BUF←CAR+L, BUFX←5
	01	1		1	3	2		A						7	IF2 : A ← BUF, ⇒ BC0
	02		7		3		4								IF3 : DA : IR ← BUF
	03	4	1	1											IF4 : ⇒ (XLATE0)
	04	5	1	1											IF5 : ⇒ (XLATE1)
	05	6	1	1											IF6 : ⇒ (XLATE2)
	06	7	1	1											IF7 : ⇒ (XLATE3)
BOUNDS CHECK	07				2	1	2							1	BC0: CAR ← CAR+1
	08		5		2	2		9	A			1			BC1 : |U≥A|→BC3; →BC2.
	09	1	C											6	BC2 : → TRAP
	0A		5		2	2		A	8			1			BC3 : |A ≥ L|→BC5; →BC4.
	0B	1	C											6	BC4:→TRAP
	0C	2	2												BC5 : ⇒

FIG. 8.17 Listing of the Instruction Fetch and Bounds Check Programs

The ROM field codes of Table 8.5 were established so that entries of 0 are very common. Such entries are shown as blanks in Fig. 8.17. Writing the correct field entries is tedious, to say the least. We will therefore discuss additional microinstructions with the modified DDL notation, as illustrated in Fig. 8.17, and assume that an assembler program is available to convert it to 38-bit ROM words.

Decoding and executing macroinstructions when 8 addressing modes for source and destination operands are available is a formidable task. Careful organization and a generalized strategy are necessary since the ROM is limited to 256 words. Instructions of form 4, such as the ADD instruction, will be decoded and executed as follows:

1.   ⇒ XLATE0
2. Place source operand in register D
3.   ⇒ XLATE1
4. Place destination operand in register C
   Place destination address, if any, in register B

5.   ⇒ XLATE2
6. Place result of execution in register *C*
7.   ⇒ XLATE3
8. Place result in file or memory
9.   → IF1

Only the actual execution step will be unique for the instructions of this form. Generally useful subroutines are used in all other cases.

Source operands are obtained by translating to address $010sss00_2$, where *sss* represents the *SM* field of the instruction. Figure 8.18 provides a SOURCE OPERAND microprogram. Absolute ROM addresses are shown in parentheses. Each addressing mode is provided four addresses by this translation. In several cases four addresses are not sufficient, and the FETCH S ADDRESS and FETCH S OPERAND programs, which are located in another part of the ROM, are required. The register mode of addressing is easily handled. The immediate mode requires a MEM reference followed by bounds checking, placing the word from MEM in the *D* register, and finally returning. Instructions SO4 and SO5 are also used for all of the other addressing modes. Direct addressing requires that the address word be obtained and then supplied to MEM as the address of the data. Instructions SO6–SO8 initiate fetching the address word. Instructions FSO1–FSO4 relocate that address and request that MEM send the data word. Since the address was not supplied by the CAR, microinstruction BC0 is not used. Then SO4 and SO5 place the data in *D* and return.

Preindexed, indirect addressing is the most involved of the addressing modes. Instructions SO22–SO24 initiate the fetch of the address word that must follow the instruction. Note that the source register contents are placed in *D* by instruction SO24. FSA1 adds register *D* to the word supplied by MEM, placing M[CAR] + F[S] in the *A* register. This address is relocated, sent to MEM, and checked by FSA2–FSA4. When the word from MEM is available, M[M[CAR] + F[S]] is relocated, sent to MEM, and checked by FSO1–FSO4. Finally the data word is placed in register *D* by SO4–SO5.

Destination operands and addresses are obtained by translating the instruction to ROM address 001*ddd*00, where *ddd* is the *DM* field of the macroinstruction. Destination operand code in ROM locations 20 to 3F and FETCH D ADDRESS and FETCH D OPERAND programs in addresses 10 through 17 are almost identical to the source operand programs of Fig. 8.18. The destination file register F[D] is used in place of F[S]. The *B* and *C* registers are used in place of the *D* register. The memory address of the destination operand is saved in the *C* register, whereas the location of the source operand is not preserved in Fig. 8.18.

With two operands available in working registers, execution of the form 4 instructions is easily accomplished. They are translated to address 10001*ccc*, where *ccc* expresses bits OP[2 :4] of the macroinstruction. Figure 8.19 provides execution instructions for some form 1 instructions as well as the form 4 instructions. In general, one or two operands are

Register	(40)	SO1 : D←F[S], ⇒
		⋮
Immediate	(44)	SO2 : IOI : BUF←CAR+L, BUFX←5
		SO3 : A←BUF, ⇒ BC0
		SO4 : DA : D←BUF
		SO5 : ⇒
Direct	(48)	SO6 : IOI : BUF←CAR+L, BUFX←5
		SO7 : A←BUF, ⇒ BC0
		SO8 : D←0, →FSO1
		.
Indexed	(4C)	SO9 : IOI : BUF ← CAR+L, BUFX←5
		SO10 : A←BUF, ⇒ BC0
		SO11 : D←F[S], →FSO1
		.
Register Indirect	(50)	SO12 : IOI : BUF←F[S]+L, BUFX←5⁻
		SO13 : A←BUF, ⇒ BC1
		SO14 : →SO4
		.
Pointer	(54)	SO15 : IOI : BUF←F[S]+L, BUFX←5
		SO16 : A←BUF, ⇒ BC1
		SO17 : F[S]←F[S]−1
		SO18 : →SO4
Indirect	(58)	SO19 : IOI : BUF←CAR+L, BUFX←5
		SO20 : A←BUF, ⇒ BC0
		SO21 : D←0, →FSA1
		.
Preindexed Indirect	(5C)	SO22 : IOI : BUF←CAR+L, BUFX←5
		SO23 : A←BUF, ⇒ BC0
		SO24 : D←F[S], →FSA1
		.
Fetch S Address	(18)	FSA1 : DA : A←D+BUF
		FSA2 : IOI : BUF←A+L, BUFX←5
		FSA3 : A←BUF, ⇒ BC1
		FSA4 : D←0
Fetch S Operand	(1C)	FSO1 : DA : A←D+BUF
		FSO2 : IOI : BUF←A+L, BUFX←5
		FSO3 : A←BUF, ⇒ BC1
		FSO4 : →SO4

FIG. 8.18   Source Operand Microprograms

combined to form a result, which is placed in register C, and a return is executed.

The compare instruction is an exception, since results are not preserved. While Fig. 8.19 does not so indicate, the AL field must indicate that the STATUS word is to be modified by these arithmetic and logic micro-instructions. A final translation to the STORE RESULTS microprogram is needed since the results must be transferred to one of two places. If the

Zero	(80)	$Z : C \leftarrow 0, \Rightarrow$
1's Comp.		$CM1 : C \leftarrow C \odot 0, \Rightarrow$
2's Comp.		$CM2 : C \leftarrow 0 - C, \Rightarrow$
Double		$DBL : C \leftarrow C + C, \Rightarrow$
Increment		$INC : C \leftarrow C + 1, \Rightarrow$
Decrement		$DEC : C \leftarrow C - 1, \Rightarrow$
Add C_x		$ADDC : C \leftarrow C + C_x, \Rightarrow$
Subtract C_x		$SUBC : C \leftarrow C - C_x, \Rightarrow$
Move	(88)	$MOV : C \leftarrow D, \Rightarrow$
Add		$ADD : C \leftarrow D + C, \Rightarrow$
Subtract		$SUB : C \leftarrow D - C, \Rightarrow$
Compare		$CMP : D - C, \rightarrow IF1$
And		$AND : C \leftarrow D \cdot C, \Rightarrow$
Or		$OR : C \leftarrow D v C, \Rightarrow$
Exclusive-or		$XOR : C \leftarrow D \oplus C, \Rightarrow$
	(8F)	$\cdot$
Store Results	(68)	$SR1 : F[D] \leftarrow C, \rightarrow IF1$
		$SR2 : IOI : BUF \leftarrow B, BUFX \leftarrow 7$
		$SR3 : DA : BUF \leftarrow C$
	(6B)	$SR4 : \rightarrow IF1$

FIG. 8.19   Execution and Storing Results

destination addressing mode is register or immediate, then the results must be placed in a file register (instruction SR1). Other addressing modes require that the result be placed in memory (instructions SR2–SR4). Translation to address $0110100x$ where

$$x = v/DM[1 : 2]$$

separates these two cases.

Figure 8.20, a simulation result, shows how this microcode fetches, decodes, and executes the following ADD instruction.

0	113142	ADD indexed to register-indirect
1	000007	
2		
$\vdots$		
10	177774	
11	000017	

Register 1 holds 000001_8; register 2 holds 000011_8. The lower bound is set to 0 while the upper limit is 777_8. The request to MEM is issued immediately. The instruction appears in IR at time 9. Incrementing the CAR and the bounds check are completed well in advance of that time. The address word is obtained from memory location 1 and added to register 1. The source operand is transferred from memory location 10_8 to the D register. Since the destination operand is addressed register-indirect, only one memory reference is required to place the destination

TIME	RAR	ROM WORD	IR	FA	FILE	STATUS	BUF	IC	TOS	ARB	IID	IIAD	STATE	MAR	MDR
1	00	0802200800051	000000			140036	0C0C00	C0	00	07			IDL	000000	000000
2	01	1010320A00070	000000			140036	0C0C00	C0	00	16	000000	05	T1	000000	000000
3	07	000212000010	000000			140036	0C0C00	C0	02	02			T2	000000	000000
4	08	0502209A10000	000000	A	000000	140036	0C0C00	C0	02	02			T3	000000	000000
5	0A	0502200A810000	000000	C	000001	140036	0C0C00	C0	02	02			T4	000000	000000
6	0C	2020000000000	000000			140036	0C0C00	C0	02	02			T5	000000	000000
7	02	0700340000000	000000			140036	113142	C0	00	42	113142	14	IDL	000000	113142
8	02	0700340000000	000000			140036	113142	C0	00	07			IDL	000000	113142
9	03	4010000000000	000000			140036	113142	C0	00	07			IDL	000000	113142
10	4C	0802200800051	000000			140036	0C0C01	C0	04	07			IDL	000000	113142
11	4D	1010320A00070	113142			140036	0C0C01	C0	04	16	000001	05	T1	000001	113142
12	07	0002120000010	113142	A	000001	140036	0C0C01	C0	4E	02			T2	000001	113142
13	08	0502209A10000	113142	C	000002	140036	0C0C01	C0	4E	02			T3	000001	113142
14	0A	0502200A810000	113142			140036	0C0C01	C0	4E	02			T4	000001	113142
15	0C	2020000000000	113142			140036	0C0C01	C0	4E	02			T5	000001	113142
16	4E	110202ED000C0	113142			140036	0C0C01	C0	04	42	000007	14	IDL	000001	000007
17	1C	070232DA00000	113142	D	000001	140036	0C0C07	C0	04	07			IDL	000001	000007
18	1D	0802200AB00051	113142	A	000010	140036	0C0C10	C0	04	16	000010	05	IDL	000010	000007
19	1E	1010320A00080	113142	A	000010	140036	0C0C10	C0	04	02			T1	000010	000007
20	08	0502209A10000	113142			140036	0C0C10	C0	1F	02			T2	000010	000007
21	0A	0502200A810000	113142			140036	0C0C10	C0	1F	02			T3	000010	000007
22	0C	2020000000000	113142			140036	0C0C10	C0	1F	02			T4	000010	000007
23	1F	1400000000060	113142			140036	0C0C10	C0	04	42	177774	14	T5	000010	177774
24	46	0700320D000000	113142	C	000010	140036	177774	C0	04	07			IDL	000010	177774
25	47	0700320000000	113142	D	177774	140036	177774	C0	04	07			IDL	000010	177774
26	04	2020000000000	113142			140036	177774	C0	00	07			IDL	000010	177774
27	30	5011000000000	113142			140036	0C0C11	C0	05	16	000011	05	IDL	000010	177774
28	31	080220F800051	113142	A	000011	140036	0C0C11	C0	32	02			IDL	000011	177774
29	08	1010320A00080	113142			140036	0C0C11	C0	32	02			T1	000011	177774
30	0A	0502209A10000	113142	A	000011	140036	0C0C11	C0	32	02			T2	000011	177774
31	0C	0502200A810000	113142			140036	0C0C11	C0	05	02			T3	000011	177774
32	32	2020000000000	113142			140036	0C0C11	C0	05	42	000017	14	T4	000011	000017
33	26	120202AB00060	113142	B	000011	140036	0C0C11	C0	05	07			T5	000011	000017
34	27	0700320C0C0070	113142	C	000011	140036	0C0C17	C0	05	07			IDL	000011	000017
35	05	0700320C00070	113142	C	000017	140036	0C0C17	C0	05	07			IDL	000011	000017
36	89	2020000C000000	113142			140036	0C0C17	C0	06	07			IDL	000011	000017
37	06	6011000000000	113142			140036	0C0C17	C0	07	07			IDL	000011	000017
38	69	202222DC80000	113142	C	000013	140036	0C0C17	C0	07	07			IDL	000011	000017
39	6A	7011000000000	113142			161036	0C0C17	C0	07	07			IDL	000011	000017
40	6B	0802200800071	113142			161036	0C0C11	C0	07	07			IDL	000011	000017
41	00	0702200C000000	113142			161036	0C0C11	C0	07	16	000011	07	IDL	000011	000017
42	6A	070200C000000	113142			161036	0C0C11	C0	07	46	000013	16	A	000011	000013
43	6B	3000000000000	113142			161036	0C0C13	C0	07	03			T1	000011	000013
44	00	0802200800051	113142			161036	0C0C13	C0	07	02			T2	000011	000013
45	01	1010320A00070	113142	A	000002	161036	0C0C02	C0	07	02			T3	000011	000013
46	07	0002120000010	113142	C	000003	161036	0C0C02	C0	02	02			T4	000011	000013
47	08	0502209A10000	113142			161036	0C0C02	C0	02	16	000002	05	IDL	000011	000013

FIG. 8.20 Results of Simulating the Fetch, Decode, and Execution of an ADD Instruction

operand (000017_8) in the C register, and the destination address of 11_8 in the B register. The sum is formed and a write request placed. Writing the sum (000013_8) in memory location 00011_8 delays initiation of the fetch of the instruction in location 2_8. In spite of attempts to place memory requests promptly, we see that MEM is still idle at times because of delays introduced by the BUF register.

Privileged instructions are not executed unless the lower limit is set to zero, or the mode of execution is privileged. The MODE CHECK subroutine of Fig. 8.21 makes these checks. While the instruction set does not appear to provide a way of attaining the privileged mode, we will see one way in the TRAP program. Instructions of form 2 have the unusual 4-bit count field. The SHIFT COUNT subroutine copies the instruction into the D register and then uses the iteration counter to shift the copy to the right 6 places. Finally, instruction SC4 loads the iteration counter with this count field and returns. Execution of the shift and TSB instruction are then rather straightforward. The destination operand is shifted repeatedly in the C register until the iteration count has been reduced to one. Storing the shifted result is accomplished with the STORE RESULTS program. Since a circulate left microinstruction is not available, the SLC instruction is executed by circulating to the right.

Only the TRAP program illustrates additional microprogramming techniques. If an illegal instruction code (or an unacceptable privileged

Mode Check	(60)	MC1 : $\|L \doteq 0\| \rightarrow$MC3; $\rightarrow$MC2. ($C = 4$)
		MC2 : $\|M \doteq 0\| \rightarrow$MC3; $\rightarrow$MC4. ($C = 2, SH = 5$)
		MC3 : $\Rightarrow$ (allow)
		MC4 : $\rightarrow$TRAP
Shift Count	(64)	SC1 : D$\leftarrow$0$\downarrow$ IR
		SC2 : IC$\leftarrow$4
		SC3 : (4)0$\downarrow$ D
		SC4 : IC $\leftarrow$0$\downarrow$ D, $\Rightarrow$
SRZ	(90)	SRZ1 : 0$\downarrow$ C, $\not\!c$IC, $\|$IC$\doteq$1$|$ $\rightarrow$SRZ2; $\rightarrow$SRZ1.
		SRZ2 : $\Rightarrow$
SRE		SRE1 : e$\downarrow$ C, $\not\!c$IC, $\|$IC$\doteq$1$|$ $\rightarrow$SRE2; $\rightarrow$SRE1.
		SRE2 : $\Rightarrow$
	$\vdots$	
SLC	(9E)	SLC1 : IC$\leftarrow$0$-$D, $\rightarrow$SLC2
	(6C)	SLC2 : c$\downarrow$ C, $\not\!c$IC, $\|$IC$\doteq$1$|$ $\rightarrow$SLC3; $\rightarrow$SLC2.
		SLC3 : $\Rightarrow$
TSB	(C0)	TSB1 : D$\leftarrow$STATUS
		TSB2 : 0$\downarrow$ D, $\|$IC$\doteq$1$|$ $\rightarrow$TSB3; $\rightarrow$TSB2.
		TSB3 : $\|\bar{S}X \doteq 1\| \rightarrow$TSB5; $\rightarrow$TSB4.
		TSB4 : $\rightarrow$IF1
		TSB5 : CAR$\leftarrow$C, $\rightarrow$IF1

FIG. 8.21   Shift and Transfer on STATUS Bit Execution

Trap   (C6) T1 : A←11
                  T2 : B←STATUS, ⇒T11
                  T3 : B←IR, ⇒T11
                  T4 : B←U, ⇒T11
                  T5 : B←L, ⇒T11
                  T6 : IR←7
                  T7 : B←F[IR], ⇒T11
                  T8 : IR←IR−1, |IR≥0| →T10; →T9.
                  T9 : CAR←12, →IF1
        (CF) T10 : L←0, →T7
                  T11 : IOI : BUF←A, BUFX←7
                  T12 : DA : BUF←B
                  T13 : A←A−1, ⇒

Fɪɢ. 8.22   The TRAP Program

instruction) or an out of bounds address is encountered, the TRAP program places the contents of file registers 0 through 9, IR and STATUS,
in memory locating 0 through 11. It also sets the lower limit to zero
(privileged mode) and sets the CAR to 12.

It is assumed that a part of the operating system begins in location 12.
That program may examine locations 0–11 to determine the encountered
difficulty and how to respond to it. The user's program may be terminated
with a message, or the operating system may perform a service for the
user's program and then reinstate it.

To conserve ROM space, a loop is written in the TRAP program of Fig.
8.22. After the IR, STATUS, U, and L registers have been transferred to
memory, the IR is used as a down-counter to provide file addresses for
the remaining file registers. The A register provides memory addresses.

Translation

The discussed microinstructions were not assigned absolute ROM
locations at random. Programs must be placed so that they can be reached
by translating macroinstructions. And, it is very desirable to simplify the
XLATE block as much as possible. Our discussions demand the instruction translation of Table 8.10. Multiplexers may constitute the bulk of the
XLATE block, although a ROM or PLA might also be used. A 4-input,
8-bit multiplexer that is controlled by the right two bits of the N field
selects the translation mode. (The left bit of the N field determines whether
the translation result is loaded in the RAR, or not.) In mode 0 translation,
either the destination modifier field or the source modifier field must be
combined with constants to form the required next ROM address. A
2-input, 8-bit multiplexer controlled by the left bit of the op-code field
(bit a in Table 8.10) provides the necessary translation. Modes 1 and 2
require 8- and 4-input multiplexers, which are controlled by more of the
op-code bits. Mode 3 translation requires only one OR gate to examine

TABLE 8.10. TRANSLATION OF MACROINSTRUCTIONS

abcd	Mode 0 (N = 4)	Mode 1 (N = 5)	Mode 2 (N = 6)	Mode 3 (N = 7)
0000	001kℓm00	1101efg0		0110100x
0001		0111efg0	0110100x	
0010		10000efg	0110100x	
0011	⋮	01100100	10010ef0	
0100		01100100	10011ef0	
0101		101efg00		
0110		01100100	11000000	
0111	001kℓm00	11100000		
1000	010efg00	001kℓm00	10001bcd	⋮
1001				
1010				
1011	⋮	⋮	⋮	
1100				
1101				
1110				
1111	010efg00	001kℓm00	10001bcd	0110100x

NOTE: Lowercase letters designate bits from the D bus.

$$D[1:16] = abcdefghijk\,\ell mnop$$
$$x = k \lor \ell$$

2 bits of the *DM* field and to provide the least significant bit of the ROM address. While this XLATE design is not unduly complex, it clearly is directed at one instruction set, and one microprogrammed realization of that instruction set.

Translation may also be accomplished with a PLA. In general an 18-input PLA is required to translate the 16 bits on the D bus and 2 bits from the *N* field of the microinstruction. The translation of Table 8.10 can be accomplished with a 12-input PLA. Bits 1–7 and 11–13 from the D bus provide the op-code, *SM*, and *DM* fields of instructions. This reduction in the size of the input variable set makes the use of a commercial 14-input, 8-output, 96-word PLA feasible. Table 8.11 shows some of the 86 PLA words that are required to accomplish the translation of Table 8.10. Note how multiplexing is programmed and how many PLA words it requires. If the use of a PLA translation were anticipated, then the microprogrammer would have greater freedom in placing microinstructions in ROM. Uniform spacing is not important, and better utilization of the ROM can be achieved. For example, the unused ROM words of Fig. 8.18 (addresses 41–43, 4B, 4F, etc.) can be avoided with the packed PLA translation of Table 8.12. The use of subroutines to avoid duplicate instructions is still recommended. With non-uniform translation the time and space of micro-instructions, such as SO18 in Fig. 8.18, which only transfer to an area of ROM where sufficient room exists to write a desired sequence of micro-instructions, can be avoided.

TABLE 8.11. PARTIAL PLA TRANSLATION

$N_2\,N_3\,D_1D_2\,D_3\,D_4D_5\,D_6\,D_7\,D_{11}D_{12}\,D_{13}$	$O_1O_2\,O_3\,O_4O_5\,O_6\,O_7O_8$	(Hexa-decimal)
INPUT	OUTPUT	
0 0 0 x x x x x x 0 0 0	0 0 1 0 0 0 0 0	(20)
0 0 0 x x x x x x 0 0 1	0 0 1 0 0 1 0 0	(24)
0 0 0 x x x x x x 0 1 0	0 0 1 0 1 0 0 0	(28)
⋮		
0 0 0 x x x x x x 1 1 1	0 0 1 1 1 1 0 0	(3C)
0 0 1 x x x 0 0 0 x x x	0 1 0 0 0 0 0 0	(40)
0 0 1 x x x 0 0 1 x x x	0 1 0 0 0 1 0 0	(44)
⋮		
0 0 1 x x x 1 1 1 x x x	0 1 0 1 1 1 0 0	(5C)
⋮		
0 1 0 0 1 1 x x x x x x	0 1 1 0 0 1 0 0	(64)
0 1 0 1 0 0 x x x x x x	0 1 1 0 0 1 0 0	(64)
⋮		
1 0 0 0 0 1 x x x 0 0 x	0 1 1 0 1 0 0 0	(68)
1 0 0 0 0 1 x x x 1 x x	0 1 1 0 1 0 0 1	(69)
1 0 0 0 0 1 x x x x 1 x	0 1 1 0 1 0 0 1	(69)
⋮		
1 0 1 0 0 0 x x x x x x	1 0 0 0 1 0 0 0	(88)
1 0 1 0 0 1 x x x x x x	1 0 0 0 1 0 0 1	(89)
1 0 1 0 1 0 x x x x x x	1 0 0 0 1 0 1 0	(8A)
⋮		

Clearly, macroinstruction formats, microprograms, and the logic of the XLATE block are highly related. To minimize the time required to fetch, decode, and execute macroinstructions, to minimize hardware costs, and to maximize macroprogramming flexibility—in short, to optimize a micro-programmed computer, these items must be developed in parallel.

8.4 PERIPHERAL EQUIPMENT

Means must be provided to convey programs and data originally recorded in human readable form to the central processing unit, and usually it is necessary to convert results generated by that unit to human readable form. Very formidable problems arise when we attempt to convert information from a human readable form to a machine readable or electronic form and vice versa. To a large extent, equipment that is able to deal with information in a form suitable for our perusal is mechanical in nature and hence performs tasks in milliseconds rather than microseconds or nanoseconds, as does the CPU. A difference in operating speed of three to six orders of magnitude commonly exists between mechanical input-output equipment and electronic data processing units. This difference gives rise to two new classes of problems. First, it is very difficult in some

TABLE 8.12. PLA TRANSLATION TO A PACKED SOURCE OPERAND MICROPROGRAM

$N_2N_3D_1D_2D_3D_4D_5D_6D_7D_{11}D_{12}D_{13}$	$O_1O_2O_3O_4O_5O_6O_7O_8$	(Hexa-decimal)	Instruction
0 0 1 x x x 0 0 0 x x x	0 1 0 0 0 0 0 0	(40)	SO1
0 0 1 x x x 0 0 1 x x x	0 1 0 0 0 0 0 1	(41)	SO2
0 0 1 x x x 0 1 0 x x x	0 1 0 0 0 1 0 1	(45)	SO6
0 0 1 x x x 0 1 1 x x x	0 1 0 0 1 0 0 0	(48)	SO9
0 0 1 x x x 1 0 0 x x x	0 1 0 0 1 0 1 1	(4B)	SO12
0 0 1 x x x 1 0 1 x x x	0 1 0 0 1 1 1 0	(4E)	SO15
0 0 1 x x x 1 1 0 x x x	0 1 0 1 0 0 1 0	(52)	SO19
0 0 1 x x x 1 1 1 x x x	0 1 0 1 0 1 0 1	(55)	SO22

cases for such "slow" mechanical equipment to supply programs and data fast enough to keep a high-speed CPU busy. Second, the time taken by a mechanical unit to perform a task can not be controlled to an accuracy of microseconds. Thus electronic and mechanical activity can not be truly synchronized. Logic close to the mechanical devices must be asynchronous in design and prepared to cope with unbounded delays introduced by the mechanical equipment.

As the computational speed of processing units has increased over the years, more and more units of mechanical input-output equipment and peripheral mass memory have been placed around the CPU in an attempt to keep the CPU busy doing useful work. As a result it is often difficult to find the processing unit in a large computing center today: it occupies one of the smaller cabinets in a room filled with peripheral devices. The cost of the CPU may likewise be a small fraction of the cost of the total installation.

To relieve an expensive CPU of the somewhat routine tasks associated with the control of a large number of peripheral units, several special purpose units are commonly placed between the CPU and the peripheral devices. This intermediate equipment solves problems other than that of connecting many peripheral units to the CPU. Peripheral units commonly have widely different characteristics. A peripheral core, disk, drum, or tape memory can usually store and retrieve data at rates that differ from the capability of the main memory by a factor of only 2 to 10. A remote terminal, on the other hand, is only capable of supplying a few bits of information per second when it is operated by one who does not type well. It can accept bits and print characters at rates on the order of 100 bits per second. But this is many orders of magnitude less than the rate at which the main memory of a computer may supply information. And the units may differ in other significant ways. The number of bits that different peripheral units treat as a unit of information may differ. Some may treat 16-bit sequences as a word; others may deal with 8-bit character codes or sequences of such code words. And the codes which they recognize often differ. In a specific installation only the special purpose equipment required to resolve the existing differences need be placed between the CPU and peripheral devices, whereas it is not usually economical or desirable to construct each CPU so that it can deal directly with the wide variety of peripheral units that are now available or will be made available for attachment to a CPU at some future date.

A *channel* is a special purpose digital computer capable of communicating with the CPU and main memory on one hand, and one or more peripheral devices on the other. Channels can be modest in complexity and cost, but in large computing systems channels often approach the size and complexity of the elementary digital computers which we have been considering. In fact, in some cases small, low cost, general purpose computers are used as channels for large high-speed processors. A channel may be dedicated to serve one peripheral device, such as a mass memory, with data rates that approach the rate of the main memory of the system. Or a channel might serve a number of high data rate units, but not simultaneously. At the other extreme, since teletypewriters, card handling equipment and line printers have data rates far below that of the main memory, a channel may be constructed to serve many peripheral units on a time-shared basis. First one unit is served, then a second, etc. until all peripheral units have been serviced once. Then the cycle repeats; if it is carefully designed such a channel is able to keep all units connected to it fully occupied.

Usually some equipment must be unique to a peripheral unit and dedicated to one specific piece of mechanical equipment. We will call such dedicated equipment the peripheral *controller*. We attempt to minimize the complexity and cost of the controllers, since many such units may be required for each CPU, and place as much costly logic as possible in the channel where it serves a number of peripheral units.

Channels may be connected to CPUs and controllers may be connected to channels in any of several ways. One organization might be called a *star* configuration. The master unit, either a CPU or channel, is designed to have several information ports. One servant unit, either a channel or controller, is connected to each port, or the port may go unused. The master unit may then present information to a specific servant by presenting it at the appropriate port. In some circumstances the star configuration lacks flexibility and requires the inclusion of data-routing circuitry which may go unused, within the master unit.

The *ring* interconnection configuration shown in Fig. 8.23 alleviates these problems. Many servant units intercept one bus connected to one port of the master unit. Each servant monitors all messages placed on the bus, but captures and responds only to those which are addressed to it. Some sort of address must accompany each word of information placed on the bus so that each servant can determine whether the word pertains to it or not. If it does not, the servant amplifies the information and passes it on to the next servant. Generally this means that servants need not be placed physically as close to the master because of the repeated amplification of the bus signals, but greater transmission delays must also be expected. If the bus is ultimately returned to the master, then a communication path

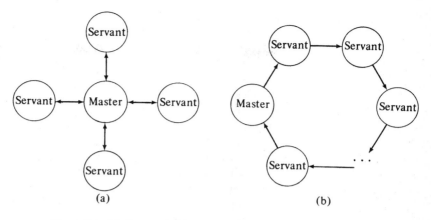

FIG. 8.23 (a) Star and (b) Ring Interconnection Configurations

from each servant to the master is established at little additional expense. We will assume the use of a ring configuration in subsequent examples.

We will look at only one example of a channel and controller, to see the function of these units and characteristics of peripheral units. Alternative designs for these units will be suggested as we proceed, but their full exploration must be left to the reader. We will attempt to provide a channel and control unit such that a number of teletypewriter terminals might be connected to the CPU of the previous section and serve as input-output units for it.

The Channel

A channel has the responsibility of passing words of information between the main memory of a computing system and a device controller. To do so the channel is required to know: (1) which peripheral device is involved and its status; (2) the direction in which the information is to be passed; (3) the location in memory where the information resides or is to reside; and (4) the number of words to be transferred. In the original version of EDC, considered in Sec. 7.1, CPU obtained this information from the input or output instruction it was executing at the time. Again the instruction set of CPU must include input-output instructions whereby a programmer can provide this information to CPU. But now CPU must present suitable information to the channel, or provisions must be made for the channel to obtain or generate the information it requires.

CPU and a channel may communicate directly via a bus such as the II bus, or indirectly via the main memory of the computing system. To illustrate CPU-channel communication, we will assume the bus structure of previous sections. Then CPU can emit words to the channels and the channels can return requested information to CPU, and issue interrupts to CPU. The INSTRUCTION FETCH microprogram can be rewritten to check for such interrupts. And, when an interrupt is found it causes CPU to return control to an executive program (perhaps with the TRAP microprogram of the previous section) that can poll the channel to find the reason for the interruption.

We will refer to a word emitted directly by CPU to a channel as a *channel command*. The channel command word might have the following format.

```
1          3 4    6 7        16
 ┌──────────┬──────┬──────────┐
 │    T     │  D   │    A     │
 └──────────┴──────┴──────────┘
```

T : bits 1–3 Command Type
D : bits 4–6 Device Address
A : bits 7–16 Memory Address

With three bits used to designate command type, any of eight different commands may be conveyed to a channel. These commands might be:

000 : Start Input
001 : Start Output
010 : End File
011 : Rewind
100 : Backspace File
101 : Advance File
110 : Backspace Record
111 : Advance Record

The actions to be taken by a channel in response to each of these command types are rather evident. The device address field is included in the command word because we assume the ring interconnection, and expect to connect more than one peripheral device to the channel. Our example of a 3-bit device address field allows the channel to be connected with up to eight peripheral units.

Field A of the channel-command word conveys a main memory address. CPU could convey word count and memory address information directly to a channel, but such direct communication would involve more than one 16-bit word. Since a channel can communicate with main memory, it is proposed here that the CPU be programmed to place word count and the starting address for data into two consecutive memory locations before it issues a command to the channel. Then it need only tell the channel where in main memory the information it requires can be found. The use of a 10-bit field in the command word requires that the word count and data address be stored in the first 1024 locations of main memory. Alternatively, the A field could be eliminated entirely from the channel command word if the channel were constructed so that it always sought word count and data address information from the same two main memory locations.

Figure 8.24 offers some detail of a channel that can respond to such channel command words addressed to it and can control any one of eight peripheral input or output devices at a time. To emphasize the main functions a channel must perform, many sophisticated details have not been included. Some possible extensions of this very basic channel will be discussed as we examine it; others are offered as problems.

Data channel DC continuously monitors the II bus for command words addressed to it. Commands from CPU are recognized only if DC is in state IDLE when the command is issued. The command type and device address fields of such commands are placed in register C; the A field is placed in register ADR along with high-order bits of 0. Then in states S1, S2, S3, and S4, DC calls on MEM for the word count and data address words that pertain to the command. These words are placed in registers WRDCNT and ADR, respectively. In state S4 the command type is decoded.

If the command is to start input, DC issues an input order to the appropriate device by emitting the order type and device address over the channel to device bus, CTOD. It continues to issue this command (by remaining

⟨AU⟩ DC : P :
    ⟨TE⟩ CTOD[22] = TYDA[6] ∘ DDATA[16],
            DTOC[22] = RESP[6] ∘ RDATA[16],
    ⟨RE⟩ C[6] = TYPE[3] ∘ DADR[3], WRDCNT[10], ADR[16],
            DATA[16], STAT.
    ⟨BO⟩ ARB[5] = STAT.
⟨ST⟩ IDLE : · /IIA\2D2 : C←IID[1 : 6], ADR←0D6 ∘ IID[7 : 16],
         STAT←0, →S1.
     S1 : ¬ARB[1] · ARB[4] : ARB[2] = 1, II = ADR ∘ 4D4,
        ¢ADR, →S2.
     S2 : ·/IIAO\8D4 : WRDCNT←IID[7 : 16], →S3.
     S3 : ¬ARB[1] · ARB[4] : ARB[2] = 1, II = ADR ∘ 4D4, →S4.
     S4 : ·/IIAO\8D4 : ADR←IID, [TYPE |0 →IN1 |1 →OUT1;
        STAT←1, →IDLE. .
     IN1 : TYDA = C, |∨/(TYDA ⊕ RESP)| →IN1;
        DATA←RDATA, →IN2. .
     IN2 : ¬ARB[1] · ARB[4] : ARB[2] = 1, II = ADR ∘ 6D4, →IN3.
     IN3 : ·/IIAO\10D4 : IID = DATA, ¢ADR, ¢WRDCNT,
        | ·/WRDCNT\1D10| →IN4; →IN1. .
     IN4 : ¬ARB[1] · ARB[6] : ARB[2] = 1, IIAO = 13D4,
        STAT←1, →IDLE.
     OUT1 : ¬ARB[1] · ARB[4] : ARB[2] = 1, II = ADR ∘ 4D4,
        →OUT2.
     OUT2 : ·/IIAO\8D4 : DATA←IID, →OUT3.
     OUT3 : CTOD = C ∘ DATA,
        |∨/(TYDA ⊕ RESP)| →OUT3; →OUT4. .
     OUT4 : ¢ADR, ¢WRDCNT, | ·/WRDCNT\1D10| →IN4;
        →OUT1. . . .

FIG. 8.24 A Very Simple Channel That Is Compatible
with the System of the Previous Section

in state IN1) until the device controller returns the order type to DC over the DTOC bus. DC accepts this message, signifying that the device controller has placed one word on the 16 RDATA lines of the DTOC bus, by loading its DATA register from these lines. Then in states IN2 and IN3 the word is transferred to MEM, the address for data is incremented, the word count decremented, and a test for completion of the command is made. If words remain to be transferred, DC returns to state IN1.

One serious shortcoming of this design is apparent. DC must wait in state IN1 for from 100 milliseconds to perhaps several minutes for a new data word if the input-output device is a terminal operated by a human. DC could and should be designed to interrogate each of its devices periodically, say every 10 ms, and respond to the device only if it is prepared to conduct a transaction. After issuing an input or output order to a device, DC could deal in turn with each of the other devices it must service. It could then return to the original device and test it asking whether it has completely responded to the previously issued order. If it has, DC must supervise the transfer of a word between MEM and the device controller before going on to service its other devices. But if the device has not yet completed its response to a previous order, DC need not wait for it to do so, but can go on directly to the interrogation of other devices. The rate at which DC must poll each device is determined by the rate at which the device can respond to commands. The teletype unit, which we will discuss, requires approximately 100 ms to generate or print a word. Polling such devices every 10 ms will certainly ensure that information is not lost and that the devices connected to a channel never need wait on their channel for service.

Other shortcomings of the design of Fig. 8.24 pertain if the input-output devices are controlled by humans. In such situations, the messages to be provided by the device operator are seldom of a fixed, known, word count. It is more desirable for CPU to request a maximum number of words to be transferred to memory, but to allow the operator to provide an "end of message" word to DC. Some keyboard character that is not otherwise used commonly serves this purpose. When DC senses this message it can terminate the input command and notify CPU that it has done so by interrupting CPU.

The human operator feels rather uncomfortable if what he types is not printed before him. The device controller or the device itself may be constructed to connect the keyboard and printing mechanism directly, but some error detection capability can be provided by having DC accept a word and then immediately return it to the input-output device for printing or electronic comparison with the word originally supplied by the device. The transmission facilities and certain of the registers of both the channel and controller may be continuously checked in this manner. If an error is detected, or the operator feels that he has sent the wrong word, he must be provided with a "correction" word that DC is prepared to detect and respond to by counting ADR down and incrementing WRDCNT.

Controller

Before we can offer details of a controller that matches a given channel to a specific input-output device, we must know the characteristics of the device as well as those of the channel. We will examine the teletype as an input-output device because it is widely used and because it is relatively simple. While a variety of teletypewriters exist, each of which may be used in a variety of ways, we will assume an instrument that can be operated in a full-duplex mode, i.e., transmission and reception of information words are independent functions. Thus the keyboard may be operated so that characters are emitted while the printing mechanism is idle or it is printing information supplied to it. This printed information need not correspond to the information being typed on the keyboard.

The teletype keyboard and associated transmitting mechanism can be viewed as a single, mechanical, normally closed switch. When a key is depressed, this switch opens for approximately 9 milliseconds and then **opens and closes to present an 8-bit ANCII code word in serial fashion with the least significant bit being presented first.** Each bit is presented for approximately 9 ms. Finally, the switch remains closed for approximately 18 ms as an "end of character" symbol. Thus each time a key is depressed, the teletype may be considered to emit an 11-bit code word over an interval of $11 \times 9 \approx 100$ milliseconds. Up to 10 characters can be transmitted per second.

The teletype printer is electrically compatible with the keyboard, and mechanically tied to it to the extent that printer and keyboard cycles of activity are synchronized. The printer may be viewed as an electric current sink for our purposes. When the printer continuously receives 20 mA (or 60 mA) of current, it is quiescent. An interruption of this current for 9 ms serves as a "start" signal that initiates a cycle of printer activity. The eight bits of a code word must be serially presented to the printer following the start bit, and two "end of character" bits must follow the information bits. Each bit must be presented for approximately 9 ms. Since the teletype is a mechanical device, only approximate rather noncritical times can be given. These times will vary slightly from one teletype unit to another, and with the function being performed by the teletype.

The teletype controller specified in Fig. 8.25 consists of two automata, one of which is synchronized with DC, while the other is approximately synchronized with the mechanical teletype cycle. These automata share an 8-bit shift register CHAR in which character codes are stored, and a number of flip-flops and terminals. Register OP, which consists of flip-flops RW (receive word) and SW (send word), holds the command that automaton X accepts from DC. The setting of one or both of these flip-flops initiates activity in automaton Y. If flip-flop SW is set, automaton Y accepts a word from the teletype transmitter and places it in register CHAR. If flip-flop RW is set, automaton Y presents the character code held by CHAR to the teletype printer mechanism. If both of these flip-

⟨SY⟩ TTY CONTROL:
　　⟨TE⟩ CTOD[21] = OTYPE[2] ∘ DEVICE[3] ∘ DDATA[16],
　　　　DTOC[21], DID[3].
　　⟨RE⟩ CHAR[8], A, B, C, D, E, F, OP[2] = RW ∘ SW.
　　⟨BO⟩ DID = iD3. (Each device is given a unique identification)
　　⟨AU⟩ X : P : | ∨ /(DEVICE ⊕ DID) | DTOC = CTOD.
　　⟨ST⟩ S1 : ↓/(DEVICE ⊕ DID) : OP ← OTYPE,
　　　　　　　　CHAR ← DDATA[9 : 16], →S2.
　　　　S2: A : →S3.
　　　　S3: A : DTOC = OP ∘ DID ∘ 0D8 ∘ CHAR, OP ← 0, →S1...
　　⟨AU⟩ Y : PP : * | RW ↓ SW | A ← B ← C ← D ← E ← 0.,
　　　　　　　　* | START | A ← 1., | B̄ ∘ D ∘ E | A ← 0.,
　　　　　　　　| C̄ · D̄ · E | B ← B̄.,
　　　　　　　　| B · D · E | C ← C̄.,
　　　　　　　　| E | D ← D̄.,
　　　　　　　　E ← Ē, F ← CHAR[8],
　　　　　　　* | SHIFT | ↓↗ S̄ · B · SW ∘ CHAR.,
　　　　　　　START = B̄ · C̄ · D̄ · (A ∨ S ∨ RW),
　　　　　　　TOTTY = START ∨ B · RW · F̄,
　　　　　　　SHIFT = B · SFT,
　　　　　　　V = U, W = Ū, RUN = A, H = TTY CONTACTS.
　　⟨EL⟩ ST(S : H), AMV(U : RUN),
　　　　MMV1(PP : V), MMV2(SFT : W)...

Fɪɢ. 8.25　The Teletypewriter Controller

flops are set, both activities are commanded. The specifications and actions
of automaton X are rather straightforward. Automaton Y is not described
in terms of state declarations, so its activities are more difficult to ascertain.
Flip-flops A, B, C, D, and E make up the control unit of automaton Y
and act essentially as a modulus-11 counter. This counter was not specified
with a state declaration, to illustrate that it is not necessary to use the state
declaration to specify a sequential network, and because a specific state
assignment was desired, to simplify the logic of the entire controller.
Specifically, flip-flop B is set when the teletype transmitter is presenting the
8 information bits and the printer is accepting them. Flip-flop A is set
whenever the teletype is executing a cycle of activity, and is reset when the
teletype is quiescent.

　　To synchronize automaton Y and the teletype unit, an astable multivibrator, AMV, with an adjustable period of approximately 9 msec is
specified with an ⟨EL⟩ declaration. This multivibrator can be stopped by
applying 0 to its input terminal, RUN. Connection specification RUN =
A indicates that the multivibrator is to run whenever flip-flop A is set, i.e.,
whenever the teletype is active. Two monostable multivibrators, MMV1
and MMV2, are used to provide very brief (say 50 nanosecond) pulses in
synchronism with the state changes of AMV. One of these sets of pulses,

PP, is used to synchronize most of the activity of automaton Y; the other set is used to shift register CHAR. Figure 8.26 shows the relative timing of many of these signals and a block diagram of the entire controller.

A Schmitt trigger, ST, is also specified as an element. This circuit is used to eliminate the contact bounce of the teletype transmitter switch and to convert switch closures and opening to electronic signals. S takes the value of 1 when the transmitter switch is open, and expresses the complement of the character code bits.

Let us assume now that DC issues an input order to the controller of Fig. 8.25. Automaton X monitors the CTOD bus and immediately passes those messages that are not addressed to it to its DTOC terminals. If the order is addressed to this controller it is stored in register OP where flip-flop SW is set. Subsequent activity of automaton X is rather straightforward.

If automaton Y is idle, as determined by $\overline{B}\overline{C}\overline{D} = 1$, when the input order is received and flip-flop SW is set, then signal START $= 0$ until the operator of the teletype depresses a key. The automaton may have to wait several seconds or minutes for this event. The teletype transmitter "start" bit causes $S = 1$, which in turn causes START $= 1$. Flip-flop A is asynchronously set and the cycle of electronic activity of automaton Y begins. When flip-flop A is set, the multivibrator begins to operate. A sequence of 12 PP pulses that causes the counter to count as shown in Fig. 8.26 is emitted. Eight SHIFT pulses cause successive values of $\overline{S} \cdot B \cdot SW$ to be shifted into the character register CHAR. Note that this shifting does not take place in synchronism with the counting activity. It is not advisable to sample signal S near the time when it may change its value, because these times can not be accurately predicted, and because severe contact bounce could cause S to take false values at these switching times. Thus S is sampled at or near the middle of each teletype bit interval.

The counter ultimately resets flip-flop A and activity stops. The period of AMV must be adjusted so that this event occurs close to or after the end of the mechanical teletype cycle, if the state of flip-flop A is to truly reflect the conditions of the mechanical teletype. Note that the 8 bits of information are placed in CHAR some 22 ms before the end of the cycle. This character code may be transmitted to DC at any time during this interval, but automaton X waits until the end of the cycle to do so.

When DC issues an output order, the 8 bits are transferred from CTOD bus to the CHAR register and the RW flip-flop is set. This in turn causes START $= 1$, which initiates activity in automaton Y. The START signal itself serves as the "start" bit passed to the teletype printer via signal TOTTY. Then the 8 information bits are presented in turn to the printer. Because register CHAR is not shifted in synchronism with the counter, and hence not in synchronism with the teletype bit boundary times, flip-flop F clocked by PP is introduced.

A number of LSI circuits are available that perform the functions of this simple controller, and more. Detailed logic design is largely avoided by using these integrated circuits, but the need to understand the charac-

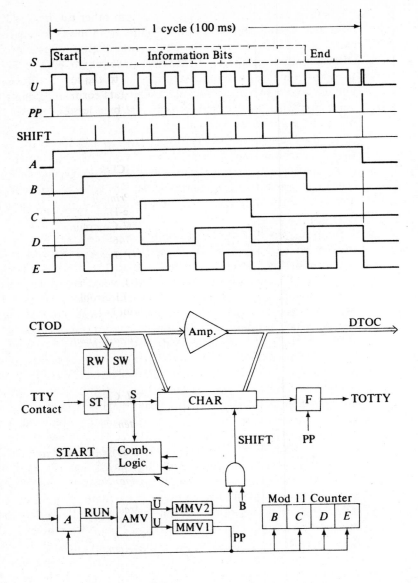

FIG. 8.26 Details of a Teletypewriter Controller

teristics of the channel and the peripheral device is not reduced.

The number of variations and improvements that one can imagine of the basic channel and controller presented here is almost without limit. A number of possibilities are explored in the problems for this section. It is not important that we be familiar with many variations in any detail. The concept of relieving the CPU of the need to remain idle while slow peripheral devices operate is fundamental. Less expensive units can be called upon to perform many of the input-output tasks that must be performed.

They may well be called upon to perform other routine tasks as well. For example, our DC transfers 16-bit words from main memory to peripheral devices. In the case of the teletype, only 8-bit character codes are used. DC might be constructed to break each 16-bit word into two 8-bit character codes, and present them successively to the teletype controller. This is almost a trivial example of the type of data formation that often must be performed, and can often be performed by a lesser unit than the CPU.

REFERENCES

1. A. M. ABD-ALLA and A. C. MELTZER, *Principles of Digital Computer Design, Vol. 1.* Englewood Cliffs, N.J.: Prentice-Hall, 1976.
2. G. H. BARNES, et al, " The ILLIAC IV Computer," *IEEE Transactions on Computers,* Vol. C-17, Aug. 1968, pp. 746–757.
3. C. G. BELL and A. NEWELL, *Computer Structures: Examples and Readings.* New York: McGraw-Hill, 1970.
4. A. W. BURKS, H. H. GOLDSTINE, and J. VON NEUMANN, " Preliminary Discussion of the Logical Design of an Electronic Computing Instrument," Institute for Advanced Study, Princeton, N.J., 1946 (reprinted: *Datamation,* Sept. 1962, pp. 24–31).
5. Y. CHU, *Introduction to Computer Organization.* Englewood Cliffs, N.J.: Prentice-Hall, 1972.
6. Y. CHU, *Computer Organization and Microprogramming.* Englewood Cliffs, N.J.: Prentice-Hall, 1972.
7. H. W. GSCHWIND, *Design of Digital Computers.* New York: Springer-Verlag, 1967.
8. H. HELLERMAN, *Digital Computer System Principles.* New York: McGraw-Hill, 1973.
9. F. J. HILL and G. R. PETERSON, *Digital Systems: Hardware Organization and Design 2nd ed.* New York: John Wiley and Sons, 1978.
10. K. E. IVERSON, *A Programming Language.* New York: John Wiley and Sons, 1962.
11. M. M. MANO, *Computer System Architecture.* Englewood Cliffs, N.J.: Prentice-Hall, 1976.
12. M. E. SLOAN, *Computer Hardware and Organization: An Introduction.* Chicago: Science Research Associates, 1976.

PROBLEMS

8.1-1. Describe in DDL the decoding and execution of an ADD instruction, assuming the file–ALU organization of Fig. 8.2.

8.1-2. The organization of Fig. 8.3(a) will support register$_1$ + register$_2$ to register$_2$, and register$_1$ + memory to register$_2$ ADD instructions. Assume that one bit of the A field distinguishes these types of ADD instructions. Use DDL to describe the hardware necessary to decode and execute these ADD instructions.

8.1-3. The EDC op-codes of Table 7.1 (ADD = 05$_8$) are used below in instructions to be fetched, decoded, and executed by the hardware of Fig. 8.4.

File registers 6 and 7 may be used as working registers, if required. Write the sequence of register transfers that must take place to fetch and execute each instruction. Indicate the mode of operation of the ALU at each step.

(a) 042003 (instruction)
    000713 (address word)
(b) 040312
(c) 050312
(d) 052020
    001577
    000033
(e) 060332
    000775

8.1-4. The instruction set of EDC, as given in Table 7.1, is extended to permit the addressing modes of Table 8.2 in instructions that do not already designate the IX bits. The hardware of EDC is that of Fig. 8.5. For each instruction, list the register transfers that must take place to fetch, decode, and execute it. The contents of selected memory locations are listed so that the numerical result of each transfer can also be given. Assume that CAR $= 000010_8$, B $= 000005_8$, and ACC $= 000111_8$ before each instruction is fetched.

| | Memory | |
(a) 044003	Address	Content
(b) 046003	0	177777
	1	000004
(c) 054000	2	000006
	3	000000
(d) 074001	4	000377
	5	000167
(e) 106001	6	000100
	7	177776

8.1-5. Assuming that a 16-register file is available, design a suitable up-down counter and control unit, as in Fig. 8.6, to create a push-down stack.

8.2-1. In Fig. 8.8, MEM gives priority to data channel requests over those of CPU. Suppose ten devices can request service, each on its own device request (DR) line. If DR[1] has highest priority, while DR[10] has lowest, design an iterative network that will identify the highest priority request of those simultaneously received, if any. Identify the "no request" case also.

8.2-2. Time-sharing a bus is a standard communication technique. The values of 25 binary signals, $A, B, \ldots, Y$ are to be transmitted in turn over line Z; Z is to have the value of each input signal for 4 millisec., every 100 msec.
(a) Specify a *multiplexer* that will generate Z.
(b) Twenty-five signals like Z are to be multiplexed and transmitted over line Z'. How must your multiplexer be changed if it is to be able to generate Z'?
(c) The narrowest pulse that may be transmitted successfully over a line has a duration $\tau = 1/f_c$ where f_c measures the *bandwidth* of the transmission line. What is the minimum bandwidth of a line to carry Z? Z'?

532   CHAPTER 8   Advanced Computer Concepts

8.2-3. The ARB bus of Fig. 8.10 requires two wires per block. To reduce this number, a single *bid* wire is to be used with blocks placed along that single line in order of priority. Develop a typical cell (of an iterative network) that may be placed in a block which determines if any unit to the left is bidding. And if not, and the block wishes to bid, the availability of the bus is indicated to the block, and all blocks to the right are notified that they may not bid.

8.2-4. In BUSEDC of Figs. 8.12 and 8.13, indicate the control signals that must be set to 1 during each clock period of a fetch and execute cycle for each of the following instruction types:

(a) LDA
(b) ADD
(c) STO
(d) SRA 5
(e) TRA

8.2-5. Modify Fig. 8.13, while preserving the architecture of Fig. 8.12, so that the ADD instruction may be addressed in the modes given in Table 8.2. Comment on the importance of the BUF register.

8.3-1. Let the BUFX register consist of six flip-flops, the right four of which are loaded from the *DATA* field when the BUFX microinstruction field is a 1 and the control unit is idle. Draw a state diagram for that control unit. It must provide the DA and IOI signals and drive the CPU bid and status lines of the ARB bus.

8.3-2. Use multiplexers in a design of a shifter for use in Fig. 8.14. The shifter must be controlled by the 3-bit SH field with the code of Table 8.5.

8.3-3. Registers 1 and 2 hold a double precision 2's complement number. The left bit of register 1 is the sign bit; register 2 holds the least significant magnitude bits. Registers 3 and 4 hold a second double precision number. Double precision results are to be placed in registers 3 and 4. Write sequences of microinstructions to perform: (a) double precision add, and (b) double precision subtract.

8.3-4. Write a microinstruction, or sequence of such instructions, to perform 1's complement (a) addition and (b) subtraction. Assume the operands are in registers 1 and 2, and the result is to be placed in register 2.

8.3-5. Write a microprogram to perform Booth multiplication. Assume the operands are in registers. Use any ROM addresses that you require, but keep the program as short as you can.

8.3-6. Write a program of macroinstructions that performs Booth multiplication. Assume the operands are in registers 1 and 2, and the results are to be placed in registers 3 and 4.

8.3-7. Alter the INSTRUCTION FETCH microprogram of Fig. 8.17 to check for an unmasked interrupt, and trap when one is found.

8.3-8. Draw a flowchart of the entire SOURCE OPERAND microprogram of Fig. 8.18.

8.3-9. The XLATE block built of multiplexers would be simpler if all instructions were translated to $001k\ell m00$ in mode 0, and instructions with operands 1000 through 1111 were translated to $010efg00$ in mode 1. What effect would this have on the macroinstructions, and how much simpler would the XLATE block really be?

8.4-1. A channel is needed to connect the main memory of Fig. 8.10 to a slower magnetic core memory with a capacity of 2^{19} 32-bit words and a cycle time of 10 μs.

(a) Respecify the automaton MEM of Fig. 8.10 so that after this channel requests its services, all other requests from CPU or other channels are ignored until this channel releases main memory MEM.

(b) Specify a controller for the 2^{19}-word, 32-bits per word, 10 μs peripheral memory, which when supplied a 10-bit address, reads or writes in sequence a block of 2^9 words. The 10-bit address supplied is to be taken as identifying a block of 512 32-bit words.

(c) Specify a channel that connects the main memory of part (a) with the peripheral memory of (b). This channel must accept commands from CPU; issue commands to both memories, break the 32-bit word supplied by the peripheral memory into two 16-bit words and supply them in turn to main memory; and pack two 16-bit words from main memory into a 32-bit word for the peripheral memory. The transfer of a block of 512 32-bit words is to take place as rapidly as possible, of course.

8.4-2. A channel is to connect at most eight teletypewriters with main memory on a time-shared basis. Each teletype is to be polled every 8 milliseconds; a clock with a 1 ms period is available to facilitate this timing. The channel is to use 16 or 24 words of main memory rather than flip-flop registers to store status, word count, and data address information for each of the teletypes. Discuss the status information that will be stored and the interrogation and response codes to be used. Specify the channel.

8.4-3. Extend the channel of Prob. 8.4-2 so that:
(a) An input command may be terminated by the operator striking the EOM (end of message) key, which provides the code word 203_8.
(b) The previous character may be deleted by the operator striking the ← key, which provides the code word 337_8.

8.4-4. The teletype controller of Fig. 8.25 includes an 8-bit shift register, which could be a part of the channel. Then communication of data between channel and controller may be serial rather than parallel.
(a) Respecify the channel and controller of Figs. 8.24 and 8.25 and the buses that interconnect them so as to use serial transfer of character cards.

(b) Respecify the time-shared channel of Prob. 8-4.2, so as to use serial transfer. Estimate the hardware saved by using this approach, assuming that eight controllers are connected to the channel.

9

Computer Manipulation of Switching Functions

Tables, maps, equations, and block diagrams provide mathematical models of binary switching circuits. These facilitate the analysis and synthesis of such circuits. Each type of model serves the logic designer according to his/her training and experience. Some are best for analysis; others may serve best when other jobs must be performed. Some models facilitate visualization; others are amenable to algorithmic procedures. We will concentrate here on the means to handle larger combinational switching circuit problems with the assistance of the general purpose digital computer. The tuple and array notation of Section 2.5 provides data formats that are well suited to efficient machine computation. Operators and a Boolean algebra of arrays are developed by generalizing set operations and algebra. This development parallels and utilizes the generalization of binary tuples that represent minterms to ternary tuples—representing product terms—in Section 2.5. Basic applications of this algebra are shown before discussing efficient computer implementations of the operators. Then the chapter turns to the order of applying operators—computational algorithms—so as to calculate desired results.

9.1 THE ALGEBRA OF BASES

When we write a switching function in sum-of-minterms form, we record those minterms associated with truth table entries of 1. The minterms may be written in algebraic form, referenced by name using the naming convention illustrated in Table 2.8, or indicated by listing the subscripts on those names. If those subscripts are expressed in binary, they reveal the input symbols for which the function takes the value of 1.

$$G = \bar{A}\bar{B}C \vee \bar{A}BC \vee AB\bar{C} \vee ABC$$
$$= m_1 \vee m_3 \vee m_6 \vee m_7$$
$$= \bigvee 1, 3, 6, 7$$

The set of minterms of a function is known as the minterm set, or true set, of the function. We call it the *base* of the function, symbolized $K^0(f)$.

$$K^0(G) = \{1, 3, 6, 7\} = \begin{Bmatrix} 001 \\ 011 \\ 110 \\ 111 \end{Bmatrix}$$

The set of input symbols expressed as decimal integers or binary n-tuples, for which a function equals 0, will be referred to as the *maxterm set*, and denoted $L^0(f)$. Thus, for example function G, we may write any of the following:

$$G = \prod M_0, M_2, M_4, M_5 \qquad L^0(G) = \{0, 2, 4, 5\} = \begin{Bmatrix} 000 \\ 010 \\ 100 \\ 101 \end{Bmatrix}$$
$$= \prod 0, 2, 4, 5$$

If the function is not completely specified, a third set of input symbols exists. The set of these symbols is the *don't care* array $D^0(f)$. Since G is a complete function, $D^0(G) = \phi$, the empty set.

Function f takes the value of 1 for every input symbol included in $K^0(f)$, value 0 for every member of $L^0(f)$, and its value is not determined for members of $D^0(f)$. Every integer (or equivalent binary n-tuple) of the universal set $U_n^0 = \{0, 1, \ldots, 2^n - 1\}$ appears in exactly one of $K^0(f)$, $L^0(f)$, and $D^0(f)$.

$$K^0(f) \cap L^0(f) = K^0(f) \cap D^0(f) = L^0(f) \cap D^0(f) = \phi$$
$$K^0(f) \cup L^0(f) \cup D^0(f) = U_n^0 \qquad \qquad \textbf{(9.1)}$$

For completely specified functions, $D^0(f) = \phi$ and

$$K^0(f) \cap L^0(f) = \phi$$
$$K^0(f) \cup L^0(f) = U_n^0 \qquad \qquad \textbf{(9.2)}$$

We assume completely specified functions in the following development.

From Eq. (9.2), we see that $K^0(f)$ and $L^0(f)$ are complementary sets. Given one, we can find the other with ease; i.e., those input symbols not in $K^0(f)$ are the members of $L^0(f)$.

$$K^0(f) = \overline{L^0(f)}$$
$$G = \bigvee 1, 3, 6, 7 = \prod 0, 2, 4, 5 \qquad \qquad \textbf{(9.3)}$$

The complement of a Boolean function may be expressed as follows:

$$\bar{f}(x_1, \ldots, x_n) = \overline{\bigvee f_d \cdot m_d} = \overline{\prod (f_d \vee M_d)} \tag{9.4}$$

where the range of d is 0 through $2^n - 1$. Using deMorgan's Theorem and the fact that each minterm is the complement of the maxterm with the same subscript gives

$$\begin{aligned} \bar{f} &= \prod (\bar{f}_d \vee \bar{m}_d) = \bigvee \bar{f}_d \cdot M_d \\ &= \prod (\bar{f}_d \vee M_d) = \bigvee \bar{f}_d \cdot m_d \end{aligned} \tag{9.5}$$

The complement of a switching function can be expressed in either canonical form for it is another switching function. The result to be noted is that the bases of f and $\bar{f}$ are complementary.

$$\begin{aligned} K^0(\bar{f}) &= \overline{K^0(f)} = L^0(f) \\ K^0(f) &= \overline{K^0(\bar{f})} = L^0(\bar{f}) \\ G &= \bigvee 1, 3, 6, 7 \qquad G = \prod 0, 2, 4, 5 \\ \bar{G} &= \bigvee 0, 2, 4, 5 \qquad \bar{G} = \prod 1, 3, 6, 7 \end{aligned} \tag{9.6}$$

If f and g are complete functions of the same input variables, then the base of their logic product is the set intersection of their bases.

$$\begin{aligned} K^0(f \cdot g) &= K^0(f) \cap K^0(g) \\ L^0(f \cdot g) &= L^0(f) \cup L^0(g) \end{aligned} \tag{9.7}$$

This result is easily derived using the fact that the logic product of different minterms is logic 0.

$$\left.\begin{aligned} m_i \cdot m_j &= 0 \\ M_i \vee M_j &= 1 \end{aligned}\right\} \quad \text{for } i \neq j \tag{9.8}$$

If:

$$E = \bar{A}B\bar{C} \vee BD \qquad K^0(E) = \{4, 5, 7, 13, 15\}$$
$$F = \bar{A}\bar{C} \vee \bar{B}\bar{C}\bar{D} \qquad K^0(F) = \{0, 1, 4, 5, 8\}$$

then

$$\begin{aligned} Z &= E \cdot F \\ &= (m_4 \vee m_5 \vee m_7 \vee \ldots) \cdot (m_0 \vee m_1 \vee m_4 \vee \ldots) \\ &= m_4 \cdot m_0 \vee m_4 \cdot m_1 \vee m_4 \cdot m_4 \vee \ldots \vee m_5 \cdot m_0 \vee \ldots \\ &= \quad 0 \quad \vee \quad 0 \quad \vee \quad m_4 \quad \vee \ldots \vee \quad 0 \quad \vee \ldots \end{aligned}$$

We see the following result emerging.

$$\begin{aligned} K^0(Z) &= K^0(E) \cap K^0(F) \\ &= \{4, 5, 7, 13, 15\} \cap \{0, 1, 4, 5, 8\} \\ &= \{4, 5\} \\ Z &= \bar{A}B\bar{C}\bar{D} \vee \bar{A}B\bar{C}D \end{aligned}$$

By Eq. (9.3),

$$L^0(Z) = \overline{K^0(Z)}$$
$$= \overline{K^0(E) \cap K^0(F)}$$
$$= \overline{K^0(E)} \cup \overline{K^0(F)}$$
$$= L^0(E) \cup L^0(F)$$
$$= \{0, 1, 2, 3, 6, 7, \dots, 15\}$$

The base of the logic sum of two complete switching functions is the union of their bases.

$$K^0(f \vee g) = K^0(f) \cup K^0(g)$$
$$L^0(f \vee g) = L^0(f) \cap L^0(g) \tag{9.9}$$

A minterm of either or both f and g is a minterm of $f \vee g$.

$$K^0(E \vee F) = \{4, 5, 7, 13, 15\} \cup \{0, 1, 4, 5, 8\}$$
$$= \{0, 1, 4, 5, 7, 8, 13, 15\}$$

When switching functions are expressed in canonical form, the Boolean algebra of switching functions reduces to set algebra. We will generalize this idea later in the chapter. For now it is interesting to look back at Section 2.3, where an algebra of truth tables was introduced. Since canonical forms are closely tied to truth tables, we may find the algebra of bases and the algebra of truth tables to be essentially the same.

9.2 THE STRUCTURE OF SWITCHING FUNCTIONS

Canonical expressions of switching functions suggest that they consist of disjoint basic objects (minterms or maxterms). Often a collection of minterms (or maxterms) forms a larger unit, and a function may be thought to have structure. The topological representation of switching functions provides a way of illustrating that structure. The related array representation of switching functions is excellent for computer manipulation.

In Section 2.5, we associated a unique binary n-tuple with each minterm of n variables. In turn each such n-tuple determines a unique point in an n-dimensional space, a unique vertex of an n-dimensional unit cube, or n-cube. Thus each minterm is associated with a unique vertex of an n-cube, a unique unit subcube of zero dimension; i.e., a 0-cube.

We measure the "distance" between minterms as the Hamming distance (Section 3.8) between their binary n-tuples. Theorem 10 in Section 2.5 allows us to combine two adjacent (Hamming distance of one) minterms. In the topological representation of switching functions, adjacent vertices

FIG. 9.1 Geometry of Th. 10

are connected by a line segment. We may take the line determined by two adjacent vertices as the geometric representation of the product term of Th. 10. Figure 9.1 illustrates this relationship for the combination of minterms m_5 and m_7. A product term of $n-1$ literals is represented by a line segment, a 1-dimensional cube, or 1-cube. Thus in Fig. 9.1, the product term AC and its n-tuple equivalent via Table 2.6, $(1, x, 1)$, label the line segment. The x in the n-tuple $(1, x, 1)$ indicates that the second variable is absent from the product term.

More generally we say that two n-tuples (two product terms) of the alphabet $\{0, 1, x\}$ are *adjacent* if they are identical in all coordinate positions except one and if a 1 (variable) appears in that position in one of the n-tuples, while a 0 (complement) appears opposite it in the other n-tuple. The algebraic manipulations that Th. 10 suggests may also be performed on n-tuples.

$$AB\bar{C} \vee ABC = AB \qquad\qquad A\bar{C}D \vee ACD = AD$$

$$\begin{Bmatrix} 110 \\ 111 \end{Bmatrix} \equiv 11x \qquad\qquad \begin{Bmatrix} 1x01 \\ 1x11 \end{Bmatrix} \equiv 1xx1$$

Because of the interchangeability of Boolean product terms and n-tuples of 0's, 1's, and x's, we refer to both product terms and corresponding n-tuples as *cubes*. We have encountered 0-cubes in which all literals appear (no x's in the n-tuples) and 1-cubes in which a single literal is missing (a single x appears in the n-tuples). We will encounter product terms with more absent literals (more x's appear). In general, if r literals are absent (r x's in the n-tuple), we refer to the cube as an r-*cube*. When we deal with an n-variable switching function, $0 \le r \le n$.

Just as $K^0(f)$ is the set of all 0-cubes of f, $K^1(f)$ is the set of all 1-cubes of that function. For the example function

$$f = \bigvee 5, 6, 7$$

we know that

$$K^0(f) = m_5, m_6, m_7 = \begin{Bmatrix} 101 \\ 110 \\ 111 \end{Bmatrix}$$

and we can find all 1-cubes of f by algebraic manipulations and by applying Th. 10 repeatedly to $K^0(f)$; thus

$$K^1(f) = \begin{Bmatrix} 11x \\ 1x1 \end{Bmatrix}$$

Figure 9.2 shows the topology of f. The 0-cubes and 1-cubes of this function are clearly seen. That some cubes are related to others is also obvious.

Fig. 9.2   Topology of $f = \bigvee 5, 6, 7$

Cover Relation

Any cube α can be decomposed to its base $K^0(\alpha)$ by using Th. 10. If α is an r-cube, then $K^0(\alpha)$ has 2^r members. They are obtained from α by replacing its r x's with 0's and 1's in all ways possible. Cube α is said to *cover* the individual members of its base. Thus, if $\alpha_1 \in K^0(\alpha)$, we write $\alpha \sqsupseteq \alpha_1$. Conversely, 0-cube $\alpha_1 \in K^0(\alpha)$ " is included in " cube α, $\alpha_1 \sqsubseteq \alpha$. Thus 1-cube AC (equivalent to n-tuple $1x1$) has the base $K^0(1x1) = \{101, 111\}$: $AC(1x1)$ covers 0-cubes $A\bar{B}C$ (101) and ABC (111); $101 \sqsubseteq 1x1$, and $111 \sqsubseteq 1x1$.

The *minimization* theorem allows us to replace the sum $A\bar{B}C \vee ABC$, set $\{101, 111\}$, with term AC, cube $1x1$, in any Boolean equation. As a result we say that cube α is *cover equivalent* to its base $K^0(\alpha)$, written $\alpha \equiv K^0(\alpha)$. For our example $\{101, 111\} \equiv \{1x1\}$. The equivalence symbol $\equiv$ is used to distinguish cover equivalence from set equality ($=$). Clearly, $\{101, 111\}$ is not composed of the same members as $\{1x1\}$. Thus $\equiv$ relates sets of cubes as $=$ relates Boolean expressions.

Cube α *covers* cube β, if $K^0(\beta) \subseteq K^0(\alpha)$. Thus cube A ($1xx$) covers cube AC ($1x1$) because $K^0(1x1)$ as given above is a subset of

$$K^0(1xx) = \{100, 101, 110, 111\}.$$

If $K^0(\alpha) = K^0(\beta)$, α and β have the same base, then they are cover equivalent to each other as well as their common base, $\alpha \equiv \beta \equiv K^0(\alpha)$. These relationships may also be defined directly in terms of cubes α and β without recourse to their bases. Cube α *covers* β if 0's (1's) appear in the β n-tuple in exact correspondence with all 0's (1's) in the α n-tuple. Any entry may appear in those positions in the β n-tuple in which the α n-tuple contains x's. Thus $1xx$ covers $1x1$ because both have 1 in the first position. Cube $1x1$ does not cover $1xx$ because the latter cube does not have 1 in the third position to correspond to the 1 which appears in that position in the former cube. When we decompose cube $1xx$ we first form set $\{10x, 1x0, 11x, 1x1\}$ from which the base of the cube may be computed with ease. Note that cube $1x1$ appears in this intermediate set. We can be sure that the base of $1x1$ will be part of the base of $1xx$.

The set and cover order relationships differ just as do the set equality and cover equivalence relationships. Thus $\{1x1\} \sqsubseteq \{1xx\}$, but clearly the first set is not a subset of the second set. The transitivity of the cover relationship is at the heart of the difference. If cube $\gamma \sqsubseteq \beta$ and $\beta \sqsubseteq \alpha$, then $\gamma \sqsubseteq \alpha$. Since $K^0(\gamma)$ is a subset of $K^0(\beta)$, which in turn is a subset of $K^0(\alpha)$, $K^0(\gamma)$ is also a subset of $K^0(\alpha)$. $K^0(\gamma) \subseteq K^0(\beta) \subseteq K^0(\alpha)$. But we cannot relate cubes α, β, γ directly with the set order relationship symbolized $\subseteq$. The more complex nature of the cover relationship was emphasized by J. P. Roth [5] when he named the mathematics of covers a " complex " rather than an algebra.

Arrays

An *array* is a set of cubes. A cube may be considered to be an array with one member. The cubes of an array are distinguished with superscripts; the components of a cube are distinguished with subscripts. Thus array $A = \{a^1, a^2, \ldots, a^p\}$ and $a^i = (a_1^i, a_2^i, \ldots, a_n^i)$, where every $a_j^i \in \{0, 1, x\}$. The base of an array is the union of the bases of its members.

$$K^0(A) = \bigcup_{i=1}^{p} K^0(a^i) \qquad (9.10)$$

An array covers the members of its base and is cover equivalent to its base. From Fig. 9.2:

$$A = \begin{Bmatrix} 1x1 \\ 11x \end{Bmatrix} \equiv K^0(A) = \begin{Bmatrix} 101 \\ 110 \\ 111 \end{Bmatrix}$$

$$101 \sqsubseteq 1x1 \sqsubseteq A$$
$$110 \sqsubseteq 11x \sqsubseteq A$$
$$111 \sqsubseteq 1x1 \sqsubseteq A$$
$$111 \sqsubseteq 11x \sqsubseteq A$$

Array A covers array B, $B \sqsubseteq A$, if $K^0(B) \subseteq K^0(A)$, and the arrays are cover equivalent, $A \equiv B$, if $K^0(B) = K^0(A)$. Thus arrays

$$A = \begin{Bmatrix} 1x1 \\ 11x \end{Bmatrix} \quad \text{and} \quad B = \begin{Bmatrix} 110 \\ 1x1 \end{Bmatrix}$$

are not the same sets, but Fig. 9.2 indicates that they have the same bases and hence are cover equivalent.

If in a sum-of-product expression of a function, one product term covers another, the *absorption* theorem allows us to simplify the expression by deleting the covered cube.

$$ABC \lor AC = AC(B \lor 1) = AC$$

Similarly, in an array of n-tuples those cubes that are covered by other cubes of the array may be deleted. Since such deletion does not alter the base of the array, the original and final arrays are cover equivalent. When we compare each cube of an array with every other member looking for the covering relationship, and then delete those cubes that are covered by other cubes, we *absorb the array*. We remove a rather obvious type of redundancy by exercising the absorption theorem.

Since every Boolean function f can be expressed as a sum of minterms, we can find the base $K^0(f)$. Theorem 10 provides us with a means of finding $K^1(f)$, the set of 1-cubes *of the function*, i.e., 1-cubes which cover only members of base $K^0(f)$. One-cube $10x$ covers one of the minterms of the function illustrated in Fig. 9.2, but it also covers 100, which is not a member of the base of the function. Hence $10x$ is not a cube of function f. Application of Theorem 10 to the members of $K^1(f)$ enables us to find $K^2(f)$, the 2-cubes of f. Then in turn $K^3(f), \ldots, K^n(f)$ may be found. If

$K^r(f)$ is found to be empty, then $K^i(f)$ where $r < i \leq n$ are also empty.

The *complex* of function f, denoted $K(f)$, is the set of all cubes of the function.

$$K(f) = \bigcup_{i=0}^{n} K^i(f) \tag{9.11}$$

The complex of a function serves much the same purpose and has the same properties as a universal set in set algebra. An *ON-array* of a function is then any subset of the complex of the function, $ON \subseteq K(f)$, that is cover equivalent to the base and to the complex of the function.

$$K^0(f) \equiv ON \subseteq K(f) \tag{9.12}$$

Any ON-array of the function f corresponds directly to a sum-of-products expression of f, and we could use Eq. (9.12) to find all sum-of-products expressions if we so desired. $K^0(f)$ is an ON-array of f, but for most functions smaller sets of larger cubes satisfy the definition of ON-array.

As an example, consider the following function.

$$K^0(f) = \left\{ \begin{matrix} 0101 \\ 0111 \\ 1010 \\ 1011 \\ 1110 \\ 1111 \end{matrix} \right\} \qquad f = \vee\ 5,\ 7,\ 10,\ 11,\ 14,\ 15$$

We can apply Th. 10 repeatedly to find all 1-cubes of f. Actually, the topology as shown in Fig. 9.3 clearly indicates that six 1-cubes exist for this function.

$$K^1(f) = \left\{ \begin{matrix} 01x1 \\ x111 \\ 1x11 \\ 111x \\ 1x10 \\ 101x \end{matrix} \right\}$$

Note that a single x appears in the cube notation for each 1-cube. One literal is absent in each product term. Figure 9.3 also suggests that this function has a 2-cube (the square). Thus:

$$K^2(f) = \{1x1x\}$$

This n-tuple with two x's may be constructed by applying Th. 10 to either of two pairs of adjacent 1-cubes. Cube $1x1x$ covers the four 1-cubes of which it is constructed and also covers the four 0-cubes which make up its base. Figure 9.4 presents the complex $K(f)$ and illustrates the covering-

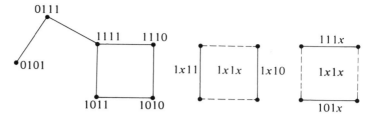

FIG. 9.3 Topology of f, and the Structure of its 2-cube

inclusion relationships of the cubes of f. Cubes appear above and are connected to the cubes that they cover.

Both the topology of Fig. 9.3 and the covering relationships of Fig. 9.4 suggest that we can express f as the sum of a small number of product terms from which some literals have been deleted. The covering

$$\text{ON} = \begin{Bmatrix} 01x1 \\ 1x1x \end{Bmatrix} \qquad f = \bar{A}BD \lor AC$$

is particularly interesting because it expresses f with a minimum number of terms, and as many literals as possible are deleted from each term. We have selected the covering that consists of a minimum number of the largest possible cubes. The circuit of Fig. 9.5 realizes this covering: it has a minimum number of connections to the OR gate (a minimum number of AND gates), and to each AND gate. Thus a minimum number of interconnections will be required to build the network. In many situations we consider this to be the " best " circuit. Methods for finding such " best " circuits are presented in Chapter 10.

The cubes of a function are also known as *implicants* of that function. Think of a function as a sum of terms. Then if any term has a value of 1, we know that the function has a value of 1. Assuming or knowing that the term has a value of 1 implies that the function has the same value. Example:

$$g = AB \lor A\bar{B}C$$

If $A = B = 1$, then AB, a term of g, has the value 1, and so does g. Those

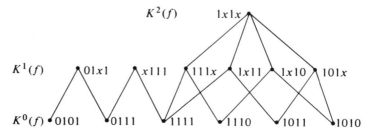

FIG. 9.4 Covering Relationships of the Cubes of f. Hasse Diagram

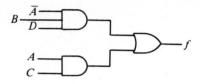

FIG. 9.5   Minimum Cost Realization of f

cubes of a complex that are covered by no other cubes of the complex are called *prime implicants*. Prime implicants are particularly interesting because they correspond to product terms of a minimum number of literals. Thus minimum cost sum-of-products expressions of a function consist of a sum of prime implicants.

Tuples represent product terms (sum terms). Arrays (sets of tuples) represent sum-of-products (product-of-sums) expressions of switching functions. In very many cases tuples and arrays are better suited for storage and processing by digital computers than the symbol sequences of Boolean equations. In fact, tuple notation and processing are often superior to Boolean notation for hand calculation. We now turn to the processing of arrays.

9.3   ARRAY OPERATORS AND ALGEBRA

The operators introduced in this section are defined in terms of *coordinate operators* and a set of rules. Thus if c^1 and c^2 are cubes, c_i^1 and c_i^2 are combined via the coordinate operator. The cube operator then gives a result determined from the set of these coordinate results. Clearly c^1 and c^2 must be tuples of the same dimension for this to be possible. This requirement will be assumed in the future, and not explicitly stated with each operator definition.

Absorb

Let a and b be two n-tuples of elements a_i, $b_i \in \{0, 1, x\}$. Then:

$$a \sqsubseteq b \quad \text{if} \quad (a_i \sqsubseteq b_i) = \varepsilon \text{ for all } i$$
$$a \not\sqsubseteq b \quad \text{if} \quad (a_i \sqsubseteq b_i) = \phi \text{ for any } i \qquad (9.13)$$

where Table 9.1 defines the coordinate covering relationship $(a_i \sqsubseteq b_i)$.

The Absorption theorem permits deleting from an array cubes that are covered by one or more other cubes of that array. The unary *absorb* operator, denoted **A**, then deletes from its operand array all cubes that are covered by other members of that array. When array A is absorbed, the resulting array, $\mathbf{A}(A)$, is cover equivalent to A, $\mathbf{A}(A) \equiv A$, and a subset of it, $\mathbf{A}(A) \subseteq A$. Consider the array and corresponding geometry shown in

TABLE 9.1 COORDINATE COVERING

$a_i \sqsubseteq b_i$	b_i		
	0	1	x,d
a_i 0	ε	ϕ	ε
1	ϕ	ε	ε
x,d	ϕ	ϕ	ε

Fig. 9.6. Cube $c^1 = 000$ does not cover $c^2 = 100$, or vice versa. In the first coordinate $(c_1{}^1 \sqsubseteq c_1{}^2) = \phi$: the 0–1 opposition always prohibits the covering relationship. But $c^1 \sqsubseteq c^3$, $000 \sqsubseteq 0x0$, and c^1 can be deleted from the array. In Fig. 9.6, 0-cube 000 is included in 1-cube $0x0$. If the array is to be a concise expression of a cover of 0-cubes, then $c^1 = 000$ is redundant and can be deleted without loss of information. Cube $c^3 = 0x0$ is covered by no other cube of array C; while one 0-cube (010) covered by c^3 is also covered by other cubes, no other cube of C entirely covers c^3. But further searching shows $c^4 = x10 \sqsubseteq c^6 = x1x$ and $c^5 \sqsubseteq c^6$.

Cube Union

If $A = \{a^1, a^2, \ldots\}$ and $B = \{b^1, b^2, \ldots\}$ are sets of cubes of the same number of variables, the union of these arrays is the *absorbed* set $A \cup B$.

$$A \bigsqcup B = \mathbf{A}\,(A \cup B) = \mathbf{A}\,(\{a^1, a^2, \ldots, b^1, b^2, \ldots\}) \qquad \textbf{(9.14)}$$

This definition varies from that of the union of sets in that an absorbed result is specified to keep that result concise. From Fig. 9.6:

$$\begin{Bmatrix} 000 \\ x11 \end{Bmatrix} \bigsqcup \begin{Bmatrix} 0x0 \\ x1x \end{Bmatrix} \equiv \begin{Bmatrix} \cancel{000} \\ \cancel{x11} \\ 0x0 \\ x1x \end{Bmatrix} \equiv \begin{Bmatrix} 0x0 \\ x1x \end{Bmatrix}$$

This operator parallels the logical sum operator and Absorption Theorem, and the same example, can be expressed algebraically.

$$\{\overline{A}\,\overline{B}\,\overline{C} \lor BC\} \lor \{\overline{A}\,\overline{C} \lor B\} = \overline{A}\,\overline{B}\,\overline{C} \lor BC \lor \overline{A}\,\overline{C} \lor B = \overline{A}\,\overline{C} \lor B$$

Thus the cube union of two arrays is a cover of the union of the bases of the original arrays.

$$\begin{matrix} 1 \\ 2 \\ 3 \\ 4 \\ 5 \\ 6 \end{matrix} \begin{Bmatrix} 000 \\ 100 \\ 0x0 \\ x10 \\ x11 \\ x1x \end{Bmatrix} \equiv \begin{Bmatrix} 100 \\ 0x0 \\ x1x \end{Bmatrix}$$

Absorbed C

Array C

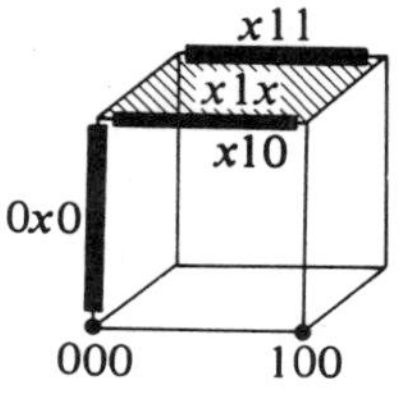

FIG. 9.6 Cubes for Purposes of Example

Cube Intersection

The cube intersection of two n-tuples, a and b, is defined through the coordinate intersection of Table 9.2 and the rules

$$a \sqcap b = \begin{cases} \phi \text{ (empty) if any } a_i \sqcap b_i = \phi \\ c \text{ otherwise, where } c_i = a_i \sqcap b_i \end{cases} \qquad (9.15)$$

Again using Fig. 9.6 for purposes of illustration:

$$000 \sqcap 0x0 = 000$$
$$000 \sqcap 100 = \phi 00 = \phi$$
$$0x0 \sqcap x1x = 010$$
$$x11 \sqcap x1x = x11$$

Examine each of these examples using Fig. 9.6 as an aid in visualizing what this operator accomplished. The first example involves a 0-cube and a 1-cube that covers it. The intersection is just the 0-cube. In terms of Boolean algebra, this example may be expressed:

$$(\overline{A}\overline{B}\overline{C}) \cdot (\overline{A}\overline{C}) = \overline{A}\overline{B}\overline{C}$$

The second example illustrates that the intersection of two 0-cubes is empty. We know that the product of minterms is logic 0 expressed here by ϕ, the empty set of cubes. The other examples above involve larger cubes: study them carefully and propose other examples of your own. One fact becomes apparent in this study—the intersection of two cubes is the largest single cube common to both of the original cubes.

The intersection of a cube b and an array $A = \{a^1, a^2, \ldots\}$ is defined as

$$A \sqcap b = \{(a^1 \sqcap b) \sqcup (a^2 \sqcap b) \sqcup \ldots\} \qquad (9.16)$$

Then the intersection of two arrays, A and B, is defined to be:

$$A \sqcap B = \{\{A \sqcap b^1\} \sqcup \{A \sqcap b^2\} \sqcup \ldots\} \qquad (9.17)$$

The cube union operators indicate that the resulting set is to be absorbed, and hence expressed somewhat compactly. Again we call on Fig. 9.6 for examples.

$$\begin{Bmatrix} 000 \\ x1x \end{Bmatrix} \sqcap \{0x0\} = \begin{Bmatrix} 000 \\ 010 \end{Bmatrix}$$

TABLE 9.2 COORDINATE CUBE INTERSECTION

	$\sqcap$	0	1	x,d
	0	0	ϕ	0
a_i	1	ϕ	1	1
	x,d	0	1	x

A 1-cube is intersected with a 0-cube which it covers, and with a 2-cube with which it shares a vertex. Note that the resulting cover could be compressed to the 1-cube $0x0$. The intersection operation does not generate this combination; it does not provide the simplifications allowed by the Minimization Theorem. The result of cube intersection is a cover of the set intersection of the bases of the original arrays, but is not necessarily a minimal cover.

Sharp Product

The base of $f_3 = f_1 \cdot \bar{f}_2$ is given by

$$K^0(f_3) = K^0(f_1) \cap \overline{K^0(f_2)} = K^0(f_1) - K^0(f_2)$$

Function f_3 consists of all vertices of f_1 that are not vertices of f_2. In set theory this relationship is known as the "relative complement." Function f_3 is the complement of f_2 relative to function f_1, rather than to the universal function. Alternatively, f_3 can be thought of as that part of f_1 left when f_2 is "removed," that part of $K^0(f_1)$ left when $K^0(f_1) \cap K^0(f_2)$ is taken from $K^0(f_1)$.

The *sharp operator* ($\#$) is the array counterpart of the relative-complement of set theory. The sharp product of two cubes, $a \# b$, is defined by the coordinate sharp product of Table 9.3 and the following rules:

$$a \# b = \begin{cases} a \text{ if } a \sqcap b = \phi, \text{ i.e. } a_i \# b_i = \phi \text{ for some } i \\ \phi \text{ if } a \sqsubseteq b, \text{ i.e. } a_i \# b_i = \varepsilon \text{ for all } i \\ \bigsqcup_i (a_1, a_2, \ldots, \bar{b}_i, \ldots, a_n) \text{ otherwise} \\ \text{where the union is for all } i \text{ for which} \\ a_i \# b_i = \alpha_i \in \{0, 1\} \end{cases} \qquad (9.18)$$

Examples based on Fig. 9.6 may clarify this definition. Consider cubes $x10$ and 000.

$$\begin{array}{r} x10 \\ \# \quad 000 \\ \hline 1\phi\varepsilon \end{array} \qquad x10 \# 000 = x10$$

These two cubes have nothing in common, i.e., they are disjoint: $x10 \sqcap 000 = \phi$. Hence when we remove from $x10$ all 0-cubes common to it and 000, we remove nothing, and $x10$ remains intact. The similarity between the coordinate sharp and cover tables is emphasized when we compute

TABLE 9.3   COORDINATE SHARP PRODUCT

$a_i \# b_i$	b_i		
	0	1	x,d
a_i 0	ε	ϕ	ε
a_i 1	ϕ	ε	ε
a_i x,d	1	0	ε

$$\# \quad \frac{\begin{array}{c} x10 \\ x1x \end{array}}{\varepsilon\varepsilon\varepsilon} \qquad x10 \,\#\, x1x = \phi$$

Cube $x1x$ entirely covers $x10$, $x1x \sqcap x10 = x10$, and hence when we remove common vertices, nothing remains. On the other hand, $x1x \,\#\, x10 = x11$. A nonempty result is obtained here since $x10$ does not completely cover $x1x$; something of $x1x$ remains when we remove $x10$. Finally, we determine

$$\# \quad \frac{\begin{array}{c} x1x \\ 010 \end{array}}{1\varepsilon1} \qquad x1x \,\#\, 010 = \{11x, x11\}$$

Fig. 9.7 details this example. One vertex is removed from 2-cube $x1x$. Two 1-cubes remain.

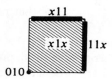

FIG. 9.7   The Sharp Product $x1x \,\#\, 010$

If $a \,\#\, b$ is not empty, then it is cube a or a set of subcubes of a of dimension one less than the dimension of cube a. The result is an absorbed array of cubes of the largest possible dimension. Thus the result is a cover of $K^0(a) \cap \overline{K^0(b)}$ expressed in a very compact fashion. If cube a happens to be the unit n-cube, then $a \,\#\, b$ expresses the complement of b relative to the universe, which we call simply the "complement" of b. Thus the complement of a minterm of three variables is:

$$\# \quad \frac{\begin{array}{c} xxx \\ 000 \end{array}}{111} \qquad xxx \,\#\, 000 = \{1xx, x1x, xx1\}$$

i.e., a maxterm.

The sharp product $A \,\#\, b$ is given by:

$$A \,\#\, b = \{\{a^1 \,\#\, b\} \sqcup \{a^2 \,\#\, b\} \sqcup \ldots\} \tag{9.19}$$

Thus b is removed from each of the cubes of A in turn. Note that the result is an absorbed array. The sharp product $a \,\#\, B$ is given by

$$a \,\#\, B = \{\ldots\{\{a \,\#\, b^1\} \,\#\, b^2\} \,\#\, \ldots\} \tag{9.20}$$

All of the vertices common to cube a and B are removed from a. First those of b^1 are taken from a; then those of b^2 are removed. Thus if $a = xxx$ and $B = \{1xx, x11\}$, corresponding to the Boolean expression $x_1 \vee x_2 x_3$,

$$a \# B = \{\{xxx \# 1xx = 0xx\} \# x11\} = \{00x, 0x0\}$$

$$\overline{x_1 \vee x_2 x_3} = \bar{x}_1 \cdot (\bar{x}_2 \vee \bar{x}_3) = \bar{x}_1 \bar{x}_2 \vee \bar{x}_1 \bar{x}_3$$

Finally, if A and B are two arrays, $A \# B$ can be defined in either of two ways.

$$A \# B = \{\ldots\{\{A \# b^1\} \# b^2\}\ldots\} \tag{9.21}$$

$$A \# B = \{\{a^1 \# B\} \bigsqcup \{a^2 \# B\}\ldots\} \tag{9.22}$$

Both accomplish the same thing—the common vertices are removed from A with the result expressed as an absorbed set of large cubes. The order of removal differs with the two definitions. Based on Fig. 9.8, let

$$A = \begin{Bmatrix} 1xxx \\ x01x \end{Bmatrix} \quad \text{and} \quad B = \begin{Bmatrix} 101x \\ 1111 \end{Bmatrix}$$

Then:

$$A \# b^1 = \begin{Bmatrix} 11xx \\ 1x0x \\ 001x \end{Bmatrix} \quad (A \# b^1) \# b^2 = \begin{Bmatrix} \cancel{110x} \\ 11x0 \\ 1x0x \\ 001x \end{Bmatrix}$$

$$\left(a_1 \# B = \begin{Bmatrix} 110x \\ 11x0 \\ 1x0x \end{Bmatrix} \right) \bigsqcup \left(a_2 \# B = \{001x\} \right) = \begin{Bmatrix} \cancel{110x} \\ 11x0 \\ 1x0x \\ 001x \end{Bmatrix}$$

Check these results with Fig. 9.8. In both cases covered cubes are deleted, as specified by the cube union operator.

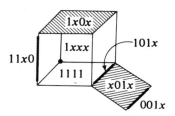

Fig. 9.8 A Four-Variable Example

Bookkeeping Operators

Array processing algorithms to be developed often require that arrays be altered in ways that are not related to Boolean algebra. Many necessary manipulations are rather straightforward in nature and can be described with words, but we gain the advantages of precision and conciseness by formalizing them. For example, we have been using the equal sign ($=$) to express set equivalence and to associate a name with a set of cubes. Such names can be thought of as variables. When we create an array and assign it a name, we assign a value to the variable. If we intersect arrays B and C, we create a new array that may be assigned an unused name, say A.

$$A = B \sqcap C$$

But if A were assigned a value previously, we must interpret this expression as a statement of set equality.

The *replacement* operator denoted with a left-going arrow ($\leftarrow$) is used to express the assignment of a new value to a previously defined variable. Absorbing an array alters the given array in general and thus accomplishes the replacement

$$A \leftarrow \mathbf{A}(A)$$

We will often need to change elements of a cube and the elements in columns of an array according to a rule. Let λ be a set of integers that identify columns of an array. Then the *change* operator, denoted $\mathbf{C}_\lambda^{b_1 b_2 b_3}(A)$ where all $b_i \in \{0, 1, x\}$, replaces all 0's with b_1's, 1's with b_2's, and x's with b_3's in the columns of array A identified by set λ. With $\lambda = \{2, 3\}$ and

$$A = \begin{Bmatrix} 0000 \\ 01x0 \\ 1x11 \end{Bmatrix}$$

$$\mathbf{C}_\lambda^{011}(A) = \begin{Bmatrix} 0000 \\ 0110 \\ 1111 \end{Bmatrix} \qquad \mathbf{C}_\lambda^{10x}(A) = \begin{Bmatrix} 0110 \\ 00x0 \\ 1x01 \end{Bmatrix}$$

The *delete* operator $\mathbf{D}_\lambda(A)$ deletes columns identified by λ from array A. Remaining columns are renumbered with consecutive integers after all deletions have been made. The *insert* operator $\mathbf{I}_i^j(A)$ inserts j new columns of x's to the right of the ith column of A. The original $(i + 1)$th, $(i + 2)$th,.. columns become the $(i + j + 1)$th, $(i + j + 2)$th, ... columns, respectively. If j is not specified, a single column is to be inserted. The *permutation* operator $\mathbf{P}_{ij}(A)$ interchanges the ith and jth columns of array A.

We will require special subcubes of U_n so often that it is to our advantage to define the following:

χ_λ^b—tuple of x's with $b \in \{0, 1\}$ in the columns identified by λ
ψ_λ—tuple of 0's with 1's in the columns identified by λ

If λ is empty or not specified, χ and ψ represent tuples of x's and 0's, respectively. Note that the dimension of these tuples is not stated explicitly; context will clearly reveal what their dimensions must be (n, m, $n + m$ are usual). These special cubes can be formed from U_n with the change operator, but the notation required to write them in that manner is verbose.

The Cartesian product $\times$, when applied to arrays, appends a fixed m-tuple to each element of an array. Thus if A is a set of n-tuples $A \times \psi$ appends an m-tuple of 0's to each element of A, and A becomes a set of $(n + m)$-tuples. This operation can be expressed in terms of the insert and change operators also, but $\mathbf{C}_\lambda^{000}(\mathbf{I}_n{}^m(A))$ is a very heavy way to describe the desired computation. Several other major array manipulating operators will be defined in subsequent sections where they are needed.

Properties of Array Operators

Both cube union and cube intersection are commutative, associative, and distributive over each other in the cover equivalence sense.

$$A \bigsqcup B \equiv B \bigsqcup A$$
$$A \bigsqcap B \equiv B \bigsqcap A$$
(9.23)

$$A \bigsqcup (B \bigsqcup C) \equiv (A \bigsqcup B) \bigsqcup C$$
$$A \bigsqcap (B \bigsqcap C) \equiv (A \bigsqcap B) \bigsqcap C$$
(9.24)

$$A \bigsqcup (B \bigsqcap C) \equiv (A \bigsqcup B) \bigsqcap (A \bigsqcup C)$$
$$A \bigsqcap (B \bigsqcup C) \equiv (A \bigsqcap B) \bigsqcup (A \bigsqcap C)$$
(9.25)

Cube union combines two sets of cubes into one and then absorbs that set. Absorbing eliminates redundancy, and does not alter the set in the sense of its being a cover of a $K^0(f)$. The order in which the sets are combined can not affect the result, and cube union is commutative in the set equality sense also.

Cube intersection provides a cover of $K^0(f)$ formed by set intersections. Symmetry of the coordinate cube intersection Table 9.2 assures us that cube intersection is a commutative operation. One may interchange the labels a^i and b^i without altering that table, and hence one may commute the cubes to be intersected without altering the result. Cube intersection is capable of dealing directly with cubes of arbitrary dimension, and breaks those cubes apart as little as possible. This specific mechanism of computing a cover results in the following theorem, which is not generally true for set intersection.

THEOREM 9.1. Fundamental Theorem of Cube Intersection.
Let c^1, c^2, ..., c^N be n-tuples. Then

$$c^1 \bigsqcap c^2 \bigsqcap \ldots \bigsqcap c^N \neq \phi$$

if and only if

$$c^i \sqcap c^j \neq \phi$$

for all $i, j \in I = \{1, 2, \ldots, N\}$.

Proof: Assume $c^i \sqcap c^j \neq \phi$ for all $i, j \in I$, and

$$c^1 \sqcap c^2 \sqcap \ldots \sqcap c^N = \phi$$

Then there exists a minimum k, $1 < k \leq N$ such that

$$\prod_{i=1}^{k-1} c^i \neq \phi \quad \text{and} \quad \prod_{i=1}^{k} c^i = \phi$$

In some coordinate, say r, c^k must contain $0(1)$ while a $1(0)$ exists in that position of $\prod_{i=1}^{k-1} c^i$. For that $1(0)$ to exist, one or more cubes, say c^l, $1 \leq l \leq k$, must have a $1(0)$ at the rth coordinate. Then

$$c^l \sqcap c^k = \phi$$

and this contradiction completes the "if" part of the theorem.

Now assume $\prod_{i=1}^{N} c^i \neq \phi$ and $c^i \sqcap c^j = \phi$ for some $i, j, \in I$. Since intersection is commutative and associative, generality is not lost by assuming $i = 1$ and $j = 2$. Then:

$$((\phi \sqcap c^3) \sqcap \ldots \sqcap c^N) = \phi$$

by repeated reduction using $\phi \sqcap c^i = \phi$.

This theorem is not generally valid for the intersection of sets or the intersection of sets of cubes, since it is possible for $A \sqcap B \neq \phi$, $A \sqcap C \neq \phi$, and $B \sqcap C \neq \phi$, but still $A \sqcap B \sqcap C = \phi$. Fig. 9.9 illustrates how this can happen with sets.

$A = \{11x\}, B = \{1x1\}, C = \begin{Bmatrix} 1x0 \\ 10x \end{Bmatrix}$

$A \sqcap B = \{111\}, A \sqcap C = \{110\}, B \sqcap C = \{101\}$

$A \sqcap B \sqcap C = \phi$

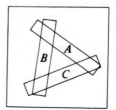

FIG. 9.9   Theorem 9.1 is not Valid for Sets

Two logic arrays are cover equivalent if they cover the same set of 0-cubes. The sharp product provides a means of determining if two arrays are so related. In general, the sharp product is not commutative: the coordinate sharp product table, Table 9.3, is not symmetric. But in one case $A \# B = B \# A$.

THEOREM 9.2. *If* $A \# B \equiv B \# A$, *then*

$$A \# B = B \# A = \phi$$

and A is cover equivalent to B, $A \equiv B$.

Proof: By definition $A \# B$ is a cover of 0-cubes covered by A and not covered by B; $B \# A$ is a cover of those 0-cubes uniquely covered by B. These two sets of 0-cubes can be equivalent only if both are empty.

The sharp operator is not associative, either.

$$A \# (B \# C) \not\equiv (A \# B) \# C \qquad (9.26)$$

Let $K^0(A)$ be the set of 0-cubes covered by array A. The left side of this inequality specifies a cover of all members of $K^0(A)$, other than those of $K^0(B)$ that are not common to $K^0(B)$ and $K^0(C)$.

$$K^0(A) \cap \overline{K^0(B) \cap \overline{K^0(C)}} = K^0(A) \cap \{\overline{K^0(B)} \cup K^0(C)\} \qquad (9.27)$$

Array C may cover members of $K^0(A)$, but these will not be removed. The right side of Eq. (9.26) specifies a cover of members of $K^0(A)$ not found in either $K^0(B)$ or $K^0(C)$.

$$\{K^0(A) \cap \overline{K^0(B)}\} \cap \overline{K^0(C)} = K^0(A) \cap \overline{\{K^0(B) \cup K^0(C)\}} \qquad (9.28)$$

Note that $K^0(B)$ and $K^0(C)$ can be commuted in Eq. (9.28): the sharp operator satisfies the following commutative-like relationship.

$$(A \# B) \# C \equiv (A \# C) \# B \qquad (9.29)$$

Both sides of this equation call for removing from $K^0(A)$ all 0-cubes covered by B or C or both. The order in which the 0-cubes are removed does not alter the set of 0-cubes formed by the relative complement operation, or the fact that the sharp product produces a cover of that set. This relationship then assures us that when computing $A \# B$ via either Eq. (9.21) or Eq. (9.22), the result does not depend on the order of the cubes in arrays A and B.

Let us examine the sharp operator further by computing $U_n \# c$ where c is a 0-cube: the n-tuple of all 1's will be used as a typical c.

$$U_n \# c = \{xx \ldots x\} \# \{11 \ldots 1\} = \begin{pmatrix} 0xx \ldots x \\ x0x \ldots x \\ xx0 \ldots x \\ \vdots \\ xxx \ldots 0 \end{pmatrix}$$

This sharp product produced a set of n $(n-1)$-cubes. Each cube is of dimension one less than U_n and each remains in the set because it covers at least one vertex covered by no other $(n-1)$-cube of the set. For example, $0xx \ldots x$ is the only $(n-1)$-cube which covers vertex $011 \ldots 1$.

Let us remove a 1-cube, say $x11 \ldots 1$, from U_n:

$$U_n \# x11 \ldots 1 = \begin{pmatrix} x0x \ldots x \\ xx0 \ldots x \\ \vdots \\ xxx \ldots 0 \end{pmatrix}$$

Now $n - 1$ $(n-1)$-cubes result. Again each provides the sole cover of a 1-cube. In general,

THEOREM 9.3. *Let c be an r-cube of n variables. Then $U_n \# c$ is a set K^{n-1} of $n - r$ $(n-1)$-cubes each of which uniquely covers an r-cube adjacent to c.*

Proof: From the definition of sharp product, $U_n \# c$ is either U_n, empty, or a set of $(n-1)$-cubes. The first case cannot occur for U_n covers c by definition. The second case occurs only if $c = U_n$ in which case $n - n = 0$ cubes exist as elements of the set. In the third case, the coordinate sharp product of U_n and c gives $\alpha \in \{0, 1\}$ in $n - r$ positions, and the definition thus provides a set of $n - r$ $(n-1)$-cubes, K^{n-1}. The ith of these provides the sole cover of the r-cube obtained by complementing the 0 or 1 in the ith position of c, which is other than an x. That r-cube is adjacent to c. None of the $(n-1)$-cubes are removed from K^{n-1} by absorbing.

Array Boolean Algebra

Set S of all n-tuples of elements from $\{0, 1, x\}$ has 3^n members. Its power set $\hat{S}$ then has 2^{3^n} members. Each member may be thought to express a switching function in *sop* (*pos*) form. If we think of arrays as representing switching functions, then the cover equivalence relation $\equiv$ partitions $\hat{S}$ into 2^{2^n} equivalence classes, one for each switching function of n variables.

Cube union and cube intersection serve as the two operators of a Boolean algebra on $\hat{S}$, with cover equivalence $\equiv$ being the equivalence relation of the algebra. Equations (9.24), (9.25), and (9.26) show these operators to be commutative, associative, and distributive. The "zero" of $\hat{S}$ is the empty set ϕ. U_n, or any array that is cover equivalent to it, serves as the "one" of the algebra. Each member of $\hat{S}$ has a complement; the sharp operator provides a means of calculating an array that describes the complementary switching function.

It is difficult to show a complete example for significant values of n

because S has so many members. The smallest example, $n = 1$, is enumerated below.

$$U_1 = x$$
$$K(U_1) = \{0 \ 1, x\}$$
$$\hat{S} = \{\phi; 0; 1; x; 0,1; 0,x; 1,x; 0,1,x\}$$
$$\langle \hat{S}, \equiv \rangle = \{\{\phi\}, \{0\}, \{1\}, \{x; 0,1; 0,x; 1,x; 0,1,x\}\}$$

Two members of each equivalence class are of particular interest to us. First, one member consists of 0-cubes only; it is the base of all members of the equivalence class. It is the canonical representative of the class. Second, some member describes the lowest cost sum-of-products realization of the switching function of the class. "Minimization" is the process of finding this member. With different cost criteria, different members must be found.

We can order switching functions in a variety of ways. Most often we define.

$$f_i \leq f_j \quad \text{if} \quad f_i \cdot f_j = f_i \quad \text{or} \quad f_i \vee f_j = f_j$$

Retaining all algebraic properties while moving to array space yields the *absorb ordering* on arrays.

$$A_i \sqsubseteq A_j \quad \text{if} \quad A_i \sqcap A_j \equiv A_i \quad \text{or} \quad A_i \sqcup A_j \equiv A_j$$

This definition of $\sqsubseteq$ is consistent with earlier ones that reveal how to detect when two cubes are $\sqsubseteq$-related, and that their bases are $\subseteq$ related.

Set $\hat{S}$ together with $\equiv$, $\sqsubseteq$, $\sqcap$, $\sqcup$, and $\#$ forms a Boolean algebra. Theorems can be developed for this array-space algebra. The simpler ones are

$$A \sqcup U_n \equiv U_n \qquad A \sqcup \phi \equiv A$$
$$A \sqcap \phi \equiv \phi \qquad A \sqcap U_n \equiv A$$

$$A \sqcup A \equiv A \qquad A \sqcup (U_n \# A) \equiv U_n$$
$$A \sqcap A \equiv A \qquad A \sqcap (U_n \# A) \equiv \phi$$
$$U_n \# (U_n \# A) \equiv A$$

$$A \sqcup (A \sqcap B) \equiv A \qquad A \sqcup (B \# A) \equiv A \sqcup B$$
$$A \sqcap (A \sqcup B) \equiv A \qquad A \sqcap ((U_n \# A) \sqcup B) \equiv A \sqcap B$$

$$(A \sqcap B) \sqcup (A \# B) \equiv A \qquad U_n \# (A \sqcap B) \equiv (U_n \# A) \sqcup (U_n \# B)$$
$$(A \sqcup B) \sqcap (A \sqcup (U_n \# B)) \equiv A \qquad U_n \# (A \sqcup B) \equiv (U_n \# A) \sqcap (U_n \# B)$$

These theorems have been written to conform to the usual theorems of Boolean algebra. The sharp operator is more general than the simple complement. Advantage can be taken of this to write "unusual" theorems. The following theorems that involve the sharp operator become very important in such development. The first two may be thought of as generalizations of deMorgan's theorems.

$$A \# (B \bigsqcup C) \equiv (A \# B) \bigsqcap (A \# C)$$
$$\equiv (A \# B) \# C$$

$$A \# (B \bigsqcap C) \equiv (A \# B) \bigsqcup (A \# C) \tag{9.30}$$

The next two are distributive-like theorems.

$$(A \bigsqcup B) \# C \equiv (A \# C) \bigsqcup (B \# C)$$

$$(A \bigsqcap B) \# C \equiv (A \# C) \bigsqcap (B \# C) \tag{9.31}$$
$$\equiv A \bigsqcap (B \# C)$$
$$\equiv B \bigsqcap (A \# C)$$

Probably the easy way to prove these theorems is by performing corresponding set operations on the bases of the arrays. Clearly the strong parallel between set algebra and array algebra is of great value.

The sharp operator is functionally complete, assuming the universal cube is available.

NOT $\bar{A} \equiv U_n \# A$

AND $A \bigsqcap B \equiv A \# (U_n \# B) \equiv B \# (U_n \# A)$

OR $A \bigsqcup B \equiv U_n \# ((U_n \# A) \# B)$

INHIBIT $A \bigsqcup \bar{B} \equiv A \bigsqcap (U_n \# B) \equiv (A \bigsqcap U_n) \# B \equiv A \# B$

9.4 FUNCTION ARRAY ALGORITHMS

Now we turn to the use of array operators in algorithms that solve specific problems. While these algorithms are interesting and valuable, our goal is to be able to develop new algorithms which generate solutions to new problems we face. Thus, in our analysis we should examine each step and ask what this step accomplishes and how it contributes to the goal of obtaining a solution.

We have represented n-input, m-output switching functions by means of truth tables with 2^n rows and $n + m$ columns. We will now call such tables, or tables that express exactly the same information, *function arrays*, and refer to each row of such an array as a cube. Let C be a function array of cubes $c^1, c^2, \ldots$ each of which is an $(n + m)$-tuple. Let $\alpha = \{1, 2, \ldots, n\}$ and $\Omega = \{n + 1, \ldots, n + m\}$. We then denote the input n-tuple part of cube c^i by $c_\alpha{}^i$, and the output m-tuple part by $c_\Omega{}^i$. Then

$$c^i = c_\alpha{}^i \circ c_\Omega{}^i \tag{9.32}$$

where $c_\alpha{}^i$ is an n-tuple of elements $x_j \in \{0, 1, x\}$ and $c_\Omega{}^i$ is an m-tuple of elements $z_k \in \{0, 1, d\}$, and $1 \le j \le n$, $n + 1 \le k \le n + m$.

A function array is characterized by (1) for two cubes c^i and c^j, if $c_\Omega{}^i \ne c_\Omega{}^j$ then

$$c_\alpha{}^i \bigsqcap c_\alpha{}^j = \phi \tag{9.33}$$

and (2) all input symbols are represented.

$$\bigsqcup_{i=1}^{|C|} c_\alpha^i \equiv U_n \qquad (9.34)$$

The input portions of cubes with different output parts are disjoint. We frequently find all c_α^i to be minterms of the input variables, and this disjointness is then true for all cubes. A function array may have only 2^n or fewer members even though it is a set of $(n + m)$-tuples. Thus $2^n \cdot 3^m - 2^n = 2^n(3^m - 1)$ $(n + m)$-tuples may never appear in a function array. Only one output m-tuple is associated with each input n-tuple; the other $3^m - 1$ output m-tuples are excluded.

Function arrays can often be compressed to achieve more compact expression. If two cubes c^i and c^j of a function array have identical output parts, $c_\Omega^i = c_\Omega^j$, then we may seek the minimum cover of $\{c_\alpha^i, c_\alpha^j\}$ and possibly replace the original pair of cubes with a single cube covering both c^i and c^j. In the uncompressed function array of Fig. 9.10, cubes 0000 011 and 0001 011 have the same output parts, and this pair of cubes can be replaced by the single cube $000x\,011$, which covers both of the original cubes and has the same output m-tuple as both original cubes. Cubes 0000 011 and 0010 011 could be replaced with $00x0\,011$, and note that $000x\,011 \sqcap 00x0\,011 \neq \phi$. But Fig. 9.10(c) suggests that greater compression is obtained by combining 0010 011 with 0110 011.

To compress a function array with don't cares:
1. Group all cubes with identical c_Ω^i parts;
2. Find a minimal cover of the group; and
3. Replace the original cubes with the minimal cover.

x_1	x_2	x_3	x_4	z_1	z_2	z_3	
0	0	0	0	0	1	1	⎧0000 011⎫
0	0	0	1	0	1	1	0001 011
0	0	1	0	0	1	1	0010 011 ⎧000x 011⎫
0	0	1	1	0	1	d	0011 01d 0x10 011
0	1	0	0	0	0	d	0100 00d 0011 01d
0	1	0	1	d	0	0	0101 d00 0100 00d
0	1	1	0	0	1	1	0110 011 0101 d00
0	1	1	1	1	1	1	0111 111 x111 111
1	0	0	0	0	0	1	1000 001 ⎨1000 001⎬
1	0	0	1	1	0	1	1001 101 1001 101
1	0	1	0	0	0	0	1010 000 1110 101
1	0	1	1	1	0	0	1011 100 1010 000
1	1	0	0	0	0	0	1100 000 1100 000
1	1	0	1	1	0	0	1101 100 1011 100
1	1	1	0	1	0	1	1110 101 ⎩1101 100⎭
1	1	1	1	1	1	1	⎩1111 111⎭

(a) Truth Table   (b) Uncompressed   (c) Compressed

FIG. 9.10 A Function Array

In the special case of $m = 1$, we can break the function array into three parts, the ON-array, the OFF-array, and the DC-array, and delete the output 1-tuple from all cubes. We are merely encoding the output information by physical separation when we do this.

ON-array: set of input n-tuples for which $f = 1$

OFF-array: set of input n-tuples for which $f = 0$

DC-array: set of input n-tuples for which $f = d$

These arrays may, but need not, be sets of 0-cubes.

We need only two of these three arrays: we can always compute the third. Let U_n denote the unit n-cube, i.e. an n-tuple of all x's. Then:

$$\text{ON} \sqcup \text{OFF} \sqcup \text{DC} \equiv U_n \tag{9.35}$$

If the OFF- and DC-arrays are known, the ON-array may be determined with the sharp product.

$$\text{ON} = U_n \, \# \, (\text{OFF} \sqcup \text{DC}) \tag{9.36}$$

Similarly,

$$\text{OFF} = U_n \, \# \, (\text{ON} \sqcup \text{DC}) \tag{9.37}$$

and

$$\text{DC} = U_n \, \# \, (\text{ON} \sqcup \text{OFF}) \tag{9.38}$$

We really have been using these relationships already. For example, we often fill 1's and d's in the Karnaugh map and understand that all other cells hold 0's, i.e. make up the OFF-array. But using these array computations may lead to surprising results. We do not necessarily get a set of minterms when we use the sharp product. We have seen that we get the largest appropriate fragments of U_n.

In some cases there is value in breaking a function array into m ON-arrays, m OFF-arrays, and m DC-arrays. This is not too difficult a task. We will need the special cube χ_i^b, an $(n + m)$-tuple of all x's except for the ith position, in which $b \in \{0, 1\}$ will appear.

ALGORITHM 9.1. *Dissect a Function Array.*
Let F be the function array we wish to dissect. Then

$$A^* = \text{ON}_i{}^* \sqcup \text{DC}_i{}^* = F \sqcap \chi_{n+i}^1 \tag{9.39}$$

$$B^* = \text{OFF}_i{}^* \sqcup \text{DC}_i{}^* = F \sqcap \chi_{n+i}^0 \tag{9.40}$$

where the * signifies that output m-tuples are still a part of each cube. We know in fact that a 1 will appear in the $(n + i)$th position of each cube of A^*, and a 0 will appear in the $(n + i)$th position of each cube of B^*.

Now let us remove the output m-tuple from all cubes of A^* and B^*.

$$A = \mathbf{D}_\Omega(A^*) = \mathbf{D}_\Omega(\text{ON}_i{}^* \bigsqcup \text{DC}_i{}^*)$$
$$B = \mathbf{D}_\Omega(B^*) = \mathbf{D}_\Omega(\text{OFF}_i{}^* \bigsqcup \text{DC}_i{}^*) \qquad (9.41)$$

A consists of on and don't care conditions for the ith output variable. We now separate these conditions.

$$\text{DC}_i = A \bigsqcap B$$
$$\text{ON}_i = A \mathrel{\#} \text{DC}_i \qquad (9.42)$$
$$\text{OFF}_i = B \mathrel{\#} \text{DC}_i$$

This may appear to be a very formal and formidable way of specifying how to perform a rather elementary task. But we are rapidly approaching the point where the assistance of a digital computer in performing our computations will be most welcome. To instruct a computer successfully, we must detail algorithms with great precision. And unless we wish to write completely new computer programs for each computation we must attempt to express all algorithms in terms of a basic set of operators which can be programmed once in the form of subroutines. And the definition and use of formal operators enhances precise communication between ourselves.

Now let us illustrate Algorithm 9.1 using the following function array, with $n = 4$ and $m = 2$.

$$F = \left\{ \begin{array}{l} 0x0x\ 11 \\ 1x00\ 10 \\ 1x01\ 01 \\ 0x11\ 1d \\ 0x10\ d1 \\ 1\ x110d \\ 1x10\ d0 \end{array} \right\}$$

Then

$$F \bigsqcap \overset{\chi_5{}^1}{\{xxxx1x\}} = \left\{ \begin{array}{l} 0x0x\ 11 \\ 1x00\ 10 \\ 0x11\ 1d \\ 0x10\ 11 \\ 1x10\ 10 \end{array} \right\} = A^*$$

$$F \bigsqcap \overset{\chi_5{}^0}{\{xxxx0x\}} = \left\{ \begin{array}{l} 1x01\ 01 \\ 0x10\ 01 \\ 1\ x110d \\ 1x10\ 00 \end{array} \right\} = B^*$$

and

$$A = \left\{ \begin{array}{l} 0x0x \\ 1x00 \\ 0x11 \\ 0x10 \\ 1x10 \end{array} \right\} \qquad B = \left\{ \begin{array}{l} 1x01 \\ 0x10 \\ 1x\,11 \\ 1x10 \end{array} \right\}$$

There are input conditions in A for which $z_1 = 1$ and others for which $z_1 = d$. To separate the on and don't care conditions we compute:

$$DC_1 = A \sqcap B = \begin{Bmatrix} 0x10 \\ 1x10 \end{Bmatrix}$$

$$ON_1 = A \ \# \ DC_1 = \begin{Bmatrix} 0x0x \\ 1x00 \\ 0x11 \end{Bmatrix}$$

Because all input parts of cubes of a function array are disjoint, no new cubes are formed by the intersection and sharp operators. Finally,

$$OFF_1 = B \ \# \ DC_1 = \begin{Bmatrix} 1x01 \\ 1x11 \end{Bmatrix}$$

Compute ON_2, OFF_2 and DC_2 to complete this example.

Variations of this algorithm, which perform a variety of interesting and useful tasks, can be formulated. For example, suppose C is a cover of a single-output switching function rather than a function array. We desire to expand the covered function, f, about some variable in the manner of the expansion theorem.

$$\begin{aligned}
f(x_1, x_2, \ldots, x_n) &= x_i \cdot f(x_1, \ldots x_{i-1}, 1, x_{i+1}, \ldots, x_n) \vee \bar{x}_i \cdot f(\ldots, 0, \ldots) \\
&= x_i(g \vee k) \vee \bar{x}_i(h \vee k) \qquad\qquad\qquad\qquad\text{(9.43)} \\
&= x_i g \vee \bar{x}_i h \vee k
\end{aligned}$$

where g, h, and k are functions of the $n - 1$ variables other than x_i.

ALGORITHM 9.2. *Expansion Theorem*
Using the operators we have available with $\lambda = \{i\}$

$$\begin{aligned}
A &= (G \sqcup K) = \mathbf{D}_\lambda(C \sqcap \chi_i^1) \\
B &= (H \sqcup K) = \mathbf{D}_\lambda(C \sqcap \chi_i^0)
\end{aligned} \qquad \text{(9.44)}$$

Then:

$$\begin{aligned}
K &= A \sqcap B \\
G &= A \ \# \ K \qquad\qquad \text{(9.45)} \\
H &= B \ \# \ K
\end{aligned}$$

gives arrays that cover functions g, h, and k, respectively.

The following 6-input variable function is very regular.

$$C = \begin{Bmatrix} 0xx110 \\ 00xx11 \\ 100xx1 \\ 1100xx \\ x1100x \\ xx1100 \end{Bmatrix}$$

If we expand about the first variable

$$A = \begin{Bmatrix} 00xx1 \\ 100xx \\ 1100x \\ x1100 \end{Bmatrix} \quad B = \begin{Bmatrix} xx110 \\ 0xx11 \\ 1100x \\ x1100 \end{Bmatrix}$$

Then

$$G = \begin{Bmatrix} 00x01 \\ 100xx \end{Bmatrix} \quad H = \begin{Bmatrix} xx110 \\ 01x11 \end{Bmatrix}$$

$$K = \begin{Bmatrix} 00x11 \\ 1100x \\ x1100 \end{Bmatrix}$$

Note that G, H, and K are not composed only of cubes of A and B. The intersect and sharp operations have broken some of the 2-cubes into 1-cubes. For example, cube $00xx1 \in A$ and cube $0xx11 \in B$ intersect to give $00x11$. This 1-cube thus belongs in the K-array of this function, and the parent cubes $00xx1$ and $0xx11$ must be broken when forming the G- and H-arrays.

$$\begin{array}{r} 00xx1 \\ \# \quad 00x11 \\ \hline 00x01 \in G \end{array} \qquad \begin{array}{r} 0xx11 \\ \# \quad 00x11 \\ \hline 01x11 \in H \end{array}$$

The solution of another very practical problem requires a similar algorithm. Suppose the 3-input, 3-output network of Fig. 9.11 were manufactured as an entity, either on a single chip of an integrated circuit or of discrete components on a small printed circuit card. (Actually this circuit is perhaps too small to yield the economy which can be obtained by such manufacture.) Now we are to use this *module* to construct switching networks. How can we make efficient use of this module? Should we develop an algebra with the very complex operator which this circuit realizes? No, we probably prefer to use our familiar Boolean algebra and its operators, and after we have specified a switching function, attempt to use the module or parts of it to implement the function. We might recognize that output w_1 is the NAND of input variables v_1 and v_2, and design NAND networks, but this certainly would not make very efficient use of the module.

Before we begin to implement a Boolean function with the module at hand, we might determine all of the submodules that can be formed from it by tying input terminals to logic 0 or 1, or together. The example module of Fig. 9.11 is so small that this would not be a big task. But if we were dealing with a 10-input, 10-output network of 50 gates, a great deal of labor would be involved, and computer assistance welcomed.

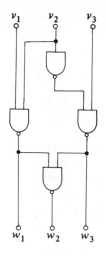

v_1	v_2	v_3	w_1	w_2	w_3
0	x	0	1	0	1
x	0	1	1	1	0
0	1	x	1	0	1
x	0	0	1	0	1
1	1	x	0	1	1

$$w_1 = v_1 \uparrow v_2$$

$$w_2 = v_1 v_2 \vee \bar{v}_2 v_3$$

$$w_3 = v_2 \vee \bar{v}_3$$

Fig. 9.11 A Module

Let F be the function array that describes a module with n input and m output variables. Because an actual circuit is described by F, don't cares will not appear in the output m-tuples of cubes of F.

ALGORITHM 9.3. *Submodule Generation*
The submodule obtained by tying the ith input terminal to logic 0 is then given by:

$$A = \mathbf{D}_i(F \sqcap \chi_i^0) \qquad (9.46)$$

The ith column is deleted since it is no longer an input terminal to which we may connect a variable signal. The submodule obtained by tying the ith input terminal to logic 1 is given by:

$$A = \mathbf{D}_i(F \sqcap \chi_i^1) \qquad (9.47)$$

If we tie v_1 of the module of Fig. 9.11 to logic 0, Eq. (9.46) gives

$$A = \begin{Bmatrix} x0\ 101 \\ 01\ 110 \\ 1x\ 101 \\ 00\ 101 \end{Bmatrix}$$

We see that it is possible to clean up this result. Observe that $w_1 = 1$ for this submodule; the first output column can be deleted; this submodule is more properly thought of as a 2-input, 2-output circuit.

If more than one input terminal is to be tied to logic 1 or 0, we can obtain the resulting submodule by intersecting successively with different appropriate mask cubes. Or we can form one, more complex, mask cube and perform a single intersection. If we desire the submodule obtained by tying v_1 to 0 and v_2 to 1, then we may use the mask cube $01xxxx$, giving

$$F \sqcap 01xxxx = \begin{Bmatrix} \cancel{010\ 101} \\ 01x\ 101 \end{Bmatrix}$$

The columns corresponding to v_1 and v_2 may be deleted, and we see that w_1, w_2, and w_3 also are trivial. This particular tying results in a trivial submodule.

Now how can we find the submodules that result when we tie input terminals together? In terms of our example module, what submodule results if we tie input terminals v_1 and v_2 together to form a new single input terminal, which we might call $v_1{}'$? Let λ denote the set of columns of variables to be tied together. Then the desired submodule is described by the array A obtained as follows.
Let

$$A^* = (\chi_\lambda{}^0 \sqcup \chi_\lambda{}^1) \sqcap F \tag{9.48}$$

All but one of the columns identified by λ may be deleted. Let λ' be the set λ with the first element deleted. Then

$$A = \mathbf{D}_{\lambda'}(A^*) \tag{9.49}$$

Now, to determine the submodule we get when v_1 and v_2 of our example module are tied together, we compute

$$A^* = \begin{Bmatrix} 000\ 101 \\ 001\ 110 \\ 000\ 101 \\ 11x\ 011 \end{Bmatrix}$$

The intersection of $\chi_\lambda{}^0$ and F gives all input conditions in which $v_1 = v_2 = 0$; the intersection of $\chi_\lambda{}^1$ and F gives all input conditions in which $v_1 = v_2 = 1$. Now the v_2 column can be deleted, and identical cubes removed.

$$\overset{\overset{\textstyle v_1{}'v_3}{\downarrow\downarrow}}{A = \begin{Bmatrix} 00\ 101 \\ 01\ 110 \\ 1x\ 011 \end{Bmatrix}} \qquad \begin{aligned} w_1 &= \bar{v}_1{}' \\ w_2 &= v_1{}' \vee v_3 \\ w_3 &= v_1{}' \vee \bar{v}_3. \end{aligned}$$

As an interesting sidelight, let us use this algorithm to examine the submodule obtained when an input and an output terminal are tied together. Let us tie v_3 and w_2, for example.

$$A^* = F \sqcap \begin{Bmatrix} xx0\ x0x \\ xx1\ x1x \end{Bmatrix} = \begin{Bmatrix} 0x0\ 101 \\ 010\ 101 \\ x00\ 101 \\ x01\ 110 \\ 111\ 011 \end{Bmatrix}$$

Here we might insist that the v_3 column be deleted and the w_2 column be retained.

$$A = \begin{Bmatrix} 0x\ 101 \\ x0\ 101 \\ x0\ 110 \\ 11\ 011 \end{Bmatrix}$$

Note that with $v_2 = 0$, either of two output m-tuples may be expected. This array is said to contain a *contradiction*; it does not describe a combinational switching network since a specific input condition (00 for example) is not associated with a unique output m-tuple. When we form loops in logic networks, sequential circuits result, as Chapter 12 will illustrate more fully. This specific sequential circuit is shown in Figure 12.9.

In summary, we see that the technique of intersecting an array with a special mask cube can often be used to identify those cubes of the original array that satisfy requirements specified by the mask cube. The sharp operator can also be used to do this job; Eqs. (9.39) and (9.40) could be written:

$$A^* = F \mathrel{\#} \chi^0_{n+i}$$
$$B^* = F \mathrel{\#} \chi^1_{n+i} \tag{9.50}$$

Note here that the roles of χ^0_{n+i} and χ^1_{n+i} are reversed.

Split Operator

In some cases a simpler operator than either the intersection or sharp operator can be employed to achieve the same or an even more desirable effect. When we form the ON, OFF, and DC-arrays from a function array, we really scan the ith output column looking for 1's ,0's, and x's. When we find a 1, we place that cube in the ON-array. Intersection separates 1's and x's from 0's and hence A^* in Eq. (9.39) is a composite of the ON and DC-arrays, and further processing is required to separate them.

Let us propose another operator denoted **S** which identifies and *transfers* to another array all cubes of a given array that are covered by a given mask cube. Thus if F is a function array, we will denote by this operator $F\mathbf{S}\mu$ the array of cubes removed from F under mask μ. Array F will be altered if cubes are removed.

ALGORITHM 9.4 *Alternative to Algorithm* 9.1

The algorithm for finding the ON- , OFF- , and DC-arrays can then be stated

$$
\begin{aligned}
&1. \quad \mathrm{ON}_i \ = \mathbf{D}_\Omega(F\mathbf{S}\chi^1_{n+i}) \\
&2. \quad \mathrm{OFF}_i = \mathbf{D}_\Omega(F\mathbf{S}\chi^0_{n+i}) \\
&3. \quad \mathrm{DC}_i \ = \mathbf{D}_\Omega(F)
\end{aligned}
\tag{9.51}
$$

Order is now very important, for we are successively removing cubes from F, and hence F is destroyed in the process of carrying out this algorithm.

Repeating a previous example,

$$\mathrm{ON}_1 = \mathbf{D}_\Omega \left(\begin{Bmatrix} 0x0x\ 11 \\ 1x00\ 10 \\ 1x01\ 01 \\ 0x11\ 1d \\ 0x10\ d1 \\ 1x11\ 0d \\ 1x10\ d0 \end{Bmatrix} \mathbf{S}\chi_5^1 \right)$$

$$= \mathbf{D}_\Omega \left(\begin{Bmatrix} 0x0x\ 11 \\ 1x00\ 10 \\ 0x11\ 1d \end{Bmatrix} \right) = \begin{Bmatrix} 0x0x \\ 1x00 \\ 0x11 \end{Bmatrix}$$

At this point

$$F = \begin{Bmatrix} 1x01\ 01 \\ 0x10\ d1 \\ 1x\ 110d \\ 1x10\ d0 \end{Bmatrix}$$

Then:

$$\mathrm{OFF}_1 = \mathbf{D}_\Omega(F\,\mathbf{S}\chi_5^0) = \mathbf{D}_\Omega \left(\begin{Bmatrix} 1x01\ 01 \\ 1\,x1101 \end{Bmatrix} \right)$$

$$= \begin{Bmatrix} 1x01 \\ 1x\ 11 \end{Bmatrix}$$

Finally, DC_1 is just what remains of the F-array after the output columns are deleted.

$$\mathrm{DC}_1 = \mathbf{D}_\Omega(F) = \begin{Bmatrix} 0x10 \\ 1x1d \end{Bmatrix}$$

This $\mathbf{S}$ (for "split") operator is easier to execute by hand than the intersect or sharp operators, and also is performed more efficiently by a computer. We examine the mask cube. Those columns in the mask cube which contain an x are of no concern, because the x covers both the 0 and 1. We need to examine only those columns in the given array, in which the mask cube contains 0 or 1, and here we must look for a precise match. When a match is found, we need only transfer the cube from one array to another. This efficiency, and the fact that this operator can often be used, justify the introduction of still another operator.

Function arrays convey important information, which is not always readily apparent from a set of $3m$ ON- , OFF- , and DC-arrays. The output section, $c_\Omega{}^i$, of each cube of a function array indicates whether the associated input section, $c_\alpha{}^i$, is a member of the ON- , OFF- , or DC-array of each output variable. In particular, if a $c_\Omega{}^i$ contains two or more 1's then the associated $c_\alpha{}^i$ is a cube of two or more ON-arrays. Such cubes lead to AND gates, which may be shared between the networks that realize the output variables. Such information can be obtained from individual ON- , OFF- , and DC-arrays, of course, but computation is required. For example, those $c_\alpha{}^i$, if any, to be associated with $c_\Omega{}^i = 111$ ($m = 3$) are given by:

$$\text{ON}_1 \sqcap \text{ON}_2 \sqcap \text{ON}_3$$

Those $c_\alpha{}^j$ associated with $c_\Omega{}^j = 110$ are given by

$$\text{ON}_1 \sqcap \text{ON}_2 \sqcap \text{OFF}_3$$

We can obtain a function array by performing all the possible intersections and appending to the results of each intersection the appropriate $c_\Omega{}^i$. One orderly way to perform the required intersections and append the necessary output columns proceeds as follows.

ALGORITHM 9.5. *Form a Function Array*
To each cube of each ON- , OFF- , and DC-array we append an m-tuple that reflects the origin of the input section.

$$
\begin{aligned}
\text{ON}_i{}^* &= \text{ON}_i \times \chi_i^1 \\
\text{OFF}_i{}^* &= \text{OFF}_i \times \chi_i^0 \\
\text{DC}_i{}^* &= \text{DC}_i \times U_m
\end{aligned}
\tag{9.52}
$$

To form a function array we begin with

$$F = \text{ON}_1{}^* \cup \text{OFF}_1{}^* \cup \text{DC}_1{}^* \tag{9.53}$$

and iteratively compute

$$F \leftarrow F \sqcap (\text{ON}_i{}^* \cup \text{OFF}_i{}^* \cup \text{DC}_i{}^*) \tag{9.54}$$

for $i = 2, \ldots, m$. The desired function array emerges.

Assume the following ON- , OFF- , and DC-arrays are given.

$$
\text{ON}_1 = \begin{Bmatrix} x111 \\ 1xx1 \\ 111x \end{Bmatrix}
\qquad
\text{ON}_2 = \begin{Bmatrix} 00xx \\ 0x1x \\ x111 \end{Bmatrix}
\qquad
\text{ON}_3 = \begin{Bmatrix} 00x0 \\ x00x \\ x11x \end{Bmatrix}
$$

$$\text{OFF}_1 = \begin{cases} 00xx \\ 0xx0 \\ x0x0 \\ xx00 \end{cases} \qquad \text{OFF}_2 = \begin{cases} 10xx \\ 1xx0 \\ x10x \end{cases} \qquad \text{OFF}_3 = \begin{cases} x10x \\ 101x \\ 110x \end{cases}$$

$$\text{DC}_1 = \{\, 0101 \,\} \qquad \text{DC}_2 = \phi \qquad \text{DC}_3 = \begin{cases} 0011 \\ 0100 \end{cases}$$

Then array F is initially assigned the value.

$$F = \text{ON}_1{}^* \cup \text{OFF}_1{}^* \cup \text{DC}_1{}^* = \begin{cases} x111\ 1xx \\ 1xx1\ 1xx \\ 111x\ 1xx \\ 00xx\ 0xx \\ 0xx0\ 0xx \\ x0x0\ 0xx \\ xx00\ 0xx \\ 0101\ xxx \end{cases}$$

$\text{ON}_1{}^*$, $\text{OFF}_1{}^*$, and $\text{DC}_1{}^*$ are easily identified in this array. The next value assigned to F is

$$F \leftarrow F \sqcap \begin{cases} 00xx\ x1x \\ 0x1x\ x1x \\ x111\ x1x \\ 10xx\ x0x \\ 1xx0\ x0x \\ x10x\ x0x \end{cases} = \begin{cases} 00xx\ 01x \\ 0x10\ 01x \\ x111\ 11x \\ 10x1\ 10x \\ 10x0\ 00x \\ 1110\ 10x \\ 1x00\ 00x \\ 1101\ 10x \\ x100\ 00x \\ 0101\ x0x \end{cases}$$

We begin to see the compressed array of Fig. 9.10 emerge. Note what intersection accomplished here. Cube $00xx$ is a member of OFF_1 and ON_2. Intersection of $00xx\ 0xx$ and $00xx\ x1x$ gives the first cube of F.

9.5 COMPUTING CONSIDERATIONS

The array model of switching functions is well suited to machine computation if an effective data structure is used. Arrays must be encoded so that they may be stored and processed efficiently. We will explore coding techniques and see samples of the associated processing in this section.

Computing considerations will be discussed in terms of the FORTRAN programming language mainly for two reasons: (1) FORTRAN is widely known and used both in educational and industrial institutions. (2) Subroutines that realize the array operators of the previous and subsequent sections have been prepared in FORTRAN for a number of general purpose digital computers. While FORTRAN dialects do differ among computing installations, the main body of the language is invariant, and programs may be made operational at new locations with relative ease.

We will assume that all program variables are of type INTEGER. The FORTRAN statement

IMPLICIT INTEGER (A–Z)

types all variables, beginning with a letter in the range A through Z as integers. Second, we assume that the host computer has a word length of **b** bits, and an integer is expressed with one such word.

DIMENSION statements are used to reserve portions of memory for storing arrays. It is obvious that a two-dimensional FORTRAN array can hold a logic array of p n-tuples, with each tuple element being stored in one word. This technique is very wasteful of memory and requires that n words be processed to process one cube; $n \times p$ words must be referenced to process the entire array. This is inefficient processing.

Since a cube element $c_j^i \in \{0, 1, x\}$ is a ternary digit or *trit*, it may be encoded with as few as two bits, using any of 24 codes. For purposes of example, we will use the following code throughout this section.

trit	code
0	01
1	10
x	11
none	00

Up to $\mathbf{b}/2$ trit codes can be packed into one computer word, or **b** trits can be stored using two words, with the left bits held in one of the words and the right bits in the other. Both of these techniques have been explored and used. We will illustrate only the first approach in the text of this section.

If $n \le \mathbf{b}/2$, then a complete cube may be stored in one computer word. With $n > \mathbf{b}/2$, successive words in memory may be used to store the cube. In general, $\lfloor (n-1)/(\mathbf{b}/2) \rfloor + 1$ words are required to store a cube of n variables, where $\lfloor \ \rfloor$ denotes "greatest integer $\le$ the enclosed quantity". Thus FORTRAN integer arithmetic is required to compute the number of words per cube.

NWPC = (N − 1)/(B/2) + 1

An array is stored in an area of memory reserved with a DIMENSION statement. If p cubes exist in the array, the reserved area must consist of $p \times nwpc$, or more words. There are good reasons to use one or more of the additional words as a header record that contains facts about the array. Storing the values of n, p, $nwpc$, and the total number of words required by an array in its header permits error detection and speeds computation within subroutines asked to process that array. As a specific example, assume that $\mathbf{b} = 24$, and that two words are used to hold these four items, each in a half-word.

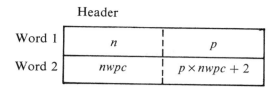

Header

Word 1	n	p
Word 2	$nwpc$	$p \times nwpc + 2$

Array A with $n = 5$ and $p = 2$ is then stored as shown below. The code 00 is used to fill unused portions of computer words; the header occupies the first two words of the array.

DIMENSION A(10)

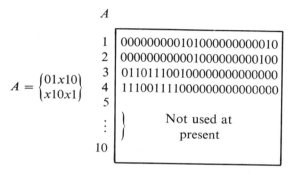

A

$$A = \begin{Bmatrix} 01x10 \\ x10x1 \end{Bmatrix}$$

1	00000000010100000000000010
2	00000000000100000000000100
3	01101110010000000000000000
4	11100111100000000000000000
5	
$\vdots$	Not used at present
10	

Coding trits and packing them into computer words uses memory efficiently, although we see parts of words going unused in the above example. While further packing is possible, it would make cube processing more complex. Using $nwpc$ words for each cube permits easy access to a cube and the parallel processing of $b/2$ trits. For example, in the absorb operation each cube must be compared with every other cube. Let c^i and c^j be the cubes being examined for the relation $c^j \sqsubseteq c^i$. Of the nine possible trit pairs (c^i_k, c^j_k), four are shown in Table 9.1 to indicate that the relation does not exist. How can we detect the presence of any of these four pairs? Examining the code combinations for all nine cases suggests that ANDing computer words on a bit-by-bit basis is valuable.

$$
\begin{array}{lllllllllll}
c^i = 000111xxx & = 01 & 01 & 01 & 10 & 10 & 10 & 11 & 11 & 11 \\
c^j = 01x01x01x & = 01 & 10 & 11 & 01 & 10 & 11 & 01 & 10 & 11 \\
c^i \cdot c^j & = 01 & 00 & 01 & 00 & 10 & 10 & 01 & 10 & 11 \\
& & \uparrow & \uparrow & \uparrow & & \uparrow
\end{array}
$$

In the permitting cases $(c^i \cdot c^j)_k = c^j_k$. In the prohibiting cases $(c^i \cdot c^j)_k \neq c^j_k$. Thus only two operations, ANDing and comparing, are needed to determine if $c^j \sqsubseteq c^i$, or not. In FORTRAN this computation is expressed for array A with:

(A(I) .AND. A(J)) .EQ. A(J)

The value of this logical expression may be used to control the deletion of cube c^j, or not. If $n > b/2$, then successive pairs of words must be tested in this manner before $c^j \sqsubseteq c^i$ is established. Covered cubes may be marked, perhaps by setting $A(J) = 0$, and all marked cubes are removed after all comparisons have been made.

We might also expect ANDing to be important in computing cube intersection. Table 9.2 indicates that two trit pairs produce the special empty result. These trit pairs give the 00 code when code words are ANDed.

$$0 \sqcap 1 = \phi \qquad 01 \cdot 10 = 00$$
$$1 \sqcap 0 = \phi \qquad 10 \cdot 01 = 00$$

Detecting the presence of 00 code words in $c^i \cdot c^j$ is not a trivial task. The OR operator gives a unique result for the 00 input combination. But the bits of a trit code word are side-by-side in a computer word, and we do not wish to tear that word apart and examine individual trit codes. Thus we shift a copy of $c^i \cdot c^j$ one bit position to the left, and then OR. Meaningful information is present in alternating positions only. Therefore, we examine only the meaningful positions with a special *mask* word that has 1's in only the meaningful positions. If we do not find a 1 in every such position, then the intersection is empty. Otherwise, it is $c^i \cdot c^j$.

$c^i = 000111xxx$	$= 01$	01	01	10	10	10	11	11	11
$c^j = 01x01x01x$	$= 01$	10	11	01	10	11	01	10	11
$c^i \cdot c^j$	$= 01$	00	01	00	10	10	01	10	11
$2 \times (c^i \cdot c^j)$	$= 10$	00	10	01	01	00	11	01	10
$t = c^i \cdot c^j \vee 2 \times (c^i \cdot c^j)$	$= 11$	00	11	01	11	10	11	11	11
mask	$= 10$	10	10	10	10	10	10	10	10
$t \cdot mask$	$= 10$	00	10	00	10	10	10	10	10
		↑		↑					

If $t \cdot mask \neq mask$, then $c^i \sqcap c^j = \phi$

In FORTRAN this processing is described:

```
R=C(I).AND.C(J)
IF(((R.OR.(R.SHIFT.1)).AND.MASK).NE.MASK) GO TO ...
```

where R is preserved as the result when the intersection is not empty.

Again, if cubes require more than one word for storage, the successive pairs of words must be ANDed and tested before the final result is found. Finally, if cubes do not exactly fill computer words, the test for 00 code words must be masked so that the 00 fill code words do not interfere.

The sharp operator includes tests for covering and intersection, and the same sort of programming can be used. If $c^i \sqcap c^j \neq \phi$, and $c^i \not\sqsubseteq c^j$, then $c^i \# c^j$ is a set of cubes formed from c^i according to Table 9.3. Forming these cubes requires that isolated trit codes of c^i be altered. An x (code 11) must be replaced with the complement of c_k^j in the kth position. To isolate that trit, a special mask with 1's only in the position of c_k^j must be established, ANDed with c^j, and the result exclusive-ORed with c^i. Assume for purposes of example that $nwpc = 1$ and the $k = 1$ position is on the left of computer words. Then the mask may be formed by shifting the binary representation of the integer 3 to the left $\mathbf{b} - 2\mathbf{k}$ positions.

$$c^i = 1xx0 = 10 \quad 11 \quad 11 \quad 01$$
$$c^j = x1x0 = 11 \quad 10 \quad 11 \quad 01$$
$$mask = 00 \quad 11 \quad 00 \quad 00$$
$$c^j \cdot mask = 00 \quad 10 \quad 00 \quad 00$$
$$c^i \oplus c^j \cdot mask = 10 \quad 01 \quad 11 \quad 01 = 10x0$$

Bit manipulation to accomplish isolated trit processing is also required in a number of the bookkeeping operations of Section 9.3. The availability of full word logic operators such as .AND., .OR. and .XOR. makes this very possible. The FORTRAN .SHIFT. and .ROTAT. operators are also very helpful, but they are not widely available. They can be replaced with multiplication, division, and logical operations, of course, but processing is slowed.

Appendix 9.1 offers descriptive information on a set of FORTRAN subroutines that implement the array operators we have studied, and others we will encounter. Their use is illustrated in subsequent appendices that list FORTRAN programs. These realize some of the algorithms presented in this chapter. The program listings offer a much more detailed presentation of the alogorithms than the corresponding text description. Thus the study of these programs may be advantageous even when their computer execution is not anticipated.

To see how the subroutines are used, let us examine the program of Appendix 9.2 which breaks a given function array into individual ON-, OFF-, and DC-arrays via Algorithm 9.1. The arrays dimensioned in this program are given the names used in describing that algorithm. First the values of n and m are read from the first data card, and the total number of variables, NV, of the function array is computed. Then the function array, F, is read into memory from data cards via the READAR subroutine. One row of this array must be punched on each successive data card, with a card bearing END OF ARRAY terminating the data deck.

Then we begin to determine the ON- , OFF- , and DC-arrays for each z_i, $i = 1, 2, \ldots, m$. First the mask cube is established by defining a unit cube of dimension NV. The $(n + i)$th element of this cube is changed to a 1 by the CNGCOL subroutine, and χ_{n+i}^1 is thus formed. The INTERX subroutine is then used to form the A^* array of Eq. (9.39). The mask cube is changed to be χ_{n+i}^0 and array B^* of Eq. (9.40) is found.

Then we delete the output variable columns for Eq. (9.41) by repeatedly calling the DELCOL subroutine, which is only able to delete one column at a time. The DC- , ON- , and OFF-arrays can then be computed using the intersection (INTERX) and sharp (SHARPS) operators in the manner of Eq. (9.42). Finally, the program prints the arrays just formed.

The statements of this FORTRAN program correspond very closely with the equations used to describe the algorithm. Additional FORTRAN statements are needed for bookkeeping and input-output functions. This is very generally true, as a comparison of the programs of other appendices and the algorithms upon which they are based will reveal.

9.6 ALGORITHMIC CONVERSION OF BOOLEAN EXPRESSIONS TO ARRAYS

Boolean expressions that involve variable names; binary operators AND, OR, and EX-OR; unary operator NOT; and parentheses can be converted to sum-of-products, parentheses-free form with the axioms and theorems of Boolean algebra.[1] When carried out by hand, such a conversion is not usually algorithmic; the manipulation of expressions via a trial-and-error application of a large set of theorems usually is not efficient computer processing.

System programmers in preparing compiler programs such as a FORTRAN compiler, have shown that arithmetic expressions can be converted efficiently to assembly language code by using an intermediate, parentheses-free representation of the expression known as *Polish string notation*. Polish strings employ *post-fix* notation, meaning that the operators appear after (to the right of) their operands. We usually use *infix* notation, in which the operators appear in place between their operands, and occasionally use *prefix* notation with operators appearing before their operands.

Polish strings are also a very valuable intermediate form when converting Boolean expressions to equivalent logic arrays. Suppose that the operator symbols of Table 9.4 are used for the logic operators. We can easily keypunch and print Boolean expressions with conventional computing equipment. The NOT unary operator symbol will be used in the prefix mode; e.g., it must appear before the variable or term enclosed in parentheses that is to be complemented. Juxtaposition of variable names will not be allowed as an expression of logic product; the asterisk must appear.

TABLE 9.4. PERMITTED LOGIC
OPERATORS AND THEIR HIERARCHY

Operator	Symbol	Precedence
NOT	−	7
AND	*	6
EX-OR	@	5
OR	+	4

Examples of Polish strings and corresponding Boolean expressions are given below. How these strings must be interpreted to see the correspondence is by no means obvious, but such discussion will be deferred until we have considered how Polish strings are generated.

---

[1] Other operators such as NAND, NOR, INHIBIT, and IMPLY could be included, but their non-associative nature imposes additional complications. The EQUIVALENCE operator can be included as easily as EX-OR; however, it illustrates nothing additional.

Infix Boolean Expression	Equivalent Polish String
$A \cdot B = A * B$	$AB*$
$A + \bar{B} = A + -B$	$AB- +$
$A \cdot (B + C) = A * -(B + C)$	$ABC+ - *$
$A \cdot B \oplus C + D = A * B @ C + D$	$AB*C@D+$

We use an implied hierarchy of operators when we write infix Boolean expressions. This hierarchy must be formalized and extended to convert infix expressions to Polish strings. A precedence number is given for each operator symbol in Table 9.4. Table 9.5 extends this list to include parentheses and the period that will be used to mark the end of the Boolean expression.

We also require a *push-down*, or *first-in, last-out* (FILO) *stack*. A stack is a storage device; however, not having a hardware stack available we use conventional random access memory and simulate a FILO stack via programming. When a new item is placed on the top of a FILO stack, all items already held by the stack are "pushed down", e.g., stored one cell more distant from the top cell. When an item is removed from a FILO stack, it is the top item, and all other items "pop up" one position toward the top.

In a conventional computer a FILO stack is achieved by reserving a block of memory for the storage of items and by maintaining a word called the *pointer*, which always gives the address of the top element in the stack. Items in the stack are thus not moved in memory when the stack is pushed or popped; the pointer is changed. Using FORTRAN, suppose that the reserved area is named B and the pointer is BP[2].

Assuming that the stack is empty initially, the following statements must appear early in the FORTRAN program:

```
DIMENSION B(500)
BP = 0
```

If X is an item to be placed in the stack, we push the stack with the statements:

TABLE 9.5. PERMITTED
PUNCTUATION

Punctuation	Precedence
)	3
(	2
.	1

[2] All variables discussed are typed INTEGER.

SECTION 9.6   *Algorithmic Conversion of Boolean Expressions to Arrays*

$$BP = BP + 1$$
$$B(BP) = X$$

If the stack is to be popped with variable X being assigned the top value, we write:

$$X = B(BP)$$
$$BP = BP - 1$$

Checks for stack overflow or underflow should be included in the FOR-TRAN code, but have not been written above so that the simplicity of simulating a FILO stack is not obscured.

Finally, we require two *first-in, first-out* (FIFO) *stacks*. We will assume that the first of these, call it A, contains the infix Boolean expression. Each time we pop this stack, we get the next variable name, operator, or punctuation mark. In effect, we will be scanning the Boolean expression from left to right. We will push the items of the Polish string into the second FIFO stack; in the first algorithm call it C. Let AP and CP be the pointers of the two FIFO stacks; both pointers will be initialized at zero.

ALGORITHM 9.6. *Infix Boolean Expression to Polish String Conversion*
The flow chart of Fig. 9.12 summarizes this algorithm. Subroutine GET pops stack A each time it is called, and translates the item into an integer. If the item is a variable name, the integer measures the position of the variable name in a table of valid names. This is also the column of the variable in the logic array to be generated ultimately. These integers are immediately transferred to FIFO stack C; e.g., variable names are immediately placed at the end of the partial Polish string.

Subroutine GET converts punctuation and operator symbols into their precedence number. If the item popped from stack A is a left parenthesis " (" with precedence 2, it is placed on the top of FILO stack B. If the item is a right parenthesis ") ", items are taken from the top of stack B and placed in stack C until a left parenthesis is encountered.[3] Neither the right nor its matching left parenthesis are placed in the C stack; they have served their purpose and are hence discarded. If a period denoting the end of the infix expression is encountered, all operators remaining in stack B, if any, are in turn popped from B and pushed into C. Checks for un-matched parentheses and other punctuation errors can be included in the programming of these transfer operations.

When GET pops a logic operator such as $-$, $*$, $@$, or $+$ from stack A, its precedence is compared with the precedence of the item on top of stack B—actually the top item. (Subroutine GET translates operators and punctuation to precedence numbers to facilitate this comparison.) If the item on top of B has greater or equal precedence, it is transferred to stack C; and the test, and perhaps transfer, is repeated. Ultimately the new operator is placed on top of stack B. Execution of this algorithm on the

---

[3] Operator integers are flagged negative to distinguish them from variable name integers in stack C.

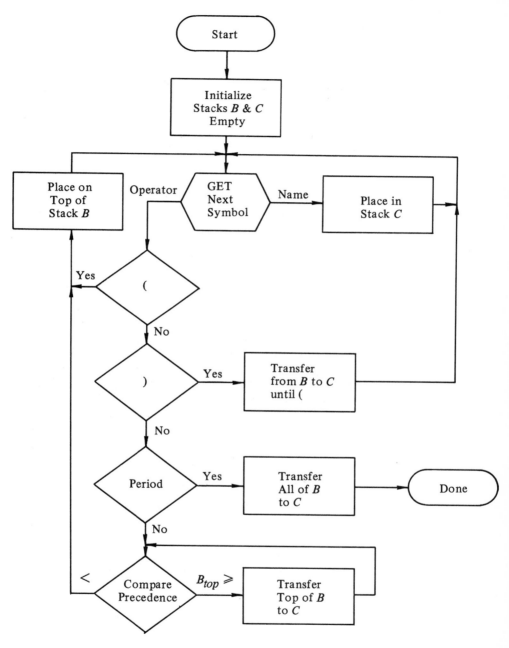

FIG. 9.12   Infix Boolean Expression to Polish String Conversion

Boolean expression, "$A\bar{B} + C \oplus D \oplus (\overline{EF}) = A* - B + C @ D @ - (E * F)$.", is illustrated below. Numerical encodings of the names and operators are not given for clarity. The tops of stacks B and C are on the right; only the item popped from A is shown at each step.

$A(AP)$	FILO stack B	FIFO stack C
A		A
$*$	$*$	A
$-$	$*-$	A
B	$*-$	AB
$+$	$+$	$AB - *$
C	$+$	$AB - *C$
$@$	$+@$	$AB - *C$
D	$+@$	$AB - *CD$
$@$	$+@$	$AB - *CD@$
$-$	$+@-$	$AB - *CD@$
$($	$+@-($	$AB - *CD@$
E	$+@-($	$AB - *CD@E$
$*$	$+@-(*$	$AB - *CD@E$
F	$+@-(*$	$AB - *CD@EF$
$)$	$+@-$	$AB - *CD@EF *$
$\cdot$		$AB - *CD@EF* - @+ \cdot$

ALGORITHM 9.7. *Polish String to Array Conversion*

We bothered to convert an infix expression to a Polish string because that string eases the problem of forming the array representation of the Boolean expression. We can see this and gain the philosophy of Algorithm 9.7 by first converting a Polish string back to infix notation. The string formed in the example above will be used. We will again require a FILO stack, but will now call it the "array stack" because arrays of cubes will be the items placed in it ultimately.

Figure 9.13 offers a flow chart of the algorithm for interpreting Polish strings. The string must be scanned from left to right; e.g., FIFO stack C must be popped successively, just as stack A was in Algorithm 9.6. When a variable name is encountered, it is placed on top of the array stack. If a

Item popped from FIFO stack C	Array stack
A	A
B	A, B
$-$	$A, \bar{B}$
$*$	$A\bar{B}$
C	$A\bar{B}, C$
D	$A\bar{B}, C, D$
$@$	$A\bar{B}, C \oplus D$
E	$A\bar{B}, C \oplus D, E$
F	$A\bar{B}, C \oplus D, E, F$
$*$	$A\bar{B}, C \oplus D, EF$
$-$	$A\bar{B}, C \oplus D, \overline{EF}$
$@$	$A\bar{B}, C \oplus D \oplus \overline{EF}$
$+$	$A\bar{B}+C\oplus D\oplus\overline{EF}$
$\cdot$	$A\bar{B}+C\oplus D\oplus\overline{EF}$

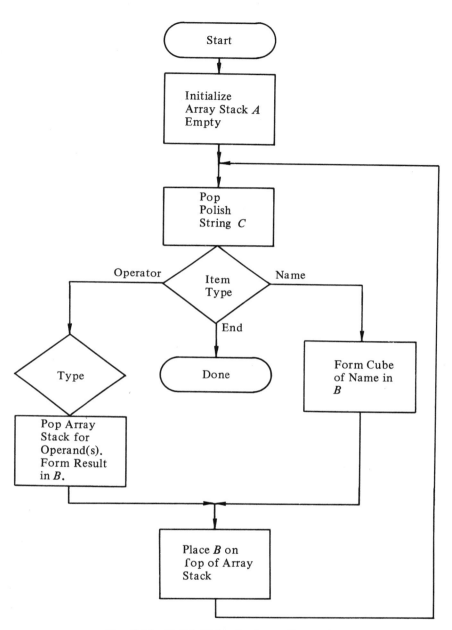

FIG. 9.13 Polish String to Array Conversion

NOT operator is encountered, the top element of the array stack is used as its operand. It is removed from the stack, complemented, and the result placed in the array stack. When a binary operator is popped from C, two items are popped from the array stack to serve as operands. The result of the operation is placed in the array stack. The execution of these simple rules on the example Polish string is shown (p.576). Again, the integer encodings of names and operators is avoided for clarity.

Programming and Manipulating the Array Stack

The array stack must provide for the management of a set of arrays of cubes. While we could establish a great number of different FORTRAN arrays, each with a unique name, the management of such a storage technique would be difficult. FORTRAN subscripts can be used to build easily a FILO stack for logic arrays in a single dimensioned area. A second small vector of pointers and a FORTRAN variable that points to the top of this pointer vector are also needed. To be specific, let A be a FORTRAN array of dimension max, P be a small array of pointers, and T be the integer that points to the last meaningful entry in P—to the top of P. $P(1)$ points to the word of A that holds the first word of the logic array in the bottom of the stack. This is usually $A(1)$. $P(T)$ gives the word of A in which the first word of the logic array on top of the stack is stored. $P(T + 1)$ points to the first word of A available should another logic array be placed in the stack. Figure 9.14 suggests these relationships.

We will begin with an empty array stack by initializing $P(1)$ and T:

```
DIMENSION  A(max), P(20)
P(1) = 1
T = 0
```

If logic array B is now to be placed on top of the array stack, we compute the number of 24-bit words occupied by B with the statements:

```
NWPC = (NV − 1)/12 + 1
NW = NWPC·NCUBES(B) + 2
```

$NWPC$ measures the number of 24-bit words required to hold a cube of NV variables. NW then is the number of words occupied by B; two words are required for its header record. To place B in the array stack and update the various pointers in anticipation of future additions, we write:

```
T = T + 1
P(T + 1) = P(T) + NW
I = P(T)
CALL COPYAR(A(I),B,max − I)
```

Note that the space left in A for placement of array B is easily expressed as $max - P(T)$. Subscripted subscripts are not permitted in many FORTRAN dialects so the variable I is introduced to accomplish computation of $A(P(T))$.

To pop the array stack transferring the top item to B, we need only write:

```
IF(T  .EQ.  0) WRITE (error message)
I = P(T)
CALL COPYAR(B,A(I),MXB)
T = T − 1
```

The IF statement is included as a reminder of the many possible error tests that can, and should be, included in a production program.

Now we will consider how logic arrays may be generated and manipulated in response to a Polish string. When executing Algorithm 9.13,

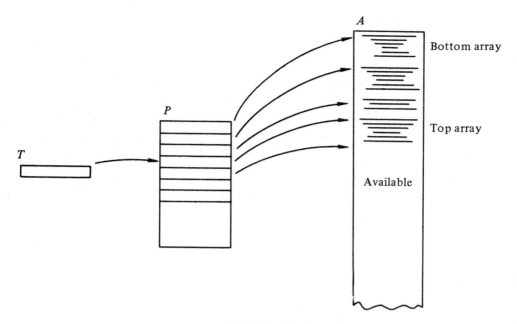

FIG 9.14   The Array Stack

suppose that the name of the *K*th variable is popped from stack *C*. The following calls establish array *B* as a single cube of $NV - 1$ *x*s and a 1 in the *K*th column; this is the array form of a Boolean expression that consists of a single variable:

```
CALL DEFINA(B,NV, 'X')
CALL CNGCOL(B,K, '111')
```

This array may now be placed on top of the array stack with the FOR-TRAN statements discussed above.

If a NOT operator is encountered in the scan of the Polish string, the top item of the array stack must be complemented. We do this with the SHARPS subroutine and an *n*-cube, U_n. The following code forms the complement in *B* (from which it can be placed on top of the array stack, as discussed above) and then pops the array stack by merely decrementing top pointer *T*.

```
CALL DEFINA(CUBE,NV, 'X')
I = P(T)
CALL SHARPS(CUBE,A(I),B,WORK,MXB,NC)
T = T—1
```

Operators AND, EX-OR, and OR require two operands. These are easily referenced with the following statements:

```
IF(T   .LT.   2) WRITE(error message)
I = P(T)
J = P(T—1)
```

The two operands are effectively removed from the array stack with the statement:

T = T − 2

(Nothing is changed in the A array or the P array by this statement; the operands really remain until another item is placed on top of the array stack.)

The AND operation is performed via the INTERX subroutine. In the following call, the result is placed in B, from which it can be placed on top of the array stack, as discussed previously.

```
CALL INTERX(A(I), A(J),B, MXB, NC)
```

The EX-OR operation can be performed in a variety of ways. Using the familiar definition $A \oplus B = A\bar{B} \lor \bar{A}B$ requires two sharp and a combine (or unite) calculations. Additional arrays are needed to store intermediate results and as work space for the SHARPS subroutine:

```
CALL SHARPS(A(I),A(J),D,WORK,MX,NC)
CALL SHARPS(A(J),A(I),B,WORK,MX,NC)
CALL COMBIN(B,D,MXB)
```

The OR operation is performed via the COMBIN subroutine. (UNITE would absorb the result and hence may be superior, but would require more time.)

```
CALL COPYAR(B,A(I),MXB)
CALL COMBIN(B,A(J),MXB)
```

In principle, $A(I)$ could be combined directly with $A(J)$ to avoid unnecessary movement of both top elements of the array stack. The standard code for adding items to the array stack could not be used then.

Transforming infix Boolean expressions to Polish strings requires a modest amount of rather simple computation. It is very worthwhile in that it makes the formation of the corresponding logic array possible with extremely straightforward and modest use of the logic subroutines. The final array generated might well be the ON-array of a complete switching function. The OFF array is then easily generated via Eq. (9.37), and the two arrays may be packed into a multiple output function array via Algorithm 9.5. If incomplete switching functions are to be accepted, then two Boolean expressions must be given and converted to arrays via the technique of this section. Then the third can be computed via Eqs. (9.36), (9.37), or (9.38), and a function array formed.

9.7 COMPUTER ASSISTED SYNTHESIS OF SYNCHRONOUS SEQUENTIAL NETWORKS

Some of the steps in synchronous sequential network design, presented in Chapter 5, are considered here because they offer additional interesting and useful examples of function array generation and processing. These steps are really very routine, even though they may have appeared rather difficult when first encountered. Specifically, we will be concerned with forming a function array of the combinational logic required to realize a given finite state machine that uses Delay flip-flops or JK flip-flops. Synthesis of that logic from such a function array is the subject of subsequent chapters.

An *input alphabet* is a set of N names, called *input symbols*, associated with n-tuples of elements from the ternary set $\{0, 1, x\}$. The tuples provide the encoding of the input symbols in terms of n binary *input variables*. The number of input symbols may be less than or equal to the number of unique binary n-tuples, e.g., $N \leq 2^n$. The set of input symbol encodings need not cover the unit n-cube; input variable n-tuples not covered by this set will be treated as don't cares in the algorithms to be developed. Thus, a finite state machine with two binary input variables might have the following input alphabet; note that the 11 encoding does not appear.

AN INPUT ALPHABET

Symbol name	Encoding $x_1 x_2$
SIT	00
ADV	01
CLR	10

An *output alphabet* is a set of M output symbol names and associated m-tuples. A *state alphabet* is a set of P state names and associated p-tuples, $M \leq 2^m$ and $P \leq 2^p$. As with input alphabets, the state and output alphabets need not be complete.

AN OUTPUT ALPHABET		A STATE ALPHABET	
Symbol name	Encoding $z_1 z_2$	State name	Encoding $y_1 y_2 y_3$
Z0	00	S0	000
Z1	01	S1	001
Z2	10	S2	010
		S3	011
		S4	$1xx$

A *state transition table* is a tabular specification of the next state and output function; these are given in terms of input symbol, output symbol, and state names. Many formats have been used for ST tables; we will use a rather popular one in which a table row exists for each state name and a column exists for each input symbol. Entries in the table are next-state—present-output-symbol pairs. Table 9.6 gives an example ST table, which may be used in conjunction with the example alphabets given above. A mod 5 counter capable of responding to ADVance and CLear commands is described. Output symbol $Z1$ is issued when the sequential network advances to state $S0$; $Z2$ is generated when the sequential network is cleared to state $S0$.

Why not present an encoded state transition table only, and forget about the alphabets and names? The answer can be found by asking what designers might be expected to do with the assistance of a computer. They may well desire to examine a variety of realizations of a fixed ST table. In particular, the state encodings are often obtained in a trial-and-error manner, especially if the remainder of the synthesis algorithm is performed automatically. Designers also may be free to modify input and output symbol encodings. The ST table is the most complicated data structure; it is advantageous to designers not to have to reprepare completely this table for each new code that they wish to try. If they expect to explore various next state and/or output functions, this advantage disappears or is reduced, of course.

Forming the DFF Function Array

Forming a function array from the information found in three alphabets and an ST table is a rather trivial task; this assumes that information is properly stored in a computer memory. One cube of a function array must be generated for each entry in the ST table; additional cubes must be established for the unspecified (don't care) possibilities. Each such cube must be an $n+p+p+m$-tuple of elements of $\{0, 1, x\}$. The first n elements of the tuple are an input symbol encoding; the next p elements give a present state encoding. These $n+p$ entries constitute input variables of the combinational logic to be designed. The next state encoding and output symbol encoding make up the remaining $p+m$ entries of the cube; these are the output variables of the combinational logic. Thus the entry in the bottom right corner of Table 9.6 is converted to the following cube.

With the generation of such cubes in mind, useful storage for the alphabets might be as follows. Assume that two dimensioned FORTRAN variables, INAME and ITUPLE, are used to hold the input alphabet[4]. Input symbol names will be packed successively in INAME(1), INAME(2), ..., INAME(N) as they are read from cards. ITUPLE will hold a logic array of cubes. The order of the cubes is important; it must correspond exactly to the order of the names in INAME. Let these cubes express the

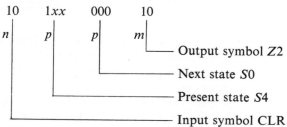

---

[4] All FORTRAN variables are assumed to be type INTEGER.

TABLE 9.6
A STATE TRANSITION TABLE

Present state	Input symbol		
	SIT	ADV	CLR
S0	S0/Z0	S1/Z0	S0/Z2
S1	S1/Z0	S2/Z0	S0/Z2
S2	S2/Z0	S3/Z0	S0/Z2
S3	S3/Z0	S4/Z0	S0/Z2
S4	S4/Z0	S0/Z1	S0/Z2

input symbol encoding in their left n coordinates and contain xs in their right $p+p+m$ coordinates. We will consider later programming techniques for filling these arrays from free-format data cards.

Output and state alphabets can be stored in similar fashion, but with $n+p+p$ xs on the left of the output symbol encodings. There are n xs on the left and $p+m$ xs on the right of the state encodings. Let ONAME, OTUPLE, SNAME, and STUPLE be the dimensioned FORTRAN arrays that hold this information. In addition, we will need logic array NSTUPL, which consists of cubes of state encodings with $n+p$ xs on the left and m xs on the right. Figure 9.15 suggests how the example alphabets would appear in storage.

Many of the cubes of a function array may now be formed by intersecting one cube from each of ITUPLE, OTUPLE, STUPLE, and NSTUPL. Intersected cubes must first be copied from these arrays with the CPYCUB subroutine so that the order of cubes in these arrays is not perturbed. As next state names of the ST table are read from data cards, they are compared with the elements of SNAME until a match is found. The position of the name in SNAME then is the number of the cube of NSTUPL to be intersected with the element of STUPLE, which corresponds to the present state, and the element of ITUPLE, which corresponds to the input symbol. Output symbol names must be compared with the elements of ONAME, and the corresponding cube of OTUPLE intersected with the result of previous intersections. (If a name is not found in an alphabet, an error message may be printed and processing terminated.)

As a specific example, suppose that logic array ICUBE contains the member of ITUPLE corresponding to the present input symbol of the table element being processed, and SCUBE contains the cube of the present state. Let X hold the name of the next state and Y hold the name of the output symbol. Assuming that valid names are held by X and Y, the following code will form the required function array cube:

```
      DO 1 I =1,NS
1     IF(X. EQ. SNAME(I)) GO TO 2
      WRITE error message
2     DO 3 J=1,NO
3     IF(Y .EQ. ONAME(J)) GO TO 4
      WRITE error message
4     CALL INTERX(ICUBE, SCUBE, A, MX, NC)
      CALL CPYCUB(NSTUPL, B, I)
```

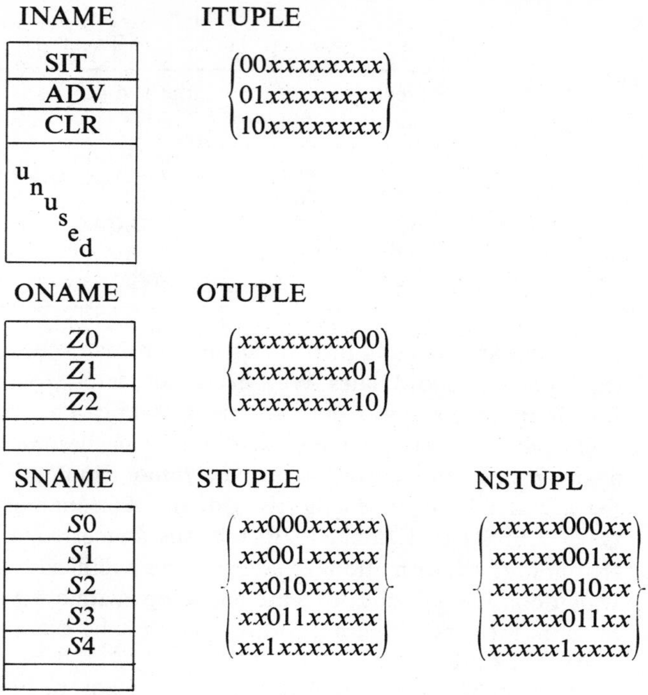

INAME   ITUPLE

SIT
ADV
CLR

$$\left\{\begin{array}{l} 00xxxxxxx \\ 01xxxxxxx \\ 10xxxxxxx \end{array}\right\}$$

u n u s e d

ONAME   OTUPLE

Z0
Z1
Z2

$$\left\{\begin{array}{l} xxxxxxx00 \\ xxxxxxx01 \\ xxxxxxx10 \end{array}\right\}$$

SNAME   STUPLE   NSTUPL

S0
S1
S2
S3
S4

$$\left\{\begin{array}{l} xx000xxxxx \\ xx001xxxxx \\ xx010xxxxx \\ xx011xxxxx \\ xx1xxxxxxx \end{array}\right\} \qquad \left\{\begin{array}{l} xxxxx000xx \\ xxxxx001xx \\ xxxxx010xx \\ xxxxx011xx \\ xxxxx1xxxx \end{array}\right\}$$

FIG. 9.15   Input, Output, and State Alphabet Storage

```
CALL INTERX(A, B, C, MX, NC)
CALL CPYCUB(OTUPLE, B, J)
CALL INTERX(C, B, A, MX, NC)
CALL COMBIN(F, A, MX)
```

The partial function array that is obtained from Table 9.6 by processing each row, in turn, from left to right is

$$F = \left\{\begin{array}{ccccc}
x_1 x_2 & y_1 y_2 y_3 & y_1' y_2' y_3' & z_1 z_2 \\
0\ 0 & 0\ 0\ 0 & 0\ 0\ 0 & 0\ 0 \\
0\ 1 & 0\ 0\ 0 & 0\ 0\ 1 & 0\ 0 \\
1\ 0 & 0\ 0\ 0 & 0\ 0\ 0 & 1\ 0 \\
0\ 0 & 0\ 0\ 1 & 0\ 0\ 1 & 0\ 0 \\
0\ 1 & 0\ 0\ 1 & 0\ 1\ 0 & 0\ 0 \\
1\ 0 & 0\ 0\ 1 & 0\ 0\ 0 & 1\ 0 \\
0\ 0 & 0\ 1\ 0 & 0\ 1\ 0 & 0\ 0 \\
0\ 1 & 0\ 1\ 0 & 0\ 1\ 1 & 0\ 0 \\
1\ 0 & 0\ 1\ 0 & 0\ 0\ 0 & 1\ 0 \\
0\ 0 & 0\ 1\ 1 & 0\ 1\ 1 & 0\ 0 \\
0\ 1 & 0\ 1\ 1 & 1\ x\ x & 0\ 0 \\
1\ 0 & 0\ 1\ 1 & 0\ 0\ 0 & 1\ 0 \\
0\ 0 & 1\ x\ x & 1\ x\ x & 0\ 0 \\
0\ 1 & 1\ x\ x & 0\ 0\ 0 & 0\ 1 \\
1\ 0 & 1\ x\ x & 0\ 0\ 0 & 1\ 0
\end{array}\right\}$$

To complete this function array, don't care input variable n-tuples and state variable p-tuples must be appended, if any exist. To generate these don't care conditions, we need only sharp arrays ITUPLE and STUPLE from a unit $n+p+p+m$-cube, and combine the results with F.

```
CALL DEFINA(CUBE, N+P+P+M, 'X')
CALL SHARPS(CUBE, ITUPLE, A, B, MX, NC)
CALL COMBIN(F, A, MX)
CALL SHARPS(CUBE, STUPLE, A, B, MX, NC)
CALL COMBIN(F, A, MX)
```

In the example, only cube $11 \circ xxx \circ xxx \circ xx$ will be added to array F.

JK Flip-Flop Function Array Generation

The function array generated to this point describes the combinational logic of a sequential machine with Delay flip-flop memory elements. If other types of flip-flops are to be used, additional processing is required. This additional processing is a tedious and error-prone task when performed by hand. It also appears to be a difficult step in the synthesis algorithm. Yet, this additional processing is almost trivial to program and requires very little computer time.

Suppose JKFFs are to be used. Table 5.9 listed the J and K signal values that are required to promote some state transitions. Table 9.7 provides a complete listing.

For a given state variable, we break array F into pieces corresponding to rows of Table 9.7, and replace the next state column y' with two columns for J and K. The pieces are then reassembled to form a function array. The splitting apart is performed with a mask cube and the SPLITA subroutine. For purposes of example we will transform the first state variable; all of the other state variables are processed in like manner.

```
NV = N+P+P+M
I = 1
PRES = N+I
NEXT = N+P+2·I−1
```

TABLE 9.7
JKFF TRANSFORMATION RULES

State change $y \rightarrow y'$		Required signals J	K
0	0	0	d
0	1	1	d
0	x	d	d
1	0	d	1
1	1	d	0
1	x	d	d
x	0	0	1
x	1	1	0
x	x	d	d

The first three rows of Table 9.7 suggest that we establish a cube of xs with a zero in the present state column, column $PRES$. If any cubes are split from F with this mask, the y' column, column $NEXT$ assuming all prior state variables have already been transformed, can serve as the J column; a column of xs must be created (by OPNCOL) for the K variable.

```
         CALL DEFINA(CUBE, NV, 'X')
         CALL CNGCOL(CUBE, PRES, '000')
         CALL SPLITA(F, A, CUBE, MX, NC)
         IF(NC .EQ. 0) GO TO 1
         CALL OPNCOL(A, NEXT, MX)
1        CONTINUE
```

Processing the example function array F with $I = 1$ gives the following A array.

$$A = \begin{Bmatrix}
x_1x_2 & y_1y_2y_3 & J_1K_1y_2'y_3' & z_1z_2 \\
0\ 0 & 0\ 0\ 0 & 0\ d\ 0\ 0 & 0\ 0 \\
0\ 1 & 0\ 0\ 0 & 0\ d\ 0\ 1 & 0\ 0 \\
1\ 0 & 0\ 0\ 0 & 0\ d\ 0\ 0 & 1\ 0 \\
0\ 0 & 0\ 0\ 1 & 0\ d\ 0\ 1 & 0\ 0 \\
0\ 1 & 0\ 0\ 1 & 0\ d\ 1\ 0 & 0\ 0 \\
1\ 0 & 0\ 0\ 1 & 0\ d\ 0\ 0 & 1\ 0 \\
0\ 0 & 0\ 1\ 0 & 0\ d\ 1\ 0 & 0\ 0 \\
0\ 1 & 0\ 1\ 0 & 0\ d\ 1\ 1 & 0\ 0 \\
1\ 0 & 0\ 1\ 0 & 0\ d\ 0\ 0 & 1\ 0 \\
0\ 0 & 0\ 1\ 1 & 0\ d\ 1\ 1 & 0\ 0 \\
0\ 1 & 0\ 1\ 1 & 1\ d\ x\ x & 0\ 0 \\
1\ 0 & 0\ 1\ 1 & 0\ d\ 0\ 0 & 1\ 0
\end{Bmatrix}$$

The fourth through sixth rows of Table 9.7 suggest the use of a mask cube with a 1 in the $PRES$ column. The complement of the $NEXT$ column can be used as the K column; a column of xs must be introduced to its left for the J variable.

```
         CALL CNGCOL(CUBE, PRES, '111')
         CALL SPLITA(F, B, CUBE, MX, NC)
         IF(NC .EQ. 0) GO TO 2
         CALL CNGCOL(B, NEXT, '10X')
         CALL OPNCOL(B, NEXT-1, MX)
         CALL COMBIN(A, B, MX)
2        CONTINUE
```

Processing what remains of the F array in this manner gives

$$B = \begin{Bmatrix}
x_1x_2 & y_1y_2y_3 & J_1K_1y_2'y_3' & z_1z_2 \\
0\ 0 & 1\ x\ x & d\ 0\ x\ x & 0\ 0 \\
0\ 1 & 1\ x\ x & d\ 1\ x\ x & 0\ 1 \\
1\ 0 & 1\ x\ x & d\ 1\ x\ x & 1\ 0
\end{Bmatrix}$$

This result is appended to A.

It is entirely possible that the F array will be empty at this point. If it is not, processing per the last three rows of Table 9.7 must be performed. The next state column can serve directly as the J variable column. To introduce complementary entries in the K variable column, without a great deal of splitting and changing columns, we make a copy of the remains of the F array, complement the next state column in the copy, and open columns in the two versions and intersect them.

```
      IF(NCUBES(F) .EQ. 0) GO TO 3
      CALL COPYAR(B, F, MX)
      CALL CNGCOL(B, NEXT, '10X')
      CALL OPNCOL(F, NEXT, MX)
      CALL OPNCOL(B, NEXT-1, MX)
      CALL INTERX(F, B, C, MX, NC)
      CALL COMBIN(A, C, MX)
3     CALL COPYAR(F, A, MX)
```

The remaining row in the example function array is transformed to:

$$x_1 x_2 \quad y_1 y_2 y_3 \quad J_1 K_1 y_2' y_3' \quad z_1 z_2$$
$$C = \{1\ 1 \quad x\ x\ x \quad d\ d\ x\ x \quad x\ x\}$$

All of the previous intermediate results are returned to F so that the process may be repeated for $I = 2$ and 3. The final function array, after all state variables have been transformed, is shown below. The order in which the cubes appear can be justified by reviewing the algorithms. The great number of don't care conditions that appear is usual with JK flip-flops and is one of their great advantages. Any synthesis program that will deal with such function arrays must take advantage of these conditions.

$$F = \begin{cases}
x_1 x_2 y_1 y_2 y_3 & J_1 K_1 J_2 K_2 J_3 K_3 z_1 z_2 \\
0\ 0\ 0\ 0\ 0 & 0\ d\ 0\ d\ 0\ d\ 0\ 0 \\
0\ 1\ 0\ 0\ 0 & 0\ d\ 0\ d\ 1\ d\ 0\ 0 \\
1\ 0\ 0\ 0\ 0 & 0\ d\ 0\ d\ 0\ d\ 1\ 0 \\
0\ 0\ 0\ 1\ 0 & 0\ d\ d\ 0\ 0\ d\ 0\ 0 \\
0\ 1\ 0\ 1\ 0 & 0\ d\ d\ 0\ 1\ d\ 0\ 0 \\
1\ 0\ 0\ 1\ 0 & 0\ d\ d\ 1\ 0\ d\ 1\ 0 \\
0\ 0\ 0\ 0\ 1 & 0\ d\ 0\ d\ d\ 0\ 0\ 0 \\
0\ 1\ 0\ 0\ 1 & 0\ d\ 1\ d\ d\ 1\ 0\ 0 \\
1\ 0\ 0\ 0\ 1 & 0\ d\ 0\ d\ d\ 1\ 1\ 0 \\
0\ 0\ 0\ 1\ 1 & 0\ d\ d\ 0\ d\ 0\ 0\ 0 \\
0\ 1\ 0\ 1\ 1 & 1\ d\ d\ d\ d\ d\ 0\ 0 \\
1\ 0\ 0\ 1\ 1 & 0\ d\ d\ 1\ d\ 1\ 1\ 0 \\
0\ 0\ 1\ x\ x & d\ 0\ d\ d\ d\ d\ 0\ 0 \\
0\ 1\ 1\ x\ x & d\ 1\ 0\ 1\ 0\ 1\ 0\ 1 \\
1\ 0\ 1\ x\ x & d\ 1\ 0\ 1\ 0\ 1\ 1\ 0 \\
1\ 1\ x\ x\ x & d\ d\ d\ d\ d\ d\ d\ d
\end{cases}$$

Processing Names

Techniques for reading names and tuples from cards punched in a rather free format are interesting and generally useful; they are not difficult to create or tailor to special situations. The FORTRAN code for techniques described here assumes a 24-bit word into which three characters may be packed. It may not be applicable to other machines because of differences in FORTRAN, machine codes for letters and digits, and machine word lengths. But modifications to the code listed here are easily made once these differences are known.

Define a *name* to be a sequence of one to three letters and/or decimal digits. The upper limit of three exists because the 24-bit word is able to hold only three 8-bit ASCII character codes. If names of more than three characters are to be allowed, more than one computer word must be used to store the name; the programming becomes slightly more complicated, as we will see when we consider storing tuples of arbitrary length. Any symbol, including the blank, other than a letter or digit will be taken as terminating the name.

In Table 1.3, notice that decimal digits are assigned contiguous encodings with a leftmost bit of zero in the ASCII code. The letters are also assigned contiguous encodings with a leftmost bit of zero. Thus, if letter or digit codes reside on the left of computer words, those words take the form of positive integers. When columns of cards are read with the A1 format of FORTRAN, the codes of individual card columns are placed on the left of computer words.

Suppose that the data card bearing a name in some column(s) is read into FORTRAN array IO with the A1 format specification.

```
      DIMENSION IO(80)
         ⋮
      READ(5,1) IO
 1    FORMAT(80A1)
```

Suppose that variable CC points to the card column of the first character of the name. We examine the character in that column and columns to the right of it until a terminating symbol is encountered, or until a full three characters have been found and packed into a single word, *NAME*. Single quote marks enclose Hollerith constants in many FORTRAN dialects.

```
      NAME = ' '
      DO 2 I = 1,3
      CHR = IO(CC)
      IF(CHR .LT. '0') GO TO 3
      IF(CHR .GT. '9' .AND. CHR .LT. 'A') GO TO 3
      IF(CHR .GT. 'Z') GO TO 3
      NAME = (NAME .SHIFT. 8) .OR. (CHR .SHIFT. −16)
      CC = CC+1
      IF(CC .LE. 80) GO TO 2
      READ(5,1) IO
      CC = 1
 2    CONTINUE
 3    CONTINUE
```

The character in card column CC is tested; if it is neither a digit nor a letter the program above proceeds to statement 3 where subsequent processing of the assembled name takes place. If a digit or letter is found, *NAME* is shifted left 8 places to make room for the new character; the word holding the character is shifted to the right 16 places. All such shifts fill vacated bit positions with zeroes. The new character may now be packed in *NAME* with the OR operator. *CC* is incremented in preparation for examining the next column. If splitting a name over a card boundary is to be permitted, the test of *CC*, READ, and resetting of *CC* may be included, as shown in the program above.

Reading Free-Formated Cubes

Reading a tuple of NV elements of $\{0, 1, x\}$ from consecutive card columns and packing it three elements per word into $NV/3 + 1$ consecutive words of memory (for use as the Hollerith string argument of the DEFINA subroutine) is a generalization of reading and packing names. In fact, the same technique and much the same programming would be used to read and pack longer names.

We assume that A is a FORTRAN array of dimension sufficient to hold the packed tuple; e.g., the dimension of A is greater than $NV/3+1$ for all expected values of NV. Again we will scan card images held in array IO with pointer CC and terminate when NV elements have been packed or when an illegal character is encountered. Assume CC is pointing at the column of the first tuple character initially.

The number of words of A to be filled is:

$$NWA = (NV-1)/3 + 1$$

We will use I to count the words of A, and L to count the number of elements packed.

```
        L = 0
        DO 13 I = 1,NWA
        X = ' '
        DO 12 J = 1,3
        IF(L .LT. NV) GO TO 10
        X = X .ROTAT. 8
        GO TO 12
10      CHR = IO(CC)
        IF(CHR .EQ. '0') GO TO 11
        IF(CHR .EQ. '1') GO TO 11
        IF(CHR .EQ. 'X') GO TO 11
        WRITE error message
        STOP
11      X = (X .SHIFT. 8) .OR. (CHR .SHIFT. —16)
        CC = CC+1
        IF(CC .LE. 80) GO TO 12
        READ(5,1) IO
        CC=1
12      L=L+1
13      A(I)=X
        A(NWA + 1) = ' '
        CALL DEFINA(CUBE, NV, A)
```

The inner DO loop packs three valid tuple characters into word X in the same manner as names were packed. The final word may not contain three tuple characters; L is used to left-justify the characters in X by rotating X to the left. An additional word of blanks is placed in $A(NWA + 1)$ to ensure that the packed Hollerith string (now in A3 format) ends with a blank; the DEFINA subroutine requires this.

Would it not be easier to read a data card bearing names or tuples with an A3 format specification, and let FORTRAN input-output programs take care of all the packing? Yes, it would be easier for the programmer. No, it would be a nightmare for users of that programmer's programs. They would be required to place names and tuples in very specific columns of data cards. Errors are more easily made, leading to ruined computer runs and frustrated, dissatisfied users.

9.8 SUMMARY

Switching functions have structure that is graphically displayed by their topological presentation. The value of such presentations declines as the number of variables increases, but the concept of "cubes" does not. The cube-array notation is as useful with many variables as it is with few, and perhaps more clearly and easily used than Boolean notation when the number of variables is large. It certainly is more amenable to computer manipulation. Therefore, covers and complexes were defined and the definiton of "ON array" was generalized.

The absorb, union, intersection, sharp, split, and bookkeeping operators provide a means of processing arrays efficiently. They form the Boolean algebra of arrays. Single output functions may be described with two Boolean expressions (giving the ON and DC conditions, for example) or with two arrays. The third expression (array) is generated by complementing their logical sum. Multiple output functions are expressed with a set of Boolean equations or a function array that is conceptually identical to a truth table.

A number of useful and interesting algorithms for processing ON and function arrays were described in terms of array operations. While algorithms may be executed by hand, the computer is a very available and able tool for doing routine processing. Hence techniques for storing and processing arrays with computers are very important. Given effective means of processing arrays, the use of computers to process switching functions in useful arrays is limited only by our imagination and resources. To stimulate the imagination, we have discussed techniques for converting Boolean expressions and state tables to function arrays. Synthesizing logic networks from such arrays is even more exciting; we now turn to that subject.

REFERENCES

1. M. A. BREUER, *Design Automation of Digital Systems*, Chapter 2, "Logic Synthesis." Englewood Cliffs: Prentice-Hall, Inc., 1972.
2. D. L. DIETMEYER and P. R. SCHNEIDER, "Identification of Symmetry, Redundancy, and Equivalence of Boolean Function," *IEEE Trans. on Electronic Computers*, Vol. EC-16, Dec. 1967, pp. 804–817.

3. E. J. MCCLUSKEY, *Introduction to the Theory of Switching Circuits*. New York: McGraw-Hill, 1965.

4. R. E. MILLER, *Switching Theory Volume I: Combinational Circuits*. New York: John Wiley and Sons, 1965.

5. J. P. ROTH, "Algebraic Topological Methods for the Synthesis of Switching Systems I," *Trans. Am. Math. Soc.*, Vol. 88, No. 2, July 1958, pp. 301–326.

6. R. H. URBANO and R. K. MUELLER, "A Topological Method for the Determination of the Minimal Forms of a Boolean Function," *IRE Trans. on Electronic Computers*, Vol. EC-5, No. 3, Sept. 1956, pp. 125–132.

7. M. E. ULUG and B. A. BOWEN, "A Unified Theory of the Algebraic Topological Methods for the Synthesis of Switching Systems," *IEEE Transactions on Computers*, Vol. C-23, Feb. 1974, pp. 255–267.

PROBLEMS

9.1-1. Find the base, K^0, and maxterm set, L^0 for:

   (a) $F_1 = \bar{A}B\bar{C}$

   (b) $F_2 = \bar{A} \vee B \vee \bar{C}$

   (c) $F_3 = (A\bar{B} \vee \bar{A}B) \cdot C$

9.1-2. The bases of two switching functions are given:

$$K^0(f) = \left\{\begin{matrix}001\\011\\101\\110\\111\end{matrix}\right\} \qquad K^0(g) = \left\{\begin{matrix}010\\011\\110\\111\end{matrix}\right\}$$

   (a) Find $L^0(f)$ and $L^0(g)$.

   (b) Find the base of each of the following new switching functions:

      (i) $r = f \cdot g$

      (ii) $s = f \vee g$

      (iii) $t = f \cdot \bar{g}$

      (iv) $u = \overline{f \vee \bar{g}}$

9.1-3. The Expansion Theorem introduced switching functions

$$x_i \cdot f(x_1, \ldots 1 \ldots, x_n) \quad \text{and} \quad \bar{x}_i \cdot f(x_1, \ldots 0 \ldots, x_n).$$

Describe the base of each of these functions in terms of $K^0(f)$.

9.1-4. Problem 2.7-8 introduced switching functions $x_i \cdot g$, $\bar{x}_i \cdot h$, and k. Describe the bases $K^0(x_i \cdot g)$, $K^0(\bar{x}_i \cdot h)$ and $K^0(k)$ in terms of $K^0(f)$.

9.1-5. (a) Find $K^0(F)$ and $K^0(G)$. Are they related, and if so, how?

$$F(A, B, C) = AB \vee \bar{B}C \qquad G = B \vee \bar{B}C$$

   (b) In an expression of $f(x_1, \ldots, x_n)$, we erase all appearances of x_i to obtain function $g(x_1, \ldots, x_n)$. Are $K^0(f)$ and $K^0(g)$ related, and if so, how?

9.2-1. (a) Use Theorem 10 to find all implicants of each of the functions below.
(b) Find K^0, K^1, K^2, etc. for each function. (c) Display the covering relationship between implicants with a Hasse diagram such as that shown

in Fig. 9.4 and with the topology of each function.

(i) $F1 = \vee\, 0, 1, 4, 7$

(ii) $F2 = \vee\, 5, 7, 9, 10, 13, 15$

(iii) $F3 = \vee\, 4, 5, 6, 7, 12, 13, 14, 15, 16, 18, 20, 22$

9.2-2. (a) How many 1-cubes does a complete 2-cube cover? 3-cube? 4-cube? n-cube? (b) How many 2-cubes does a complete 3-cube cover? 4-cube? n-cube? (c) How many r-cubes does an n-cube cover where $r \leq n$?

9.2-3. (a) Find the smallest (fewest members) subset C of each $K(f_i)$ which covers the corresponding $K^0(f_i)$.

$$K^0(f_1) = \left\{ \begin{matrix} 010 \\ 100 \\ 101 \\ 110 \\ 111 \end{matrix} \right\} \qquad K(f_1) = \left\{ \begin{matrix} 010 \\ 100 \\ 101 \\ 110 \\ 111 \\ x10 \\ 10x \\ 1x0 \\ 1x1 \\ 11x \\ 1xx \end{matrix} \right\}$$

$$K^0(f_2) = \left\{ \begin{matrix} 0011 \\ 0100 \\ 0101 \\ 0111 \\ 1001 \\ 1101 \\ 1110 \\ 1111 \end{matrix} \right\} \qquad K(f_2) = \left\{ \begin{matrix} K^0(g) \\ 010x \\ 01x1 \\ 0x11 \\ x101 \\ x111 \\ 1x01 \\ 11x1 \\ 111x \\ x1x1 \end{matrix} \right\}$$

(b) Repeat part (a), attempting to find the smallest subset of the largest cubes that covers each minterm set.

9.2-4. Use Karnaugh maps to verify that $K(f_1)$ and $K(f_2)$ as given in Prob. 9.2-3 are complete. Review your answers to that problem with the Karnaugh maps in mind.

9.2-5. Use Karnaugh maps to find $K(f_i)$ for each of the following functions. Select a smallest covering subset of $K(f_i)$.

(a) $f_1 = \vee\, 2, 4, 7, 9, 11, 12, 13, 14$

(b) $f_2 = \vee\, 1, 4, 5, 7, 8, 9, 11, 12, 15$

9.2-6. Use Karnaugh maps to find the base, the complex, and a smallest covering set for each function.

(a) $f = A\bar{C}\bar{D} \vee B\bar{C}D \vee \bar{A}BD \vee \bar{A}\bar{C}D \vee A\bar{B}\bar{C}$

(b) $g = \bar{A}\bar{C} \vee A\bar{B}\bar{C}\bar{D} \vee CD(\overline{AB})$

9.2-7. The cover relationship $\sqsubseteq$ can be used to order Boolean functions. (a) Using $\sqsubseteq$, find all order relationships between members of the set of all switching functions of two variables $\{f_0, f_1, \ldots, f_{15}\}$. See Table 2.4 (b) Which pairs of elements are not related by $\sqsubseteq$? (If none exist the set is said to be *totally ordered*.)

(c) Display these ordering relationships on a *Hasse* diagram (Fig. 9.4 is an example) in which covered items are placed below and connected to all items that cover them. Describe the topology of your Hasse diagram.

9.2-8. (a) Find the set of all implicants of $f(x_1, x_2) = 1$.

(b) Find an order relationship which partially orders this set. Draw a Hasse diagram.

9.2-9. (a) Find the set of all implicants of $f = C(\overline{B\overline{D}}) \vee \overline{C}\overline{D}(\overline{AB}) \vee BD$.

(b) Order the implicants of f using $\sqsubseteq$ and draw a Hasse diagram that displays the ordering.

(c) Determine from the Hasse diagram which implicants are prime.

(d) From the Hasse diagram select a minimal cover of f.

(e) Could the partially ordered set of implicants serve as the basis for a lattice?

9.2-10. Assume the sole prime implicant $11xxxx$ of a switching function is found from its base. How many 0-cubes, 1-cubes, 2-cubes, etc., are generated in the process of arriving at this prime implicant?

9.2-11. A switching function of six variables consists of all of the unit 6-cube except for the 1-cube $x11111$. Describe the sum-of-products expression of this function as a set of cubes.

9.2-12. In a *n*-cube how many vertices are distance $1, 2, 3, \ldots,$ from a given vertex?

9.2-13. Call a vertex and all vertices adjacent to it a "code set." How many disjoint code sets exist in an *n*-cube? If a Hamming code (Section 3.10) is to be used so that single errors may be detected and corrected, how many parity bits, $n-k$, must be introduced if k measures the number of message bits?

9.2-14. The structure shown below is a part of a complete *n*-cube. All adjacencies are shown as lines. Vertices not connected by lines are distance 2 or greater apart. Find a labeling of the vertices that satisfies these rules. What is the smallest value which *n* may assume? Show a consistent labeling for this value of *n*.

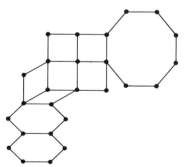

9.2-15. (a) Define the off complex $L(f)$ and don't care-complex $DC(f)$ for a switching function.

(b) Give generalized definitions for "OFF-array" and "DC-array."

9.2-16. Product-of-sums expressions may also be represented with arrays. Let each sum term be expressed with a tuple by the following code: 1 for true variable, 0 for complement, and x for absent variable. Write arrays for each of the following switching functions.

(a) $F_1 = (A \vee B \vee C)(A \vee B \vee \bar{C})(A \vee C)$
(b) $F_2 = A(B \vee \bar{C} \vee D)(\bar{B} \vee C \vee E)$

9.2-17. Does the cover relation $\sqsubseteq$ apply to cubes that represent sum terms as in Prob. 9.2-16?

9.3-1. (a) Absorb the following arrays.

$$A = \left\{ \begin{array}{l} 11x1 \\ x0x1 \\ 1001 \\ 0x1x \\ x1xx \\ 0000 \end{array} \right\} \qquad B = \left\{ \begin{array}{l} xx11 \\ 11x1 \\ 00xx \\ 1x11 \end{array} \right\}$$

(b) Form $A \sqcap B$, $A \# B$, and $A \sqcup B$.

9.3-2. Problem 9.2-11 calls for the computation of $A = xxxxxx \# x11111$. Compute array A by using the sharp operator, and compare with your answer to Prob. 9.2-11.

9.3-3. If a, b, and c are subcubes of a common cube, which of the following expressions are cover equivalent? Justify your answer.

1. $(a \sqcap b) \# c$
2. $(a \# c) \sqcap (b \# c)$
3. $a \sqcap (b \# c)$
4. $(a \# c) \sqcap b$
5. $(a \sqcap b) \# (c \sqcap b)$

9.3-4. Equations (9.21) and (9.22) offer two ways of computing the sharp product $A \# B$. Examine these two ways and test them on example arrays, attempting to determine which is, in general, the more efficient means of computing $A \# B$.

9.3-5. Show that the following distributive-like laws are valid.

$$(A \sqcup B) \# C \equiv (A \# C) \sqcup (B \# C)$$
$$(A \sqcap B) \# C \equiv (A \# C) \sqcap (B \# C)$$

9.3-6. C is a cover of Boolean function f; $c^i \in C$ and $C - c^i$ denotes array C with member c^i removed.
In general, is $C - c^i \equiv C \# c^i$?

9.3-7. Which, if any, of the following arrays are cover equivalent?

$$A = \left\{ \begin{array}{l} x10x \\ 00x0 \\ 11x1 \\ 101x \end{array} \right\} \qquad B = \left\{ \begin{array}{l} 0x00 \\ 110x \\ 0101 \\ 1x11 \\ x010 \end{array} \right\}$$

$$C = \left\{ \begin{array}{l} 0x00 \\ 11xx \\ x101 \\ 1x1x \\ 0010 \end{array} \right\}$$

$$D = \begin{cases} 0000 \\ 0010 \\ 0100 \\ 0101 \\ 1010 \\ 1011 \\ 1100 \\ 1101 \\ 1111 \end{cases} \qquad E = \begin{cases} 00x0 \\ 1x1x \\ 11xx \\ x10x \end{cases}$$

9.3-8. (a) Prove that if an r-cube is removed from an n-cube, where $r \le n$, then $n - r$ $(n-1)$-cubes remain.

(b) Show that $\mathcal{U}_n \# c$, where c is an r-cube, is a set of $n - r$ $(n-1)$-cubes.

9.4-1. Develop an algorithm that uses the array operators to form the base $K^0(f)$ from a cover $C(f)$. Program your algorithm and test your program on a number of covers of the same and different functions. Minimize the number of statements used to actually do the conversion; i.e., do not count input, output, dimension, etc., statements.

9.4-2. We can use n-tuples to represent other than product terms. Suppose the cubes c^i of C each represent a sum term and C describes a product-of-sums expression of a Boolean function.

(a) Develop an algorithm for converting C into array D, which expresses the function in sum-of-products form. Example: convert $f = (A \lor B)$ $(\bar{C} \lor D)(\bar{A} \lor \bar{D} \lor E)$ to sum-of-products form.

(b) Develop an algorithm for converting D to C.

(c) Prepare and test FORTRAN programs that implement your algorithms.

9.4-3. We know that any Boolean function can be expressed as the exclusive-sum of product terms. For example: $A \lor B = A \oplus B \oplus AB$, $A \cdot \bar{B}$ $\oplus B \cdot C = A\bar{B} \lor BC$.

(b) If C is an array that represents a sum-of-products expression of a Boolean function and D is an array of product terms of the exclusive-sum-of-products expression of the same function, develop an algorithm for converting D to C.

(c) Develop an algorithm for generating D from C.

(d) Prepare and test FORTRAN programs that implement your algorithms. Test with more complex functions than those given as examples in part (a).

9.4-4. We know that a Boolean function can be expressed as the exclusive-sum-of-products terms of variables only. For example: $A \cdot \bar{B} = A \oplus AB$.

(b) Repeat parts (b) and (c) of Prob. 9.4-3, but now complements will not appear in D.

(c) Prepare and test FORTRAN programs that implement your algorithms.

9.4-5. Array C represents a sum-of-products expression of a switching function. Develop an algorithm for converting C into a cover D of the same function in which all cubes are disjoint; i.e., $d^i \sqcap d^j = \phi$, for all $i \ne j$. Do not submit an algorithm given to answer a previous problem.

9.4-6. Array F is not a complete function array. Only those input conditions for which all output variables have value 0 are missing. Develop and program an algorithm to convert F to a function array. Test your program on a variety of incomplete, and one complete, function arrays.

9.4-7. A 4-input, 3-output combinational circuit is described by the truth table below.

(a) What submodules can be obtained by tying a single input to logic 0? to logic 1? Eliminate all redundancy in the list of submodules.

(b) What submodules can be obtained by tying a pair of input to 0? to1?

(c) What input—output terminal ties result in combinational submodules?

0000	000	1000	100
0001	010	1001	110
0010	010	1010	110
0011	001	1011	101
0100	100	1100	001
0101	110	1101	011
0110	110	1110	011
0111	101	1111	001

9.4-8. Prepare and test a FORTRAN program that executes Algorithm 9.5. What limits does your program place on n, m, or a combination of n and m?

9.4-9. Can the split operator **S** be used in place of the intersect operator $\sqcap$ in Algorithm 9.2? Algorithm 9.3? If not, why not?

9.5-1. A single-output, completely specified Boolean function of n variables can be recorded with 2^n bits using a 2^n-tuple. If the base of the function consists of k minterms, the function may also be expressed via k n-tuples with kn bits. For $2 \le n \le 6$ compute the value of k such that the n-tuple expression of the function requires fewer bits. What fraction of the Boolean functions of 2 and 3 variables are expressed equally or more concisely with n-tuples.

9.5-2. The following code permits the easy identification of xs in cubes: Just examine the left bit.

0	0 0
1	0 1
x	1 0, 1 1

Either the 10 or 11 code may be found for the x in a cube. Develop $\sqsubseteq$ and $\sqcap$ algorithms for the computer processing of words that hold cubes encoded with this code.

9.5-3. Cube elements are encoded as in Prob. 9.5-2, but the left bits of elements of cube c^i are packed into computer word c^{i1}, while the right bits are packed into c^{i2}. Cube c^j is similarly stored. Develop $\sqsubseteq$ and $\sqcap$ algorithms for the computer processing of these four words.

9.5-4. Left and right shifting with vacated positions filled with 0's can be accomplished with integer multiplication and division, respectively. (Ignore sign and overflow problems.) Show how a circular shift (rotate) can be programmed in FORTRAN by using arithmetic operations.

9.6-1. Convert each of the Boolean expressions to a Polish string.

(a) $A \cdot B \vee \bar{A} \cdot C \cdot (E \vee F)$

(b) $A \cdot (B \oplus C) \cdot (\overline{\overline{D \vee E}}) \oplus F$

(c) $A \cdot (B \cdot C \cdot (\overline{\overline{D \vee E}}))$

9.6-2. Convert each Polish string to an infix Boolean expression. Introduce parentheses, if necessary, so that the expression is correctly read.

(a) $AB @ - A - C + - *$

(b) $ABC - * +$

9.6-3. Complete Boolean equations are to be read with the dependent variable written on the left of an $=$ sign that expresses the "connect" operation. The Boolean expression appears to the right of the equal sign.

(a) What precedence number would you assign to the equal sign? (Renumber other symbols if you feel it to be necessary.)

(b) If completely specified multiple-output switching functions are described by sets of such equations, what array operations would you perform in response to the equal sign and its operands in preparation for forming a function array? Detail your answer with a segment of FORTRAN code.

9.6-4. The binary NAND operator is to be included in the set of accepted logic operators and keypunched via the dollar sign ($).

(a) Where would you place $ on the precedence list?

(b) What array operations would you perform in response to a $?

Illustrate your answer with a segment of FORTRAN code that is compatible with the segments given in the text.

9.7-1. Prepare a DFF function array from the following alphabets and state transition table.

Input			State			Output				YES	NO
YES	0	1	A	0	x	X	0		A	B/X	X/Y
NO	1	0	B	1	0	Y	1		B	C/X	A/X
			C	1	1				C	A/Y	A/X

9.7-2. How must a DFF function array be transformed if (a) RSFFs and (b) TFFs are to be used?

9.7-3. Write and exercise a FORTRAN program that reads a card in A1 format, packs the characters to An format (where n is the maximum number of characters that may be packed into one word in your computer system), and then prints the card image using An format. Extra characters should not be printed on the right.

9.7-4. Write and exercise a FORTRAN program that reads names of 6 or fewer characters from one or more cards. Names are separated by a comma or blank followed by any number of blanks. The last name is followed by a period. All 80 columns of cards may be used. Names are to be stored, left justified in the words of a symbol table, and that table is to be printed after all names have been read.

9.7-5. Extend the program of Prob. 9.7-4 to print error messages and (a) truncate long names and (b) delete duplicate names.

9.7-6. Assume that names are listed in a symbol table in the order of input variables for an ON-array. Develop and program an algorithm for printing the ON-array as a sum-of-products Boolean expression.

ANSWERS

9.1-1. (c) $K^0(F_3) = \{3, 5\}$, $L^0(F_3) = \{0, 1, 2, 4, 6, 7\}$

9.1-2. (iii) $K^0(t) = K^0(f) \cap L^0(g) = \begin{Bmatrix} 001 \\ 101 \end{Bmatrix}$

9.1-3. $K^0(x_i \cdot f(x_1, \ldots 1 \ldots, x_n))$ consists of those members of $K^0(f)$ with a 1 in the ith position. Thus $K^0(x_i \cdot f_i) \subseteq K^0(f)$.

9.2-1. $F_1 = \bar{A}\bar{B}\bar{C} \vee \bar{A}\bar{B}C \vee A\bar{B}\bar{C} \vee ABC \vee \bar{A}\bar{B} \vee \bar{B}\bar{C}$

$$K^0 = \begin{Bmatrix} 000 \\ 001 \\ 100 \\ 111 \end{Bmatrix} \quad K^1 = \begin{Bmatrix} 00x \\ x00 \end{Bmatrix}, \quad K^2 = \varnothing$$

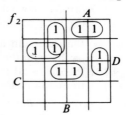

9.2-3. $C(f_1) = \begin{Bmatrix} 1xx \\ x10 \end{Bmatrix} \quad C(f_2) = \begin{Bmatrix} 010x \\ 0x11 \\ 111x \\ 1x01 \end{Bmatrix}$

9.2-5. Minimal covers. Find another minimal cover for function f_2.

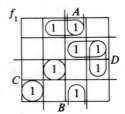

9.2-6.

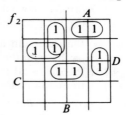

$K^0(f) = \{1, 5, 7, 8, 9, 12, 13\}$

$$C(f) = \begin{Bmatrix} 1x0x \\ xx01 \\ 01x1 \end{Bmatrix}$$

9.2-10. An n-cube covers $\binom{n}{r}2^{n-r}$ r-cubes.

9.2-12. $\binom{n}{d}$ where d is the distance. Any of d bits may be changed in a 0-cube to form a distant cube.

9.2-17. Yes. We are treating a maxterm as a point (0-cube). Two adjacent points define a line (1-cube).

$$(A \vee B \vee C)(A \vee B \vee \bar{C}) = A \vee B$$

In a space where minterms appear as points, the topology of $A \vee B$ lies within that of $A \vee B \vee C$, but we are not using that space here.

9.3-1. $11x1 \sqsubseteq x1xx$, $1001 \sqsubseteq x0x1$ in A

$$A \sqcap B = \begin{cases} 11x1 \\ x011 \\ 00x1 \\ 0x11 \\ 001x \\ x111 \\ 0000 \end{cases}$$

9.3-6. No

$$\begin{cases} 00x \\ 0x1 \end{cases} - 0x1 = 00x; \qquad \begin{cases} 00x \\ 0x1 \end{cases} \# 0x1 = 000$$

9.3-8. See Theorem 9.3.

9.4-2. Only one program is necessary. Sharp and complement columns.

9.4-4. Remove duplicate cubes in pairs. $A \oplus A = 0$.

9.4-6. Shape the given input tuples from U_n.

9.5-2. Two cases where $c_k^i = x$ cause problems in evaluating $c^j \sqsubseteq c^i$, if the 10 cube is permitted.

9.5-4. $B = A*(2**n) + A/(2**(b - n))$.

9.6-1. (a) $AB \cdot A - C \cdot EF \vee \cdot \mathbf{v}$

9.6-2. (b) $A \vee (B\overline{C})$

9.7-2. "Don't know" present states cannot be transformed to specific next states with TFFs.

APPENDIX 9.1
A Combinational Logic Subroutine Set

Many of the logic array manipulating operators that are defined and utilized in this and subsequent chapters have been implemented as FORTRAN subroutines for a number of digital computers. Details of the subroutine set differ between computers. Only a brief description of the UNIVAC 1108 subroutines is given here; subsequent appendices illustrate their use.

Each element of a cube is encoded with two or more bits, and code words are packed into one or consecutive words of memory. The first word or words of each array is called the *header* and stores the following facts about the array:

> Number of words used to store each cube
> Number of variables
> Number of cubes in the array
> Number of words used to store array

The number of bits used to express each item limits its range.

In the subroutine descriptions that follow, the FORTRAN variables have the following significance.

A,B,C, ...	Name of an array. These variables *must* be dimensioned and are treated as type INTEGER by the subroutines.
N	Number of Input Variables
M	Number of Output Variables
NV	Total number of variables (usually $NV = N + M$)
MXA, MXB, ...	Dimensioned size of array $A, B, \ldots$
ICUBE	Integer indentifying a cube of an array

ICOL, JCOL	Integer indentifying a column of an array
NCC	Number of cubes in array C

0. SYSTEM SUBROUTINES

These subroutines are utilized by other subroutines and are not generally available to the user, but they do emit error messages.

0.1 ERRORS

This subroutine is called whenever other subroutines encounter illegal data. It causes the UNIVAC 1108 executive system to print:

A COMPUTED GO TO THAT WAS OUT OF RANGE WAS DETECTED AT SEQUENCE NUMBER 174

and then the subroutine that called ERRORS and the subroutine calling that subroutine, if any, etc., until finally the source statement of the main program is identified.

0.2 WHEAD

This subroutine writes the header record associated with each logic array.

0.3 RHEAD

This subroutine reads the header record for an array. If some field is not correct, the subroutine prints the error message:

INVALID ARRAY HEADER RECORD

and called the ERRORS routine to provide a walkback to the source statement of the main program which promoted the reference to the illegal array.

1. INPUT-OUTPUT SUBROUTINES

1.1 READAR   CALL READAR (A, NV, MXA)

An array is transferred from punched cards to array A. One cube must be punched per card; only 0, 1, 2, X, and D are taken as valid characters (2, X, and D are all logically equivalent). Blank columns of punched cards are ignored. A card with other than one of these characters in the leftmost position is taken as an end-of-array indicator. Only the NV leftmost legal characters are read from each card. MXA indicates the amount of storage available.

Error Messages

READAR—ILLEGAL CHARACTER ON CARD
—   card image   —
  n CARDS SKIPPED

The message is printed if an illegal character is encountered in other than the leftmost position. The subroutine searches for the end of array indicator, indicates the number of cards skipped, and allows processing to continue.

READAR—INSUFFICIENT VARIABLES ON CARD
—   card image   —
n CARDS SKIPPED

Less then NV legal characters were found on the card.

READAR—ARRAY OVERFLOW
n CARDS SKIPPED

The complete array could not be placed in the MXA memory locations provided. Again the end of array card is located and processing allowed to continue.

1.2 PRNTAR   CALL PRNTAR (A, N)

Array A is printed with a blank placed between the Nth and $(N+1)$st character. If N is outside the range $0 < N \leq NV$ where NV is the number of variables of array A, then N is taken as NV. If illegal characters are encountered in an array, they are printed as question marks (?). 0's, 1's, and X's are printed in the N leftmost columns; 0's, 1's, and —'s are printed in the right columns.

1.3 DEFINA   CALL DEFINA (A, NV, CHAR)

Array A is generated as a single cube of NV variables. CHAR, a Hollerith string such as '$0X1D$' of six or fewer characters from the set {0, 1, 2, X, D} determines the cube. If $NV \leq$ the number of characters in CHAR, the NV leftmost characters of CHAR are employed. If NV exceeds the number of characters in CHAR, the rightmost character of CHAR is extended as needed.
Error Messages

DEFINA—ILLEGAL CHARACTER IN–char–TAKEN AS X

Any nonblank character other than 0, 1, 2, X, D is taken as equivalent to an X. Processing continues.

2. ARRAY BOOKKEEPING SUBROUTINES

2.1 COMBIN $A \leftarrow A \cup B$ CALL COMBIN (A, B, MXA)

The unabsorbed union of arrays A and B replaces array A, if this union requires less than MXA storage locations and if the arrays are of the same number of variables.

COPYAR $A \leftarrow B$ CALL COPYAR (A, B, MXA)

Array A is replaced with a copy of array B, if array B occupies less than MXA locations.

Error Messages

COMBIN—UNEQUAL NUMBER OF VARIABLES

Arrays A and B have differing numbers of variables.

COMBIN–ARRAY OVERFLOW

$A \cup B$ or B will not fit into MXA locations.

2.2 NCUBES N = NCUBES (A)

CUBCNT   CALL CUBCNT (A, NCUBE)

The number of cubes in array A is determined.

2.3   KUBCST (c^i) KOST = KUBCST (A, N, ICUBE)

      CUBCST   CALL CUBCST (A, N, ICUBE, KOST)

The number of 0's and 1's in the left N positions of the ICUBEth cube of array A is returned as KOST, if array A contains at least ICUBE cubes. If N falls outside the range $0 < N \leq NV$, N is taken as NV, the number of variables of array A.
Error Message
    CUBCST—NCUBE OUT OF RANGE
    Array A contains fewer than ICUBE cubes.

2.4   CPYCUB $B \leftarrow a^i$ CALL CPYCUB (A, B, ICUBE)

      MOVCUB $B \leftarrow a^i$, $A \leftarrow A - a^i$ CALL MOVCUB (A, B, ICUBE)

The ICUBEth cube of array A is copied as array B, or moved from array A to array B, if the cube existed in the array. Array B consists of a single cube in both cases.
Error Message
    CPYCUB—NONEXISTENT CUBE, B ARRAY SET EMPTY
    NCUBE ≤ 0 or greater than the number of cubes in array A cause this message. Processing continues.

2.5   IDTCOL   CALL IDTCOL (A, ICOL, NRET)

The ICOLth column of array A is examined, and the results returned via the value of NRET.

NRET = 0	Array A is empty or ICOL is out of range
1	Column of all 0's
2	Column of all 1's
3	Column of 0's and 1's
4	Column of all x's
5	Column of 0's and x's
6	Column of 1's and x's
7	Column of 0's, 1's, and x's

Error Messages
    IDTCOL—ILLEGAL CHARACTER
    Other than a 0, 1, or x was encountered in the array examined.
    IDTCOL—ICOL OUT OF RANGE
    ICOL does not fall within $0 < ICOL \leq NV$ for array A.
    NRET = 0 is returned and processing continues.

2.6   CNGCOL $A \leftarrow \mathbf{C}_t^{b_1 b_2 b_3}(A)$ CALL CNGCOL (A, ICOL, CNG)

The ICOLth column of array A is altered in a manner determined by CNG, if the column exists. CNG is a 3-character Hollerith string of characters 0, 1, 2, X, and D. If CNG is $b_1 b_2 b_3$, then 0's in the ICOLth column are replaced by b_1, 1's are replaced by b_2, and x's are replaced by b_3. If array A is empty, it is not altered.
Error Messages
    CNGCOL—ILLEGAL CHARACTER IN CNG
    Other than a 0, 1, 2, X, or D appears in the first three characters of CNG.

CNGCOL—ICOL OUT OF RANGE
ICOL is not in the range $0 < \text{ICOL} \leq NV$.

2.7 DELCOL $A \leftarrow \mathbf{D}_i(A)$ CALL DELCOL (A, ICOL)

The ICOL*th* column of array A is deleted. If A is empty, no change is made.
Error Message
DELCOL—ICOL OUT OF RANGE
ICOL is outside the range $0 < \text{ICOL} \leq$ the number of variables of array A.

2.8 OPNCOL $A \leftarrow \mathbf{I}_i(A)$ CALL OPNCOL (A, ICOL, MXA)

A column of x's is inserted in array A to the right of the ICOL leftmost column if ICOL falls within the range $0 \leq \text{ICOL} \leq NV$ and the expanded array can be stored in MXA words of storage.
Error Messages
OPNCOL—ICOL OUT OF RANGE
OPNCOL—ARRAY OVERFLOW

2.9 PRMCOL $A \leftarrow \mathbf{P}_{ij}(A)$ CALL PRMCOL (A, ICOL, JCOL)

Columns ICOL and JCOL of array A are interchanged, if ICOL and JCOL fall in the range $0 < \text{ICOL}, \text{JCOL} \leq$ number of variables of array A. If array A is empty, no change is made. If $\text{ICOL} = \text{JCOL}$, no change is made.
Error Message
PRMCOL—ICOL OR JCOL OUT OF RANGE

2.10 COLHGT   CALL COLHGT (A, IHGT)

The number of 0's and the number of 1's in the ith column of array A are counted and the greater count placed in cell IHGT(i) of integer array IHGT for each column of A. Thus IHGT must have a dimension which equals or exceeds the number of variables of array A. If A is empty, NV cells of IHGT are set to 0.

3. LOGIC OPERATORS

3.1 ABSORB $A \leftarrow \mathbf{A}(A)$ CALL ABSORB (A)

Array A is absorbed if it originally consists of more than one row.

3.2 UNITE $A \leftarrow A \sqcup B$ CALL UNITE (A, B, MXA)

The absorbed union of arrays A and B replaces array A, if that union can be stored in MXA locations or less. Array B will be absorbed if the unabsorbed union $A \cup B$ requires more than MXA locations. Otherwise it will not be altered.

Error Messages
UNITE—ARRAY OVERFLOW
UNITE—UNEQUAL NV

3.3 INTERX $C \leftarrow A \sqcap B$ CALL INTERX (A, B, C, MXC, NCC)

The intersection of arrays A and B is placed in array C, if the intersection requires fewer than MXC locations. NCC returns the number of cubes in array C. Array C must not utilize the same memory locations as array A or array B.
Error Messages
INTERX—ARRAY OVERFLOW
INTERX—UNEQUAL NV

3.4 SHARPS $C \leftarrow A \# B$ CALL SHARPS (A, B, C, D, MXC, NCC)

The sharp product of arrays A and B is placed in array C, if that product can be stored in MXC locations. Array D is a work array used by SHARPS and assumed to have the same dimension as array C, i.e., MXC. NCC returns the number of cubes placed in array C. Arrays C and D may not utilize the same memory locations as arrays A and B.
Error Messages
SHARPS—UNEQUAL NV
SHARPS—ARRAY OVERFLOW
Overflow may refer to either array C or array D.

3.5 SPLITA $B \leftarrow A\, \mathbf{S}\, \text{MASK}$ CALL SPLITA (A, B, MASK, MXB, NCB)

All cubes of array A that subsume the first cube of array $MASK$ are removed from array A and collected in array B, if MXB or fewer memory locations suffice to hold all. NCB returns the number of cubes transferred to array B.
Error Messages
SPLITA—UNEQUAL NV
SPLITA—ARRAY OVERFLOW

3.6 CNSSUS $A \leftarrow * (A)$ CALL CNSSUS (A, B, C, MXA, NIV)

The cubes of array A are converted into prime implicants using the generalized consensus algorithm. The left NIV variables are treated as input variables. If NIV falls outside the range $0 < NIV \leq NV$, all variables are treated as input variables. If $NIV < NV$, Bartee coding of the output variables is assumed; cubes with an all-0 output part are not generated. Arrays B and C are work arrays and are assumed to have the same dimension as array A, i.e., MXA.
Error Messages
CNSSUS—ARRAY OVERFLOW

3.7 FACTOR $B \leftarrow \mathbf{F} (A)$ CALL FACTOR (A, B, NFM)

The common factor cube of array A is placed in array B. NFM measures the figure-of-merit of the factor cube, and will be 0 if A contains one or fewer cubes, or if the common factor cube placed in B in the universal cube U_n.

3.8 SUBCOL CALL SUBCOL (A, ICOL, JCOL, NRET)

Columns ICOL and JCOL of array A are compared for the covering relationship, and the result returned via the value of NRET.

$$
\begin{aligned}
\text{NRET} &= 1 & &\text{No relationship between columns} \\
&= 2 & &\text{Column ICOL} \sqsubseteq \text{column JCOL} \\
&= 3 & &\text{Column ICOL} = \text{column JCOL} \\
&= 4 & &\text{Column ICOL} \sqsupseteq \text{column JCOL}
\end{aligned}
$$

Error Messages
SUBCOL—ICOL OR JCOL OUT OF RANGE

3.9 CONTRD CALL CONTRD (A, N, NRET)

Array A is examined for contradictions, i.e., a single input n-tuple associated with disjoint output m-tuples, and the results are returned via the value of *NRET*.

$$
\begin{aligned}
\text{NRET} &= 1 & &\text{No contradiction exists} \\
\text{NRET} &= 2 & &\text{Contradiction exists}
\end{aligned}
$$

Error Message
CONTRD—NO OUTPUT VARIABLES
The given value of N equals or exceeds the total number of variables of array A.

4. SUPPLEMENTARY SUBROUTINES

4.1 SELECT CALL SELECT (ISEL, N, M, IND)

All selections of M integers from the set $\{1, 2, \ldots, N\}$ are successively generated in integer array *ISEL*. With $IND = 1$, the subroutine defines $ISEL(I) = I$ for $I = 1, 2, \ldots, M$, and sets $IND = 2$. With $IND = 2$, the next selection is placed in array *ISEL*, if a next selection exists. If the final selection has previously been generated, *IND* is set to 3.
Error Messages
SELECT—IND = 3
The subroutine was called with IND = 3.
SELECT—M .GT. N

4.2 NANDS NGATES = NANDS (I, L)

The number of L-fan-in limited NAND gates required to achieve an I-fan-in NAND gate is calculated by evaluating

$$
\text{NANDS} = \begin{cases} 0 & \text{if } I \leq 1 \\ 2^* \left\lfloor \dfrac{L - I}{L - 1} \right\rfloor + 1 & \text{if } I > 1 \end{cases}
$$

where $\lfloor \; \rfloor$ denotes "integer less-than" $\dfrac{L - I}{L - 1}$

APPENDIX 9.2

A program based upon Algorithm 9.1: Dissect a Function Array. This program splits a function array into m ON, m OFF, and m DC-arrays.

```
      INTEGER A(100), B(100), F(100), MASK(5)
      INTEGER ON(100), OFF(100), DC(100), WORK(100)
      MAX = 100
C
C     READ PARAMETERS AND FUNCTION ARRAY
C
      READ 1, N, M
    1 FORMAT(1X, 3I2)
      NV = N + M
      CALL READAR(F, NV, MAX)
C
C     FOR EACH OUTPUT VARIABLE—
C     ESTABLISH MASKS AND SPLIT FUNCTION ARRAY
C
      DO 6 I = 1,M
      CALL DEFINA(MASK, NV, 'XXXXXX')
      CALL CNGCOL(MASK, N + I, '111')
      CALL INTERX(F, MASK, A, MAX, NCA)
      CALL CNGCOL (MASK, N + I, '000')
      CALL INTERX(F, MASK, B, MAX, NCB)
C
C     DELETE OUTPUT COLUMNS
C
      DO 2 J = 1,M
      CALL DELCOL(A, N + 1)
    2 CALL DELCOL(B, N + 1)
C
C     FORM DC, ON, AND OFF—ARRAYS
C
      CALL INTERX(A, B, DC, MAX, NCDC)
      CALL SHARPS(A, DC, ON, WORK, MAX, NCON)
      CALL SHARPS(B, DC, OFF, WORK, MAX, NCOFF)
C
C     PRINT RESULTS
C
      PRINT 3, I
    3 FORMAT (4H0ON(, I2, 7H)—ARRAY)
      CALL PRNTAR(ON, N)
      PRINT 4, I
    4 FORMAT(5H OFF(, I2, 7H)—ARRAY)
      CALL PRNTAR(OFF, N)
      PRINT 5, I
    5 FORMAT(4H DC(, I2, 7H)—ARRAY)
    6 CALL PRNTAR(DC, N)
      CALL EXIT
      END
```

Annotations alongside the code:

χ^1_{n+i}

$A^* = F \sqcap \chi^1_{n+i}$

$B^* = F \sqcap \chi^0_{n+i}$

$A = \mathbf{D}_\Omega(A^*)$

$B = \mathbf{D}_\Omega(B^*)$

$DC_i = A \sqcap B$

$ON_i = A \,\#\, DC_i$

$OFF_i = B \,\#\, DC_i$

APPENDIX 9.3

A program based upon Algorithm 9.3, in part. All submodules of a given module with 2 inputs tied together are formed.

```
      INTEGER F(100), A(100), MASK(5), ISEL(10)
      MAX = 100
C
C     READ PARAMETERS AND MODULE ARRAY
C
      READ 1, N, M
    1 FORMAT(1X, 3I2)
      NV = N + M
      CALL READAR(F, NV, MAX)
      IND = 1
C
C     SELECT THE TWO INPUTS TO BE TIED
C
    2 CALL SELECT(ISEL, N, 2, IND)                         Form λ
      GO TO (2, 3, 10), IND
C
C     FORM MASK ARRAY
C
    3 CALL DEFINA(MASK, NV, 'XXXXXX')
      DO 4 I = 1,2
    4 CALL CNGCOL(MASK, ISEL(I), '000')                    χλ^0
      CALL COPYAR(A, MASK, MAX)
      DO 5 I = 1,2
    5 CALL CNGCOL(MASK, ISEL(I), '111')                    χλ^1
      CALL COMBIN(MASK, A, 5)                              M = χλ^0 ∪ χλ^1
C
C     FORM SUBMODULE ARRAY
C
      CALL INTERX(F, MASK, A, MAX, NCA)                    A* = F ⊓ M
      IF(NCA .EQ. 0) GO TO 2
C
C     PRINT RESULTS
C
      PRINT 6, ISEL(1), ISEL(2)
    6 FORMAT(15H0MODULE WITH V(, I2, 4H) = V(, I2, 1H))
      CALL PRNTAR(A, N)
C
C     CHECK FOR TRIVIAL OUTPUT VARIABLES
C
      DO 9 I = 1,M
      CALL IDTCOL(A, N + I, NRET)
      GO TO (7, 7, 9, 7, 7, 7, 9), NRET
    7 PRINT 8, I
    8 FORMAT(10H OUTPUT W(, I2, 12H) IS TRIVIAL)
    9 CONTINUE
      GO TO 2
   10 CALL EXIT
      END
```

APPENDIX 9.4

A subroutine based upon Algorithm 9.5: Form a Function Array. This subroutine forms a function array from m ON and m DC-arrays. Variables and arrays are in COMMON storage so that this subroutine can be used in conjunction with the subroutines of subsequent appendices.

```
          SUBROUTINE FORMFA
          IMPLICIT INTEGER(A–Z)
          COMMON N, M, P, NV, IOC, MX, NC, X(3)
          COMMON F(200), ON(200), E(200), A(200), B(200), C(200), D(200)
          COMMON CUBE(5), SAVE(200)
          MX = 200
          READ 1, N, M
     1    FORMAT(1X, 3I2)
          NV = N + M
          IOC = N + 1
          IF(NV .EQ. 0) CALL EXIT
          CALL DEFINA(F, NV, 'XXXXXX')
          CALL DEFINA(ON, 0, 0)
          CALL DEFINA(CUBE, N, 'XXXXX')
          DO 4 I = 1,M
     C
     C    READ ON AND DC–ARRAYS, MINIMIZE ON, FORM OFF–ARRAY
     C
          CALL READAR(A, N, MX)
          CALL CNSSUS(A, B, C, MX, N)
          NPI = NCUBES(A)
          DO 2 J = 1,NPI
          CALL MOVCUB(A, B, 1)
          CALL SHARPS( B, A, C, D, MX, NC)
          IF(NC .GT. 0) CALL COMBIN(A, B, MX)
     2    CONTINUE
          CALL READAR(B, N, MX)
          CALL COPYAR(C, A, MX)
          CALL COMBIN(C, B, MX)
          CALL SHARPS(CUBE, C, E, D, MX, NC)
     C
     C    OPEN OUTPUT COLUMNS, ADD OUTPUT CODING
     C
          DO 3 J = 1,M
          CALL OPNCOL(A, N, MX)
          CALL OPNCOL(B, N, MX)
     3    CALL OPNCOL(E, N, MX)
          CALL CNGCOL(A, N + I, '111')
          CALL CNGCOL(E, N + I, '000')
     C
     C    ADD TO ON–ARRAY, UPDATE FUNCTION ARRAY
     C
          CALL COMBIN(ON, A, MX)
          CALL COMBIN(E, A, MX)
          CALL COMBIN(E, B, MX)
          CALL INTERX(F, E, A, MX, NC)
     4    CALL COPYAR(F, A, MX)
          DO 5 J = IOC,NV
     5    CALL CNGCOL(ON, J, '010')
          RETURN
          END
```

10

Minimization of Switching Functions

Minimization is the process of obtaining that expression of a switching function which is optimum under some criterion. Usually a *cost* criterion is established and the optimum expression is that which dictates a minimum cost realization of a function. Thus the type of circuits we will use when actually constructing a network influences the cost criterion, and the equation-to-circuit transformation must be clearly established and kept in mind when we establish a minimization procedure.

Cost criteria can be simple or very complex. At one time diodes were expensive, and reducing the number of diodes required in AND and OR gates was very worthwhile. Costs of resistors, other electronic components, and fabrication of the circuit were neglected. Then transistors in NAND and NOR gates became the expensive items, it was important to minimize transistor count, ignoring diode, resistor, and fabrication cost. Simple cost criteria like these apply today when integrated circuits are being designed. For the designer who interconnects available integrated circuits, IC count is a meaningful criterion for purposes of comparing alternative realizations of a switching function.

Integrated circuit gates present a new problem. The cost of making a connection to a gate or mounting the gate on supporting material can equal or exceed the cost of the gate itself. Cost of design may well exceed the cost of the hardware components if only a few systems are to be fabricated: then the first expression written is optimum in that it requires the least design effort. A valid cost criterion must be very complicated if such facts are included.

10.1 MANUAL MINIMIZATION

To simplify this section, we will consider a gate to be most expensive and an interconnection to cost less, and we will not emphasize attaining circuits of absolutely minimum cost. Thus first emphasis will be placed on minimizing the number of gates employed; within that criterion the number of interconnections required will be minimized. Further, we will consider only two-level realizations of switching functions, and speak of AND and OR gates constructed of diodes. The discussion applies equally well to two-level NAND gate networks in which we attempt to minimize the number of required gates and gate interconnections.

The axioms and theorems of Boolean algebra can be used to minimize a Boolean function. But use of algebra to minimize other than simple functions has serious disadvantages. First, in what order should we apply the axioms and theorems? No answer can be given. With experience, insight, luck, and trial and error we may find a minimal circuit, but an orderly procedure is certainly to be desired. Second, when do we stop attempting to apply the axioms and theorems to minimize further? With few exceptions it is difficult to tell when a Boolean function has in fact been expressed in optimum fashion.

Several orderly procedures (algorithms) have been developed which, if followed, will give a minimal sum-of-product (by duality: product-of-sum) expression of a Boolean function. Some of these procedures require rather complex computation and are presented later. This section will be concerned with procedures that can give a minimum or near-minimum cost realization of a Boolean function with relatively small effort on our part. All these procedures are essentially the same and can be divided into two parts. First, all prime implicants are found. Then a subset of the set of all prime implicants is selected and the Boolean function expressed as a sum of the selected prime implicants.

Generation of Prime Implicants

We saw in Section 9.1 how to form the larger cubes of a function by combining smaller cubes via Theorem 10. From the minterm set $K^0(f)$, we also found the 1-cube set $K^1(f)$. From $K^1(f)$ we found $K^2(f)$, and continued until we eventually found an empty cube set. All cubes of a function were named "implicants" in Section 9.1. We now wish to use this procedure, and identify those implicants which deserve special attention.

To begin, let us consider the function f.

$$f = 5, 7, 10, 11, 14, 15$$

$$= \bar{A}B\bar{C}D \vee \bar{A}BCD \vee A\bar{B}C\bar{D} \vee A\bar{B}CD \vee ABC\bar{D} \vee ABCD$$

Some of the possible combinations of minterms that Th. 10 provides are:

$$\bar{A}B\bar{C}D \vee \bar{A}BCD = \bar{A}BD$$

$$\bar{A}BCD \lor ABCD = BCD$$

$$A\bar{B}CD \lor ABCD = ACD$$

We proceed to form new terms from and add them to the sum-of-products expression of the function until we can form no new terms via Th. 10, and then delete all terms that are covered by other terms.

$$f = \bar{A}\bar{B}\bar{C}D \lor \bar{A}\bar{B}CD \lor \bar{A}\bar{B}C\bar{D} \lor \bar{A}\bar{B}CD \lor \bar{A}BC\bar{D} \lor \bar{A}BCD$$
$$\lor \bar{A}BD \lor BCD \lor A\bar{C}\bar{D} \lor A\bar{B}C \lor AC\bar{D} \lor ABC \lor AC$$

In terms of Section 9.1, we have found the complex of the function, and then absorbed the complex. Exhaustive application of Th. 10 produces all implicants of f; prime implicants remain after applying the *absorption* theorem.

This procedure may also be performed using the truth table representation of a function. Using the same example as above, Fig. 10.1 gives the truth table of f and a division of that table into two sets of 0-cubes, an ON-array of f and an OFF-array.

If we deal with the members of $K^0(f)$ just as we dealt with the minterms of the Boolean-equation expression of f, all pairs of adjacent 0-cubes are found; each produces a 1-cube of the function. In Fig. 10.2 those 0-cubes that are members of such a pair(s) are checked to remind us that they are included in a larger cube(s) of the function. Since a 0-cube may be adjacent to several other 0-cubes and we seek all 1-cubes of the function, we must not ignore a cube after it has been checked, but continue to evaluate its distance from other cubes of the set.

A	B	C	D	f
0	0	0	0	0
0	0	0	1	0
0	0	1	0	0
0	0	1	1	0
0	1	0	0	0
0	1	0	1	1
0	1	1	0	0
0	1	1	1	1
1	0	0	0	0
1	0	0	1	0
1	0	1	0	1
1	0	1	1	1
1	1	0	0	0
1	1	0	1	0
1	1	1	0	1
1	1	1	1	1

(a) Truth Table

(b) ON-array $K^0(f)$

0	1	0	1
0	1	1	1
1	0	1	0
1	0	1	1
1	1	1	0
1	1	1	1

(c) OFF-array

0	0	0	0
0	0	0	1
0	0	1	0
0	0	1	1
0	1	0	0
0	1	1	0
1	0	0	0
1	0	0	1
1	1	0	0
1	1	0	1

FIG. 10.1 Array Representations of a Switching Function

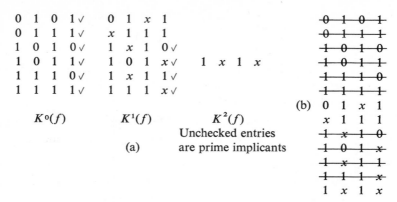

FIG. 10.2 Finding Prime Implicants

From $K^0(f)$, the minterm set, we form the 1-cube set $K^1(f)$ and check all 0-cubes that are included in members of $K^1(f)$. Then we form $K^2(f)$ by locating all pairs of adjacent 1-cubes, writing the 2-cube each pair dictates via Th. 10, and checking the 1-cubes which are covered by one or more 2-cubes of $K^2(f)$. Actually, we need only compare 1-cubes which lack the same literal (x in same column) for Th. 10 can combine only such cubes. The procedure is repeated until an empty cube set is encountered, members of $K^i(f)$ being formed from adjacent members of $K^{i-1}(f)$ via Th. 10. Those cubes of $K^0(f)$, $K^1(f)$, etc., that remain unchecked are the prime implicants of f.

Quine [14] originally proposed this method for finding prime implicants in 1952. In 1956 McCluskey [9] proposed a modification for increasing the efficiency of obtaining prime implicants. The modified method is thus known as the *Quine-McCluskey algorithm* (Q-M, for short).

McCluskey's modification consists of ordering the minterms of the function according to the number of 1's in their binary representation. All of the 0-cubes, if any, with zero 1's are written first, then the 0-cubes with a single 1, etc. Associate with each group an integer index that indicates the number of 1's in the minterms of that group. Now cubes of one index need be compared only with cubes of the next higher index because adjacent cubes can appear only in groups with adjacent indices. Implicant 0000 with index 0 can not be combined with 0110 (index 2); it is a waste of time to examine such a pair.

With experience the Quine-McCluskey procedure can be performed on the decimal integer representation of minterms. Again order minterms as suggested by McCluskey. The decimal representation of adjacent minterms differ by a power of two. The minterm of small index must be subtracted from that of higher index. Cube 0111 is adjacent to 0101. In decimal, $7 - 5 = 2$, which is a power of two; the difference (2) indicates the coordinate in which the two minterms differ. This difference must be recorded, as implicants with different literals deleted cannot be combined and should not even be compared. Thus the decimal pair (5, 2) is a shorthand notation for the implicant $01x1$. The 5 denotes the lowest numbered minterm enter-

2	0101 ✓		01x1	1x1x
	1010 ✓	2	101x ✓	
	0111 ✓		1x10 ✓	
3	1011 ✓		x111	
	1110 ✓	3	1x11 ✓	
4	1111 ✓		111x ✓	

Prime implicants are unchecked

FIG. 10.3   Quine-McCluskey Generation of
Prime Implicants

ing into the combination and also the position of the 1's in the implicant. The decimal 2 indicates the position of the x.

$$\left.\begin{array}{l} 5 = 0101 \\ 2 = 0010 \end{array}\right\} \Rightarrow 01x1 \qquad (5, 2)$$

During subsequent iterations we compare only cubes with "x's in the same columns," or, in decimal, "cubes with the same second integer." The first integers must again differ by a power of two. Thus $(10, 4) = 1x10$ and $(11, 4) = 1x11$ satisfy the requirements and are combined to $(10, 5)$. The first integer is again the smallest minterm number; the second integer is the sum of the original (4) and the new differences ($11 - 10 = 1$).

Figure 10.4 illustrates generation of the prime implicants of f (again) with this procedure. Study it carefully for all of the details of this modification of the basic Quine-McCluskey procedure.

The topology of example function f is shown in Fig. 10.5(a). All implicants of a function are displayed by the topology of a function. We need only recognize 0-cubes, 1-cubes, etc., to obtain all implicants. Once a 1-cube has been recognized a tendency exists to ignore the 0-cubes of which it is composed. This is desirable since we seek prime implicants. We see three prime implicants in Fig. 10.5(a). One-cubes $01x1$ and $x111$ are covered by no larger cubes of the function, nor is 2-cube $1x1x$. Let us denote the set of all prime implicants of function f by $PI(f)$. Then for our example function

$$PI(f) = \left\{\begin{array}{l} 01x1 \\ x111 \\ 1x1x \end{array}\right\}$$

5 ✓	5, 2	10, 5		
10 ✓	10, 1 ✓		$5, 2 = \left.\begin{array}{l} 0101 \\ 0010 \end{array}\right\}$	$01x1$
7 ✓	10, 4 ✓			
11 ✓	7, 8		$7, 8 = \left.\begin{array}{l} 0111 \\ 1000 \end{array}\right\}$	$x111$
14 ✓	11, 4 ✓			
15 ✓	14, 1 ✓		$10, 5 = \left.\begin{array}{l} 1010 \\ 0101 \end{array}\right\}$	$1x1x$

FIG. 10.4   Decimal Quine-McCluskey

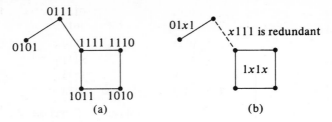

FIG. 10.5 Topology of *f*, and Its Minimum Cover

Selection of a Prime Implicant Cover

We know that a Boolean function may be expressed as a sum of min-terms. Minterms may be combined, in general, to form implicants, and hence switching functions can be expressed as a sum of implicants, but each of the original minterms of the function must be covered by one or more of these implicants. Using prime implicants is desirable in that a maximum number of literals have been deleted from each. Hence in the AND-feeding-OR-gate realization of a function, each AND gate will require a minimum number of input signal connections. In general, a switching function can be expressed as a sum of fewer than all prime implicants. If we can find the smallest set of prime implicants

$$M(f) \subseteq PI(f),$$

such that each of the minterms of the function is covered by one or more of the members of $M(f)$, we can express a switching function as a minimum sum, and build a circuit that requires a minimum number of AND gates and connections to both AND and OR gates. We seek the smallest set of prime implicants that satisfies

$$K^0(f) \equiv M(f) \subseteq PI(f) \tag{10.1}$$

Finding a smallest covering set is not always easy. We begin with the set of all prime implicants of the function $PI(f)$, the base of the function $K^0(f)$, and an empty set $M(f) = \varnothing$. Then we examine and compare prime implicants until we decide to discard one or transfer it from $PI(f)$ to $M(f)$. With each such transfer we delete those members of $K^0(f)$ that the transferred prime implicant covers. Finally, $K^0(f)$ becomes empty. How we examine and compare prime implicants is the crucial part of this procedure.

Graphical representations are very useful when comparing the prime implicants of a function. The topology of some functions goes beyond revealing their prime implicants by pointing out rather clearly which prime implicants, if any, are redundant. In Fig. 10.5(b) the prime implicants of *f* are emphasized. But note that prime implicant $x111$ covers only vertices covered by other prime implicants: prime implicant $x111$ is redundant, and f may be most concisely expressed as a sum of two of its three prime implicants.

$$M(f) = \begin{Bmatrix} 01x1 \\ 1x1x \end{Bmatrix}$$

$$= \bar{A}BD \vee AC$$

By not including prime implicant $x111$ in $M(f)$, we save one 3-input AND gate and one connection to the final OR gate (four diodes are saved).

One or more vertices of a function may be unusual, or *distinguished*, in that each is covered by only one prime implicant. Vertices 0101 and 1010 are examples. Prime implicants that cover distinguished vertices *must* appear in minimal expressions of a function and hence have been given a special name: *extremals* (core prime implicants, or essential prime implicants are also used). Prime implicants that cover only vertices also covered by extremals are then redundant and will not appear in a minimal expression of a function. We will call them *irrelevant*.

Selecting the prime implicants to appear in a minimal expression of a function is not always as simple as in the above example. Consider the function of Fig. 10.6, for example. This function, let us call it g, is covered by a set of 8 prime implicants, $\pi^1, \pi^2, \ldots, \pi^8$.

$$PI(g) = \begin{cases} \pi^1 = 10x0 \\ \pi^2 = x000 \\ \pi^3 = 000x \\ \pi^4 = 00x1 \\ \pi^5 = 0x11 \\ \pi^6 = 011x \\ \pi^7 = 01x0 \\ \pi^8 = 0x00 \end{cases}$$

Prime implicant π^1 is the only extremal of this function. It is the only prime implicant that covers vertex 1010, and all other vertices are covered

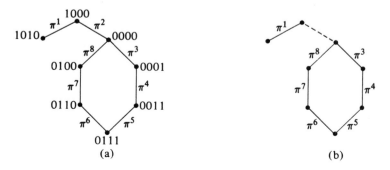

FIG. 10.6 An Example of Less-Than and a Cyclic Structure

by two or more prime implicants. We must include π^1 in the minimal cover, M, that we seek. So to this point

$$M(g) = \{10x0\}$$

and vertices 1010 and 1000 are covered by $M(g)$. We need select no other prime implicant for inclusion in $M(g)$ because it covers either of these vertices, although some of the prime implicants that are selected as we proceed may cover one or another of them. In general, when a prime implicant is selected for membership in $M(g)$ the vertices that it covers lose their importance. They do not influence subsequent selections of prime implicants. We say that vertices become *inactive* when a covering prime implicant is included in $M(g)$. As we proceed fewer and fewer 0-cubes of the function remain active. When the entire base of the function is inactive a minimal cover of the function has been found.

None of the members of $PI(g)$ are irrelevant. For example, π^2 covers vertex 0000 as well as 1000. While 1000 is covered by an extremal, 0000 is not, and hence π^2 is not irrelevant. But there are other prime implicants (π^3 and π^8) that also cover vertex 0000. These are 1-cubes, as is π^2, and hence dictate the same number of AND gate input terminals as π^2. Further, π^3 and π^8 cover other vertices, not covered by π^2. It would seem wise to select either π^3 or π^8 or both rather than π^2 to cover vertex 0000. Prime implicant π^i is *less than* π^j, $\pi^i \leq \pi^j$, if (1) π^i is a smaller cube or the same size as π^j and (2) π^j covers all of the active vertices that π^i covers. Less-than prime implicants need never be included in a minimal cover; in fact, it is clearly necessary to include the greater-than prime implicant. It covers as many and possibly more active vertices and may dictate fewer AND gate connections. Thus we discard prime implicants when we recognize that they are less than other prime implicants. In our example, $\pi^2 \leq \pi^3$ and $\pi^2 \leq \pi^8$. Thus π^2 warrants no further consideration.

But even with π^2 deleted, no minimum cost selection of the remaining prime implicants is apparent in Fig. 10.6(b). Each of the active vertices is covered by two prime implicants. Such a configuration is known as a *cyclic* structure. A general procedure for obtaining a minimal cover of such structures can be extremely complex, so at this point a shortcut will be proposed, which gives a low but not necessarily minimum cost cover.

Select the largest cube remaining in PI when a cyclic structure is encountered.

In Fig. 10.6(b) all remaining prime implicants $\pi^3-\pi^8$ are of the same size: they are all 1-cubes. This rule suggests that we can arbitrarily select any prime implicant. If we select π^3, for example, the cycle is broken: π^4 and π^8 become less-than cubes. Deleting them makes π^5 and π^7 appear to be extremals. We refer to them as *secondary extremals*. We must select them to cover 0011 and 0100, respectively, and when we do, M covers all vertices of the function, and we have completed the task of finding one minimal sum-of-product expression of g.

$$M(g) = \begin{cases} \pi^1 = 10x0 \\ \pi^3 = 000x \\ \pi^5 = 0x11 \\ \pi^7 = 01x0 \end{cases}$$

What if we don't pick π^3, but break the cycle by picking π^4? Then π^3 and π^5 become less-than π^8 and π^6, respectively, and π^6 and π^8 become secondary extremals. Function g has two minimal covers, and we have just found the second.

$$M'(g) = \begin{cases} \pi^1 = 10x0 \\ \pi^4 = 00x1 \\ \pi^6 = 011x \\ \pi^8 = 0x00 \end{cases}$$

Circuits based upon M and M' will have exactly the same number (4) of AND gates. Each AND gate must have 3-input terminals. Thus the same number of diodes are required to construct both circuits.

All these things can be accomplished without drawing the topology of the function. If we are minimizing by hand and have found prime implicants with the Quine-McCluskey algorithm, the *prime-implicant table* will help us to select a minimal cover. Each row of such a table is labeled by a prime implicant, and each column by a minterm. Entries then indicate which minterms each prime implicant covers or conversely, which prime implicant(s) cover each minterm.

The tables of Fig. 10.7 can be established after we find prime implicants as in Figs. 10.2 through 10.4. We must keep track of, or now go back and find out, which minterms each prime implicant covers so that we can place the entries in these tables.

We begin to make our selection of prime implicants by looking for columns that contain a single x. There are several in Fig. 10.7. Such columns correspond to distinguished vertices and the rows in which the x's lie correspond to extremals. Thus in Fig. 10.7, $\bar{A}BD$ and AC are extremals and must be included in the minimum expression of function f. Then we cross out all minterms covered by these selected prime implicants.

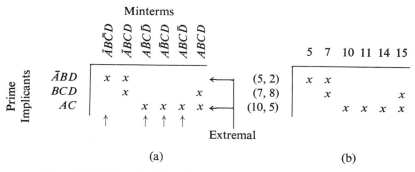

FIG. 10.7 Prime Implicant Tables for the Function of Fig. 10.1

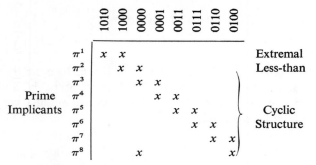

FIG. 10.8   A Less-Than and Cyclic Structure

These are easily identified, for they identify columns containing x's in selected rows. In Fig. 10.7 the extremals cover all minterms and our selection is therefore complete.

From Fig. 10.6 we can form the prime implicant table of Fig. 10.8 for function g. Here we find π^1 to be the only extremal. The first two columns, minterms 1010 and 1000, can be marked as inactive.

All of the active minterms are covered by two or more prime implicants. We see that π^2 has little to offer in comparison with π^3 and π^8 and rule it out as a less-than prime implicant. Then we are left with the cyclic structure. Select any remaining prime implicant, and mark the minterms that it covers inactive. Rule out less-than prime implicants which appear. Two prime implicants become secondary extremals. These must be picked to complete the selection and obtain either $M(g)$ or $M'(g)$.

Karnaugh Maps

Functions of five or fewer variables are most efficiently minimized by hand using the foregoing procedure, but with the map representation of switching functions. A map from which prime implicants can be recognized and selected, often by observation, is most easily prepared. Two adjacent cells containing 1's form an area corresponding to a 1-cube. Encircling those two cells is the graphic equivalent of applying Theorem 10. We have also seen that 2-cubes appear as 4 adjacent cells on the Karnaugh map.

Prime implicants are the largest groupings of 1, 2, 4, 8, or 16 adjacent cells covered by no larger groupings. Figure 10.9 shows the maps of f and g. We can find all the prime implicants found earlier, although it is tempting to overlook redundant prime implicants and encircle only members of the minimum cover.

Extremals are very easily recognized. Usually it is not even necessary to encircle all prime implicants before extremals can be spotted. Any cell that falls into only one grouping is distinguished. The grouping which covers that cell is thus an extremal. In Fig. 10.9(a). $\bar{A}B\bar{C}D$ is a distinguished vertex and $\bar{A}BD$ is therefore an extremal. Prime implicant AC is also an

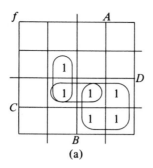 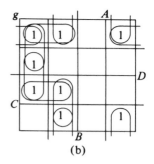

FIG. 10.9 All Prime Implicants of f and g

extremal in that figure; it covers 3 disinguished cells. One extremal can be found in Fig. 10.9(b).

Usually we will not find it necessary to identify formally extremals, irrelevant prime implicants, and less-than prime implicants. The human brain is very good at pattern recognition, and with some experience prime implicants can be recognized and selected simultaneously.

Don't cares offer us some freedom when designing a combinational network. We can elect to have the network produce either 0 or 1 for each of the don't care input *n*-tuples. Advantage should be taken of this freedom and a more economical circuit designed. Suppose a circuit is required to accept the four bits of an 8421 encoded digit, and generate an output signal of 1 when a digit in the range 6-8 is presented and an output of 0 when any other digit is presented. Let A, B, C, and D be the four input bits and F be the output variable of the circuit. A truth table of F is shown in Fig. 10.10. A value of d is placed in the column for F for the six unused code words.

A	B	C	D	F
0	0	0	0	0
0	0	0	1	0
0	0	1	0	0
0	0	1	1	0
0	1	0	0	0
0	1	0	1	0
0	1	1	0	1
0	1	1	1	1
1	0	0	0	1
1	0	0	1	0
1	0	1	0	d
1	0	1	1	d
1	1	0	0	d
1	1	0	1	d
1	1	1	0	d
1	1	1	1	d

$$\begin{pmatrix} 0 & 1 & 1 & 0 \\ 0 & 1 & 1 & 1 \\ 1 & 0 & 0 & 0 \end{pmatrix}$$
ON-array
$K^0(F)$

$$\begin{pmatrix} 1 & 0 & 1 & 0 \\ 1 & 0 & 1 & 1 \\ 1 & 1 & 0 & 0 \\ 1 & 1 & 0 & 1 \\ 1 & 1 & 1 & 0 \\ 1 & 1 & 1 & 1 \end{pmatrix}$$
DC-array
$D^0(F)$

$$\begin{pmatrix} 0 & 0 & 0 & 0 \\ 0 & 0 & 0 & 1 \\ 0 & 0 & 1 & 0 \\ 0 & 0 & 1 & 1 \\ 0 & 1 & 0 & 0 \\ 0 & 1 & 0 & 1 \\ 1 & 0 & 0 & 1 \end{pmatrix}$$
OFF-array
$L^0(F)$

$F = \bigvee 6, 7, 8 \vee \bigvee_d 10, 11, 12, 13, 14, 15$

don't care

FIG. 10.10   A Function with Don't Cares

A subtle difference exists between d, and the x used previously. Where the x represented *both* 0 and 1 in the expression of a large cube composed of smaller cubes, the d represents either the 0 or 1, but not both, the specific value being as yet undetermined. A value will be assigned in the design procedure, for our circuit must present either 0 or 1 in all cases, but as yet the advantage of one value over the other is not clear.

The Karnaugh map expression of F is most revealing. If d_{10}, d_{12}, d_{14}, and d_{15} are set to 1 and the others replaced by 0's, two large groupings appear. Figure 10.11(b) shows this most desirable assignment of the don't cares: from it we write

$$F = BC \lor A\bar{D}$$

Other don't care assignments are possible. We might propose solving the don't care assignment problem by setting all don't cares to 0 or to 1. The following respective expressions for F, while easily obtained are not minimal in terms of the cost function of the previous section.

$$F_0 = \bar{A}BC \lor A\bar{B}\bar{C}\bar{D}$$
$$F_1 = BC \lor AC \lor AB \lor A\bar{D}$$

Note that

$$F \neq F_0 \neq F_1$$

Actually three different Boolean functions have been derived from the same map. Different assignments of the don't cares lead to different switching functions. But all the members of the set of functions specified by a truth table with don't cares have one thing in common—they satisfy the "care" specifications. If s d's appear (and each may be assigned a value of 0 or 1) then 2^s different complete switching functions may be derived from a partial specification. At least one minimal expression exists for each; our task then is to find the most minimal of these expressions. In Karnaugh maps we seek the smallest number of the largest groupings of 1's and d's such that each 1's cell is included in some grouping. Thus each grouping

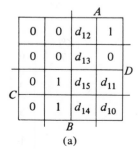

(a)

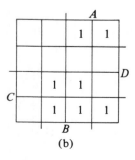

(b)

FIG. 10.11 Minimization of F

must provide unique coverage of at least one 1's cell, and groupings of d's only are avoided.

The Quine-McCluskey minimization algorithm with suitable modification may be used to select the most desirable don't care assignment just as the Karnaugh map was so used. First we find prime implicants not from $K^0(F)$ alone but rather from the union of $K^0(F)$ and $D^0(F)$. Then we select prime implicants to obtain a minimal cover of $K^0(F)$.

$$K^0(F) \sqsubseteq M(F) \subseteq PI\,(K^0(F) \cup D^0(F)) \qquad (10.2)$$

Some don't care terms may be covered by selected prime implicants: others may not be covered. But all members of $K^0(F)$ are covered by $M(F)$. Figure 10.12 shows this more generalized Quine-McCluskey algorithm for example function F.

All ON and DC minterms are used to obtain the four prime implicants of F. Both $1xx0$ and $x11x$ are extremals and must be selected. Together they cover $K^0(F)$.

$$M = \begin{Bmatrix} 1xx0 \\ x11x \end{Bmatrix}$$

$$F = A\bar{D} \vee BC$$

Prime implicants $1x1x$ and $11xx$ cover only don't cares. Usually don't care minterms are not placed in the prime implicant table. They are included in Fig. 10.12 only to show that the selected prime implicants cover don't care as well as care minterms: $1xx0$ covers don't care minterms 10, 12, and 14. We selected it because it is the only cover of minterm 8. Comparison with the Karnaugh maps of Fig. 10.11 reveals that the same minimal cover and assignment of don't cares can be obtained from the map representation of F.

Considerable time and effort may be required to minimize a switching function. As a result, many nonminimum switching circuits have been

1	1000 √		10x0 √	1	1xx0
	$\overline{0110}$ √	1	1x00 √		x11x
2	1010 √		$\overline{011x}$ √	2	1x1x
	1100 √		x110 √		11xx
	$\overline{0111}$ √	2	101x √		
3	1011 √		1x10 √		
	1101 √		110x √		
	1110 √		11x0 √		
4	$\overline{1111}$ √		$\overline{x111}$ √		
		3	1x11 √		
			11x1 √		
			111x √		

	$K^0(F)$			$D^0(F)$				
	6	7	8	10	11	12	13	14	15
$1xx0$			x	x		x		x	
$x11x$	x	x						x	x
$1x1x$				x	x			x	x
$11xx$						x	x	x	x
				\multicolumn{6}{l}{Ignore when selecting}

FIG. 10.12 Q-M Minimization of F

designed. There is no excuse for this if functions are small enough so that the Karnaugh map may be used; everyone has time to draw and minimize with the Karnaugh map. The savings must be substantial if manual minimization of larger switching functions is to be economically justified. Much smaller savings can be justified by computer minimization of switching functions.

10.2 ALGORITHMIC MINIMIZATION OF SINGLE-OUTPUT SWITCHING FUNCTIONS

In this section we seek more efficient procedures for finding minimal two-level realizations of switching functions. Our main concern will be with efficiency of computation and we will not restrict ourselves to procedures easily carried out by hand.

We seek a switching circuit that is optimum in some respect. To be able to compare circuits that realize the same switching function, we establish a measure of the "cost" of each circuit, and search for the circuit of minimum cost. Several measures of cost have traditionally been employed. They may or may not reflect dollar costs today, but these costs can all be determined rather easily from the Boolean expression that represents a circuit, or from the equivalent array. Let C be a cover consisting of $p = |C|$ n-tuples. Cube c^i, an r-cube of C, is said to have a *cube cost* of

$$\$c^i = n - r \qquad (10.3)$$

Then

1. $$\$_1(C) = \sum_{i=1}^{p} \$c^i \qquad (10.4)$$

    This cost measures the number of literals that appear in the sum-of-product expression of a function. It thus reflects the number of connections that must be made to AND gates in a two-level network.

2. $$\$_2(C) = \sum_{i=1}^{p} (\$c^i + 1) \qquad (10.5)$$

    This cost function measures the total number of inputs to AND and OR gates in a two-level network. It reflects the number of diodes required in a diode pyramid, although exceptional cases are easily found. In general, $\$c^i$ measures the diodes required in the AND gate which realizes c^i, and p is the number of diodes required in the OR gate. But if $\$c^i = 1$, then an AND gate is not required. A true diode cost function recognizes this exception and hence is a bit more complicated.

3. $$\$_3(C) = p + 1 \qquad (10.6)$$

    This function measures the number of logic blocks required. It is independent of cube cost and assumes that each cube can be realized with one AND gate of fixed cost.

Other cost functions may be more realistic in a specific situation, but often are too complicated to deal with efficiently by hand.

In general we seek a minimum (according to some cost function) cost cover $\underline{C}(ON)$, of the ON-array that uses the 0-cubes covered by the DC-array to advantage. We seek:

$$ON \sqsubseteq \underline{C}(ON) \sqsubseteq (ON \bigsqcup DC) \qquad (10.7)$$

An r-cube of $K^r(f)$ is a *prime implicant* of switching function f if it is covered by no member of $K^{r+1}(f)$. The absorbed complex of f, $PI(f) = \mathbf{A}[K(f)]$ then is the set of all prime implicants of f. We usually seek prime implicants, for they are cubes of minimum cost. $\underline{C}(f)$ will consist of prime implicants if $\$_1(C)$ and $\$_2(C)$ are used as measures of cost. It may or may not be necessary to construct $\underline{C}(f)$ of prime implicants if $\$_3(C)$ is employed, but we can never obtain a lower $\$_3(C)$ cost by not using prime implicants than by using them.

A first step in most minimization processes then consists of obtaining the *PI*-array. The Q-M algorithm offers one means of obtaining the *PI*-array of a switching function; let us look a bit more carefully at that procedure. It requires that we first obtain $K^0(f)$. This is often a most undesirable step to take since we know that $K^0(f)$ may have as many as 2^n members. The function of eleven variables,

$$f = A \vee \bar{A}BCDEFGHIJK$$

has 1025 0-cubes but is trivial to minimize. This is a very long list to write by hand, and is even undesirably long to form and store within a digital computer. If $15 \le n \le 20$, we may encounter K^0 sets that are too large to store in most current digital computer memories.

The Quine-McCluskey procedure next calls for the formation of

$$K^0(f) \bigsqcup K^1(f)$$

$K^1(f)$ can have many more members than $K^0(f)$. It takes a great deal of time to form them and memory capacity to store them. Continuing, the Quine-McCluskey algorithm asks that we find

$$PI(f) = \bigsqcup_{i=0}^{n} K^i(f) \qquad (10.8)$$

We may be unable to do so because we lack sufficient time, money, or computer memory capacity. Methods of finding PI that are more efficient of time and memory are required.

Generation of Prime Implicants—Sharp Algorithm

We have seen that the ON-array may be computed from the OFF- and DC-arrays. Use of the sharp product in Eq. (9.36) produces a very interesting ON-array. For the moment assume that $DC = \phi$: we have a completely specified switching function. Further assume that the OFF-array covers exactly one r-subcube. We saw in a previous section that removing an r-cube from U_n produced $n - r$ $(n - 1)$-cubes. We now recognize that these $(n - 1)$-cubes are prime implicants. The sharp operation gives us the set of *all* prime implicants

$$PI(f) = U_n \# OFF(f) \tag{10.9}$$

where $OFF(f)$ covers an r-cube. Further, Theorem 9.3 indicates that each $(n - 1)$-cube is the sole cover of some member of $K^0(f)$: hence $PI(f)$ is the minimum cost cover of f with the three cost functions defined by Eqs. (10.4), (10.5), and (10.6).

In general, the sharp product does not give a minimum cost cover, but a complete set of prime implicants.

THEOREM 10.1

Let $C(\bar{f})$ be a cover of $\bar{f}$, a function of n variables. Then:

$$PI(f) = U_n \# C(\bar{f})$$

is the set of all prime implicants of f.

Proof: Let $c^1, c^2, \ldots$ be members of $C(\bar{f})$. First compute $U_n \# c^1$ and obtain a set of $(n - 1)$-cubes all of which are prime implicants of a function g with base $K^0(g) = K^0(U_n) \cap \overline{K^0(c^1)}$. All prime implicants of g are given.

Now assume $U_n \# \{c^1, c^2, \ldots, c^{i-1}\}$ gives all prime implicants of function h with base

$$K^0(h) = K^0(U_n) \cap \overline{K^0(c^1 \cup c^2 \cup \ldots \cup c^{i-1})}$$

Let $d^1, d^2, \ldots$ be the cubes of $U_n \# \{c^1, c^2, \ldots, c^{i-1}\}$. Then form

$$\{d^1, d^2 \ldots\} \# c^i.$$

If $d^j \# c^i = d^j$, that cube was a prime implicant of h and now is a prime implicant of a new function k defined similarly to g and h. If c^i covers some 0-cube of d^j, the sharp operation as defined gives all maximal-sized fragments of $d^j \# c^i$. Those which are not absorbed are prime implicants of k.

As a simple example, let

$$OFF = \begin{Bmatrix} 000 \\ 111 \end{Bmatrix}$$

Then:

$$xxx \# 000 = \begin{Bmatrix} 1xx \\ x1x \\ xx1 \end{Bmatrix}$$

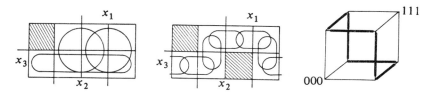

FIG. 10.13 Sharping on the Karnaugh Map

$$
\begin{Bmatrix} 1xx \\ x1x \\ xx1 \end{Bmatrix} \# 111 = \begin{Bmatrix} 10x \\ 1x0 \\ 01x \\ x10 \\ 0x1 \\ x01 \end{Bmatrix} = PI
$$

The equivalent construction can be performed on the Karnaugh map as in Fig. 10.13, which reveals this function to have the topology of a hexagon.

ALGORITHM 10.1. *Prime Implicants via the Sharp Operator*
If we begin with an ON-array of a complete switching function f, the OFF-array can always be obtained with the sharp product.

$$OFF(f) = U_n \# ON(f) \qquad (10.10)$$

The fact that $OFF(f)$ consists of all prime implicants of $\bar{f}$ may be of value. But if we seek $PI(f)$, we use the sharp operation twice.

$$PI(f) = U_n \# (U_n \# ON(f)) \qquad (10.11)$$

If the don't care-array is not empty, appropriate modifications to Eqs. (10.10) and (10.11) must be made.

$$OFF = U_n \# (ON \cup DC) \qquad (10.12)$$

$$PI = U_n \# (U_n \#(ON \cup DC)) \qquad (10.13)$$

Now PI consists of all prime implicants of ON $\cup$ DC. Some of these prime implicants may cover only don't care vertices; ultimately we will want procedures to detect and eliminate such members of PI.
    As an example of this algorithm, let

$$
ON = \begin{Bmatrix} 00xx \\ 0xx0 \\ xx00 \\ x00x \end{Bmatrix} \qquad DC = \begin{Bmatrix} 1101 \\ 1011 \end{Bmatrix}
$$

SECTION 10.2   **Algorithmic Minimization of Single-Output Switching Functions**

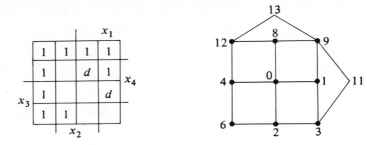

FIG. 10.14 Karnaugh Map of Example Partial Switching Function

Then:

$$\text{OFF} = \left\{ \begin{matrix} \cancel{1xxx} \\ \cancel{x1xx} \\ \cancel{11xx} \\ \cancel{x1x1} \\ \cancel{1x1x} \\ \cancel{1xx1} \\ \cancel{111x} \\ \cancel{11x1} \\ \cancel{1x11} \end{matrix} \quad \begin{matrix} 01x1 \\ x111 \\ \cancel{1111} \\ 111x \\ 1x10 \end{matrix} \right\} = \left\{ \begin{matrix} 01x1 \\ x111 \\ 111x \\ 1x10 \end{matrix} \right\}$$

and

$$\text{PI} = \left\{ \begin{matrix} \cancel{1xxx} \\ \cancel{x0xx} \\ \cancel{xxx0} \\ \cancel{10xx} \\ 1x0x \\ \cancel{1xx0} \\ 0xx0 \\ \cancel{x0x0} \\ xx00 \end{matrix} \quad \begin{matrix} 00xx \\ x00x \\ x0x1 \\ \cancel{100x} \\ \cancel{10x1} \\ 1x00 \\ 00x0 \\ \cancel{x000} \end{matrix} \right\} = \left\{ \begin{matrix} 1x0x \\ 0xx0 \\ xx00 \\ 00xx \\ x00x \\ x0x1 \end{matrix} \right\}$$

Intermediate results are shown above as an indication of the amount of computation required. An attempt to reproduce these results by hand demonstrates that the sharp operator is not one of the easier operations for the human to perform, although with practice one *is* able to extract the sharp product efficiently. Fig. 10.14 shows the Karnaugh map and topology of the function considered: check that all prime implicants of OFF and of ON ∪ DC are formed.

Generation of Prime Implicants—Star Product

A second well-known procedure for finding prime implicants is based on what Quine [14, 15] first called the *consensus* of implicants. Implicant A is said to be the consensus of implicants AB and $A\bar{B}$, as in the Minimization

TABLE 10.1 COORDINATE *-product

	b_i		
*	0	1	x
a_i 0	0	ϕ	0
1	ϕ	1	1
x	0	1	x

Theorem. But more generally, AC is said to be the consensus of implicants AB and $\bar{B}C$. The equivalent *-*product* of cubes was defined by Roth as the coordinate *-product Table 10.1 and the rules:

$$a * b = \begin{cases} \phi \text{ if } a_i * b_i = \phi \text{ for } more \text{ than one } i \\ c \text{ where } c_i = \begin{cases} a_i * b_i \neq \phi \\ x \text{ when } a_i * b_i = \phi \end{cases} \end{cases} \quad (10.14)$$

Note first that Table 10.1 is identical to Table 9.2, but the rule of Eq. (10.14) differs slightly from that of Eq. (9.15) for the intersection operation. We allow one ϕ coordinate *-product: no ϕ gives cube intersection, which may be considered to be a degenerate case of the *-product. For one ϕ to appear, cubes a and b must have adjacent faces. The *-product is the coface of those adjacent faces; it's the largest cube *between* two cubes!

Examples illustrate this result. First we repeat those expressed above.

$$
\begin{array}{l}
\quad 11xx \\
* \underline{\quad 10xx} \\
\quad 1\phi xx
\end{array}
\qquad
\begin{array}{l}
AB * A\bar{B} = A \\
11xx * 10xx = 1xxx
\end{array}
$$

Here the consensus cube covers both parent cubes.

$$
\begin{array}{l}
\quad 11x \\
* \underline{\quad x01} \\
\quad 1\phi 1
\end{array}
\qquad
\begin{array}{l}
AB * \bar{B}C = AC \\
11x * x01 = 1x1
\end{array}
$$

The consensus cube covers neither of the parent cubes. Fig. 10.15 shows the geometry of this example.

$$
\begin{array}{l}
\quad x1x \\
* \underline{\quad x00} \\
\quad x\phi 0
\end{array}
\qquad
x1x * x00 = xx0
$$

Here we find with the *-operator that a 2-cube lies between a 2-cube and a 1-cube adjacent to one edge of the 2-cube. The consensus cube covers one of the parent cubes but not the other. If we compute $x1x * xx0$ we obtain

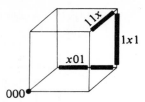

FIG. 10.15 Examples of the ∗–product

a consensus cube, $x10$, that is covered by its parents. Finally, if two cubes have no adjacent faces, their ∗-product is empty.

$$
\begin{array}{r}
000 \\
* \quad 11x \\
\hline
\phi\phi0
\end{array}
$$

More than one ϕ appears, so $000 * 11x = \phi$. (See Fig. 10.15.)

The ∗-product does not give prime implicants by itself. But an algorithm involving ∗-product and absorbing can be shown to give all prime implicants.

ALGORITHM 10.2. *Iterative Consensus*
  For all cubes c^i and c^j of the given cover C, if

$$c^i * c^j \neq \phi$$

replace C with $C \bigsqcup (c^i * c^j)$.

$$C \leftarrow C \bigsqcup (c^i * c^j)$$

Continue until
(a) no consensus terms can be found, or
(b) all new consensus terms found are immediately deleted when the absorbed union is formed.

Each prime implicant lies between, and is the largest cube between, their faces, and hence will be found before the algorithm terminates. All cubes included in prime implicants are deleted from C in the absorbing operation.

Using the same ON- and DC-arrays as were used to illustrate Algorithm 10.1, we might begin by comparing $c^1 = 00xx$ with all other cubes of ON $\cup$ DC.

$$
\begin{array}{lll}
00xx * 0xx0 = 00x0 \sqsubseteq 00xx & \quad \text{do not add to list} \\
00xx * xx00 = 0000 \sqsubseteq 00xx & \quad \text{do not add to list} \\
00xx * x00x = 000x \sqsubseteq 00xx & \quad \text{do not add to list} \\
00xx * 1101 = \phi & \\
00xx * 1011 = x011 \sqsupseteq 1011 & \quad \text{add } x011, \text{ delete } 1011.
\end{array}
$$

At this point our list of cubes is

$$
\text{ON} \begin{cases} 00xx \\ 0xx0 \\ xx00 \\ x00x \end{cases}
$$

$$
\text{DC} \begin{cases} 1101 \\ \cancel{1011} \\ x011 \end{cases}
$$

Continuing, we eventually arrive at the following PI-array.

$$
\left.
\begin{array}{l}
\text{ON} \begin{cases} 00xx \\ 0xx0 \\ xx00 \\ x00x \end{cases} \\
\text{DC} \begin{cases} \cancel{1101} \\ \cancel{1011} \\ \cancel{x011} \\ \cancel{001x} \\ \cancel{110x} \\ x0x1 \\ 1x0x \end{cases}
\end{array}
\right\} \text{PI}
$$

When we compute this result we find that we must compare cubes as with the Quine-McCluskey algorithm (which is a special case of the iterative consensus algorithm). While the comparisons to be made are more complicated than with the Q-M algorithm, they are far fewer in number, and we need not start with a 0-cube cover of a function. This is a very great advantage.

While the iterative consensus algorithm is far more efficiently performed by hand or computer than the Q-M algorithm, both algorithms require an excessive number of cube comparisons. In the iterative consensus algorithms we compare cubes searching for a nontrivial *-product, and when one is found compare it with many, perhaps all, cubes for inclusion . With proper organization, the amount of comparison required can be further reduced.

Generation of Prime Implicants—Generalized Consensus

Tison [20] recently generalized the concept of consensus, and suggested a most efficient algorithm for generating prime implicants, which takes advantage of the close relationship between the star and intersection operators. Again let C be a cover of a function for which we wish to find all prime implicants. But now rather than pick one cube and compare it with others, let us concentrate on a column of C. We might begin with the first column which, in general, will contain 0's, 1's, and x's. Let us split C into three parts based upon the first element of each cube of C.

$$A = C\,\mathbf{S}\,\chi_1^1$$
$$B = C\,\mathbf{S}\,\chi_1^0$$

(10.15)

Now all cubes of A have a 1 in column 1. If we are looking in column 1 for the single $0 - 1$ opposition required of a nontrivial *-product, we see that we need not compare cubes of A with each other. Similarly, in array B 0 is found in the first position of each cube, and we need not compare the cubes of B with each other.

But any cube of A and any cube of B are known to have a 0–1 opposition in column 1. If the remainder of these cubes intersect, then a nontrivial *-product exists, which is the concatenation of an x and the result of the intersection.

$$a^i = 1011 \qquad\qquad\qquad 011 \;\bigcap\; 0xx = 011$$
$$b^j = 00xx$$
$$a^i * b^j = x \circ 011 \;= x011$$

Now we can generalize the formation of a *-product illustrated above. Given the arrays A and B of Eq. (10.15), let us compute

$$\mathbf{I}\;_0(\mathbf{D}_1(A) \;\bigcap\; \mathbf{D}_1(B))$$

The result will be called the " generalized consensus with respect to column 1." It consists of all of the *-product cubes with an x in column 1 that can be obtained by comparing only cubes of the original array C. Let us use the same example function as before.

$$C = \begin{cases} 00xx \\ 0xx0 \\ xx00 \\ x00x \\ 1101 \\ 1011 \end{cases} \qquad
\begin{aligned}
B &= \begin{cases} 00xx \\ 0xx0 \end{cases} & \searrow \text{ delete col.} \\
& & \text{intersect } = \{x011\} \\
A &= \begin{cases} 1101 \\ 1011 \end{cases} & \nearrow \text{ insert col.}
\end{aligned}$$

Thus the generalized consensus with respect to column 1 consists of a single cube $\{x011\}$ in this case.

ALGORITHM 10.3. *Generalized Consensus*

Given an array C of n variables, in turn for $i = 1, 2, \ldots, n$ compute:

1. $A = C\,\mathbf{S}\,\chi_i^1$
    $B = C\,\mathbf{S}\,\chi_i^0$
    $C \leftarrow A \cup B \cup C$ (This simply restores array C.)
2. $C \leftarrow C \;\bigsqcup\; \mathbf{I}_{i-i}(\mathbf{D}_i(A) \;\bigcap\; \mathbf{D}_i(B))$

(10.16)

This algorithm calls for computation of the generalized consensus with respect to each column in turn, and formation of the absorbed union of each generalized consensus and the original array. It is thus very similar to the iterative consensus algorithm, but can be far faster since a large number of futile and repetitious cube comparisons are eliminated. Many comparisons are required to verify that the iterative consensus procedure can be terminated. There is no question of when the generalized consensus algorithm terminates.

Finding Extremals

Any prime implicant of $PI(f)$ that is the sole cover of a member of $K^0(f)$ is known as an *extremal*, and is of interest because it must be included in any cover that is minimal under the cost functions discussed earlier. With the methods of generating the PI–array of a function discussed above, it is not necessary to generate $K^0(f)$; it would be undesirable to have to do so now. We do not wish to be forced to form a prime implicant table, and "look for columns containing a single x."

An alternative procedure works directly with PI. Let $\pi^1, \pi^2, \ldots$ be the cubes of PI, and let "$\mathrm{PI} - \pi^i$" denote the array PI less cube π^i. Suppose $DC = \phi$ and we compute

$$e = \pi^i \mathbin{\#} (\mathrm{PI} - \pi^i) \tag{10.17}$$

Cube π^i covers some members of $K^0(f)$; array $\mathrm{PI} - \pi^i$ covers the remaining members of $K^0(f)$, but may also cover members covered by π^i. A 0-cube may be covered by more than one prime implicant. Now if $e = \phi$, then all 0-cubes covered by π^i are also covered by other prime implicants, members of $\mathrm{PI} - \pi^i$, and π^i is not an extremal. But if $e \neq \phi$, then one or more 0-cubes of π^i are not covered by $\mathrm{PI} - \pi^i$ and π^i is an extremal.

If don't care vertices exist, 0-cubes uniquely covered by π^i, if any, may be don't care vertices. We may eliminate this possibility when testing π^i by computing

$$e = (\pi^i \mathbin{\#} (\mathrm{PI} - \pi^i)) \mathbin{\#} DC \tag{10.18}$$

For the example function of Fig. 10.14 and its PI-array:

$$00xx \mathbin{\#} \begin{Bmatrix} 0xx0 \\ xx00 \\ x00x \\ x0x1 \\ 1x0x \end{Bmatrix} = \phi \qquad 00xx \text{ is not an extremal}$$

$$0xx0 \mathbin{\#} \begin{Bmatrix} 00xx \\ xx00 \\ x00x \\ x0x1 \\ 1x0x \end{Bmatrix} = 0110 \mathbin{\#} \underset{DC}{\begin{Bmatrix} 1101 \\ 1011 \end{Bmatrix}} = 0110$$

$0xx0$ is an extremal. It is the only prime implicant which covers vertex 0110. (See Fig. 10.14.)

To find all extremals of a prime implicant array, we need only test each member of that array in turn. But if we are seeking a minimal cover, some additional activity is necessary, either after we have found all extremals or as we find each one. Suppose the test of Eq. (10.18) reveals that π^i is an extremal; it must then be set aside in an array E in which we collect the members of a minimal cover.

$$E \leftarrow E \cup \pi^i$$

We now have included in E a prime implicant that covers some members of the original ON-array of the function. In future computation we need never select another prime implicant because it covers one or another of these members of the ON-array; as far as future computation is concerned, these ON conditions have all of the properties of don't care conditions. Thus when we find π^i to be an extremal, we can also include it in the DC-array.

$$DC \leftarrow DC \cup \pi^i$$

If π^i is not found to be an extremal, it must be returned to the PI-array; it may yet appear in the minimal cover, but we do not yet know that it *must* appear.

A computer program based upon these ideas is given in Appendix 10.1. In brief it:
1. Computes PI by the generalized consensus algorithm
2. For each i = 1, 2, $\cdots$, |PI| in turn
   a. $PI \leftarrow PI - \pi^i$
   b. $A \leftarrow \pi^i \,\#\, PI$
   c. If $A = \phi$, $PI \leftarrow PI \cup \pi^i$ and i is advanced
      If $A \neq \phi$, $B \leftarrow A \,\#\, DC$
   d. If $B = \phi$, $PI \leftarrow PI \cup \pi^i$ and i is advanced
      If $B \neq \phi$, $E \leftarrow E \cup \pi^i$, $DC \leftarrow DC \cup \pi^i$, and i is advanced.

Nonredundant Covers

With minor modifications the program of Appendix 10.1 can provide a *nonredundant cover* of a switching function, a set of prime implicants in which no member is covered by the logical sum of two or more other members of the set. Consider again the function of Fig. 10.13 which has a hexagon topology. Fig. 10.16 below repeats that topology and the set of all prime implicants. This function is interesting, because it has two equally minimal two-level covers.

$$\pi^1 \lor \pi^5 \lor \pi^4 = A\bar{B} \lor \bar{A}C \lor B\bar{C}$$
$$\pi^2 \lor \pi^3 \lor \pi^6 = A\bar{C} \lor \bar{A}B \lor \bar{B}C$$

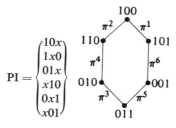

$$PI = \begin{Bmatrix} 10x \\ 1x0 \\ 01x \\ x10 \\ 0x1 \\ x01 \end{Bmatrix}$$

FIG. 10.16 A Function with Several Nonredundant Covers

These are nonredundant covers, but are not the only nonredundant covers of this function. While the following covers are not minimal, they are non-redundant; no prime implicant can be removed without uncovering a vertex of the function.

$$\pi^1 \vee \pi^2 \vee \pi^3 \vee \pi^5 = A\bar{B} \vee A\bar{C} \vee \bar{A}B \vee \bar{A}C$$
$$\pi^2 \vee \pi^4 \vee \pi^5 \vee \pi^6 = A\bar{C} \vee B\bar{C} \vee \bar{A}C \vee \bar{B}C$$
$$\pi^1 \vee \pi^3 \vee \pi^4 \vee \pi^6 = A\bar{B} \vee \bar{A}C \vee B\bar{C} \vee \bar{B}C$$

Nonredundant covers are of interest because they are of low if not minimum cost, and can be obtained more easily than absolute minimum cost covers. For example, in the program of Appendix 10.1, when we find a prime implicant that is not a unique cover of any vertex, i.e., when the computation of Eq. (10.18) gives an empty result, rather than replacing that prime implicant in the PI-array from which it was extracted, *throw it away!*

Using the PI-array of Fig. 10.16 for purposes of example, we examine the first cube, $\pi^1 = 10x$.

$$\pi^1 \# (PI - \pi^1) = \phi$$

Thus we delete this cube from further consideration, and proceed to examine π^2. That cube *now* is found to be an extremal: of the remaining prime implicants, only π^2 covers 0-cube 100. Thus it must be placed in the E- and DC-arrays. When π^3 is examined, it is found to provide sole cover-age of no 0-cubes. Throw it out! Then when π^4 is examined, it is an extremal (cube 010). Similarly π^5 and π^6 are extremals since we have discarded π^1 and π^3. We have a nonredundant cover that is not a minimum cost cover, but one obtained by a straightforward algorithm.

Note that the order of the cubes in PI influences the result. This procedure will give the minimal covers if the prime implicants are encountered in the correct order, but unfortunately we cannot predict that order in advance, in general.

Less-than Cubes

If we seek a minimum cost cover, then the algorithm for extracting extremals may be followed by a procedure which declares some of the remaining prime implicants to be less desirable than others and deletes these "less-than" prime implicants. Cube π^i is *less than* π^j if

(a) $\$(\pi^i) \geq \$(\pi^j)$, and $\qquad\qquad\qquad\qquad\qquad\qquad$ **(10.19)**

(b) $(\pi^i \# \text{DC}) \# \pi^j = \phi$

The cost of π^i must exceed or equal that of π^j (if $\$_3(C)$ is being used, all cubes are of the same cost), and the care vertices covered by π^i must also be covered by π^j. Keep in mind that any extremals found were added to the DC-array. The second requirement would seldom be satisfied if this were not the case.

With the function of Fig. 10.17, π^1 is an extremal. Placing it in E and DC makes π^2 and π^3 less than π^4 and π^5, respectively. The only vertex covered by π^2 which is not in the DC-array is vertex "α." But π^4 covers it and additional as yet uncovered vertices. And the cost of π^4 equals that of π^2 if the cube cost function proposed earlier is employed.

The deletion of π^2 from the prime implicant list for the function of Fig. 10.17, makes π^4 an extremal. It is the only remaining prime implicant that covers vertex α.

In general, we are able alternately to find extremals, and cast out less-than cubes until (1) the prime implicant array becomes empty, in which case a minimal cover has been found, or (2) no new extremals or less-than cubes can be found. In this second case, we have encountered a cyclic structure. Finding the absolute minimum cost cover becomes substantially more difficult.

If we are willing to forego a guarantee of obtaining a minimum cost cover, we can avoid a great deal of work when a cyclic structure is encountered simply by placing the least costly prime implicant of those remaining into the E- and DC-arrays. This straightforward rule, first proposed in Section 10.1, often leads to the minimum cost cover.

Branching

If, while finding the prime implicants of a minimum cost cover, all active vertices are each covered by two or more prime implicants, and no prime implicant can be deleted as less desirable than another, a cyclic structure has been encountered. The hexagon of Fig. 10.16 is an example. One thing can be said about each remaining prime implicant: "it either is or is not a part of the minimum cost cover." This safe statement gives rise to a means of breaking the cyclic structure.

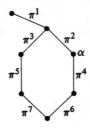

FIG. 10.17   Less-Than Cubes

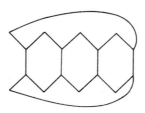

FIG. 10.18   Cycles within cycles

Branching consists of picking one of the remaining prime implicants and

(a) Assuming it is a part of the minimum cost cover, and proceeding to recognize less-than and extremals to find the remainder of the cover, and

(b) Assuming it is not a part of the minimum cost cover and proceeding to find a cover based upon this assumption.

Then the costs of the two covers are compared and the lower cost cover selected. Two covers must be obtained so branching doubles the amount of computation required. With the hexagon of Fig. 10.16 branching leads to two equally minimal covers.

Branching would not be an undesirable technique if it had to be employed only once per function. But cyclic structures can exist within cyclic structures. Hexagons and octagons can be interconnected in a loop, as in Fig. 10.18. Now we must branch each original branch, or in other words, find four covers and compare their costs. And cyclic structures within cyclic structures, etc., ten levels deep have been demonstrated. Finding and comparing 1024 covers, as the branching algorithm demands, may be too large a problem to solve economically even with digital computers.

The Extraction Algorithm

J. P. Roth first assembled these techniques into an overall switching function minimization algorithm, which was named the "extraction algorithm." While the various steps can vary in their details, we can obtain a minimal cover of a switching function by

ALGORITHM 10.4.   *Extraction Algorithm*

1. Compute all prime implicants of ON $\cup$ DC.
2. Extract extremals and place them in a separate array, adding them to the DC-array. Delete them from the array of prime implicants.
3. Delete all less-than prime implicants from the array of prime implicants.
   a. If the prime implicants array is empty, a minimal cover has been found.
   b. If PI $\neq \phi$, and one or more less-than prime implicants have been deleted, go back to step 2.
   c. If PI $\neq \phi$, but no less-than prime implicants exist, a cyclic structure exists.
4. Perform branching and retain the cover with the lowest cost.

Appendix 10.2 lists a program written as a FORTRAN subroutine that follows the extraction algorithm with the exception of the branching step. Branching is replaced by the rule of thumb introduced in Section 10.1. The costs of all remaining prime implicants are compared, and the least expensive prime implicant is arbitrarily picked to be part of the final cover. Thus when a cyclic structure is encountered and a cube chosen, we can no longer be certain of being supplied with a minimum cost cover. Branching requires a much more complicated program, which performs a great deal of bookkeeping, but does not require any logic manipulations not already examined.

10.3 MINIMIZATION OF MULTIPLE OUTPUT SWITCHING FUNCTIONS

Usually a logic designer is faced not with the problem of designing a single-output combinational logic network, but with the need for a network which realizes many, say m, Boolean functions of subsets of a set of n input variables.

$$f : P_2^n \rightarrow P_2^m \tag{10.20}$$

Such multiple-output problems may be treated either as many single-output problems, or as a single many-input, many-output problem. We have been taking the first approach but this approach often leads to poorer results than simultaneously considering all of the output variables.

Karnaugh maps for three functions are shown in Fig. 10.19. If we individually minimize each, we obtain:

$$z_1 = AD \lor BD \lor ABC$$
$$z_2 = \bar{A}\bar{B} \lor \bar{A}C \lor BCD$$
$$z_3 = \bar{B}\bar{C} \lor BC \lor \bar{A}\bar{B}$$

If individual networks are constructed for each function, 9 AND gates and 3 OR gates are required. A second glance at the three functions reveals that the implicant $\bar{A}\bar{B}$ is common to both z_2 and z_3. If technology permits, and it commonly does, one of the two AND gates which realize $\bar{A}\bar{B}$ may be eliminated and the other used as a common part of both networks (fan-out).

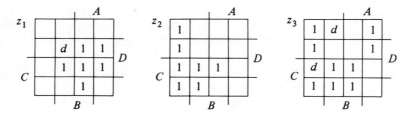

FIG. 10.19 Maps of Three Functions

Additional AND gates can be eliminated if we express these three functions in the following nonminimal fashion.

$$z_1 = BCD \lor ABC \qquad\qquad \lor AD$$
$$z_2 = BCD \qquad\qquad \lor \overline{A}\overline{B} \lor \overline{A}C$$
$$z_3 = \qquad\qquad ABC \lor \overline{A}\overline{B} \lor \overline{A}C \qquad \lor \overline{B}\overline{C}$$

Now we see that the AND gate to realize BCD can be shared by z_1 and z_2, another to realize ABC can be shared by z_1 and z_3, and two AND gates may be shared by z_2 and z_3. A total of 6 AND gates and 3 OR gates is required.

How can we obtain these nonminimal expressions which reduce the cost of the overall network at the expense of individual networks? Boolean algebra and searching can be used on problems the size of our example. We can also use Karnaugh maps on this size problem.

We must find low-cost implicants that can be shared by output variables, as well as traditional prime implicants. Figure 10.20(a) shows the Karnaugh map for the multiple output example above. The selected prime implicants are shown. If we look very carefully, we will find that $\overline{A}\overline{B}$ is redundant in the expression of z_3 given above. One AND to OR gate connection can be eliminated and the minimum diode circuit drawn as in Fig. 10.20(b).

As long as m and n are quite small, manipulation of equations and map techniques can lead to a satisfactory if not minimal solution with an expenditure of a modest amount of effort. But when the number of equations or input variables is greater than four or five, more orderly computer-oriented algorithms are required if we desire optimum two-level networks.

We will now find how minimal, two-level, multiple-output networks may be obtained using a generalized form of the Extraction Algorithm, and how

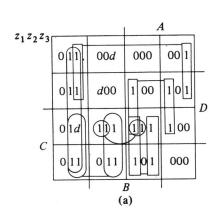

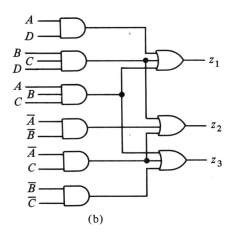

(a)   (b)

Fig. 10.20 Multiple Output Map Minimization and the Resulting Circuit. Values of (z_1, z_2, z_3) are placed in each cell.

economical networks may be obtained using somewhat simpler algorithms. We will attempt to minimize the number of gates required to realize a function, and will reduce, if not minimize, the number of necessary connections. While true minimization of connections (diodes) is possible in principle, we will see that the effort required to obtain such results may be so great that the value of known algorithms is questionable.

A switching function may be specified by a set of 3m ON-, OFF-, and DC-arrays, or as a single function array. The function array immediately indicates implicants common to two or more output variables. This information may lead to the sharing of AND gates by the networks which realizes the output variables. This type of information must be fully developed if we seek minimal networks, so let us assume that we begin minimization algorithms with a function array F available. We are able to obtain ON-, OFF-, and DC-arrays from the F-array if it becomes necessary to have them.

The Extraction Algorithm (Section 10.2) begins by generation of the prime implicants of a function. In a parallel development, we now seek what we call "multiple-output prime implicants." Let us refer to

$$ON_i \cup DC_i$$

as A_i for $i = 1, 2, \ldots, m$. We are familiar with and have methods for generating the prime implicants of each output variable, z_i, from its corresponding A_i. But what might we say of prime implicants of $A_i \cap A_j$ where $i \neq j$? These are of interest since they are implicants common to z_i and z_j (and possibly other output variables); they are the largest cubes z_i and z_j have in common. They each suggest a minimum input AND gate, which can be shared between networks that realize z_i and z_j. Generalizing, we can make corresponding statements about the prime implicants of $A_i \cap A_j \cap \ldots \cap A_k$.

Let $\lambda \subseteq \{1, 2, \ldots, m\}$, c_α be an n-tuple, and c_Ω be an m-tuple which identifies λ in any of several possible ways. Then cube $c_\alpha \circ c_\Omega$ will be said to be a *multiple output prime implicant* if:

(1) c_α is a prime implicant of $A_\lambda = \prod_{\text{all } i \in \lambda} A_i$, and

(2) c_α is covered by no prime implicants of $A_{\lambda'}$, where $\lambda \subseteq \lambda'$.

The first requirement insures that we are dealing with the largest cubes common to all output variables z_i identified by λ. The second requirement insures that we have included *all* of the output variables in λ to which a cube is common.

Given a function array description of a function, how do we find its multiple-output prime implicants? We might try to use the algorithms of the previous section for finding prime implicants. Consensus techniques appear to fail. For example, the *-product

$$0000 \circ 111 * 0001 \circ 110 = \phi$$

does not reveal the obvious implicant of z_1 and z_2, $000x$, because two

0–1 oppositions exist in these cubes. The opposition in the output part prevents formation of the meaningful ∗-product. We could examine only the input parts of cubes with the ∗-product and examine the output parts in some other manner, but such techniques are to be avoided if possible because they require computations of greater complexity.

Consensus techniques can be used to find multiple-output prime implicants if the encoding of output variable information is altered in one of several ways. Bartee [2, 3] has shown that if each 1 in the c_Ω part of each cube of a function array is replaced with an x, then the c_Ω parts

(1) Will never prohibit the formation of a ∗-product,
(2) Keep account of the output variables to which each c_α applies, and
(3) Prevent the loss of multiple-output prime implicants through absorbing.

The example shown above becomes:

$$0000 \circ xxx * 0001 \circ xx0 = 000x \circ xx0$$

Here the resulting cube covers $0001\,xx0$ and this parent cube can be deleted since it is not a multiple-output prime implicant. Cube $0000\,xxx$ is not covered by the resulting cube even though $0000 \sqsubseteq 000x\colon c_\alpha = 0000$ is a cube of z_1, z_2, and z_3 whereas $000x$ is a cube of only z_1 and z_2.

When we transform the output variable encoding by replacing 1's with x's we lose information. It is not possible to distingish between ON and DC entries after this transformation. This distinction must be made later in the Extraction Algorithm, so we must retain a copy of the original function array, the ON-arrays of the output variables, or the DC-arrays of the output variables. We have some choice here; we can write the extremal test in several ways.

To be specific, let us elect to retain the ON-arrays, which are easily obtained from a given function array by a simplified version of either Algorithm 9.1 or 9.4. To be able to compare multiple-output prime implicants and ON information, we will append ψ_i, an m-tuple of all 0's except for the ith position, in which a 1 appears, to every cube of each ON_i-array and form one array of ON information.

$$\text{ON} = \bigcup_{i=1}^{m} \text{ON}_i \times \psi_i \tag{10.21}$$

One final point requires attention before we examine an example. Any cube of a function array in which the c_Ω part is all 0's describes input conditions for which no output variables have the value of 1. Such input conditions are of no interest when we seek sum-of-product expressions of functions, and we will delete all such cubes from the function array, and neglect them if they appear in a search for prime implicants.

A 4-input, 4-output variable function is defined by the function array F below. Cubes with an all-0 c_Ω part have already been deleted. Array F' is determined from F by replacing all 1's with x's in the output variable columns. Array ON is obtained from F using Algorithm 9.1 to find ON_i, the Extraction Algorithm to obtain a minimal cover of ON_i, and Eq.

(10.21) to form the ON-array.

$$
F = \left\{
\begin{array}{l}
0000\ x011 \\
0001\ 001x \\
0010\ 0111 \\
0011\ 1100 \\
0111\ 1111 \\
1000\ 00x1 \\
1001\ 1011 \\
1011\ 0010 \\
1100\ 01x0 \\
1101\ 101x \\
1110\ 01xx \\
1111\ 10x0
\end{array}
\right\}
\qquad
F' = \left\{
\begin{array}{l}
0000\ x0xx \\
0001\ 00xx \\
0010\ 0xxx \\
0011\ xx00 \\
0111\ xxxx \\
1000\ 00xx \\
1001\ x0xx \\
1011\ 00x0 \\
1100\ 0xx0 \\
1101\ x0xx \\
1110\ 0xxx \\
1111\ x0x0 \\
000x\ 00xx \\
x00x\ 00xx \\
00x1\ 0000
\end{array}
\right\}
\qquad
\mathrm{ON} = \left\{
\begin{array}{l}
1x01\ 1000 \\
0x11\ 1000 \\
x111\ 1000 \\
11x0\ 0100 \\
001x\ 0100 \\
0x11\ 0100 \\
000x\ 0010 \\
00x0\ 0010 \\
0111\ 0010 \\
1x01\ 0010 \\
10x1\ 0010 \\
00x0\ 0001 \\
100x\ 0001 \\
0111\ 0001
\end{array}
\right\}
$$

Iterative consensus may now be performed on F' to obtain all multiple-output prime implicants. A few steps of this process are shown below F'. Note how cubes with all-0 c_Ω parts are encountered. The final array of multiple-output prime implicants is given below.

$$
\mathrm{PI} = \left\langle
\begin{array}{l}
x00x\ 00xx \\
0111\ xxxx \\
1x01\ x0xx \\
00x0\ 00xx \\
1110\ 0xxx \\
0010\ 0xxx \\
0000\ x0xx \\
11xx\ 00x0 \\
1x0x\ 00x0 \\
11x1\ x0x0 \\
1xx1\ 00x0 \\
x111\ x0x0 \\
11x0\ 0xx0 \\
001x\ 0x00 \\
0x11\ xx00
\end{array}
\right\rangle
$$

The definition of "multiple-output prime implicants" is well illustrated here. Minterm $0000 \circ x0xx$ remains in PI even though large cubes such as $x00x \circ 00xx$ and $00x0 \circ 00xx$ appear. These larger cubes do not pertain to all of the output variables z_1, z_3, and z_4. No larger cube which covers 0000 and is common to these output variables exists. This can most easily be seen by examining simultaneously the Karnaugh maps of all output variables. Locate each of the members of PI on the maps of Fig. 10.21 to see that each is in fact a multiple-output prime implicant.

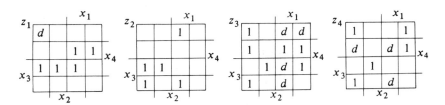

FIG. 10.21 Maps of the Output Variables

Iterative consensus is not a particularly efficient procedure for finding multiple-output prime implicants. We are dealing with $(n + m)$-tuples and with arrays that usually contain more cubes than we encounter in single-output function minimization. Substantial searching may be required to find a nontrivial consensus term. Absorbing then requires that a great number of cubes be compared.

The generalized consensus algorithm of the previous section (Algorithm 10.3) may be used for greater efficiency. We need compute only the generalized consensus with respect to the n input variables, of course. While Algorithm 10.3 can quickly give multiple-output prime implicants, one problem appears if we attempt to use it on very large arrays. After splitting a transformed function array on an input variable, the intersection of the fragments is formed. Many cubes with an all-0 c_Ω part may be formed in the computation of the intersection. While most of these cubes are removed when the array of intersection results is absorbed, they must be generated and temporarily stored. Storage may be the greater problem. Table 10.2 illustrates this by listing statistics pertaining to the example function. Note that for $i = 4$, 24 cubes are formed, nine of which have an all-0 c_Ω part. Twenty of these 24 cubes are covered by the four new multiple-output prime implicants found in this last step of the algorithm.

We can substantially reduce the number of trivial cubes formed if we remove all cubes with an all-0 c_Ω part at each step. This is most easily accomplished by forming a mask cube with an all-0 output part, $U_n \circ \psi$, and then removing the undesirable cubes with the **S** operator.

$$A \leftarrow PI \, \mathbf{S} \, U_n \circ \psi \tag{10.22}$$

Selection of the Minimal Cover

If we seek a minimum gate network that realizes a given multiple-output switching function, we can proceed to select a subset of the array of all

TABLE 10.2 ARRAY SIZES IN FORMING PI FROM F'-ARRAY

	$i = 1$	2	3	4
No. of cubes in array to be split	12	13	13	16
No. of intersection product cubes	4	8	9	24
No. with all-0 c_Ω part	1	2	3	9
No. of new cubes which are covered	0	3	4	20

multiple-output prime implicants. We will find that redundant gate connections may be specified by a cover composed of multiple-output prime implicants, and desire to eliminate such redundancy. But if we wish to find the true minimum connection (diode) network, we must substantially extend the PI-array to obtain the array from which the minimal cover is selected. We will briefly consider this case after we find how to obtain minimum gate, reduced connections networks.

To obtain a minimum gate network we must select the smallest set of multiple-output prime implicants that covers all of the ON_i-arrays. These arrays have been collected into one ON-array via Eq. (10.21) so that we can proceed as though we were concerned with a single-output minimization problem.

Extremals may be found using much the same method as was used for single-output minimization. Let π^i be the prime implicant under test. Then

$$A = ON \bigcap \pi^i \tag{10.23}$$

is the array of cubes (or subcubes) of the ON-array that π^i covers. If $A = \phi$, π^i covers no active members of any ON_i-array and it may be discarded. It either covers only don't cares, or the cubes of an ON_i-array that it originally covered are now inactive as a result of a prior multiple-output prime implicant selection. When $A \neq \phi$, to determine if π^i is an extremal, we compute

$$B = A \,\#\, (PI - \pi^i) \tag{10.24}$$

If $B = \phi$, then π^i is not a unique cover of A, is not an extremal, and may be returned to PI. But if $B \neq \phi$, then π^i provides the sole cover of one or more vertices of one or more output variables. It must be included in a minimal cover,

$$E \leftarrow E \cup \pi^i \tag{10.25}$$

and the ON-array entries that π^i covers must be removed from the ON-array.

$$ON \leftarrow ON \,\#\, A \quad \text{or} \quad ON \leftarrow ON \,\#\, \pi^i \tag{10.26}$$

The specific format of the ON-array prescribed by Eq. (10.21) was selected to facilitate the test of Eq.(10.23) and this updating step. The removed cubes subsequently may be covered by other selected prime implicants, but we need not select additional prime implicants specifically to cover these cubes. Their status is now that of a don't care. Of course, extremal π^i is deleted from PI.

Using the PI- and ON-arrays for our example function, we begin by defining E to be an empty array. The first prime implicant to be examined is $\pi^1 = x00x\,00xx$.

$$\pi^1 \bigcap \text{ON} = \begin{Bmatrix} 000x\ 0010 \\ \overline{0000\ 0010} \\ 1001\ 0010 \\ \overline{1001\ 0010} \\ 0000\ 0001 \\ 100x\ 0001 \end{Bmatrix} \neq \phi$$

π^1 covers vertices of z_3 and z_4 which are not covered by prime implicants already selected. Hence, we determine array B via Eq. (10.24).

$$B = \begin{Bmatrix} 0001\ 0010 \\ 1000\ 0001 \end{Bmatrix}$$

This result may also be easily obtained using the Karnaugh maps of Fig. 10.21. We must add π^1 to the set of selected prime implicants.

$$E = \{x00x\ 00xx\}$$

and remove π^1 from the ON-array. Cubes $000x \circ 0010$ and $100x \circ 0001$ will be completely removed from ON. Cube $00x0 \circ 0010$ must be replaced with $0010 \circ 0010$.

$$00x0 \circ 0010 \ \# \ x00x \circ 00xx = 0010 \circ 0010$$

Similar replacements must also be made in the ON-array.

$$1x01 \circ 0010 \ \# \ \pi^1 = 1101 \circ 0010$$
$$10x1 \circ 0010 \ \# \ \pi^1 = 1011 \circ 0010$$
$$00x0 \circ 0001 \ \# \ \pi^1 = 0010 \circ 0001$$

Evaluation of the remaining cubes of PI-array in the same manner gives the following extremal- and ON-arrays.

$$E = \begin{Bmatrix} x00x\ 00xx \\ 0111\ xxxx \\ 1x01\ x0xx \\ 1xx1\ 00x0 \\ 11x0\ 0xx0 \\ 0x11\ xx00 \end{Bmatrix} \qquad \text{ON} = \begin{Bmatrix} 1111\ 1000 \\ 0010\ 0100 \\ 0010\ 0010 \\ 0010\ 0001 \end{Bmatrix}$$

The ON-array is not empty so a complete cover has not yet been found and we must perform a less-than test.

To identify undesirable multiple-output prime implicants which may remain in the PI-array after the extremal test has been completed, we determine what cubes of the current ON-array each prime implicant covers.

$$A = \pi^i \bigcap \text{ON} \tag{10.27}$$

If $A = \phi$, and this is now very possible, then π^i is no longer of interest, and it may be discarded immediately. When $A \neq \phi$, we must compare π^i with every other remaining multiple-output prime implicant π^j. If the cost of π^i equals or exceeds that of π^j, and π^j covers A, then we say π^i is *less-than* π^j, and delete π^i in favor of π^j. The second requirement is straightforward.

$$B = A \# \pi^j \tag{10.28}$$

We ask whether or not B is an empty array. Comparing the cost of π^i and π^j is a bit more involved since we have not established what we mean by the cost of a cube with encoded output information. We can use either $\$\pi^i$, the cube cost evaluated over all coordinates, or $\$\pi_\alpha^i$, cube cost evaluated only over the input variables. The use of $\$\pi^i$ places a small input cube that applies to a number of output variables on a par with a large input cube that is restricted in its range of applicability. For example, $1110 \circ 0xxx$ and $11xx \circ 00x0$ have the same cost under this definition. This might seem to be a reasonable approach, in that the smaller cube is applicable to a number of output variables and thus AND gate sharing should result.

Experience contradicts this. Very often the prime implicant with the high-cost π_α^i part is redundant with respect to several of the output variables. Networks with lower connection costs if not lower gate costs are obtained by evaluating cube cost over the input variables only.

Returning to our example, the following prime implicants remain after the extremal test is completed.

$$\text{PI} = \begin{cases} 00x0\ 00xx & 1 \\ 1110\ 0xxx & 2 \\ 0010\ 0xxx & 3 \\ 0000\ x0xx & 4 \\ 11xx\ 00x0 & 5 \\ 1x0x\ 00x0 & 6 \\ 11x1\ x0x0 & 7 \\ x111\ x0x0 & 8 \\ 001x\ 0x00 & 9 \end{cases}$$

$$A = \pi^1 \prod \text{ON} = \begin{cases} 0010\ 0010 \\ 0010\ 0001 \end{cases}$$

π^3 covers A but $\$\pi_\alpha^3 > \$\pi_\alpha^1$ so we can not disregard π^1 yet. No other prime implicant covers A so π^1 must remain. $\pi^2 \prod \text{ON} = \phi$; thus π^2 may be thrown away.

$$\pi^3 \prod \text{ON} = \begin{cases} 0010\ 0100 \\ 0010\ 0010 \\ 0010\ 0001 \end{cases}$$

No other prime implicant covers this array so π^3 must remain. π^4, π^5, and

π^6 cover no part of the current ON-array and hence are removed from the PI-array.

$$\pi^7 \sqcap ON = \{1111\ 1000\}$$

π^8 covers this cube and $\$\pi_\alpha{}^7 = \$\pi_\alpha{}^8$ so we say π^7 is less-than π^8, and eliminate π^7 from further consideration. π^8 must clearly be retained in the PI-array.

$$\pi^9 \sqcap ON = \{0010\ 0100\}$$

Only π^3 covers this array, but $\$\pi_\alpha{}^3 > \$\pi_\alpha{}^9$ so that we cannot delete π^9.

Another extremal test is now in order since we have deleted a less-than cube. That extremal test will locate π^8. It must be added to the E-array, and cube $1111 \circ 1000$ deleted from the ON-array. Now we attempt another less-than test.

$$PI = \begin{Bmatrix} 00x0\ 00xx \\ 0010\ 0xxx \\ 001x\ 0x00 \end{Bmatrix} \qquad ON = \begin{Bmatrix} 0010\ 0100 \\ 0010\ 0010 \\ 0010\ 0001 \end{Bmatrix}$$

We find no prime implicant to be less than any other. Each element of the ON-array is covered by two prime implicants whose cost is such that none may be deleted.[1] We must resort to a branching procedure or a rule for breaking the deadlock.

In previous work we have replaced branching with the rule of picking the lowest cost cube. We see an example here where that rule is a poor one. Clearly the single prime implicant $0010 \circ 0xxx$ should be chosen over the other two. Another rule is suggested, which is generally better than the one we have been using.

Pick the prime implicant which covers the greatest number of elements of the ON-array.

Use of this rule give the final E-array.

$$E = \begin{Bmatrix} x00x\ 00xx \\ 0111\ xxxx \\ 1x01\ x0xx \\ 1xx1\ 00x0 \\ 11x0\ 0xx0 \\ 0x11\ xx00 \\ x111\ x0x0 \\ 0010\ 0xxx \end{Bmatrix}$$

---

[1] Use of $\$\pi^i$ in the less-than test rather than $\$\pi_\alpha{}^i$ would resolve the problem in this case, but not in general.

Appendix 10.3 lists a FORTRAN subroutine, MOMIN, which generates and selects multiple-output prime implicants in the manner which we have discussed. Note that the program that calls MOMIN must establish the function array, F, ON-array, values of N, M, $N + M$, etc. in the COMMON area of storage. Subroutine FORMFA of Appendix 9.4 may be used to do these things.

Redundancy in Connection Arrays

Conversion of all the x's in output columns of the E-array to 1's will facilitate a later computation. We will refer to an array with only 0's and 1's in the output variable columns as a *connection array*. A connection array is not necessarily a consistent function array, but it does describe the connections to be made in a two-logic-level network. Zero in an output coordinate indicates that no AND gate to OR gate connection is to be made. One indicates that such a connection is required. To see that a connection array need not be a function array let us consider the example.

$$E = \begin{pmatrix} x00x\ 0011 \\ 0111\ 1111 \\ 1x01\ 1011 \\ 1xx1\ 0010 \\ 11x0\ 0110 \\ 0x11\ 1100 \\ x111\ 1010 \\ 0010\ 0111 \end{pmatrix}$$

Looking at array E we might ask if z_1 is to have the value 0 or 1 under the input n-tuple 1001. Cube $1x01 \circ 1011$ suggests the value is to be 1; cube $1xx1 \circ 0010$ argues that $z_1 = 0$, if we view E as a function array. Clearly E is not a function array.

A connection array must be consistent with the care specifications of the function array from which it is derived, and must describe a completion of the original function. Let F be a function array with members $f^1, f^2, \ldots$; and $E = \{e^1, e^2, \ldots\}$ be a connection array for the same function. For every 0-cube, f^0 such that $f^0 \sqsubseteq f^i_\alpha$ and $(f^i_\Omega)_j = 1$; i.e., the jth component of the output part of f^i is a 1, there must exist a cube $e^k \in E$ such that $f^0 \sqsubseteq e^k_\alpha$ and $(e^k_\Omega)_j = 1$. In reverse, for every $e^0 \sqsubseteq e^i_\alpha$ and $(e^i_\Omega)_j = 1$, there must exist a cube $f^k \in F$ such that $e^0 \sqsubseteq f^k_\alpha$ and $(f^k_\Omega)_j \neq 0$. "Not equal to zero" means $(f^k_\Omega)_j \in \{1, x\}$. Therefore, the 1's in the output parts of the cubes of a connection array do reveal the ON-arrays of the selected completions of the output variables.

Array E does describe a network of 8 AND gates and 4 OR gates that realizes the given switching function. Each cube of E dictates an AND gate (unless the c_α portion of the cube has unit cube cost). The number of 1's in the c_Ω part of each cube determines the fan-out required of the corresponding AND gate. The number of 1's in an output variable column determines the fan-in of the OR gate that generates that output variable.

If the number of logic-block interconnections (diode fan-in) is to be reduced, one task remains to be performed after prime implicant selection. Often we are required to select a multiple-output prime implicant that is applicable to a number of output variables because it is an extremal with respect to one or more but not necessarily all of these variables. Other selected multiple-output prime implicants may cover the same input conditions for some of the output variables. When this occurs, redundant AND to OR gate connections are specified in the connection array.

In the example E-array above, cube $0111 \circ 1111$ appears because it is an extremal with respect to z_4. Cubes $0x11 \circ 1100$ and $x111 \circ 1010$ cover minterm 0111 for output variables z_1, z_2, and z_3. Thus three connections can be eliminated if $0111 \circ 1111$ is replaced by $0111 \circ 0001$.

ALGORITHM 10.5. *Eliminate Redundant Connections*

Most redundancy can be quickly eliminated by selecting a nonredundant cover of each output variable from the prime implicants which appear in the E-array. For each i in turn, $1 \le i \le m$, let E_i be the cover of z_i obtained · from the E-array in the following manner.

$$E_i^* = E S \chi_{n+i}^1 \qquad (10.29)$$
$$E_i = \mathbf{D}_\Omega(E_i^*)$$

We must retain E_i^* so that the E array can be reformed. Now we test each of the cubes of E_i, say e^j, for redundancy with either of the following extremal tests.

$$A = (e^j \# (E_i - e^j)) \# \mathrm{DC}_i \qquad (10.30)$$
$$A = (e^j \bigcap \mathrm{ON}_i) \# (E_i - e^j) \qquad (10.31)$$

The DC_i-array can be obtained from a copy of the original function array, and the ON_i-array may be obtained from a copy of either the function array or the ON-array.

$$\mathrm{ON}_i = \mathbf{D}_\Omega(F S \chi_{n+i}^1) = \mathbf{D}_\Omega(\mathrm{ON} \bigcap \chi_{n+i}^1) \qquad (10.32)$$

If $A \ne \phi$, e^j is not redundant in this cover of z_i, and it must be returned to E_i. Its associated output coding in E_i^* must not be altered. But if $A = \phi$, e^j is redundant. It is not returned to the E_i-array and in array E_i^* the 1 in the $(n+i)$th coordinate of the corresponding cube must be replaced with a 0.

After each member of E_i is tested and E_i^* altered appropriately, E may be formed again.

$$E \leftarrow E \cup E_i^* \qquad (10.33)$$

Each output variable is examined in this manner. It is possible that cubes with an all-0 c_Ω part will now appear in E. These are easily removed with the mask cube $U_n \circ \psi$ as in Eq. (10.22).

The final connection array for our example function requires 6 fewer connections than the original array. Maximum OR gate fan-in is reduced from 7 to 4.

$$E = \begin{cases} x00x\ 0011 \\ 0111\ \mathbf{0001} \\ 1x01\ 1000 \\ 1xx1\ 0010 \\ 11x0\ 0100 \\ 0x11\ 1100 \\ x111\ 1010 \\ 0010\ 0111 \end{cases}$$

This is not a unique solution; other nonredundant covers of some output variables may be selected.

Appendix 10.4 lists FORTRAN subroutine EREDCN, which eliminates redundancy in a connection array. The program that calls this subroutine must establish connection array E, function array F, and values of N, M, $N + M$, etc., in the proper COMMON storage locations. Subroutine MOMIN of Appendix 10.3 does these things as do other programs to be developed.

Connection Minimization

While the Extraction Algorithm gives a minimum gate, two-level realization of a multiple-output switching function, which is often also a minimum connection network, this algorithm can not guarantee that the solution it gives requires a minimum number of connections. In fact, for some functions the algorithm is not capable of giving a minimum connection realization. We can see this clearly with an example.

$$\begin{aligned} z_1 &= x_1 x_2 \\ z_2 &= x_1 x_2 \bar{x}_3 \\ z_3 &= x_1 x_2 x_3 x_4 \\ z_4 &= x_1 x_2 x_3 \bar{x}_4 \end{aligned} \qquad PI = \begin{cases} 110x\ xx00 & 1 \\ 1111\ x0x0 & 2 \\ 1110\ x00x & 3 \\ 11xx\ x000 & 4 \end{cases}$$

The first three prime implicants of PI must be selected because they are extremals with respect to z_2, z_3, and z_4, respectively. The fourth prime implicant is then redundant. The network demanded by the Extraction Algorithm consists of three AND gates, each feeding an output terminal, and an OR gate that provides z_1. The three connections to the OR gate exceed the two connections that would be required to realize z_1 with the AND gate suggested by the Boolean equation above.

Minimum connection networks can be obtained with an extraction algorithm if suitable modifications are made to the prime implicant-array before we begin to select the minimal cover. The example above suggests the nature of these modifications: we were forced to select the first three prime implicants because they are extremals; in doing so, we selected an undesirable cover of z_1. To avoid this undesirable cover we may expand the prime implicant set to provide for the choice of *not* sharing an AND gate. We may include entries such as $110x \circ x000$ and $1111 \circ 00x0$ in the PI-array, for example.

In general, if a minimum connection network is desired, for each multiple-output prime implicant with two or more x's in its output part we must add to that array entries which have the same c_α part, but fewer x's in their c_Ω part. If p x's appear, then p entries with $p - 1$ x's each may be added. If $p - 1 > 1$, then each of these entries may be supplemented in the same manner. An entry need not be added to the PI-array if its c_α part subsumes the c_α part of some other entry with the same c_Ω part. We can not simply absorb the supplemented PI-array, for all of the added cubes are covered by their parents.

For the example function we must add $110x\, 0x00$, $1111\, 00x0$, and $1110\, 000x$ to the PI-array. Cubes $110x\, x000$, $1111\, x000$, and $1110\, x000$ need not be added because of the presence of $11xx\, x000$.

$$\text{PI} = \begin{Bmatrix} 110x\ xx00 \\ 110x\ 0x00 \\ 1111\ x0x0 \\ 1111\ 00x0 \\ 1110\ x00x \\ 1110\ 000x \\ 11xx\ x000 \end{Bmatrix}$$

We can now see some of the disadvantages of this algorithm. First, we have expanded the PI-array substantially. Greater storage and processing will be required when we select a cover. Second, when we begin to select a cover, we may find that no extremals exist (as in this case). If we demand a truly minimal cover, we are forced immediately to the branching procedure. This happens quite generally and is not a peculiarity of the example function.

Thus while we have a means of obtaining minimum connection two-level AND-to-OR gate networks, the computation called for is great and may be considered excessive in most cases. The savings gained by eliminating a few connections may be far less than the cost of designing that network. If integrated circuits are to be used, the cost of connections made on the chip is totally negligible.

Nonredundant Covers

The cost of executing the Extraction Algorithm to obtain a minimum gate network may be undesirably high when large arrays must be processed. We saw in the previous section that a nonredundant cover of a single-output function can be obtained with far less effort than a guaranteed minimal cover. We can also select a nonredundant cover of a multiple-output function with less computation.

Nonredundant covers can be selected in several ways. We could examine prime implicants in the order in which they appear in the PI-array, as we did in the previous section. Or, if we feel that slightly more processing is justified, we might examine prime implicants in an order that favors selecting lower cost cubes over higher cost cubes.

Appendix 10.5 offers a program that selects a nonredundant cover by generating all multiple output prime implicants, and examining them in order from lowest to highest cost over the c_α part. When a low cost prime implicant is found to cover a vertex or vertices of the ON-array, it is selected, and the vertices it covers are removed from the ON-array. When the ON-array becomes empty, the program converts the E-array to a connection array; redundant connections may then be eliminated with the EREDCN subroutine of Appendix 10.4.

10.4 FUNCTION REDUCTION

Multiple-output function minimization would seem to offer a way of finding the programming for the PLAs of Section 3.3. The cost function must be changed drastically: Cost is very high if the final connection array has more cubes than the PLA has AND gates, and very small otherwise. Prime implicants and large cubes are important only if their selection reduces the size of the final connection array. Also, rather modest switching functions can be found that have very large prime implicant arrays. Minimization procedures that do not require all prime implicants to be found initially are thus of interest and value. We will consider one procedure which gives a reduced, if not minimum cost connection array for a multiple-output function, and does not require that we generate multiple-output prime implicants as such.

Let F be a function array of cubes f^i. All cubes with an all-0 Ω-part are to be deleted. We will now refer to the set of all $f_\alpha{}^i$ as F_α, and to the set of all $f_\Omega{}^i$ as F_Ω. Let λ be a set of output column designators.

$$\lambda \subseteq \Omega = \{n + 1, n + 2, \ldots, n + m\} \tag{10.34}$$

Array F can be thought of as the union of $2^m - 1$ or fewer subarrays, each of which consists of cubes that have 1's in all of the coordinate positions specified by λ and have 0's or d's in all other positions. Let F^λ denote such a subarray. Then

$$F = \bigcup_{k=1}^{2^m-1} F^{\lambda_k} \tag{10.35}$$

where

$$F^{\lambda_k} = \left\{ f^i \mid f_j{}^i = \begin{cases} 1 \text{ for all } j \in \lambda_k \\ 0 \text{ or } d \text{ for } j \notin \lambda_k \end{cases} \right\} \tag{10.36}$$

As with F we let F_α^λ and F_Ω^λ denote the input and output subarrays of F^λ, respectively.

$$F^\lambda = F_\alpha^\lambda \circ F_\Omega^\lambda \tag{10.37}$$

A function array may be thought of as the union of $3^m - 1$ or fewer subarrays, each of which consists of cubes with *identical* output m-tuples. It can be expressed more compactly if a minimal cover of the input part of each of these subarrays is found. Now we ask if greater compression is possible when we assign d's in some optimum fashion.

Suppose that the Ω-part of $f^i \in F$ consists of a mixture of 0's, 1's, and d's, and that μ is the set of positions of the 1's in f_Ω^i. Then $f_\alpha^i \in F_\alpha^\mu$. But if one or more of the d's in f_Ω^i were set to 1, then $f_\alpha^i \in F_\alpha^\lambda$, where $\mu \subseteq \lambda$. We could consider f_α^i to be a "don't care" term when forming a minimal (or at least a nonredundant) cover of F_α^λ. Greater compression of F_α^λ is often possible.

If we do not insist on retaining a consistent function array, but attempt to convert a function array into a connection array by optimum assignment of Boolean values to the d's, then every $f_\alpha^i \in F_\alpha^\lambda$ for all $\lambda \supset \mu$ can be used as "don't care" cubes in arriving at a minimal cover of F_α^μ.

To see these points, consider the following function array.

$$F = \begin{Bmatrix} 000\ 11 \\ 001\ 1d \\ 01x\ 10 \\ 100\ dd \\ 101\ d1 \\ 11x\ 01 \end{Bmatrix}$$

With $\lambda = \{4, 5\}$ we might replace cube $000 \circ 11$ with $00x \circ 11$, which assigns the value of 1 to the d in the second cube, or with $x00 \circ 11$, which assigns the value 1 to both d's in cube $100 \circ dd$. We might replace cube $01x \circ 10$ with $0xx \circ 10$, a multiple-output prime implicant that can be located by using cubes $000 \circ 11$, $001 \circ 1d$, $100 \circ dd$, and $101 \circ d1$ as a don't care array in conjunction with $F^{\{4\}} = 01x \circ 10$. If we make all of these replacements, we no longer have a consistent function array. Cube $00x \circ 11$ contradicts cube $0xx \circ 10$ for input condition 001 because we have assigned a d the value of 1 in the first case and the value of 0 in the last case.

Let $|\lambda|$ denote the number of elements in set λ, and

$$D1^\lambda = \{f^i \mid f_j^i = d \text{ for all } j \in \lambda\}$$
$$D2^\lambda = \{f^i \mid f_j^i = 1 \text{ or } d \text{ for all } j \in \lambda\}, \text{ and}$$
$$D^\lambda = D1^\lambda \cup D2^\lambda \tag{10.38}$$

Let C_α^λ be a minimal (or nonredundant) cover such that

$$F_\alpha^\lambda \sqsubseteq C_\alpha^\lambda \sqsubseteq F_\alpha^\lambda \cup D_\alpha^\lambda \tag{10.39}$$

If we begin computation with $|\lambda| = m$, and find C_α^λ and set it aside in array E, we may change f_j^i to a d for all $j \in \lambda$ for every $f^i \in F^\lambda$. This corresponds to the transfer of covered cubes from the ON- to the DC-array in minimization procedures. These covered cubes may be used to advantage as don't cares when finding C_α^μ where $\mu \subset \lambda$. This process may be repeated for $|\lambda| = m - 1, m - 2, \ldots, 1$ until no 1's remain in F_Ω. The union of all $C_\alpha^\lambda \circ \psi_\lambda$ describes a network with a minimum or near minimum of AND gates and network interconnections. Redundancy in this connection array may then be removed with Algorithm 10.5.

ALGORITHM 10.6. *Reduction of Function Arrays.*
Let F be a given function array with all f^i deleted for which $f_j^i = 0$ for all $j \in \Omega$.

1. Set $E = \phi$, $|\lambda| = m$.

2. Select the next λ. For a given value of $|\lambda|$, $p = \binom{m}{|\lambda|}$ such selections will be made in turn. If all selections for a specific value of $|\lambda|$ have been made, reduce $|\lambda|$ by 1, make a selection and proceed to step 3. When $|\lambda|$ is reduced to zero, go to step 7.

3. $F^\lambda = F\, \mathbf{S}\, \chi_\lambda^{\ 1}$
   If $F^\lambda = \phi$, go to step 2, otherwise obtain D^λ by:
   $D^\lambda = F \prod \chi_\lambda^{\ 1}$
   For all $j \in \lambda$ and all $f^i \in F^\lambda$, change f_j^i from 1 to d: $F^\lambda \leftarrow \mathbf{C}_\lambda^{0dx}(F^\lambda)$.
   Then: $F \leftarrow F \cup F^\lambda$.

4. $F_\alpha^\lambda = \mathbf{D}_\Omega(F^\lambda)$, $D_\alpha^\lambda = \mathbf{D}_\Omega(D^\lambda)$
   Find C_α^λ using the single output extraction algorithm or an algorithm which gives a nonredundant cover.

5. $E \leftarrow E \cup C_\alpha^\lambda \times \psi_\lambda$

6. In F for every $f^i \in D_\alpha^\lambda \prod C_\alpha^\lambda$, set $f_j^i = d$ for all $j \in \lambda$. This step may require that f^i be broken into subcubes if a part of f_α^i is covered by C_α^λ. Go to step 2.

7. Eliminate redundancy in the connection array E, with the EREDCN algorithm.

The program of Appendix 10.6 is based upon this algorithm, which can be performed by hand with a modest expenditure of effort in comparison with the extraction algorithm, and also is executed efficiently by a computer. To see this, let us consider the function array that was minimized earlier in this section. With $|\lambda| = 4$, $\lambda = \{5, 6, 7, 8\}$, $F^\lambda = 0111 \circ 1111$, and $D^\lambda = \phi$. Thus we place this cube in array E and alter F by replacing cube $0111 \circ 1111$ with $0111 \circ dddd$.[2]

---

[2] While d and x are equivalent, we use both here to point out changes made in the F-array.

$$E = \{0111 \circ 1111\}$$

$$F = \begin{cases} 0000\ x011 \\ 0001\ 001x \\ 0010\ 0111 \\ 0011\ 1100 \\ 0111\ dddd \\ 1000\ 00x1 \\ 1001\ 1011 \\ 1011\ 0010 \\ 1100\ 01x0 \\ 1101\ 101x \\ 1110\ 01xx \\ 1111\ 10x0 \end{cases}$$

Now we select: $\quad \lambda = \{5, 6, 7\} \quad$ and find $\quad F^\lambda = \phi$

$\qquad\qquad\quad\ \lambda = \{5, 6, 8\} \quad$ and find $\quad F^\lambda = \phi$

$\qquad\qquad\quad\ \lambda = \{5, 7, 8\} \quad$ and find $\quad F^\lambda = \{1001 \circ 1011\}$

$$D^\lambda = \{0111 \circ dddd,\ 1101 \circ 101x,$$
$$0000 \circ x011\}$$
$$C_\alpha{}^\lambda = \{1x01\}$$

Thus we add $1x01 \circ 1011$ to E and replace $1001 \circ 1011$ with $1011 \circ d0dd$, and $1101 \circ 101x$ with $1101 \circ d0dx$ in F.

With $\lambda = \{6, 7, 8\}$ we find $F^\lambda = \{0010 \circ 0111\}$ and $D^\lambda = \{0111 \circ dddd,\ 1110 \circ 01xx\}$. We must add $0010 \circ 0111$ to E and change its output portion in F. Then we set $|\lambda| = 2$ and for $\lambda = \{5, 6\}$ find $F^\lambda = \{0011 \circ 1100\}$ and $D^\lambda = \{0111 \circ dddd\}$. Cube $0x11 \circ 1100$ must be added to E and the Ω part of $0011 \circ 1100$ altered in F. At this point we have the following arrays.

$$E = \begin{pmatrix} 0111\ 1111 \\ 1x01\ 1011 \\ 0010\ 0111 \\ 0x11\ 1100 \end{pmatrix}$$

$$F = \begin{cases} 0000\ x011 \\ 0001\ 001x \\ 0010\ 0ddd \\ 0011\ dd00 \\ 0111\ dddd \\ 1000\ 00x1 \\ 1001\ d0dd \\ 1011\ 0010 \\ 1100\ 01x0 \\ 1101\ d0dx \\ 1110\ 01xx \\ 1111\ 10x0 \end{cases}$$

We can see that the next nonempty F^λ array will be obtained for $\lambda = \{7, 8\}$. Here $F^\lambda = \{0000 \circ x011\}$ and D^λ consists of seven cubes. $C_\alpha{}^\lambda = x00x$ and the following replacement must be made in F.

$$0000 \circ x011 \Rightarrow 0000 \circ x0dd$$

$$0001 \circ 001x \Rightarrow 0001 \circ 00dx$$

$$1000 \circ 00x1 \Rightarrow 1000 \circ 00xd$$

Continuing with $|\lambda| = 1$ leads to the following E-array; elimination of redundancy changes the boldface 1's to 0's.

$$E = \begin{cases} 0111\ \mathbf{1111} \\ 1x01\ 10\mathbf{11} \\ 0010\ 0111 \\ 0x11\ 1100 \\ x00x\ 0011 \\ 11x1\ 1000 \\ 11x0\ 0100 \\ 1xx1\ 0010 \end{cases}$$

This connection array specifies a network of 8 AND gates and 4 OR gates, which could be constructed of 37 diodes. The Extraction Algorithm was used previously to obtain another network of the same number of gates and diodes. Thus in this case, the reduction algorithm led to a minimum gate network. This will not generally happen. If we compare the effort which the two algorithms demand, we must contrast generating and selecting multiple-output prime implicants with finding minimal covers of many arrays of n-tuples. In the example, finding most of these required minimal covers was a trivial task.

Experience with this straightforward reduction algorithm reveals several problems. First, nearly all of 2^m λ-sets must be established, and masks formed and tested. This number grows very rapidly with m; the number of masks for which nonempty ON-arrays exist is usually well below 2^n. Thus for $m > n$, a great deal of time is spent forming and testing nonproductive masks. Computation speed can be greatly increased by selecting productive masks from the current F array at each step. We may use the output parts of the cubes of F in turn as masks or, with greater effort, we may select the output part with the greatest number of 1's (or that applies to the greatest number of cubes, etc.,). Using the first approach on the example function array gives the same input parts as listed above, with only one redundant connection after applying only six masks.

Cubes with an output part of all ds appear in all DC arrays. If the number of such cubes is substantial, as it can be especially in switching functions that involve binary encoded decimal digits, then prime implicants of these cubes are formed over and over again at great expense. A single-output reduction algorithm that searches the DC-array for the necessary minterms to complete a prime implicant from the few cubes of the ON-array avoids this expense. An operator that finds the smallest cube to cover an array is defined and used extensively in the next chapter.

If the reduced algorithm is to be used to determine programming for PLAs, then the cost function of PLAs should be incorporated. One approach to minimizing the number of cubes placed in the connection array consists of first finding and using those mask cubes that apply to the largest number of cubes in the function array. Such masks are most able to replace a number of cubes with one cube. Again, this topic is identical to topics of the next chapter.

Deciding if an output variable of a PLA should be realized in a *sop* or AOI form is an open problem. Clearly the decision can be significant. The following compressed function array for a complete 3-input, 3-output switching function can be viewed as a connection array. Five product terms appear.

$$\begin{pmatrix} 00x\ 001 \\ 01x\ 010 \\ 100\ 100 \\ x01\ 001 \\ 11x\ 110 \end{pmatrix}$$

If the first output variable is realized in AOI form, then only four product terms are needed. (Complement the z_1 column to find the cube of $\bar{z}_1$, or draw Karnaugh maps for the three output variables.)

Four product terms remain in the connection array if z_2 is complemented. Finally, if z_3 is realized in AOI form, then only three product terms are needed.

$$\bar{z}_3$$
$$\begin{pmatrix} 01x\ 001 \\ 100\ 101 \\ 11x\ 111 \end{pmatrix}$$

Complementing pairs and all three output variables do not result in even fewer product terms in this example.

We might double m by including the complement of each output variable in the function array, minimize or reduce the function array, and then choose either z_i or $\bar{z}_i$ for each i, based upon the results. Exhaustive searches such as this require a great deal of computation. We resort to them only when the possible results can justify the expense.

10.5 MINIMIZATION OF FUNCTIONS WITH CONSTRAINED DON'T CARES

Sometimes restrictions exist on the don't cares of the switching functions to be realized. In the Karnaugh map of Fig. 10.22 d's represent traditional don't cares and α and β represent constrained don't cares. If one α is set to 1 then both must be, i.e., α may take either the 0 or 1 value, but both α

entries must be the same. If β is set to 1, then $\bar{\beta}$ must be set to 0. Even minimization by inspection of the Karnaugh map becomes a more difficult task when a few constrained don't cares are introduced.

We will refer to α, β, etc., used to express don't care constraints, as *parameters* of the function. The cubical notation and algorithms and techniques of previous sections can be used to find a minimal expression of a function with parameters if a column is added to the function array for each parameter. In the case of single-output functions $(n + l)$-tuples must thus be considered, where l measures the number of parameters of the function. The encoding in these added columns reflects the constraint placed on the input symbol portion of the $(n + l)$-tuples. Thus the function of Fig. 10.22 may be expressed

$$
\text{ON} = \begin{Bmatrix} 0000\ xx \\ 0001\ xx \\ 0010\ xx \\ 0100\ xx \end{Bmatrix}
\qquad
\text{DC} = \begin{Bmatrix} \overset{\alpha\beta}{0011\ xx} \\ 0101\ 1x \\ 1001\ 1x \\ 0110\ x1 \\ 1011\ x0 \end{Bmatrix}
$$

Using this coding, don't care 0011 is independent of the parameters. Cube 0101 is to be a minterm of the function only if α is set to 1. Minterm 0110 is a cube of the function if $\beta = 1$, but then 1011 is not. With $\beta = 0$, 1011 is a cube of the function and 0110 is not.

If iterative or generalized consensus is performed on $\text{ON} \cup \text{DC}$, the following prime implicant array is obtained.

$$
\text{PI}^* = \begin{Bmatrix} 0x00\ xx \\ x001\ 1x \\ x011\ x0 \\ 00xx\ xx \\ 0x0x\ 1x \\ 0xx0\ x1 \\ x0x1\ 10 \end{Bmatrix}
$$

Calling the cubes of this array prime implicants can be questioned. Cube $0x00 \circ xx$ is not really a prime implicant, regardless of parameter assignment. If α is set to 1, then cube $0x0x$ $(\alpha\beta = 1x)$ is a prime implicant which

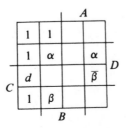

Fig. 10.22 A Function with Constrained Don't Cares

covers $0x00$; if β is set to 1 then $0xx0$ ($\alpha\beta = x1$) is a larger cube of the function which covers $0x00$. Thus $0x00 \circ 00$ should replace $0x00 \circ xx$ because it indicates the parameter assignment under which the input cube is a true prime implicant.

One way to obtain a correct prime implicant array is to expand PI* on the parameter columns by replacing each x in those columns with 0 and with 1 duplicating the rest of the cube as often as necessary, and then absorbing the resulting array. For the example function,

$$
\text{PI} = \left\{
\begin{array}{l}
0x00\ 00 \\
x001\ 11 \\
x011\ 00 \\
00xx\ 00 \\
00xx\ 01 \\
00xx\ 10 \\
00xx\ 11 \\
0x0x\ 10 \\
0x0x\ 11 \\
0xx0\ 01 \\
0xx0\ 11 \\
x0x1\ 10
\end{array}
\right\}
=
\left\{
\begin{array}{l}
0x00\ 00 \\
x001\ 11 \\
x011\ 00 \\
00xx\ xx \\
0x0x\ 1x \\
0xx0\ x1 \\
x0x1\ 10
\end{array}
\right\}
$$

Clearly this array can be compressed if we reintroduce x's in the parameter columns. It may be undesirable to do so however. For example, cube $00xx$ is an extremal with parameter assignments $\alpha\beta = 00$, 01, or 10, but is redundant with $\alpha\beta = 11$. If we pick cube $00xx \circ xx$ without knowing the optimum parameter assignment and later find $\alpha\beta = 11$ to be optimum, we will have introduced a redundant term into the solution. And this will happen if we use the extremal tests of the previous sections. If the compressed array above is used as the PI array and $\pi^i = 00xx \circ xx$, then

$$
(\pi^i \bigcap \text{ON}) \# (\text{PI} - \pi^i) = \left\{
\begin{array}{l}
0001\ 0x \\
0010\ x0
\end{array}
\right\}
$$

Since the sharp product is not empty, we conclude that π^i is an extremal. Examining the sharp product reveals that π^i is the sole cover of 0001 if $\alpha = 0$, and the sole cover of 0010 if $\beta = 0$. Thus we can only call $00xx \circ xx$ an extremal if we agree to one or both of these parameter assignments.

If we work with the expanded PI-array above, then the parameter l-tuples separate the prime implicants into disjoint subsets. Single-output minimization of this array of prime implicants gives the result

$$E = \begin{Bmatrix} 0x00\ 00 \\ 00xx\ 00 \\ x011\ 00 \\ 0xx0\ 01 \\ 00xx\ 01 \\ 0x0x\ 10 \\ 00xx\ 10 \\ x0x1\ 10 \\ 0x0x\ 11 \\ 0xx0\ 11 \\ x001\ 11 \end{Bmatrix} \qquad \begin{aligned} \alpha\beta = 00:\ &\overline{A}\overline{C}\overline{D} \vee \overline{A}\overline{B} \vee \overline{B}CD \\ 01:\ &\overline{A}\overline{D} \vee \overline{A}\overline{B} \\ 10:\ &\overline{A}\overline{C} \vee \overline{A}\overline{B} \vee \overline{B}D \\ 11:\ &\overline{A}\overline{C} \vee \overline{A}\overline{D} \vee \overline{B}\overline{C}D \end{aligned}$$

which differs from PI only by the absence of $00xx \circ 11$. This result is not a minimum cost cover of the ON-array, but a collection of the minimum cost covers for each parameter assignment. We may now select the optimum parameter assignment by comparing the costs of these covers. Clearly the assignment $\alpha\beta = 01$ leads to the lowest cost circuit.

Alternatively, we might have split the expanded prime implicant array on each possible parameter assignment, and arrived at the minimum cost cover for each assignment in turn. In either case the procedure is exhaustive, in the sense that all possible parameter assignments are examined in detail. This is very undesirable even when the number of parameters is as modest as 4 or 5. Unfortunately, a nonexhaustive procedure for finding the absolute minimum cost cover is not known to exist.

Nonredundant covers can be found more directly, of course. And now we might work with the uncorrected prime implicant array, PI*. A cube may be selected because of its cost (measured over the input n-tuple), or because of the number of members of the ON-array that it covers. Suppose from array PI* we select cube $00xx \circ xx$ for both reasons and also because it is the first 2-cube we encounter. Its parameter l-tuple gives no indication of the optimum parameter assignment. But selecting this cube reduces the ON-array to $\{0001 \circ xx\}$. The next cube in PI*, $0x0x \circ 1x$, covers this remaining ON-cube and introduces the parameter assignment $\alpha = 1$. If we include $0x0x \circ 1x$ in the nonredundant cover, the remaining original member of the ON-array may be deleted, but cubes $0101 \circ 1x$ and $1001 \circ 1x$ must now be transferred from the DC- to the ON-array. When we elect to set $\alpha = 1$, we insist that minterms 0101 and 1001 be covered. The prime implicant $0x0x \circ 1x$ just selected covers $0101 \circ 1x$, so only $1001 \circ 1x$ remains in the ON-array to be covered. The lowest cost member of PI* which covers this entry is $x0x1 \circ 10$. Selecting this prime implicant completes the parameter assignment by setting $\beta = 0$. Cube $1011 \circ x0$ must now be transferred to the ON-array and immediately deleted because $x0x1 \circ 10$ covers it. Thus the ON-array vanishes and the selection is complete.

Now suppose that the cubes of PI* were rearranged so that $0xx0 \circ x1$ is the first cube encountered which has minimum cost *and* covers a maximum number of entries of the ON-array. When $\beta = 1$ is selected, cube $0110 \circ x1$

must be transferred from the DC- to the ON-array and immediately deleted because the selected prime implicant covers it. Array PI* may now be reduced. Cubes $x011 \circ x0$ and $x0x1 \circ 10$ may be deleted because they are prime implicants only if $\beta = 0$. This point would have been encountered much earlier if a PI-array had been used rather than PI*.

The only cube remaining in the ON-array is $0001 \circ xx$. It is covered by several of the remaining prime implicants. If $00xx \circ xx$ is selected for inclusion in the nonredundant cover, the value of α has not been fixed, but the ON-array has vanished. If $\alpha = 0$ is selected, no members of the DC-array need be transferred to the ON-array and a nonredundant cover, actually the minimal cover, has been found. If α is arbitrarily set to 1, then $0101 \circ 1x$ and $1001 \circ 1x$ must be transferred to the ON-array and the selection procedure continued. Again the order of cubes in the PI-array influences the cost of the cover selected. But now, that order influences the parameter assignment as well. If the wrong parameter assignment is made, then there is no possibility that the nonredundant cover is of minimum cost.

The multiple-output minimization algorithms of previous sections may be modified to handle $(n + l + m)$-tuples with the above discussion in mind. Don't cares of different output variables may be constrained. Chapter 5 presented a source of particularly interesting and difficult constraints. Assume that we have a two-output Boolean function and input symbol δ is a don't care for both z_1 and z_2, but constrained in the following fashion. If δ is covered by z_1 then δ must also be covered by z_2. If δ is not covered by z_1, then δ is an unconstrained don't care of z_2.

If we use α as the parameter, this complicated constraint can be expressed with the following function array entries.

$$
\begin{array}{cccc}
x_1 \ldots x_n & z_1 & z_2 & \alpha \\
\delta_1 \ldots \delta_n & 1 & 1 & 1 \\
\delta_1 \ldots \delta_n & 0 & d & 0
\end{array}
$$

These rows would appear to introduce a contradiction, in that z_1 is to take both values for input symbol $\delta_1 \ldots \delta_n$. But once parameter α has been assigned a value, one of these two function array entries must be dropped because it does not satisfy the parameter assignment. Alternatively, the same constraint may be expressed: if z_2 covers δ then δ is an unconstrained don't care with respect to z_1; if δ is not covered by z_2, then it must not be covered by z_1. This expression of the constraint suggests the following function array entries.

$$
\begin{array}{cccc}
\delta_1 \ldots \delta_n & d & 1 & 1 \\
\delta_1 \ldots \delta_n & 0 & 0 & 0
\end{array}
$$

A second type of constraint, introduced in Chapter 5, may be described as follows: If input symbol ∂, a constrained don't cares of z_1 and z_2, is

	∂^1	∂^2	∂^3	∂^4
	1	1	δ^2	1
	δ^1	d	d	δ^3
x_3	1	d	d	1

z_1 x_1 , x_4 , x_2

	∂^1	∂^2	∂^3	∂^4
	0	0	δ^2	0
	δ^1	d	d	δ^3
x_3	1	d	d	1

z_2 x_1 , x_4 , x_2

x_1	x_2	x_3	x_4	z_1	z_2	δ_1	δ_2	δ_3	∂_1	∂_2	∂_3	∂_4
0	0	0	0	1	0	x	x	x	1	x	x	x
0	0	0	0	0	d	x	x	x	0	x	x	x
0	x	0	1	1	0	x	x	x	x	x	x	x
x	0	0	1	1	0	x	x	x	x	x	x	x
x	0	1	0	1	1	x	x	x	x	x	x	x
0	0	1	1	1	1	1	x	x	x	x	x	x
0	0	1	1	0	d	0	x	x	x	x	x	x
0	1	0	0	1	0	x	x	x	x	1	x	x
0	1	0	0	0	d	x	x	x	x	0	x	x
x	1	1	x	d	d	x	x	x	x	x	x	x
1	0	0	0	1	0	x	x	x	x	x	x	1
1	0	0	0	0	d	x	x	x	x	x	x	0
1	0	1	1	1	1	x	x	1	x	x	x	x
1	0	1	1	0	d	x	x	0	x	x	x	x
1	1	0	0	1	0	x	x	x	x	x	1	x
1	1	0	0	0	d	x	x	x	x	x	0	x
1	1	0	1	1	1	x	1	x	x	x	x	x
1	1	0	1	0	d	x	0	x	x	x	x	x

FIG. 10.23   A Function with δ and ∂ Constraints

not covered by z_1, then it is a true don't care for z_2, but if it is covered by z_1 then it must not be covered by z_2. The following entries may be placed in a function array to express this constraint where β is the parameter.

$$
\begin{array}{cccc}
x_1 \dots x_n & z_1 & z_2 & \beta \\
\partial_1 \dots \partial_n & 1 & 0 & 1 \\
\partial_1 \dots \partial_n & 0 & d & 0
\end{array}
$$

The alternative expression of this constraint is given by:

$$
\begin{array}{cccc}
\partial_1 \dots \partial_n & d & 0 & 1 \\
\partial_1 \dots \partial_n & 0 & 1 & 0
\end{array}
$$

Fig. 10.23 shows an example function which illustrates these types of constraints.

The map for z_1 is unusual in that no zeros appear.[3] Thus if all parameters are assigned to 1, $z_1 = 1$. However, with this parameter assignment the function array of Fig. 10.23 reduces to:

$$F = \begin{cases} 0000\ 10 \\ 0x01\ 10 \\ x001\ 10 \\ x010\ 11 \\ 0011\ 11 \\ 0100\ 10 \\ x11x\ dd \\ 1000\ 10 \\ 1011\ 11 \\ 1100\ 10 \\ 1101\ 11 \end{cases} \qquad z_1 = \{xxxx\} \qquad z_2 = \begin{cases} 11x1 \\ xx1x \end{cases}$$

This expression of z_2 requires two gates and five diodes. While this is not an expensive circuit, it is not the minimum diode cost realization of this switching function. With the parameter assignment

$$(\delta_1\ \delta_2\ \delta_3\ \partial_1\ \partial_2\ \partial_3\ \partial_4) = (1011101),$$

the following connection array may be found.

$$\begin{pmatrix} 0xxx\ 10 \\ x0xx\ 10 \\ xx1x\ 01 \end{pmatrix}$$

which requires only two diodes (and two complements).

Minimum cost covers may be found by a suitably modified extraction algorithm. The necessary modifications were pointed out earlier. Thus the output coding of a function array such as that of Fig. 10.23 may be changed to the Bartee encoding and an ON-array that extends Eq. (10.21) may be formed. Prime implicants may be formed with the generalized consensus procedure. If the general consensus is formed only over the input variables, a partially corrected array results. If a nonredundant cover is desired, the general consensus may be found over the parameter columns as well. Usually prime implicant arrays are very large. Generalized consensus over the input variables of the function array of Fig. 10.23 produces 56 prime implicants. General consensus over all columns produces an array of 50 prime implicants.

Prime implicants may be selected in two ways. Selection on the full, corrected prime implicant array will produce not a single minimal cover,

---

[3] This is true of all functions with a similar origin. (See Section 5.3.)

but a cover for each parameter assignment. Alternatively, the prime implicant array may be split on each possible parameter assignment and a minimal cover selected from the subset of prime implicants that satisfy the parameter assignment. In either case redundancy must be eliminated (as with multiple output minimization), the cost of covers for various parameter assignments compared, and the least cost cover selected. Both procedures are exhaustive searches over all possible parameter assignments and hence require a great deal of computation.

Minimizing functions that contain constraints of the complexity of these, or even greater, is not an easy task even when the functions are modest in other respects. The Karnaugh map is much less useful as a tool because the user must keep these constraints in mind or use some sort of notation that makes the map much less tractable. Computer minimization is not attractive if the absolute minimum cost circuit is required. The exhaustive search of parameter assignments becomes very expensive when each don't care input symbol introduces a parameter. We are almost forced to accept nonredundant, near-minimum cost circuits by practical considerations, such as the cost and availability of designers and computers.

10.6 SUMMARY

Let us think of creating a software system for use by designers of combinational logic networks. To describe the switching function to be realized, we might permit a logic designer to write the following: (1) a truth table, (2) a set of ON-and DC-arrays, (3) a set of Boolean equations, or (4) a state transition table and associated alphabets. The truth table is a function array; we need merely read it into computer memory. The ON- and DC-arrays must be used to form OFF-arrays for the output variables, and all arrays must be packed into a function array via Algorithm 9.5. Algorithms for the conversion of Boolean equations and state transition tables to function arrays were shown in the last chapter. Hence we are able to arrive at a function array from all of these forms for describing a switching function.

If a ROM is to be used to realize the function, the input parts of function array cubes must be converted to 0-cubes. This is an easy task with array operators available (See Prob. 9.4-1). Also, don't cares must be assigned values. Since no ROM simplification is possible, the values assigned are of no importance. The function array then provides the necessary ROM programming.

The ROM simplification point is not entirely valid. Sometimes, especially for function arrays with many don't cares like those derived from state tables for JK flip-flops, some output variables can be eliminated by proper don't care assignment. They may be made: (1) trivial (logic 0 or 1), (2) equal to an input variable or its complement, or (3) equal to another output variable or its complement. The existence of such conditions can simplify a ROM realization or a gate, or PLA realization. Since multiple-output minimization and reduction algorithms often fail to identify such simple relations, we may wish to check for their existence with column testing programs, i.e., the SUBCOL operator of Appendix 9.1, or similar but more general operators. Since making two output variables equal or complementary may require assigning don't cares of both, we should check to be certain that the network cost is not increased by establishing the relation through don't care assignments. Single-output minimization of the functions with and without the relation establishing don't care assignments can give an estimate of the cost savings.

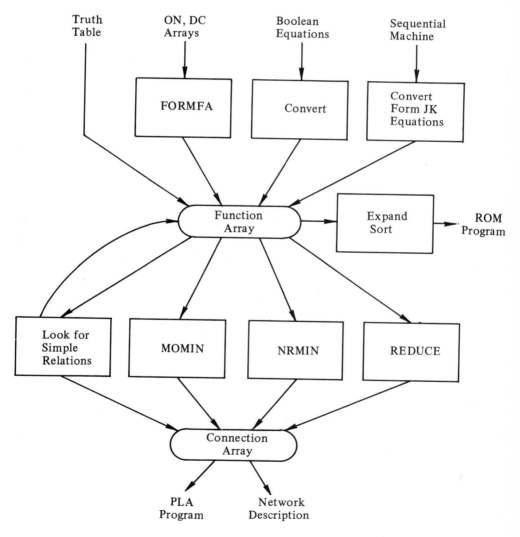

FIG. 10.24   Software Help for the Logic Designer

If we have found single-output covers of the output variables, these can be packed to a connection array for the multiple-output function. If we recognize and merge cubes with identical input parts, very respectable connection arrays can be obtained. If we demand better or the best connection arrays, then multiple-output minimization (MOMIN), non-redundant minimization (NRMIN), reduction (REDUCE), or variations of these can be applied. We use Algorithm 10.5 (subroutine EREDCN) to eliminate redundant connections in each case. The resulting connection array can be printed directly (PLA program), converted to Boolean equations (Prob. 9.7-6), or published as a list of gates and their interconnections.

Sometimes network cost can be further reduced by converting the network to three or more levels. The next chapter considers techniques for finding economical multiple-level networks.

REFERENCES

1. A. BARNA and D. I. PORAT, *Integrated Circuits in Digital Electronics*. New York: John Wiley and Sons, 1973.
2. T. C. BARTEE, "Computer Design of Multiple-output networks," *IRE Trans. on Electronic Computers*, Vol. EC-10, March 1961, pp. 21–30.
3. T. C. BARTEE, I. L. LEBOW, and I. S. REED, *Theory and Design of Digital Machines*. New York: McGraw-Hill Book Co., 1962.
4. H. FLEICHER and L. I. MAISSEL, "An Introduction to Array Logic," *IBM Jour. of Research and Development*, Vol. 19, March, 1975, pp. 98–109.
5. S. J. HONG, R. G. CAIN, and D. L. OSTAPKO, "MINI: A Heuristic Approach to Logic Minimizations," *IBM Jour. of Research and Development*, Vol. 18, Sept., 1974, pp. 443–458.
6. J. W. JONES, "Array Logic Macros," *IBM Jour. of Research and Development*, Vol. 19, March, 1975, pp. 120–126.
7. R. M. KARP, F. E. MCFARLIN, J. P. ROTH and J. R. WILTS, "A Computer Program for the Synthesis of Combinational Switching Circuits," *Proc. Second Annual AIEE Symposium on Switching Circuit Theory and Logical Design*, Oct. 1961, pp. 182–194.
8. J. C. LOGUE, et al., "Hardware Implementation of a Small System in Programmable Logic Arrays," *IBM Jour. of Research and Development*, Vol. 19, March, 1975, pp. 110–119.
9. E. J. MCCLUSKEY, "Minimization of Boolean Functions," *Bell System Tech. Journal*, Vol. 35, No. 6, Nov. 1956, pp. 1417–1444.
10. E. J. MCCLUSKEY, Jr. and T. C. BARTEE, et al, *A Survey of Switching Circuit Theory*. New York: McGraw-Hill, 1962.
11. R. E. MILLER, *Switching Theory Volume I: Combinational Circuits*. New York: John Wiley and Sons, 1965.
12. D. L. OSTAPKO and S. J. HONG, "Generating Test Examples for Heuristic Boolean Minimization," *IBM Jour. of Research and Development*, Vol. 18, Sept., 1974, pp. 459–464.
13. S. R. PETRICK, "A Direct Determination of the Irredundant Forms of a Boolean Function from the Set of Prime Implicants," Air Force Cambridge Research Center, Cambridge, Mass., *Tech Report AFCRC-TR-56-110*, 1956.
14. W. V. QUINE, "The Problem of Simplifying Truth Functions," *Am. Math.*

Monthly, Vol. 59, 1952, pp. 521–531.

15. W. V. QUINE, "A Way to Simplify Truth Functions," *Am. Math. Monthly*, Vol. 62, 1955, pp. 627–631.

16. J. P. ROTH, and E. G. WAGNER, "Algebraic Topological Methods for the Synthesis of Switching Systems, Part III. Minimization of Nonsingular Boolean Trees," *IBM Journal*, Vol. 4, No. 4, Oct. 1959, pp. 326–345.

17. J. P. ROTH, "Minimization over Boolean Trees," *IBM Journal*, Vol. 4, No. 5, Nov. 1960, pp. 543–558.

18. J. P. ROTH and R. M. KARP, "Minimization over Boolean Graphs," *IBM Journal*, Vol. 6, No. 2, April 1962, pp. 227–238.

19. Y-H SU and D. L. DIETMEYER, "Computer Reduction of Two-Level, Multiple Output Switching Circuits," *IEEE Trans. on Computers*, Vol. C-18, Jan. 1969, pp. 58–63.

20. P. TISON, "Generalization of Consensus Theory and Application to the Minimization of Boolean Functions," *IEEE Trans. on Electronic Computers*, Vol. EC-16, Aug. 1967, pp. 446–456.

21. P. WEINER and T. F. DWYER, "Discussion of Some Flaws in the Classical Theory of Two-Level Minimization of Multiple-Output Switching Networks," *IEEE Trans. on Computers*, Vol. C-17, Feb. 1968, pp. 184–186.

PROBLEMS

10.1-1. Apply the axioms and theorems of Boolean algebra to simplify each of the following expressions.

(a) $x_1\bar{x}_2x_3 \vee x_1\bar{x}_2 \vee x_1\bar{x}_2x_4$

(b) $x_1x_2 \vee x_1\bar{x}_3 \vee \bar{x}_1x_2x_4 \vee \bar{x}_2x_3x_4$

(c) $x_1x_2x_3 \vee x_1x_2x_4 \vee x_1x_2\bar{x}_3\bar{x}_4$

10.1-2. Use Theorem 10 to find all implicants of the following functions. Indicate which implicants are prime implicants.

(a) $F_1 = \vee\ 0, 1, 4, 7$

(b) $F_2 = \vee\ 1, 2, 3, 5, 7$

(c) $F_3 = \vee\ 0, 1, 5, 7, 10, 14, 15$

(d) $F_4 = \vee\ 0, 2, 8, 9, 10, 12, 13$

10.1-3. For each of the functions in Prob. 10.1-2, (a) write an ON-array and use it to find prime implicants; (b) draw the topology and recognize all prime implicants; and (c) draw the Karnaugh map and recognize all prime implicants.

10.1-4. Identify the extremals of the functions of Problem 10.1-2.

10.1-5. Define a *nonredundant* cover of a function as a sum-of-prime implicant expression of the function in which no prime implicant is covered by a sum of other prime implicants. For example, in the sum-of-prime implicants $A\bar{B} \vee AC \vee BC$ while AC is not covered by either $A\bar{B}$ or BC alone, it is covered by their sum $A\bar{B} \vee BC$ and can therefore be removed from the sum.

(a) Find a minimal sum-of-products cover of each function below.

(b) Find all nonredundant covers of each function.

$$F_1 = \vee\ 0, 2, 3, 4, 5, 7 \quad \text{(2 minimum, 5 nonredundant)}$$
$$F_2 = \vee\ 0, 1, 3, 7, 8, 12, 14, 15$$

10.1-6. Use the Karnaugh map to find a minimal sum-of-products expression of each of the following functions. Use the same maps to find minimal product-of-sums expressions.

(a) $F_1 = \bar{A}C\bar{D} \vee \bar{A}B\bar{D} \vee \bar{A}B\bar{C} \vee B\bar{C}D \vee A\bar{B}D$

(b) $F_2 = \bar{B}\bar{D} \vee AB\bar{C} \vee \bar{A}C\bar{D} \vee \bar{A}BCD \vee BC\bar{D}$

(c) $F_3 = \bar{A}\bar{B}CDE \vee \bar{A}B\bar{E} \vee AB\bar{D}E \vee A\bar{B}CE \vee \bar{A}\bar{B}DE \vee A\bar{B}\bar{C}\bar{D}E$

10.1-7. Use the Quine-McCluskey algorithm to find a minimal expression of the following functions.

(a) $F_1 = \vee$ 0, 2, 4, 6, 7, 8, 10, 12, 13, 14

(b) $F_2 = \vee$ 7, 9, 10, 12, 14, 17, 19, 21, 23, 24, 28

(c) $F_3 = \vee$ 8, 9, 10, 11, 13, 15, 29, 33, 35, 41, 42, 43, 47, 49, 51, 57, 59

(d) $F_4 = \vee$ 0, 1, 3, 4, 6, 7, 14, 15, 22, 30, 31, 47, 55, 63, 94, 95, 127, 128, 255, 257, 311, 384, 385, 439, 503, 511

10.1-8. Select the prime implicants of a minimal cover from each primˈ cant table. The topology of each function is interesting.

F_1	0	1	3	7	15	16	18	19	23	31	F_2	0	1	3	5	7	15	19	21	31
$0000x$	x	x									$0000x$	x	x							
$000x1$	x	x									$x0011$			x				x		
$x0x11$			x	x				x	x		$x0101$				x				x	
$xx111$				x	x				x	x	$0x111$					x	x			
$1001x$						x	x				$x1111$						x			x
$100x0$						x	x				$00xx1$	x	x	x	x					
$x0000$	x					x														

10.1-9. Find a minimal sum-of-products and a minimal product-of-sums expression of each function.

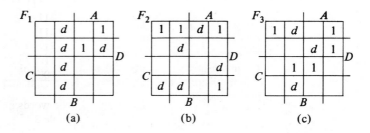

(a)   (b)   (c)

10.1-10. Use the Quine-McCluskey algorithm to obtain a minimal sum-of-products expression of the following functions with don't cares.

(a) $F_1 = \vee$ 0, 1, 2, 3, 4, 6, 8, 9, 12 $\vee$ $\vee_d$ 11, 13

(b) $F_2 = \vee$ 0, 1, 3, 5, 7, 15, 19, 21 $\vee$ $\vee_d$ 31

(c) $F_3 = \vee$ 1, 4, 6, 14, 20, 28, 30, 60, 61 $\vee$ $\vee_d$ 0, 3, 7, 33, 35, 62

10.1-11. The ON and DC arrays as given below are somewhat compressed. Expand each to a minterm array and then use the Q-M algorithm to find a minimal sum-of-products expression of each function. Write the minimal expression as an array.

$$\left.\begin{array}{c} 00xx \\ 01x0 \\ 100x \\ 1x00 \end{array}\right\} \qquad \left.\begin{array}{c} 10x1x0 \\ 001xx0 \\ 10x0x0 \\ 00x1x0 \\ 00xx1x \end{array}\right\}$$

$$\text{ON}_1 \qquad\qquad\quad \text{ON}_2$$

$$\left.\begin{array}{c} 1011 \\ 1101 \end{array}\right\} \qquad\qquad \{xx000x\}$$

$$\text{DC}_1 \qquad\qquad\qquad \text{DC}_2$$

10.1-12. A circuit is required to accept an 8421 code word and generate the 8421 representation of the 9's complement of the encoded input digit, e.g., with an input 4-tuple 0000 the output 4-tuple should be 1001.

(a) Write the truth table for the required switching function.

(b) Find a minimal sum-of-products expression of each individual output variable.

(c) Find a minimal 2-level, multiple output network.

In parts (b) and (c) minimize (i) gate count and (ii) diode count.

10.1-13. A 5-bit creeping (or walking) code for the decimal digits is given below. A minimal network is required which accepts words of this code and represents an output of 1 if the input digit, D, satisfies the requirement (a) $D > 5$; (b) $D \geq 5$; (c) $D = 5$; (d) $3 \leq D \leq 7$; (e) $D_1 \leq D \leq D_2$. (You pick D_1 and D_2.)

Digit	Creeping Code
0	00000
1	00001
2	00011
3	00111
4	01111
5	11111
6	11110
7	11100
8	11000
9	10000

10.1-14. Design a network that accepts 8421 code words and generates the corresponding Hamming code words of Table 3.11.

10.1-15. Design a network that accepts 7-bit code words, views them as Hamming code words of Table 3.11, and generates the following output variables: $z_1 = 1$ if no single errors are detected; $z_1 = 0$ otherwise. z_2, z_3, z_4 encode in binary the bit position found to be in error when $z_1 = 0$. When $z_1 = 1$ (no single error), z_2, z_3, and z_4 may have any value.

(a) Design using your familiarity with the Hamming code and EX-OR gate.

(b) Design separate minimal, 2-level networks, taking advantage of don't cares, for z_1, z_2, z_3, and z_4.

(c) Attempt to share AND gates in your design of part (b).

10.2-1. Two completely specified Boolean functions are specified by arrays below. Find the prime implicants of each by (a) the Quine-McCluskey algorithm, (b) the sharp algorithm, (c) the iterative consensus algorithm, and (d) the generalized consensus algorithm. In each case carefully count *all* of the cube comparisons you make. Compare the counts. (Do not overlook those comparisons for which the results are "obvious" and on which you spend very little time and perhaps take no action.)

$$\begin{pmatrix} x111 \\ 10x1 \\ 0100 \\ x001 \\ 11x1 \end{pmatrix} \qquad \begin{pmatrix} 1x11 \\ 0x0x \\ 1x01 \\ x100 \\ 100x \end{pmatrix}$$

10.2-2. Reexamine your work of Prob. 10.2-1 to find the algorithm(s) in which the order of cubes in the original cover and intermediate arrays greatly affects the amount of computation. If order is important for an algorithm, reorder the original arrays so that computation will be minimized. Can you predict the optimum order in advance of executing the algorithm? If order is not of great importance, develop a formula that measures the computation an algorithm entails.

10.2-3. Write and run FORTRAN programs that implement the sharp algorithm, iterative consensus, and the generalized consensus algorithms. Use arrays substantially larger than those of Prob. 10.2-1. Measure the time each program requires to compute a prime implicant array. Do the running times correlate with the results of Prob. 10.2-1?

10.2-4. (a) Given a cover of a completely specified Boolean function and a vertex v of that function, write an algorithm that will produce only all prime implicants that cover v.

(b) How might this algorithm be incorporated in a minimization program?

(c) For what sorts of Boolean functions would this minimization algorithm be more efficient of either computer time or memory than the extraction algorithm?

10.2-5. If the extremal test of Eq. (10.18) is to be a part of a minimization algorithm, the DC-array of a function must be retained and maintained.

(a) Write an extremal test which utilizes the ON-array of a function rather than the DC-array.

(b) How must the ON-array be maintained as extremals are found and selected for inclusion in the minimal cover of a function?

10.2-6. Find all nonredundant prime implicant covers of the following functions.

$$\lor\ 0, 2, 3, 4, 5, 7, 8, 9, 13$$
$$\lor\ 0, 2, 3, 4, 5, 7, 8, 10, 14, 15$$

10.2-7. In order to design reliable asynchronous sequential circuits (Chapter 12) prime implicants must be selected so that each 1-cube of a function is covered.

(a) Develop an extraction-like algorithm that will select a minimum cost cover to satisfy this additional requirement.

(b) Program, run, and test the algorithm on the functions of Prob. 10.2-6 as well as $\lor\ 0, 1, 4, 5, 6, 7, 9, 11, 13, 15$.

10.2-8. $f = \vee\ 3, 5, 6, 7, 9, 10, 11, 12, 13, 14, 15, 17, 18, 19, 20, 21, 22, 23, 24, 25,$
26, 27, 28, 29, 30
(a) Find the complex of f.
(b) Find the set of all prime implicants of f.
(c) Find a minimum diode cost cover of f.
(d) Estimate the number of equally minimum cost covers of f.
(e) Rigorously execute the extraction algorithm to find a minimum cost cover. How many levels of branching are encountered? Does the cube upon which branching is based influence the number of subsequent branches which must be examined?

10.2-9. Write a FORTRAN main program that calls subroutine XTRACT of Appendix 10.2 and prints the given and resulting arrays. Test your program with the arrays of Prob. 10.2-1 and more extensive arrays of 5 to 10 variables.

10.2-10. The following algorithm is effective on some single-output, complete switching functions.
Given a cover C of the function:

(1) For each $c^i \in C$,
    For each $c_j^i, j = 1, 2, \ldots, n$
      If $c_j^i = x$, advance to the next value of j
      If $c_j^i \in \{0, 1\}$ replace c_j^i with x forming cube c^*
      If $c^* \sqsubseteq C$, replace c^i with c^*, and then advance j.
    (Each $c^i \in C$ is replaced with a prime implicant.)

(2) Eliminate redundant prime implicants, if any.

(a) Exercise this algorithm faithfully on the arrays of Prob. 10.2-1 and the functions of Prob. 10.2-6. Carry out all steps in the prescribed order. Note the paper required to arrive at each solution. For what type of cover is this algorithm best suited?
(b) Program this algorithm and compare its execution times with those of the XTRACT program.

10.3-1. Multiple-output function f is described by:

$$z_1 = \vee\ 1, 6, 7, 9, 11, 14, 15$$
$$z_2 = \vee\ 4, 5, 6, 7, 10, 12, 13, 14, 15$$
$$z_3 = \vee\ 1, 4, 5, 9, 10, 11, 12, 13, 14, 15$$

(a) Minimize each of the output variables individually. Evaluate the total gate and diode cost of the three networks.
(b) What AND gates, if any, can be shared to reduce the total network gate and diode cost?
(c) Write a compressed function array for f.
(d) Find a minimum gate cost realization of f.
(e) Find a minimum diode cost realization of f.

10.3-2. Find minimum gate two-level networks that realize the following function arrays. Cubes with all-0 output parts have been deleted from the function arrays. Compute network gate and diode costs.

$$\left\{\begin{array}{ll} 0000 & d0dd \\ 0011 & d0dd \\ 0101 & 0dd0 \\ 0110 & 1110 \\ 1001 & 01dd \\ 1010 & dd00 \\ 1011 & d001 \\ 1100 & 0dd1 \\ 1101 & dd10 \\ 1110 & 10d0 \\ 1111 & 11dd \end{array}\right\} \quad \left\{\begin{array}{ll} 0001 & 1111 \\ 0010 & 1000 \\ 0011 & 000d \\ 0100 & 0001 \\ 0110 & 1100 \\ 0111 & 1dd0 \\ 1000 & 1d11 \\ 1001 & 1000 \\ 1010 & 0101 \\ 1011 & 0101 \\ 1100 & 0d11 \\ 1101 & 0011 \\ 1111 & 0010 \end{array}\right\}$$

10.3-3. (a) Detail an algorithm of the generalized consensus philosophy that will generate all multiple-output prime implicants of a given function array. The algorithm should be efficient with respect to both memory required and execution time.

(b) Program your algorithm and test using the arrays of previous problems.

10.3-4. The ON-array described by Eq. (10.21) is particularly inefficient of storage. Could two entries such as 0000∘1000 and 0000∘0100 be replaced with a single entry such as 0000∘1100 or 0000∘xx00? If yes, demonstrate that the extraction algorithm will be successful by reworking the example problem of Sec. 10.3 with a compressed ON-array. If no, why not?

10.3-5. Rather than retaining the ON-array of Eq. (10.21) could the don't care information be retained and utilized in an extraction algorithm? If yes, specify an extraction algorithm which will utilize the don't care information and the form in which that information is to be retained. If no, why not?

10.3-6. How might Algorithm 10.5 be altered to give lower diode cost networks?

10.3-7. Find nonredundant covers of the multiple-output functions of Probs. 10.3-1 and 10.3-2. Compute network gate and diode costs and compare with the results of these problems.

10.3-8. A random number generator RANDOM (A) is available. Each time this subroutine is called it assigns a new value to A in the range $0 \leq A < 1$.

(a) Use this subroutine in a FORTRAN program which will generate a non-trivial n-input, m-output function array. An array is to be rejected if any column does not contain at least one 0 and one 1.

(b) Generate several function arrays of modest size and minimize them by hand and with computer programs.

(c) Will your program be "successful" if $n + m > 15$? If not, in what way(s) will it fail, or be inefficient? How might you alter your program to overcome these difficulties?

10.3-9. A comment card in Appendix 9.4 claims that each ON_i-array (read into array A) is minimized. (a) In what sense is each ON_i-array minimized? (b) Why is it desirable to process each ON_i-array in this manner? Why aren't the DC_i-and OFF_i-arrays processed in the same way?

10.3-10. (a) Write a FORTRAN main program that calls subroutines FORMFA (Appendix 9.4), MOMIN (Appendix 10.3), and then EREDCN (Appendix 10.4). Use your program to solve Prob. 10.3-2. Print the resulting arrays.
(b) Substitute NRMIN (Appendix 10.5) for MOMIN. Solve Prob. 10.3-7.

10.4-1. (a) Reduce the arrays of Prob. 10.3-2 with Algorithms 10.6 and 10.5. Compare your results with those of Probs. 10.3-2 and 10.3-7.
(b) Repeat (a), selecting masks from the output parts of cubes of the function array.

10.4-2. Alter the program of Appendix 10.6 to select masks from the output parts of the function array. Run your revised program and compare its execution times with those of the original program.

10.4-3. (a) Convert the function array below to a minimum diode cost connection array.
(b) Show how to realize the function with an $n = m = p = 4$ PLA. Draw a logic diagram of the programmed PLA. (Maps may be helpful.)

$$F = \begin{pmatrix} 0 & 0 & 0 & & 0 & 0 & 1 \\ 0 & 0 & 1 & & 1 & 0 & 0 \\ 0 & 1 & 0 & & 0 & 1 & 0 \\ 0 & 1 & 1 & & 1 & 1 & 0 \\ 1 & 0 & 0 & & 1 & 1 & 1 \\ 1 & 0 & 1 & & 1 & 1 & 0 \\ 1 & 1 & 0 & & 0 & 0 & 1 \\ 1 & 1 & 1 & & 0 & 0 & 0 \end{pmatrix}$$

10.5-1. Find the minimum literal count, sum-of-products expression of each of the following switching functions in which α, β, etc., denote constrained don't cares.

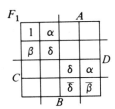

10.5-2. Calculate a prime implicant array PI* and a corrected prime implicant array PI for the functions of Prob. 10.5-1.

10.5-3. Write a formal algorithm for correcting a PI*-array to form either a compressed or uncompressed PI-array.

10.5-4. Use the extraction algorithm to select a minimal cover from the PI-arrays of Prob. 10.5-2. If branching is required, use a rule of thumb to pick a "desirable" cube. Compare results with those of Prob. 10.5-1.

10.5-5. (a) A multiple-output network is to realize the function array given below and take the form of Fig. P10.5-5(a). Draw Karnaugh maps for signals y_1, y_2, and y_3 recognizing constrained don't cares. Complete the design of the network so that y_1, y_2, and y_3 are realized with 2-level networks and the overall network involves a minimum number of gates.

A	B	C	D	z_1	z_2
0	0	0	x	0	d
0	x	1	1	d	0
x	0	1	0	0	0
x	1	1	0	1	1
1	x	x	1	1	1
1	x	0	0	0	0
0	1	0	0	1	0
0	1	0	1	0	1

Fig. P10.5-5 (a)

(b) Repeat for the functions in Figs. P10.5-5(a) and (b) and the assumed form in Fig. P10.5-5(b). No restrictions are placed on the forms of the y_1, y_2, and y_3 networks.

A	B	C	D	E	z_1	z_2
1	x	x	x	x	1	1
x	1	x	x	x	1	1
x	x	1	x	x	1	1
0	0	0	1	1	1	0
0	0	0	0	0	0	1
all others					0	0

Fig. P10.5-5 (b)

10.5-6. The Quine-McCluskey algorithm and Prime Implicant Table can be generalized to handle constrained don't cares. Examine these algorithms and determine what alterations must be made. Test your altered algorithms on the functions of Prob. 10.5-1.

10.5-7. For the following function array, in which the δ's and ∂ are the types of constraints discussed in the text, find the don't care and parameter assignments which lead to a minimum: (a) z_1 network, (b) z_2 network, (c) overall network.
In all cases determine the cost of the overall network.

A	B	C	D	z_1	z_2
x	0	0	0	1	1
0	0	1	x	1	0
x	0	1	1	1	0
x	1	1	x	d	d
1	0	1	0	δ type constraint	
x	1	0	x	∂ type constraint on each minterm	

ANSWERS

10.1-3.

$$K(F_1)=\begin{Bmatrix}000\\001\\100\\111\\00x\\x00\end{Bmatrix} \qquad PI(F_2)=\begin{Bmatrix}00x\\x00\\111\end{Bmatrix}$$

$$K(F_2)=\begin{Bmatrix}001\\010\\011\\101\\111\\0x1\\x01\\01x\\x11\\1x1\\xx1\end{Bmatrix} \qquad PI(F_2)=\begin{Bmatrix}xx1\\01x\end{Bmatrix}$$

10.1-5.
$$\begin{aligned}
F_1 &= \bar{A}\bar{C} \lor BC \lor A\bar{B} \\
&= \bar{A}B \lor AC \lor \bar{B}\bar{C}
\end{aligned} \Bigg\}\text{minimum}$$
$$\left.\begin{aligned}
&= \bar{A}\bar{C} \lor \bar{B}\bar{C} \lor AC \lor BC \\
&= \bar{A}\bar{C} \lor \bar{A}B \lor A\bar{B} \lor AC \\
&= \bar{B}\bar{C} \lor A\bar{B} \lor \bar{A}B \cdot \lor BC
\end{aligned}\right\}\text{nonredundant}$$

10.1-6.
$$\begin{aligned}
F_1 &= \bar{A}\bar{D} \lor B\bar{C}D \lor A\bar{B}D \\
&= (\bar{A} \lor D)(\bar{B} \lor \bar{C} \lor \bar{D})(A \lor B \lor \bar{D})
\end{aligned}$$

10.1-7.

$$K^0(F_1) = \left\{ \begin{array}{l} 0000 \\ 0010 \\ 0100 \\ 0110 \\ 0111 \\ 1000 \\ 1010 \\ 1100 \\ 1101 \\ 1110 \end{array} \right\}$$

$0000\ \checkmark$	$00x0\ \checkmark$	$0xx0\ \checkmark$	$xxx0\ \gamma$
$\overline{0010}\ \checkmark$	$0x00\ \checkmark$	$x0x0\ \checkmark$	
$0100\ \checkmark$	$x000\ \checkmark$	$0xx0\ \checkmark$	
$1000\ \checkmark$	$\overline{0x10}\ \checkmark$	$xx00\ \checkmark$	
$\overline{0110}\ \checkmark$	$x010\ \checkmark$	$\overline{xx10}\ \checkmark$	
$1010\ \checkmark$	$01x0\ \checkmark$	$x1x0\ \checkmark$	
$1100\ \checkmark$	$x100\ \checkmark$	$1xx0\ \checkmark$	
$\overline{0111}\ \checkmark$	$10x0\ \checkmark$		
$1101\ \checkmark$	$1x00\ \checkmark$		
$1110\ \checkmark$	$\overline{011x}\ \alpha$		
	$x110\ \checkmark$		
	$1x10\ \checkmark$		
	$110x\ \beta$		
	$11x0\ \checkmark$		

	0	2	4	6	7	8	10	12	13	14	
α					x	x					
β								x	x		
γ	x	x	x	x		x	x	x		x	All are extremals.

$$F = \alpha \lor \beta \lor \gamma = \bar{A}BC \lor AB\bar{C} \lor \bar{D}$$

10.1-8. $F_1 = xx111 \lor x0x11 \lor 0000x \lor 100x0$

$$= CDE \lor \bar{B}DE \lor \bar{A}\bar{B}\bar{C}\bar{D} \lor A\bar{B}\bar{C}\bar{E}$$

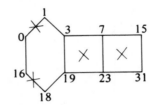

10.1-9. $F_1 = A\bar{B}\bar{C} \lor A\bar{C}D$ (or $B\bar{C}D$)

$$F_1 = A\bar{C}(\bar{B} \lor D)$$

10.1-11.

$$ON_1 = \left\{ \begin{array}{l} 0000 \\ 0001 \\ 0010 \\ 0011 \\ 0100 \\ 0110 \\ 1000 \\ 1001 \\ 1100 \end{array} \right\}$$

	$0\ \checkmark$	$0,1\ \checkmark$	$0,3$
	$\overline{1}\ \checkmark$	$0,2\ \checkmark$	$0,9$
	$2\ \checkmark$	$0,4\ \checkmark$	$0,6$
	$4\ \checkmark$	$0,8\ \checkmark$	$0,12$
	$8\ \checkmark$	$\overline{1,2}\ \checkmark$	$\overline{1,10}$
	$\overline{3}\ \checkmark$	$1,8\ \checkmark$	$8,5$
	$6\ \checkmark$	$2,1\ \checkmark$	
	$9\ \checkmark$	$2,4\ \checkmark$	
	$12\ \checkmark$	$4,2\ \checkmark$	
DC	$\overline{11}\ \checkmark$	$4,8\ \checkmark$	
	$13\ \checkmark$	$8,1\ \checkmark$	
		$8,4\ \checkmark$	
		$\overline{3,8}\ \checkmark$	
		$9,2\ \checkmark$	
		$9,4\ \checkmark$	

$$F = 0,6 \ \text{v} \ 1,10 \ \text{v} \ 0,12$$
$$= 0xx0 \ \text{v} \ x0x1 \ \text{v} \ xx00$$

$$\begin{pmatrix} 0xx0 \\ x0x1 \\ xx00 \end{pmatrix}$$

	0	1	2	3	4	6	8	9	12	
0, 3	x	x	x	x						LT(1, 10)
0, 9	x	x					x	x		LT
0, 6	x				x	x				E
0, 12	x				x		x		x	←Pick either
1, 10		x			x			x		E
8, 5							x	x	x	

10.1-13. Let $b_1b_2b_3b_4b_5$ be the bits of a creeping code word.
  (a) $z = b_1\bar{b}_5$
  (b) $z = b_1$
  (c) $z = b_1b_5$

10.2-1. (c)

$$x111$$
$$\cancel{10x1}$$
$$0100$$
$$x001$$
$$\cancel{11x1}$$

$$\cancel{1x11}$$
$$\cancel{1111}$$
$$\cancel{1001}$$
$$1xx1$$

10.2-2. Order is important for iterative consensus.
10.2-5. $e = (\pi^i \ \# \ (\text{PI} - \pi^i)) \bigcap \text{ON}$
  $\text{ON} \leftarrow \text{ON} \ \# \ e$, if e is an extremal.

10.2-8.

$x_1 = 0$

		x_2			
			1		
	1	1	1		x_5
x_4	1	1	1	1	
	1	1	1		
		x_3			

$x_1 = 1$

		x_2			
		1	1	1	
	1	1	1	1	x_5
x_4	1	1		1	
	1	1	1	1	
		x_3			

10.3-1. (a)

	Gates	Diodes	
z_1	4	11	Share at least ACD, $\bar{B}\bar{C}D$
z_2	2	5	
z_3	4	9	
	10	25	

10.3-4. No, unless some new scheme is developed for updating the ON-array as extremals are selected.

10.3-6. EREDCN does a fast nonredundant selection of the available prime implicants. A modified extraction algorithm could be used.

10.4-3. (a) $\begin{cases} 10x & 110 \\ 0x1 & 100 \\ 01x & 010 \\ 1x0 & 001 \\ x00 & 001 \end{cases}$

(b) $z_3 = \overline{C} \cdot (\overline{\overline{A}B})$
$\quad = A\overline{C} \vee \overline{C}z_2$

10.5-1. $F_1 = \overline{A}\,\overline{C} \vee ACD$

10.5-5. $y_1 = A \vee \overline{D},\ y_3 = C \vee D$

APPENDIX 10.1

A program to find extremals, or a nonredundant cover.

```
      INTEGER ON(100), DC(100), E(100)
      INTEGER A(100), B(100), C(100), CUBE(5)
      MAX = 100
      READ 1, N
    1 FORMAT(1X,3I2)
C
C     SET EXTREMAL ARRAY EMPTY, READ ON AND DC-ARRAYS
C
      CALL DEFINA(E, 0, 0)                          E = ϕ
      CALL READAR(ON, N, MAX)
      CALL READAR(DC, N, MAX)
C
C     FORM PRIME IMPLICANTS OF ON U DC
C
      CALL COMBIN(ON, DC, MAX)                      ON ← ON ∪ DC
      CALL CNSSUS(ON, A, B, MAX, N)                 ON ← *(ON)
      NPI = NCUBES(ON)
C
C     EXAMINE EACH PRIME IMPLICANT
C
      DO 3 I = 1,NPI
      CALL MOVCUB(ON, CUBE, 1)                      CUBE = πⁱ,
      CALL SHARPS(CUBE, ON, A, B, MAX, NCA)         ON ← ON − πⁱ
      IF(NCA .EQ. 0) GO TO 2                        A = πⁱ # (ON)
      CALL SHARPS(A, DC, B, C, MAX, NCB)            B = A # DC
      IF(NCB .EQ. 0) GO TO 2
C
C     PRIME IMPLICANT IS AN EXTREMAL
C
      CALL COMBIN(E, CUBE, MAX)                     E ← E ∪ πⁱ
      CALL COMBIN(DC, CUBE, MAX)                    DC ← DC ∪ πⁱ
      GO TO 3
C
C     PRIME IMPLICANT IS NOT AN EXTREMAL
C
    2 CALL COMBIN(ON, CUBE, MAX) ←                  ON ← ON ∪ πⁱ
    3 CONTINUE                                      Replace with
C                                                   2 CONTINUE
C     PRINT RESULTS                                 to obtain a
C                                                   nonredundant cover
```

```
      PRINT 4
  4   FORMAT(10H0EXTREMALS)
      CALL PRNTAR(E, N)
      CALL EXIT
      END
```

APPENDIX 10.2

A program based upon Algorithm 10.4: The EXTRACTION Algorithm. Note that this program is written as a subroutine which accepts an ON and DC-array and values of MAX and N. The minimum cover is returned as array E. All arrays are assumed to be of dimension MAX. This subroutine picks the lowest cost cube rather than branching when a cyclic structure is encountered.

```
      SUBROUTINE XTRACT(ON, DC, E, A, B, C, MAX, N)
      INTEGER MASK(5), CUBE(5)
      CALL DEFINA(E, 0, 0)
      CALL COMBIN(DC, ON, MAX)
      CALL CNSSUS(DC, A, B, MAX, N)
C
C     EXTREMAL TEST
C
  1   NPI = NCUBES(DC)
      DO 3 I = 1,NPI
      CALL MOVCUB(DC, CUBE, 1)
      CALL INTERX(ON, CUBE, A, MAX, NCA)
      IF(NCA .EQ. 0) GO TO 3
      CALL SHARPS(A, DC, B, C, MAX, NCB)
      IF(NCB .EQ. 0) GO TO 2
      CALL COMBIN(E, CUBE, MAX)
      CALL SHARPS(ON, CUBE, A, B, MAX, NCA)
      CALL COPYAR(ON, A, MAX)
      IF(NCA) 10, 10, 3
  2   CALL COMBIN(DC, CUBE, MAX)
  3   CONTINUE
C
C     REMOVE LESS–THAN CUBES
C
  4   NPI = NCUBES(DC)
      K = NPI − 1
      DO 7 I = 1,NPI
      CALL MOVCUB(DC, CUBE, 1)
      CALL INTERX(ON, CUBE, A, MAX, NCA)
      IF(NCA .GT. 0) GO TO 5
      K = K − 1
      IF(K) 10, 1, 7
  5   ICOST = KUBCST(CUBE, N, 1)
      DO 6 J = 1,K
      JCOST = KUBCST(DC, N, J)
      IF(ICOST .LT. JCOST) GO TO 6
      CALL CPYCUB(DC, MASK, J)
      CALL SHARPS(A, MASK, B, C, MAX, NCB)
      IF(NCB .GT. 0) GO TO 6
      K = K − 1
      IF(K) 10, 1, 7
  6   CONTINUE
      CALL COMBIN(DC, CUBE, MAX)
  7   CONTINUE
      IF(K + 1 .NE. NPI) GO TO 1
```

```
C
C     PICK LOW COST CUBE
C
      LOW = 999
      DO 8 I = 1,NPI
      ICOST = KUBCST(DC, N, I)
      IF(LOW .LE. ICOST) GO TO 8
      LOW = ICOST
      J = I
    8 CONTINUE
      CALL MOVCUB(DC, CUBE, J)
      PRINT 9
    9 FORMAT('0XTRACT PICKED')
      CALL PRNTAR(CUBE, N)
      CALL COMBIN(E, CUBE, MAX)
      CALL SHARPS(ON, CUBE, A, B, MAX, NCA)
      CALL COPYAR(ON, A, MAX)
      IF(NCA .GT. 0) GO TO 4
   10 RETURN
      END
```

APPENDIX 10.3

A subroutine which minimizes multiple output switching functions via the Extraction Algorithm. The subroutine picks a desirable cube rather than branching when a cyclic structure is encountered.

```
      SUBROUTINE MOMIN
      IMPLICIT INTEGER(A–Z)
      COMMON N, M, P, NV, IOC, MX, NC, X(3)
      COMMON F(200), ON(200), E(200), A(200), B(200), C(200), D(200)
      COMMON CUBE(5), SAVE(200)
      CALL DEFINA(E, 0, 0)
      CALL COPYAR(SAVE, F, MX)
      CALL DEFINA(CUBE, NV, 'XXXXXX')
      DO 1 I = IOC,NV
    1 CALL CNGCOL(CUBE, I, '000')
      CALL SPLITA(F, A, CUBE, MX, NC)
      DO 2 I = IOC,NV
    2 CALL CNGCOL(F, I, '0XX')
      CALL CNSSUS(F, A, C, MX, N)

C
C     EXTREMAL TEST
C
    3 NPI = NCUBES(F)
      DO 5 I = 1,NPI
      CALL MOVCUB(F, CUBE, 1)
      CALL INTERX(ON, CUBE, A, MX, NC)
      IF(NC .EQ. 0) GO TO 5
      CALL SHARPS(A, F, B, C, MX, NC)
      IF(NC .EQ. 0) GO TO 4
      CALL COMBIN(E, CUBE, MX)
      CALL SHARPS(ON, CUBE, A, C, MX, NC)
      CALL COPYAR(ON, A, MX)
      IF(NC) 13, 13, 5
    4 CALL COMBIN(F, CUBE, MX)
    5 CONTINUE
```

```
C
C      LESS–THAN TEST
C
     6  NPI = NCUBES(F)
        K = NPI − 1
        DO 10 I = 1,NPI
        CALL MOVCUB(F, CUBE, 1)
        CALL INTERX(ON, CUBE, A, MX, NC)
        IF(NC .GT. 0) GO TO 8
     7  K = K − 1
        IF(K) 13, 3, 10
     8  ICOST = KUBCST(CUBE, N, 1)
        DO 9 J = 1,K
        IF(ICOST .LT. KUBCST(F, N, J)) GO TO 9
        CALL CPYCUB(F, D, J)
        CALL SHARPS(A, D, C, B, MX, NC)
        IF(NC .EQ. 0) GO TO 7
     9  CONTINUE
        CALL COMBIN(F, CUBE, MX)
    10  CONTINUE
        IF(K .LT. NPI−1) GO TO 3
C
C      PICK BEST CUBE
C
        JCOST = 0
        DO 11 I = 1,NPI
        CALL CPYCUB(F, CUBE, I)
        CALL INTERX(ON, CUBE, A, MX, NC)
        IF(NC .LT. JCOST) GO TO 11
        JCOST = NC
        J = I
    11  CONTINUE
        CALL MOVCUB(F, CUBE, J)
        PRINT 12
    12  FORMAT ('0MOMIN PICKED')
        CALL PRNTAR(CUBE, N)
        CALL COMBIN(E, CUBE, MX)
        CALL SHARPS(ON, CUBE, A, B, MX, NC)
        CALL COPYAR(ON, A, MX)
        IF(NC .GT. 0) GO TO 6
C
C      CHANGE OUTPUT CODING
C
    13  DO 14 I = IOC,NV
    14  CALL CNGCOL(E, I, '011')
        CALL COPYAR(F, SAVE, MX)
        RETURN
        END
```

APPENDIX 10.4

A subroutine that eliminates redundancy in a connection array in the manner of Algorithm 10.5.

```
        SUBROUTINE EREDCN
        IMPLICIT INTEGER(A–Z)
        COMMON N, M, P, NV, IOC, MX, NC, X(3)
        COMMON F(200), ON(200), E(200), A(200), B(200), C(200), D(200)
        COMMON CUBE(5), SAVE(200)
        CALL DEFINA(CUBE, NV, '1XXXXX')
        CALL PRMCOL(CUBE, 1, N)
```

```
C
C          EXAMINE EACH OUTPUT VARIABLE
C
           DO 4 I = IOC, NV
           CALL PRMCOL(CUBE, I–1, I)
           CALL SPLITA(E, A, CUBE, MX, NPI)
           IF(NPI.EQ. 0) GO TO 4
           CALL COPYAR(B, A, MX)
           CALL SPLITA(F, ON, CUBE, MX, NC)
           CALL COMBIN(F, ON, MX)
           DO 1 J = 1,M
           CALL DELCOL(ON, IOC)
        1  CALL DELCOL(B, IOC)
C
C          SELECT A NONREDUNDANT COVER
C
           DO 3 J = 1,NPI
           CALL MOVCUB(A, C, 1)
           CALL MOVCUB(B, C(10), 1)
           CALL INTERX(ON, C(10), D, MX, NC)
           CALL SHARPS(D, B, C(20), SAVE, MX, NC)
           IF(NC .GT. 0) GO TO 2
           CALL CNGCOL(C, I, '10X')
           CALL COMBIN(E, C, MX)
           GO TO 3
        2  CALL COMBIN(A, C, MX)
           CALL COMBIN(B, C(10), MX)
        3  CONTINUE
        4  CALL COMBIN(E, A, MX)
C
C          ELIMINATE USELESS CUBES
C
           DO 5 I = IOC,NV
        5  CALL CNGCOL(CUBE, I, '000')
           CALL SPLITA(E, A, CUBE, MX, NC)
           RETURN
           END
```

APPENDIX 10.5

A subroutine that selects a nonredundant, prime implicant cover of a multiple output switching function. Prime implicants are examined in order of increasing cost of their input parts.

```
           SUBROUTINE NRMIN
           IMPLICIT INTEGER(A–Z)
           COMMON N, M, P, NV, IOC, MX, NC, X(3)
           COMMON F(200), ON(200), E(200), A(200), B(200), C(200), D(200)
           COMMON CUBE(5), SAVE(200)
           CALL DEFINA(E, 0, 0)
           CALL COPYAR(SAVE, F, MX)
C
C          FORM PRIME IMPLICANTS
C
           CALL DEFINA(CUBE, NV, 'XXXXXX')
           DO 1 I = IOC,NV
        1  CALL CNGCOL(CUBE, I, '000')
           CALL SPLITA(F, A, CUBE, MX, NC)
           DO 2 I = IOC, NV
        2  CALL CNGCOL(F, I, '0XX')
           CALL CNSSUS(F, A, C, MX, N)
```

```
C
C       SELECT PRIME IMPLICANTS, EXAMINE CUBES IN
C       ORDER OF INCREASING COST
C
        NPI = NCUBES(F)
        DO 5 I = 1,N
        K = 0
        DO 4 J = 1,NPI
        IF(KUBCST(F, N, J–K) .GT. I) GO TO 4
        CALL MOVCUB(F, CUBE, J–K)
        CALL INTERX(ON, CUBE, A, MX, NC)
        IF(NC .EQ. 0) GO TO 3
        CALL SHARPS(ON, CUBE, B, C, MX, NC)
        CALL COPYAR(ON, B, MX)
        CALL COMBIN(E, CUBE, MX)
        IF(NC .EQ. 0) GO TO 6
      3 K = K + 1
      4 CONTINUE
      5 NPI = NPI – K
C
C       CHANGE OUPUT CODING, RESTORE FUNCTION ARRAY
C
      6 DO 7 I = IOC,NV
      7 CALL CNGCOL(E, I, '011')
        CALL COPYAR(F, SAVE, MX)
        RETURN
        END
```

APPENDIX 10.6

A subroutine that generates a reduced cost connection array from a function array via Algorithm 10.6.

```
        SUBROUTINE REDUCE
        IMPLICIT INTEGER(A–Z)
        COMMON N, M, P, NV, IOC, MX, NC, X(3)
        COMMON F(200), ON(200), E(200), A(200), B(200), C(200), D(200)
        COMMON CUBE(5), SAVE(200)
        COMMON G(200), H(200), MASK(5), ISEL(20)
        CALL DEFINA(E, 0, 0)
        CALL COPYAR(SAVE, F, MX)
C
C       SELECT LAMBDA SET
C
        DO 3 K = 1,M
        I = M + 1 – K
      1 L = 1
      2 CALL SELECT(ISEL, M, I, L)
        GO TO (1, 4, 3), L
      3 CONTINUE
        CALL COPYAR(F, SAVE, MX)
        RETURN
C
C       FIND COVER OF SUBARRAY
C
      4 CALL DEFINA(MASK, NV, 'XXXXXX')
        DO 5 J = 1,I
      5 CALL CNGCOL(MASK, N + ISEL(J), '111')
        CALL SPLITA(F, A, MASK, MX, NC)
        IF(NC .EQ. 0) GO TO 2
```

```
      CALL INTERX(F, MASK, B, MX, NC)
      CALL COPYAR(G, B, MX)
      DO 6 J = 1,I
      CALL CNGCOL(G, N + ISEL(J), 'XXX')
    6 CALL CNGCOL(A, N + ISEL(J), 'XXX')
      CALL COMBIN(F, A, MX)
      DO 7 J = 1,M
      CALL DELCOL(A, IOC)
    7 CALL DELCOL(B, IOC)
      CALL COMBIN(B, A, MX)
      CALL CNSSUS(B, C, D, MX, N)
      NCB = NCUBES(B)
      DO 8 J = 1,NCB
      CALL MOVCUB(B, CUBE, 1)
      CALL INTERX(A, CUBE, C, MX, NC)
      IF(NC .EQ. 0) GO TO 8
      CALL SHARPS(C, B, D, H, MX, NC)
      IF(NC .NE. 0) CALL COMBIN(B, CUBE, MX)
    8 CONTINUE
C
C     UPDATE FUNCTION AND CONNECTION ARRAY!
C
      DO 9 J = IOC,NV
      CALL OPNCOL(B, N, MX)
    9 CALL CNGCOL(MASK, J, '010')
      CALL INTERX(B, MASK, A, MX, NC)
      CALL COMBIN(E, A, MX)
      CALL INTERX(B, G, A, MX, NC)
      IF(NC .EQ. 0) GO TO 2
      CALL SHARPS(F, A, B, C, MX, NC)
      CALL COPYAR(F, B, MX)
      CALL COMBIN(F, A, MX)
      GO TO 2
      END
```

11

Multiple Level Synthesis

Algorithms for deriving low cost, multiple-level realizations of switching functions usually require considerable searching. Knowledge of the form of a function, redundancies in its expression, and its symmetries can often be used to reduce or eliminate the searching. After we look at these aspects of switching functions, we will examine factoring as a means of deriving multiple-level networks. This technique obtains its name from the idea upon which it is based—factoring terms in Boolean equations. Finally, we will explore the idea of decomposing a switching function into a composition of simpler functions.

11.1 FORM, SYMMETRY, AND REDUNDANCY

In Chapter 2, we saw that knowledge of the form of a switching function could be used to avoid repetitious optimization of the same network. Our familiarity with topology now permits us to examine forms more effectively. Functions can be classified in a number of ways. Two functions are said to be of the same form if one can be derived from the other after:

(1) complementing selected input variables, or
(2) (1) above and permuting selected input variables, or
(3) (2) above and complementing the output variable.

Optimum realizations for three variable functions have been derived and published for a number of different technologies. Tables of optimum networks are organized according to functional form; usually only input variable permutations are permitted. Let us find a smaller set of forms by using definition (2) above.

TABLE 11.1 ASSIGNMENTS

a	b	c
l_1	l_2	l_3
l_1	l_3	l_2
l_2	l_1	l_3
l_2	l_3	l_1
l_3	l_1	l_2
l_3	l_2	l_1

Let l_1, l_2, and l_3, be the three literals of Boolean functions of 3 variables. Let a be one of these three literals, b be another, and c be the third. Six assignments of l_1, l_2, and l_3 to a, b, and c exist, and are shown in Table 11.1.

We can find the forms of 3-variable functions in a number of ways. One way is based upon the fact that all functions of a given form are the sum of the same number of minterms. If we sum zero minterms we obtain the first functional form of Table 11.2. This form represents only one Boolean function and is not particularly interesting. If we sum one minterm, we obtain functions such as $\bar{x}_1 \cdot \bar{x}_2 \cdot \bar{x}_3$, $\bar{x}_1 \bar{x}_2 x_3$, etc., the minterms themselves. All of these functions are of the form $l_1 \cdot l_2 \cdot l_3$: we do not need a, b, and c as yet, but we will soon, so let us express this form as $a \cdot b \cdot c$.

Summing two minterms can lead to functions such as $x_1(x_2 x_3 \vee \bar{x}_2 \bar{x}_3)$, $\bar{x}_3(\bar{x}_1 x_2 + x_1 \bar{x}_2)$, etc. Here we must use the form expression $a(bc \vee \bar{b}\bar{c})$ rather than $l_1(l_2 l_3 + l_2 l_3)$ for any one of the three literals can appear outside of the parentheses. If we continue in this manner until all 8 minterms are summed, we have the data presented in Table 11.2.

Alternative expressions for many of the functional forms exist. Entries in Table 11.2 were selected to emphasize that many forms are the duals of others. Thus form 1 is the dual of form 22, and form 2 is the dual of form 21. This duality exists with the exception of those forms arising from four minterms: forms 9–14. All functions of forms 2 and 21 are of the same form under definition (3) above.

All functions of two variables are functions of three variables also, and the forms of functions of two variables are, therefore, included in Table 11.2 as entries 1, 5, 9, 14, 18, and 22.

Functions of form 10 have been given a special name—MAJORITY. If two or three of the literals of the function equal 1, then the function equals 1; in a sense a vote is taken to determine what value the function should take.

Table 11.2 also lists the number of functions that each form represents. How can we determine these numbers without writing out all 256 functions and counting those of a specific form? Topology is a very useful tool in such a determination. For the topology of the function

$$F = A \vee B \cdot \bar{C}$$

see Fig. 11.1(a). If we write down a few other functions of form 15 such

TABLE 11.2 FORMS OF BOOLEAN FUNCTIONS OF THREE VARIABLES

Form No.	No. of Minterms	No. of Functions Represented	Functional Form
1.	0	1	0
2.	1	8	$a \cdot b \cdot c$
3.	2	4	$abc \lor \bar{a}\bar{b}\bar{c}$
4.	2	12	$a(bc \lor \bar{b}\bar{c})$
5.	2	12	ab
6.	3	8	$ab\bar{c} \lor a\bar{b}c \lor \bar{a}bc$
7.	3	24	$a\bar{b}\bar{c} \lor bc$
8.	3	24	$a(b \lor c)$
9.	4	6	a
10.	4	8	$ab \lor ac \lor bc$ MAJORITY
11.	4	24	$ab \lor \bar{a}c$
12.	4	24	$\bar{a}bc \lor a(\bar{b} \lor \bar{c}) = a \oplus (bc)$
13.	4	2	$\bar{a}\bar{b}\bar{c} \lor ab\bar{c} \lor a\bar{b}c \lor \bar{a}bc = a \oplus (b \equiv c)$
14.	4	6	$a\bar{b} \lor \bar{a}b = a \oplus b$
15.	5	24	$a \lor bc$
16.	5	24	$(a \lor \bar{b} \lor \bar{c})(b \lor c)$
17.	5	8	$(a \lor b \lor \bar{c})(a \lor \bar{b} \lor c)(\bar{a} \lor b \lor c)$
18.	6	12	$a \lor b$
19.	6	12	$(a \lor b \lor c)(a \lor \bar{b} \lor \bar{c})$
20.	6	4	$(a \lor b \lor c)(\bar{a} \lor \bar{b} \lor \bar{c})$
21.	7	8	$a \lor b \lor c$
22.	8	1	1

as $A \lor BC$, $\bar{C} \lor \bar{A}\bar{B}$, etc., find the minterms of each, and plot each on a 3-dimensional cube, we find the square-leg structure of Fig. 11.1(b). The predominant features of that unlabeled structure are (1) a square and (2) a line attached to one corner of the square. That the vertices of this structure can be labeled in 24 different ways, each of which satisfies the adjacencies implied by the structure, can be argued in several ways. The easiest is:

(1) The square can be any one of the six faces of a 3-dimensional cube;
(2) The line can be attached to any of the four corners of the square;
(3) Thus $6 \times 4 = 24$ different sets of minterms are represented by this unlabeled structure, and each such set of minterms specifies a different Boolean function.

It is a relatively easy task to find the topologies of the functional forms of 2 variables shown in Fig. 11.2. Each functional form has a unique topology and minterm set size. We might use this fact to find more easily than by the exhaustive search procedure suggested above, that 3-variable functions fall into 22 different forms. We see that the number of minterms, and the distances between them are invariant properties of each form.

With $n = 4$, use of functional forms begins to break down although all 402 forms have been listed along with an electronic circuit to generate

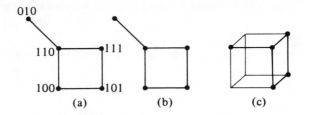

FIG. 11.1   Topology of Functions of Form 15

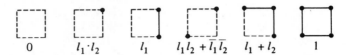

FIG. 11.2   Topology of the 2-Variable Forms

each. With $n = 5$ over one million forms exist; their enumeration is of little value to us.

Symmetry

If we look more closely at why it is necessary to use a, b, and c in Table 11.2, we find a familiar concept of practical as well as mathematical importance. The AND and OR operators were assumed to be commutative (Axiom 1). In terms of circuits, if we have two wires bearing signals A and B, and two input terminals on one AND gate, we can connect either wire to either terminal, and obtain the same output signal. The AND and OR functions are *totally symmetric*—the function does not change (is invariant) when we interchange two variables in the expression of the function.

Many switching functions are totally symmetric. Let us examine the majority function, for example.

$$f = AB \lor AC \lor BC$$

If we replace A ($\bar{A}$) wherever it appears with B ($\bar{B}$) and replace B ($\bar{B}$) with A ($\bar{A}$) we permute A and B, $A \leftrightarrow B$.

$$f = BA \lor BC \lor AC$$

Has f changed as a result of this permutation? Does f change if we permute $A \leftrightarrow C$, or $B \leftrightarrow C$? No, it does not, and thus f is a totally symmetric function.

Incidentally, but not accidentally, the topology of the majority function does not change if we graphically permute $A \leftrightarrow B$ by labeling the axes of the 3-dimension space with A and B interchanged. Figure 11.3 shows this.

Try other topological permutations to see that the topology does not vary with $A \leftrightarrow C$, and $B \leftrightarrow C$. Totally symmetric functions have totally symmetric topologies.

Another function of form 10 presents more of a challenge.

$$g = A\bar{B} \vee AC \vee \bar{B}C$$

Again, g is a totally symmetric function (if one function is totally symmetric, all functions of that form are symmetric), but it is symmetric with respect to the set of variables $\{A, \bar{B}, C\}$. Thus if we permute $A \leftrightarrow \bar{B}$—replace A ($\bar{A}$) with $\bar{B}$ (B)—g is invariant.

$$g = \bar{B}A \vee \bar{B}C \vee AC$$

The functions of form 4 (Table 11.2) are not totally symmetric.

$$h = A\,(BC \vee \bar{B}\bar{C})$$

If we permute $A \leftrightarrow B$, we obtain

$$h' = B\,(AC \vee \bar{A}\bar{C})$$

which is not the same function as h. (Find the minterm sets of the two functions.) But h is not totally asymmetric either, for if we permute $B \leftrightarrow C$ we obtain h.

$$h = A\,(CB \vee \bar{C}\,\bar{B})$$

Symmetry is more involved than we might expect; let us look at it more formally.

A single-output switching function f is said to be *pairwise symmetric* with respect to two input variables (literals, in general) x_i and x_j, if it is possible to assign don't care conditions (if they exist) such that the function is not altered when all appearances of x_j are replaced with x_i and all appearances of x_i are replaced with x_j. More briefly, f is symmetric with respect to x_i and x_j if it is invariant under their permutation.

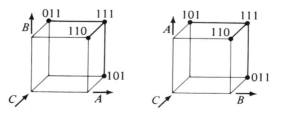

FIG. 11.3 Invariant Topology of the Majority Function

SECTION 11.1   **Form, Symmetry, and Redundancy**

$$f(x_1, \ldots, x_i, \ldots, x_j, \ldots) = f(x_1, \ldots, x_j, \ldots, x_i, \ldots) \qquad \textbf{(11.1)}$$

For example, the Boolean function

$$f = x_1 x_2 \ \lor \ x_3 \bar{x}_4 \ \lor \ x_5$$

is pairwise symmetric with respect to $\{x_1, x_2\}$ and $\{x_3, \bar{x}_4\}$. To test, rewrite f with x_2 replacing x_1 and vice versa.

$$f' = x_2 x_1 \ \lor \ x_3 \bar{x}_4 \ \lor \ x_5$$

Does $f' = f$? Then permute x_3 and $\bar{x}_4$: where x_3 appears write $\bar{x}_4$; where $\bar{x}_3$ appears write x_4; where $\bar{x}_4$ appears write x_3: etc.

$$f'' = x_2 x_1 \ \lor \ \bar{x}_4 x_3 \ \lor \ x_5$$

Now since $f'' = f$, f is pairwise symmetric with respect to $\{x_3, \bar{x}_4\}$.

A Boolean function f is said to be *partially* symmetric with respect to a subset, λ, of the input literals if an assignment of the don't cares exists such that f remains invariant under all permutations of the variables of λ. Pairwise symmetry is a special case where two variables appear in λ. *Total symmetry* is another special case in which λ includes all of the input literals.

$$f = x_1 x_2 x_3 \bar{x}_4 x_5 \ \lor \ x_1 x_2 x_3 x_4 \bar{x}_5$$

is partially symmetric in the set $\lambda = \{x_1, x_2, x_3\}$ and also in the set $\{x_4, \bar{x}_5\}$. It is not symmetric with respect to x_1 and x_4, for example, and hence is not totally symmetric. Interchanging x_1 and x_4 gives:

$$f' = x_4 x_2 x_3 \bar{x}_1 x_5 \ \lor \ x_4 x_2 x_3 x_1 \bar{x}_5 \neq f$$

On the other hand the function

$$f = \bar{x}_1 x_2 \ \lor \ \bar{x}_1 x_3 \ \lor \ x_2 x_3$$

is a totally symmetric function. Permute x_2 and x_3; the function is invariant. Permutation of x_1 and x_2 does not give a function equivalent to f, but permutation of $\bar{x}_1$ and x_2 does.

$$f' = x_2 \bar{x}_1 \ \lor \ x_2 x_3 \ \lor \ \bar{x}_1 x_3 = f$$

Similarly, permute $\bar{x}_1$ and x_3.

$$f'' = x_3 x_2 \ \lor \ x_3 \bar{x}_1 \ \lor \ x_2 \bar{x}_1 = f$$

Thus the function is totally symmetric with respect to $\lambda = \{\bar{x}_1, x_2, x_3\}$.

A special notation and algebra were developed by Shannon [6, of Ch. 2] for totally symmetric, single-output functions.

THEOREM 11.1. *A function* $f(x_1, x_2, \ldots, x_n)$ *is totally symmetric in the literal set* $\lambda = \{l_1, l_2, \ldots, l_n\}$ *if and only if a set of integers* $a = \{a_1, a_2, \ldots, a_k\}$ *exists such that* $f(x_1, x_2, \ldots, x_n) = 1$ *if and only if exactly* a_i *for* $i = 1, 2 \ldots k$ *of the literals equal 1 and all others equal 0.*

The totally symmetric function above equals 1, whenever any 2 members of $\lambda = \{\bar{x}_1, x_2, x_3\}$ equal 1 or whenever all 3 equal 1. Thus the " a " numbers for this function are $a = \{2, 3\}$ and this function may be expressed with the following notation.

$$S_a(\lambda) = S_{2,3}(\bar{x}_1, x_2, x_3) = \bar{x}_1 x_2 \ \text{v} \ \bar{x}_1 x_3 \ \text{v} \ x_2 x_3$$

It may also be expressed with a different λ and a.

$$S_{2,3}(\bar{x}_1, x_2, x_3) = S_{0,1}(x_1, \bar{x}_2, \bar{x}_3)$$

If none or exactly one of the literals x_1, $\bar{x}_2$, and $\bar{x}_3$ equals 1, then $f = 1$.

$$f = \underbrace{\bar{x}_1 x_2 x_3}_{0 \text{ literals}} \ \text{v} \ \underbrace{x_1 x_2 x_3 \ \text{v} \ \bar{x}_1 \bar{x}_2 x_3 \ \text{v} \ \bar{x}_1 x_2 \bar{x}_3}_{\text{exactly 1}}$$

With the exception of this S notation for totally symmetric functions, the above definitions could be rephrased to treat multiple-output functions. Finding a partial symmetry set is the difficult problem in either case, so we will prepare for algorithms to solve this problem. Assume that we have a function array F of cubes $c_\alpha{}^i \circ c_\Omega{}^i$. Let us partition the set

$$\alpha = \{x_1, x_2, \ldots, x_n\}$$

into two disjoint subsets λ and μ such that

$$\lambda \cup \mu = \alpha$$
$$\lambda \cap \mu = \phi \qquad\qquad\qquad \textbf{(11.2)}$$

Suppose F is not compressed so that all $c_\lambda{}^i$ are 0-cubes, and suppose that $\lambda \subseteq \alpha$ is a subset of p members, where $p \geq 2$ and $\mu = \alpha - \lambda$.[1] We can always reorder the columns of F so that its ith row may be expressed

$$c_\lambda{}^i \circ c_\mu{}^i \circ c_\Omega{}^i$$

although this reordering is not really necessary.

Now we partition F into many subarrays, F_i, of rows of F such that
(1) All rows of a subarray have identical c_μ parts, and
(2) All rows of a subarray have the same number of 1's appearing in their c_λ parts.

[1]Symmetries involving x_i and $\bar{x}_j$ will not be treated explicitly. To do so requires only that the x_j column in F be complemented.

Let $c^{i(1)}, c^{i(2)}, \ldots, c^{i(q)}$ be the cubes of subarray F_i. In Fig. 11.4(a), assume $\lambda = \{x_1, x_2\}$ and hence $\mu = \{x_3, x_4\}$. Then the function array given is partitioned by these rules into the 12 subarrays of Fig. 11.4(b). What is the significance of these subarrays? Consider F_1; if F is symmetric with respect to $\lambda = \{x_1, x_2\}$, then permutation of the x_1 and x_2 columns of F_1

x_1	x_2	x_3	x_4	z_1	z_2
0	0	0	0	1	0
0	0	0	1	1	0
0	0	1	0	d	0
0	0	1	1	1	0
0	1	0	0	1	0
0	1	0	1	d	0
0	1	1	0	0	1
0	1	1	1	1	1
1	0	0	0	1	0
1	0	0	1	1	d
1	0	1	0	0	1
1	0	1	1	1	1
1	1	0	0	0	1
1	1	0	1	1	1
1	1	1	0	0	0
1	1	1	1	1	1

(a) Function Array

	$\overbrace{\lambda}$		$\overbrace{\mu}$		$\overbrace{\Omega}$	
	x_1	x_2	x_3	x_4	z_1	z_2
$F_1 =$	{0	0	0	0	1	0}
$F_2 =$	(0	1	0	0	1	0)
	(1	0	0	0	1	0)
$F_3 =$	{1	1	0	0	0	1}
$F_4 =$	{0	0	0	1	1	0}
$F_5 =$	(0	1	0	1	d	0)
	(1	0	0	1	1	d)
$F_6 =$	{1	1	0	1	1	1}
$F_7 =$	{0	0	1	0	d	0}
$F_8 =$	(0	1	1	0	0	1)
	(1	0	1	0	0	1)
$F_9 =$	{1	1	1	0	0	0}
$F_{10} =$	{0	0	1	1	1	0}
$F_{11} =$	(0	1	1	1	1	1)
	(1	0	1	1	1	1)
$F_{12} =$	{1	1	1	1	1	1}

(b) Partitioned Function Array

$\lambda = \{x_1, x_3\}$

λ		μ		Ω	
x_1	x_3	x_2	x_4	z_1	z_2
(0	1	0	0	d	0)
(1	0	0	0	1	0)
(0	1	0	1	1	0)
(1	0	0	1	1	d)
(0	1	1	0	0	1)
(1	0	1	0	0	1)
(0	1	1	1	1	1)
(1	0	1	1	1	1)

$\lambda = \{x_2, x_3\}$

x_2	x_3	x_1	x_4	z_1	z_2
(0	1	0	0	d	0)
(1	0	0	0	1	0)
(0	1	0	1	1	0)
(1	0	0	1	d	0)
(0	1	1	0	0	1)
(1	0	1	0	0	1)
(0	1	1	1	1	1)
(1	0	1	1	1	1)

$\lambda = \{x_1, x_2, x_3\}$

x_1	x_2	x_3	x_4	z_1	z_2
(0	0	1	0	d	0)
(0	1	0	0	1	0)
(1	0	0	0	1	0)
(0	1	1	0	0	1)
(1	0	1	0	0	1)
(1	1	0	0	0	1)
(0	0	1	1	1	0)
(0	1	0	1	d	0)
(1	0	0	1	1	d)
(0	1	1	1	1	1)
(1	0	1	1	1	1)
(1	1	0	1	1	1)

(c) Verification that F is Partially Symmetric with Respect to Three Additional Variable Sets

Fig. 11.4 Symmetry Test Example Function

must leave that subarray invariant. It does so because the c_λ part of the single cube in F_1 consists of all 0's. Subarray F_1 and F_3, F_4, F_6, F_7, F_9, F_{10}, and F_{12} may be disregarded when determining whether F is in fact symmetric with respect to x_1 and x_2, because they will always be invariant under permutation of the members of λ.

Now we consider F_2, which consists of cubes $c^{2(1)}$ and $c^{2(2)}$. If we permute the x_1 and x_2 columns in these cubes, $c_\lambda^{2(1)}$ becomes $c_\lambda^{2(2)}$ and vice versa. If F is to remain invariant under the permutation, then $c_\Omega^{2(1)}$ must equal $c_\Omega^{2(2)}$, or more precisely:

$$c_\Omega^{2(1)} \bigcap c_\Omega^{2(2)} \neq \phi \tag{11.3}$$

To see why we need this definition, consider F_5, which consists of two cubes with output parts $c_\Omega^{5(1)} = d0$ and $c_\Omega^{5(2)} = 1d$ that are not identical, but which can be made so by assigning appropriate values to don't cares. The intersection $d0 \bigcap 1d = 10$ gives the appropriate values.

THEOREM 11.2.
A function F is partially symmetric with respect to a λ set if and only if, for every nontrivial subarray F_i of $g \geq 2$ cubes,

$$c_\Omega^{i(1)} \bigcap c_\Omega^{i(2)} \bigcap \dots \bigcap c_\Omega^{i(q)} \neq \phi \tag{11.4}$$

Proof. Pick a nontrivial subarray and assume F is partially symmetric with respect to the variables of λ. Permutation of the variables of λ will not alter the function, but any such permutation is now equivalent to permutation of the $c_\Omega^{i(j)}$ associated with the $c_\alpha^{i(j)}$ of F_i. Since F is symmetric, F_i must not change as a result of such a permutation: there must be an assignment of don't cares that makes all $c_\Omega^{i(j)}$ of F_i equal. Thus the intersection of all $c_\Omega^{i(j)}$ must be nonempty.

Now assume that the nonempty intersection exists. Then there is an assignment of don't cares that forces all $c_\Omega^{i(j)}$ to be equal. Since any permutation of the variables of λ is again equivalent to a permutation of the $c_\Omega^{i(j)}$ of cubes of each F_i and since all such $c_\Omega^{i(j)}$ can be made equivalent, the function is invariant under such a permutation.

From Fig. 11.4(b) and Theorem 11.2, the example function is known to be partially symmetric with respect to $\lambda = \{x_1, x_2\}$. Fig. 11.4(c) shows the subarrays that must be examined to show that F is partially symmetric with respect to $\lambda = \{x_1, x_3\}$, $\{x_2, x_3\}$, and $\{x_1, x_2, x_3\}$. We could look for other symmetry sets; in particular, we might well be interested in the largest symmetry sets.

Must we test each and every possible λ in an attempt to find the symmetry sets of a function, or, having found some partial symmetry sets and non-symmetry sets, can we use such information to advantage in locating larger symmetry sets? The following theorems offer the answer that small symmetry sets do suggest the larger sets of a function.

THEOREM 11.3.

A necessary and sufficient condition for $\lambda = \{x_1, x_2, x_3\}$ to be a partial symmetry set is that $\{x_1, x_2\}$, $\{x_1, x_3\}$, and $\{x_2, x_3\}$ be partial symmetry sets.

Proof: For every c_μ, two subarrays are nontrivial.

$$F_1 = \begin{Bmatrix} 001 \circ c_\mu^1 \circ c_\Omega^1 \\ 010 \circ c_\mu^2 \circ c_\Omega^2 \\ 100 \circ c_\mu^3 \circ c_\Omega^3 \end{Bmatrix} \qquad F_2 = \begin{Bmatrix} 011 \circ c_\mu^4 \circ c_\Omega^4 \\ 101 \circ c_\mu^5 \circ c_\Omega^5 \\ 110 \circ c_\mu^6 \circ c_\Omega^6 \end{Bmatrix}$$

From pairwise symmetry we know

$$\{x_1, x_2\} : c_\Omega^2 \bigcap c_\Omega^3 \neq \phi \quad c_\Omega^4 \bigcap c_\Omega^5 \neq \phi$$
$$\{x_1, x_3\} : c_\Omega^1 \bigcap c_\Omega^3 \neq \phi \quad c_\Omega^4 \bigcap c_\Omega^6 \neq \phi$$
$$\{x_2, x_3\} : c_\Omega^1 \bigcap c_\Omega^2 \neq \phi \quad c_\Omega^5 \bigcap c_\Omega^6 \neq \phi$$

These nonempty intersections together with Theorem 9.1 ensure that

$$c_\Omega^1 \bigcap c_\Omega^2 \bigcap c_\Omega^3 \neq \phi$$
$$c_\Omega^4 \bigcap c_\Omega^5 \bigcap c_\Omega^6 \neq \phi$$

This theorem allows us to declare the existence of a larger symmetry set from knowledge that smaller symmetry sets exists. In general, all three pairwise symmetries must exist as the theorem states; in special cases weaker conditions can be given. The following function is pairwise symmetric with respect to $\{x_1, x_2\}$ and $\{x_1, x_3\}$, but is not symmetric with respect to $\{x_2, x_3\}$ and hence not symmetric with respect to $\{x_1, x_2, x_3\}$.

x_1 x_2 x_3 x_4	x_1 z_2	$\overbrace{x_2 \ x_3}^{\lambda}$ $\overbrace{x_1 \ x_4}^{\mu}$ $\overbrace{z_1 \ z_2}^{\Omega}$	
1 0 0 0	d 0	$\begin{Bmatrix} 1 & 0 & 0 & 0 & 0 & 0 \\ 0 & 1 & 0 & 0 & 1 & 0 \end{Bmatrix}$	not symmetric
0 1 0 0	0 0		
0 0 1 0	1 0		
all other	d d		

It is also interesting to note that if the don't care of the first cube is assigned so as to make the function symmetric with respect to $\{x_1, x_2\}$, then the function will not be symmetric with respect to $\{x_1, x_3\}$.

$$\begin{matrix} x_1 & x_2 & x_3 & x_4 & z_1 & z_2 \\ \begin{Bmatrix} 1 & 0 & 0 & 0 & d & 0 \\ 0 & 1 & 0 & 0 & 0 & 0 \end{Bmatrix} \end{matrix} \longleftrightarrow \begin{matrix} x_1 & x_3 & x_2 & x_4 & z_1 & z_2 \\ \begin{Bmatrix} 1 & 0 & 0 & 0 & d & 0 \\ 0 & 1 & 0 & 0 & 1 & 0 \end{Bmatrix} \end{matrix}$$

THEOREM 11.4.

For the variables of $\lambda = \{x_1, x_2, \ldots, x_p\}$, $p \geq 3$, to constitute a partial symmetry set of F, it is necessary that all subsets of λ be partial symmetry sets.

Proof: Denote the subarray of F based upon a specific c_μ and with r 1's in each c_λ as $F_i(\mu, r)$. Let λ_s be a subset of λ and let μ_s contain the remaining input variables. Now any subarray $F_j(\mu_s, k)$ is a subarray of some $F_j(\mu, r + k)$ and since the intersection of all $c_\Omega{}^i$ is not empty, the intersection of a subset of these $c_\Omega{}^i$ must also be nonempty. Hence λ_s is a partial symmetry set.

Unfortunately the conditions of this theorem, while necessary, are not *sufficient* to ensure that λ is a partial symmetry set. To see this consider the function shown in Fig. 11.5. All subsets of 2 or 3 inputs constitute partial symmetry sets. But this knowledge fails to reveal the empty intersection of $c_\Omega{}^1$ and $c_\Omega{}^2$. When investigating symmetry sets of 2 or 3 variables, these two rows are always placed in different subarrays.

A sufficient condition is given by

THEOREM 11.5.

For $\lambda = \{x_1, x_2, \ldots, x_p\}, p \geq 3$, to be a partial symmetry set it is sufficient that

1. *A subset λ_s of $p - 1$ elements of λ is a symmetry set, and*

2. *A pairwise symmetry exists between the remaining element and some element of λ_s after assigning the don't cares necessary to make λ_s a partial symmetry set.*

Proof: Pick any $F_i(\mu, r)$. After making the don't care assignments necessary for λ_s to be a partial symmetry set, the $c_\Omega{}^i$ of F_i take on at most two distinct values. One value of $c_\Omega{}^i$ is associated with all $c_\lambda{}^i$ of F_i that have a 1 in the coordinate of the element not in λ_s. The other value is associated with the $c_\lambda{}^i$ that have a 0 in that coordinate. The required pairwise symmetry guarantees that these two values have a nonempty intersection.

Fig. 11.4(c) shows $\lambda = \{x_1, x_2, x_3\}$ to be a partial symmetry set of the function of Fig. 11.4. Three don't care assignments are required. If these are made, the array of Fig. 11.6 results. Then x_4 is not pairwise symmetric with any other variable. The sufficient requirements of Theorem 11.5 are not satisfied. In fact, the necessary requirements of Theorem 11.4 are not met, and this function is not totally symmetric.

	x_1	x_2	x_3	x_4	z_1
$c^1 =$	0	0	1	1	1
$c^2 =$	1	1	0	0	0
	0	0	1	0	0
	0	0	0	1	0
	1	0	1	1	0
	0	1	1	1	0
	all other				d

FIG. 11.5 All Subsets of 2- or 3-Input Variables are Symmetry Sets

It is possible to weaken the sufficiency conditions if a function is completely specified. Let $\lambda = \{x_1, x_2, \ldots, x_p\}$ and $J = \{1, 2, \ldots, p\}$ be a set of integers. A set of $p - 1$ distinct pairwise symmetry sets involving the members of λ is *complete* if and only if

1. Each $x_j \in \lambda$ appears in at least one pairwise symmetry set; and
2. It is not possible to write a sequence of pairwise symmetry sets in the manner $\{x_{j(1)}, x_{j(2)}\}, \{x_{j(2)}, x_{j(3)}\}, \ldots, \{x_{j(t)}, x_{j(1)}\}$ where all $j(k) \in J$.

THEOREM 11.6.

If a complete set of pairwise symmetry sets based upon

$$\lambda = \{x_1, x_2, \ldots, x_p\}$$

exists for a function F without don't cares, then F is pairwise symmetric with respect to every pair of members of λ.

Proof: The permutation of any two columns of F that correspond to members of λ can be accomplished by successively permuting pairs of columns dictated by known pairwise symmetry sets. All of these latter permutations leave the function invariant.

For example, if a complete switching function is known to be symmetric with respect to $\{x_1, x_2\}$ and $\{x_1, x_3\}$, then permutation of the first and second columns of its function array does not alter that array; nor does permutation of the first and third columns and then the first and second columns alter the array. But this sequence of permutations has the effect of permuting x_2 and x_3, and hence the function is symmetric with respect to $\{x_2, x_3\}$.

x_1	x_2	x_3	x_4	z_1	z_2
0	0	0	0	1	0
0	0	0	1	1	0
0	0	1	0	1	0
0	0	1	1	1	0
0	1	0	0	1	0
0	1	0	1	1	0
0	1	1	0	0	1
0	1	1	1	1	1
1	0	0	0	1	0
1	0	0	1	1	0
1	0	1	0	0	1
1	0	1	1	1	1
1	1	0	0	0	1
1	1	0	1	1	1
1	1	1	0	0	0
1	1	1	1	1	1

$$\begin{array}{cccc} x_1 & x_4 & x_2 & x_3 \\ \end{array} \quad \begin{array}{cc} z_1 & z_2 \end{array}$$
$$\left. \begin{array}{cccccc} 0 & 1 & 0 & 1 & 1 & 0 \\ 1 & 0 & 0 & 1 & 0 & 1 \end{array} \right\} \begin{array}{l} \{x_1, x_4\} \text{ is not a symmetry set.} \end{array}$$

$$\begin{array}{cccc} x_2 & x_4 & x_1 & x_3 \\ \end{array} \quad \begin{array}{cc} z_1 & z_2 \end{array}$$
$$\left. \begin{array}{cccccc} 0 & 1 & 0 & 1 & 1 & 0 \\ 1 & 0 & 0 & 1 & 0 & 1 \end{array} \right\} \{x_2, x_4\} \text{ is not}$$

$$\begin{array}{cccc} x_3 & x_4 & x_1 & x_2 \\ \end{array} \quad \begin{array}{cc} z_1 & z_2 \end{array}$$
$$\left. \begin{array}{cccccc} 0 & 1 & 0 & 1 & 1 & 0 \\ 1 & 0 & 0 & 1 & 0 & 1 \end{array} \right\} \{x_3, x_4\} \text{ is not}$$

FIG. 11.6 Example Function with Don't Cares Assigned

THEOREM 11.7.

For $\lambda = \{x_1, x_2, \ldots, x_p\}$ to be a partial symmetry set of a completely specified function, it is sufficient that a complete set of pairwise symmetry sets exist.

Proof: Since there are no don't cares, knowledge of pairwise symmetry sets ensures the equality of pairs of the $c_\Omega{}^{i(j)}$ of every F_i. Since the set of pairwise symmetry sets is complete, all $c_\Omega{}^{i(j)}$ of each F_i are identical and hence have a nonempty intersection.

Knowledge of the pairs of variables in which a function is symmetric is thus of considerable interest. That a function is pairwise symmetric with respect to a set λ can be verified by examining subarrays as we have been doing, but often this is not the most efficient procedure. For example, single-output functions are not usually expressed in the function array format, but rather in terms of their ON, OFF, and DC-arrays. Pairwise symmetry can be determined easily from these arrays also.

Let $\mathbf{P}_{ij}(c^k)$ denote a cube derived from cube c^k by permutation of $c_i{}^k$ and $c_j{}^k$. This *permutation* operation will also be applied to entire arrays, in which case the ith and jth columns are to be interchanged.

Now assume we wish to know whether $\lambda = \{x_i, x_j\}$ is a partial symmetry set of a single-output function expressed in terms of its ON- and OFF-arrays. If $c^k \in \text{ON}$, $\mathbf{P}_{ij}(c^k)$ must be included in either the ON- or the DC-array of the function if λ is to be a symmetry set of that function. If $\mathbf{P}_{ij}(c^k) \in \text{OFF}$, permutation of x_i and x_j alters the function. This same requirement must be satisfied for all cubes of the ON-array. Thus, for λ to be a partial symmetry set

$$\mathbf{P}_{ij}(\text{ON}) \bigcap \text{OFF} = \phi \qquad (11.5)$$

By complementing one test column, variable–complement symmetry pairs may be detected. While this algorithm is very straightforward, it may not be suitable if the number of variables is large. Formation of the OFF-array via the sharp operator can require a great deal of computing time and storage, and the complete intersection of Eq. (11.5) must be formed. Testing individual cubes of $\mathbf{P}_{ij}(\text{ON})$ via Eq. (11.5) may be faster and can be terminated as soon as a nonempty intersection is encountered. If formation of the OFF-array is to be avoided, a test such as $\mathbf{P}_{ij}(\text{ON}) \# (\text{ON} \cup \text{DC})$ may be used.

If the single-output function examined with the above algorithm is a partial switching function, that algorithm indicates which pairwise symmetries may exist. In general, for a specific pairwise symmetry to exist don't cares must be assigned appropriate values. Such an assignment based upon one symmetry set may in fact prevent other pairwise symmetries from existing as was shown in the example associated with Theorem 11.3. To establish the symmetry set $\{x_i, x_j\}$ it is necessary to compute

$$\text{ON} \leftarrow \text{ON} \bigsqcup (\mathbf{P}_{ij}(\text{ON}) \bigsqcap \text{DC})$$
$$\text{DC} \leftarrow \text{DC} \# \mathbf{P}_{ij}(\text{ON}) \tag{11.6}$$

which place appropriate don't care vertices in the ON-array, i.e., assign 1's to certain don't cares. It is also necessary to compute

$$\text{OFF} \leftarrow \text{OFF} \bigsqcup (\mathbf{P}_{ij}(\text{OFF}) \bigsqcap \text{DC})$$
$$\text{DC} \leftarrow \text{DC} \# \mathbf{P}_{ij}(\text{OFF}) \tag{11.7}$$

which assign 0's to appropriate d's of the function.

If we seek larger partial symmetry sets, don't care assignment may be deferred until the existence of such larger sets has been established. Two algorithms for finding larger symmetry sets may be proposed. First, if the function is completely specified, $\text{DC} = \phi$, Theorem 11.7 and a list of all pairwise symmetries may be employed to suggest which larger sets should be considered, and verify that a specific λ is a partial symmetry set. The ON-array of a complete switching function is shown in Fig. 11.7 together with all the pairwise symmetry sets of that function. These pairwise symmetries suggest that only $\lambda = \{x_1, x_3, x_4\}$ need be investigated as a partial symmetry set. Theorem 11.7 assures us that it is a symmetry set.

If the switching function is partially specified, knowledge of pairwise symmetries and Theorem 11.3 are used to establish three-element symmetry sets. These sets, together with Theorem 11.4, propose candidate four-element symmetry sets, which may be investigated using Theorem 11.5.

We can arrive at maximal candidate symmetry sets in another fashion. Suppose a function of five variables $f(x_1, x_2, x_3, x_4, x_5)$ is *not* pairwise symmetric with respect to $\{x_2, x_4\}$ and $\{x_2, x_5\}$. Then these pairs of variables may not appear together in any symmetry set. Let

$$S_0 = \{x_1, x_2, x_3, x_4, x_5\}$$

be the set of all variables. Since x_2 and x_4 may not appear together, delete x_2 from one copy of S_0, and delete x_4 from another copy.

$$\text{ON} = \left\{ \begin{matrix} x000x \\ 0x00x \\ 00x0x \\ 000xx \\ xx000 \\ 0xx00 \\ 0x0x0 \\ 11111 \end{matrix} \right\} \quad \begin{matrix} \{x_1, x_3\} \\ \\ \\ \{x_1, x_4\} \\ \\ \{x_3, x_4\} \\ \\ \{x_2, x_5\} \end{matrix}$$

FIG. 11.7 A Completely Specified Function and its Pairwise Symmetries

$$S_{11} = \{x_1, x_2, x_3, x_5\} \qquad S_{12} = \{x_1, x_3, x_4, x_5\}$$

As far as we know to this point, S_{11} and S_{12} constitute symmetry sets. But we also know that x_2 and x_5 may not appear together, so we break the S_{ij} subsets in which both x_2 and x_5 appear.

$$S_{21} = \{x_1, x_2, x_3\} \qquad S_{22} = \{x_1, x_3, x_5\} \qquad S_{12} = \{x_1, x_3, x_4, x_5\}$$

Since S_{22} is a proper subset of S_{12}, S_{22} may be deleted. Then S_{21} and S_{12} are potential partial symmetry sets of maximum size. Theorem 11.3 assures us that S_{21} *is* in fact a symmetry set. We must investigate S_{12} further to verify whether it is or is not a symmetry set. As an example, consider the function specified by the following ON- and OFF-arrays.

$$\text{ON} = \{000xx\} \qquad \text{OFF} = \{01000\}$$

All pairwise symmetries exist, with the exception of $\{x_2, x_4\}$ and $\{x_2, x_5\}$, although some pairwise symmetries will require that don't care assignments be made. For this function to be symmetric with respect to $\{x_1, x_2, x_3\}$, don't care vertices 10000 and 00100 must be assigned to the OFF-array. For $\{x_1, x_3, x_4, x_5\}$ to be a symmetry set, don't care cubes $x0x00$, $x00x0$, $x000x$, $00x0x$, and $00xx0$ must be assigned to the ON-array. Note that these don't care assignments conflict. Vertex 10000 can not be assigned to both the OFF- and ON-arrays. Hence this function can not be made to be simultaneously symmetric with respect to both $\{x_1, x_2, x_3\}$ *and* $\{x_1, x_3, x_4, x_5\}$.

While detection of pairwise symmetry sets in multiple-output functions is similar in philosophy to the algorithm given above for the single-output case, the details must differ. Suppose we are investigating $\lambda = \{x_i, x_j\}$. For each row c^k of a function array, form $\mathbf{P}_{ij}(c^k)$. When $\mathbf{P}_{ij}(c^k)$ and c^k are identical we are dealing with a trivial subarray and may advance k to its next value. Otherwise examine $\mathbf{P}_{ij}(c_\alpha{}^k) \bigcap c_\alpha{}^l$ where $l > k$. If this intersection is empty, advance l. When a nonempty intersection is encountered, cubes of the same subarray have been located. The intersection $c_\Omega{}^k \bigcap c_\Omega{}^l$ must be examined. If this intersection is empty, the output sections of the two cubes of a subarray differ and the pairwise symmetry can not exist. If the intersection is not empty, the search must continue until all pairs of cubes have been examined.

Larger symmetry sets may be located from pairwise symmetry sets in the same manner as discussed above for single output functions. To make a multiple-output function symmetric with respect to a set λ will require assignment of don't cares in general. This can be accomplished by making the don't care assignments required by each pairwise symmetry of a complete set of such symmetry sets. An algorithm for making don't care assignments to satisfy a pairwise symmetry, say $\lambda = \{x_i, x_j\}$, may proceed in a manner similar to that given above to verify the existence to a pairwise symmetry. Examine each cube c^k of the function array and compare it with all cubes c^l below it in the array. This effects comparison of all pairs of cubes. Only when $\mathbf{P}_{ij}(c^k) \neq c^k$ and $\mathbf{P}_{ij}(c_\alpha^k) \sqcap c_\alpha^l \neq \phi$ must c_Ω^k and c_Ω^l be examined. If $c_\Omega^k = c_\Omega^l$, no don't cares require assignment. When $c_\Omega^k \neq c_\Omega^l$ row c^k must be replaced by (1) the row

$$(c_\alpha^k \sqcap \mathbf{P}_{ij}(c_\alpha^l)) \circ (c_\Omega^k \sqcap c_\Omega^l)$$

which reflects the vertices of c_α^k that are members of a subarray together with vertices of c_α^l and (2) the array

$$(c_\alpha^k \# \mathbf{P}_{ij}(c_\alpha^l)) \times c_\Omega^k$$

which covers the remaining vertices of c_α^k. The output coding for these vertices is not altered. Cube c^l is similarly replaced with the row

$$(c_\alpha^l \sqcap \mathbf{P}_{ij}(c_\alpha^k)) \circ (c_\Omega^l \sqcap c_\Omega^k) \quad \text{(don't cares assigned)}$$

and the array

$$(c_\alpha^l \# \mathbf{P}_{ij}(c_\alpha^k)) \times c_\Omega^l$$

The function of Fig. 11.8(a) is symmetric with respect to $\{x_1, x_2, x_3\}$ and $\{x_2, x_3, x_4\}$. Let us first make the don't care assignments required by $\{x_2, x_3\}$ which is a subset of both larger symmetry sets. A number of don't cares must be assigned and one cube, $00x1d0$, broken into $0001d0$, which retains the original output part, and 001100, in which the don't care has been assigned. Only the cubes that are altered are shown, to emphasize the change. If we now make the don't care assignments dictated by the pair-

x_1	x_2	x_3	x_4	z_1	z_2	(b)	(c)	(d)
0	0	0	0	1	1			
0	1	1	1	1	1			
0	1	0	0	1	0			
0	0	1	0	1	d	0 0 1 0 1 0		
1	1	1	x	0	1			
0	1	0	1	0	0			
1	0	0	1	0	0			
0	1	1	0	0	d		0 1 1 0 0 0	0 1 1 0 0 0
1	0	1	1	d	1		1 0 1 1 1 1	1 0 1 1 0 1
0	0	x	1	d	0	0 0 1 1 0 0		
						0 0 0 1 d 0		0 0 0 1 1 0
1	0	x	0	d	0		1 0 0 0 1 0	1 0 1 0 0 0
							1 0 1 0 0 0	
1	1	0	0	d	d	1 1 0 0 d 0	1 1 0 0 0 0	1 1 0 0 0 0
1	1	0	1	d	d	1 1 0 1 d 1	1 1 0 1 1 1	1 1 0 1 0 1
		(a)				(b)	(c)	(d)

FIG. 11.8 (a) Function Array; (b) Don't Care Assignment to Satisfy $\{x_2, x_3\}$; (c) Additional Assignments to Satisfy $\{x_1, x_2, x_3\}$; (d) Additional Assignments to Satisfy $\{x_2, x_3, x_4\}$.

wise symmetric set $\{x_1, x_2\}$, we achieve a function array symmetric in $\{x_1, x_2, x_3\}$. Again, assigning don't cares requires splitting cubes. Alternatively, we can follow the $\{x_2, x_3\}$ assignment by assigning don't cares to satisfy $\{x_3, x_4\}$. Different don't care assignments are required as shown in Fig. 11.8(d), and the function can not be simultaneously symmetric in both $\{x_1, x_2, x_3\}$ and $\{x_2, x_3, x_4\}$.

Redundant Input Variables

A *contradiction* is said to exist in a function array if two cubes of that array, c^i and c^j, can be found such that

$$c_\alpha{}^i \bigcap c_\alpha{}^j \neq \phi, \quad \text{and} \quad c_\Omega{}^i \bigcap c_\Omega{}^j = \phi \tag{11.8}$$

In Section 10.3 we saw that connection arrays contain contradictions. We define an input variable to be *redundant* if the array remaining after that variable column is deleted is free of contradictions. We do not expect redundant input variables in Boolean functions, and tend to delete them as soon as they are discovered. But in the next section we will examine procedures that can cause input variables to become redundant. In preparation, we now will be concerned with the problem of showing that a variable, or set of input variables, is redundant.

Input variable x_1 is redundant in f, which is described by the function array F given below. Deleting the first column of F gives array F', an array that can be made free of contradictions by assigning appropriate values to the three don't cares. If variable x_3 is deleted from F, array F'' results; this array is also free of contradictions if the appropriate, but different, don't care assignments are made.

$$
F = \begin{pmatrix}
000 & d \\
001 & 0 \\
010 & 0 \\
011 & d \\
100 & 1 \\
101 & d \\
110 & 0 \\
111 & 0
\end{pmatrix}
\quad
F' = \begin{pmatrix}
00 & d = 1 \\
01 & 0 \\
10 & 0 \\
11 & d = 0 \\
00 & 1 \\
01 & d = 0 \\
10 & 0 \\
11 & 0
\end{pmatrix}
\quad
F'' = \begin{pmatrix}
00 & d = 0 \\
00 & 0 \\
01 & 0 \\
01 & d = 0 \\
10 & 1 \\
10 & d = 1 \\
11 & 0 \\
11 & 0
\end{pmatrix}
$$

Both x_1 and x_3 are redundant input variables: we may express f in terms of x_2 and x_3, or in terms of x_1 and x_2. The subarray that results when the first (third) column is deleted describes a subset of the set of switching functions specified by F after don't care assignments are made so as to eliminate possible contradictions. But x_1 and x_3 are not simultaneously redundant. We can not assign values to the don't cares so as to make them simultaneously redundant.

More generally, a number of sets of input variables may exist such that the members of any one set can be made to be simultaneously redundant by assigning appropriate values to the don't cares. If a function is completely specified, then only one such *redundancy set* can exist.

The most direct means of testing a proposed redundancy set λ of input variables consists of deleting the variables of λ and verifying that the resulting array is free of contradictions. Because deletion and testing for contradictions are rather easily performed, such an algorithm is also very efficient.

ALGORITHM 11.1. *Redundancy Test and Don't Care Assignment for Single Output Functions.* Let a single output function be expressed in terms of its ON-, OFF-, and DC-arrays. Set λ of q input variables constitutes a redundancy set if and only if

$$\mathbf{D}_\lambda(\text{ON}) \bigcap \mathbf{D}_\lambda(\text{OFF}) = \phi \qquad (11.9)$$

If λ is found to be a redundancy set, then $\mathbf{D}_\lambda(\text{ON})$, $\mathbf{D}_\lambda(\text{OFF})$, and $[\mathbf{D}_\lambda(\text{DC}) \# \mathbf{D}_\lambda(\text{ON})] \# \mathbf{D}_\lambda(\text{OFF})$ constitute the ON-, OFF-, and DC-arrays, respectively, of a function of $n - q$ input variables that satisfies all the care conditions of the original function.

ALGORITHM 11.2. *Redundancy Test for Multiple Output Functions.*
Let N be the number of rows in a given function array F, and λ be a set of q input variables. Let $C = \mathbf{D}_\lambda(F)$. Then for each cube c^k, $c_\beta^k \circ c_\Omega^k$ where $1 \le k \le N$, and c_β^k is an $(n - q)$-tuple, compare c_β^k and c_β^l where $k < l \le N$. If $c_\beta^k \bigcap c_\beta^l \ne \phi$ and $c_\Omega^k \bigcap c_\Omega^l = \phi$, a contradiction exists and the members of λ cannot be simultaneously redundant. If testing is completed without encountering a contradiction, then λ is a redundancy set.

ALGORITHM 11.3. *Redundancy Don't Care Assignment for Multiple Output Functions.*
Assume λ is known to be a redundancy set. Again let $C = \mathbf{D}_\lambda(F)$, and examine the cubes of C in the manner of Algorithm 11.2. When

$$c_\beta^k \bigcap c_\beta^l \ne \phi \quad \text{and} \quad c_\Omega^k \ne c_\Omega^l,$$

don't care assignment is required. Cube $c_\beta^k \circ c_\Omega^l$ must be replaced by the array $(c_\beta^k \# c_\beta^l) \times c_\Omega^k$ and the row $(c_\Omega^k \bigcap c_\beta^l) \circ (c_\Omega^k \bigcap c_\Omega^l)$. Cube $c_\beta^l \circ c_\Omega^l$ must be replaced by the array $(c_\beta^l \# c_\beta^k) \times c_\Omega^l$. Array C may be absorbed to eliminate repetition of cubes.

To illustrate this algorithm and the cube replacements specified, suppose cubes $x110x \circ d1$ and $111x1 \circ 0d$ are part of array C, from which q redundant variables have already been eliminated. The β-parts of these cubes cover 11101; to this β-part output coding $d1 \bigcap 0d = 01$ must be appended. Don't cares have been assigned to eliminate a contradiction. But without knowledge of the other cubes of C, we have no reason to assign a value to the d associated with the remainder of the two original cubes. Thus $d1$ may be retained with $x110x \neq 111x1 = \{0110x, x1100\}$, and $0d$ may continue to serve as the c_Ω-part associated with $111x1 \neq x110x = 11111$.

Redundancy sets can be found in a manner similar to that in which symmetry sets were established. Individual variables can be shown to be redundant, and then pairs, triplets, etc. of these variables tested for redundancy. But when function arrays contain a substantial number of redundant variables and the largest redundancy set is sought, a more efficient algorithm consists merely of testing λ-sets of size $n - 1$ then $n - 2$, etc. until a set is shown to be redundant, or λ becomes the empty set. Such an exhaustive search is not feasible for detecting symmetry sets because of the complexity of symmetry tests.

11.2 FACTORING

Most "minimization" procedures are oriented toward two logic level networks. Attempts to generalize such procedures to obtain minimum cost multiple-level networks have yielded algorithms that demand excessive amounts of computation. Yet some switching functions can be realized at lower cost with multiple (more than two) level networks, and practical constraints such as fan-in limit may force the use of networks of more than two levels. In this section we will emphasize fan-in limitation as motivation for designing low if not minimum cost multiple-level networks.

Consider Boolean function f expressed below in both minimum sum-of-product and product-of-sum forms.

$$f = AB \lor ACD$$
$$= A(B \lor C)(B \lor D)$$

Both expressions suggest 7-diode networks, but use of the distributive axioms on these expressions—factoring the equations—leads to a Boolean equation that describes a 6-diode, three-level network.

$$f = A(B \lor CD)$$

Counting diodes in multiple-level networks lacks practical significance, since three or more level diode networks are not used. However, the number of connections to input terminals of amplifying gates is significant, or, if the number of input terminals on such gates is fixed, then the number of gates is a meaningful measure of network cost.

Suppose that function f is to be realized with 2 input NAND gates. Two NAND gates can be used in tandem to form an AND gate and fan-in extended by this means. Fig. 11.9(a) shows how the sum-of-products expression of f might be realized with five 2-input NAND gates. The product-of-sums expression of f also demands a 3-input AND gate, and hence a total of six NAND gates are required to satisfy this expression, as shown in Fig. 11.9(b). In the factored expression of f all operators combine two variables and we need not extend fan-in through the use of AND gates. Fig. 11.9(c) shows the 4 NAND-gate network that this expression of f suggests. Note that complements are shown at input terminals in Fig. 11.9. We will assume that complements are as available as variables and that no additional cost is associated with a complement.

We can factor Boolean equations with the distributive laws, and to a limited extent the Karnaugh map may be used as a tool to find factored forms of Boolean equations. But such procedures are not particularly effective if the number of variables is large and NAND gates are to be used to construct a realization, the situation we assume.

Suppose that we have an array C that covers a Boolean function. Array C may be a minimal cover or a nonredundant cover, or any other cover, for that matter. We normally think of C as a sum-of-products expression of the function, with each cube c^i describing one product term The 0's and 1's of c^i describe the complements and variables, respectively, that are to be connected to input terminals of an AND gate. Each of those variables (complements), say x_j, can be described by an $(n-1)$-cube d with $d_j = 1$ (0). Cube c^i can be thought of as the cube-by-cube intersection of those $(n-1)$-cubes just as the product term is the logical product of individual literals.

Each AND gate realizes a Boolean function of the literals applied to its input terminals. We can speak of the signal produced by each AND gate as a "single line" realization of a Boolean function. But the *bundle* of lines into the AND gate can be thought of collectively as realizing or representing the same Boolean function in certain circumstances. One such circumstance is illustrated in Fig. 11.10. Either the single line representation of the function $x_1 \cdot x_2$ or the bundle $\{x_1, x_2\}$ may be applied to input terminals of the NAND gate.

In a like manner, a single line or a bundle may represent a Boolean function described by an array of cubes. In a two-level AND-to-OR network, the bundle of signals generated by the AND gates represents the function, and in some cases can be used in place of the single signal generated by the OR gate.

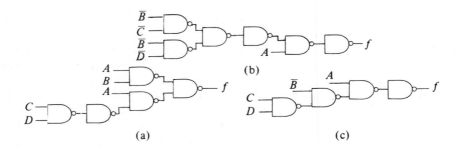

FIG. 11.9 Realizing a Function with Fan-in Limited Gates

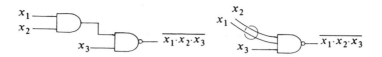

FIG. 11.10 The Use of a Bundle

A NAND gate can be thought of as an AND gate followed by an inverter. Thus, if the bundle of signals described by a cube c^i are connected to the input terminals of a NAND gate, the output signal is described by the array $U_n \# c^i$. If array C consists of a number of cubes $c^1, c^2, \ldots, c^p$, and the bundle described by each cube is applied to a separate NAND gate, then p signals described by $U_n \# c^1, \ldots, U_n \# c^p$ are generated. If these signals are now connected to the input terminals of another NAND gate, the signal produced by that gate is described by:

$$U_n \# ((U_n \# c^1) \sqcap (U_n \# c^2) \sqcap \ldots \sqcap (U_n \# c^p)) \equiv$$
$$U_n \# (U_n \# (c^1 \sqcup c^2 \sqcup \ldots \sqcup c^p)) \equiv$$
$$U_n \# (U_n \# C) \equiv C \tag{11.10}$$

Thus the signal generated by a two-level NAND tree represents the function described by C. But we are familiar with the fact that a two-level AND-to-OR gate network can be replaced with a NAND tree. The bundle of signals into the output NAND gate is a representation of $\bar{C} = U_n \# C$.

Suppose that the literals specified by two cubes c^1 and c^2 are applied to two NAND gates, which are interconnected in the manner of Fig. 11.11. The resulting circuit produces a signal described by

$$U_n \# (c^2 \sqcap (U_n \# c^1)) \equiv U_n \# (c^2 \# c^1)$$
$$\equiv c^1 \sqcup (U_n \# c^2) \tag{11.11}$$

This array consists of cube c^1 and $\$c^2$ $(n-1)$-cubes, if none are absorbed.

Any literal appearing in both c^1 and c^2 is redundant in c^1, in the sense that its deletion from c^1 will not alter the function realized by this network. If $c_j^1 = c_j^2 \neq x$, then one of the cubes described by $U_n \# c^2$ will consist of $n-1$ x's with $\bar{c}_j^2$ appearing in the jth coordinate position. A single 0–1 opposition will exist between this cube and c^1, and we could replace c^1 with a larger cube that has an x in the jth position.

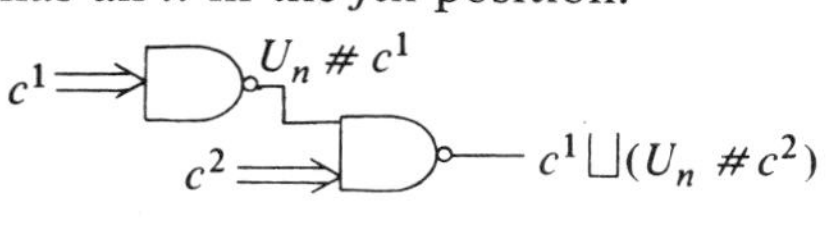

$$c^1 \Rrightarrow \quad U_n \# c^1$$
$$c^2 \Rrightarrow \quad c^1 \sqcup (U_n \# c^2)$$

FIG. 11.11 An Interesting NAND Network
The double lines represent a bundle of signal lines

On the other hand, if in some position $c_j{}^1 = \bar{c}_j{}^2 \neq x$, then c^1 is entirely redundant. The $(n-1)$-cube with $(\bar{c}_j{}^2) = c_j{}^1$ in position j will cover c^1, and it may be deleted from the final array.

As examples consider the following pairs of cubes.

$$c^1 = 0011xx \qquad\qquad\qquad c^1 = 0011xx$$
$$c^2 = 0x1x0x \qquad\qquad\qquad c^2 = 1x1x0x$$

$$c^1 \bigsqcup (U_n \# c^2) = \begin{Bmatrix} 0011xx \\ 1xxxxx \\ xx0xxx \\ xxxx1x \end{Bmatrix} \qquad\qquad c^1 \bigsqcup (U_n \# c^2) = \begin{Bmatrix} \cancel{0011xx} \\ 0xxxxx \\ xx0xxx \\ xxxx1x \end{Bmatrix}$$

Using $c^1 = x0x1xx$ would give an equivalent array $\qquad\qquad$ c^1 is redundant

In summary, the network of Fig. 11.11 gives significant results only if $c^1 \bigcap c^2 \neq \phi$. Assuming c^1 and c^2 intersect, $c_j{}^1$ may be changed to an x (if it is not already an x) for all j for which $c_j{}^2 \neq x$.

Now let us return to the realization of a single signal described by array C. If this signal is to be generated by a NAND gate, that gate must be supplied with (1) a single signal described by $\bar{C} = U_n \# C$, or (2) a bundle of signals described by cubes whose intersection is $\bar{C}$. Many bundles may be used. If we partition C into two subarrays $C1$ and $C2$ and achieve two signals described by $\overline{C1}$ and $\overline{C2}$ respectively, then this bundle of two signals may be used.

$$\overline{C1} \bigcap \overline{C2} \equiv U_n \# (C1 \bigsqcup C2) \equiv \bar{C} \qquad\qquad (11.12)$$

We may further partition either or both of $C1$ and $C2$, thus partitioning C into more than two subarrays. In the extreme case, C may be partitioned into p parts, each of which consists of a single cube of C. Then the literals of each cube may be applied to a NAND gate. The bundle of signals provided by these p gates describes $\bar{C}$. This is just the situation described in Eq. (11.10).

Suppose it is possible to select subarray $C1 = \{c^1, c^2, \ldots, c^q\}$ from array C so that a cube $\gamma \neq U_n$ exists, and satisfies

$$c^i = c^i \bigcap \gamma \qquad\qquad (11.13)$$

for all $i = 1, 2, \ldots, q$. Array $C1$ is a *common factor subarray* of C; cube γ is a *common factor* of $C1$. We derive cube d^i from cube c^i in the following manner:

$$d_j{}^i = \begin{cases} c_j{}^i & \text{if} \quad \gamma_j = x \\ x & \text{if} \quad \gamma_j \neq x \end{cases} \qquad\qquad (11.14)$$

Let $D = \{d^1, d^2, \ldots, d^q\}$. A bundle representation of $C1$ may then consist of the literals of γ and one or more signals that represent D. A single line representation of $\overline{C1}$ may then be formed with the right NAND gate of Fig. 11.11, with γ serving as c^2 and either a single line or bundle representation of D serving as $U_n \# c^1$. Realization of $C1$ ($\overline{C1}$) has been reduced to the problem of realizing D, an array of $\$\gamma$ fewer variables than $C1$.

As an example, in the following function the product $x_1 \cdot x_2$ can clearly be factored from the first two terms.

$$f = x_1 x_2 x_3 \bar{x}_4 \lor x_1 x_2 \bar{x}_3 x_5 \lor x_3 x_4 x_5$$
$$= x_1 x_2 (x_3 \bar{x}_4 \lor \bar{x}_3 x_5) \lor x_3 x_4 x_5$$

Let

$$f_1 = x_1 x_2 (x_3 \bar{x}_4 \lor \bar{x}_3 x_5)$$
$$g = x_3 \bar{x}_4 \lor \bar{x}_3 x_5$$

Product term $x_1 \cdot x_2$ is a common factor of f_1; g is a function that remains to be synthesized after the network of Fig. 11.12 has been established. For this same function:

$$C = \begin{Bmatrix} 1110x \\ 110x1 \\ xx111 \end{Bmatrix} \qquad \begin{aligned} C1 &= \begin{Bmatrix} 1110x \\ 110x1 \end{Bmatrix} \\ C2 &= \{xx111\} \end{aligned} \qquad \begin{aligned} \gamma &= 11xxx \\ D &= \begin{Bmatrix} xx10x \\ xx0x1 \end{Bmatrix} \end{aligned}$$

Array D remains to be realized.

We are forced to abandon the two-level NAND tree and consider multiple-level network structures if the NAND gates we expect to use are fan-in limited. Let Λ measure that fan-in limit. Any cube with cost $\$c^i > \Lambda$ or array with $p > \Lambda$ prohibits two-level synthesis.

Let $\kappa > \Lambda$ be the number of signals we wish to apply to a Λ-limited NAND gate. We can partition all but $\kappa - \Lambda G$ of these signals into a minimum number G of groups of Λ or fewer each, such that

$$G + (\kappa - \Lambda G) \le \Lambda$$

or

$$G \ge \frac{\kappa - \Lambda}{\Lambda - 1} \qquad\qquad \textbf{(11.15)}$$

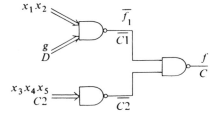

FIG. 11.12 Partial Realization of Example Function

Members of each group may be connected to their respective AND gates, which in turn are connected to the input terminals of the original NAND gate. Fig. 11.9(a) illustrates this for $\kappa = 3$ and $\Lambda = 2$.

Define:

$$\mathscr{G}(\kappa) = \begin{cases} 0 & \text{if} \quad \kappa = 1 \\ 1 & \text{if} \quad 1 < \kappa \le \Lambda \\ 1 + 2\left\lceil \dfrac{\kappa - \Lambda}{\Lambda - 1} \right\rceil & \text{if} \quad \kappa > \Lambda \end{cases} \tag{11.16}$$

where $\left\lceil \dfrac{\kappa - \Lambda}{\Lambda - 1} \right\rceil$ denotes "minimum non-negative integer greater than or equal to $\dfrac{\kappa - \Lambda}{\Lambda - 1}$. Then $\mathscr{G}(\kappa)$ measures the number of Λ-limited NAND gates required to realize a κ-limited NAND gate. The multiplier of 2 appears because two NAND gates are required to achieve an AND gate.

Blind use of AND gates to solve NAND gate fan-in problems often result in an excessive number of logic levels and gates. Fig. 11.9 illustrates that the recognition and application of factors can result in lower cost networks with the same number or fewer levels of logic.

The Value of Factoring

The factor γ common to two cubes c^i and c^j is found with the *factor product* operator **F** defined by Table 11.3 and the rule

$$\gamma = (\gamma_1, \gamma_2, \ldots, \gamma_n) \quad \text{where} \quad \gamma_k = c_k{}^i \mathbf{F} c_k{}^j \tag{11.17}$$

If no common factor exists, $\gamma = U_n$. Given an array of cubes, the common factor of the array is then given by

$$\gamma = (\ldots ((c^1 \mathbf{F} c^2) \mathbf{F} c^3) \ldots \mathbf{F} c^q) = \mathbf{F}(C) \tag{11.18}$$

Factor γ is said to have a width of $w = \$\gamma$, and a height of $q = |C|$, the number of cubes of C. The *figure-of-merit* of factor γ of array C is defined as

$$\mathscr{M}(C, \gamma) = w \cdot q \tag{11.19}$$

If $C1$ is a subarray of cubes of C, and γ is a common factor of $C1$, then

TABLE 11.3 COORDINATE FACTOR PRODUCT

		$c_k{}^j$		
	F	0	1	x
	0	0	x	x
$c_k{}^i$	1	x	1	x
	x	x	x	x

$\overline{C1}$ may be realized as a bundle of q lines or as a single line (Fig. 11.12). The gate cost of the bundle realization using AND gates to solve second level fan-in problems is:

$$\$_g^1(C1) = \sum_{i=1}^{q} \mathscr{G}(\$c^i) \tag{11.20}$$

This approach clearly aggravates the fan-in problem at the first logic level, which we must face eventually. In fact, if $q > \Lambda$ one or more AND gates will obviously be required to solve this problem.

In an iterative synthesis procedure where the necessity of applying N_f signals to the first-level gate has already been established and $p - q$ cubes remain in $C - C1$, the cost of the bundle realization of $C1$ is better estimated as

$$\$_g^{1e}(C1) = \sum_{i=1}^{q} \mathscr{G}(\$c^i) + \mathscr{G}(N_f + p) - 1 \tag{11.21}$$

The cost of AND gates required to solve first-level fan-in problems is included and perhaps overestimated in $\$_g^{1e}(C1)$.

A single line realization of $\overline{C1}$ alleviates the first-level fan-in problem. The structure of Fig. 11.12 recognizes and utilizes the common factor of $C1$, but leaves us with array D to realize. We can look for common factor subarrays of D in an iterative procedure, which will ultimately terminate because D is an array of $\$\gamma$ fewer variables than C, and common factor subarrays of D will be of still fewer variables. But for the moment we will consider only two realizations of D.

First, a two-level NAND tree may be used to achieve a single line representation of D. Fan-in problems will arise unless $q \leq \Lambda$ and $\$d^i \leq \Lambda$ for all $1 \leq i \leq q$. The number of input connections to the second- and third-level gates will be $\$\gamma + 1$ and q, respectively, and $\mathscr{G}(\$\gamma + 1)$ and $\mathscr{G}(q)$ NAND gates are thus required at these positions. For each cube d^i of D, $\mathscr{G}(\$d^i)$ gates are required. The total gate cost of this realization of $\overline{C1}$ is thus

$$\$_g^2(C1) = \mathscr{G}(\$\gamma + 1) + \mathscr{G}(q) + \sum_{i=1}^{q} \mathscr{G}(\$d^i) \tag{11.22}$$

This cost may be less than $\$_g^1(C1)$.

Alternatively, D may be realized with a bundle. Let $\Delta = \{\delta^1, \delta^2, \ldots, \delta^s\}$ be a minimal or nonredundant cover of $\overline{D}$.

$$\overline{D} \sqsubseteq \Delta \subseteq (U_n \# D) \tag{11.23}$$

The s cubes of Δ then specify the literals to be applied to NAND gates whose output lines form a bundle representation of D. $\mathscr{G}(\$\gamma + s)$ gates will be required in place of the single second-level gate shown in Fig. 11.12. The total gate cost of this realization of $\overline{C1}$ is then given by

$$\$_g^3(C1) = \mathscr{G}(\$\gamma + s) + \sum_{i=1}^s \mathscr{G}(\$\delta^i) \tag{11.24}$$

This cost may be less than that of the others and the number of logic levels may be less than that of the single line realization of D. Again factors of Δ might be used in a multiple-level factoring procedure.

C is a nonredundant, prime-implicant cover of a 10-variable switching function. Assume a fan-in limit $\Lambda = 4$.

$$C = \left\{ \begin{matrix} x0001x11x0 \\ x0111x11x0 \\ x0101x10x0 \\ x0010x10x0 \\ 0x1001xxxx \\ x1xxxx010x \\ 1x111xxxxx \\ 1x0x1xx0xx \\ 1x0001xxxx \\ x110011xx1 \end{matrix} \right\}$$

Partition C with

$$C1 = \left\{ \begin{matrix} x0001x11x0 \\ x0111x11x0 \\ x0101x10x0 \\ x0010x10x0 \end{matrix} \right\}$$

The bundle realization of $\overline{C1}$ has a cost of $\$_g^1(C1) = 3 + 3 + 3 + 3 = 12$ NAND gates. $\$_g^{1e}(C1) = 12 + \mathscr{G}(10) - 1 = 11 + 5 = 16$ NAND gates. (See Fig. 11.13). But $C1$ is a common factor subarray of C.

$$\gamma = \{x0xxxx1xx0\}, \qquad \$\gamma = 3$$

$$D = \left\{ \begin{matrix} xx001xx1xx \\ xx111xx1xx \\ xx101xx0xx \\ xx010xx0xx \end{matrix} \right\} \qquad \Delta = \left\{ \begin{matrix} xx11xxx0xx \\ xx10xxx1xx \\ xx01xxx1xx \\ xxx00xxxxx \\ xxxx0xx1xx \\ xx0x1xx0xx \end{matrix} \right\}$$

$$q = 4 \qquad\qquad s = 6$$

Thus a single line realization of both $\overline{C1}$ and D has a gate cost of

$$\$_g^2(C1) = 1 + 1 + (1 + 1 + 1 + 1) = 6 \text{ NAND gates}$$

The bundle realization of D has a cost of

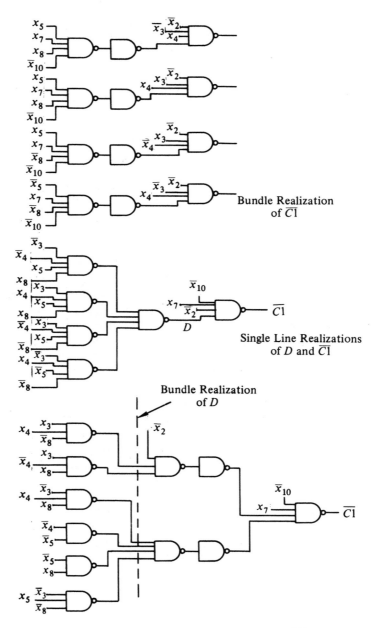

FIG. 11.13 Three Realizations of $\overline{C1}$

$$\$_g{}^3(C1) = 11 \text{ NAND gates.}$$

The use of a NAND tree to realize D is clearly advisable in this case

Finding Desirable Factors

How do we find a desirable factor of an array C? In general we will search for factors with a maximum figure-of-merit, subject to certain restrictions. We will assume C to be a nonredundant, prime-implicant cover of a Boolean function f of n variables. Experience suggests that little or no advantage is gained by beginning with an absolute minimal cover of f, and hence the extra processing required to obtain that cover will be avoided.

When examining C in search of a factor of width w, cubes with cost less than or equal to w may be excluded. Such cubes do not represent the product of sufficient literals. Thus we may form and deal with array C^*,

$$C^* = \{c^i \mid c^i \in C, \$c^i > w\} \qquad (11.25)$$

Let p^* be the number of members of C^*. Other arguments help to minimize the effort of locating a most desirable factor.

The common factor γ of array C with the maximum figure-of-merit may be found by searching. The amount of searching to be performed may be reduced successfully by limiting the range of factor widths examined. Theoretically, factors of width 1 and 2 give maximum figure-of-merit, but in a specific case a wider factor may maximize the figure-of-merit. But when searching for wider factors, at what point should the search terminate and which of the many factors found should be chosen?

Let k be the largest cube cost of the cubes of given array C. We need never seek a factor wider than $k - 1$, as none can exist. The costs of bundle realizations of a common factor subarray are independent of the width of the factor of that subarray. Thus the single line realization of array D primarily determines the width of factors worth finding. For example, if a factor of width $w = k - \Lambda$ exists and is chosen, neither $\$d^i$ nor $\$\delta^i$ can exceed Λ, and fourth and third level fan-in problems are avoided in realizations of D and Δ. Use of the factor of this width will create a second-level fan-in problem if $k - \Lambda > \Lambda - 1$, but it is more economical of gates and logic levels to use a few AND gates to realize the factor rather than many AND gates to realize the cubes of D.

A wider factor, of width $w \leq \Lambda - 1$, may be chosen when $k - \Lambda < \Lambda - 1$. Such a factor will not induce a second-level fan-in problem, and it does reduce $\$d^i$ for each cube, thus reducing the loading of the input variables and the number of interconnections in the network. It may permit the use of NAND gates with fan-in less than Λ to realize the cubes of D.

Thus we will search for the factor with the greatest figure-of-merit that satisfies

$$w \leq \min(k - 1, \max(k - \Lambda, \Lambda - 1)) \qquad (11.26)$$

If factors with the same figure-of-merit are encountered, we choose the widest subject to $w \leq \Lambda - 1$. Thus we may retain a narrower factor giving a higher common factor subarray to avoid second level fan-in problems and include more cubes of C in $C1$.

For the array C of the previous example, $n = 10$ and $\Lambda = 4$.

$$k - 1 = 6$$
$$k - \Lambda = 3$$
$$\Lambda - 1 = 3$$

Thus we consider only factors of width 3 or less. One such factor has been shown with the maximum figure–of–merit of 12. Another factor of width 4 leads to the same figure–of–merit.

$$\gamma = \{x0xx1x1xx0\} \qquad C1 = \begin{pmatrix} x0001x11x0 \\ x0111x11x0 \\ x0101x10x0 \end{pmatrix}$$

If this factor were utilized, a second-level fan-in problem would be encountered and a greater portion of C would remain to be synthesized.

In forming C^* we eliminate rows that can not enter a common factor subarray with factor of width w. Columns can also be identified as being unworthy of inclusion in a search. Let $h_i^{\,0}$ measure the number of 0's in the ith column of C^* and $h_i^{\,1}$ measure the number of 1's. Let

$$h_i = \max (h_i^{\,0}, h_i^{\,1}) \tag{11.27}$$

If we have previously found a factor with figure-of-merit M and are now searching for a factor of width w with a greater figure-of-merit, only those columns with $w \cdot h_i > M$ are of interest.

The following algorithm utilizes all of these facts about factors to reduce the amount of searching that must be performed before the most desirable factor within an array is located. The given array is repeatedly searched in a binary fashion, i.e. the array is split into two parts and each of the parts examined. If one part reveals a superior factor, it and its figure–of–merit are recorded. If no factor is found in a part, but its dimensions are such that a better factor could exist within it, the other part is set aside in array $SAVE$ and the first part is split into two parts each of which are examined, etc. Eventually the parts set aside in the $SAVE$ array are examined in the same fashion.

ALGORITHM 11.4. *Find the Factor of Maximum Figure–of–Merit.*
Let C be the n-variable array to be factored and w the width of factors under consideration. Let M be the largest figure of merit of all previously found factors.
Step 1. $M = 1$, $\gamma = U_n$. Perform the following steps, through 9, for $w = 1$, 2, etc., up to the maximum value dictated by Eq. (11.26), unless the procedure is terminated earlier.
Step 2. Select $C^* = \{c^i \mid c^i \in C \text{ and } \$c^i > w\}$. Let p^* be the height of C^*.
    If $w \cdot p^* \leq M$ advance w to its next value and repeat.
Step 3(*a*). Find h_i for each column of C^*.
Step 3(*b*). Find set $\Theta = \{i \mid w \cdot h_i > M\}$

If $|\Theta| < w$, advance w and return to step 2.

Step 4(a). Select a new $\theta \subseteq \Theta$ where $|\theta| = w$. If all such selections have already been made, advance w and return to step 2; otherwise $A \leftarrow C^*$.

Step 4(b). $SAVE = \phi$.

Step 5(a). For each $i \in \theta$ in turn,

$$X0 \leftarrow A\mathbf{S}\chi_i^0$$
$$X1 \leftarrow A\mathbf{S}\chi_i^1$$

Step 5(b). $\gamma' = \mathbf{F}(X0)$

If $\mathcal{M}(X0, \gamma') > M$ go to step 9.

If $|X0| < 2$ discard $X0$ by proceeding to step 6.

If $w \cdot |X0| < M$ discard $X0$ by proceeding to step 6.

Otherwise $X0$ can still provide a desirable common-factor subarray. Before continuing, determine if $X1$ can contain a factor of interest.

If $w \cdot |X1| \geq M$, $SAVE \leftarrow X1 \cup SAVE$.

$A \leftarrow X0$ and repeat this step for the next i. If the last column has already been selected, go to step 7.

Step 6. $\gamma' = \mathbf{F}(X1)$. Test as in step 5.

If $\mathcal{M}(X1, \gamma') > M$ go to step 9.

If $|X1| < 2$ discard $X1$ by proceeding to step 8.

If $w \cdot |X1| < M$ discard $X1$ by proceeding to step 8.

Otherwise $A \leftarrow X1$ and repeat step 5 for the next i.

Step 7. Splitting on all columns of θ has been completed. If $\mathcal{M}(X, \gamma') = M$ and $\$\gamma' < \Lambda$, then $\gamma \leftarrow \gamma'$.

Step 8. If $SAVE \neq \phi$, $A \leftarrow SAVE$ and go to step 4(b). Otherwise go to step 4(a).

Step 9. A new factor has been found. Set $M = \mathcal{M}(X, \gamma')$ and $\gamma \leftarrow \gamma'$. Go to step 3(b).

The following example will clarify this algorithm; the FORTRAN program of Appendix 11.1 implements this program and offers greater detail.

Example

We seek the factor of array C of previous examples with maximum figure–of–merit and width ≤ 3.

1. $M = 1$. (A larger initial value prevents finding factors in very small arrays.)

       $\gamma = \{xxxxxxxxxx\}$ (This is an indication that no true factor has as yet been found, or may be our end result if no factor exists.)

$$\boxed{w = 1}$$

2. $C^* = C$ since no cubes of C have unit cost.

      $p^* = 10$ and $w \cdot p^* > M$.

3.   Column heights are as shown below. Column 9 need not be examined for a factor of width 1.

$$C* = \begin{cases} x0001x11x0 \\ x0111x11x0 \\ x0101x10x0 \\ x0010x10x0 \\ 0x1001xxxx \\ x1xxxx010x \\ 1x111xxxxx \\ 1x0x1xx0xx \\ 1x0001xxxx \\ x110011xx1 \end{cases}$$

$h_i = \quad 3455535314$

4. Select column 1 as the test column, and let $SAVE = \phi$.

5(a) $X0 = \{0x1001xxxx\}$

5(b) $\gamma' = \{xxxxxxxxxx\}$

$\mathcal{M}(X0, \gamma') < M$, and $|X0| < 2$

6.
$$X1 = \begin{cases} 1x111xxxxx \\ 1x0x1xx0xx \\ 1x0001xxxx \end{cases} \qquad |X1| = 3$$

$\gamma' = \{1xxxxxxxxx\}$

$\mathcal{M}(X1, \gamma') = 3 > M$

9. $\gamma = \{1xxxxxxxxx\}$, $M = 3$

3(b) $\theta = \{2, 3, 4, 5, 7, 10\}$. Only these columns can lead to more desirable factors of unit width.

4. Select $\theta = \{2\}$. $i = 2$.

5(a)
$$X0 = \begin{cases} x0001x11x0 \\ x0111x11x0 \\ x0101x10x0 \\ x0010x10x0 \end{cases} \qquad X1 = \begin{cases} x1xxxx010x \\ x110011xx1 \end{cases}$$

5(b) $\gamma' = \{x0xxxx1xx0\}$ Note that a factor of width greater than 1 has been found.

$\mathcal{M}(X0, \gamma') = 12 > M$

9. $\gamma = \{x0xxxx1xx0\}$, $M = 12$

3(b) No columns are of further interest with $w = 1$.

2. $\boxed{w = 2}$ $C* = C$, $p* = 10$, and $w \cdot p* > M$.

3. Column heights are as previously shown, since $C*$ has not been altered. But no column heights are sufficient to satisfy $w \cdot h_i > M$.

2. $\boxed{w = 3}$ $C* = C, \ldots$

3. $\theta = \{3, 4, 5, 7\}$

4. Select columns $\theta = \{3, 4, 5\}$, $i = 3$, $SAVE = \phi$.

5(a)
$$X0 = \begin{cases} x0001x11x0 \\ x0010x10x0 \\ 1x0x1xx0xx \\ 1x0001xxxx \end{cases} \qquad X1 = \begin{cases} x0111x11x0 \\ x0101x10x0 \\ 0x1001xxxx \\ 1x111xxxxx \\ x110011xx1 \end{cases}$$

5(b) $\gamma' = \{xx0xxxxxxx\}$ $\qquad\qquad$ $\gamma' = \{xx1xxxxxxx\}$

$\mathscr{M}(X0, \gamma') = 4 < M$, $|X0| > 2$, but $w \cdot |X0| = M$.

$\mathscr{M}(X1, \gamma') = 5 < M$, $|X1| > 2$, and $w \cdot |X1| > M$. $SAVE \leftarrow X1$.

$A \leftarrow X0$.

5(a) Split A on column 4.

$$X0 = \begin{Bmatrix} x0001x11x0 \\ 1x0001xxxx \end{Bmatrix} \qquad X1 = \{x0010x10x0\}$$

5(b) Since $w \cdot |X0| < M$, discard $X0$.

6. $\quad$ Since $|X1| < 2$, discard $X1$.

8. $\quad$ $A \leftarrow SAVE$.

4(b) Set SAVE empty.

5(a) Splitting on column 3 gives

$X0 = \phi$

$$X1 = \begin{Bmatrix} x0111x11x0 \\ x0101x10x0 \\ 0x1001xxxx \\ 1x111xxxxx \\ x110011xx1 \end{Bmatrix}$$

6. $\quad$ For $X1$, $|X1| = 5$, $\gamma' = \{xx1xxxxxxx\}$ but $\mathscr{M}(X1, \gamma') = 5 < M$.

Therefore, $A \leftarrow X1$.

5(a) Splitting on column 4:

$$X0 = \begin{Bmatrix} x0101x10x0 \\ 0x1001xxxx \\ x110011xx1 \end{Bmatrix} \qquad X1 = \begin{Bmatrix} x0111x11x0 \\ 1x111xxxxx \end{Bmatrix}$$

5(b) $w \cdot |X0| < M$, discard $X0$.

6. $\quad$ $w \cdot |X1| < M$, discard $X1$.

8. $\quad$ Since $SAVE = \phi$, go to step 4(a).

4(a) All columns of interest for $\theta = \{3, 4, 5\}$ have been tested. The above steps are repeated for $\theta = \{3, 4, 7\}$; with $\theta = \{4, 5, 7\}$ no better factor is found. Since $w = 3$ is the maximum width factor sought, execution of the algorithm is complete.

This example suggests that Algorithm 11.4 requires a rather large amount of computation and many complex decisions. It is not the way that we would find a factor by hand. (By hand we often fail to find the best factor.) Could we not more simply establish each possible factor of width 1, 2, and 3, and test each? Only the split operator and a simple computation would be required to evaluate a factor. We could, but a total of

$$\binom{10}{1} \cdot 2 + \binom{10}{2} \cdot 2^2 + \binom{10}{3} \cdot 2^3 = 1160$$

such tests would have to be made. In the example we performed only six (6) pairs of splits!

It has been observed when executing this algorithm that the desirable factor can usually be found by splitting an array on each column in turn and examining the fragments with the factor operator. The preceding example illustrates this. An algorithm based upon this observation is extremely simple and efficient.

ALGORITHM 11.5. *Find a Factor Rapidly*. See Appendix 11.2.

Step 1. $M = 1$, $\gamma = U_n$. Perform the following steps for each column of C. Let i denote the column under test.

Step 2. $A \leftarrow C$

Step 3. For $b = 0$, then $b = 1$

$$X \leftarrow A\mathbf{S}\chi_i^b$$
$$\gamma' = \mathbf{F}(X)$$

If $\mathcal{M}(X, \gamma') \geq M$, $M = \mathcal{M}(X, \gamma')$, $\gamma \leftarrow \gamma'$, and if $|X| = |C|$ terminate the search.

The final test has been found to be a desirable addition. It may cause the algorithm to terminate with a very tall common-factor subarray, but less than maximum figure-of-merit.

Extensive testing of these two algorithms in conjunction with the synthesis algorithm to be presented next reveals that most often they find factors that lead to networks of the same cost. Algorithm 11.5 leads to lower cost circuits when, after $\Lambda - 1$ factors have already been selected, it selects one factor common to all remaining cubes, thus avoiding a first-level fan-in problem. Algorithm 11.4 has produced more economical networks when it selected one factor rather than another of equal figure-of-merit. Algorithm 11.5 requires substantially less computer time, as might be expected.

Synthesis Algorithm for Single-Output Functions

We now utilize the facts and techniques of this section in an algorithm to synthesize single-output networks of fan-in-limited NAND gates. This algorithm is not particularly powerful or general since it finds only one factor of each subarray and then uses AND gates to solve remaining fan-in problems. But it provides the techniques needed in a more general iterative factoring algorithm.

ALGORITHM 11.6. *Synthesis of Fan-In-Limited NAND Gate Networks*.

Assume we are given a nonredundant, prime-implicant cover C of a Boolean function f.

Step 1. Form $C' = \{c^i | c^i \in C, \$c^i = 1\}$. Let $C \leftarrow C - C'$. Use $c^s = U_n \# C'$ as input literals to a first-level gate. We thus eliminate $(n - 1)$-cubes from further considerations.

Step 2. Find the factor γ with the largest figure-of-merit, using Algorithm 11.4 (or find a factor with Algorithm 11.5). Let $C1$ be the subarray with common factor γ. $C1 \leftarrow C\mathbf{S}\gamma$. Compute the D and Δ arrays. If at some point there is no common-factor subarray of C, Algorithms 11.4 and 11.5 provide a factor $\gamma = U_n$ indicating that $C1 = C$.

Step 3. Calculate $\$_g^{1e}(C1)$, $\$_g^2(C1)$ and $\$_g^3(C1)$, and compare as follows:

(a) $\$_g^{1e}(C1) \leq \$_g^2(C1)$

If $\$_g^{1e}(C1) \leq \$_g^3(C1)$, use a bundle realization of $\overline{C1}$.

If $\$_g^{1e}(C1) > \$_g^3(C1)$, use a bundle realization of D.

(b) $\$_g^{1e}(C1) > \$_g^2(C1)$

If $\$_g{}^2(C1) < \$_g{}^3(C1)$, use a single line realization of D.

If $\$_g{}^2(C1) \geq \$_g{}^3(C1)$, use a bundle realization of D.

Step 4. If necessary, use AND gates to solve the first logic level fan-in problem.

We have already found the first factor of the example 10-variable array, and the costs of the three realizations of the corresponding common factor subarray in previous examples. A 6-NAND gate, single line realization of D is clearly suggested.

Delete $C1$ from C and repeat the process. Another factor is found.

$$C1 = \begin{Bmatrix} 0x1001xxxx \\ 1x0001xxxx \\ x110011xx1 \end{Bmatrix} \qquad D = \begin{Bmatrix} 0x1xxxxxxx \\ 1x0xxxxxxx \\ x11xxx1xx1 \end{Bmatrix}$$

$$\gamma = \{xxx001xxxx\} \qquad \Delta = \begin{Bmatrix} 0x0xxxxxxx \\ 1x1xxx0xxx \\ 101xxxxxxx \\ 1x1xxxxxx0 \end{Bmatrix}$$

Again $\$_g{}^2(C1)$ is less than $\$_g{}^{1e}(C1)$ and $\$_g{}^3(C1)$, and a single line realization of this new D-array is used.

A third common-factor subarray leads to a bundle realization of D. While $\$_g{}^1(C1) < \$_g{}^3(C1)$, the first-level fan-in problem causes $\$_g{}^{1e}(C1) > \$_g{}^3(C1)$.

$$C1 = \begin{Bmatrix} 1x111xxxxx \\ 1x0x1xx0xx \end{Bmatrix} \qquad D = \begin{Bmatrix} xx11xxxxxx \\ xx0xxxx0xx \end{Bmatrix}$$

$$\gamma = \{1xxx1xxxxx\} \qquad \Delta = \begin{Bmatrix} xx10xxxxxx \\ xx0xxxx1xx \end{Bmatrix}$$

The remaining cube of C is realized directly, as shown in Fig. 11.14. Note that only 16 gates and 4 levels are needed. Transforming the 2-level AND-OR network directly to a 2-level NAND network, and then using AND gates to solve the fan-in problem leads to a 29-gate, 6-level network.

A FORTRAN program that implements this algorithm is given in Appendix 11.3. With minor modifications this algorithm can place less emphasis on gate count and give networks of minimum or near-minimum logic levels. A number of additional considerations lead to techniques efficacious in special circumstances, which might be incorporated into more sophisticated NAND synthesis algorithms. A number of these are introduced in the problems.

Factoring Multiple-Output Networks

The types of factoring discussed to this point apply to a single-output switching function, or to a single output variable of a multiple-output switching function. The use of AND gates to resolve fan-in problems was generally discouraged. When synthesizing multiple-output switching

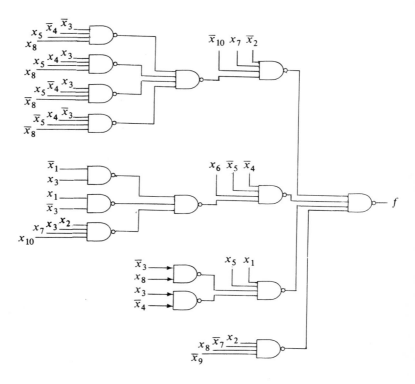

FIG. 11.14 The Synthesized Network

functions, the use of AND gates to realize common factors can be desirable.
  Consider the following function array F and connection array E derived
from F via one of the two-level minimization algorithms.

$$F = \begin{pmatrix} 0000 \; 0d0d0 \\ 0001 \; 0dd00 \\ 0010 \; d11d1 \\ 0011 \; d0d00 \\ 0100 \; 0d1d1 \\ 0101 \; 0dd00 \\ 0110 \; d11d1 \\ 0111 \; d0d01 \\ 1000 \; 1d1d1 \\ 1001 \; 0dd00 \\ 1010 \; d01d0 \\ 1011 \; d1d01 \\ 1100 \; 1d0d0 \\ 1101 \; 0dd00 \\ 1110 \; d00d0 \\ 1111 \; d1d10 \end{pmatrix} \qquad E = \begin{pmatrix} 1xx0 \; 10000 \\ 1xx1 \; 01000 \\ 10xx \; 00100 \\ 01xx \; 00100 \\ 111x \; 00010 \\ 1011 \; 00001 \\ 1000 \; 00001 \\ 01x0 \; 00001 \\ 011x \; 00001 \\ 0x10 \; 01101 \end{pmatrix}$$

In the input variable portion of the E array factor $\gamma_1 = 10xx\ xxxxx$ with figure-of-merit of 6 is common to three cubes.

$$\begin{cases} 10xx\ 00100 \\ 1011\ 00001 \\ 1000\ 00001 \end{cases}$$

The fan-in requirement of two second-level NAND gates can be reduced from 4 to 3 if this factor is realized with an AND structure. This, in fact is a very attractive factor; half of the AND structure must already be a part of the two-level realization dictated by E because of the first cube in the common factor array above. Thus the reduced fan-in may be purchased at a cost of one additional inverter.

Note that γ_1 can not be used in the manners previously discussed in this section because cube $10xx \circ 00100$ of the common factor array pertains to output variable z_3 while the other two cubes pertain to z_5. We do not seek a single line realization of a $C1$ but must realize $10xx$ as part of the z_3 subnetwork, and 1011 and 1000 as part of the z_5 network.

Realizing γ_1 with an AND structure does not reduce the fan-in to the first-level gate which provides z_5. If we examine the cubes which describe the z_5 subnetwork, several factors with figure-of-merit of 4 are found.

x_1	x_2	x_3	x_4	γ_1	z_5
x	x	1	1	1	1
x	x	0	0	1	1
0	1	x	0	x	1
0	1	1	x	x	1
0	x	1	0	x	1

These factors may be found, evaluated, and utilized via Algorithm 11.6. Figure 11.15 shows a realization of this switching function in which factor $01xxx$ is used to reduce the fan-in of the first-level gate of z_5 from 5 to 4. Factors $0xxxx$ or $xx1xx$ could be used to reduce this fan-in to 3, but the realization of their D-arrays is expensive. And since cube $0x10x \circ 01101$ is common to z_2, z_3, and z_5, a second-level gate which realized $0x10x$ must be included in the network for purposes of generating z_2 and z_3, even if this cube is realized in a factored structure in the z_5 subnetwork.

Finding AND structure factors to reduce second level fan-in and finding traditional factors to reduce first level fan-in require some modification of Algorithms 11.4 and 11.5. Deciding which of many found factors to actually use requires modification of Algorithm 11.6. The cost function can be changed from a simple gate count to a proportion of the fan-in of gates, which more accurately reflects the cost of IC gates or the area required by a gate in an LSI realization. The use of NOR gates in place of AND gates

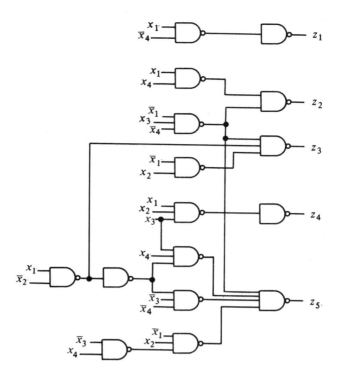

FIG. 11.15 A Factored Multiple-output Network

can be permitted. While all of these modifications are interesting, they add little that is fundamentally different, and are of little value if we cannot represent a multiple-level network in array form.

Array Expression of Multiple-Level Networks

Arrays are excellent for representing two-level networks. When we factor, we replace a single 2-level network with a network of 2-level networks. Each of the subnetworks and their interconnections are easily expressed in an array by naming the output variables of the subnetworks and introducing columns for them.

Two factors are shown in Fig. 11.15. Let us call the functions found u and v.

$$u = x_1 \bar{x}_2$$
$$v = x_3 \vee \bar{x}_4$$

If two columns are introduced for each intermediate variable, the connections of Fig. 11.15 may be expressed as follows

$$
E' = \left\{
\begin{array}{cccccc|ccccccc}
x_1 & x_2 & x_3 & x_4 & u & v & u & v & z_1 & z_2 & z_3 & z_4 & z_5 \\
1 & x & x & 0 & x & x & 0 & 0 & 1 & 0 & 0 & 0 & 0 \\
1 & x & x & 1 & x & x & 0 & 0 & 0 & 1 & 0 & 0 & 0 \\
0 & x & 1 & 0 & x & x & 0 & 0 & 0 & 0 & 1 & 0 & 1 \\
0 & 1 & x & x & x & x & 0 & 0 & 0 & 0 & 1 & 0 & 0 \\
1 & 1 & 1 & x & x & x & 0 & 0 & 0 & 0 & 0 & 1 & 0 \\
x & x & 1 & 1 & 1 & x & 0 & 0 & 0 & 0 & 0 & 0 & 1 \\
1 & 0 & x & x & x & x & 1 & 0 & 0 & 0 & 1 & 0 & 0 \\
x & x & 0 & 0 & 1 & x & 0 & 0 & 0 & 0 & 0 & 0 & 1 \\
x & x & 1 & x & x & x & 0 & 1 & 0 & 0 & 0 & 0 & 0 \\
x & x & x & 0 & x & x & 0 & 1 & 0 & 0 & 0 & 0 & 0 \\
0 & 1 & x & x & x & 1 & 0 & 0 & 0 & 0 & 0 & 0 & 1
\end{array}
\right\}
$$

The left u and v columns indicate how these variables appear in product terms, serve as input variables to AND gates. The right u and v columns give connection information. They identify the product terms that are summed to realize u and v, respectively.

Using Algorithm 9.1 to find ON-arrays only gives

$$\text{ON}(u) = \{10xxxx\} \qquad x_1 \cdot \bar{x}_2$$

$$\text{ON}(v) = \begin{Bmatrix} xx1xxx \\ xxx0xx \end{Bmatrix} \qquad x_3 \vee \bar{x}_4$$

$$\text{ON}(z_3) = \begin{Bmatrix} 0x10xx \\ 01xxxx \\ 10xxxx \end{Bmatrix} \qquad \bar{x}_1 x_3 \bar{x}_4 \vee \bar{x}_1 x_2 \vee x_1 \bar{x}_2$$

$$\text{ON}(z_5) = \begin{Bmatrix} 0x10xx \\ xx111x \\ xx001x \\ 01xxx1 \end{Bmatrix} \qquad \bar{x}_1 x_3 \bar{x}_4 \vee x_3 x_4 u \vee \bar{x}_3 \bar{x}_4 u \vee \bar{x}_1 x_2 v$$

An extended connection array may be processed in the same way as a conventional connection array. We might look for further factors in the input part of array E', for example; and those factors could now include variables u and v to achieve multiple level factoring.

In the function given by

$$z_1 = A\bar{B} \vee \bar{A}C$$
$$z_2 = \bar{A}B \vee A\bar{B}$$
$$z_3 = A\bar{B} \vee A\bar{C}$$

z_3 may also be expressed as a function of z_2.

$$z_3 = \bar{z}_2 \bar{C} \vee A\bar{C}$$

At times there is real value in expressing output variables as functions of others (see Prob. 10.4-3). The same sort of extended connection array may be used to record such expressions. For this example, z_2 must appear as an input, as well as an output variable.

$$ABCz_2 \quad z_1z_2z_3$$

$$\begin{pmatrix} 1\,0\,x\,x & 1\,1\,0 \\ 0\,x\,1\,x & 1\,0\,0 \\ 0\,1\,x\,x & 0\,1\,0 \\ x\,x\,0\,0 & 0\,0\,1 \\ 1\,x\,0\,x & 0\,0\,1 \end{pmatrix}$$

Factoring is a technique for altering *sop* (*pos* by duality) Boolean expressions to compound forms so as to achieve economy or meet fan-in limitations. It does not provide means for expressing dependent variables as functions of other dependent variables, as is occasionally advantageous. How then do we derive such expressions?

11.3 FEED-FORWARD NETWORK SYNTHESIS

A network in which one output signal serves as an input signal for another output signal is a *feed-forward* network. If the second output signal also serves as an input signal to the first, a loop exists and the network is an example of a *feed-back* network. Feedback seldom appears in combinational networks; the 1's complement adder with the end-around carry is a marginal example since that network does exhibit the behavior of sequential networks. Ho [7] has shown how feed-forward and feed-back connections that reduce NAND network costs can be found algorithmically. While his full treatment of the subject is too extensive to include here, some interesting extensions of array theory are presented in this section.

Cube c of function array F is an $(n + m)$-tuple where the n-tuple c_α and the output part c_Ω convey different types of information. c_α describes a product term. An *extended cube* $c = c_\alpha \circ c_\varepsilon$, with $c_\varepsilon \neq U_m$, is an $(n + m)$-tuple description of a product of input and at least one output variable, or their complements. The m-tuple c_ε extends the traditional cube using the same code.

Since an extended cube encodes a product term, it describes a switching function on the primary input variables. ON-, OFF-, and DC-arrays exist for that function. They are found with the following algorithm.

ALGORITHM 11.7. ON-, OFF-, *and* DC-*arrays of an extended cube.*
For a given function array F and extended cube c:

Step 1. Replace each cube $f^i \in F$ with $K^0(f_\alpha^i) \times f_\Omega^i$.
Step 2. $F^1 = F\mathbf{S}c$
Step 3. $A = U_{n+m} \,\#\, c$ and $F^0 = \phi$
Step 4. For each $a \in A$ compute $F^0 \leftarrow F^0 \bigsqcup (F\mathbf{S}a)$
Step 5. ON$(c) = \mathbf{D}_\Omega(F^1)$
$\quad\quad$ OFF$(c) = \mathbf{D}_\Omega(F^0)$
$\quad\quad$ DC$(c) = \mathbf{D}_\Omega(F)$
Step 6. $F \leftarrow F \cup F^1 \cup F^0$

An example clarifies and justifies this algorithm; let

$$c = 1xx \circ 1xxx \qquad \text{and } F = \begin{cases} 000 & 1101 \\ 001 & 0110 \\ 01x & 0011 \\ 100 & 1100 \\ 101 & x011 \\ 110 & 0100 \\ 111 & 1xx0 \end{cases}$$

$$(x_1 z_1)$$

Step 1 requires that the input parts of cubes of F be reduced to 0-cubes. Thus the third cube must be replaced with

$$\begin{cases} 010 & 0011 \\ 011 & 0011 \end{cases}$$

This is one of the few points where an x representing both the 0 and 1, and the d representing don't cares must be treated differently. The split operator used in subsequent steps cannot distinguish between the two so the x's must be removed.

Step 2 filters from F all cubes for which all variables of the product term described by c have value 1. The product term has value 1 in these cases. For the product $x_1 z_1$

$$F^1 = \begin{cases} 100 & 1100 \\ 111 & 1xx0 \end{cases}$$

A product term has value 0 if any of its components has value 0. The A array describes the complement of the product term. Step 4 filters all cases in which the product is 0 by finding cases in which a component is 0.

$$A = \begin{cases} 0xx & xxxx \\ xxx & 0xxx \end{cases}$$

$$F^0 = \begin{cases} 000 & 1101 \\ 001 & 0110 \\ 010 & 0011 \\ 011 & 0011 \\ 110 & 0100 \end{cases} \qquad \begin{aligned} & FSa^1 \\ \\ & FSa^2 \end{aligned}$$

At this point

$$F = \{101 \quad x011\}$$

Step 5 deletes the output variable columns, giving the desired arrays.

$$ON(c) = \begin{cases} 100 \\ 111 \end{cases} \qquad OFF(c) = \begin{cases} 000 \\ 001 \\ 010 \\ 011 \\ 110 \end{cases} \qquad DC(c) = \{101\}$$

To see the validity of these results, prepare a Karnaugh map of z_1 from the given function array; then prepare a map for $x_1 z_1$.

Two definitions of the cover relation were given in Chapter 9. While the absorb operator of Section 9.3 indicates that $xxx1xx0 \not\sqsubseteq 1xx1xxx$ and

$1xx1xxx \not\sqsubseteq xxx1xx0$, the spirit of the definition, "$c^i \sqsubseteq c^j$ if and only if $K^0(c^i) \subseteq K^0(c^j)$", is violated when those cubes are viewed as extended cubes of the example function array above. They have the same ON-, OFF-, and DC-arrays, and therefore, express the same switching function and should be related. Ho found it to be very valuable to define the *extended cover relation* $c^i \leq c^j$ if $ON(c^i) \sqsubseteq ON(c^j)$, which extends the base definition. Since extended cubes generally describe partial switching functions, we cannot conclude that $OFF(c^i) \sqsupseteq OFF(c^j)$ from $c^i \leq c^j$. While $c^i \sqsubseteq c^j$ implies that $c^i \leq c^j$, the converse is not true, as the example cubes show.

Members of connection arrays are also $(n + m)$-tuples. The coding in the output part of a connection array is simpler than in a function array; in this section we will use 1 to represent a connection and an x to represent the absence of a connection. An *extended connection* array (EC-array) is a set of $(n + m + m)$-tuples, $c = c_\alpha \circ c_\varepsilon \circ c_\Omega$, that describes a realization of the function array from which it was derived. If the extension part $c_\varepsilon = U_m$, then cube c describes a product of input literals only and the connection of an AND gate to one or more OR gates.

An extended connection array does not necessarily describe a lower cost network than a traditional connection array. Algorithms for finding absolute minimum cost, extended connection arrays are not available, but network cost can sometimes be reduced in the following manner. First, extend the connection array by extending each cube with $c_\varepsilon = U_m$. Then, repeatedly replace subsets of the extended connection array with a single extended prime implicant such that (1) the cost of the network is reduced and (2) the new network realizes the original function. This rather obvious algorithm requires us to find extended prime implicants that are potentially suitable substitutes and maximal subsets of EC arrays that extended prime implicants can replace.

Finding potentially useful extended prime implicants is very involved. Ho argues that only those with a single 0 or 1 in c_ε have significant probability of reducing network cost. Of course, if $c_\alpha \neq U_n$, then a 1 should appear in c_ε; if $c_\alpha = U_n$, then a 0 should appear in c_ε for NAND network cost reduction. These later cubes describe one output signal that drives directly the NAND gate, which produces another or other output signals. The extended prime implicant $xxx \circ 0xxx \circ x1x1$ indicates that $\bar{y}_1$ is a valid product term of y_2 and y_4, which could be written

$$y_2 = \bar{y}_1 \vee \cdots$$
$$y_4 = \bar{y}_1 \vee \cdots$$

To realize such sums, variable y_1 must be supplied to a first level NAND operator.

Extended prime implicants with a single 0 in c_ε may be generated with the following algorithm.

ALGORITHM 11.8. *Unit cube cost, extended prime implicants.*
For a given function array F:

Step 1. $EPI(F) = \phi$.

Step 2. For each $h = 1, \ldots, m$ in turn:

    (a) Set $S = U_m$.

    (b) For each $i = 1, \ldots, m$, but $i \neq h$ in turn:

        If $ON(\bar{y}_i \bar{y}_h) = \phi$, then $S \leftarrow \mathbf{C}_i^{111} S$

    (c) If $S \neq U_m$, then $EPI(F) \leftarrow EPI(F) \cup (U_n \circ \chi_h^0 \circ S)$

Step 3. If extended prime implicants with a single 1 in c_ε are desired, then repeat step 2, substituting y_h for $\bar{y}_h$ in step 2(b) and χ_h^1 in place of χ_h^0 in step 2(c).

We will not prove that the resulting cubes are extended prime implicants. Their extended input parts of $(n + m - 1)$-cubes suggests their desirability. The test of $ON(\bar{y}_i \bar{y}_h)$ ensures that the ON-array of $\bar{y}_i$ and the OFF-array of y_h have nothing in common. If this test is satisfied, then $\bar{y}_h$ can appear as a product term in the expression of y_i. With $h = 1$ and $i = 2, 3, 4$ for the example function array examined earlier

$$ON(\bar{y}_2 \bar{y}_1) = \{01x\} \neq \phi$$

Therefore $\bar{y}_1$ is not an implicant of y_2.

$$ON(\bar{y}_3 \bar{y}_1) = \{110\}$$

$$ON(\bar{y}_4 \bar{y}_1) = \begin{Bmatrix} 001 \\ 110 \end{Bmatrix}$$

and $S = U_4$ so no cubes are added to $EPI(F)$ in step 2(c). For $h = 2$, $S = xx11$ and

$$EPI(F) = \{xxx \quad x0xx \quad xx11\}$$

For $h = 3$, $S = x1xx$.
For $h = 4$, $S = x1xx$.

Thus after step 2 is completed

$$EPI(F) = \begin{Bmatrix} xxx & x0xx & xx11 \\ xxx & xx0x & x1xx \\ xxx & xxx0 & x1xx \end{Bmatrix}$$

and $\bar{y}_2$ may appear in the expressions for y_3 and y_4, and $\bar{y}_3$ and $\bar{y}_4$ may appear in the expressions of y_2. We must be certain to avoid a feedback loop that leads to a sequential rather than combinational network.

If we test these extended prime implicants against the connection array

$$C(F) = \begin{Bmatrix} 1x1 & 1xxx \\ x00 & 11xx \\ 00x & x1xx \\ 11x & x1xx \\ xx1 & xx1x \\ 01x & xx11 \\ 101 & xxx1 \\ 0x0 & xxx1 \end{Bmatrix}$$

we find the *bundle equivalent* of each, the subset of $C(F)$ that can be replaced by the extended implicant.

$$BE(xxx \circ x0xx \circ xx11) = \begin{pmatrix} 01x & xx11 \\ 101 & xxx1 \end{pmatrix}$$

Replacing two cubes of $C(F)$ with this entended prime implicant would save 6 connections and 2 gates.

$$BE(xxx \circ xx0x \circ x1xx) = \begin{pmatrix} 00x & x1xx \\ 11x & x1xx \end{pmatrix}$$

This replacement would save 5 connections and 2 gates.

$$BE(xxx \circ xxx0 \circ x1xx) = \{11x \quad x1xx\}$$

Only two connections and one gate can possibly be saved.

If we make the first replacement to gain the greatest cost reduction, the connection array becomes

$$EC(F) = \begin{cases} xxx & x0xx & xx11 \\ 1x1 & xxxx & 1xxx \\ x00 & xxxx & 11xx \\ 00x & xxxx & x1xx \\ 11x & xxxx & x1xx \\ xx1 & xxxx & xx1x \\ 0x0 & xxxx & xxx1 \end{cases}$$

Making additional replacements results in a feedback network with sequential properties, and therefore those replacements must not be made.

11.4 DECOMPOSITION

Network synthesis based on common factors of a cover proceeds from the output terminals of the ultimate network towards the input terminals. We now investigate a synthesis philosophy that proceeds from the input terminals towards the output terminals, and allows the introduction of practical constraints such as fan-in and type of available gates or modules. This type of synthesis, known as *decomposition*, is based upon the idea of breaking a function into small parts, each of which can be realized more easily than the original function, if not directly, by available gates or assemblies of gates, and hence is particularly appropriate to integrated circuit technology, in which assemblies of permanently interconnected gates can be fabricated at low cost.

Let X be the set of input variables, $X = \{x_1, x_2, \ldots, x_n\}$, of a single output function $f(X)$. Further let λ and μ be subsets of X such that

$$\lambda \cup \mu = X \tag{11.28}$$

If $f(X)$ can be expressed in the composite form

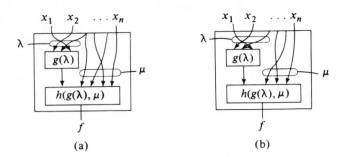

FIG. 11.16 Decomposition of $f(X)$.
(a) Nondisjoint Decomposition
(b) Disjoint Decomposition

$$f(X) = h(g(\lambda), \mu) \tag{11.29}$$

that expression of $f(X)$ is a *simple decomposition* of $f(X)$. When λ and μ are disjoint subsets of X we refer to this expression as a *simple disjoint decomposition*; otherwise it is known as a *nondisjoint decomposition*.

The function

$$F = \bar{A}C \lor AB\bar{C} \lor \bar{B}C$$

may be rewritten

$$F = \bar{A}C \lor G\bar{C} \lor \bar{B}C$$

where $G = AB$

which expresses a nondisjoint decomposition in which $\lambda = \{A, B\}$ and $\mu = \{A, B, C\}$. Alternatively, again with $G = AB$, we may write

$$F = G \oplus C$$

which expresses a disjoint decomposition, since A and B, the variables of G, are not explicit variables of F.

Each type of simple decomposition has advantages and disadvantages. The simple disjoint decomposition eliminates fan-out of input variables and reduces the synthesis problem to one of synthesizing two functions each of less than n variables. Thus we might expect synthesis of the " pieces " to be an easy matter. Unfortunately, disjoint decompositions do not always exist and do not necessarily lead to lowest cost networks. For example, if we attempt to decompose example function F above in the manner $F(G(B, C), A)$, we find that no single output Boolean function $G(B, C)$ exists to make this possible.

More generally, a *decomposition* of a switching function $f(X)$ is a sequence of functions $g_1(\lambda_1, G_1), g_2(\lambda_2, G_2), \ldots, g_p(\lambda_p, G_p)$ where $\lambda_i \subseteq X$

and $G_i \subseteq \{g_1, g_2, \ldots, g_{i-1}\}$ such that g_p is defined and equal to $f(X)$ for every input symbol for which $f(X)$ is defined. Thus $f(X)$ need not be completely specified, and the individual functions $g_i(\lambda_i, G_i)$ may correspond either to disjoint or nondisjoint decompositions. Synthesis may then consist of finding a sequence of $g_i(\lambda_i, G_i)$ such that each corresponds to an available logic gate or assembly of gates and g_p expresses $f(X)$. We may seek to minimize p as well, or introduce a cost function that permits one sequence of g_i's to be chosen over another.

A decomposition may be found by repeated simple decompositions— an iterative procedure. To obtain an efficient synthesis procedure, we will want not merely to find a simple decomposition, but the "best" simple decomposition at each step, where "best" means the simple decomposition that leads to the overall decomposition optimum under a given cost function. In general, this implies that for each known decomposition, we must be able to look ahead to the set of final networks that may be derived from that simple decomposition. An exhaustive examination of all decompositions of the function is suggested. The expense of an exhaustive search may prevent our finding the optimum realization of a function or knowing that a network we have found is in fact optimum.

Disjoint Decompositions

Assume that we seek a disjoint decomposition of $f(X)$ such that $g(\lambda)$ describes a logic gate available to us. A number of problems arise. How do we select $\lambda \subseteq X$? We can always test all the subsets of X; known partial symmetry sets of $f(X)$ can be used to limit the list of λ's we test, or they can be used to suggest subsets of X to examine first as likely to lead to a disjoint decomposition. The number of members of λ must correspond to the number of input terminals on some available logic gate, in any case. Once we have arrived at a set λ in some manner we must ask: "Does a disjoint decomposition of $f(X)$ exist for this λ?" In answering this question we will also determine the gate specified by $g(\lambda)$, if any, and the manner in which the members of λ must be connected to the input terminals of that gate.

Let λ have r members; then $\mu = X - \lambda$ has $n - r$ members. For conceptual purposes we will reorder the columns of the function array expressing $f(X)$ so that the members of λ appear on the left. Then the columns of that function array may be labeled $\lambda_1, \lambda_2 \ldots, \lambda_r, \mu_{n-r}, \ldots, \mu_n, z_1, \ldots, z_m$, and the ith cube of the array may be expressed as

$$c_\alpha{}^i \circ c_\Omega{}^i = c_\lambda{}^i \circ c_\mu{}^i \circ c_\Omega{}^i.$$

The λ-parts of two cubes c^i and c^j are said to be *incompatible*, expressed $c_\lambda{}^i \sim c_\lambda{}^j$, if:

$$\begin{aligned} &1.\ c_\mu{}^i \textstyle\bigcap c_\mu{}^j \neq \phi, \text{ and} \\ &2.\ c_\Omega{}^i \textstyle\bigcap c_\Omega{}^j = \phi. \end{aligned} \qquad \textbf{(11.30)}$$

Further, all binary r-tuples that $c_\lambda{}^i$ covers are incompatible with those covered by $c_\lambda{}^j$. Those binary r-tuples that are not incompatible are *compatible*.

Let $\lambda = \{x_1, x_2\}$ and $\mu = \{x_3, x_4\}$ with the following function array.

| | λ | μ | |
	$x_1 x_2$	$x_3 x_4$	z_1
c^1	1 0	1 x	1
c^2	1 x	1 0	1
c^3	0 1	x x	1
c^4	1 x	0 0	0
c^5	x 0	0 x	0
c^6	1 1	x 1	d
c^7	0 0	1 x	d

As we compare cubes of this function array with the above requirements in mind, we find $c_\mu{}^1 \sqcap c_\mu{}^2 = 10 \neq \phi$, but $c_\Omega{}^1 \sqcap c_\Omega{}^2 \neq \phi$. Therefore we can not say that the λ-parts of these cubes are incompatible. But $c_\mu{}^3 \sqcap c_\mu{}^4 \neq \phi$ and $c_\Omega{}^3 \sqcap c_\Omega{}^4 = \phi$. Therefore, $c_\lambda{}^3 = 01$ and $c_\lambda{}^4 = 1x$ are incompatible. And $c_\lambda{}^3 = 01$ and $c_\lambda{}^5 = x0$ are incompatible. Thus for this function, 01 is not compatible with 00, 10, and 11. These last three 2-tuples are compatible with each other. We may summarize these findings with the following sets of mutually compatible λ-parts: $\{01\}$, $\{00, 10, 11\}$.

A simple disjoint decomposition exists for a given $f(X)$ and λ set if a function $g(\lambda)$ can be found such that $g(c_\lambda{}^i) \neq g(c_\lambda{}^j)$ for every $c_\lambda{}^i \sim c_\lambda{}^j$. To see this, suppose $c_\lambda{}^i$ and $c_\lambda{}^j$ have been shown to be incompatible. Let $c_\mu{}^i \sqcap c_\mu{}^j = a$. Then $c_\lambda{}^i \circ a \circ c_\Omega{}^i$ and $c_\lambda{}^j \circ a \circ c_\Omega{}^j$ could be listed as cubes of the function array if they do not already appear. But $c_\Omega{}^i \sqcap c_\Omega{}^j = \phi$. Therefore, either $c_\Omega{}^i$ or $c_\Omega{}^j$ equals 0; the other equals 1. Now if a function g is selected such that $g(c_\lambda{}^i) = g(c_\lambda{}^j) = b$, then since $f(c_\lambda{}^i, a) \neq f(c_\lambda{}^j, a)$ it is necessary for $h(b, a) \neq h(b, a)$. This is clearly not possible. In terms of Fig. 11.16(b), it is not possible to find a function $h(g, \mu)$ that equals $f(X)$ for all input symbols for which $f(X)$ is defined unless block g passes information to block h to enable h to distinguish between input symbols with incompatible λ-parts.

For the example above, a decomposition can exist only if we can find a gate such that

$$g(0, 1) \neq g(0, 0)$$
$$g(0, 1) \neq g(1, 0)$$
$$g(0, 1) \neq g(1, 1)$$

These conditions are satisfied by the following complementary gate functions:

x_1	x_2	$g' = \bar{x}_1 x_2$
0	0	0
0	1	1
1	0	0
1	1	0

x_1	x_2	$g'' = x_1 \vee \bar{x}_2$
0	0	1
0	1	0
1	0	1
1	1	1

If we use either of these gate functions, we can describe the example function without recourse to input variables x_1 and x_2.

g'	x_3	x_4	h'
0	1	x	1
0	1	0	1
1	x	x	1
0	0	x	0
0	0	0	0
0	x	1	d
0	1	x	d

g''	x_3	x_4	h''
1	1	x	1
1	1	0	1
0	x	x	1
1	0	x	0
1	0	0	0
1	x	1	d
1	1	x	d

These truth tables are derived directly from the original function array and the truth tables for g' and g'', respectively. Clearly, absorption may be performed to compress these arrays. And note that the don't cares must be assigned to make these truth tables consistent function arrays.

g'	x_3	x_4	h'
0	1	x	1
1	x	x	1
0	0	x	0

$$h' = g' \vee x_3$$

g''	x_3	x_4	h''
1	1	x	1
0	x	x	1
1	0	x	0

$$h'' = \bar{g}'' \vee x_3$$

In fact, we now see that x_4 is a redundant input variable.

Compatibility is not a transitive relationship for incomplete switching functions. The following array has as mutually compatible sets {00, 10}, {00, 11}, and {01} with $\lambda = \{x_1, x_2\}$ and $\mu = \{x_3, x_4\}$.

x_1	x_2	x_3	x_4	f	
1	0	1	x	1	
1	x	1	0	1	$10 \sim 11$
0	1	x	x	1	$01 \sim 11$
1	x	0	0	0	$01 \sim 1x$
x	0	0	x	0	$01 \sim x0$
1	1	x	1	0	
0	0	1	x	d	

In order for $00 \sim 10$ the don't care must be assigned the value of 1 (first and last rows). For $00 \sim 11$ the don't care must be assigned the value of 0 (sixth and last rows). Since these don't care assignments can not be simultaneously satisfied, 10 can not be compatible with 11 even though each is separately compatible with 00.

For this example, if a simple disjoint decomposition is to exist we must find a gate function g such that:

$$g(0, 0) \neq g(0, 1)$$

$$g(1, 0) \neq g(0, 1)$$

$$g(1, 1) \neq g(0, 1)$$

$$g(1, 1) \neq g(1, 0)$$

We are allowed either $g(0, 0) = g(1, 0)$ or $g(0, 0) = g(1, 1)$, but not both. Or putting it still another way, the set of input symbols of g may be partitioned into either of two sets of mutually compatible elements.

$$\{\{00, 10\}, \{11\}, \{01\}\}$$

$$\{\{10\}, \{00, 11\}, \{01\}\}$$

These requirements can not be satisfied if g is a single-output function. If β mutually compatible, disjoint subsets of r-tuples exist for a given $f(X)$ and λ, then $g(\lambda)$ must be an α-output function where $\beta \leq 2^\alpha$ so that a unique α-tuple may be associated with the members of each mutually compatible set. For example function f and set λ, many 2-output functions may be used as $g(\lambda)$, two of which are given below.

x_1	x_2	$g_1{}'$	$g_2{}'$
0	0	0	0
0	1	0	1
1	0	0	0
1	1	1	1

$$g_1{}' = x_1 x_2$$

$$g_2{}' = x_2$$

x_1	x_2	$g_1{}''$	$g_2{}''$
0	0	0	0
0	1	0	1
1	0	1	0
1	1	0	0

$$g_1{}'' = x_1 \bar{x}_2$$

$$g_2{}'' = \bar{x}_1 x_2$$

This first 2-output gate function suggests the use of a nondisjoint decomposition, since $g_2{}' = x_2$. Again since the gate functions above are based on the compatibilities of the given $f(X)$ and λ set, a realizable function h can be found.

$g_1{}'$	$g_2{}'$	x_3	x_4	h'
0	0	1	x	1
1	1	1	0	1
0	1	x	x	1
1	1	0	0	0
0	0	0	x	0
1	1	x	1	0
1	0	x	x	d

$g_1{}''$	$g_2{}''$	x_3	x_4	h''
1	0	1	x	1
0	0	1	0	1
0	1	x	x	1
0	0	0	0	0
1	0	0	x	0
0	0	x	1	0
1	1	x	x	d

In both cases the original don't care was assigned, but because the output alphabets of functions g' and g'' are incomplete, new don't care input symbols of h' and h'' appear. Since h' and h'' are both functions of four variables, as was the original function, it is not clear that the use of these simple decompositions simplifies the problem of synthesizing the original function.

A synthesis algorithm based on these ideas might proceed as follows.

ALGORITHM 11.9. *Synthesis via Disjoint Decomposition.*

Step 1. Given a function f to be realized and a set of available logic elements to be employed, in turn select each possible λ set.

Step 2. For a selected λ set calculate the incompatible r-tuples and the mututally compatible, disjoint subsets of λ-variable r-tuples.

Step 3. Determine the gate function that each set of such subsets prescribes. If the gate function corresponds to an available logic element,

(i) arbitrarily elect to employ the disjoint decomposition, or
(ii) continue to find all disjoint decompositions that prescribe available logic elements and then pick the one that is "best" according to some criterion.

Otherwise, select the next λ set at step 1.

Step 4. If a decomposition has been selected, record it, determine function h which results, and repeat the algorithm for this new function.

This algorithm is often ineffective and usually inefficient. A great number of λ sets exist and must be tested. The calculation of incompatibilities and mutually compatible subsets for each λ set is lengthy, and many disjoint, mutually compatible subsets may exist for a given λ set, all of which should be investigated. Further, for each set of disjoint, mutually compatible subsets several gate functions exist each of which may or may not correspond to an available logic element. Thus a great deal of searching is prescribed by this algorithm, all of which may not lead to a disjoint decomposition utilizing an available logic element, in which case the algorithm fails. When a number of disjoint decompositions exist, much more searching is implied since we can not look ahead to the effect of choosing a particular decomposition without completing the synthesis for that decomposition. Thus an exhaustive examination of each disjoint decomposition is specified by this algorithm.

However, the algorithm is not entirely without merit. If we do not insist that the gate function for a given λ set correspond to an available simple gate but accept gate functions that can be realized from such gates at modest cost, or if we have a large variety of logic elements available, including multiple output elements, then the algorithm can lead to realizations of a function. The realizations may not be built of truly disjoint decompositions. Again the problem of finding the "best" realization can be solved by extensive searching. If we have reason to insist that a decomposition based on a specific λ set be used, then most of this searching can be avoided; we need examine only the gate functions that the incompatibilities of the λ set allow.

Module Application

Before we examine other synthesis algorithms that utilize the philosophy of decomposition, let us generalize and formalize several procedures, including that of forming an array description of function h from arrays for given f and g functions. Let F be the function array for an n-input, m-output function f. Rows with an all-d output part need not be included in F, but rows with an all-0 output part must be included. Let G be the function array for an r-input, s-output logic element or *module* where $r \leq n$. The input terminals of the module will be $v_1, v_2, \ldots, v_r$; the output terminals will be $w_1, w_2, \ldots, w_s$. Then v^i and w^j will be referred to as input and output symbols of the module, just as x^i and z^j are input and output symbols, respectively, of the function. Let τ be a rule that associates a unique x_i with every v_j, i.e., it indicates which function input variable is to be connected to each module input terminal. If a set $\lambda \subseteq X$ of r function input variables is connected to module g, then h is a description of f in terms of input variables $x_1, \ldots, x_n, w_1, \ldots w_s$. Some of these input variables may be redundant; clearly the s w's are redundant. But we will not be concerned with this point immediately.

ALGORITHM 11.10. *Apply Module.*
To find H, the function array specification of function h, from given function arrays F and G, set λ, and mapping τ:

Step 1. Insert s columns of x's in F to the immediate right of the nth column.

$$F \leftarrow \mathbf{I}_n^s(F)$$

Cubes of F are now $(n + s + m)$-tuples.

Step 2. Insert $n - r$ columns of x's in G between the input and output variable columns. Append m columns of x's on the right of G.

$$G \leftarrow \mathbf{I}_{n+s}^m \left(\mathbf{I}_r^{n-r}(G) \right)$$

Cubes of G are now also $(n + s + m)$-tuples. The output (z) columns of F

oppose columns of x's in G; the output (w) columns of G oppose columns of x's in F.

Step 3. Permute the first n columns of G according to the mapping τ. If x_i is to be connected to terminal v_j, then the original jth column of G must become the ith column of the permuted G array. A means of expressing τ and performing the necessary permutations is described below.

Step 4. $H = G \bigsqcap F$.

Given the module described by G and the function array F below, assume τ specifies the connections of x_3 to v_1 and x_1 to v_2.

	F						G			
x_1	x_2	x_3	z_1	z_2		v_1	v_2	w_1	w_2
0	0	x	0	1		x	0	0	1
x	0	1	0	1		0	1	1	0
0	1	x	1	0		1	1	0	0
1	x	0	0	0					
1	1	x	0	0					

We open the appropriate columns in F and G.

$$F = \begin{Bmatrix} 00x\,xx\,01 \\ x01\,xx\,01 \\ 01x\,xx\,10 \\ 1x0\,xx\,00 \\ 11x\,xx\,00 \end{Bmatrix} \qquad G = \begin{Bmatrix} x0x\,01\,xx \\ 01x\,10\,xx \\ 11x\,00\,xx \end{Bmatrix}$$

Columns of G must now be permuted so that the first column (v_1) becomes the third column (opposite the x_3 column), and the second column (v_2) is placed opposite the first column of F.

$$G = \begin{Bmatrix} 0xx\,01\,xx \\ 1x0\,10\,xx \\ 1x1\,00\,xx \end{Bmatrix}$$

Then intersecting F and G gives

$$H = \begin{matrix} x\ \ w\ \ z \end{matrix}$$
$$H = \begin{Bmatrix} 00x\,01\,01 \\ 101\,00\,01 \\ 01x\,01\,10 \\ 1x0\,10\,00 \\ 111\,00\,00 \end{Bmatrix}$$

F associates an output symbol z^k with every input symbol x^i. Under a given mapping τ, G may be said to associate an output symbol w^j with every x^i. Array H presents both the original F array and this permuted G array as one compressed array. Alternatively, if we view $x^i \circ w^j$ as an input symbol of H, then H associates an output symbol z^k with input symbol $x^i \circ w^j$.

Not all $x^i \circ w^j$ are represented in the H array formed by Algorithm 11.10. We might look for input symbol $000 \circ 00$ in the example above. It is not present. With $x_1 x_2 x_3 = 000$ and the mapping used above, only $w_1 w_2 = 01$ is possible. Thus input symbols $000 \circ 00$, $000 \circ 10$, and $000 \circ 11$ can not occur and may be associated with an all don't care output part. All the following rows could be included in the H array for similar reasons.

$$
H' = \begin{cases}
xx0 \; 00 \; dd \\
0xx \; 1x \; dd \\
0xx \; x0 \; dd \\
xx1 \; 1x \; dd \\
1xx \; x1 \; dd \\
xxx \; 11 \; dd
\end{cases}
$$

It is clearly desirable to omit these rows from H. Then we can in fact be sure that H will consist of 2^n or fewer rows, regardless of the value of s.

If we mentally form $H \cup H'$ for a moment, we can see that the d's of H' can always be assigned so that $H \cup H'$ associates the same output symbol z^k with every input symbol $x^i \circ w^j$ that array F associates with input symbol x^i. Consider $x_1 x_2 x_3 = 000$ again. H associates $z_1 z_2 = 01$ with $000 \circ 01$. If the 2-cubes of H' are broken into their subcubes, the d's of H' may be assigned so that $z_1 z_2 = 01$ is associated with $000 \circ 00$, $000 \circ 10$, and $000 \circ 11$. Thus the H array may be made to express the same switching function expressed by the F array.

The mapping τ may be expressed in several ways, but the permutations of Step 3 of Algorithm 11.10 are easily made if τ is expressed in a natural way as an ordered set $(\tau_1, \tau_2, \ldots, \tau_r)$ where τ_1 reflects the member of $\lambda \subseteq X$ which is to be connected to v_1, τ_2 reflects the member of λ to be connected to v_2, and so on. Thus for the example above: $\tau_1 = x_3$, $\tau_2 = x_1$, or, using only the subscripts, $\tau = (3, 1)$. To this ordered set we add $n - r$ 0's corresponding to the $n - r$ columns of x's introduced in the G array. Thus $\tau = (3, 1, 0)$ for the example function.

ALGORITHM 11.11. *Permuting the Module Array.*

For $i = 1, 2, \ldots, n$ in turn:
   Step 1. Let $j = \tau_i$. If $i = j$, or $j = 0$, advance i.
   Step 2. Otherwise: $G \leftarrow \mathbf{P}_{ij}(G)$, $\tau_i \leftarrow \tau_j$, $\tau_j \leftarrow j$, and repeat Step 1.

Both τ, the ordered set of integers, and the columns of array G are permuted by this algorithm. For the example, the computation proceeds as follows:

i	$j=\tau_i$	τ_1	τ_2	τ_3	G_1	G_2	G_3
		3	1	0	v_1	v_2	—
1	3						
		0	1	3	—	v_2	v_1
1	0						
2	1						
		1	0	3	v_2	—	v_1
2	0						
3	3						

As a more comprehensive example, suppose the mapping desired is given by $\tau = \{2, 3, 4, 0\}$.

i	$j=\tau_i$	τ_1	τ_2	τ_3	τ_4	G_1	G_2	G_3	G_4
		2	3	4	0	v_1	v_2	v_3	—
1	2								
		3	2	4	0	v_2	v_1	v_3	—
1	3								
		4	2	3	0	v_3	v_1	v_2	—
1	4								
		0	2	3	4	—	v_1	v_2	v_3
1	0								
2	2								
3	3								
4	4								

Now we can better see the philosophy of the algorithm. We begin by asking: "Is the correct v_i in column 1 of G?" If not, we put the v_i in column 1 (v_1 at first) where it belongs, according to τ. This is repeated until either the correct v_i or all x's reside in column 1. Then the remaining columns are treated in a similar manner. With each permutation of G at least one v_i is put in its proper place.

Now what does the apply module algorithm accomplish as far as a synthesis procedure is concerned? First, if we are interested in disjoint decompositions, we may now test the H array with the algorithms of the previous section to determine if the members of set λ are redundant variables. Deletion of the x_1 and x_3 columns of H gives

$$\mathbf{D}_{1,\,3}(H) = \begin{array}{c} x_2 w_1 w_2 \quad z_1 z_2 \\ \begin{pmatrix} 0\ 0\ 1 & 0\ 1 \\ 0\ 0\ 0 & 0\ 1 \\ 1\ 0\ 1 & 1\ 0 \\ x\ 1\ 0 & 0\ 0 \\ 1\ 0\ 0 & 0\ 0 \end{pmatrix} \end{array}$$

This is a consistent function array and hence the decomposition is disjoint. But $\{x_1, w_1\}$, $\{x_3, w_2\}$, and $\{w_1, w_2\}$ are also sets of redundant variables.

TABLE 11.4   COLUMN COVER

	z_i^k		
$w_j^k \sqsubseteq z_i^k$	0	1	d
w_j^k 0	ε	ϕ	ε_d
w_j^k 1	ϕ	ε	ε_d
w_j^k x	ϕ	ϕ	ε

For example,

$$
\begin{array}{c}
x_1 x_2 w_1 \quad z_1 z_2 \\
\mathbf{D}_{3,\,5}(H) = \begin{pmatrix}
0 & 0 & 0 & 0 & 1 \\
1 & 0 & 0 & 0 & 1 \\
0 & 1 & 0 & 1 & 0 \\
1 & x & 1 & 0 & 0 \\
1 & 1 & 0 & 0 & 0
\end{pmatrix}
\end{array}
$$

is a consistent function array. Which, if any, of these four redundancy sets should be recognized in a synthesis procedure is not clear. The disjoint decomposition has no obvious advantages over the others.

Second, from the H array generated by the apply module algorithm, we can determine if a module application realizes one or more of the output variables of the function. If a w_j column is covered by an output variable column z_i, then w_j realizes z_i.

ALGORITHM 11.12. *Outputs Realized Test.*
For each w_j for each z_i in turn, if $w_j^k \sqsubseteq z_i^k = \varepsilon$ or ε_d, per Table 11.4, for all k, then w_j realizes z_i. Record the fact and delete the z_i column. If a ε_d is encountered, don't cares of z_i are assigned by w_j.

To see how these algorithms might be used, we apply the example module G to the example function F according to all possible λ sets and τ mappings. Table 11.5 summarizes the findings. Note that $\{w_1,\ w_2\}$ is always found to be a redundancy set, as we might expect.

The mapping $\tau = (1, 2, 0)$ is of great interest, because w_1 realizes z_1 with it. Since the realization of all output variables is the ultimate goal of synthesis, we might very well select this application so that further decom-

TABLE 11.5 RESULTS OF APPLYING MODULE G TO FUNCTION F

λ	τ	output realized	largest redundancy sets
1, 2	1, 2, 0	$w_1 = z_1$	$\{x_2, w_1\}, \{w_1, w_2\}$
	2, 1, 0	none	$\{x_1, w_1\}, \{w_1, w_2\}$
1, 3	1, 3, 0	none	$\{x_3, w_1\}, \{w_1, w_2\}$
	3, 1, 0	none	$\{x_1, x_3\}, \{x_1, w_1\}, \{x_3, w_2\}, \{w_1, w_2\}$
2, 3	2, 3, 0	none	$\{x_3, w_1\}, \{x_3, w_2\}, \{w_1, w_2\}$
	3, 2, 0	none	$\{x_2, w_1\}, \{w_1, w_2\}$

positions may be directed to realizing z_2. We can delete x_2 and w_1, or w_1 and w_2 (reject the decomposition), or any one of these variables, from further consideration because they are redundant. But it is not clear that it is desirable to do so. Module application with mapping $\tau = (3, 1, 0)$ is also interesting because it is the only application that leads to a disjoint decomposition.

A simple if not effective synthesis algorithm might proceed much as we have above.

ALGORITHM 11.13. *Single Module Synthesis.*

Given a function array F description of an n-input, m-output switching function f and an array G which describes an r-input, s-output module to be used as the sole type of logic element of a network which realizes f:

Step 1. In turn select each possible λ set of r members. When all have been selected go to step 6.

Step 2. For the selected λ set, in turn form each of the possible τ mappings. When all have been selected go to step 1. If partial symmetry sets of G are known, then less than all $r!$ possible permutations of the members of λ need be formed.

Step 3. Apply module array G to function array F according to τ.

Step 4. Test to determine if any w_j realizes a z_i. If none do, proceed to step 5. If this module application realizes more output variables than any previously examined application, record it and the first of its largest redundancy set which does not include the module output variables. If this module application realizes the same number of output variables as a previous application, but has a larger redundancy set, record it. In either case go to step 2.

Step 5. Find the size of the largest redundancy set (not including module output variables) of this application. If the size exceeds the size of the largest redundancy set previously found, record the size, τ, and the redundancy set. Go to step 2.

Step 6. Select the application that realizes the greatest number of output variables, or if none are realized, the application with the largest redundancy set. Print the τ mapping and the redundancy variables. Delete the output variables realized, if any, and the members of the maximum redundancy set from the H array which corresponds to the selected mapping. If output variables remain to be realized, call this deleted H array F and go to step 1.

To synthesize the example function array F with modules G this algorithm requires that we compute the information of Table 11.5, but retain only mapping $\tau = (1, 2, 0)$ and redundancy set $\{x_2\}$. That w_1 realizes output variable z_1 is to be recorded. Applying module G to F with this mapping, deleting the x_2 and z_1 columns from the resulting H array, and relabeling the w_1 and w_2 columns as x_4 and x_5 respectively then gives the following array, for which the algorithm is to be repeated.

$$F \leftarrow \mathbf{D}x_2, z_1(H) = \begin{matrix} x_1 & x_3 & x_4 & x_5 & z_2 \\ \begin{pmatrix} 0 & x & 0 & 1 & 1 \\ x & 1 & 0 & 1 & 1 \\ 0 & x & 1 & 0 & 0 \\ 1 & 0 & 0 & 1 & 0 \\ 1 & x & 0 & 0 & 0 \end{pmatrix} \end{matrix}$$

Mapping $\tau = (x_3, x_1)$ is the first encountered to give a redundancy set of three members. Hence it is selected, variables x_1, x_3, and x_4 are deleted from the corresponding H array, and the output variables of this second module are named x_6 and x_7 respectively.

$$F \leftarrow \mathbf{D}x_1, x_3(H) = \begin{matrix} x_5 & x_6 & x_7 & z_2 \\ \begin{pmatrix} 1 & 0 & 0 & 1 \\ 1 & 0 & 1 & 1 \\ 0 & 0 & 0 & 0 \\ 0 & 0 & 1 & 0 \\ 0 & 1 & 0 & 0 \\ 1 & 1 & 0 & 0 \end{pmatrix} \end{matrix}$$

Applying module G in all possible ways to this function array reveals that z_2 may be realized by applying module G with $\tau = (x_6, x_5)$. This is the only output variable which remains, so the synthesis is complete. Fig. 11.17 shows the network which the algorithm has generated. The numerals indicate the order in which the module applications were selected.

Unfortunately the successful synthesis of this example cannot be taken as typical. The algorithm is not generally able to find low cost realizations and may fail to find any realization at all.

Algorithm 11.13 can be extended to become much more efficacious. A nonredundant library of all submodules of the given module(s) must be

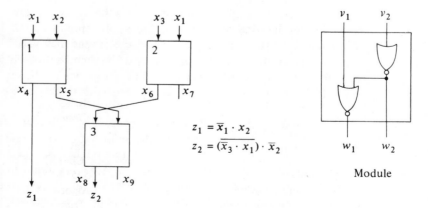

$$z_1 = \overline{x}_1 \cdot x_2$$
$$z_2 = (\overline{x_3 \cdot x_1}) \cdot \overline{x}_2$$

Module

FIG. 11.17 A Realization of F

formed. Then all modules of the nonredundant library must be applied under all τ mappings that are meaningful, as determined by the symmetry sets of the modules (looking first for the realization of function output variables). If a number of applications realize the same maximum number of variables, then this list must be screened further to find those module applications that lead to the maximum size redundancy set. If a number of applications produce redundancy sets of the same maximum size, then further criteria must be introduced to select the single application upon which the ultimate network will be based. The cost of the module, the delay that an application introduces, and the loading that the application imposes might serve to help make this final selection. Limited experience with an algorithm of this type indicates that these criteria do lead to networks of reasonable cost and form, with a reasonable amount of computation [10].

REFERENCES

1. R. L. ASHENHURST, "The Decomposition of Switching Functions," *Proc. International Symposium on the Theory of Switching*, April 2–5, 1957, Vol. 29 of *Annals of Computation Laboratory of Harvard University*, 1959, pp. 74–116.

2. M. A. BREUER, "*Design Automation of Digital Systems*", Ch. 2, Logic Synthesis, Englewood Cliffs, N.J.: Prentice-Hall, 1972.

3. H. A. CURTIS, *A New Approach to the Design of Switching Circuits*. Princeton: Van Nostrand, 1962.

4. D. L. DIETMEYER and P. R. SCHNEIDER, "Identification of Symmetry, Redundancy, and Equivalence of Boolean Functions," *IEEE Trans. on Electronic Computers*, Vol. EC-16, Dec. 1967, pp. 804–817.

5. D. L. DIETMEYER and Y-H SU, "Logic Design Automation of Fan-In Limited NAND Networks," *IEEE Trans. on Computers*, Vol. C-18, Jan. 1969, pp. 11–22.

6. J. F. GIMPEL, "The Minimization of TANT Networks," *IEEE Transactions on Electronic Computers*, Vol. EC-16, Feb. 1967, pp. 18–38.

7. B-W Ho, "NAND Synthesis of Multiple-Output Combinational Logic Using Implicants Containing Output Variables," Ph.D. thesis, University of Wisconsin—Madison. 1976.

8. R. M. KARP, F. E. MCFARLIN, J. P. ROTH and J. R. WILTS, "A Computer Program for the Synthesis of Combinational Switching Circuits," *Proc. Second Annual AIEE Symposium on Switching Circuit Theory and Logical Design*, Oct. 1961, pp. 182–194.

9. A. MUKHOPADHYAY, "Detection of Total or Partial Symmetry of a Switching Function with the Use of Decomposition Charts," *IEEE Trans. on Electronic Computers*, Vol. EC-12, Oct., 1963, pp. 553-557.

10. P. R. SCHNEIDER and D. L. DIETMEYER, "An Algorithm for Synthesis of Multiple Output Combinational Logic," *IEEE Trans. on Computers*, Vol. C-17, Feb. 1968, pp. 117–128.

PROBLEMS

11.1-1. Draw the topology of each of the 22 forms of functions of three variables. Compare the forms which are duals of each other.

11.1-2. From the topology of each form argue that the number of functions of that form is correct.
(a) Form 3, 4 functions; (b) Form 5, 12 functions; (c) Form 8, 24 functions.

11.1-3. If the *class* of a function contains all functions of the same form and the dual form, how many classes of functions of three variables exist? Which forms make up each class?

11.1-4. Single-output function f is known to be partially symmetric with respect to $\{x_1, x_2\}$. The base of f includes minterms 0010, 0110, and 1110. What other minterms must that base include? What other minterms may it include?

11.1-5. Cube $10x$ is a prime implicant of a totally symmetric function. Find the other prime implicants of the function. Repeat, if $11x$ is the known prime implicant.

11.1-6. Plot the Karnaugh map of the following totally symmetric function

$$f = S_a(\lambda)$$

for each of the following λ sets and $a = 0, 1,$ and 2 in turn.

$$\lambda_1 = \{A, B, C, D\}$$
$$\lambda_2 = \{A, B, C, \bar{D}\}$$
$$\lambda_3 = \{\bar{A}, \bar{B}, \bar{C}, D\}$$

Deduce from these maps how one might examine an arbitrary map to determine if the function is totally symmetric, the a numbers, and the symmetry set λ.

11.1-7. A 3-input, 2-output module is described by the following function array.

$$M = \begin{Bmatrix} v_1 & v_2 & v_3 & w_1 & w_2 \\ 0 & 0 & 0 & 1 & 0 \\ 0 & x & 1 & 0 & 1 \\ x & 1 & 0 & 1 & 1 \\ 1 & 0 & x & 0 & 1 \\ 1 & 1 & 1 & 1 & 0 \end{Bmatrix}$$

Find all symmetry sets of w_1, of w_2, and of M.

11.1-8. Find all symmetry sets of z_1, of z_2, and of the function f described by array F below. For each, indicate the don't care assignments which must be made, if any.

$$F = \begin{cases} 0x00 & 11 \\ 00x1 & 10 \\ 0011 & 1d \\ 0x01 & 10 \\ 0110 & d1 \\ 0111 & 01 \\ 1x00 & d1 \\ 1001 & d0 \\ 1010 & 01 \\ 1011 & 11 \\ 1101 & dd \\ 111x & 00 \end{cases} \qquad z_1 z_2$$

11.1-9. Which input variables, if any, are redundant in each of the following arrays. All input symbols not present in these arrays are associated with an all–don't-care output 2-tuple.

$$\begin{cases} 0011 & 10 \\ 0110 & 01 \\ 10x1 & 00 \\ 1x01 & 00 \\ 1110 & 00 \\ x001 & 00 \end{cases} \qquad \begin{cases} 00111 & 10 \\ 01110 & 01 \\ 11101 & 00 \\ 10x11 & 00 \\ xx011 & 00 \end{cases} \qquad \begin{cases} x0001 & 01 \\ 1x001 & 01 \\ x0110 & 10 \\ 01001 & 00 \\ x1100 & 00 \end{cases}$$

11.1-10. Prepare and test a FORTRAN program that will identify all redundant variables in a given array. Show how this program can be used to determine sets of redundant variables. Test on the arrays of Prob. 11.1-9.

11.2-1. Only 2-input NAND gates are available. Find a low gate count realization of the function described by the following cover.

$$\begin{cases} 1 & 0 & 1 & 1 & 0 & x \\ 1 & 0 & 0 & 1 & x & 1 \\ 0 & 0 & x & 0 & x & x \\ 0 & 1 & 1 & 1 & 1 & 1 \\ 1 & 0 & x & 1 & 0 & 0 \end{cases}$$

11.2-2. The following function is to be realized with a minimum number of 3-input NAND gates. Complements are NOT available as input signals. Show a logic block diagram.

$$f = \begin{cases} x101 \\ 101x \\ 01x1 \\ 1x10 \end{cases}$$

11.2-3. The output signal of an AND gate is equivalent to the bundle of its input lines when connected to a NAND gate. See Fig. 11.10. The output signal of what types of gates, if any, are equivalent to the bundle of their input lines when connected to (a) an AND gate, (b) an OR gate, (c) a NOR gate, (d) an Exclusive-OR gate?

11.2-4. Write a FORTRAN subroutine for evaluating $\mathscr{G}(\kappa)$, Eq. (11.16), for values of κ and Λ supplied as arguments. Do not use division in your subroutine.

11.2-5. (a) Given an array of subcubes of U_n, define an operator that produces the smallest, single subcube of U_n that covers all of the 0-cubes covered by the given array (and possibly other 0-cubes as well).

(b) Discuss real or potential uses of your operator.

11.2-6. Array $C1$ is a common factor subarray of C with common factor γ. Factor γ' is a narrower common factor of $C1$, i.e., $\$\gamma' < \$\gamma$. Array D is generated from $C1$ by deleting common factor literals. Complements are available at no additional cost.

(a) What effect, if any, does the use of γ' rather than γ have on the single-line NAND network realization of $\overline{C1}$, if a single-line realization of D is employed?

(b) What effect, if any, does the use of γ' rather than γ have on the single-line NAND network realization of $\overline{C1}$, if a bundle realization of D is employed?

11.2-7. The following four types of NAND-gate packs are available and have equal cost. Fan-in can not be extended on any of these packs.

Pack 1 contains four 2-input NAND gates
Pack 2 contains three 3-input NAND gates
Pack 3 contains two 4-input NAND gates
Pack 4 contains one 8-input NAND gate

Find a low (minimum if possible) pack count realization of the function described by the following cover. Realize any cover of the function that you like.

$$\begin{pmatrix} 0 & 0 & 1 & 1 & 0 & 0 & 1 \\ 0 & 0 & 1 & 1 & 0 & 1 & 0 \\ 0 & 1 & 0 & 0 & 0 & 0 & x \\ 0 & 1 & 0 & 1 & 1 & 1 & x \\ 0 & 1 & 1 & 0 & 1 & 0 & 1 \\ 0 & 1 & 1 & 0 & 1 & 1 & 0 \\ 1 & 0 & 0 & 1 & 1 & 0 & 1 \\ 1 & 0 & 0 & 1 & 1 & 1 & 0 \\ 1 & 0 & 1 & 0 & 1 & 0 & x \\ 1 & 1 & 0 & 0 & 0 & 1 & x \\ 1 & 1 & 0 & 1 & 0 & 0 & 1 \\ 1 & 1 & 0 & 1 & 0 & 1 & 0 \end{pmatrix}$$

11.2-8. If C is a nonredundant, prime-implicant cover of a function of n variables,

(a) prove that (i) every factor of C has width $w \leq n - 2$, (ii) C consists of 2^{n-1} or fewer cubes, and (iii) if subarray $C1$ has q members with a common factor of width w, then $q \leq 2^{n-w-1}$.

(b) These results indicate that the figure–of–merit of a factor is $\leq w \cdot 2^{n-w-1}$. What value of w maximizes this limit?

11.2-9. Find a cover for which Algorithm 11.5 will not find the factor of maximum figure–of–merit but for which Algorithm 11.4 will find it.

11.2-10. Subarray $C1$ with common factor γ, $\$\gamma \geq \Lambda$, may be realized with a single line realization of the D array derived from $C1$ and γ at a NAND gate cost $\$_g{}^2(C1)$ given by Eq. (11.22).

(a) Derive an expression for $\$_g{}^4(C1)$, the cost of the network obtained by using common factor γ' rather than γ where $\$\gamma' = \Lambda - 1 \leq \$\gamma$.

(b) In what sort of circumstances will $\$_g^4 (C1) < \$_g^2 (C1)$?

(c) Find a cover C that demonstrates these circumstances.

11.2-11. Single output function z has been factored extensively and the following connection array obtained. Draw a NAND network realization of z from this array.

$$
\begin{array}{cccccccc c cccc}
A & B & C & D & E & F & p & q & r & & p & q & r & z \\
\end{array}
$$

$$
\left(
\begin{array}{cccccccc}
1 & x & x & x & x & x & 1 & x & x \\
x & 1 & x & x & x & x & x & 1 & x \\
x & x & x & x & x & x & x & x & 0 \\
0 & x & x & x & x & x & x & 1 & 1 \\
x & x & 0 & x & x & x & x & x & x \\
x & x & x & 0 & x & x & x & x & x \\
x & x & x & x & 0 & x & x & x & x \\
x & x & x & x & x & 0 & x & x & x \\
\end{array}
\right.
\left.
\begin{array}{cccc}
0 & 0 & 0 & 1 \\
1 & 0 & 0 & 0 \\
1 & 0 & 0 & 0 \\
0 & 0 & 0 & 1 \\
0 & 1 & 0 & 0 \\
0 & 1 & 0 & 0 \\
0 & 0 & 1 & 0 \\
0 & 0 & 1 & 0 \\
\end{array}
\right)
$$

11.3-1. A 3-input, 3-output function is specified by the function array F. Find the ON-, OFF-, and DC-arrays for each of the extended cubes.

(a) $x0x\ 1xx$

(b) $x1x\ 0xx$ $\qquad F = \begin{cases} 00x & 101 \\ 0x0 & 101 \\ x11 & 110 \\ 10x & 011 \\ 1x0 & 011 \end{cases}$

(c) $xxx\ xx0$

11.3-2. Draw Karnaugh maps of the output variables of function F of Prob. 11.3-1. Can z_1 or $\bar{z}_1$ be used to reduce the costs of realizing z_2? z_3? Write Boolean equations for z_2 and z_3, and draw a NAND realization of F based upon these equations. Repeat using z_3 to realize z_1 and z_2.

11.3-3. Does a NAND network based upon the following equations realize the function of Prob. 11.3-1? Justify your answer.

$$z_1 = \bar{z}_2 \vee BC$$
$$z_2 = \bar{z}_1 \vee BC$$
$$z_3 = \overline{z_1 z_2}$$

11.3-4. Which, if any, of the following relations are valid for extended cubes of array F in Prob. 11.3-1?

(a) $00x\ xxx \sqsubseteq xxx\ xx1$

(b) $00x\ xxx \leq xxx\ xx1$

(c) $xx1\ 1x1 \leq x1x\ x1x$

(d) $xxx\ 0xx \leq xxx\ x11$

11.3-5. For the final example of the section, use Karnaugh maps or Boolean equations to show that extended cube $xxx\ x0xx$ can be used to replace $01x$ in the realization of z_3, and $\{01x, 101\}$ in the realization of z_4, at a cost savings of six connections and two gates. Repeat for the other two extended prime implicants and their listed BE-arrays.

11.3-6. If a different cover of z_2 is selected, the connection array $C(F)$ for the final example of the section is altered; that is, $11x\ x1xx$ is replaced by $1x0\ x1xx$. Repeat Problem 11.3-5 using this altered connection array.

11.4-1. Function f is described by the function array below.

(a) With $\lambda = \{x_1, x_3\}$ find all incompatibilities. Note that the array has been rearranged for your convenience.

x_1	x_3	x_2	x_4	f
x	x	0	0	1
1	x	1	x	1
0	0	x	x	1
x	1	1	x	1
1	x	0	1	0
x	1	0	1	0

(b) Does a disjoint, 2-input AND, OR, NAND, or NOR decomposition exist with $\lambda = \{x_1, x_3\}$? If so, express f in terms of x_2, x_4, and g, the output of the gate selected.

(c) Complete the synthesis using only the 2-input blocks mentioned above.

11.4-2. For the function f given below with $\lambda = \{x_1, x_2\}$:
(a) Find all incompatibilities.
(b) Find all sets of mutually compatible subsets of 2-tuples.
(c) Find all gate functions which each set of (b) permits.
(d) Express f in terms of the output variable of each permitted gate and x_3 and x_4. Make a judgment as to which decomposition is "best."

x_1	x_2	x_3	x_4	f
0	0	x	0	0
0	x	0	1	1
x	0	0	1	1
0	0	1	1	d
0	1	0	0	d
0	1	1	x	0
x	1	1	0	0
1	0	0	0	d
1	0	1	x	0
1	1	0	x	1
1	1	1	1	d

11.4-3. Repeat Prob. 11.4-2 for the following function with $\lambda = \{x_2, x_3, x_4\}$.

x_1	x_2	x_3	x_4	f
x	0	0	x	0
x	0	x	0	0
x	x	0	0	0
x	0	1	1	1
x	1	0	1	1
x	1	1	0	1
0	1	1	1	0
1	1	1	x	1

11.4-4. Switching function F described below is to be realized using module M. As a first attempt at decomposition, x_1 is connected to v_1, x_2 to v_2, and x_3 to v_3.

(a) Describe F in terms of x_1, x_2, x_3, x_4, w_1, and w_2. Absorb your result; call it H.

<table>
<tr><th colspan="4"></th><th colspan="2">F</th></tr>
<tr><th>x_1</th><th>x_2</th><th>x_3</th><th>x_4</th><th>z_1</th><th>z_2</th></tr>
<tr><td>0</td><td>0</td><td>0</td><td>0</td><td>0</td><td>0</td></tr>
<tr><td>0</td><td>0</td><td>0</td><td>1</td><td>1</td><td>d</td></tr>
<tr><td>x</td><td>0</td><td>1</td><td>0</td><td>0</td><td>d</td></tr>
<tr><td>x</td><td>0</td><td>1</td><td>1</td><td>d</td><td>1</td></tr>
<tr><td>0</td><td>1</td><td>0</td><td>x</td><td>d</td><td>1</td></tr>
<tr><td>0</td><td>1</td><td>1</td><td>x</td><td>0</td><td>1</td></tr>
<tr><td>1</td><td>0</td><td>0</td><td>0</td><td>1</td><td>d</td></tr>
<tr><td>1</td><td>0</td><td>0</td><td>1</td><td>1</td><td>1</td></tr>
<tr><td>1</td><td>1</td><td>0</td><td>0</td><td>1</td><td>1</td></tr>
<tr><td>1</td><td>1</td><td>0</td><td>1</td><td>1</td><td>d</td></tr>
<tr><td>1</td><td>1</td><td>1</td><td>x</td><td>0</td><td>0</td></tr>
</table>

<table>
<tr><th colspan="5">M</th></tr>
<tr><th>v_1</th><th>v_2</th><th>v_3</th><th>w_1</th><th>w_2</th></tr>
<tr><td>0</td><td>0</td><td>0</td><td>1</td><td>0</td></tr>
<tr><td>0</td><td>x</td><td>1</td><td>0</td><td>1</td></tr>
<tr><td>x</td><td>1</td><td>0</td><td>1</td><td>1</td></tr>
<tr><td>1</td><td>0</td><td>x</td><td>0</td><td>1</td></tr>
<tr><td>1</td><td>1</td><td>1</td><td>1</td><td>0</td></tr>
</table>

(b) Does either w_1 or w_2 realize either z_1 or z_2?

(c) Find all redundancy sets of H (include w_1 and w_2).

(d) Repeat for $\tau = (x_2, x_1, x_4)$.

ANSWERS

11.1-3. 14 classes. Class 1—forms 1 and 22
Class 14—form 14

11.1-4. 1010 must be included.

11.1-6. Look for rings around the cell of the product of the variables of symmetry.

11.1-7. w_2 is totally symmetric.

11.2-2. 5 gates with fan-in $= 3$: $BD(\overline{AC}) \vee AC(\overline{BD})$
4 gates with fan-in $= 4$: $BD(\overline{ABCD}) \vee AC(\overline{ABCD})$

11.2-5. Factor product, of course.

11.2-8. The widest possible prime implicants are minterms. Two minterms have $n - 2$ or fewer variables in common. The 2^{n-1} minterms of exclusive-or form functions are all prime implicants.

11.3-1. $ON(x0x \circ 1xx) = 00x$: $OFF = \begin{Bmatrix} 1xx \\ x1x \end{Bmatrix}$

11.3-3. The equations describe a sequential network.

11.3-6. $\bar{z}_2$ cannot be used to realize z_4.

11.4-1. $00 \sim 1x$, $00 \sim x1$
OR and NOR decompositions exist.

$$f = x_2 \vee \bar{x}_4 \vee \bar{g}_{OR}$$
$$= x_2 \vee \bar{x}_4 \vee g_{NOR}$$

11.4-2. 00 $\sim$ 11, therefore the following compatible subsets may be used for disjoint decompositions.

$$\{\{00\}, \{01, 10, 11\}\} \quad \text{OR, NOR}$$
$$\left.\begin{array}{l} \{\{00, 01\}, \{10, 11\}\} \quad x_1 \text{ or } \bar{x}_1 \\ \{\{00, 10\}, \{01, 11\}\} \quad x_2 \text{ or } \bar{x}_2 \end{array}\right\} \text{redundant variable?}$$
$$\{\{00, 01, 10\}, \{11\}\} \quad \text{AND, NAND}$$

Using x_1 or x_2 avoids a gate.

APPENDIX 11.1

A subroutine based upon Algorithm 11.4 to find a factor with maximum figure of-merit.

```
      SUBROUTINE MAXFCT
      IMPLICIT INTEGER(A–Z)
      COMMON N, M, P, NV, IOC, MX, NC, IFAN, X(2)
      COMMON F(200), C(200), GAM(5), A(200), B(200), X0(200), X1(200)
      COMMON SAVE(200), MASK(5), HGT(20), COLS(20), ISEL(20)
      MFM = 1
      CALL DEFINA(GAM, N, 'XXXXXX')
      CALL COPYAR(B, C, MX)
      NCB = NCUBES(B)
      K = 0
      DO 1 I = 1,NCB
    1 K = MAX(K, KUBCST(B, N, I))
      MFW = MIN(K—1, MAX(K–IFAN, IFAN–1))
      DO 14 IW = 1,MFW
C
C     FIND CUBES AND COLUMNS WORTHY OF EXAMINATION
C
      DO 2 I = 1,NCB
      J = NCB + 1 – I
      IF(KUBCST(B, N, J) .LE. IW) CALL MOVCUB(B, A, J)
    2 CONTINUE
      NCB = NCUBES(B)
      IF(IW*NCB .LE. MFM) GO TO 14
      IF(IW*2**(N–IW–1) .LE. MFM) RETURN
      CALL COLHGT(B, HGT)
    3 K = 0
      DO 4 I = 1,N
      IF(HGT(I)*IW .LE. MFM) GO TO 4
      K = K + 1
      COLS(K) = I
    4 CONTINUE
      IF(K .LT. IW) GO TO 14
C
C     SELECT TEST COLUMNS
C
      IND = 1
    5 CALL SELECT(ISEL, K, IW, IND)
      CALL COPYAR(A, B, MX)
      GO TO (5, 7, 14), IND
    6 CALL COPYAR(A, SAVE, MX)
    7 CALL DEFINA(SAVE, 0, 0)
C
C     SPLIT ON SELECTED COLUMNS
C
      ISAVE = 1
```

```
                DO 10 I = 1,IW
                CALL DEFINA(MASK, N, 'XXXXXX')
                J = ISEL(I)
                CALL CNGCOL(MASK, COLS(J), '000')
                CALL SPLITA(A, X0, MASK, MX, NX0)
                CALL CNGCOL(MASK, COLS(J), '111')
                CALL SPLITA(A, X1, MASK, MX, NX1)
                IF(NX0 .LT. 2) GO TO 9
                IF(IW*NX0 .LT. MFM) GO TO 9
                CALL FACTOR(X0, MASK, NFM)
                IF(NFM .GT. MFM) GO TO 13
                CALL COPYAR(A, X0, MX)
                IF(IW*NX1 .LT. MFM) GO TO 10
                CALL COMBIN(SAVE, X1, MX)
                ISAVE = 2
                GO TO 10
         9 IF (NX1 .LT. 2) GO TO 12
                IF (IW*NX1 .LT. MFM) GO TO 12
                CALL FACTOR(X1, MASK, NFM)
                IF (NFM .GT. MFM) GO TO 13
                CALL COPYAR(A, X1, MX)
        10 CONTINUE
    C
    C       SAVE DESIRABLE FACTOR
    C
                IF(NFM-MFM) 12, 11, 13
        11 IF(KUBCST(MASK, N, 1) .LT. IFAN) CALL COPYAR(GAM, MASK, MX)
        12 GO TO (5, 6), ISAVE
        13 CALL COPYAR(GAM, MASK, MX)
                MFM = NFM
                GO TO 3
        14 CONTINUE
                RETURN
                END
```

APPENDIX 11.2

A subroutine based upon Algorithm 11.5 to find a good factor rapidly.

```
                SUBROUTINE FASTFT
                IMPLICIT INTEGER(A–Z)
                COMMON N, M, P, NV, IOC, MX, NC, IFAN, X(2)
                COMMON F(200), C(200), GAM(5), A(200), B(200), GAMP(5), MSK(5)
                NCC = NCUBES(C)
                MFM = 1
                CALL DEFINA(GAM, N, 'XXXXXX')
                DO 3 I = 1,N
                CALL COPYAR(A, C, MX)
                CALL DEFINA(MSK, N, 'XXXXXX')
                CALL CNGCOL(MSK, I, '000')
                DO 3 J = 1,2
                CALL SPLITA(A, B, MSK, MX, NCB)
                CALL FACTOR(B, GAMP, NFM)
                IF(NFM-MFM) 3, 1, 2
         1 IF(KUBCST(GAMP, N, 1) .GE. IFAN) GO TO 3
         2 MFM = NFM
                CALL COPYAR(GAM, GAMP, MX)
                IF(NCB .GE. NCC) RETURN
         3 CALL CNGCOL(MSK, I, '111')
                RETURN
                END
```

APPENDIX 11.3

A program based upon Algorithm 11.6 to find a low cost, fan-in limited NAND gate realization of a switching function. This program finds only one factor of each subarray and then uses AND gates to solve remaining fan-in problems. Subprogram KOST evaluates $S_g{}^1(A)$ for given array A and parameters.

```
        SUBROUTINE NANDSY
        IMPLICIT INTEGER(A–Z)
        COMMON N, M, P, NV, IOC, MX, NC, IFAN, X(2)
        COMMON F(200), C(200), GAM(5), A(200), B(200), C1(200), D(200)
        COMMON DEL(200), MSK(5)
        CALL COPYAR(C, F, MX)
        NCC = NCUBES(C)
        NETCT = 0
        NFLA = 0
C
C       GET FACTOR, FORM D-ARRAY
C
    1   CALL MAXFCT
        CALL SPLITA(C, C1, GAM, MX, NCD)
        CALL COPYAR(D, C1, MX)
        NCC = NCC − NCD
        DO 2 I = 1,N
        CALL IDTCOL(GAM, I, NRET)
        IF(NRET .NE. 4) CALL CNGCOL(D, I, 'XXX')
    2   CONTINUE
C
C       FORM DEL–ARRAY
C
        CALL DEFINA(MSK, N, 'XXXXXX')
        CALL SHARPS(MSK, D, DEL, A, MX, NCDL)
        DO 3 I = 1,NCDL
        CALL MOVCUB(DEL, MSK, 1)
        CALL SHARPS(MSK, DEL, A, B, MX, NCA)
        IF(NCA .GT. 0) CALL COMBIN(DEL, MSK, MX)
    3   CONTINUE
C
C       COMPUTE COSTS
C
        NCDL = NCUBES(DEL)
        KOST1 = KOST(C1, NCD, N, IFAN)
        KST1E = NANDS(NCD + NFLA + NCC, IFAN) + KOST1
        KST = KUBCST(GAM, N, 1)
        KOST2 = NANDS(NCD, IFAN) + NANDS(KST + 1, IFAN) +
    1       KOST(D, NCD, N, IFAN)
        KOST3 = NANDS(KST + NCDL, IFAN) + KOST(DEL, NCDL, N, IFAN)
        IF(KST1E .LE. KOST2) GO TO 4
        IF(KOST2 − KOST3) 6, 9, 9
    4   IF(KST1E .GT. KOST3) GO TO 9
C
C       PRINT RESULTS
C
        K = 1
        PRINT 5, KOST1, K
    5   FORMAT(' 0USE ',I3, ' NAND GATES IN A FORM', I2, ' REALIZATION')
        CALL PRNTAR(C1, N)
        NFLA = NFLA + NCD
        NETCT = NETCT + KOST1
```

```
              GO TO 11
         6 K = 2
              PRINT 5, KOST2, K
              PRINT 7
         7 FORMAT('OFACTOR')
              CALL PRNTAR(GAM, N)
              PRINT 8
         8 FORMAT ('OD–ARRAY')
              CALL PRNTAR(D, N)
              NFLA = NFLA + 1
              NETCT = NETCT + KOST2
              GO TO 11
         9 K = 3
              PRINT 5, KOST3, K
              PRINT 7
              CALL PRNTAR(GAM, N)
              PRINT 10
        10 FORMAT('ODEL–ARRAY')
              CALL PRNTAR(DEL, N)
              NFLA = NFLA + 1
              NETCT = NETCT + KOST3
        11 IF(NCC .GT. 0) GO TO 1
              K = NANDS(NFLA, IFAN)
              IF(K .EQ. 0) K = 1
              PRINT 12, K
        12 FORMAT('OUSE', I5, 2X, 'NAND GATES AT FIRST LEVEL')
              NETCT = NETCT + K
              PRINT 13, NETCT, IFAN
        13 FORMAT('ONET REQUIRES', 2I4, '–INPUT NAND GATES')
              RETURN
              END

              FUNCTION KOST(A, NR, N, LM1)
              J = 0
              DO 10 I = 1,NR
              CALL CUBCST(A, N, I, K)
        10 J = J + NANDS(K, LM1)
              KOST = J
              RETURN
              END
```

12

Asynchronous Sequential Logic

In Chapter 4 a clock signal was introduced to equalize delays "in effect," and synchronous sequential networks were modeled, analyzed, and synthesized with relative ease because transient signals arising from varying gate delays are ignored by synchronous memory elements that are under the control of a clock signal. In some situations synchronous operation of a network is undesirable. It may be too expensive, too slow, or synchronization of incoming signals may not be possible. In such cases we must turn to asynchronous sequential switching circuits in which no attempt is made to equalize delays. In this chapter we will model and analyze networks that are to operate in a specified manner even though the precise magnitudes of delays are not known.

12.1 ANALYSIS OF ASYNCHRONOUS SEQUENTIAL NETWORKS

Various network elements introduce delays of differing magnitudes and characteristics. Wires conveying electrical signals can usually be considered to introduce pure delay even though they also attenuate and distort the waveform of a signal. Each foot of wire introduces approximately one nanosecond of delay. While delays of such magnitude are totally negligible in many networks, some logic gates introduce delays of the same order of magnitude. Wiring delays are significant in networks of such gates, and in networks in which slower logic elements are separated by great distances.

Logic gates also introduce delays that are measured in nanoseconds or tens of nanoseconds. Such delays may be very complex. The signals presented to different input terminals of a gate may encounter slightly different delays. Temperature may influence the magnitudes of these delays; delay of the leading edge of a waveform may differ from delay of the trailing edge (particularly true of DTL gates). Again, we will ignore such complications for the most part, and model each gate by placing a single delay element in its output lead. For many purposes, we may consider that the delay element translates in time the signal presented to it: the output signal of the delay element is an exact duplicate of the input waveform. In other situations, it may be appropriate to endow the delay element in our model of a gate with *inertial* characteristics. If we attempt to pass shorter and shorter pulses through a gate, we will encounter a pulse width so short that the output signal of the gate will not change sufficiently so that other gates can recognize the excursion as a change of logic level. Thus gates may eliminate very brief transients that appear on their input line. We will assume that an inertial delay element does not respond to two or more changes of its input signal if those changes occur within a period of time that is less than the delay of the element. If an input signal changes infrequently, the inertial delay acts as a pure delay.

No two gates can be expected to offer exactly the same delay, even if they are manufactured at the same time of nominally equivalent components. But if the delays they introduce are known to fall between a lower and an upper limit, the delays are *bounded*. When an electronic switching system must interact with mechanical devices or people, then delays may be without bound. An elevator control must wait until someone presses a button. If a high speed electronic computer signals that an operator is to supply another deck of punched cards, or mount another reel of magnetic tape, a great number of nanoseconds elapse before signals indicating that the operator has complied are presented to waiting logic circuits.

In the first sections of this chapter we will assume that all delays are inertial and bounded. For the most part, wire delays will be lumped with the delay of the gate driving the wire. No precise delay magnitudes or bounds will be assumed. Relative magnitudes of delays that can lead to network failures will be determined. Whether such magnitudes fall within bounds and might therefore actually occur in a network depends on the

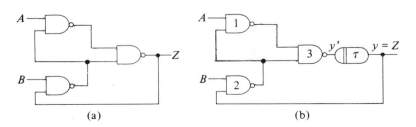

(a)   (b)

FIG. 1₂.1   An Asynchronous Network and its Fundamental Model

SECTION 12.1   **Analysis of Asynchronous Sequential Networks**

specific hardware of which a network is to be constructed and thus cannot enter our discussion.

Figure 12.1(a) offers a first asynchronous network for our consideration. We will model such networks in two ways. A *fundamental model* consists of a given network in which a minimum number of delay elements have been inserted such that a delay element appears in each loop of gates.[1] All gates are considered to be delayless in such a model. Figure 12.1(b) shows a fundamental model of the network of Fig. 12.1(a). The magnitude of delay, τ, cannot be directly related to gate delay since one loop of two gates and another of three gates exist. A *detailed model* in which a delay element is inserted in the output line of each gate is more realistic. With both models we assume that only true variables appear on input signal lines. If complements are required, they must be explicitly generated by the network under investigation so that the delay introduced in their generation will be included in the study.

The fundamental model has the advantage of simplicity. A minimum number of delay elements is involved, so bookkeeping details are minimized. The detailed model can give more accurate results (closer to actual network behavior) but with much more effort, as we will find in this section by analysis of a network using both models. As a result we will use the fundamental model whenever it can give sufficient information, and allude to the detailed model only when refinement is required.

Both the fundamental and detailed models of a network are special cases of the very general model presented in Fig. 4.5. And the general definitions of Section 4.1 may be used to describe mathematically both types of models. Signals on the output lines of delay elements are known as *state variables* and denoted y_i; those on their input lines are known as *next-state* or *excitation variables*, y'. If the ith delay element of a model introduces delay τ_i, its state and excitation variables are related by

$$y_i(t + \tau_i) = y_i'(t)$$

The p-tuple of valued state variables, $y^i = (y_1, y_2, \ldots, y_p)$, measures the *state* of a network at any instant of time. In a fundamental model, p will be small; in a detailed model, p will equal the number of gates of the network. The $(n + p)$-tuple formed by concatenating an input symbol, x^i, and a network state, y^j, i.e., $x^i \circ y^j$, is known as a *total state* of the network.

If at some instant $y_i'(t) \neq y_i(t)$, then state variable y_i is said to be *excited*. When $y_i'(t) = y_i(t)$, y_i is said to be *quiescent*. When one or more state variables of a network are excited, the state of the network is about to change and is said to be *unstable with respect to the input symbol* being applied. The total state at such times is said to be *unstable*. Conversely, when all state variables are quiescent, none of the state variables are tending to change and the network state will not vary as long as the current

---

[1] Existence of a loop is a necessary but not sufficient condition for a network of gates to be sequential. We need not place delay elements in combinational loops.

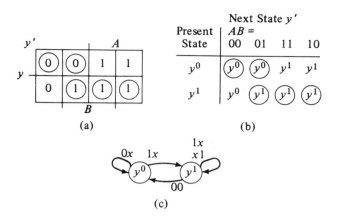

FIG. 12.2   State Descriptions of the Network of Fig. 12.1

input symbol persists. The network state is *stable with respect to the applied input symbol*; the total state is said to be *stable* in such cases. Rather than speaking of stable or unstable total states, we abbreviate and say that a network is in an *unstable state* when at least one state variable is excited, and in a *stable state* when all state variables are quiescent. Stable states are usually emphasized by encircling them as in Fig. 12.2.

Only one state variable y appears in Fig. 12.1(b); its excitation variable may be expressed as a function of the input variables and itself.

$$y' = A(\bar{B} \lor \bar{y}) \lor By$$

From the excitation map of y' in Fig.12.2(a) we see that y' may be more simply expressed as

$$y' = A \lor By$$

In Fig. 12.2(a), state $y = 0$ is stable with respect to input n-tuples $AB = 00$ and 01, and is unstable with respect to $AB = 11$ and 10. State $y = 1$ is unstable only with respect to $AB = 00$. Thus total states $ABy = 000$, 010, 011, 111, and 101 are stable, and 001, 110, and 100 are unstable total states.

We will view an inertial delay as follows: if the excitation is removed from an excited state variable before that variable changes its value in response to the excitation, then the brief excitation is ignored. This model of delay, while not perfect, does introduce the inertial aspects of delay and is manageable.

A state-transition table can be prepared directly from the excitation map if input variables are used to label the columns of the map and state variables are used to label its rows. A unique state name must be assigned to each row. That name is also placed in each column of the row in which a circled (stable state) entry appears. Uncircled entries are replaced with the state name associated with the row identified by the uncircled entry. This transformation is particularly simple for the example network, and leads to the table of Fig. 12.2(b), in which the state names of Section 5.1 are used. A more extensive example to be considered soon offers a more general example of this transformation.

The state-transition diagram of Fig. 12.2(c) can be prepared directly from the state-transition table. From the diagram, we see that a state is stable with respect to an input symbol if that symbol labels a loop transition arrow[2] (from the state to itself). A state is unstable with respect to all input symbols that label transitions from that state to others.

Do the tabular and graphic models of a network offer a description of its behavior? We must trace signals in Fig. 12.1(b) and compare the results with Fig. 12.2. First, two stable states appear in the $AB = 01$ column of Fig. 12.2. Tracing signals in Fig. 12.1(b) reveals that with $AB = 01$, y may have either value and the network maintains whichever value y has. With $AB = 01$, $y' = 0 \vee 1 \cdot y = y$. Gates 2 and 3 of Fig. 12.1(b) form a latch circuit, treated in the previous chapter. It is essential that two or more stable states appear in one or more columns of a state-transition table. If only one (or no) stable condition appears in each column, then the table describes a combinational network, since the output m-tuple of the network is uniquely determined by the input symbol (with the exception of transients).

Suppose that $y = 1$ and $AB = 01$ and then the value of A is changed to 1. Examining Fig. 12.1(b) reveals that when $y = 1$ and $B = 1$, a change in the value of A causes very little activity in the network. The state variable is not excited by such changes. Fig. 12.2(b) indicates that the machine is stable in state y^1 with $AB = 01$. Changing A to 1 requires that we move horizontally in the y^1 row from the 01 to the 11 column. We find that state y^1 is also stable with respect to $AB = 11$. Hence the state table predicts the lack of activity, which we observed in the network.

If $y = 1$, $AB = 01$ and B is changed to 0, the entry in the y^1 row and $AB = 00$ column of Fig. 12.2(b) is not circled, indicating that the state variable is excited. The entry in that cell of the table indicates that the next state of the machine is to be y^0 ($y = 0$). If the $AB = 00$ input symbol persists for longer than τ seconds, the value of y will change to 0. In the y^0 row and 00 column a stable entry appears. The total state has changed from (011) to 001 to (000). The change in input symbol requires that we scan the state table horizontally in the row of the original stable state, to the column of the new input symbol. The change in the state variable corresponds to vertical motion in the new input-symbol column of the state table. Analysis of the network also reveals that its state is changed when the input symbol is changed from 01 to 00.

Further analysis of the network and state-transition table gives more evidence that the state table (or diagram) does summarize activity of a fundamental model of a network. Time is not considered in the state table

---

[2] The loop should not suggest that a network leaves the state and then returns to it. Loop transitions could be omitted, but are usually drawn to indicate that the behavior of the network under the labeling input symbol is specified as opposed to being unspecified (don't cares).

summary. We cannot determine how long a network will remain in an unstable condition before it jumps to a stable condition. All we can be sure of is that if the input symbol persists, the jump will eventually be made.

In a detailed model of an asynchronous network, we insert an inertial delay element in the output lead of each gate. This delay represents both the gate delay and delay introduced by the interconnection. Again gates are thought of as ideal after the delay elements have been inserted.

Figure 12.3(a) shows the detailed model of the network of Fig. 12.1. The excitation equations for the three state variables may be written directly from the model.

$$y_1' = \bar{A} \vee \bar{y}_2$$
$$y_2' = \bar{B} \vee \bar{y}_3$$
$$y_3' = \bar{y}_1 \vee \bar{y}_2$$

Entries in a state table are calculated from these equations. A map is a less useful intermediate step here because of the larger number of variables. In Fig. 12.3(b) stable network conditions are again circled for emphasis, and asterisks have been placed after those next state-variable values that differ from their corresponding present state entries, so that we may more easily identify the excited state variables.

Now we are in a position to examine activity of the network of Fig. 12.1 in greater detail. But first let us determine that Fig. 12.3(b) prescribes the same activity as Fig. 12.2. Suppose that $AB = 00$ and the network has attained stable condition $y_1 y_2 y_3 = 110$. If we now change the input symbol to $AB = 01$, we move horizontally in Fig. 12.3(b) and find that state 110 is also stable with respect to $AB = 01$. But we discovered this when we examined the first row of Fig. 12.2(b).

If with $y_1 y_2 y_3 = 110$, the input symbol is changed from 00 to 10, we move in the 110 row to the fourth next-state column of Fig. 12.3(b). The entry 0*10 indicates that y_1 is excited. After τ_1 seconds have elapsed y_1 will change to 0 and the network state will be 010. But the entry in the 010 row, 10 column of Fig. 12.3(b) indicates that 010 is not a stable network condition with respect to $AB = 10$; y_3 is now excited and will attain the value of 1 after τ_3 seconds elapse. Then the network state will be 011, which is stable with respect to $AB = 10$. In brief, change of the input symbol from 00 to 10 causes the network state to change from 110 to 010 to 011. The values of y_1 and y_3 must change in sequence before the state transition is completed. Analysis of the network model reveals exactly the same activity.

Suppose that $AB = 00$ and $y_1 y_2 y_3 = 000$. The state-transition table indicates that all three state variables are excited. When two or more state variables are simultaneously excited, a *race* exists. We cannot predict the results of a race because we do not know the exact magnitudes of the delays in the network. We may only assume relative delay magnitudes and determine the activity that will take place if our assumptions happen to be valid. For example, if $\tau_1 < \tau_2$ and $\tau_1 < \tau_3$, then state 000 will be followed

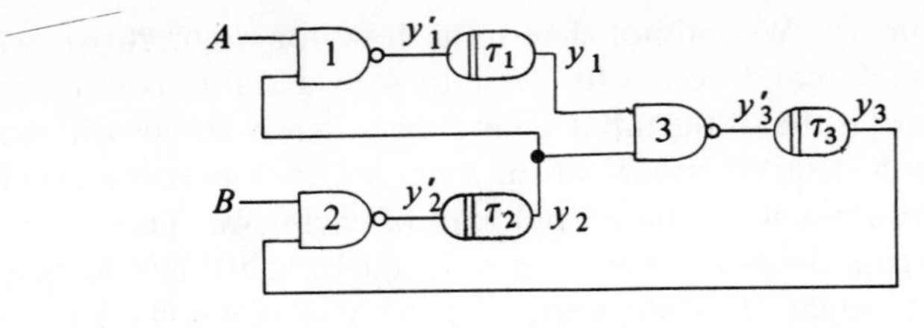

(a)

Present state			$AB =$	(Excitation $y_1'y_2'y_3'$)		
y_1	y_2	y_3	00	01	11	10
0	0	0	1*1*1*	1*1*1*	1*1*1*	1*1*1*
0	0	1	1*1*1	1*0 1	1*0 1	1*1*1
0	1	0	1*1 1*	1*1 1*	0 1 1*	0 1 1*
0	1	1	1*1 1	1*0*1	0 0*1	0 1 1
1	0	0	1 1*1*	1 1*1*	1 1*1*	1 1*1*
1	0	1	1 1*1	1 0 1	1 0 1	1 1*1
1	1	0	1 1 0	1 1 0	0*1 0	0*1 0
1	1	1	1 1 0*	1 0*0*	0*0*0*	0*1 0*

(b)

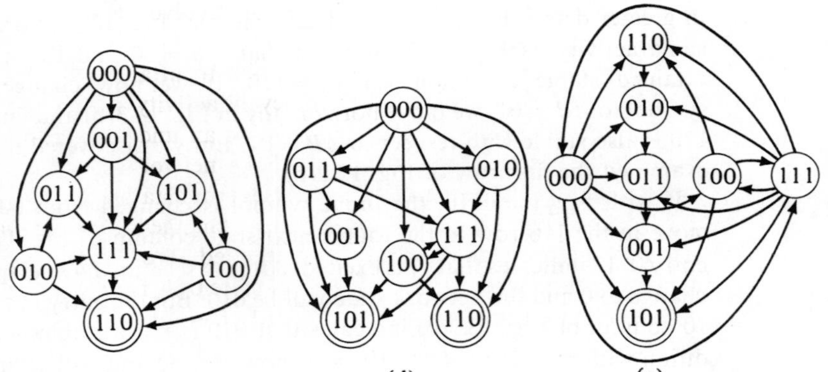

(c) (d) (e)

FIG. 12.3 (a) Detailed Network Model. (b) State Transition Table. (c) State Diagram for $AB = 00$. (d) State Diagram for $AB = 01$. (e) State Diagram for $AB = 11$.

by state 100. This is one of seven possible delay distributions; each of the other six promotes a different state transition.

Figure 12.3(c) summarizes all the state transitions that may occur with $AB = 00$. The transitions are not labeled since the entire diagram applies for only this one input symbol. Where more than one transition arrow leaves a state, a race exists. Network activity may follow any one of the arrows in such cases. This then is a *nondeterministic* state diagram; we cannot predict the exact path of network activity, and two identical interconnections of logic gates do not necessarily follow the same paths. The diagram indicates not only the transitions that will actually take place in a network, but all transitions that might occur.

All paths lead to state 110 in Fig. 12.3(c). No transitions emerge from that state, and it is therefore the only state that is stable with respect to $AB = 00$. Since this one stable state is always attained regardless of the magnitudes of the delays if $AB = 00$ is maintained for a sufficient time, the network is *speed independent* with respect to $AB = 00$, and races are *noncritical*.

Two states are stable with respect to $AB = 01$. Figure 12.3(d) illustrates all possible state transitions for this input symbol. If the network state is 000, 010, 011, 100 or 111, then either of the two stable states may be reached. Races from these five states are *critical*: one state variable changing first rather than another makes a lasting difference. If the network state is 100 and $\tau_3 < \tau_2$, stable state 101 will be attained; if $\tau_2 < \tau_3$ the network will reach stable state 110. And if $\tau_2 = \tau_3$, the network may oscillate between states 111 and 100.

Transitions form closed paths in the state diagrams of Figs. 12.3(d) and (e). States 100 and 111 are included in one such loop. If the network were to reach one of these states with $AB = 11$ and $\tau_2 = \tau_3$, then the network state would alternate between 100 and 111 without ever reaching a stable state. Such activity is *cyclic*, and it may involve more than two unstable states.

Is it possible for the example network to attain the states mentioned above, which lead to critical races and cyclic activity? Yes it is, if the network is operated in what we will consider to be an undesirable manner. To see this, let us assume that $AB = 01$ and the network state is 110. Then we change AB to 11; the network progresses from state 110 to 010 to 011, etc. according to Fig. 12.3(e). Suppose that we change the input symbol back to $AB = 01$ before stable state 101 is attained. We might make this input symbol change when the network is in state 011, for example. Fig. 12.3(d) describes state transitions with $AB = 01$; from it we see that the network may ultimately attain one of the two states that are stable with respect to $AB = 01$, or it may cycle between states 100 and 111.

To avoid this critical race and all others that may be encountered by too-rapid alteration of the input n-tuple, we place a restriction on the manner in which we operate asynchronous networks. We insist that the input n-tuple never change twice in a period of time shorter than the longest time the network requires to perform a stable-state-to-stable-state transition. In the example network we have seen that $\tau_1 + \tau_3 + \tau_2 + \tau_1$ seconds elapse when the network moves from state 110 to 101. If this were the slowest transition of the network, we would insist that the input n-tuple never change twice in any interval of time shorter than this. Clearly this is a conservative restriction. Some transitions require less time than others. What we really must insist, if we are to be able to predict network performance, is that the network be allowed to attain a state that is stable with respect to an applied input symbol before that input symbol is replaced with the next one. When this requirement is satisfied, we operate a network in its *fundamental mode*.

If this restriction is met, both the fundamental and detailed models of a network may be used to perform meaningful analysis. While state descriptions of a network are easily developed from a fundamental model they offer minimum detail of network activity. State descriptions based upon a detailed model are developed and explored with the expenditure of much more effort.

Other models with more or fewer state variables than called for by the detailed model may be developed, of course. In all cases the models are described by the very general model of Fig. 4.5 and state tables and diagrams have much the same form and significance as those of the previous chapter.

We have also seen some of the ways in which an asynchronous sequential network can fail to perform as expected and have placed one restriction on how we operate such networks to avoid these failure modes. Unfortunately, asynchronous sequential networks may fail for other reasons, as we will see in the next sections. We will have to place additional restrictions on the operation of asynchronous networks, and develop design methods that suppress these other failure modes.

2.2 HAZARDS IN COMBINATIONAL NETWORKS

A hazard is an "actual or potential circuit malfunction as a result of signals encountering various delays in the paths of a network." This very general definition includes races, as defined in the previous section. Thus races are hazards for which a very descriptive name has been found. In this and the next section a variety of malfunctions that arise when two or more changing signals race toward a common point will be considered.

Signals generated by some network structures may temporarily disagree with the predictions of Boolean algebra, which do not take delay into account. Such networks produce *transitory hazards*. We have already seen that combinational switching networks may temporarily produce false signal values. In the next section we will examine hazardous conditions in asynchronous sequential networks and find that they can cause signals to assume false values permanently. Then the behavior of a network differs from the behavior described by a state table. Such conditions are classified as *steady-state hazards*.

Again we assume that only uncomplemented input variables are presented to a network. The requirement that needed complements be generated within a network ensures that the delay of the complementing circuitry will not be overlooked. Often such delay is the origin of a hazard.

Static Hazards

Figure 5.4 presents a Karnaugh map of a switching function and a minimum AND-to-OR gate realization of that function. Suppose that

$$ABCD = 0110;$$

we expect $F = 1$. Now we change A so that $ABCD = 1110$; again we expect $F = 1$. But the new value of A propagates toward the output terminal along two paths. One path includes blocks 1, 3, and 6; the second path includes only gates 4 and 6. If the total delay introduced by blocks 1 and 3 is less than the delay of gate 4, $\tau_1 + \tau_3 < \tau_4$, then for a period of time $[\tau_4 - (\tau_1 + \tau_3)]$ zeros will be present at all input terminals of the OR gate. Zero will appear on the output line if these input signals persist long enough so that gate 6 can respond, i.e., $\tau_6 < \tau_4 - (\tau_1 + \tau_3)$. We have found that it is possible for a 0 to appear temporarily on the output line in contradiction to ideal Boolean algebra. Whether a 0 will actually appear in a given network depends on the specific delay magnitudes of that network.

Figure 12.5(a) summarizes the activities that can take place in the network of Fig. 12.4(b) when the input symbol changes from 0110 to 1110. While we are dealing with a combinational network, the detailed model and notation developed in the previous section are used. Only one stable state must appear for each input symbol, of course. The delay magnitude relationship that causes the network to make a transition labels the arrow that symbolizes that transition. We see again that if the network is in state $y_1 y_2 y_3 y_4 y_5 y_6 = 111001$, which is stable with respect to $ABCD = 0110$, and if $\tau_1 + \tau_3 < \tau_4$, then y_6 will be excited and may take the value of 0 for a short time when the input symbol is changed to 1110. If $\tau_1 \geq \tau_4$, then y_6 will not vary as a result of this change of input n-tuple.

If x^i and x^j are adjacent input symbols (one input variable differs) and a switching function has the same value for both x^i and x^j, i.e.,

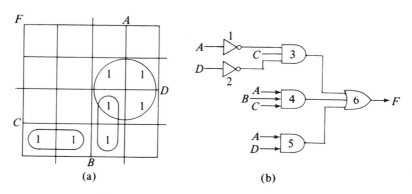

(a)   (b)

FIG. 12.4 A Switching Function and A Hazardous Realization

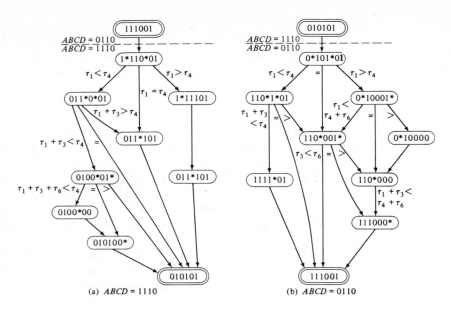

FIG. 12.5 Transition of Network Conditions for Two Input *n*-tuples

$$f(x^i) = f(x^j) = 1 \quad \text{(or 0)},$$

then a network that realizes that function contains a *static* hazard if a 0 (1) can appear temporarily on its output line when the input symbol is changed from x^i to x^j, or vice versa. The adjective "static" was chosen by D. Huffman [4, 5 (Chapter 13)] because the output value is supposed to remain constant.

The condition $\tau_1 + \tau_3 < \tau_4$ may not be satisfied in the network of Fig. 12.4, and we could increase the delay of block 1 to ensure that it will not be satisfied. But this will not remove the hazardous condition. If $\tau_1 + \tau_3 > \tau_4$, then the false value will not accompany the input-symbol change from 0110 to 1110, but it will appear when the reverse change from 1110 to 0110 is made. Figure 12.5(b) shows possible network activity following this change. Comparison of the two parts of that figure shows that if the false value does not appear when A is changed from 0 to 1 because of assumed delay magnitudes, then it may appear when A is changed from 1 to 0. The figure also indicates that no false value will appear if $\tau_1 + \tau_3 = \tau_4$, or if the difference between $\tau_1 + \tau_3$ and τ_4 is so small that the inertia of gate 6 removes the transient.

Fortunately Huffman not only identified static hazards, but also showed that they may be eliminated by proper design. One prime implicant selected when designing the network of Fig. 12.4(b) covers minterm 0110. Another covers minterm 1110. AND gates 3 and 4 correspond directly to these prime implicants. When we change the input symbol as discussed above, we cause the output signal of one of these gates to change from

0 to 1, and the output signal of the other AND gate to change from 1 to 0. If these changes do not take place at the same time, the false value of F appears. Huffman's solution to this problem is the addition of a redundant AND gate that maintains an output value of 1 when A is changed. It is always possible to find a suitable AND gate (prime implicant) since we are considering changes of a single input variable only (adjacent minterms). If we add to the network of Fig. 12.4(b) an AND gate that realizes prime implicant $BC\bar{D}$ (the prime implicant that covers minterms 0110 and 1110), the hazard we have been discussing is removed. This additional AND gate, redundant in the sense of network minimization, continuously presents a 1 to the OR gate when A is changed. Thus F may not vary as a result of these changes of A.

Static hazards are eliminated from combinational logic networks by basing such networks on a prime implicant cover of a switching function such that every pair of adjacent minterms (every 1-cube of the function) is covered by at least one selected prime implicant. It is a simple matter to identify the prime implicants that must be selected, if the Karnaugh map of a function is available. When the number of input variables is large, the prime implicant selection portion of Quine-McCluskey or other minimization algorithms must be altered so that each 1-cube of a function is covered by at least one selected prime implicant. Usually more than the minimum set but fewer than all prime implicants must be selected to meet this requirement.

Minimization procedures lead to sum-of-products expressions that suggest two-level networks. Static-hazard–free multiple-level networks can be designed by manipulating sum-of-products expressions, but such manipulations must not reintroduce hazards. Theorems such as $x \cdot \bar{x} = 0$, $x \lor \bar{x} = 1$, $x \lor \bar{x}y = x \lor y$, which involve both a variable and its complement and cancel one or both must not be used. If we were to split a prime implicant, say BCD, into a sum of terms, $BCD(x \lor \bar{x}) = BCDx \lor BCD\bar{x}$, and perhaps combine one of these terms with other terms of the function, we would introduce two paths along which a change in x propagates, a static hazard.

Hazards are easily introduced if the logic diagram, rather than the equation, is manipulated. Suppose that we desire a static-hazard–free NAND realization of

$$F = AB \lor \bar{B}C \lor AC$$

Prime implicant AC has been included only to eliminate static hazards. To reduce the number of required gates, we might factor,

$$F = AB \lor C(A \lor \bar{B})$$

and arrive at the network of Fig. 12.6. Now we recognize that gate 3 can be used to provide $\bar{A}$ and thereby remove the need for an inverter. But analyzing the resulting network yields

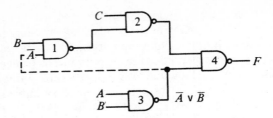

FIG. 12.6 Reintroducing a Static Hazard

$$F = AB \lor C(\bar{B} \lor AB)$$

A prohibited theorem was used and a static hazard reintroduced. This last expression of F does not include a product term that covers adjacent minterms 101 and 111. Thus when we change the input symbol from 101 to 111 or vice versa, we may observe $F = 0$ for a short time.

Dynamic Hazards

Huffman also pointed out that transients may appear when the value of the output variable of a network should change in response to a change of input symbol and does, but changes an odd number $N > 1$ times. Thus when an output signal should change from 0 to 1, we might observe it changing from 0 to 1 to 0 before it changes and remains at 1. Such activity constitutes a *dynamic hazard*.

For a dynamic hazard to appear in a gate network in response to a change of a single input variable, that change must propagate toward the output terminal along 3, 5, etc. paths. Suppose the output is to change from 1 to 0. For a dynamic hazard to appear, the input variable change must propagate along a path with minimum delay, to cause the output variable to go to 0; along a path of intermediate delay to cause the output variable to go to 1; and along a path with maximum delay to cause the output variable to return to and remain at 0. These last two paths then must also give rise to static hazards.

Dynamic hazards are rare in well-designed gate networks. Inertial delays mask dynamic hazards, and gates added to eliminate static hazards also prohibit dynamic hazards.

Multiple Input-Change Hazards

E. B. Eichelberger [2 (Chapter 13)] has shown that if two or more input variables are allowed to change simultaneously, then false values of the output variable may arise in either of two ways. One is a generalization of the static hazard and can be prevented by a generalization of the technique for preventing static hazards. Figure 12.7 provides a map and realization of a switching function that illustrates the generalized static hazard. The network is free of static hazards if only a single input-variable change is allowed. But if the input symbol is changed from $ABCD = 1111$ to 0101, i.e., A and

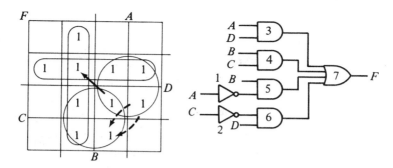

FIG. 12.7   A Network which is not Static Hazard-free
if Two Input Variables Change

C are changed at the same time, then the output signals of all four AND gates must change. If AND gates 3 and 4 provide 0's to the OR gate before AND gates 5 and 6 provide 1's (due to the delay introduced by the inverters), then a false value of F may appear.

Generalizing to include this type of activity, a *static hazard* exists if $f(x^i) = f(x^j) = 1$ (or 0) and $f = 0$ (1) can appear temporarily when the input symbol is changed from x^i to x^j. This static hazard may be eliminated by including in the sum-of-products expression of f a prime implicant that covers x^i and x^j, if such prime implicant exists. In the example above, prime implicant BD must be included to prevent the static hazard mentioned. Eliminating static hazards in general requires that all prime implicants be included in the sum-of-products expression of a function. Networks based upon the sum of all prime implicants can be very expensive and unfortunately will not be free of all hazards if arbitrary input-symbol changes are allowed. No prime implicant covers minterms 1011 and 1110 in Fig. 12.7(a). Thus input-symbol changes from 1011 to 1110, or vice versa, are not necessarily free of hazards.

If an input symbol is allowed to change from x^i to x^j and the smallest cube which covers minterms x^i and x^j is not an implicant of the function, then a *function hazard* exists. In Fig. 12.7 if the input symbol changes from 1011 to 1110, i.e. both B and D change, then the output signal of AND gate 3 is to change from 1 to 0, and the output signal of AND gate 4 is to change from 0 to 1. AND gates 5 and 6 present 0's to the OR gate. The delays of gates 3 and 4 may well be such that all 0's are temporarily presented to the OR gate and F may assume the false value of 0. Differences in gate delays have the effect of changing B and D in sequence. But because the exact magnitudes of these delays are unknown, we cannot say which sequence, B then D, or D then B, the OR gate will see. The network can take either of the two paths marked on the Karnaugh map of Fig. 12.7.

Function hazards cannot be eliminated by adding redundant gates to a network; and function hazards are so common if multiple input-variable changes are allowed that input symbol changes must be restricted when it is necessary to have hazard-free network operation, as when a combinational network is embedded in an asynchronous sequential network. Thus a second restriction is commonly placed upon the manner in which asynchronous sequential networks are operated: input-symbol changes are restricted to single input-variable changes.

Hazard Detection

The Karnaugh map is a most useful tool for detecting hazards in two-level combinational networks. But if the number of input variables is large or the network has many levels of gates, other tools are needed. Let x^i and x^j be two input symbols for which $f(x^i) = f(x^j)$, and x^i and x^j are at distance $n - q$.

$$x^i = (l_1, l_2, \ldots, l_q, l_{q+1}, \ldots, l_n)$$
$$x^j = (l_1, l_2, \ldots, l_q, \bar{l}_{q+1}, \ldots, \bar{l}_n)$$

To determine whether a hazard exists under this input-symbol change, in which q input variables remain constant, label input terminals $x_1, x_2, \ldots, x_q$ with $l_1, l_2, \ldots, l_q$, respectively. Then label the output lines of all gates whose output value is uniquely determined by $l_1, l_2, \ldots, l_q$ with that unique value. Continue this labeling as long as possible. If the network output line remains unlabeled, a hazard exists for this input-symbol change.

Whether the output signal of a gate is uniquely determined by one or more valued input variables depends on the logic of the gate and the values of the input variables. A 0 labeling a single input terminal of an AND gate fixes the output signal of that gate at 0. A 1 on any input terminal of an OR gate fixes the output signal of that gate. But a 1 (0) on one of several input terminals of an AND (OR) gate does not uniquely determine the value of the output signal of that gate.

With this procedure we attempt to find a path from the input terminals through the network to its output terminal along which all signals are determined by the nonvarying input variables. If such a path exists, the output signal of the network will not vary with the input-symbol change since it is determined by the input variables which do not vary.

In Fig. 12.8(a) we examine the network of Fig. 12.6 with $A = B = 1$ and C changing. A path through gates 3 and 4 fixes the value of F so no hazard

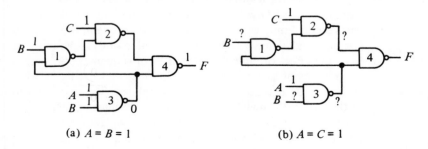

(a) $A = B = 1$ (b) $A = C = 1$

FIG. 12.8   Analysis of the Network of Fig. 12.6 for Hazards Accompanying Changes of (a) C and (b) B

accompanies changes of C. But with $A = C = 1$ and B changing as in Fig. 12.8(b), the output-signal levels of all gates are uncertain and a hazard exists.

12.3 HAZARDS IN SEQUENTIAL NETWORKS

If a hazardous combinational network is used to generate a set or reset signal for an asynchronous flip-flop, then it is easy to imagine that the flip-flop might be set or reset erroneously when the combinational network generates a false signal value. But hazards in combinational networks also cause false state transitions when they are incorporated in feedback networks. To see this we will ignore the warning and form a hazardous network that realizes the state-transition equation of the gated-latch:

$$F' = AB \vee \bar{B}F$$

Note that this function was discussed in open-loop form in connection with Fig. 12.6.

Figure 12.9(a) gives a fundamental model of a network derived directly from the above equation; no redundant term has been introduced to eliminate hazards. The state-transition table of Fig. 12.9(c) reveals that if $B = 0$, then the network is stable with either $F = 0$ or $F = 1$. Variable B then corresponds to the gate variable of the gated-latch; A serves as the data variable. The existence of a static hazard in the network is quickly ascertained from the Karnaugh map of Fig. 12.9(b): when the input symbol is changed from $AB = 11$ to 10 or vice versa, F' may temporarily assume a false value.

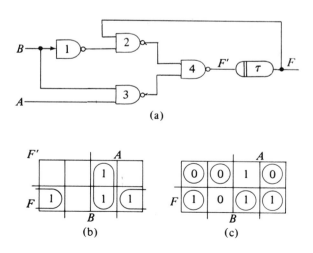

(a)

(b)   (c)

FIG. 12.9   A Hazardous Combinational Network is used to form
an Asynchronous Sequential Network

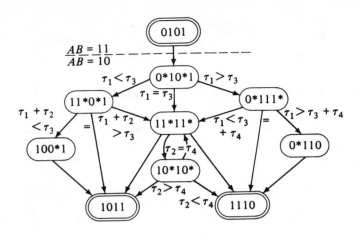

FIG. 12.10   Activity in the Network of Fig. 12.9

Figure 12.10 summarizes the results of the analysis of a detailed model of the network for this input-symbol change. State is measured by the 4-tuple $y_1 y_2 y_3 y_4$ where $y_4 = F$. After the input-symbol change from 11 to 10, the network may attain either of two stable states. The value of y_4 is different for these states. A possibility of endless cycling is also indicated. If $\tau_1 > \tau_3 + \tau_4$ or $\tau_1 + \tau_2 > \tau_3 + \tau_4$, then because of the static hazard (temporary false value of y_4) the network attains the "wrong" state with $F = 0$ as opposed to the $F = 1$ state which Fig. 12.9 indicates should be attained. Thus the asynchronous sequential network contains a steady-state hazard. The temporary false value of y_4 created by the static hazard races with the change in B toward gate 2. If gate 1 is particularly slow, 0 will reach gate 2 before the change in B, and the false value of F will be maintained by the NAND-latch, gates 2 and 4. If all gates introduce approximately the same delay, then it is probable that $\tau_1 + \tau_2 > \tau_3 + \tau_4$ and the network will display this steady-state hazard by malfunctioning.

Static hazards do not always result in steady-state hazards, but we see that they may do so. Even when static hazards do not cause false state transitions, they may result in the appearance of temporary false values at the output terminals of a network, which in turn can cause the network driven by these signals to malfunction. Thus we will usually eliminate static, dynamic, and function hazards from combinational networks destined for asynchronous sequential applications by agreeing to:
1. Operate the network in its fundamental mode, i.e., not change the input n-tuple until the network has attained a stable state;
2. Not change more than one input variable at a time; and
3. Add logic gates to the network to eliminate static hazards that remain after 1 and 2 are satisfied.

Alas, these three conditions do not guarantee perfect performance of asynchronous sequential networks; other types of hazards exist.

Critical State-Variable Races

Suppose we desire a network that behaves according to the state-transition table of Fig. 12.11(a). This is a very practical desire because the table describes an asynchronous TFF, which has many applications. Each change of input variable x from 1 to 0 is to be accompanied by a change in the output variable z. Because synthesis is the subject of the next section, here we simply make the state assignment of Fig. 12.11(b), write state transition equations, and draw the network of Fig. 12.11(c). For the moment consider the delay elements to be explicit delay circuits with τ_1 and τ_2 much larger than gate and wire delays but of otherwise unknown magnitude.

A probable source of malfunction can be identified by examination of the state-transition table. Suppose that the network is in state $S1$ and x changes from 1 to 0. Both y_1 and y_2 are then excited; the race between these state variables is critical as we can see by examining Fig. 12.11(b) or Fig. 12.12, which summarizes possible network activity. If $\tau_1 > \tau_2$ or $\tau_2 = 2\tau_1$, y_2 changes first and the network attains state $S0$ rather than the expected state $S2$. If $\tau_2 \gg \tau_1$ the value of y_1 will alternate until y_2 changes and the network attains either state $S0$ or $S2$. Another critical race takes place when the network leaves state $S3$.

If we think of τ_1 and τ_2 as representing the wire and gate delays of the network of Fig. 12.11(c), then we can be almost certain of a steady-state hazard. In order for the network to operate in the manner of the original state table, a change in x must propagate through gates 1, 2, and 3 in less time than it takes to reach terminal y_2.

Present State	Next State $x = 0$	1	Z
$S0$	$S0$	$S1$	0
$S1$	$S2$	$S1$	0
$S2$	$S2$	$S3$	1
$S3$	$S0$	$S3$	1

(a)

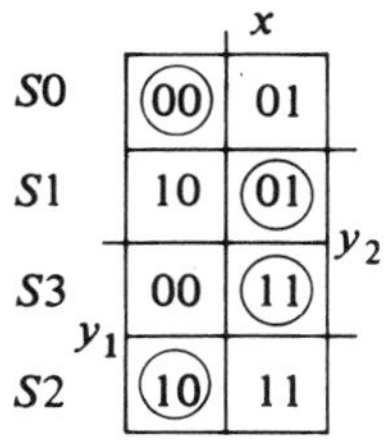

$y_1' = xy_1 \vee y_1\bar{y}_2 \vee \bar{x}\bar{y}_1 y_2$

$y_2' = x \qquad Z = y_1$

(b)

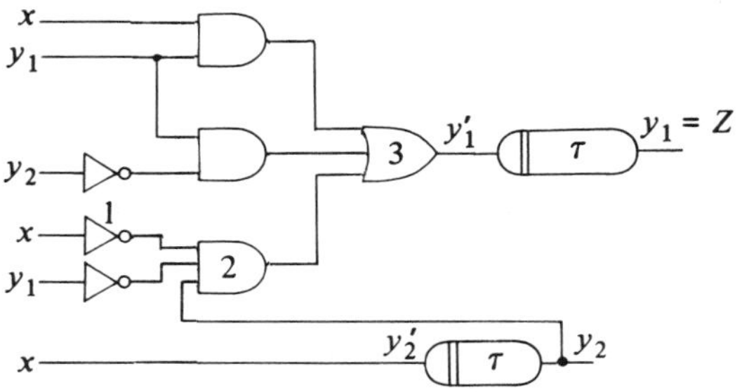

(c)

FIG. 12.11 Binary Trigger with Critical Races

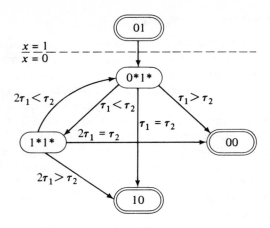

FIG. 12.12   A Critical Race

Critical races between state variables in a fundamental model can be avoided by proper encoding of the states of the machine. This then is a significant difference between design procedures for synchronous and asynchronous sequential networks: any state assignment is acceptable when designing synchronous networks; asynchronous sequential machines must be encoded so as to prevent critical races. As a result the state-assignment problem will receive a great deal of attention in Section 13.2.

Essential Hazards

Critical state-variable races are avoided in the TFF with the state encoding of Fig. 12.13(a). The network derived from this assignment has been made free of static-hazards by the inclusion of redundant prime implicants in the excitation equations. We will assume that only wire and gate delays are present in this network.

The network may malfunction in spite of these efforts to avoid hazards. Assume that the network state is $S0$, $y_1 y_2 = 00$, with $x = 0$, and x is changed to 1. Figure 12.13(a) indicates that state $S1$, $y_1 y_2 = 01$, is to be attained. But if $\tau_3 > \tau_4 + \tau_8 + \tau_2 + \tau_7 + \tau_1 + \tau_6$, then the network activity will be as in Fig. 12.13(c), which shows the network attaining stable state $S3$, $y_1 y_2 = 10$. The order in which these delays are listed indicates the sequence of state-variable changes leading to this erroneous state. We can see the change in x propagate through gates 4, 8, 2 (to change y_2), 7, 1 (to change y_1), and 6 (to lock y_1 at the wrong value). This activity is racing with the change of x propagating through gate 3.

S. H. Unger [17] first recognized this source of malfunction as an *essential hazard*. A state table contains an essential hazard if it specifies that a single change in input variable x_i is to cause a state change from state S_j to state S_k, but that three consecutive changes in x_i are to cause state changes from S_j, which terminate in state $S_l \neq S_k$. The state table of Fig. 12.11

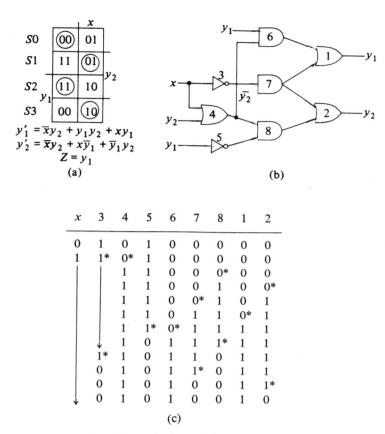

$$y'_1 = \overline{x}y_2 + y_1y_2 + xy_1$$
$$y'_2 = \overline{x}y_2 + x\overline{y}_1 + \overline{y}_1y_2$$
$$Z = y_1$$

(a)

(b)

x	3	4	5	6	7	8	1	2
0	1	0	1	0	0	0	0	0
1	1*	0*	1	0	0	0	0	0
	1	1	1	0	0	0*	0	0
	1	1	0	0	1	0	0*	
	1	1	0	0*	1	0	1	
	1	1	0	1	1	0*	1	
	1	1*	0*	1	1	1	1	
	1	0	1	1	1*	1	1	
	1*	1	0	1	1	0	1	1
	0	1	0	1	1*	0	1	1
	0	1	0	1	0	0	1	1*
	0	1	0	1	0	0	1	0

(c)

Fig. 12.13 An Essential Hazard

specifies that starting from state $S0$ with $x = 0$, a single change of x is to place the machine in $S1$, but three consecutive changes of x are to place the machine in $S3$. Thus this table contains an essential hazard. Every network that realizes this table could display the essential hazard by making false state transitions, unless the network is specifically designed to mask the essential hazard. Thus in every network that realizes a state table containing an essential hazard, we can find a delay distribution that will cause the network to display a steady-state hazard. The delay distributions will vary with the network structure, and in some networks it may be highly unlikely that they will be satisfied. The steady-state hazard discussed above will only appear if τ_3 exceeds the delay of six other gates. Satisfaction of the delay distributions that will cause state $S0$ rather than $S2$ to follow state $S1$, $S1$ rather than $S3$ to follow $S2$, and $S2$ rather than $S0$ to follow $S3$ may be more probable.

The presence of an essential hazard in a state table ensures that two paths of propagation of changes in x_i exist. One path is directed toward

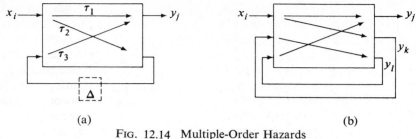

(a)   (b)

FIG. 12.14 Multiple-Order Hazards

state variable y_j; the other leads to y_k.[3] If the change in x_i propagates rapidly to change y_k and this change reaches the network that generates y_j before the change in x_i, a steady-state hazard may result. Thus an essential hazard introduces a critical race between an input variable and a state variable. If the changes in x_i and y_k reach the y_j network at the same time, the possibility of a function hazard exists despite our efforts to avoid function hazards by changing only one input variable at a time. In Fig. 12.13(c) the change in x caused y_2 to change. This change in y_2 reached the y_1 network before the change in x, and a steady-state hazard resulted.

Figure 12.14(a) illustrates these two paths. Assume $\Delta = 0$ for a moment. If $\tau_1 > \tau_2 + \tau_3$, then the essential hazard manifests a steady-state hazard. One way to avoid the steady-state hazard consists of inserting delay Δ such that $\tau_1 < \tau_2 + \tau_3 + \Delta$. Then the change in x_i will reach the y_j network before the change in y_k and the network will perform as expected. In general essential hazards can be masked by inserting massive delays in all feedback paths so that changes in the input variables propagate through all combinational logic before any state variables change. While this is a straightforward solution to the essential hazard problem, it greatly increases the time required by a network to attain its next stable state and hence reduces the rate at which input symbol changes may be provided to the network.

Multiple-Order Hazards

Steady-state hazards may be ordered in a very natural way. The steady-state hazard caused by the static hazard, Figure 12.9, may be viewed as a *first-order hazard* in that only a single state variable was involved. The essential hazard of Fig. 12.13 induces a *second-order hazard*: two state variables are involved. Figure 12.14(b) suggests the nature of a *third-order hazard*. If, in response to a change in input variable x_i, state variable y_k changes, which causes y_l to change, and if the change in y_l reaches the y_j network before the change in x_i, then a steady-state hazard may result. Still higher order steady-state hazards require that a greater number of variables change before the change in x_i reaches the y_j network. The probability of occurrence of a high-order hazard is usually very small.

---

[3] It may be necessary to assume different delays in the branches of a fan-out to find these two paths.

Essential hazards are multiple-order hazards, that can be detected by examining the uncoded state-transition table. Unfortunately not all multiple-order hazards can be detected by such an examination because the state encoding may induce a multiple-order hazard in a state table that is free of essential hazards. Figure 12.15 shows a state table free of essential hazards, which has been encoded in a very straightforward manner. Note that critical races are avoided by causing the network to pass from one unstable condition to another on its way to a stable state. Thus a transition from state $S1$ to state $S3$ is to be accomplished by the following sequence of state variable changes: $\boxed{01} \to 11 \to 10 \to \boxed{10}$.

Rather than develop and examine a network from this encoded state table, let us examine the table directly for possible malfunctions. Assume that the network is in state $S0$, $y_1 y_2 = 00$, and x changes to 1. State variable y_2 is excited and will eventually change its value to 1. Now suppose that this change takes place and reaches the y_1 network before the change in x. Then the y_1 circuitry is led to believe that the overall network state is $S1$, $y_1 y_2 = 01$, with $x = 0$, i.e., the second row and first column of Fig. 12.15(b). As a result, y_1 is excited. If the change in x has not reached the y_1 circuitry when y_1 changes, the network finds state 11 to be stable (y_1 circuitry thinks $x = 0$, y_2 circuitry thinks $x = 1$). If the change in x now reaches the y_1 circuitry, it causes no further action.

This examination of the encoded state-transition table is a means of detecting steady-state hazards without designing and analyzing networks, and can save a great deal of time and effort. Whether a network derived from a table will actually exhibit false state transitions or not depends on its structure and the magnitudes of its delays. Thus we may be required to analyze the network eventually, but knowledge derived from the coded state table indicates what state transitions must be examined at that time. We may also reject a state encoding because it introduces the possibility of multiple-order hazards, thus saving the effort of designing a marginal network. Analysis of a coded state table may also reveal unexpected sequences of states that will give rise to temporary false values of the output variables, even though the expected stable state is ultimately reached.

Unger has shown that a state table free of essential hazards can always be realized by a network that does not include explicit delay elements. But

Present state	Next state $x = 0$	1
$S0$	$\boxed{S0}$	$S1$
$S1$	$S3$	$\boxed{S1}$
$S2$	$S3$	$\boxed{S2}$
$S3$	$\boxed{S3}$	$S1$

(a)

	x	
$S0$	$\boxed{00}$	01
$S1$	11	$\boxed{01}$
$S2$	10	$\boxed{11}$
$S3$	$\boxed{10}$	00

(b)

FIG. 12.15 The state table is free of essential hazards, but a network based upon the coded table may display a second-order hazard

such a network may be entirely uneconomical because of the state encoding required to achieve this result. Unger has also shown that the network that realizes a state table containing essential hazards must contain at least one explicit delay element if steady-state hazards are to be avoided and no upper bound is placed on the magnitudes of wire delays in the network. He further argued that only one explicit delay need be added, but again it is often more economical to insert more than one delay element than to add the logic required to ensure proper network performance with only one explicit delay element.

We have seen that multiple-order hazards can be eliminated by inserting explicit delay elements of the right magnitude at the right point in a network. Massive delay can always be inserted in all state variable lines, of course, at the expense of network speed. More intelligent decisions as to whether to add delay, where to insert it, and how much delay to introduce to maximize network speed but ensure correct network performance can be made only after detailed network analysis. Such analysis is extremely expensive of time and effort, but it is systematic and can be performed by a suitably programmed digital computer. Whether a network will actually display a hazard depends on its structure and the magnitudes of its delays. Only analysis of a detailed model of the network takes these points into consideration, and can give the delay magnitude relationships that result in network malfunction.

Many asynchronous networks have been constructed without prior detailed analysis and found to perform satisfactorily. The failures, often expensive, of other networks when first constructed, after they have been in service for a time, or occasionally as temperature, humidity, or supply voltages vary have promoted the study of hazards and means of avoiding them. The next chapter presents such means.

PROBLEMS

12.1-1. (a) Draw a fundamental model of the circuit of Problem 4.1-3.
    (b) Write the excitation equation(s) for the fundamental model.
    (c) Develop the state-transition table and diagram for this circuit.
    (d) From the state-transition table and diagram determine how this circuit responds to the input symbol sequence $x_1x_2 = 01, 00, 10, 11, 10, 00, 10, 00, 01$.

12.1-2. Repeat Problem 12.1-1 for the circuit of Problem 4.1-8, using the input symbol sequence $x = 0, 1, 0, 1, 0, 1$.

12.1-3. Form a detailed model and state-transition table for the circuit of Prob. 4.1-3. Does the circuit response to input symbol change as predicted by this model differ from that predicted by the model of Prob. 12.1-1? If so, how?

12.1-4. The excitation equations for a network are:

$$y_1' = x_2 \vee \bar{x}_1 y_1$$
$$y_2' = x_1 \vee x_2 \vee y_1 y_2$$

(a) Develop a state-transition table and diagram for the network.

(b) The network is in a stable state with $x_1 = x_2 = y_1 = y_2 = 0$. Then x_2 changes to 1. Describe the activity predicted by the state table.

(c) The network is in a stable state with $x_1 = x_2 = y_1 = y_2 = 1$. How must the input variables be changed to get the network to total stable state $x_1 x_2 y_1 y_2 = 0000$? Describe the network activity that accompanies this change of the input variables.

(d) With respect to what input symbols is this network speed independent?

12.1-5. A network is described by the following excitation equations:

$$y_1' = \bar{x}_2 y_1 \vee y_1 y_2 \vee x_1 x_2 \bar{y}_2$$
$$y_2' = x_2$$

(a) Form the state-transition table for the network.

(b) Which, if any, transitions between stable states are accompanied by noncritical races? by critical races?

(c) Let $y_1(t + \tau_1) = y_1'(t)$ and $y_2(t + \tau_2) = y_2'(t)$. Develop state-transition tables and diagrams for this network for each of the following cases.

  (i) $\tau_1 < \tau_2$
  (ii) $\tau_1 = \tau_2$
  (iii) $\tau_1 > \tau_2$

12.1-6. (a) With reference to Fig. 12.3, how much time does the network require to achieve the following transitions? (Express time in terms of τ_1, τ_2, and τ_3.)

  (i) $x_1 x_2 y_1 y_2 y_3 = 00110$ to 10011
  (ii) $x_1 x_2 y_1 y_2 y_3 = 10011$ to 11101

(b) What is the minimum interval during which each input symbol must be maintained so that the network will be operated in its fundamental mode?

12.2-1. Only one input variable may change at a time.

$$f = \vee \ 0, 1, 2, 5, 6, 7$$

(a) Show that a network based upon a minimum cost, sum-of-products expression of f is not free of static hazards.

(b) Is a network based upon a minimum cost, product-of-sums expression of f free of static hazards?

(c) What relationship exists between a static-hazard–free, sum-of-products expression of f and a static-hazard–free, product-of-sums expression of f, if any?

12.2-2. Repeat Prob. 12.2-1 for the function of Fig. 12.4. Does the same relationship of part (c) exist?

12.2-3. The prime implicants for a function are given below. Find a minimum cost sum-of-products expression of the function that is static-hazard free if only one input variable is changed at a time.

$$\left.\begin{cases} 00x0 \\ 0x00 \\ x000 \\ 001x \\ 010x \\ 100x \\ 0x11 \\ 01x1 \\ x011 \\ 10x1 \\ 1x01 \\ x101 \end{cases}\right.$$

12.2-4. Two networks based on the following equations realize the same function.

$$f_1 = AB \vee AC$$
$$f_2 = A \cdot (B \vee C)$$

Which of these networks can generate a false value during the following input symbol changes?

(a) $ABC = 101$ to 110 or vice versa

(b) $ABC = 100$ to 010 or vice versa

12.2-5. Function $f = \vee 3, 4, 5, 6$ is realized with the 3-level, 4 NAND-gate network described by

$$f = [A \uparrow (A \uparrow B \uparrow C)] \uparrow [B \uparrow C \uparrow (A \uparrow B \uparrow C)]$$

Determine the input-symbol changes for which f is to maintain the same value, but for which the network may generate a false value if: (a) only a single input variable may change at a time, and (b) any number of input variables may change simultaneously.

12.3-1. The sum-of-products excitation equation for y_i' is factored on y_i:

$$y_i' = y_i \cdot g \vee \bar{y}_i \cdot h \vee k$$

where g, h, and k are functions of the input variables and state variables other than y_i.

(a) If y_i' is not to assume false values, what must be true of g, h, and k during state transitions during which (i) y_i is to remain at 0; (ii) y_i is to change from 0 to 1; (iii) y_i is to change from 1 to 0; (iv) y_i is to remain at 1.

(b) If g is free of static hazards, will y_i' be?

(c) Must g, h, and k all be static-hazard–free in order for y_i' to be?

12.3-2. In the excitation equation of Prob. 12.3-1 suppose function $h \neq 0$. When can the term $\bar{y}_i \cdot h$ take the value of 1?

12.3-3. Direct implementations of the following excitation equations are proposed as replacements for the network of Fig. 12.9.

(i) $F' = (A \uparrow B) \uparrow [F \uparrow (B \uparrow (A \uparrow A))]$ 5 gates

(ii) $F' = (A \uparrow B) \uparrow [F \uparrow (B \uparrow (A \uparrow B))]$ 4 gates

(a) Is the open-loop combinational logic of either of these networks free of static hazards if only one input variable is changed at a time?

(b) Is the closed-loop form of either or both of these networks free of steady-state hazards?

(c) What activity can possibly occur in these networks if with $AB = 00$ $F = 1$, and then A and B are simultaneously changed to 1? If with $AB = 11$ and $F = 1$, A and B are simultaneously changed to 0? Is the activity predicted by Fig. 12.9?

12.3-4. The sequential function of Fig. 12.9(a) is to be realized with a NOR-gate network based upon the following excitation function.

$$F' = (A \lor \bar{B})(B \lor F)$$

(a) Is the open-loop combinational logic free of static hazards?

(b) Is the closed-loop asynchronous network free of steady-state hazards?

12.3-5. A network is based upon the following excitation equations. Note the shared gate.

$$y_1' = \bar{x}_1 y_1 \lor \bar{x}_2 y_2 \lor x_1 \bar{x}_2 \quad y_2' = \bar{x}_1 y_1 \lor \bar{y}_1 y_2 \lor x_1 y_2$$

This expression of y_2' is not free of static hazards.

(a) Derive a state-transition table for this network.

(b) Which total-state transitions might not be properly performed by the network?

(c) What gate delay distribution must exist if the network is to fail to properly perform each of these transitions?

(d) Repeat part (c) for the network described by:

$$y_1' = (x_1 \downarrow \bar{y}_1) \lor (x_2 \downarrow \bar{y}_2) \lor x_1 \bar{x}_2$$
$$y_2' = (x_1 \downarrow \bar{y}_1) \lor \bar{y}_1 y_2 \lor x_1 y_2$$

12.3-6. The network of Fig. 12.13 is modified slightly to correspond to the following excitation equations.

$$y_1' = (x \downarrow \bar{y}_2) \lor y_1(x \lor y_2) \quad y_2' = (x \downarrow \bar{y}_2) \lor \bar{y}_1(x \lor y_2)$$

The essential hazard remains in the state table. What distribution of delays must exist for this modified network to malfunction?

12.3-7. What delay distributions cause the following false state transitions in the network of Fig. 12.13?

(a) State S0 follows state S1.

(b) State S1 follows state S2.

(c) State S2 follows state S3.

12.3-8. A network is a direct realization of the following equations.

$$y_1' = x_1 y_1 \lor x_2 y_1 \bar{y}_2 \lor x_1 x_2 \bar{y}_2$$
$$y_2' = x_1 \bar{x}_2 y_1 \lor \bar{y}_1 y_2 \lor x_2 y_2 \lor \bar{x}_1 x_2 \bar{y}_1$$
$$z = y_1 \bar{y}_2$$

(a) Are these equations free of static hazards?

(b) Develop a state-transition table for the network. Does it contain essential hazard(s)? If so, during what total state transition will critical input-variable–state-variable races take place? Identify the input variable and state variable in each case, if any, and determine the gate-delay distributions that must exist for a steady-state hazard to appear.

(c) How does the network accomplish a transition from total state $x_1 x_2 y_1 y_2 = 1011$ to 0000?

12.1-1. (b) $z' = x_1x_2 \lor z \cdot (x_1x_2 \lor \bar{x}_2)$

(c)

	x_1x_2			
z'	00	01	11	10
z 0	0	0	1	0
1	1	0	1	1

(d) This gated-latch sets with $x_1x_2 = 11$ and resets with 01.

12.1-3. $\alpha' = x_1 \uparrow x_2 \qquad \gamma' = \beta \uparrow z$
$\beta' = \alpha \uparrow x_2 \qquad z' = \alpha \uparrow \gamma$

The network is stable in total states:

$$x_1x_2\,\alpha\beta\gamma z = 001101,\ 001110$$
$$011010$$
$$110101$$
$$101101,\ 101110$$

Detailed gate switching can be seen. Possibilities of races are shown by the full ST table. For example, with $x_1x_2 = 00$ or 11, state 1100 may be followed by either stable state or oscillation.

12.1-5. Critical races from $x_1x_2\,y_1y_2 = 1000$ to 1111, 0010 to 0101.

12.2-1. (a)

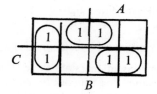 or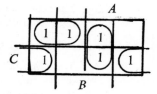

(b) All adjacent cells holding 0 are encircled by:

$$f = (\bar{A} \lor B \lor C)(A \lor \bar{B} \lor \bar{C})$$

Suppose $f = 1$ before and after a variable is changed. Pick A as the changing variable, for example. Then either B or $C = 1$ and the other $= 0$. These values ensure that the OR gates give 1's throughout the transition.

(c) One can be generated from the other without using theorems based upon the complement axiom of Boolean algebra.

12.2-3. Use all prime implicants.

12.2-4. (a) f_1 if the AND gates have unequal delays.

(b) f_2 because of OR gate delay.

12.3-1. (a) (i) h and k must be static–hazard-free.

(ii) $g = k = 1$.

(iii) $g = k = 0$.

(iv) g must be static-hazard–free.

(b) No.

k must be static-hazard–free when $y_i = 0$. g must be static-hazard–

free when $y_i = 1$. h is not useful; $h = 0$ in practical circuits unless an oscillator is desired.

12.3-4. (a) No.   (b) No.   Change AB between 00 and 01.

12.3-7. (a) The delay between gates 7 and 2 must be very large. Gate signals 3, 7, and 1 go up; then 5, 8, 2, and 4 drop.

(b) The delay between x and gate 4 must be very large. When x goes to 1, gate signals 3, 7, 2, 4, and 6 drop, breaking the y_1 loop.

13

Synthesis of Asynchronous Sequential Networks

This chapter presents an overall synthesis algorithm. That the steps of asynchronous and synchronous sequential network design are quite parallel is emphasized in the first section of this chapter. Emphasis in asynchronous design is shifted to eliminating hazards and gaining reliability. A state assignment must be selected that avoids critical races. This selection problem is difficult enough to require the full second section of the chapter. Combinational logic must be free of static and some function hazards. Rather than using massive delays that slow network response, the third section shows how gate delays can be used to advantage in gaining reliable networks. The single-input-change restriction is relaxed in the final section.

13.1 THE SYNTHESIS ALGORITHM

Design procedures for asynchronous and synchronous sequential networks are similar in many respects. Thus the outline of a synthesis procedure for asynchronous networks given below parallels the procedure of Section 5.2. One new step is introduced, and a shift in emphasis will be noted in the state assignment step.

1. Convert the word specification of a desired network to a state-transition table. This first state table is also known as a *primitive flow table*.

2. Detect and eliminate equivalent states. The basic procedure given in Section 5.2 must be generalized because the primitive flow table is incomplete: some state transitions and output symbols may be unspecified.

3. *Merge* the rows of the primitive flow table. This new step can be used to minimize the number of states to ultimately be realized, or endow the final network with other desirable properties. This step may convert an initial Moore model state table into a Mealy model.

4. Encode the state variables. While any state assignment will do for synchronous machines, asynchronous machines must be encoded so as to avoid critical races. This step not only influences the cost of a network, but also the speed with which it operates. Section 13.2 treats this step more fully.

5. Write excitation equations for the state variables and output variable equations from the encoded state-transition table.
   (a) If a gate network is to be constructed, the excitation and output equations must be free of hazards.
   (b) If asynchronous flip-flops are used for the state variables, then flip-flop input equations must be derived as in Section 5.3 and made free of certain hazards.

6. Use available technology to implement the excitation or flip-flop input equations and the output variable equations, but avoid reintroducing hazards. Section 13.3 will treat these last two steps more fully.

A network that satisfies the following specifications will be designed to illustrate these steps.

> Periodic clock signal T is used to synchronize a synchronous digital system. For purposes of testing and maintaining that system a network is needed which generates a single, complete positive portion of T each time switch S is closed. Switch S is free of contact bounce. S and T will not change simultaneously.

Figure 13.1 illustrates a typical input-output sequence for the desired network. Note that only a single pulse is to be emitted even if S is closed for thousands of periods of T, and that a complete pulse is to be emitted even if S is closed in the middle of a T pulse. The numbers that appear in the figure label important combinations of values of S, T, and z.

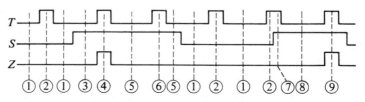

FIG. 13.1 An Input-Output Sequence for the Desired Network

Primitive Flow Table

We begin to obtain a state-transition table model of a desired network by specifying a first stable state. One network condition may appear as a most natural first state to specify; perhaps the network is usually in that state and rarely in others, but this is a matter of individual experience and taste. Here, for example, we might begin by naming as state "1" the network condition when $S, T = (0, 0)$. This input symbol is very common; we know that only one output symbol is to accompany this input symbol. We summarize our specification of a first state in the manner of Fig. 13.2(a). Note also that the network is in state 1 at the extreme left of Fig. 13.1.

Now we assume that the machine is in this first defined state and that the input symbol changes, and ask how the machine is to respond. Suppose ST is changed from 00 to 01 in correspondence with Fig. 13.1. The machine must attain another stable condition; we name it state "2." We specify this transition by placing an uncircled 2 in the first row and the $ST = 01$ column, and a circled 2 in that column and a new, second row. Each new stable state will be allotted a row in the primitive flow table; because of this it is not necessary to have a "present state" column as in all previous state-transition tables. The output symbol that the network is to provide when in state 2 is 0, and this is entered as in Fig. 13.2(b).

If T returns to 0 before S changes, we specify that the machine is to return to state 1; it is not necessary or desirable to have the machine "remember" that T has gone to 1 and returned to 0 while S has remained at 0. The uncircled 1 in the second row of the primitive flow table indicates that state 1 is to follow state 2 under this input-symbol change. It is important that previously specified states be specified as next states whenever it is possible. Development of the primitive flow table will not terminate if new states are introduced *ad infinitum*.

Now let us again assume that the machine is in state 1, but that S changes rather than T. This is an interesting and important input-symbol change, since $ST = 10$ is the first in a series of input symbols that causes the machine to generate a nonzero symbol. We propose another stable state, state 3, and indicate that state 3 is to follow state 1 under this input-symbol change, in the manner shown in Fig. 13.2(c).

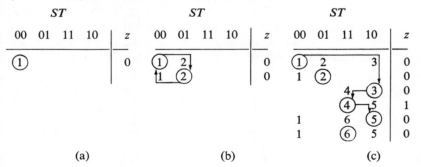

(a)　　　　　　　(b)　　　　　　　(c)

Fig. 13.2 Development of the Primitive Flow Table

To obtain $z = 1$, ST must change from 00 to 10 to 11. The corresponding state sequence includes state 4, as shown in Fig. 13.2. When T returns to 0 a complete pulse has been emitted and the machine must attain state 5 rather than returning to state 3. If it returned to state 3 and $S = 1$ were maintained, then eventually T would again˜ equal 1, and the machine would go to state 4 and emit another pulse. Such activity would not satisfy the given specifications. When the machine is in state 3 it is "remembering" that S has been closed but no pulse has as yet been emitted for this closure. When in state 5 the machine is remembering that a pulse has already been generated for this closure of S, and S must be opened before generation of another pulse is allowed. State 6 in the $ST = 11$ column must also be introduced to prevent the generation of a second pulse. Now we see in Fig. 13.2(c) that we are concerned with a sequential problem: with $ST = 11$, $z = 1$ when the machine is in state 4, and $z = 0$ when it is in state 6.

Several other entries may be placed in the primitive flow table. Again suppose that the machine is in state 1, but that S and T simultaneously take the value of 1. What is the machine to do? The specifications do not say, but indicate that this and all other double-variable, input-symbol changes will never occur. As designers we gain freedom by interpreting this specification to mean that the next state is of no concern should this change actually occur; we indicate the next state as a "don't care" as in Fig. 13.3(a). The reader may feel uneasy about this because it is not difficult to imagine the rare event in which the button is pressed in synchronism with the clock pulse. He is encouraged to place a definite entry in a duplicate table and process his table in parallel with the table of Fig. 13.3. Function hazards must be avoided, of course, if network performance is to be described by the machine he develops.

The number of different input sequences of finite length that can be presented to a machine is finite, though it may be very large. When developing a primitive flow table, all possible input sequences must be considered. In

| ST | | | | |
00	01	11	10	z
①	2	–	3	0
1	②		–	0
	–	4	③	0
		④	5	1
1	–	6	⑤	0
		⑥	5	0

(a)

| ST | | | | |
00	01	11	10	z
①	2	–	3	0
1	②	7	–	0
1	–	4	③	0
–		④	5	1
1	–	6	⑤	0
–		⑥	5	0
–		⑦	8	0
–		9	⑧	0
–		⑨	10	1
–		6	⑩	0

(b)

| ST | | | | |
00	01	11	10	z
①	2	–	3	0
1	②	7	–	0
1	–	4	③	0
–	11	④	5	1
1	–	6	⑤	0
–	2	⑥	·5	0
–	2	⑦	8	0
1	–	9	⑧	0
–	11	⑨	10	1
1	–	6	⑩	0
1	⑪	4	–	1

(c)

FIG. 13.3 Completing the Primitive Flow Table

practice we consider only the subsequences of which longer sequences may be composed. The partially developed flow table helps to identify these subsequences. Every cell of such a table must contain an entry. To complete a partially developed table, we may locate an empty cell and consider what entry must be placed in that cell. Some entries are easily determined; others require consideration of the desired response to a subsequence.

Looking at the third row of the table of Fig. 13.3(a), we might quickly decide that should $S = 1$ and return to 0 before $T = 1$, no pulse will be emitted and the machine will return to state 1. This decision is in violation of the specifications, and again the reader may wish and is encouraged to disagree and develop his own machine. But if T is truly a high-speed electronic clock with a period measured in microseconds or less, it is difficult to imagine depressing a mechanical switch for a fraction of this period.

The blank in the second row of Fig. 13.3(a) requires greater deliberation. The machine attains state 2 with $ST = 01$. If switch S is closed in the middle of a $T = 1$ pulse, the machine must wait and present the next full T pulse. State 7 is introduced to remember that while $ST = 11$ no pulse has been emitted and none is being emitted now: a pulse is to accompany the next occurrence of $T = 1$. When $T = 0$ the machine attains state 8 where it remembers similar facts. State 9 is reached, and the full pulse emitted, when $T = 1$ again. Then when T returns to 0 we may have the machine go to state 10 where it remembers that a full pulse has been emitted. Should $T = 1$ again while $S = 1$, we recognize that state 6 represents the condition of the machine, and no additional states need be introduced. See Fig. 13.3(b).

The blank entry in the fourth row of the flow table corresponds to the switch being opened while the machine is emitting a pulse. The machine must complete that pulse by attaining state 11 and then returning to state 1 when T goes to 0. The other blanks in the flow table may be quickly filled, as shown in Fig. 13.3(c). Again the reader may not agree with all of these entries. In fact, if T is a high-frequency signal and S is controlled by a slow finger, the input-symbol sequence that leads to state 11 may not be considered to be feasible: state 11 will never be entered and the reader may prefer to delete it. Specifications are seldom complete and free of ambiguity. Either the specifications must be clarified or assumptions must be made by the designer.

Compatible States

Don't care state transitions and output symbols may appear in a primitive flow table. Such entries require the generalization of the definition of equivalent states, and procedure for finding them, given in Section 5.2. A very general definition and procedure for identifying and removing equivalent states are beyond our need and interest. We will use a more restricted, simpler definition of state equivalence, which satisfies our immediate needs.

We will say that two stable states in the same column of a flow table are *compatible* if:

1. Identical output symbols are associated with the states, and

2. In every column of the rows of the two stable states
    (a) identical or compatible state names appear, or
    (b) a state name appears opposite a don't care entry, or
    (c) two don't care entries appear.

Note that this definition is recursive. Item 2(a) indicates that two states may be compatible if they are followed by states that have been shown to be compatible. We will take this to mean that if the compatibility of states S_i and S_j depends only on the compatibility of states S_k and S_l, and the compatibility of S_k and S_l depends only on the compatibility of S_i and S_j, then both S_i and S_j are compatible and S_k and S_l are compatible.

State 1 is the only stable entry in the first column of Fig. 13.3(a), and hence can be compatible to no other state. In the second column, states 2 and 11 are not compatible because different output symbols are associated with them. In the third column, state pairs (4, 6), (4, 7), (6, 9), and (7, 9) are not compatible sets for the same reason. But (4, 9) and (6, 7) cannot yet be ruled out as pairs of compatible states. Examination of the states that follow states 4 and 9 reveals identical entries in the first and second columns. In the fourth column state 4 is followed by state 5, while state 9 is followed by state 10. If we can show that states 5 and 10 are compatible, then states 4 and 9 are compatible. Similarly, the compatibility of states 6 and 7 is dependent upon the compatibility of states 5 and 8.

In the fourth column, states 3 and 5 are not compatible because for $ST = 11$ they are followed by states 4 and 6, which have been shown to be incompatible. States 5 and 8 are not compatible because states 6 and 9 are not. Now we also know that states 6 and 7 are not compatible. States 5 and 10 have the same output symbol and identical next-state entries. Hence they are compatible: as a result states 4 and 9 are also compatible. And since states 4 and 9 are compatible, states 3 and 8 are compatible. Thus three pairs of compatible states, (5, 10), (4, 9), and (3, 8) exist in the primitive flow table of Fig. 13.3(c).

The search for compatible states may be performed in a more organized fashion using the notation and techniques of Section 5.2. The state set is first partitioned on the columns of the flow table. All stable entries in a column constitute a block of the partition.

$$\{1\}, \{2, 11\} \{4, 6, 7, 9\}, \{3, 5, 8, 10\} \quad \text{partition on columns}$$

Then these blocks are further divided on the output symbols associated with the states of a block.

{1}, {2}, {11}, {4, 9}, {6, 7}, {3, 5, 8, 10}   partition on output symbol

Further division of the blocks is based upon consideration of next-state entries. A don't care entry does not rule out compatibility, and hence is not justification for dividing a block.

$$\{1\}, \{2\}, \{11\}, \{4, 9\}, \{6, 7\}, \{3, 8\}, \{5, 10\}$$
$$\{1\}, \{2\}, \{11\}, \{4, 9\}, \{6\}, \{7\}, \{3, 8\}, \{5, 10\}$$

There is no need to retain compatible states in a flow table (we will find it desirable to reintroduce compatible states in some cases). To remove redundant rows, we delete all but one of a set of compatible rows for each compatibility set. Any uncircled references to the deleted rows must then be replaced by references to the remaining compatible row. We may delete rows 8, 9, and 10 from Fig. 13.3(c). The reference to state 8 in the seventh row must be replaced by 3, the retained row compatible to row 8. Figure 13.4 shows the table that results when redundant states are eliminated.

Compatible states exist in a primitive flow table because already defined states have not been specified as next states when it is appropriate to do so. Specifically, we argued that state 7 was necessary to ensure that a full-width pulse would be generated should S be closed while $T = 1$. When T went to 0 we argued for state 8, in which the machine is remembering that it has yet to emit a pulse. But the machine is remembering exactly the same thing when in state 3. We could have avoided compatible states in the primitive flow table by recognizing this fact when we were forming that table.

Merging

Merging is a process of replacing two or more rows of a primitive flow table with a single equivalent row containing more than one circled entry. It has the effect of reducing the number of states of a machine. A number

ST

00	01	11	10	z
①	2	–	3	0
1	②	7	–	0
1	–	4	③	0
–	11	④	5	1
1	–	6	⑤	0
–	2	⑥	5	0
–	2	⑦	3	0
1	⑪	4	–	1

FIG. 13.4   Compatible States Have Been Removed

of primitive states, each stable for one input symbol, are replaced with a single state that is stable for all input symbols for which the members of the collection were stable.

Two rows of a flow table may be merged, regardless of the output symbols associated with the rows (really with the stable entries in those rows), if in every column of the two rows

1. Identical state entries (circled or not, and including don't cares) appear, or
2. A state entry appears opposite a don't care entry.

Members of a set of more than two rows may be merged if and only if all pairs of rows of the set satisfy these conditions.

Rows 1 and 2 of Fig. 13.4 may be merged. In the first two columns of these rows we find identical state entries. In the last two columns we find a state name opposing a don't care. Similarly, rows 1 and 3 may be merged. But rows 2 and 3 cannot be merged since different state entries appear in the third columns of these rows. And therefore we cannot merge all three rows into one.

The results of a comparison of all pairs of rows of a primitive flow table are usually summarized in a *merger diagram*. Such a diagram consists of a vertex for each row and connecting lines that reflect possible mergers. The merger diagram of Fig. 13.5 results from an examination of Fig. 13.4.

The merger diagram displays the many sets of mutually mergeable states. We see in Fig. 13.5 that for the example machine, rows 1, 2, and 7 may be merged. Row 11 may be merged with either row 3 or row 4, but all three may not be merged. Row 5 may be merged only with row 6. Some of the partitions of the set of states of the primitive flow table that describe allowed simultaneous mergers are:

$$\{1, 2, 7\}, \{3\}, \{4, 11\}, \{5, 6\}$$
$$\{1, 3\}, \{2, 7\}, \{4, 11\}, \{5, 6\}$$
$$\vdots$$
$$\{1\}, \{2\}, \{3\}, \{4\}, \{5\}, \{6\}, \{7\}, \{11\}$$

This last partition suggests that we may choose not to merge any rows.

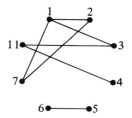

FIG. 13.5 Merger Diagram

The state table that results from the merging procedure has as many rows as there are blocks in the merger partition employed. This tells the number of states that must be encoded in the next step of the synthesis procedure. The number of state variables of an encoding may be reduced if a merger partition with a small number of blocks is chosen. Only two state variables may be required to encode a 3- or 4-row table. If a 5- to 8-block merger partition is employed, 3 or more state variables will be required. Thus using a merger partition with a small number of blocks tends to lead to fewer state variables and a lower cost network.

In some cases it is possible to select a merger partition such that all rows of a block are associated with the same output symbol. The choice of such a partition reduces and in some cases eliminates explicit combinational logic to generate the output variables. In the example it may be undesirable to merge rows 3 and 11 since they have different associated output symbols.

The merger partition employed also influences the speed with which the network responds to input symbol changes. If we elect to merge rows 1 and 3, for example, then no state variables need change when ST changes from 00 to 10. If rows 1 and 3 are not merged, then at least one state variable change must accompany the 00 to 10 change of ST. Network response is necessarily slower.

No algorithmic means of isolating the optimum merger partition is available. We take the above considerations and perhaps others into account, weighing each as we feel to be appropriate, and select some partition. The exhaustive design of networks, based upon each available merger partition, may be employed to find the best merger, of course. For purposes of further examples we will use the four-block partition

$$\{1, 3\}, \{2, 7\}, \{4, 11\}, \{5, 6\}$$

although we have no firm reason to expect that it is substantially better or worse than other available 4-block partitions.

To merge two rows:

1. Associate the output symbol of each row with the circled entry in that row.
2. For each column
   (a) move the circled entry and output symbol of the second row to the first row, and
   (b) replace don't care entries in the first row with the entry in the second row.
3. Delete the second row.

Thus to merge rows 1 and 3 of Fig. 13.4, we alter row 1 to

$$①^p \quad 2 \quad 4 \quad ③^p$$

where the superscripts denote the output symbol associated with the stable state, and then delete the third row of the flow table. Employing the merger partition selected, Fig. 13.4 merges to the table of Fig. 13.6(a). Figure 13.6(b) shows an equivalent state-transition table using a more familiar format. Since a single output symbol is associated with all stable-state entries in each row, a single output column is shown (Moore model). If rows 3 and 11 had been merged, this would not be possible and an output column would be required for each input symbol (Mealy model). In either case, the given output symbols are associated only with stable entries; the output symbols to be associated with the unstable entries remain to be determined.

Merging has eliminated all don't care state transitions in the example. Merging is facilitated and often possible only because don't care entries exist in a primitive flow table. Merging assigns many if not all of these don't cares in an optimum manner under the criteria used to pick the merger partition. Thus we now see that if the machine is in state 1 (a) and S and T change simultaneously to 1, state 1 will be followed by state 4 (c), and a pulse emitted (assuming no function hazard interferes). Assignments of other don't cares can be determined by comparison of Figs. 13.4 and 13.6(a).

The state-transition table of Fig. 13.6(b) could be developed directly from the original specifications much as state tables were developed directly in the previous chapter. But whether it is desirable or advantageous to do so and whether the very same table would be developed are difficult questions to answer. Development of the primitive flow table requires that transitions from all stable conditions for all possible input symbols be considered. Direct development of merged state tables provides an opportunity to overlook some of these transitions. The merger diagram displays the many ways in which a primitive flow table may be merged. Each available merger partition leads to a different state-transition table and network. If the merging process is avoided, the option of introducing desirable properties by selecting a particular merger partition is waived.

$ST =$ 00	01	11	10		Present State	Next State $ST =$ 00	01	11	10	z
(1)°	2	4	(3)°		a	(a)	b	c	(a)	0
1	(2)⁰	(7)⁰	3		b	a	(b)	(b)	a	0
1	(11)¹	(4)¹	5		c	a	(c)	(c)	d	1
1	2	(6)⁰	(5)⁰		d	a	b	(d)	(d)	0
	(a)						(b)			

FIG. 13.6   Merged State-Transition Table

State Assignment

The absolute minimum number of state variables that may be used to realize an r-row state table is:

$$p_{\min} = \lfloor \log_2 r \rfloor \qquad (13.1)$$

where $\lfloor \ \ \rfloor$ denotes "integer greater than or equal to" $\log_2 r$. Often $p > p_{\min}$ state variables must be used in the realization of asynchronous machines to avoid critical races. Determination of the value of p is not always an easy matter.

Four rows appear in Fig. 13.6(b), and two or more state variables must therefore be used. Figure 13.7(a) shows an abbreviated state diagram for this machine. All transitions are shown but their direction and labels are omitted. We see from this diagram that transitions between states a and b are specified in the state table. If a critical race is to be avoided, states a and b must be assigned adjacent p-tuples. Similarly, states a and c, a and d, c and d, and b and d must be assigned adjacent code words.

It is not possible to assign 2-tuples to the four states so that all of these requirements are satisfied. If the assignment for state a is adjacent to that of c and d, then the codes for states c and d will not be adjacent. Triangular patterns appear in Fig. 13.7(a). We can not encode the vertices of a triangle or odd-sided polygon so as to satisfy Hamming distance requirements. In general, the state diagram and table must be altered so that triangles do not appear. Such alterations usually result in a need for more than the absolute minimum number of state variables, as we will find in the next section.

In a *normal state table* all unstable conditions lead directly to stable states. The state table of Fig. 13.6 is a normal state table: no uncircled entry points to another uncircled entry. Thus state variables need not change in sequence to accomplish a state transition, and normal state tables lead to networks that respond most rapidly to input-symbol changes.

We will treat this point, and the state assignment problem in general, more fully in the next section. For now, we can and will encode the example machine with two state variables and still avoid critical races. Only a's appear in the first column of the state table. Thus a race when $ST = 00$ cannot be critical, and any state assignment will be satisfactory as far as the first column is concerned. We must avoid an assignment that would cause the machine to cycle endlessly between unstable states, of

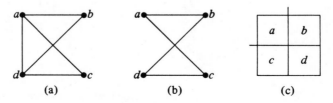

(a)   (b)   (c)

FIG. 13.7 Transition Diagrams

course. If the transitions specified in the first column are ignored, the transition diagram reduces to Fig. 13.7(b), which can be encoded with 2-tuples. Any of the eight possible labelings of Fig. 13.7(c) will produce an acceptable code, and each such code will lead to a slightly different network. We have no efficient means of determining which of these encodings leads to the minimum cost network; we can always resort to exhaustive search if network cost must really be minimized.

For purposes of example we will arbitrarily choose the following straightforward assignment.

State	y_1	y_2
a	0	0
b	0	1
c	1	0
d	1	1

Each of the state names of Fig. 13.6(b) may now be replaced with its assigned 2-tuple.

Additional attention may be directed to the first column of the encoded state-transition table. All entries in that column will be 00, the encoding for state a. A noncritical race would then accomplish the transition from state d to state a. If this race is felt to be undesirable, it may be avoided by making the state table *abnormal*. The transition from state d to state a may be performed indirectly by passing through state b. Fig. 13.8(a)

Present state		Next state ($y_1'y_2'$) ST				z
y_1	y_2	00	01	11	10	
a 0	0	(00)	01	10	(00)	0
b 0	1	00	(01)	(01)	00	0
d 1	1	01	01	(11)	(11)	0
c 1	0	00	(10)	(10)	11	1

(a)

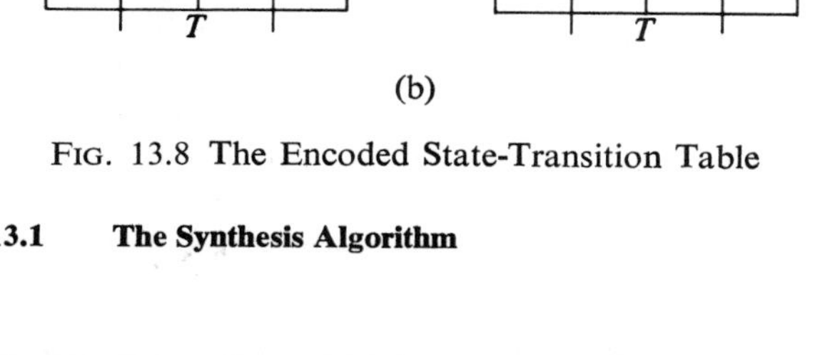

(b)

Fig. 13.8 The Encoded State-Transition Table

illustrates the encoding of this column that will cause the machine state changes $\circledd \rightarrow d \rightarrow b \rightarrow a \rightarrow \circled a$ when ST changes from 10 to 00. With this encoding we can be sure that the network will require more time to complete the transition from state d to state a than to complete any other transition.

Excitation Equations

Excitation equations for the state variables may be written directly from the encoded state-transition table if the number of variables is modest and the rows and columns are ordered as in a Karnaugh map. But separate Karnaugh maps for y_1' and y_2' are shown in Fig. 13.8(b) to facilitate writing the following static hazard-free state-transition equations.

$$y_1' = Sy_1 \vee Ty_1 \bar{y}_2 \vee ST\bar{y}_2$$
$$y_2' = S\bar{T}y_1 \vee y_1 y_2 \vee Ty_2 \vee \bar{S}T\bar{y}_1$$

These equations may be implemented directly or manipulated to obtain a lower cost or otherwise more desirable network. The y_1 and y_1', y_2 and y_2' terminals may be directly connected or separated by a delay element if hazards prove to be a problem. See Fig. 13.10(a).

Asynchronous NAND-latches or NOR-latches may be used to store state variables in much the same fashion as flip-flops were used in synchronous networks. The NOR-latch of Fig. 4.3 acts in the manner of an RSFF. The output variables of the two NOR gates are complements except during switching and when both R and S equal 1. Transient false 1's (0's) on an input line of NOR (NAND)-latches may cause them to assume the wrong state. Thus latch input equations must be free of only 0-1-0 (1-0-1) transient hazards.

Latch input equations are derived from an encoded state-transition table with the procedure of Section 5.3. Transformation of the maps of Fig. 13.8(b) via that procedure leads to the following input equations.

$$S_1 = ST\bar{y}_2 \qquad S_2 = \bar{S}T\bar{y}_1 \vee ST y_1$$
$$R_1 = \bar{S}\bar{T} \vee \bar{S}y_2 \qquad R_2 = T\bar{y}_1$$

If NAND-latches are to be used, the complements of these variables must be realized.

Output Equations

Output symbols are associated only with the stable entries in the merged state-transition table. The output symbol to be generated when the machine is in an unstable condition remains to be determined. If the original and final stable states of a state transition provide the same output symbol, then temporary false values on the output lines are avoided by associating the same output symbol with intermediate unstable entries in the state-transition table. But if different output symbols are associated with the original and final stable states, then the symbol of either state may be associated with intermediate unstable states. If the output symbol of the original stable state is assigned to the intermediate state, change of the output symbol is delayed until the final stable state is reached.

Figure 13.9(a) shows the known output symbols for the example machine. Figure 13.9(b) shows the output-symbol assignment that results in the widest possible output pulses. Figure 13.9(c) shows the assignment that results in minimum output logic.

$$z = y_1 \bar{y}_2$$

This assignment causes the output symbol $z = 1$ to be maintained while the machine is in the process of leaving state c ($y_1 y_2 = 10$).

Implementation

The output equations, and either the excitation equations or the latch input equations, must be realized with available logic gates. Thus we face a multiple-output combinational logic design problem and all the pertinent design techniques of previous chapters as well as those to be presented in Section 13.3 may be employed. The merger partition and state encoding previously selected both influence these equations directly, and hence the network we generate. Unfortunately, it is not generally possible to predict the nature of the influence of merging and encoding without designing the network.

Figure 13.10 shows two networks that may be drawn directly from the equations derived above. Further manipulation of those equations might well lead to faster or lower-cost networks. Analysis of these networks for hazards is also in order. For example, the merger partition used introduces essential hazards (third and fourth columns of Figs. 13.6 or 13.8). Thus it may be necessary to insert delay elements in the networks of Fig. 13.10, or redesign the networks, perhaps using a different merger partition. A search for alternative networks may also be motivated by economic or speed considerations. Both of the networks of Fig. 13.10 are rather expensive, and also slow, since signals must propagate through many gates before they

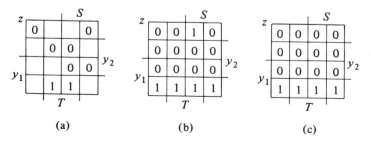

FIG. 13.9   Output Symbol Logic

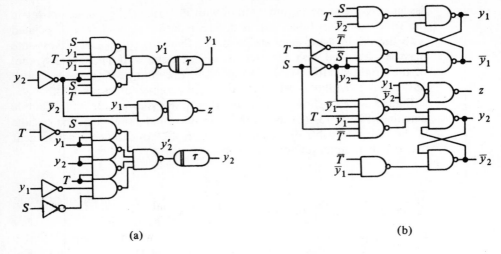

(a)   (b)

FIG. 13.10   A Gate and a Latch Realization
of the Example Machine

affect the output signal. If the delay introduced by these gates is of approximately the same order of magnitude as the positive portion of T, then a detailed analysis of the network is also most appropriate. It may reveal that the network substantially shortens or lengthens the T pulse that it presents at its output terminal.

13.2   STATE ASSIGNMENT TECHNIQUES

The determination of an encoding of the states of an asynchronous machine that avoids critical races is one of the more important and difficult steps of asynchronous network design. A number of methods that always produce a critical-race–free assignment have been developed. These approaches lead to networks with different costs, potential operating speeds, and other properties. Therefore, in a specific design situation one or another of these general methods may be quickly adopted or summarily rejected because of the network properties they induce. And in a specific situation the ingenious application of elements of these approaches can often lead to the network that is considered to be optimum. We saw an example of this in the previous section, where recognition of the unique nature of one column of a state table made it possible to use two rather than three state variables.

These various approaches will be illustrated for the state table of Fig. 13.11, which can not be encoded in a critical-race–free manner with two state variables. The triangle that appears in the state diagram can not be avoided in the manner illustrated in the previous section. If more than two, say p, state variables are used to encode the table, then a state table with

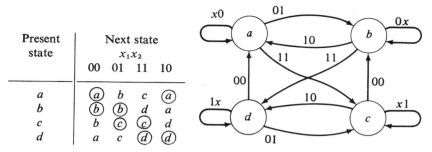

Present state	Next state $x_1 x_2$			
	00	01	11	10
a	ⓐ	b	c	ⓐ
b	ⓑ	ⓑ	d	a
c	b	ⓒ	ⓒ	d
d	a	c	ⓓ	ⓓ

FIG. 13.11   A State Table to be Encoded

2^p rows, which does not contradict the original state table, must be developed. The added states must be specified so as to satisfy the given state transitions (and output-variable specifications, which are not presented here in order to reduce the amount of detail to be considered), and avoid critical races and other undesirable network activity such as remaining indefinitely in one of the added states or cycling endlessly between unstable conditions.

Map to the p-Cube

A state table may be encoded with p state variables if the corresponding state diagram may be drawn as a p-dimensional unit cube or a part thereof. Often it is relatively easy to alter a state diagram so that this is possible. Triangles (pentagons, etc.) that appear are converted to rectangles (hexagons, etc.) by introducing an unstable state in one leg of each triangle. The diagram of Fig. 13.11 may be transformed to a 3-cube by adding additional states in a number of ways. Unstable states must be placed in a number of the original transition paths. Figure 13.12 illustrates one way and suggests a variety of other ways in which the state diagram may be altered to the topology of a 3-cube. Unstable states e, f, g, and h have been introduced. If the states are encoded to satisfy the Hamming distance relationships of this figure, critical races are avoided.

Transitions between states a and b (c and d) will involve only one state-variable change, and hence will be accomplished in minimum time, because states a and b (c and d) are associated with adjacent vertices of the 3-cube in Fig. 13.12. Transitions from state d to state a (c to b) are accomplished by passing through unstable state e (f). Two successive state-variable changes are required and as a result, these transitions will take a longer time than the direct transitions. Finally, a to c (b to d) transitions are accomplished by three successive state-variable changes, ⓐ$\rightarrow g \rightarrow h \rightarrow$ⓒ, and will be the slowest transitions. These transitions thus determine the rate at which the input symbol may be allowed to change.

Any one of the 48 possible labelings of the 3-cube that satisfied all Hamming distance requirements provides a state assignment free of critical races. One code is shown in Fig. 13.12; the others can be derived from it by complementing any one or more of the given columns, permuting any two or more of the columns, or combinations of complementing and permuting the columns. Three 2-block partitions of the state set describe all of these 48 codes.

$$\pi_1 = \overline{\{a, d, e, g; b, c, f, h\}}$$
$$\pi_2 = \overline{\{a, b, g, h; c, d, e, f\}}$$
$$\pi_3 = \overline{\{a, b, e, f; c, d, g, h\}}$$

In this abbreviated notation the bars (and semicolons) do not refer to logic complementation, rather they delimit blocks of a partition. In some column of any of these 48 codes, states a, d, e, and g will be assigned the same value, either 0 or 1; states b, c, f, and h will be assigned the other value in that column. In another column the state variable will be assigned the value of 0 (1) for states a, b, g, and h, and 1 (0) for states c, d, e, and f. Partition π_3 indicates that the remaining state variable will be assigned one value for states a, b, e, and f, and the other value for states c, d, g, and h. Each block of these partitions also lists the states assigned to the vertices of a face of the 3-cube; see Fig. 13.12. Since states a and b appear in different blocks of π_1 and in the same blocks of π_2 and π_3, the code words assigned to these states will be adjacent. This is also true of many other pairs of states. States a and d appear in the same block of π_1 and in different

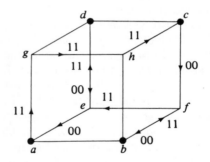

Present state	Next state x_1x_2				One possible assignment		
	00	01	11	10	y_1	y_2	y_3
a	$ⓐ$	b	g	$ⓐ$	0	0	0
b	$ⓑ$	$ⓑ$	f	a	1	0	0
c	f	$ⓒ$	$ⓒ$	d	1	1	1
d	e	c	$ⓓ$	$ⓓ$	0	1	1
e	a	–	d	–	0	1	0
f	b	–	e	–	1	1	0
g	–	–	h	–	0	0	1
h	–	–	c	–	1	0	1

Fig. 13.12   One Mapping of a State Diagram to a 3-cube

blocks of π_2 and π_3. Thus the code words assigned to these two states will be separated by a Hamming distance of 2. Continuing in this manner reveals the complete structure of the labeled 3-cube of Fig. 13.12.

With larger state tables it may not be practical to draw a complete p-cube and associate states with vertices. Additional rows can be appended to the state table and a tentative state assignment made. In terms of this assignment, critical races can be located and eliminated by altering the table to call for successive transitions through the added rows of the table. It may be necessary to alter the tentative assignment to accomplish this. For example, if states a, b, c, and d are tentatively assigned vertices of a 2-cube, then difficulty is encountered when we attempt to specify the a to c and b to d transitions, since they appear in the same column of the state table. We are forced to change our tentative assignment, if we propose to use only three state variables.

The don't cares that appear in the state table of Fig. 13.12 may be used to advantage to minimize network cost.

One-Hot Codes

While the previous technique for finding a state assignment allows one to be ingenious, it can lead to networks with varying transition times, as in the example, and finding an acceptable use of unstable states can be time consuming. A straightforward assignment that equalizes but does not minimize transition times can always be made. But such a state assignment utilizes one state variable for each row of the given state table. For the example state table, four state variables must be used.

To illustrate this type of state assignment, we will assign a 1-hot code word to each state.

State	1-hot code
a	1000
b	0100
c	0010
d	0001

Now every transition between these states can be accomplished in a race-free manner by passing through the unstable state with an assignment that is adjacent to the code of both the original and final stable states. The transition from state a to b is accomplished by passing through state 1100.

$$\textcircled{a} \rightarrow e \rightarrow \textcircled{b}$$
$$1000 \rightarrow 1100 \rightarrow 0100$$

The encoded state table for the example machine is then as given in Fig. 13.13.

Present	Next state			
state	00	01	11	10
a (1000)	Ⓐ	e	f	Ⓐ
b (0100)	Ⓑ	Ⓑ	g	e
c (0010)	h	Ⓒ	Ⓒ	i
d (0001)	j	i	Ⓓ	Ⓓ
e (1100)	–	b	–	a
f (1010)	–	–	c	–
g (0101)	–	–	d	–
h (0110)	b	–	–	–
i (0011)	–	c	–	d
j (1001)	a	–	–	–
all others	–	–	–	–

FIG. 13.13　A 1-hot State Assignment

Other codes of this type can be derived, again by complementing and permuting columns of the given state-assignment table. The state set partitions of these codes are particularly simple to generate and interpret.

$$\pi_1 = \overline{\{a, e, f, j;} \quad \overline{b, c, d, g, h, i\}}$$
$$\pi_2 = \overline{\{b, e, g, h;} \quad \overline{a, c, d, f, i, j\}}$$
$$\pi_3 = \overline{\{c, f, h, i;} \quad \overline{a, b, d, e, g, j\}}$$
$$\pi_4 = \overline{\{d, g, i, j;} \quad \overline{a, b, c, e, f, h\}}$$

One of the original states appears in the first block of each partition; the other three appear in the second block. The remaining states of the first block are those unstable states that are adjacent to the stable state of the first block.

While this type of state assignment is easily made, it does require a large number of state variables, and leads to networks with equal, but greater than minimum, state-transition times.

Equivalent States

A state diagram may also be altered to form a p-cube by adding stable states that are equivalent[1] to original states. The added states may be introduced in a variety of ways so as to avoid critical races, but two basic methods can be used to assign any state table. Figure 13.14 offers examples of these methods, and the corresponding expanded state tables for the example machine. States with upper case names are equivalent to their lower case counterparts in both examples.

---

[1] Equivalent in the sense that it will not be possible to distinguish between them by performing input-output tests on the machine.

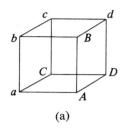
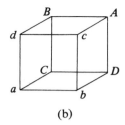

| (a) | | | | | (b) | | | |

Present state	Next state					Present state	Next state			
	00	01	11	10			00	01	11	10
a	ⓐ	b	C	ⓐ		a	ⓐ	b	C	ⓐ
b	ⓑ	ⓑ	B	a		b	ⓑ	ⓑ	D	a
c	b	ⓒ	ⓒ	d		c	b	ⓒ	ⓒ	d
d	D	c	ⓓ	ⓓ		d	a	c	ⓓ	ⓓ
A	Ⓐ	B	a	Ⓐ		A	Ⓐ	B	c	Ⓐ
B	Ⓑ	Ⓑ	d	A		B	Ⓑ	Ⓑ	d	A
C	c	Ⓒ	Ⓒ	D		C	B	Ⓒ	Ⓒ	D
D	A	C	Ⓓ	Ⓓ		D	A	c	Ⓓ	Ⓓ

| (c) | | | | | (d) | | | |

FIG. 13.14 Use of Equivalent States to Avoid
Critical Races

In the first of these methods, a set of equivalent states replaces each original state. Each member of a set of equivalent states is assigned a p-tuple that is adjacent to the encoding of one or more other members of the set. And the encoding of at least one member of each set must be adjacent to that of a member of another set for all other sets. In Fig. 13.14(a, c) we see that states a and A have been assigned adjacent code words, and that state a is adjacent to b and C while state A is adjacent to state D. Thus the requirements are satisfied for set $\{a, A\}$.

Because each member of a set of equivalent states is adjacent to some other member, transitions between *members* of a set may be free of races. And since some member of each set is adjacent to a member of any other selected set, transitions between *sets* may be free of races. Suppose the example machine is in state d of Fig. 13.14(c). A transition to state c requires only a single state-variable change. But the state d to state a transition of the original state table is accomplished by first attaining state D and then state A. Since state A can not be distinguished from state a, it is not necessary to continue to state a. States B and D may be reached directly from state A. But the state a to c transition of the original machine is now accomplished by first attaining state a and then state C.

$$ⓓ→ D →Ⓐ→ a →Ⓒ$$

The state-set partitions dictated by the cube of Fig. 13.14(a) are:

$$\pi_1 = \overline{\{a, A, b, B;} \ \overline{c, C, d, D\}}$$
$$\pi_2 = \overline{\{a, A, C, D;} \ \overline{b, B, c, d\}}$$
$$\pi_3 = \overline{\{a, b, c, C;} \ \overline{A, B, d, D\}}$$

States to be given adjacent assignments must appear in the same block in all but one partition. Thus a and A appear in the same block in π_1 and π_2, and are in different blocks of π_3. States a and b appear together in π_1 and π_3 and hence will be assigned adjacent code words in any state assignment based upon these partitions.

Other partitions than those above may also satisfy these requirements. For example, the sets of equivalent states need not all be of the same size and partitions

$$\pi_1 = \overline{\{a, \alpha, A, c;} \ \overline{b, \gamma, C, d\}}$$
$$\pi_2 = \overline{\{a, \alpha, b, d;} \ \overline{A, c, C, \gamma\}}$$
$$\pi_3 = \overline{\{a, A, b, C;} \ \overline{\alpha, c, \gamma, d\}}$$

based upon triplicating states a and c and having no states equivalent to b and d, also dictate critical-race–free state assignments for any state table with four rows.

In general,

$$p = 2 \lfloor \log_2 r \rfloor - 1 \qquad (13.2)$$

state variables must be used to encode a state table with r rows. Thus this type of assignment requires nearly twice the minimum number of state variables. While transition times are not equal, some control is exerted by the choice of equivalent state sets, and transition times may be less than those obtained by inserting unstable states to map the state diagram to the p-cube.

Figure 13.14(b) illustrates the second general method for utilizing equivalent states to achieve a critical-race-free state assignment. This method requires

$$p = 2^{\lfloor \log_2 r \rfloor} - 1 \qquad (13.3)$$

state variables, in general. In Fig. 13.14(b) we find that state a is adjacent to a member of every other set of equivalent states. Thus transitions from state a to states b, C, or d are accomplished with a single state variable change. This is generally true: state b is adjacent to states a, c, and D, etc. To be able to assign states in such a manner, the dimension of the code words used must be equal or exceed the number of rows less one, $p \geq r - 1$. If r is a power of two, then Eq. (13.3) indicates that the number of state variables required is one less than the number of rows in the given state table.

This type of state assignment leads to networks with the fastest possible response time. Only one state-variable change accomplished every transition of a normal state table. Thus such state assignments are known as *single transition time* (STT) encodings.

The state set partitions of Fig. 13.14(b) are

$$\pi_1 = \overline{\{a, b, c, d;\ A, B, C, D\}}$$
$$\pi_2 = \overline{\{a, b, C, D;\ A, B, c, d\}}$$
$$\pi_3 = \overline{\{a, B, C, d;\ A, b, c, D\}}$$

Here we find that one member of each set of equivalent states appears in each block, and that the members of an equivalence set never appear together in a block. In the more general case where equivalence sets are larger, every member of an equivalence set must be at distance 3 from some other member of that set. Thus two equivalent states must appear together in all but three partitions. If state b is adjacent to state a, and c is adjacent to b, then c must also be adjacent to some state A, which is equivalent to state a; states a and A are distance 3 apart.

While the idea of an STT state assignment is very attractive, the number of state variables that this method of obtaining an STT assignment requires is very large for more than 4-row state tables. Before we examine other

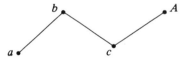

ways of obtaining STT assignments that lead to more acceptable numbers of state variables, we must formalize the manipulation of state set partitions.

Partition Algebra

Let π_i and π_j be two partitions on some set of elements S.

$$\pi_i = \{B_{i1};\ B_{i2};\ \ldots;\ B_{is}\}$$
$$\pi_j = \{B_{j1};\ B_{j2};\ \ldots;\ B_{jt}\}$$

We may think of S as the set of states of a machine. Then the blocks of a partition, B_{kl}, are disjoint subsets of the set of states. If the set union of all blocks of a partition is not S, then the partition is said to be *incomplete*.

The *product* of two partitions is the partition

$$\pi_k = \pi_i \cdot \pi_j = \{B_{i1} \cap B_{j1};\ B_{i1} \cap B_{j2};\ \ldots;\ B_{is} \cap B_{jt}\} \qquad \textbf{(13.4)}$$

where $\cap$ denotes set intersection and empty intersections are not recorded as blocks of π_k. If either π_i or π_j or both are incomplete partitions, then π_k will also be incomplete. The element(s) absent from π_i or π_j will not appear in π_k.

Using the last set of state-assignment partitions given above

$$\pi_1 \cdot \pi_2 = \{\overline{a, b, c, d};\ \overline{A, B, C, D}\} \cdot \{\overline{a, b, C, D};\ \overline{A, B, c, d}\}$$
$$= \{\overline{a, b};\ \overline{c, d};\ \overline{C, D};\ \overline{A, B}\}, \text{ and}$$
$$\pi_3 \cdot (\pi_1 \cdot \pi_2) = \{\bar{a};\ \bar{b};\ \bar{c};\ \bar{d};\ \bar{A};\ \bar{B};\ \bar{C};\ \bar{D}\}$$

The partition in which each element of S appears alone as a block is the finest possible division of S. It serves as the "0" of partition algebra and will therefore be denoted π^0. In the example above, the product of the three state-assignment partitions is π^0. The single block partition with all elements of S in one block is the "1" of partition algebra and is denoted π^1.

If a set of partitions is to specify a unique code word for each state, the product of those partitions must be π^0. Suppose that the product of three 2-block partitions is not π^0, but the following partition.

$$\pi_1 \cdot \pi_2 \cdot \pi_3 = \{\bar{a};\ \bar{b};\ \bar{c};\ \bar{d};\ \bar{A};\ \bar{B};\ \overline{C, D}\}$$

Then states C and D must appear together in a block of π_1 and their encodings will have the same value of y_1. They must also appear together in π_2 and π_3, and hence their encodings will have identical values of y_2 and y_3. These two states are given the same encoding.

Partition π_i *covers* partition π_j, $\pi_j \sqsubseteq \pi_i$, if every block of π_j is a subset of some block of π_i. Every block of $\pi_i \cdot \pi_j$ must be a subset of some block of π_i and π_j, since $B_{ik} \cap B_{jl} \subseteq B_{ik}, B_{jl}$. We can be sure that

$$\pi_i \cdot \pi_j \sqsubseteq \pi_i \quad \text{and} \quad \pi_i \cdot \pi_j \sqsubseteq \pi_j.$$

Partition π_j is *less than* π_i, $\pi_j \leq \pi_i$, and π_i is *greater than* π_j if and only if every block of π_j is a subset of a unique block of π_i. Note the difference between this definition and that of partition cover. Partition π^1 covers every partition of the same set of elements, but it is not greater than any of them. Partition π^0 is covered by all other partitions of the same set of elements, but it is not less than any of these other partitions. Given the partitions

$$\pi_1 = \{\overline{a, d};\ \overline{b, e};\ \bar{c}\}$$
$$\pi_2 = \{\overline{a, c};\ \overline{b, e}\}$$
$$\pi_3 = \{\overline{a, c, d};\ \overline{b, e}\}$$

we can say that π_3 covers both π_1 and π_2, but π_3 is greater than π_2 only. Blocks $\{a, d\}$ and $\{c\}$ of π_1 are not subsets of different blocks of π_3.

The *sum* of partitions π_i and π_j, $\pi_i + \pi_j$, is the partition with the greatest number of blocks such that $\pi_i \sqsubseteq \pi_i + \pi_j$ and $\pi_j \sqsubseteq \pi_i + \pi_j$. The blocks of $\pi_i + \pi_j$ may be constructed inductively. Suppose block B_{i1} of partition π_i and block B_{j1} of π_j have members in common, i.e., $B_{i1} \cap B_{j1} \neq \phi$. Then B_{i1}, B_{j1}, and $B_1 = B_{i1} \cup B_{j1}$ must all be subsets of the block of $\pi_i + \pi_j$ that we seek, and B_1 may be thought of as a first approximation to the unknown block of $\pi_i + \pi_j$. Any other block of either π_i or π_j, say B_k, that intersects B_1 must also be covered by this block of $\pi_i + \pi_j$. Thus we form a second approximation to this block of $\pi_i + \pi_j$, $B_2 = B_1 \cup B_k$, and continue in the same manner until no additional blocks of π_i and π_j that intersect the final approximation of the block of $\pi_i + \pi_j$ can be found. This final approximation is then the block we seek. Other blocks of $\pi_i + \pi_j$ are found in the same manner by examining the remaining blocks of π_i and π_j, if any.

Using the three partitions immediately above, $\pi_1 + \pi_2$ is formed as follows. Since block $\{a, d\}$ of π_1 intersects block $\{a, c\}$ of π_2 we form $\{a, c, d\}$. The third block of π_1 intersects this block, and is covered by it. So $\{a, c, d\}$ is one block of the sum. Block $\{b, e\}$ of π_1 intersects $\{b, e\}$ of π_2, and $\{b, e\}$ is a block of the sum. All blocks of π_1 and π_2 are covered by one or the other of these two blocks of $\pi_3 = \pi_1 + \pi_2$.

To this point, state assignments have been summarized by a set of 2-block partitions with each partition dictating the assignment of one state variable. Any state assignment can be described by a set of 2-block partitions, but partitions with more than two blocks can also be used to express state assignments. If partition π_i has $|\pi_i|$ blocks, then it dictates the assignment of $\lfloor \log_2 |\pi_i| \rfloor$ state variables. The assignment of state variables y_i and y_j

State	$y_i\, y_j$
a	0 0
b	0 1
c	0 1
d	1 1

can be described by two 2-block partitions $\{a, b, c; \overline{d}\}$ and $\{\overline{a}; b, c, \overline{d}\}$, or one 3-block partition $\{\overline{a}; b, c; \overline{d}\}$. Note that this third partition is the product of the first two. Conversely $\lfloor \log_2 |\pi_i| \rfloor$ 2-block partitions can be formed from the $|\pi_i|$ blocks of partition π_i by combining them in different ways so that the product of the 2-block partitions formed is π_i. Thus partition $\{\overline{a, d}; b, c\}$ may also be formed from the 3-block partition above and used in conjunction with either of the 2-block partitions to specify a state assignment. The 3-block partition summarizes all the assignments that can be described by pairs of 2-block partitions that cover it, and thus is more general than any one pair of 2-block partitions.

STT State Assignments

J. H. Tracey [16] developed several methods for obtaining STT state assignments that require most reasonable numbers of state variables. Noncritical races between state variables are allowed. A first method requires that a *column partition* be constructed from each column of a given state-transition table. Each block of such a partition consists of those present states that have the same next-state entries (circled or not) in the selected column. Thus the first next-state column of Fig. 13.11 yields the partition

$$\pi_1 = \{\overline{a, d};\ \overline{b, c}\}$$

The remaining columns respectively yield the following partitions.

$$\pi_2 = \{\overline{a, b};\ \overline{c, d}\}$$
$$\pi_3 = \{\overline{a, c};\ \overline{b, d}\}$$
$$\pi_4 = \{\overline{a, b};\ \overline{c, d}\}$$

Since $\pi_4 = \pi_2$, we have only three unique partitions, but their product is π^0 so they can produce a state assignment. (Since $\pi_1 \cdot \pi_2 = \pi^0$, π_1 and π_2 alone can produce a unique encoding for each state, but not one free of critical races.) One state assignment based upon these three partitions is shown in Fig. 13.15.

Given this encoded state table, suppose the input symbol is changed from 00 to 01 with the machine initially in state a. Then a transition from state a to state b is to be executed. A race between y_1 and y_3 is specified in Fig. 13.15. If the added states are properly encoded, this race will not be critical; the possible intermediate states 001 and 100 may be specified so as to carry the machine to state b (101) by leaving y_2 unexcited. Partition π_2 is based upon the column of the transition, column 2, and ensures that states a and b are encoded with the same value of y_2. Thus it will not be possible for the machine to attain either state c or d as long as y_2 remains quiescent. The input symbol 01 together with $y_2 = 0$ fix the next state of the machine as b regardless of the values of y_1 and y_3.

All the other possible transitions in the encoded state table of Fig. 13.15 are also performed by a noncritical racing of two state variables, but are determined by the state variable that does not change, together with the input symbol. The manner in which the partitions that lead to the state

Present state $y_1 y_2 y_3$	Next state $(y_1' y_2' y_3')$ $x_1 x_2$			
	00	01	11	10
a 000	ⓐ 000	b 101	c 110	ⓐ 000
b 101	ⓑ 101	ⓑ 101	d 011	a 000
c 110	b 101	ⓒ 110	ⓒ 110	d 011
d 011	a 000	c 110	ⓓ 011	ⓓ 011
001	000	101	011	000
010	000	110	110	011
100	101	101	110	000
111	101	110	011	011

FIG. 13.15   A Column-Partition State Assignment

assignment of Fig. 13.15 were constructed ensures that one or more state variables will be quiescent, with a unique value for every transition. While two state variables change with each transition, they do so simultaneously and this is therefore an STT state assignment.

Other aspects of this procedure for obtaining an STT state assignment are illustrated by the state table of Fig. 13.16. This table is incomplete in that don't care next state entries appear. As a result some of the column partitions are incomplete.

$$\pi_1 = \{\overline{a, b, c}; \overline{d, e}\}$$

$$\pi_2 = \{\overline{a, d}; \overline{b, e}; \overline{c}\}$$

$$\pi_3 = \{\overline{a, c}; \overline{b, e}\}$$

$$\pi_4 = \{\overline{b, c}; \overline{d, e}\}$$

Because π_3 and π_4 do not partition the complete state set, a state assignment based upon these partitions will not be complete. For example, π_4 does not specify the value of a state variable in the encoding of state a. Either 0 or 1 may be used to complete the assignment, or state a may be duplicated by using both 0 and 1.

Present state	Next state			
	00	01	11	10
a	ⓐ	d	ⓐ	–
b	a	ⓑ	e	c
c	a	ⓒ	a	ⓒ
d	ⓓ	ⓓ	–	e
e	d	b	ⓔ	ⓔ

FIG. 13.16   An Incomplete State-Transition Table

The following state assignment is based directly on the four-column partitions of the table of Fig. 13.16.

State	π_1 π_2 π_3 π_4 $y_1\overline{y_2}\overline{y_3}y_4y_5$
a	0 0 0 0 –
b	0 0 1 1 0
c	0 1 1 0 0
d	1 0 0 – 1
e	1 0 1 1 1

If the don't care in the assignment for state a is replaced by 0, then $y_5 = y_1$, and y_5 is redundant. Partition π_4, which determines the assignment of y_5, provides a variable that has the same value for states b and c and the other value for states d and e, thus ensuring that transitions in the fourth column will be performed without critical races. But π_1 fulfills this requirement through state variable y_1. Since $\pi_4 \leq \pi_1$, π_4 contributes nothing to a state assignment. In general we need not include in the set of partitions upon which a state assignment is to be based a partition that is less than some other member of that set.

New partitions may be formed from those given by computing the sum and products of the given partitions. From the column partitions for Fig. 13.16.

$$\pi_1 + \pi_2 = \pi^1 \qquad\qquad \pi_1 \cdot \pi_2 = \pi^0$$

$$\pi_1 + \pi_3 = \pi^1 \qquad\qquad \pi_1 \cdot \pi_3 = \{\overline{a, c};\ \overline{b};\ \overline{e}\}$$

$$\pi_5 = \pi_2 + \pi_3 = \{\overline{a, c, d};\ \overline{b, e}\} \qquad\qquad \pi_2 \cdot \pi_3 = \{\overline{a};\ \overline{d};\ \overline{b, e}\}$$

Column-partition products are not usually of value when making STT assignments. In $\pi_1 \cdot \pi_3$, state b appears in a different block than states a and c. Hence $\pi_1 \cdot \pi_3$ can not be used in place of π_1 to ensure that the transitions of column 1 of the state table will be performed without critical races. Neither can the product $\pi_1 \cdot \pi_3$ replace π_3.

Partition sums are often useful. Only one nontrivial sum, π_5, appears above. While π_5 covers both π_2 and π_3, it is greater than only π_3, $\pi_3 \leq \pi_5$. Since π_5 will ensure transitions between states a and c and between b and e it can replace π_3. But π_5 is not able to ensure that state a to state d transitions will not end in state c, and therefore π_5 can not replace π_2.

Any set of partitions such that each column partition is less than or equal to some member of that set can be used to form an STT assignment. If the product of all partitions of such a set is not π^0, then mergeable states exist. These states may be merged or an artificial partition introduced, so that each state has a unique encoding. For Fig. 13.16, we might use partitions π_1, π_2, and π_5 as such a set to arrive at a four-state-variable STT assignment. Thus we see the value of searching for such a set: it may require fewer state variables than the set of column partitions.

While writing column partitions, forming partition sums, and selecting a set of partitions such that every column partition is less than some member of the selected set is an easy procedure to carry out, it does not necessarily lead to STT assignments with a minimum number of state variables. Such STT assignments can be found with greater effort, however. If entries in a column of a normal state-transition table specify a direct transition from state S_i to S_j, and other entries in that same column specify a direct transition from state S_k to state S_l, we form the *transition partition* (*t*-partition for short),

$$\{\overline{S_i, S_j};\ \overline{S_k, S_l}\}$$

If an S_i-to-S_j transition and stable state S_k all appear in the same column, we form the *t*-partition

$$\{\overline{S_i, S_j};\ \overline{S_k}\}$$

The collection of all *t*-partitions that may be formed from a state table is known as the *t-partition list*. The *t*-partition list for the state table of Fig. 13.16 is

$$
\begin{aligned}
t_1 &= \{\overline{a, b};\ \overline{d, e}\} &\qquad& \text{column 1} \\
t_2 &= \{\overline{a, c};\ \overline{d, e}\} &\qquad& \text{column 1} \\
t_3 &= \{\overline{a, d};\ \overline{b, e}\} &\qquad& \text{column 2} \\
t_4 &= \{\overline{a, d};\ \bar{c}\} &\qquad& \text{column 2} \\
t_5 &= \{\overline{b, e};\ \bar{c}\} &\qquad& \text{column 2} \\
t_6 &= \{\overline{a, c};\ \overline{b, e}\} &\qquad& \text{column 3} \\
t_7 &= \{\overline{b, c};\ \overline{d, e}\} &\qquad& \text{column 4}
\end{aligned}
$$

T-partitions are less complete than column partitions, but are more basic. Each *t*-partition is related to a particular state transition in a particular column of a state-transition table. If the *t*-partitions are used to form a state assignment, each leads to a state variable that ensures that its specific transition will be free of critical races by remaining quiescent during that transition. Above, t_1 would ensure that transitions between states a and b never lead to either state d or e (ignoring multiple-order hazards). The other six *t*-partitions would insure other transitions of the state table from which they were derived. While it is clearly undesirable to base a state assignment on the *t*-partition list, the set of partitions upon which any STT state assignment is based must satisfy all *t*-partitions of that list. Some member of such a set must be greater than each *t*-partition to insure the transition of each *t*-partition.

To find all partitions greater than two or more t-partitions, we form the sums of all pairs of t-partitions. Only nontrivial sums need be recorded: π^1 and incomplete single-block partitions are not greater than the t-partitions from which they are formed. All sums of sums of t-partitions, etc, are then formed. The process terminates when no new nontrivial partitions can be found. If t-partitions are thought of as minterms, then the sum-of-t-partitions may be thought of as implicants and prime implicants. From the t-partition list above we form

$$\pi_1 = \{\overline{a, b, c}; \overline{d, e}\} = t_1 + t_2$$
$$\pi_2 = \{\overline{a, b, d, e}; \overline{c}\} = t_1 + t_4$$
$$\pi_3 = \{\overline{a, c}; \overline{b, d, e}\} = t_2 + t_5$$
$$\pi_4 = \{\overline{a, d}; \overline{b, e, c}\} = t_3 + t_4$$
$$\pi_5 = \{\overline{a, c, d}; \overline{b, e}\} = t_3 + t_6$$
$$\pi_6 = \{\overline{a, d, e}; \overline{b, c}\} = t_4 + t_7$$

Many of these sum partitions are encountered when t-partition sums other than those shown are formed. Note that every t-partition is less than one of these six partitions.

Now we must select the subset of the set of all sum-partitions that will give a minimum state-variable STT assignment. This is a covering type of problem; by analogy we may use any of the prime-implicant selection techniques used to minimize combinational networks. We may form the "prime implicant" table of Fig. 13.17. Entries in this table indicate that a t-partition is less than the sum-partition that labels the row of the entry.

Partition π_1 is an "extremal." Only π_1 is greater than t_1, and thus must be used to form an STT assignment. Partition π_4 is to be avoided if possible since it is a 3-block partition and will introduce two state variables if it is selected. Its cost is thus twice that of other sum-partitions. If π_4 is avoided, π_5 and then π_2 must be selected to complete the set of sum-partitions such that every t-partition is less than some member of the selected set.

t-partitions

	t_1	t_2	t_3	t_4	t_5	t_6	t_7
π_1	x	x					x
π_2				x	x		
π_3		x			x	x	
π_4			x	x	x		
π_5			x		x	x	
π_6				x			x

sum-partitions $\pi_1 \dots \pi_6$

FIG. 13.17   Selection of Sum-partitions

State	y_1	y_2	y_3
a	0	0	0
b	0	0	1
c	0	1	0
d	1	0	0
e	1	0	1
—	0	1	1
—	1	1	0
—	1	1	1

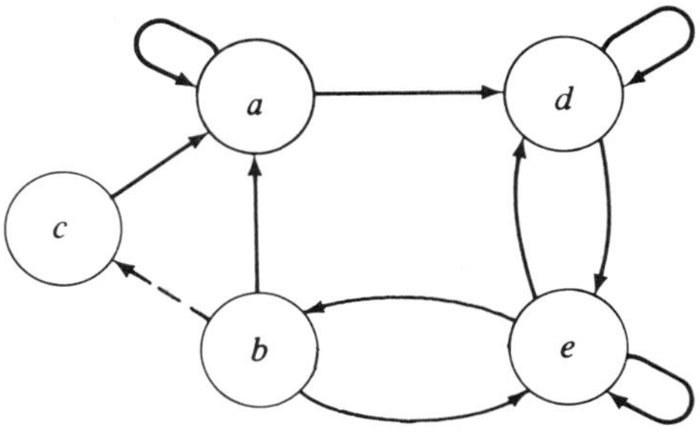

FIG. 13.18   A Minimum State Variable STT Assignment

Figure 13.18 shows a state assignment based upon partitions π_1, π_2, and π_5. The state diagram of the machine is also shown so that the virtues of this assignment can be seen more easily. Only the transition from state b to state c is accompanied by a noncritical race. All other transitions are direct. We have obtained a very desirable state assignment in this case.

13.3 IMPLEMENTATION OF ASYNCHRONOUS NETWORKS

The implementation of excitation, latch input, and output variable equations is a problem of combinational logic design, and all the techniques developed in previous chapters may be employed. But unlike our work in previous chapters, we must now place great emphasis on arriving at implementations free of hazards; the reduction of network cost and operating speed must be a secondary consideration. Suppression of hazards may require the insertion of redundant logic and explicit delay elements; it may require a very specific network structure, which may not be altered. Reduction of network cost and speed is achieved by minimization of the number of logic gates and logic levels, avoidance of explicit delay elements, and use of an STT state assignment. These requirements conflict and network optimization is thus a very difficult problem. This section will be concerned primarily with approaching optimum networks by avoiding the use of explicit delay elements to suppress hazards. Network cost and logic level minimization will be performed only to the extent that it does not interfere with the requirement that the resulting network be free of hazards.

The implementation of many asynchronous sequential networks need not include the rather detailed effort prescribed in this section. We saw in Section 13.1 that excitation or latch input equations and output equations can be implemented directly. Static hazards are eliminated easily. Some of the techniques of this section can be incorporated as easily. We have means of detecting multiple-order hazards and determining the conditions under which they will result in network malfunction. If analysis reveals that no essential hazards exist, or that the probability of malfunction is very small, we may agree to accept a direct design without further ado. This section is directed at situations in which we wish to be certain that multiple-order hazards are suppressed or change input variables simultaneously.

D. B. Armstrong, et al. [1] have shown how the ill-effects of essential hazards can be avoided without insertion of explicit delay elements in networks that meet certain requirements. An *input-level gate* of a network has only input variables and state variables supplied to its input terminals. A *loop* is then a signal path in a network from an input terminal of an input-level gate to an input terminal of the same or another input-level gate. The accumulation of wire and gate delays encountered in touring a loop is its *loop-delay*. The conditions under which explicit delay elements are not required are then:

1. The network is operated in its fundamental mode.
2. Input symbol changes are restricted to single input variable changes.
3. Only uncomplemented input variables are available.
4. Normal state-transition tables are encoded with a STT state assignment.
5. No wire delay magnitude may exceed the smallest loop-delay of the network.
6. All wire and gate delays are bounded.

The assumption of a normal state-transition table encoded with a STT assignment ensures that (a) each state transition is directed by the new input symbol, which promotes the transition, and state variables that do not vary during the transition, and (b) each transition will take a minimum amount of time. We will find the primary problem of synthesis to be that of ensuring that those state variables that should remain invariant during a transition do in fact do so. Condition 5 is very important: it permits us to argue that a new value of an input variable will reach all input-level gates before new values of state variables reach those gates. This argument is of great value when we attempt to ensure theoretically that invariant variables do not change value during a transition. It is also a very reasonable assumption: often all the components of a network are physically close to each other so that interconnecting wire lengths are short.

The normal state-transition table of Fig. 13.11 will be used for purposes of illustration and discovery because it contains essential hazards, and because the STT state assignment of Fig. 13.15 presents us with a five-variable problem, which motivates the use of array techniques of Chapter 9 but does not entirely exclude the use of Karnaugh maps. An encoded state-transition table may be presented as an array of $(n + 2p + m)$-tuples. Each input-symbol–present-state $(n + p)$-tuple must be concatenated with the appropriate next-state–output-symbol $(p + m)$-tuple to form the cubes of such an array. Each input-symbol–present-state $(n + p)$-tuple portion of such cubes is a *total state* of the network.

Figure 13.19 repeats Fig. 13.11 for our convenience and shows Karnaugh maps and minimized excitation equations for the three state

Present state			Next state ($y_1'\,y_2'\,y_3'$) $x_1 x_2$			
y_1 y_2 y_3			00	01	11	10
a 0 0 0			a 000	b 101	c 110	a 000
b 1 0 1			b 101	b 101	d 011	a 000
c 1 1 0			b 101	c 110	c 110	d 011
d 0 1 1			a 000	c 110	d 011	d 011
0 0 1			000	101	011	000
0 1 0			000	110	110	011
1 0 0			101	101	110	000
1 1 1			101	110	011	011

$$y_1' = \bar{x}_1 y_1 \vee \bar{x}_1 x_2 \vee x_2 \bar{y}_3$$

$$y_2' = x_2 y_2 \vee x_1 x_2 \vee x_1 y_2$$

$$y_3' = \bar{x}_1 \bar{x}_2 y_1 \vee \bar{x}_1 x_2 \bar{y}_2 \vee x_1 \bar{x}_2 y_2 \vee x_1 x_2 y_3$$

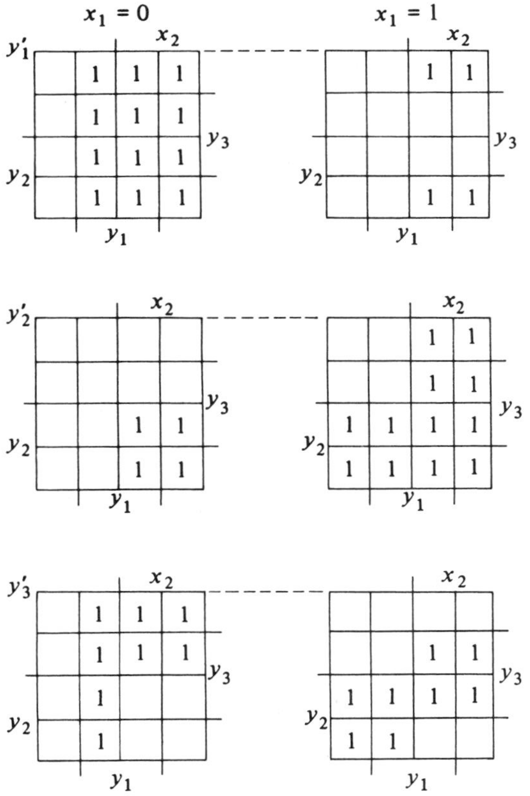

FIG. 13.19 Realizing the Encoded State Table of Fig. 13.15

variables. While detailed analysis might show that a network based upon these equations would perform in the manner of the state-transition table, the presence of static hazards in the equation for y_3' and essential hazards in the original table suggests that such a network is likely to fail. We will not analyze a questionable network with the hope of finding it to be reliable, but rather will design so that there can be no question of network failures arising from the presence of hazards.

Product terms exist in excitation equations for one or both of the following reasons. When the value of a state variable is to change from 0 to 1, all terms of its excitation equation must initally have the value 0. Then during the transition the value of some term or terms must change to 1. Such terms are said to "excite" the state variable. Other terms "hold" the state variable at 1 during other transitions.

State-Variable Changes

As a result of our use of a normal state table and an STT state assignment, each total-state transition of a machine is accomplished ideally by each state variable either remaining fixed at 0, fixed at 1, or changing once from 0 to 1 or from 1 to 0. If those variables that are to remain invariant do so, then the transition ultimately will be completed successfully. For a specific state transition, those variables that are to remain invariant can be determined easily. If S_i and S_f are the original and final total states of a transition, then the factor product[2] $S_i F S_f$ indicates which variables change and which remain invariant. For the total state transition from

$$(x_1, x_2, y_1, y_2, y_3) = 00000 \text{ to } 10000, \quad 00000 \textbf{F} 10000 = x0000$$

Thus x_2, y_1, y_2, and y_3 are to remain invariant at 0; the state of the machine does not change. No terms of the excitation equations cover either the initial or final total state of this transition, and none of the state variables may become excited. The factor product $S_i \textbf{F} S_f$ is the smallest single cube that covers both the initial and the final total states of a transition. We will refer to this cube as the *transition cube*.

Variables x_1 and y_3 are to remain invariant throughout the transition from state a to state c, total state 10000 to 11110.

$$10000 \textbf{F} 11110 = 1xxx0$$

But the machine may enter any of the eight total states covered by this transition cube as x_2, y_1, and y_2 change from 0 to 1; y_3 may be falsely excited when the machine is in one of these states. In a network based directly upon the excitation equations of Fig. 13.19, the change in x_2 will quickly excite y_1 via the term $x_2 \bar{y}_3$ in the y_1' equation; state variable y_2 will be excited via term $x_1 x_2$. If y_2 changes and that change reaches the circuit that generates y_3 before the change in x_2 reaches that circuit, then y_3 will be excited via term $x_1 \bar{x}_2 y_2$, which is present in the expression of

---

[2] Factor product is defined in Section 11.2.

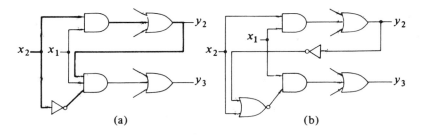

F<small>IG</small>. 13.20 Fixing Input-Variable State-Variable Races

y_3' to excite y_3 during other transitions. Fig. 13.20(a) shows paths of the racing signals that the essential hazard of the original state-transition table introduces. This race is critical: state variable y_3 is to remain quiescent at 0 to ensure that state a is followed by state c under the given input-symbol change; we see now that it may be excited. If y_3 is allowed to change, then the network performance may not correspond to the state table.

Input-variable versus state-variable races can be "fixed" by inserting a delay element in the state-variable line so that the input-variable change will reach state-variable-generating circuitry before the state-variable change. Alternatively, we may propagate the input-variable change more rapidly. In Fig. 13.20(a) the change in x_2 is delayed by an inverter as well as by several wires. While we can not remove the requirement of inverting x_2, we can introduce y_2 and x_2 at the same point in the y_3 circuit as shown in Fig. 13.20(b). The crucial term in the y_3' equation can be rewritten to correspond to this gate structure: $x_1 \bar{x}_2 y_2 = x_1 \cdot \overline{(x_2 \vee \bar{y}_2)}$.

Figure 13.20(b) might suggest that we are really delaying y_2 by introducing an inverter and presenting $\bar{y}_2$ to the y_3 circuitry. While an additional inverter is shown in Fig. 13.20(b), it is not always necessary to add an inverter, and the delay of this added inverter is not of prime significance. Because we assume that all wire delays are shorter than the least loop delay, the change in x_2 will reach the NOR gate before the change in y_2: the race between these two variables is thus resolved.

If π is a product term in a sum-of-products excitation equation y_i' and π is to maintain the same value during some total state transition, and the literals of π include complemented input variables and state variables that are to change value during the transition, then π may be rewritten to remain invariant by partitioning its literal set into two subsets. In the first subset we place all complemented input variables and all state variables; in the second set we place all uncomplemented input variables, if any. Product term π may then be expressed as the product of members of the second set and the complement of the sum of complements of the members of the first set. Then all input variables are introduced directly.

$$x_i \cdot \bar{x}_j \cdot y_k \cdot \bar{y}_l = x_i \cdot \overline{(x_j \downarrow \bar{y}_k \downarrow y_l)}$$

In terms of the network corresponding to the excitation equation for y_i, an input-level AND gate must be replaced with a NOR-AND gate pair, and with just a NOR gate if the second set is empty. If only uncomplemented input variables originally appear in π, then no critical race can exist and the NOR gate is unnecessary. We will soon encounter the need for a more general structure, of which the NOR-AND-gate pair is a part.

State variable y_1 is to remain invariant during the state c to state b (total state 01110 to 00101) transition.

$$01110 \, \mathbf{F} \, 00101 = 0x1xx$$

A term or terms of the y_1' equation must cover the original total state 01110, the final total state 00101, and if false values of y_1' are to be avoided, all possible intermediate total states. Transition cube $0x1xx$ covers all of these states; term $\bar{x}_1 y_1$ described by it appears in the y_1' equation of Fig. 13.19. The presence of this term ensures that y_1 will not assume a false value of 0 at any time during the state c to b transition.

State variable y_3 is to remain invariant during the state b to state d (total state 01101 to 11011) transition. See Fig. 13.21(a).

$$01101 \, \mathbf{F} \, 11011 = x1xx1$$

But in this case the product term $x_2 y_3$ suggested by the transition cube is not an implicant of the excitation function y_3': we will refer to such transition cubes as "incomplete." Thus no single term of y_3' can cover the initial total state, the final total state, and all of the possible intermediate total states. In the y_3' equation of Fig. 13.19, terms $\bar{x}_1 x_2 \bar{y}_2$ and $x_1 x_2 y_3$ cover the initial state and final state, respectively, and some of the possible intermediate states. If the AND gates that realize these product terms exhibit different delays, then $\bar{x}_1 x_2 \bar{y}_2$ could go to 0, and y_3' along with it, before $x_1 x_2 y_3 = 1$. The sum of these two terms is not free of static hazards. While a redundant term could be included to eliminate the static hazard, a more serious problem exists.

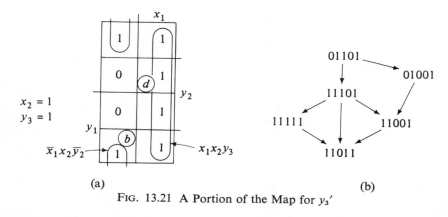

(a)   (b)

FIG. 13.21 A Portion of the Map for y_3'

Suppose that in some machine, state variable y_i is to be invariant throughout a transition from total state $I_1 \circ \alpha$ to total state $I_2 \circ \beta$, but that the transition cube is not an implicant of y_i'. Figure 13.22(a) suggests the entries that must appear in the map of y_i'. If the state-transition table of this machine is normal in form, as we are assuming, the 0 entry for total state $I_1 \circ \beta$ indicates that a transition from state β to some other stable state γ for which $y_i = 0$ is specified by the original state table. A transition from state γ to some stable state in the I_2 column must also be specified. That state may be state β as shown in Fig. 13.22(b): then the state table is said to contain a *d-trio*. If a transition from γ to state δ is specified as in Fig. 13.22(c), the state-transition table contains an essential hazard. Therefore, if a transition cube is not an implicant of an excitation variable, the state table must contain either a *d*-trio or an essential hazard. Conversely, if a state table contains an essential hazard, then an incomplete transition cube can be expected for the state variable(s) that distinguish state α from state γ.

An essential hazard exists in the $x_1 x_2 = 01$ and $x_1 x_2 = 11$ columns of the example state-transition table of Fig. 13.19. With the STT assignment of that table, this essential hazard manifests itself as a function hazard in the circuitry that generates y_3. With x_1, y_1, and y_2 all changing at the same time, it is possible for the machine to attain either total state 01011 or 01111 for which $y_3' = 0$. See Fig. 13.21(a). To avoid these states, the total state of the machine must be forced to follow one of the paths suggested by Fig. 13.21(b) when executing the state b to state d transition.

A direct implementation of $x_2 y_3 (x_1 \vee \bar{y}_2)$ avoids the static hazard, that arises when two disjoint prime implicants are used in the y_3' equation, and directs the total state along the acceptable paths above. The change in x_1 must reach this circuit before changes in y_1 and y_2, because of our wire delay assumption. Thus the OR-gate output must remain invariant as must the AND-gate signal during the transition. As far as the y_3' circuitry is concerned, the total state changes first from 01101 to 11101. Subsequent changes in y_1 and y_2 cause the y_3' circuitry ultimately to view the total state as 11011; the race between y_1 and y_2 will not be critical with respect

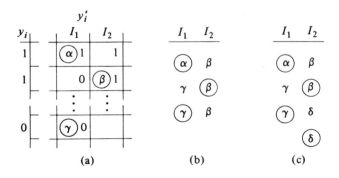

(a) (b) (c)

FIG. 13.22   The *d*-trio and Essential Hazard Lead to
Incomplete Transition Cubes

to the y_3 circuitry, and the function hazard is avoided. Thus it is possible to avoid the ill effects of function hazards in asynchronous machines (not in general) because not all transitions between total states are permitted and because the change of the state variables is dependent on the change of the input variables.

The term $x_2 y_3(x_1 \vee \bar{y}_2)$ may be generated from the Karnaugh map in Fig. 13.23 by "removing" the total states to be avoided from the cube that is universal to the transition, term $x_2 y_3$. The sharp operator of Chapter 9 may be used.

$$x_2 y_3 \;\#\; (\bar{x}_1 y_2) = x_2 y_3(\overline{\bar{x}_1 y_2}) = x_2 y_3(x_1 \vee \bar{y}_2)$$

To avoid input-variable–state-variable races, we avoid the NAND gate suggested by the intermediate result in favor of the OR gate, which combines the uncomplemented input variable with the changing state variable.

In general, if state variable y_i is to remain invariant at 1 during the transition from total state S_j to S_k, then $S_j\,\mathbf{F}S_k$ is the smallest cube that covers S_j and S_k and all of the other total states that may be entered during the transition. Let OFF_i be the OFF-array of y_i. Then

$$A = (S_j\,\mathbf{F}S_k) \bigcap \mathrm{OFF}_i$$

covers the total states to be avoided during the transition, if any, and

$$B = (S_j\,\mathbf{F}S_k) \;\#\; \mathrm{OFF}_i$$

covers the total states that may be entered, including the initial and final total states. If $A = \phi$, then $S_j\,\mathbf{F}S_k$ is an implicant of y_i' which may be realized with a single AND gate as in the first example above. When $A \neq \phi$, B must be realized, not as a sum of prime implicants, but by an AND gate that combines the invariant variables given by $S_j\,\mathbf{F}S_k$ and the

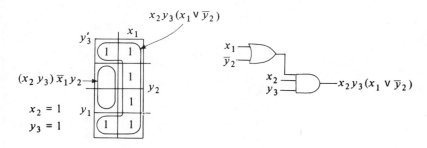

FIG. 13.23 Factoring to Avoid a Function Hazard

output of a gate that realizes the complement of a cover of A which is disjoint from B. This is much like factoring of Section 11.2. Since the variables of $S_j \mathbf{F} S_k$ are common factors of a set of cubes that covers B, they need not appear in the expression of the complement of A. Cube $\bar{x}_1 y_2$, used in the second example above, covers the area $\bar{x}_1 x_2 y_2 y_3$ shown in the Karnaugh map. In general, if the cover of A is expressed as a sum-of-products, an AND-gates-to-NOR-gate structure realizes the complement of A. Each AND gate of such a structure that combines complemented input variables with state variables, must be replaced by a NOR-AND-gate pair to resolve the race between these changing variables. Usually this general NOR-AND-NOR-AND-gate structure may be simplified. In the first example immediately above, it reduced to just an AND gate; in the second it reduced to a NOR-INVERT-AND, or just OR-AND, structure.

During the state a to c (total state 10000 to 11110) transition, state variables y_1 and y_2 are to change their value from 0 to 1. Some term or terms of the y_1' and y_2' excitation equations must cover the final total state 11110 and must be independent of y_1 and y_2, respectively, since a product term that includes variable y_i can not cause y_i' to change its value from 0 to 1. The use of an STT state assignment ensures that the transition is directed by the new input symbol and invariant state variable(s). Thus the state assignment ensures that a state variable that is to go to 1, will be excited whenever the machine is in a total state covered by the concatenation of the new input symbol and the factor product of the original- and final-state p-tuples. (This factor product indicates which state variables remain invariant.) For the state a to c transition of the example machine, both y_1 and y_2 must be excited whenever the network is in any total state covered by $11xx0$. An examination of the third next-state column of Fig. 13.19 reveals this to be the case. Thus the term $x_1 x_2 \bar{y}_3$ or a prime implicant that covers it must appear in the excitation equations for y_1' and y_2'. In the y_1' equation of Fig. 13.19, the term $x_2 \bar{y}_3$ appears; term $x_1 x_2$ of the y_2' equation covers $x_1 x_2 \bar{y}_3$.

Because a state variable is excited does not necessarily mean that it will change its value without transients. During the transition from state a to b (total state 00000 to 01101), $00000 \mathbf{F} 01101 = 0xx0x$, and y_2 is to remain invariant at 0 while y_1 and y_3 change to 1. The term $\bar{x}_1 x_2 \bar{y}_2$ appears in the y_3' equation of Fig. 13.19 and ensures that y_3 will change its value. The term $\bar{x}_1 \bar{x}_2 y_1$ also appears in that equation. If the y_3 circuitry detects the change in y_1 before the change in x_2, then y_3 may change to 1 and then back to 0 before returning and remaining at 1 since these two terms are disjoint. This static hazard could be avoided by adding redundant terms to the y_3' equation to make it static-hazard–free or by replacing the product terms with NOR-AND terms so that the input-variable changes will be seen first. But this static hazard can not result in a steady-state hazard. With many signals changing their values, function hazards can exist and also cause such transients. We need be concerned with such transients only if they can excite a state variable that should remain invariant at 1. Then we must make use of the NOR-AND–gate pair.

Transients may also appear when state variables change from 1 to 0, but again they can not result in steady-state hazards. We exert no special effort to avoid them.

Implementation without Delay Elements

Excitation equations upon which to base a reliable, delay-element-free network may be obtained by examining each of the state transitions of the encoded state-transition table. If state variable y_i is to remain at 0 or drop to 0 during a transition, no term need be added to the expression of y_i', but any term that can falsely excite the state variable must be replaced with a NOR-AND term. If y_i is to remain at 1 during a transition, then a NOR-AND-NOR-AND term must be added to the Boolean expression of y_i'. If y_i is to change to 1 during a transition then a prime implicant that covers the final state of the transition and does not involve any of the changing state variables must be included in the y_i' equation. If the term can falsely excite the state variable during some other transition, it must be replaced with a NOR-AND term. The added terms may be based upon prime implicants. The prime-implicant arrays of the excitation variables of the example machine of Fig. 13.19 are:

$$PI_1 = \begin{pmatrix} 0x1xx \\ 01xxx \\ x1xx0 \end{pmatrix} \quad PI_2 = \begin{pmatrix} 11xxx \\ 1xx1x \\ x1x1x \end{pmatrix} \quad PI_3 = \begin{cases} 001xx \\ 0x10x \\ 01x0x \\ 10x1x \\ 1xx11 \\ 11xx1 \\ x1x01 \\ x011x \end{cases}$$

Note that prime implicants of y_i' can not involve $\bar{y}_i$.

The examination of all transitions may be performed in an orderly manner. Table 13.1 lists all the possible transitions of the example machine. The transition cube is computed for each. Then the necessary "hold" terms of each excitation equation may be selected from the prime implicants of each excitation variable. If the transition cube for a transition indicates that a state variable is to remain at 1, the prime implicant that covers that transition cube must be included in the excitation equation of the state variable. This prime implicant must be the product of invariant variables only. If no single prime implicant covers the transition cube, then a function hazard must be avoided by the selection of a NOR-AND-NOR-AND term. This may be accomplished by first intersecting the transition cube and the prime-implicant array for the state variable in question. The resulting array then consists of the subcubes of the transition cube that cover the intermediate states we can allow the machine to enter when performing the transition. The invariant variables may be factored from these subcubes and the remainder of the cubes (the D array in Section 11.2) may be realized directly, or with NOR–AND–gate pairs if a subcube combines the complement of a changing input variable with changing state variables.

Table 13.1 shows the prime implicants selected to hold the state variables at 1. The b to d transition introduces transition cube $x1xx1$. No prime implicant of y_3' covers this cube. Intersection of this transition cube and PI_3 gives

$$\begin{Bmatrix} \text{~~01101~~} \\ \text{~~01x01~~} \\ \text{~~11x11~~} \\ 11xx1 \\ x1x01 \end{Bmatrix}$$

where the deleted terms are covered by other members of this array. Note that $x1xx1$ is a common factor of this array, as it must be. If x_2 and y_3 are deleted, $\{1xxxx, xxx0x\}$ remains. This array suggests that x_1 and $\bar{y}_2$ be combined in an OR gate, the structure we developed earlier.

To determine the terms that must be included in each excitation equation to excite the state variable, we first form the concatenation of the new input symbol and the factor product of the initial- and final-state p-tuples for each transition. We then search the list of transitions for those that involve a state-variable change from 0 to 1. The transition from state a to state b is accomplished by a change to 1 of both y_1 and y_3. Thus for each of these state variables we must select a prime implicant that covers the new input symbol concatenated with the invariant state variables and thus covers some of the intermediate total states as well as the final total state of the transition. For this transition we must find a prime implicant that covers $01x0x$. State-variable change is then directed by the input variables and invariant state variables. Table 13.1 again summarizes the prime implicants that excite the state variables. Note that some of these prime implicants also hold the state variables at 1 during different state transitions.

Finally, we must determine if any of the terms of the excitation equations which excite can cause a state variable to take the value of 1 when it should remain at 0. Again we examine all of the transitions, but now searching for transitions during which a state variable is to remain at 0. During the transition from state a to state a, all three state variables are to remain at 0. During the transition from state a to state b, y_2 is to remain at 0. For this transition, the machine may enter any of the eight total states covered by transition cube $0xx0x$. Can any of the excite terms for y_2' take the value of 1 in any of these states? Since the intersection of the transition cube and the one excite term of y_2' is empty, the answer is *no*.

TABLE 13.1. TERMS OF THE EXCITATION EQUATIONS

Transition	$S_i \rightarrow S_f$	$S_i \mathbf{F} S_f$	Hold terms y_1'	y_2'	y_3'	New input Symbol ∘ Invariant state variables	Excite terms y_1'	y_2'	y_3'
$a \rightarrow a$	00000 → 10000	x0000	—	—	—	10000	—	—	—
$a \rightarrow b$	00000 → 01101	0xx0x	—	—	—	01x0x	01xxx	—	01x0x
$a \rightarrow c$	10000 → 11110	1xxx0	—	—	—	11xx0	x1xx0	11xxx	—
$b \rightarrow b$	00101 → 01101	0x101	0x1xx	—	0x10x	01101	—	—	—
$b \rightarrow a$	00101 → 10000	x0x0x	—	—	—	10x0x	—	—	—
$b \rightarrow d$	01101 → 11011	x1xx1	—	—	$x_2 y_3 (x_1 \vee \bar{y}_2)$	11xx1	—	11xxx	—
$c \rightarrow c$	01110 → 11110	x1110	x1xx0	x1x1x	—	11110	—	—	—
$c \rightarrow b$	01110 → 00101	0x1xx	0x1xx	—	—	001xx	—	—	001xx
$c \rightarrow d$	11110 → 10011	1xx1x	—	1xx1x	—	10x1x	—	—	10x1x
$d \rightarrow d$	11011 → 10011	1x011	—	1xx1x	1xx11	10011	—	—	—
$d \rightarrow a$	10011 → 00000	x00xx	—	—	—	000xx	—	—	—
$d \rightarrow c$	11011 → 01110	x1x1x	—	x1x1x	—	01x1x	01xxx	—	—

State variable y_3 is to remain invariant at 0 during the transition from state a to state c. The transition cube for this transition, $1xxx0$, does intersect one of the excite terms for $y_3{}'$.

$$1xxx0 \ \sqcap \ 10x1x = 10x10$$

Thus state variable y_3 will be falsely excited during the state a to c transition if the machine is allowed to enter total states $10x10$. To enter one of these states, y_2 must change before x_2 as far as the y_3 circuitry is concerned. This is the input-variable–state-variable race we examined earlier. We now fix the race as we did then, by replacing the prime implicant with a NOR-AND term. This term is underlined in Table 13.1; it is the only excite term that requires correction.

We may now gather together all of the hold and excite terms of Table 13.1 into the excitation equations for a reliable network.

$$y_1{}' = \bar{x}_1 y_1 \ \vee \ x_2 \bar{y}_3 \ \vee \ \bar{x}_1 x_2$$
$$y_2{}' = x_2 y_2 \ \vee \ x_1 y_2 \ \vee \ x_1 x_2$$
$$y_3{}' = \bar{x}_1 y_1 \bar{y}_2 \ \vee \ x_2 y_3 (x_1 \ \vee \ \bar{y}_2) \ \vee \ x_1 y_2 y_3 \ \vee \ \bar{x}_1 x_2 \bar{y}_2$$
$$\vee \ \bar{x}_1 \bar{x}_2 y_1 \ \vee \ x_1 \overline{(x_2 \ \vee \ \bar{y}_2)}$$

The network of Fig. 13.24 is a direct implementation of these equations. Note that gates that combine complemented input and state variables exist in this network. The variables combined never race, and hence it is not necessary to replace these AND gates with NOR-AND–gate pairs. While it is not necessary to insert delay elements in this network, it *is*

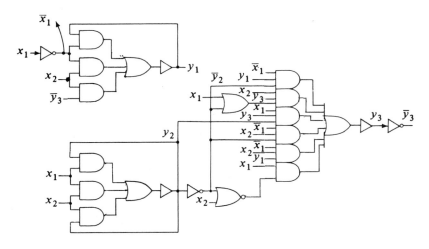

FIG. 13.24   A Delay-Element-Free Implementation

necessary to have electronic amplification present in every loop of the network. If passive diode gates are to be used, then amplifiers must be inserted as shown in Fig. 13.24. If the AND and OR gates are replaced with NAND gates that provide amplification, then explicit amplifiers need not be present

This implementation algorithm is rather exhausting. It can be abbreviated, but network costs usually will be increased if we do so. For example rather than perform the search for excite terms that require correction, we might simply correct all excite terms. Every excite term that combines complements of input variables and state variables may be replaced with NOR-AND terms. And rather than search for the required excite terms, we might include all prime implicants of y_i' that do not involve y_i in the sum-of-products expression of y_i'. Any prime implicants serving as hold terms need not be repeated, of course. These shortcuts provide the following excitation equations for the example machine.

$$y_1' = \bar{x}_1 y_1 \ \vee x_2 \bar{y}_3 \vee \ \bar{x}_1 x_2$$

$$y_2' = x_2 y_2 \ \vee x_1 y_2 \vee x_1 x_2$$

$$y_3' = \bar{x}_1 y_1 \bar{y}_2 \ \vee x_2 y_3 (x_1 \vee \bar{y}_2) + x_1 y_2 y_3 \vee \overline{x_1 \vee x_2 \vee \bar{y}_1}$$
$$\vee \ x_2 \overline{(x_1 \vee y_1)} \vee x_1 \overline{(x_2 \vee \bar{y}_2)} \vee \overline{x_2 \vee \bar{y}_1 \vee \bar{y}_2}$$

The equations for y_1' and y_2' have not been changed. One additional term appears in the expression for y_3': two AND gates are replaced with NOR gates and two additional NOR terms are introduced. Thus the network obtained with less design effort is more expensive.

Can we further abbreviate the synthesis algorithm by selecting hold terms without examining all transitions in detail? Those transitions for which the transition cube is complete do not present any problems. The hold terms required for such transitions are prime implicants and will be included in the excitation equations because we agreed earlier to include all prime implicants. But those transitions for which the transition cube is incomplete present a more difficult problem. We can not eliminate all function hazards in a combinational network, in general. The list of transitions of a machine indicates the few function hazards that must be removed. Thus we must use the list of transitions or reconstruct it within a design algorithm. Such reconstruction is possible and might be done in an algorithm to be executed by a computer, but can not be considered a simplification of the algorithm given above.

Latch Implementation

The need to develop and realize specially factored hold terms can be eliminated by the use of latches for the state variables, since transient, false values of 0 on the input lines of such latches can not cause them to change their state. False values of 1 must be avoided by the use of NOR-AND–gate pairs as before. The encoded state-transition table may be transformed to determine the latch input equations in the manner presented in Section 5.3. Actually, the excite terms of Table 13.1 must make up the terms of the S-input equations.

The following latch input equations may be written directly from the transformed excitation maps of Fig. 13.25.

$$S_1 = \bar{x}_1 x_2 \vee x_2 \bar{y}_3 \qquad\qquad S_2 = x_1 x_2$$
$$R_1 = x_1 \bar{x}_2 \vee x_1 y_3 \qquad\qquad R_2 = \bar{x}_1 \bar{x}_2$$

$$S_3 = \bar{x}_1 \bar{x}_2 y_1 \vee \bar{x}_1 x_2 \bar{y}_2 \vee x_1 \bar{x}_2 y_2$$
$$R_3 = \bar{x}_1 \bar{x}_2 \bar{y}_1 \vee \bar{x}_1 x_2 y_2 \vee x_1 \bar{x}_2 \bar{y}_2$$

Some of the terms of S_3 and R_3 combine complemented input and state variables and hence offer the possibility of a false 1 as a result of an input-variable versus state-variable race. Table 13.1 indicates that only one term of S_3 requires correction. We could examine the transitions of that table for which $y_3{}'$ is to remain at 1, and determine if any of the terms of R_3 can take the value of 1 while any of these transitions is in progress. Or we might simply replace all of the terms of R_3 with equivalent NOR-AND

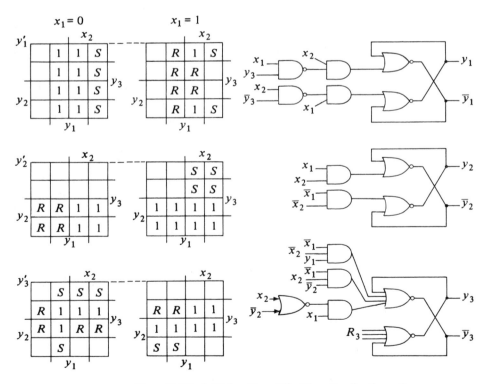

Fig. 13.25 A Delay-Free Flip-Flop Realization of the Example Machine

terms. In the following equations, which are realized in Fig. 13.25, only the terms that must be corrected have been rewritten.

$$S_3 = \bar{x}_1 \bar{x}_2 y_1 \vee \bar{x}_1 x_2 \bar{y}_2 \vee x_1 \overline{(x_2 \vee \bar{y}_2)}$$

$$R_3 = \bar{x}_1 \bar{x}_2 \bar{y}_1 \vee x_2(x_1 \vee \bar{y}_2) \vee x_1 \bar{x}_2 \bar{y}_2$$

13.4 MULTIPLE-INPUT-VARIABLE CHANGES

We now relax the restriction that only one input variable change at a time; an arbitrary number of input variables will be allowed to change "simultaneously." It is not practical to insist that input variables change at exactly the same instant. Even the signal change emanating from a single source will reach and be sensed by two input-level gates at slightly different times because of different wire delays and variations in the components of the gates. Thus we will take *simultaneous* to mean that all changes reach all input-level gates before any gate has had time to respond to any change. If the times of arrival of input variable changes at input-level gates are spread over an interval Δ and the minimum delay offered by any gate in a network is τ_{min}, then the input variables change simultaneously if $\Delta < \tau_{min}$. We will assume that if more than one input variable changes, all change simultaneously in the sense of this definition and that no further changes occur until the network has reached a stable state. The other conditions presented at the beginning of this section are also assumed to be satisfied.

With this definition of a simultaneous input-variable change, all state-variable networks have the opportunity to sense the new input symbol before any state variable changes its value. Only proper design ensures that these networks do in fact respond to a simultaneous input-variable change in the manner predicted by a horizontal transition in a state table from the column of the original input symbol to the column of the new input symbol. To see how the network might fail to respond in the ideal fashion to a multiple input-variable change, assume that the network of Fig. 13.24 (or 13.25, for that matter) is in state a of Fig. 13.19 with $x_1 = x_2 = 0$, and that x_1 and x_2 change simultaneously to 1. State variables y_1 and y_2 are to change to 1 while y_3 remains at 0. But the network that realizes y_3 includes an AND gate that realizes the term $\bar{x}_1 x_2 \bar{y}_2$. Both $\bar{x}_1$ and $\bar{y}_2$ have the value of 1 when the network is in its initial state. When x_1 and x_2 change, the change in x_2 must reach this AND gate before the change in x_1 because of our assumption that $\Delta < \tau_{min}$. The inverters providing $\bar{x}_1$ and $\bar{y}_2$ must hold $\bar{x}_1 = 1$ and $\bar{y}_2 = 1$ for a time. During that interval $\bar{x}_1 x_2 \bar{y}_2 = 1$ and hence $y_3' = 1$; y_3 is falsely excited. As far as the y_3 network is concerned, the input variables are changing in the order x_2, then x_1, rather than simultaneously.

Since state-variable networks sense the new input symbol before any state variable changes, we need only be concerned with transitions within a row of a state table. We know how to resolve input-variable versus state-variable races in favor of input variables, and transitions between rows involve state-variable changes. What we are now concerned with then, are races between input variables and function hazards in the combinational circuitry dictated by a row of a state table. Further, in the following discussion we will be concerned only with transitions between nonadjacent input symbols.

Function-Hazard-Free Rows

Two types of row can be excluded immediately from further consideration. First, note that a transition within a row must originate at a stable state. It may terminate either at another stable state in the same row or at an unstable state of that row. A state table can not prescribe a transition from an unstable state to another unstable or stable state in the same row. Thus a row of all unstable states need not receive attention. A row of all stable states can not lead to malfunctions either. In such a row, all state variables have the same value for all input symbols. The state variables are independent of the input variables in such a case, and can not falsely be excited by any input-symbol change if we take care to avoid static hazards. Such a row has rather limited value in a state table, of course.

Rows that present a mixture of stable and unstable states require attention. Transitions between two stable states of the same row of a state transition table are *reversible transitions* since they may occur in either direction. A transition from a stable state to an unstable state of the same row is *nonreversible*: the reverse transition may not occur. The following row illustrates one reversible and one nonreversible transition under a multiple input-variable change.

a	00	01	11	10
a	ⓐ	ⓐ	ⓐ	b

States a and b may appear together in a block of the partitions used to encode one or more state variables of the complete state table. For those state variables, represented by y_i below, this row takes the form of all 0's or all 1's.

$\ldots y_i \ldots$	y_i'				$\ldots y_i \ldots$	y_i'			
	00	01	11	10		00	01	11	10
0	0	0	0	0	1	1	1	1	1

Such state variables are independent of the input variables as long as the other state variables remain invariant, and thus can not assume a false value for transitions within the row if static hazards are eliminated. But for some one or more state variables, states a and b must appear in different blocks of their state-assignment partition. For these state variables the example row takes one of the following forms.

$\ldots y_j \ldots$	y_j'			
	00	01	11	10
0	0	0	0	1

$\ldots y_j \ldots$	y_j'			
	00	01	11	10
1	1	1	1	0

y_j', x_1

	0	0
x_2	1	0

y_j', x_1

	1	1
x_2	0	1

The Karnaugh maps suggested by these rows reveal the presence of function hazards if x_1 and x_2 are allowed to change simultaneously from 00 to 11 or vice versa. A great number of other rows can be shown to contain function hazards. Some of these will be examined subsequently.

We can show that other rows are free of function hazards for certain state assignments. Fortunately, such rows can be detected from the state table and state-variable partitions without examination of the fully encoded state table. By definition, a function hazard can only exist if a switching function is to maintain the same value for some change of more than one input variable. This is exactly what is to happen for all state variables during transitions from one stable state to another in the same row. The factor products of the initial and final input symbols for such transitions, $x^i \mathbf{F} x^f$, cover all the input symbols that may be seen briefly by the state-variable networks. If no state variable is to be falsely excited, then each excitation variable must be assigned the same value for all these possible input symbols. This in turn means that a stable entry must appear in the row under examination for all of the input symbols covered by the factor product $x^i \mathbf{F} x^f$. If all multiple input-variable changes are allowed, exactly 2^k, where k is an integer, stable entries must appear in a row if that row is to be entirely free of function hazards for all state variables. The example row given above contains three stable entries. This simple count is thus sufficient evidence to ensure that the row is not free of function hazards for all state variables. The row below contains $2^k = 4$ stable states and the input symbols of the stable states form a k-cube in the n-dimensional space of input symbols. Each state variable must be assigned the same value for all of the input symbols that may be presented when a reversible transition is being executed.

	000	001	011	010	100	101	111	110
a	ⓐ	ⓐ	b	b	ⓐ	ⓐ	c	d

Thus reversible transitions within the row will be performed without malfunction arising from function hazards. But we can not yet be sure that the nonreversible transitions will be free of function hazards. For a row to be entirely free of function hazards, it is necessary but not sufficient that the row contain 2^k stable states and that their input symbols form a k-cube. Finally, in the following row the input symbols of the two stable states do not form a 1-cube, and we can expect function hazards in one or more state variable maps as a result.

	00	01	11	10
a	ⓐ	b	ⓐ	c

A sufficient condition for a row to be free of function hazards entirely requires knowledge of the state assignment for the state-transition table of which the row is a part. Such an assignment may be described by p 2-block partitions of the state set. Each such partition may now be used to partition the set of next state entries of a row. Let $S1$ be the block of such a partition that contains all of the stable states and possibly some of the unstable states, and $S2$ be the block that contains the remaining unstable states, if any. If a state assignment is based upon partitions $\pi_1 = \{\overline{a, b}; \overline{c, d}\}$ and $\pi_2 = \{\overline{a, d}; \overline{b, c}\}$, then π_1 partitions the row

a	ⓐ	ⓐ	b	c

into $S1 = \{\text{ⓐ ⓐ } b\}$ and $S2 = \{c\}$, while π_2 partitions this row into $S1 = \{\text{ⓐ ⓐ}\}$ and $S2 = \{b, c\}$. The entry ⓐ is repeated in these blocks because the two appearances represent different total states.

A row will necessarily be free of function hazards if every state-assignment partition divides the next-state entries of the row into sets $S1$ and $S2$ such that $S1$ consists of 2^k members where k is an integer and the input symbols associated with these members form a k-cube. The expression of y_i' may contain a prime implicant that covers this k-cube. Then y_i must remain invariant for transitions between members of $S1$. Transitions from a member of $S1$ to a member of $S2$, necessarily an unstable state, require that y^i change value, and discussion of function hazards is not appropriate.

In the example row above, partition π_1 induces block $S1 = \{\text{ⓐ ⓐ } b\}$ with three members. Therefore, this row is not entirely free of function hazards for the state assignments dictated by partitions π_1 and π_2. The function hazard will appear in the map for y_1'. The following rows are free of function hazards if states are assigned with these two partitions.

	00	01	11	10	π_1			π_2	
a	ⓐ	ⓐ	c	c	$S1 = \{\text{ⓐ ⓐ}\}$	$S2 = \{c, c\}$	$S1 = \{\text{ⓐ ⓐ}\}$	$S2 = \{c, c\}$
b	ⓑ	a	d	c	$\{\text{ⓑ ⓐ}\}$	$\{c, d\}$	$\{\text{ⓑ } c\}$	$\{c, d\}$
c	ⓒ	ⓒ	d	d	$\{\text{ⓒ ⓒ } d, d\}$	$\{\phi\}$	$\{\text{ⓒ ⓒ}\}$	$\{d, d\}$

If a state table contains one or more rows that are not free of function hazards, then either multiple input-variable changes must be prohibited, or the network that realizes the state table must be designed so that the function hazard can not cause a false state transition. The manner in which the design of the network must proceed can again be determined by an examination of the rows of the state table and the state-assignment partitions.

Assume that the following row is part of a state table to be encoded with the partitions shown.

	00	01	11	10		
a	Ⓐ	Ⓐ	Ⓐ	b		$\pi_1 = \{a, b; \overline{c, d}\}$

$\pi_1 = \overline{\{a, b; c, d\}}$

$\pi_2 = \{a, d; \overline{b, c}\}$

This row then might be encoded in the following manner.

		y_1'							y_2'			
y_1	y_2	00	01	11	10		y_1	y_2	00	01	11	10
0	0	0	0	0	0		0	0	0	0	0	1

y_2' map:

$y_2' \backslash x_1$		
	0	1
x_2	0	0

Both y_1 and y_2 are to remain invariant in response to an input-symbol change from $x_1 x_2 = 00$ to $x_1 x_2 = 11$ or vice versa, and we see that y_1 will indeed remain invariant; this row satisfies the conditions stated above when the next-state entries of the row are partitioned by π_1. But π_2 gives blocks $S1 = \{Ⓐ, Ⓐ, Ⓐ\}$ and $S2 = \{b\}$, and the presence of a function hazard is indicated by the size of $S1$. The Karnaugh map for y_2' above also suggests that y_2 may be falsely excited when this input-symbol change is made.

If we can lead the network that realizes y_2 to believe that the input-symbol change from 00 to 11 is accomplished by changing first x_2 and then x_1, i.e., $00 \rightarrow 01 \rightarrow 11$, then y_2 will not be falsely excited. We circumvent the input symbol 10, which falsely excites y_2. Some term in the expression for y_2' must cover $x_1 \bar{x}_2 \bar{y}_1 \bar{y}_2$. For purposes of example, suppose that term is $x_1 \bar{x}_2 \bar{y}_1$. If we realize this term with a NOR gate rather than with an AND gate, the change in x_2 will reach the NOR gate before the change in x_1.

$$x_1 \bar{x}_2 \bar{y}_1 = \overline{\bar{x}_1 \lor x_2 \lor y_1}$$

False excitation of y_2 is avoided by using the delay of the inverter that provides $\bar{x}_1$ to resolve the critical race between x_1 and x_2 in favor of x_2.

Since we are discussing a reversible transition within a row, we must also be concerned with avoiding false excitation of y_2 when the input symbol is changed from 11 to 00. Now we would like the network to believe that x_1 changes before x_2. The NOR gate suggested above guarantees that y_2 will be falsely excited in this case. What we require for this transition is the AND gate that realizes $x_1 \bar{x}_2 \bar{y}_1$ and senses the change in x_1 first. Considering both directions of this reversible transition, we see the need for a network that resolves the race between x_1 and x_2 in favor of x_2 when both x_1 and x_2 change from 0 to 1, and in favor of x_1 when both input variables change from 1 to 0. Figure 13.26 shows such a network: gates 1 and 2 ensure that f remains at 0 when both input variables change to 1; gates 2 and 3 resolve input-variable races to 0 in favor of x_1. State variable y_1 is introduced at gate 3 rather than gate 1 to resolve input-variable versus state-variable races in favor of the input variables.

Avoidable Function Hazard

Suppose more generally that set $S1$ for state-assignment partition π_i and some row of a state table does not satisfy the condition necessary to make the row entirely free of function hazards. If the input symbols associated with the members of $S1$ form subcubes all linked together, i.e., each subcube has a nonempty intersection with some other subcube, then a function hazard may be *avoided*. As a first example of how we can design a network so that the input-symbol change is forced to follow the linked set of input symbols, consider the following row.

y_i	y_i'			
	00	01	11	10
1	1	1	1	0

Two linked subcubes of input symbols, $0x$ and $x1$, are associated with the members of $S1$. A sum of terms must appear in the expression of y_i' that covers $y_i(\bar{x}_1 \vee x_2)$. We will concentrate on realizing the sum $\bar{x}_1 \vee x_2$, which can then be ANDed with y_i, which is to remain invariant. We are

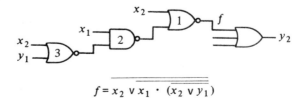

$$f = \overline{x_2 \vee x_1 \cdot \overline{(x_2 \vee y_1)}}$$

FIG. 13.26 A Network Which Avoids Exciting a State Variable Falsely

primarily concerned with multiple input-variable changes from $x_1 x_2 = 00$ to 11 and vice versa.

Figure 13.27 shows waveforms provided by various networks when x_1 and x_2 change simultaneously. Each gate is assumed to introduce approximately the same delay. The times of switching are not precisely indicated because gate and wire delays will differ slightly. The direct implementation of $\bar{x}_1 \vee x_2$ generates a false value when both x_1 and x_2 drop to 0 and thus is not suitable if the state transition for input symbols 00 and 11 is reversible. This network is suitable if the transition from 00 to 11 is nonreversible, however. During the 00 to 11 input-symbol change, the NOT gate that provides $\bar{x}_1$ holds this sum term at 1 until $x_2 = 1$ is felt by the OR gate, even if x_1 changes slightly sooner than x_2.

If the term $\bar{x}_1 \vee x_2$ is realized with a NAND gate, as suggested by $x_1 \uparrow \bar{x}_2$, a false value is generated when both x_1 and x_2 take the value of 1. This network is suitable if a transition from input 11 to input symbol 00 is nonreversible. If a reversible transition is specified by the state table providing the example row, then the sum $(\bar{x}_1 \vee x_2) \vee x_1 \uparrow \bar{x}_2$ suggests a network that will hold y_i invariant in all cases. The term $\bar{x}_1$ of this sum contributes little, and $x_2 \vee x_1 \uparrow \bar{x}_2$ provides a network free of function hazards. When both x_1 and x_2 switch value from 0 to 1, the NAND gate holds this expression at 1 until the change of x_2 reaches the OR gate, even if the change in x_1 reaches the NAND gate at a slightly earlier time. When x_1 and x_2 simultaneously drop to 0, the inverter providing $\bar{x}_2$ holds the NAND-gate output signal at 1 until the change in x_1 reaches the NAND gate, even if the change in x_2 reaches the inverter slightly earlier.

The waveforms of Fig. 13.27(b) pertain to terms of a network to realize the following row of a state table.

| y_i | y_i' | | | | | | | |
	000	001	010	011	100	101	110	111
...1...	1	0	0	0	1	1	0	1

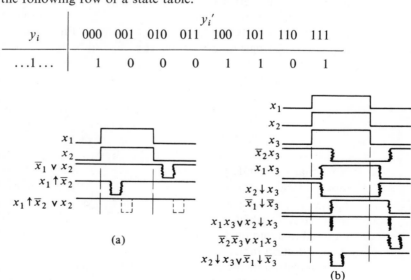

(a)

(b)

FIG. 13.27 Waveforms provided by various terms in response to multiple input variable changes

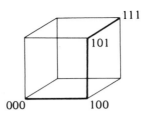

Here again the members of set $S1$ are associated with input symbols that form linked subcubes. A minimum cover of the input symbols for which y_i is to remain invariant at 1 is $\bar{x}_2\bar{x}_3 \lor x_1x_3$. In the worst case — all three input variables changing simultaneously — this sum produces a false value after the input symbol changes from 111 to 000, as shown near the bottom of Fig. 13.27(b). Again each of the sum terms may be realized with a NOR gate rather than an AND gate: $x_2 \downarrow x_3 \lor \bar{x}_1 \downarrow \bar{x}_3$ provides a false value following a 000 to 111 input-symbol change. Thus each of these networks may be used if a nonreversible transition is specified in the state-table row upon which our example is based. But if a reversible transition is specified, then the sum of these two sums, i.e., $(\bar{x}_2\bar{x}_3 \lor x_1x_3) \lor (x_2 \downarrow x_3 \lor \bar{x}_1 \downarrow \bar{x}_3)$, must be used. The waveforms suggest that a simpler network based upon $x_1x_3 \lor x_2 \downarrow x_3$, where all input variables are introduced at the same level, might be used. The inertial properties of the OR gate that combines the AND and NOR terms might remove brief false values, which can appear because of the slightly different arrival times of the input-variable changes and the difference in the delays provided by the AND and NOR gates, but this is clearly a questionable approach to the development of reliable networks.

To see how these ideas may be incorporated in a design situation, we consider the state table and Karnaugh maps of Fig. 13.28. If we assume first that only single input-variable changes will be allowed and execute the design procedure presented earlier in this section we obtain the following excitation equations.

$$y_1' = x_2y_1 \lor x_1 y_1 \bar{y}_2 \lor \bar{x}_1x_2 \lor x_1 \downarrow \bar{y}_2$$
$$y_2' = y_2(\overline{x_1x_2 y_1}) \lor \bar{x}_1x_2 \lor x_1 \bar{y}_1$$

Note that one NOR term appears in the expression of y_1' to resolve an input-variable versus state-variable race. The complex hold term, $y_2(\overline{x_1x_2 y_1})$, in the expression for y_2' offers a very economical cover of all total states for which y_2 is to remain invariant at 1; a complex hold term is required to avoid the function hazard induced by the d-trio of the original state-transition table.

Examination of the rows of the Karnaugh maps for y_1' and y_2' reveals rows free of function hazards, and other rows that contain avoidable function hazards. To fix critical input-variable races, we can examine each row individually and correct the excitation equations, or we may

Present state $y_1 y_2$ / Next state $x_1 x_2$

$y_1 y_2$		00	01	11	10
0 0	a	(a)	c	b	b
0 1	b	c	c	(b)	(b)
1 1	c	(c)	(c)	d	b
1 0	d	a	c	(d)	(d)

y_1' map (x_1 across, y_2 columns, y_1 rows, x_2 below):

0	1	0	0
1	1	0	0
1	1	1	0
0	1	1	1

y_2' map:

0	1	1	1
1	1	1	1
1	1	0	1
0	1	0	0

Transition		S_i **F** S_f	Hold Terms y_1'	y_2'	Required sequence of input variable changes y_1'	y_2'
$a \to b$	$0000 \to 1101$	x x 0 1			x_1, x_2	
$b \to c$	$1101 \to 0011$	x x x 1		$\bar{y}_1 y_2$		
$b \to c$	$1001 \to 0111$	x x x 1		$\bar{y}_1 y_2$		
$c \to d$	$0011 \to 1110$	x x 1 x	$y_1 y_2(\bar{x}_1 \vee x_2)$		x_2, x_1	
$c \to b$	$0111 \to 1001$	x x x 1		$y_1 y_2(\bar{x}_1 \vee \bar{x}_2)$		x_2, x_1
$d \to a$	$1110 \to 0000$	x x x 0				x_2, x_1
$d \to c$	$1010 \to 0111$	x x 1 x	$y_1 \bar{y}_2(x_1 \vee x_2)$		x_2, x_1	

FIG. 13.28 An Examination of Multiple Input-Variable Changes

examine each possible multiple input-variable-change transition in much the same manner as Table 13.1. Figure 13.28 summarizes all such transitions. The factor product S_i **F** S_f for each transition reveals that one state variable is to remain invariant for each transition. We must make certain that such variables do in fact remain invariant. For example, y_1' is to remain at 1 for the nonreversible transition from state (c) to d. Examination of the third row of the y_1' map indicates that a function hazard is avoided by forcing the change of x_2 to be sensed before the change in x_1. Terms $\bar{x}_1 y_2$ and $x_2 y_1$ of the excitation equations above cover the 1's of this third row; a realization of $\bar{x}_1 y_2 \vee x_2 y_1$ will sense the change of x_2 first, so the y_1' excitation equations need not be altered.

But the transition from state (d) to c in the fourth row of the state table will cause a malfunction unless the change in x_2 is sensed prior to the change in x_1. Terms $x_2 y_1$ and $x_1 y_1 \bar{y}_2$ of the original excitation equations cover the 1's of this fourth row. The realization of $x_2 y_1 \vee x_1 y_1 \bar{y}_2$ does not clearly sense the change in x_2 first. If the delays offered by the two AND gates that realize these terms differ substantially, then y_1' may take a false value. Inertial characteristics of the OR gate that sums these two terms might suppress this temporary false value, but the design is clearly questionable. We may wish to introduce an additional AND gate and realize the second term with $y_1(x_1 \bar{y}_2)$ to ensure that the network will respond correctly to this multiple input variable change.

Those transitions for which a state variable is to remain at 0 require that we examine the way in which excite terms are realized. In the first row of the y_1' map, the change in x_1 must be sensed first when the transition from total states 0000 to 1101 is being made. The excite term $\bar{x}_1 x_2$, which covers the 1 in this first row, will sense the change in x_2 first and must therefore be replaced with $x_1 \downarrow \bar{x}_2$. This new term may be combined with $x_1 \downarrow \bar{y}_2$ to give $x_1 \downarrow (x_2 \downarrow y_2)$. In the fourth row of the y_2' map and the state $\textcircled{d}$ to a transition, the change in x_2 must be sensed first. The term $\bar{x}_1 x_2$ must therefore remain intact in the y_2' equation and we can not share this AND gate between the y_1 and y_2 networks.

In summary, if we expect the network to respond properly to multiple input changes, we may realize the network by direct implementation of the following excitation equations.

$$y_1' = x_1 \downarrow (x_2 \downarrow y_2) \ \vee \ x_2 y_1 \ \vee \ y_1(x_1 \bar{y}_2)$$
$$y_2' = \bar{x}_1 x_2 \ \vee \ x_1 \bar{y}_1 \ \vee \ y_2(\overline{x_2(x_1 y_1)})$$

Some variations on these equations also meet all the requirements for a network that avoids all function hazards involving state- versus input-variable races, and input- versus input-variable races.

Unavoidable Function Hazards

Rows that contain unavoidable function hazards remain to be considered. The first two rows of the state table of Fig. 13.29 fall into this category. States a and d must appear in different blocks of some state assignment partition, as must states a and b, and a and c. For purposes of example we will assume that this state table is encoded with an STT state assignment based upon partitions $\pi_1 = \{\overline{a, b}; \overline{c, d}\}$ and $\pi_2 = \{\overline{a, d}; \overline{b, c}\}$. Then π_1 divides the first row into sets $S1 = \{\textcircled{a}, \textcircled{a}\}$ and $S2 = \{d, d\}$. The input symbols associated with the members of $S1$ are disjoint. The Karnaugh map and excitation equations for y_1' shown in Fig. 13.29 also reveal that y_1 will be falsely excited if the input symbol is changed from 00 to 11. Both $x_1 \bar{x}_2$ and $\bar{x}_1 x_2 \bar{y}_2$ in the equation for y_1' take the value of 1 for the length of time required by the inverters supplying $\bar{x}_2$ and $\bar{x}_1$ respectively to respond to the input symbol change. Partition π_2 induces an unavoidable function hazard in the second row of the state table. From the Karnaugh map for y_2' we see that y_2 is to be held at 1 when the input symbol is changed from 01 to 10, but that its value will be 0 if input symbol 00 or 11 is sensed. The equation for y_2' does not clearly reveal the source of the malfunction since y_2' is expressed in a highly factored form. If the delays of the inverters providing $\bar{x}_1$ and $\bar{x}_2$ differ, then term $\overline{x}_1 \overline{x}_2 \bar{y}_1$ might well take the value of 0. If the changes of x_1 and x_2 arrive at the NAND gate providing $\overline{x_1 x_2}$ at different times, then this term could also take the value

0; however, the inertia of the NAND gate might well eliminate this transient value, since we assume that the times of arrival are not widely separated.

If a network is to circumvent states in which a state variable is falsely excited, when the network is responding to a multiple input-variable change, we must provide a sequence of states for which the state variable has a constant value. We will do this by introducing an additional state variable. This will be an unusual state variable in that it never retains the value of 1 for any length of time, i.e., there are no hold terms in its excitation equation. The introduction of this variable will also make our state table abnormal. Not all transitions will be direct transitions from one stable total state, possibly through one unstable state, to the final total stable state. Thus the resulting network will not execute all transitions in the same period of time. In a sense we will be encoding the state table with a non-STT assignment.

We begin to form the additional state variable by locating the total states of the original state table in which unavoidable function hazards could result in a wrong value of a state variable. These total states have been marked with asterisks in Fig. 13.29. We will call these total states *hazard states*. Our additional state variable will take the value 1 for all hazard states and the value 0 for all other total states. Thus the Karnaugh map shown in Fig. 13.30 for this additional variable, h, is easily prepared.

To re-encode the state table to include this additional variable, we prepare two copies of the Karnaugh maps for y_1' and y_2'; $h = 0$ is associated with one of each pair of maps and $h = 1$ labels the other. In the maps labeled $h = 0$, we complement the entries that represent potential false values for multiple input-variable changes. These entries appear in the cells of hazard states. Thus the y_1' maps of Fig. 13.30 differ from the y_1' map of Fig. 13.29 in the first row of the $h = 0$ submap. The second row of the y_2' map for $h = 0$ differs from its counterpart in Fig. 13.29. The maps for $h = 1$ are

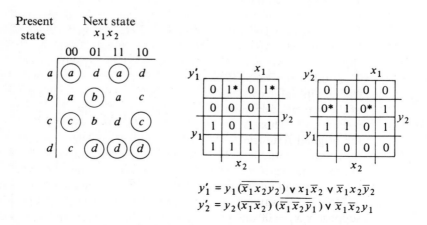

$$y_1' = y_1(\overline{x_1}x_2y_2) \vee x_1\overline{x}_2 \vee \overline{x}_1x_2\overline{y}_2$$
$$y_2' = y_2(\overline{x_1}\overline{x}_2)(\overline{x_1}\overline{x}_2\overline{y}_1) \vee \overline{x}_1\overline{x}_2y_1$$

FIG. 13.29 Unavoidable Function Hazards

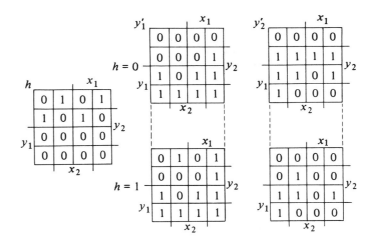

FIG. 13.30 Introducing the Hazard Variable

Present state hy_1y_2		Next state x_1x_2			
		00	01	11	10
000	a	ⓐ	e	ⓐ	e
001	b	f	ⓑ	f	c
011	c	ⓒ	b	d	ⓒ
010	d	c	ⓓ	ⓓ	ⓓ
100	e	a	j	a	j
101	f	e	b	e	c
111	g	c	b	d	c
110	j	c	d	d	d

duplicates of the maps of Fig. 13.29. If we now reform an unencoded state table from these maps we obtain the 8-state table of Fig. 13.30.

Examination of Fig. 13.30 reveals the philosophy of introducing variable h in the manner described above. First, let us examine the way in which the transition from state a to state d is made when a single input variable changes. Assume $x_1x_2 = 00$ and $y_1y_2 = 00$; $h = 0$ is the normal condition. If x_2 takes the value of 1, neither y_1 nor y_2 is excited, but h is. When h changes its value to 1, state e is attained and y_1 becomes excited. The state of the network advances to state j and then finally to state d as variable h returns to 0. Those state transitions that passed through hazard states of the original state-transition table are now performed indirectly. For these transitions, state variables are not excited until variable $h = 1$. After the excited state variables attain their new value, h returns to 0. Thus these transitions will not be performed as rapidly as the more direct transitions.

Now let us examine multiple input-variable changes in Fig. 13.30. Again suppose $x_1x_2 = 00$ and $hy_1y_2 = 000$. Now assume x_1 and x_2 simultaneously change to 1. Variable h may be falsely excited, but the maps for y_1' and

y_2' of Fig. 13.30 indicate that neither of these state variables will be falsely excited. If the new values of the input variables are sensed by the y_1 and y_2 networks before h can change its value, then subsequent changes of h can have no effect since y_1' and y_2' are independent of h for all nonhazard total states.

To clarify this point further let us examine the following excitation equations derived from the maps of Fig. 13.30.

$$h = \bar{y}_1(\bar{x}_1\bar{x}_2 y_2 \vee \bar{x}_1 x_2 \bar{y}_2 \vee x_1\bar{x}_2 \bar{y}_2 \vee x_1 x_2 y_2)$$

$$y_1' = y_1(\overline{\bar{x}_1 x_2 y_2}) \vee x_1\bar{x}_2 y_2 \vee \bar{x}_1 x_2 \bar{y}_2 h$$

$$y_2' = y_2(\overline{x_1 x_2 y_1}) (\overline{x_1 x_2 h}) (\overline{\bar{x}_1\bar{x}_2 \bar{y}_1 h}) \vee \bar{x}_1\bar{x}_2 y_1$$

Circuitry based upon these equations might malfunction for any of several reasons. If the inverter that provides $\bar{x}_1$ introduces more delay than the total network that generates h, then the term $\bar{x}_1 x_2 \bar{y}_2 h$ could take the value of 1 when the state a to state a transition is attempted. Here we see an input-variable versus state-variable race. It must be resolved in favor of x_1 by replacing the term with $x_2(\overline{x_1 \vee h \vee y_2})$. The term $\bar{x}_1\bar{x}_2 \bar{y}_1 h$ should be replaced with $\overline{x_1 \vee x_2 \vee y_1 \vee h}$ to resolve a similar race in the circuitry which realizes y_2. Finally, avoidable function hazards exist in the third and fourth rows of the original state table of Fig. 13.29. To avoid the malfunctions these hazards can cause, further terms of the equations for y_1' and y_2' must be altered.

$$y_1' = y_1(\overline{x_1 \vee \bar{x}_2 \vee \bar{y}_2}) \vee x_1\bar{x}_2 y_2 \vee x_2(\overline{x_1 \vee h \vee y_2})$$

$$y_2' = y_2(\overline{x_1 x_2 y_1}) (\overline{x_1 x_2 h}) (\overline{x_1 \vee x_2 \vee y_1 \vee h}) \vee \overline{x_1 \vee (x_2 \vee \bar{y}_1)}$$

In this section we have concentrated upon achieving reliable asynchronous networks by using the delay introduced by logic gates to resolve races. This is certainly not the only approach to obtaining reliable networks nor is it always the best approach. If speed is not of prime concern the use of delay elements in the feedback paths may be a simpler if not more economic approach. If the assumptions of this section concerning wire and loop delays and "simultaneous" changes of input variables can not be satisfied, then some other approach must be taken. An examination of the networks designed suggests that the methods of this section do not result in networks of much greater cost than the marginal network one obtains by ignoring essential hazards, d-trios, and multiple input-symbol change hazards.

13.5 SUMMARY

Every digital system is asynchronous and therefore subject to the failure modes discussed in this and the previous chapter. In a synchronous system we attempt to eliminate many failure modes by using synchronized flip-flops. The flip-flops are asynchronous networks, of course; delays in the clock distribution system may invalidate the synchronization we assume to exist. By using carefully designed flip-flops, observing their setup, hold, and minimum pulse width restrictions, and carefully distributing clock signals, we gain reliability with ease of design. The cost is slower system response time. The clock period must be longer than the maximum delay through the combinational logic. When operations that do not involve maximum delay are performed, the system waits idly for the next clock pulse. Set-up times might be placed in the same category.

Asynchronous networks respond promptly to input signal changes but are subject to many failure modes. Delay, which gives us memory, now is a culprit. The most prevalent failure modes are called hazards and can be described by introducing delay elements into systems models. In the fundamental model we introduced the minimum number of delay elements so that every loop contained delay. In the detailed model we associated a delay element with every gate. Other models could be proposed, but are seldom necessary. Each delay element added compounds analysis; so we tend to use the minimum number that provide the information we seek.

Combinational networks can not be freed of function hazards in general. But, combinational networks in sequential networks are exposed to a restricted set of input symbol changes. Only a restricted set of function hazards need be eliminated. We do this by adding logically redundant gates and perhaps massive delays in the feedback path and by choosing the state assignment carefully. Single Transition Time assignments are found with modest effort, and lead to fast networks. If a variety of gate types are available and gate delays exceed wire delays, which is often the case, then combinational networks may be rearranged so that gate delays insure reliable networks.

This chapter concluded with an investigation of multiple-input-variable changes. While this is a very practical topic and a small variety of alternative approaches for designing reliable networks appears in the literature, it is a very difficult topic, and we can hope that superior design techniques will emerge.

Most asynchronous networks are designed by "hooking" together latches and gates. Very frequently they work. When they do not and when we must have great confidence in a design, then we must resort to detailed analysis and synthesis algorithms. G. G. Landon [7] provides perspective.

REFERENCES

1. D. B. ARMSTRONG, A. D. FRIEDMAN, and P. R. MENON, "Realization of Asynchronous Sequential Circuits Without Inserted Delay Elements," *IEEE Trans. on Computers*, Vol. C-17, Feb. 1968, pp. 129–134.
2. E. B. EICHELBERGER, "Hazard Detection in Combinational and Sequential Switching Circuits," *IBM Jour. of Research and Development*, March 1965, pp. 90–99.

3. A. D. FRIEDMAN and P. R. MENON, *Theory and Design of Switching Circuits.* Woodland Hills, Calif.: Computer Science Press, 1975.

4. D. A. HUFFMAN, "The Synthesis of Sequential Switching Circuits," *Jour. Franklin Inst.*, Vol. 257, March 1954, pp. 161–190, and April 1954, pp. 275–303.

5. D. A. HUFFMAN, "The Design and Use of Hazard-Free Switching Networks," *Jour. ACM*, Vol. 4, Jan. 1957, pp. 47–62.

6. G. G. LANGDON, Jr., "Analysis of Asynchronous Circuits Under Different Delay Assumptions," *IEEE Transactions on Computers*, Vol. C-17, Dec. 1968, pp. 1131–1143.

7. G. G. LANGDON, Jr., *Logic Design, A Review of Theory and Practice.* New York: Academic Press, 1974.

8. C. N. LIN, "A State Variable Assignment Method for Asynchronous Sequential Switching Circuits," *Jour. ACM*, Vol. 10, April 1967, pp. 202–216.

9. G. K. MAKI and D. H. SAWIN, "Fault-Tolerant Asynchronous Sequential Machines," *IEEE Trans. on Computers*, Vol. C-23, July 1974, pp. 651–657.

10. G. A. MALEY and J. EARLE, *The Logic Design of Transistor Digital Computers.* Englewood Cliffs, New Jersey: Prentice-Hall, 1963.

11. M. P. MARCUS, *Switching Circuits for Engineers.* Englewood Cliffs, New Jersey: Prentice-Hall, 1962; 1967.

12. E. J. MCCLUSKEY, Jr., "Transients in Combinational Logic Circuits," *Redundancy Techniques for Computing Systems.* Spartan Book Co., 1962, pp. 9–46.

13. E. J. MCCLUSKEY, Jr., *Introduction to the Theory of Switching Circuits.* New York: McGraw-Hill, 1965.

14. R. E. MILLER, *Switching Theory Vol. II: Sequential Circuits and Machines.* New York: John Wiley, 1965.

15. C-J TAN, "State Assignment for Asynchronous Sequential Machines," *IEEE Trans. on Computers*, Vol. C-20, April 1971, pp. 382–391.

16. J. H. TRACEY, "Internal State Assignment for Asynchronous Sequential Machines," *IEEE Trans. on Electronic Computers*, Vol. EC-15, Aug. 1966, pp. 551–560.

17. S. H. UNGER, "Hazards and Delays in Asynchronous Sequential Switching Circuits," *IRE Trans. on Circuit Theory*, Vol. CT-6, March 1959, pp. 12–25.

18. S. H. UNGER, *Asynchronous Sequential Switching Circuits.* Wiley-Interscience, New York, 1969.

19. S. H. UNGER, "Asynchronous Sequential Switching Circuits with Unrestricted Input Changes," *IEEE Tran. on Computers*, Vol. C-20, Dec. 1971, pp. 1437–1444.

PROBLEMS

13.1-1. A network is needed to determine the winner of model car races. If switch $S1$ closes as a car passes over it, before switch $S2$, then z_1 is to go to 1 and remain at 1 and z_2 is to remain at 0. If $S2$ closes first, then z_2 is to go to 1 while z_1 remains at 0. Assume that $S1$ and $S2$ will not close simultaneously. The switches are free of contact bounce.

(a) Develop a primitive flow table for the network.

(b) Merge, assign, and implement the network with AND and OR gates.

(c) Qualitatively discuss the operation of your network. How would you extend it to resolve three-car races? How would you extend it so that it can be reset in preparation for another race?

(d) What will your network do if S1 and S2 close simultaneously?

13.1-2. A combination lock is to open, $z=1$, only if two switches are closed in the following sequence: (i) S1 is closed and held, (ii) S2 is closed (while S1 is held), (iii) S2 is opened (while S1 is held), (iv) S1 is released and the lock opens. If S2 is closed and then opened following this sequence, the lock is to reset. The switches can not change simultaneously.

(a) Develop a suitable primitive flow table. Eliminate equivalent states, if any.

(b) Draw a merger diagram. List all merger partitions. Select a merger partition with a small if not minimum number of blocks and, if possible, one that leads to a Moore-model machine.
Merge your primitive flow table using the partition you selected.

(c) Is it possible to reset your lock with other than the prescribed resetting input sequence? List all other resetting input sequence, if any.

(d) Does your merged table contain essential hazards?

(e) Assume a network performs exactly as prescribed by your merged state-transition table. Will an input sequence other than the one pre-scribed above open the lock?

13.1-3. For the merged table of Prob. 13.1-2, draw a transition diagram and select a state-variable assignment that avoids critical races. Implement the assigned state-transition table using only NAND gates.

13.1-4. Develop primitive flow tables and merged, minimum state, state-transition tables for a leading edge-triggered (a) JKFF, (b) TFF, and (c) DFF. The clock input signal will not change at the same time as the control input signals.

13.1-5. For the primitive flow table of Fig. 13.4:

(a) Find all 4-block merger partitions; show a merged state-transition table for each.

(b) Merge rows, 1, 2, 7; rows 1, 3; rows 4, 11; and rows 5 and 6. Note that row 1 is used twice. From this example are there any advantages or disadvantages apparent to this type of merging?

13.1-6. Find and eliminate any compatible states, which exist in the following primitive flow table. Then merge to the minimum number of rows.

$x_1x_2 = 00$	01	11	10	z
①	2	–	3	0
1	②	7	–	0
4	–	7	③	1
④	6	–	5	0
1	–	7	⑤	1
1	⑥	9	–	0
–	2	⑦	8	1
4	–	9	⑧	0
–	6	⑨	10	1
1	–	9	⑩	0

13.1-7. Alternative encodings of the state-transition table of Fig. 13.6(b) are given below. Point out the disadvantage of each, if one exists.

Present state	Next state 00	01	11	10	z
a 00	00	01	10	00	0
b 01	00	01	01	00	0
d 11	10	01	11	11	0
c 10	00	10	10	11	1

Present state	Next state 00	01	11	10	z
a 00	00	01	11	00	0
b 01	00	01	01	00	0
d 10	00	01	10	10	0
c 11	00	11	11	10	1

Present state	Next state 00	01	11	10	z
a 00	00	01	10	00	0
b 01	10	01	01	00	0
d 11	01	01	11	11	0
c 10	11	10	10	11	1

13.1-8. (a) To evaluate the effectiveness of the output symbol logic of Fig. 13.9(c), determine the duration Δ of the pulse provided by the network of Fig. 13.10(a), assuming that each gate introduces 50 ns delay and T assumes the value of 1 for N nanoseconds.

(b) Repeat, assuming $z = Ty_1\,\bar{y}_2$, and compare the two values of Δ. Which output network results in the more faithful reproduction of the T pulse?

13.1-9. What gate-delay distribution for the network of Fig. 13.10 will cause a steady-state hazard to appear when ST changes from 10 to 11 with the machine originally in state a?

13.2-1. The state table below can not be reliably realized with two state variables, but the five following 3-state-variable assignments may be used. For each, develop a complete 8-row encoded table for the machine, indicating stable total states by encircling. Label the vertices of a 3-cube with the state assigned to each, and determine which state transition will require the greatest amount of time to complete.

State	Encodings				
	i	ii	iii	iv	v
a	000	000	000, 010	000	000, 111
b	001	001	001, 011	001	001, 110
c	011	010	110, 111	011, 010	011, 100
d	111	100	101, 100	110, 111, 101, 100	010, 101

Present state	Next state 00	01	11	10
a	ⓐ	ⓐ	c	b
b	ⓑ	a	d	ⓑ
c	b	d	ⓒ	ⓒ
d	a	ⓓ	ⓓ	c

13.2-2. Write the set of state-set partitions suggested by each of the five assignments of Prob. 13.2-1. Find two encodings of the machine of that problem that are not given by any of these state-set partitions.

13.2-3. Encode the following state-transition tables with three state variables by mapping them to a 3-cube. For each show the transition diagram, a 3-cube with labeled vertices, the encoded state table, and the state-set partitions that describe your encoding. Do not introduce equivalent states.

	00	01	11	10		00	01	11	10		00	01	11	10
a	ⓐ	b	—	d	a	ⓐ	d	ⓐ	e	a	ⓐ	d	ⓐ	c
b	ⓑ	ⓑ	c	ⓑ	b	ⓑ	e	a	ⓑ	b	a	ⓑ	ⓑ	d
c	—	d	ⓒ	d	c	ⓒ	d	—	d	c	d	ⓒ	b	ⓒ
d	b	ⓓ	e	ⓓ	d	b	ⓓ	e	ⓓ	d	ⓓ	ⓓ	e	ⓓ
e	a	ⓔ	ⓔ	f	e	c	ⓔ	ⓔ	ⓔ	e	f	c	ⓔ	c
f	ⓕ	e	c	ⓕ	f	a	ⓕ	ⓕ	b	f	ⓕ	b	a	ⓕ

13.2-4. (a) By using equivalent states and three state variables, attempt to encode the state tables of Prob. 13.2-3 to obtain faster networks. Assume only one input variable changes at a time and allow indirect transitions, if these help.

(b) If you can not encode a table with three state variables, use four or five and sets of equivalent states with encodings that are linked; i.e., each member is adjacent to some other member. You need not show the final encoded state table, but show the transition diagram with each vertex labeled with the n-tuples used to encode the equivalent states.

13.2-5. Given an arbitrary state table with eight states, $2^3 - 1 = 7$ state variables must be used, per Eq. (13.3), to arrive at an STT assignment without critical or noncritical races. Each state must be duplicated $2^7/8 = 16$ times. To arrive at the sets of 16 7-tuples to encode each state, proceed as follows:

(a) Extend the Hamming code of Table 3.11 to encode all 16 4-bit message words, i.e., all hexadecimal digits. Show that each of these 7-tuples is distance 3 from some other one by writing them in an order such that each 7-tuple is distance 3 from the one that precedes it. Start with 0000000.

(b) Re-encode the hexadecimal digits, computing k_1 to give odd parity over positions 1, 3, 5, and 7. Compute k_2 and k_4 to give even parity as before. Show that each of these 16 new 7-tuples is distance 3 from some other new 7-tuple and adjacent to one of the 7-tuples of part (a).

(c) Repeat part (b), computing k_2 to give odd parity over positions 2, 3, 6, and 7; as k_1 and k_4 to give even parity.

(d) Summarize how you would continue to find the 5 remaining sets of 16 7-tuples.

13.2-6. Find Tracey STT state assignments for the state-transition tables of Prob. 13.2-3.

13.2-7. (a) Using column partitions, find an STT assignment for the state table below.

(b) Find a minimum state variable STT state assignment.

Present state	Next state 00	01	11	10
a	(a)	(a)	i	k
b	(b)	f	j	(b)
c	(c)	a	–	k
d	(d)	f	–	b
e	c	(e)	(e)	l
f	a	(f)	i	–
g	c	(g)	e	–
h	d	g	(h)	(h)
i	–	e	(i)	l
j	–	g	(j)	h
k	b	–	j	(k)
l	d	–	h	(l)

13.2-8. Let A be the set of all column partitions for a given state-transition table. Let B be a set of state-set partitions such that every member of A is less-than or equal to some member of B and the product of all members of B is not equal to π^0. The text suggests that mergeable rows exist in the ST table; demonstrate this with an example. What information does the non-π^0 product provide?

13.3-1. Merge the primitive flow table below so that the resulting state table (a) contains many d-trios; and (b) is free of d-trios.

x_1x_2 00	01	11	10	z
(1)	4	–	8	0
(2)	3	–	8	1
2	(3)	6	–	0
1	(4)	6	–	1
–	4	(5)	8	0
–	4	(6)	7	1
2	–	6	(7)	0
2	–	5	(8)	1

13.3-2. (a) Identify d-trios and essential hazards in the state table below, if any exist.

(b) Using the state assignment given, write the sum-of-products excitation equation, y_1' and y_2', and draw the corresponding network. Assuming only single input-variable changes, for each transition identify which gate(s) is exciting, possibly falsely exciting, or holding each state variable.

(c) Show how the essential hazard(s) found in part (a) results in an input-variable versus state-variable race that can result in a false transition.

(d) Alter the logic of your network so that state variables will not be falsely excited and input-versus state-variable races will be resolved.

Present state	Next state $x_1 x_2$ 00	01	11	10	state encoding y_1	y_2
a	ⓐ	d	b	b	0	0
b	ⓑ	c	ⓑ	ⓑ	0	1
c	ⓒ	ⓒ	d	ⓒ	1	1
d	a	ⓓ	ⓓ	c	1	0

13.3-3. What changes, if any, would you make to the networks of Fig. 13.10 to ensure that the essential hazard in the state-transition table does not result in a circuit malfunction? Does the fact that the encoded state table of Fig. 13.8 is abnormal cause any difficulty?

13.3-4. Find an STT state assignment and realization of the state table below which is free of steady-state hazards and will not generate any temporary false values on the output line in response to single input variable changes.

Present state	Next state $x_1 x_2$ 00	01	11	10	Output z $x_1 x_2$ 00	01	11	10
a	ⓐ	c	ⓐ	ⓐ	0	0	0	1
b	e	–	c	ⓑ	1	–	0	0
c	e	ⓒ	ⓒ	a	1	0	0	1
d	a	–	c	ⓓ	0	–	0	1
e	ⓔ	ⓔ	ⓔ	b	1	1	1	0

13.3-5. What will your network of Prob. 13.3-4 do if the input symbol is changed from 10 to 01, if it is initially in state a? b? d? If you can not say exactly what the final states of each transition will be, indicate the possible final states and the conditions under which each will be reached.

13.3-6. A gate is driven by signals $\bar{x}$ and y. Replace this gate with a logically equivalent network that is driven directly by x,
(a) AND gate   (c) NAND gate
(b) OR gate   (d) NOR gate

13.3-7. (a) Manually transcribe the encoded state-transition table of Fig. 13.8 to the form of a state-transition array of $(n + 2p + m)$-tuples. (b) With the aid of Table 5.9 transcribe the ST-array to an $(n + p)$-input, $(2p + m)$-output function array description of the combinational logic required to realize the machine with RS latches. (c) Write SET, RESET, and output equations from the function array, using Karnaugh maps to minimize each, and compare with the equations given in Section 13.1.

13.4-1. A variation of the RS latch that retains its present state when $RS = 11$ as well as when $RS = 00$ is required. Design such a latch; it is to operate properly even when the input variables are changed simultaneously.

13.4-2. Figure 13.26 presents a realization of an excite term that avoids falsely exciting the state variable.
(a) Develop corresponding networks for the following encoded state table rows.

	y_i	y_i' x_1x_2 00	01	11	10
i	0	0	0	1	0
ii	0	0	1	0	0
iii	0	1	0	0	0

(b) Develop networks that realize hold terms for the following individual rows of encoded state tables, and will not fail to hold even when the input variables change simultaneously.

	y_i	y_i' x_1x_2 00	01	11	10
i	1	1	1	1	0
ii	1	1	1	0	1
iii	1	1	0	1	1
iv	1	0	1	1	1

13.4-3. The state table of Fig. 13.28 is to be realized using two RS latches for the state variables. Derive input equations for the latches such that the network will not fail even when the input variables change simultaneously.

13.4-4. The state table of Fig. 13.19 contains a number of avoidable function hazards.

(a) For state transitions resulting from a multiple input-variable change, list the state variable(s) that may be falsely excited, and the order in which the input variable changes must be sensed if the false value is to be avoided.

(b) How must the networks of Figs. 13.24 and 13.25 be changed if the network is to respond in the manner of the original state table when input variables change simultaneously?

13.4-5. The JK latch described by the following state-transition table is to operate as described by this table when the input variables change simultaneously. The output variable z is to be free of false transient values. Design a suitable network.

Present state	Next state x_1x_2 00	01	11	10	z
a	ⓐ	ⓐ	c	b	0
b	ⓑ	a	d	ⓑ	1
c	b	a	ⓒ	b	1
d	a	a	ⓓ	b	0

ANSWERS

13.1-1. (a)

	S1 S2			z_1	z_2
00	01	11	10		
①	—	—	2	0	0
5	—	3	②	1	0
—	4	③	2	1	0
5	④	3	—	1	0
⑤	4	—	2	1	0
9	⑥	7	—	0	1
—	6	⑦	8	0	1
9	—	7	⑧	0	1
⑨	6	—	8	0	1

(b)

	S1 S2			z_1	z_2
00	01	11	10		
ⓐ	c	—	b	0	0
ⓑ	ⓑ	ⓑ	ⓑ	1	0
ⓒ	ⓒ	ⓒ	ⓒ	0	1

Use z_1 and z_2 as state variables

$$z_1' = z_1 \vee S1 \cdot \bar{z}_2$$
$$z_2' = z_2 \vee S2 \cdot \bar{z}_1$$

13.1-4. (c)

	DC			z
00	01	11	10	
①	2	—	4	0
1	②	3	—	0
—	2	③	4	0
1	—	7	④	0
⑤	—	—	8	1
5	⑥	7	—	1
—	6	⑦	8	1
5	—	7	⑧	1

{1, 4; 2, 3; 5, 8; 6, 7}
{1, 2, 3; 4; 5; 6, 7, 8}

13.1-7. (a) d to a transition is indirect.

(b) a to c is critical race; d to b is critical race.

(c) b to a transition is very indirect.

13.1-9. Both y_1 and y_2 must reach 1 before the T inverter switches.

13.2-1. (i) d to a requires 3 variables to change.

(ii) c to b and other transitions require 2 variable changes.

(v) An STT assignment.

13.2-3. (a) $\{\overline{a, e, d}; \overline{b, c, f}\}$

$\{\overline{a, b}; \overline{c, d, e, f}\}$

$\{\overline{a, e, f}; \overline{b, c, d}\}$

13.2-6. (c) Column partitions suggest 8 state variables; t-partitions give 5 variable state assignments.

13.2-7. (a) Column partitions suggest 8 state variables.

(b) $\{\overline{afi}; \overline{bjk}; \overline{ceg}; \overline{dhl}\}$

$\{\overline{ack}; \overline{bdf}; \overline{eil}; \overline{ghj}\}$

13.3-1. (a) $\{\overline{1, 4}; \overline{2, 3}; \overline{5, 8}; \overline{6, 7}\}$

(b) $\{\overline{1, 5}; \overline{2, 8}; \overline{3, 7}; \overline{4, 6}\}$

13.3-3. Problem 13.1-9 shows that the T inverter must be very slow for the malfunction to occur. We could replace the AND gate that realizes $S\overline{T}y_2$ with a NOR-AND pair that realized $S \cdot (T \downarrow \bar{y}_2)$.

13.3-6. (a) NOR gate.   (b) NAND gate.

13.4-2. (a) (i) $y_i' = x_2(x_1(x_2 \bar{y}_i)) \vee \cdots$

(iii) $y_i' = \overline{x_1 \vee (x_2 \vee (x_1 \vee y_i))} \vee \cdots$

(b) (i) $y_i' = y_i(x_2 \vee \overline{x_1 \cdot \bar{x}_2}) \vee \cdots$

(ii) $y_i' = y_i\overline{(x_2(x_1(x_2)))} \vee \cdots$

13.4-3. $S_1 = \bar{x}_1 x_2 \vee \bar{x}_1 y_2$

$= x_2(x_1 \downarrow (\bar{x}_2)) \vee x_1 \downarrow \bar{y}_2$

$R_1 = x_1 \bar{x}_2 y_2 \vee \bar{x}_1 \bar{x}_2 \bar{y}_2$

$= x_2 \downarrow (x_1 \uparrow (x_2 \downarrow \bar{y}_2)) \vee x_1 \downarrow (x_2 \vee (x_1 \vee y_2))$

Index

Modulus arithmetic, 196–201
MOMIN, 678
Monostable multivibrators, 289–291
Moore model, 308
m output signals, 183–184
Multiple-input-change hazards, 762–764
Multiple-input-variable changes, 832–844
Multiple-level networks, array expression of, 719–721
Multiple-level synthesis, 683–749
Multiple-order hazards, 780–782
Multiple-output factoring networks, 716–719
Multiple-output switching functions, 636–650
Multiplexers, 16, 180–182
  3-way, 133
  4-input, 181
  8-input, 181
Multiplication, 31–32, 213–223
  complement, 219–223
  repeated, 28
  sign-magnitude, 215–219
Multipliers, Booth, 222
Multivibrators:
  astable, 291–292
  monostable, 289–291

N

NAND gates, wiring open-collector, 113
NAND-latches, 262
NAND network synthesis, 172–180
NANDSY, 748
Next-state variables, 752
n input signals, 184
Noise margins, 56
Nonredundant covers, 649–650
Nonreversible transitions, 833–835
NOR-latches, 261–262
NOR network synthesis, 172–180
NOT gates, 11, 13
NRMIN, 680
n-tuple, 6
Numbers:
  bit, 196–197

left-to-right comparison, 195
negative, 201
positional systems, 22–32
positional value, 22–23
right-to-left comparison, 193
signed, complement codes for, 201–210
unitary system, 22

O

OFF-array, 88
ON-array, 88
One-hot codes, 227, 805–806
Operation code (*opcode*), 435
OPerator declarations, 135–137
OR gates, 12–13
  diode, 50, 51
OR operation, 156
Output:
  alphabet, 84, 307, 581
  common emitter amplifiers, 53–54
  equations, 800–801
  multiple switching functions, 636–650
  signals, 13
  single switching functions, 622–636
  symbols, 84, 279, 307
OUTPUT, 435–437
OVF terminals, 136–137

P

Parallel adders, 191
Parallel gating, 375, 377
Parallel switches, 48
Parameters of dynamic behavior, 56–57
Partition:
  algebra, 809–812
  column, 812
  transition, 815
Place markers, 94
Pointers, 486–488, 573
Polish string notation, 572–577
Pos equations, 99
Positional number systems, 22–32
Prefix notation, 572
Primary input signals, 13

Prime implicants:
  generation of, 610–614, 623–631
  selection of a cover, 614–618
  tables, 617
Primitive flow tables, 790–792
Processing elements, 11–19
Product terms, 74
Programmable counters, 380
Programmable logic arrays (PLA), 185–189
  sequential, 405–409
Programmable Read-Only-Memory (PROM), 185
Programming:
  arrays, 578–580
  of elementary digital computers, 461–468
Propagation delay, 56–57
  of the gate, 12
Push-down, 573

Q

Quine-McCluskey algorithm (Q-M), 612–613, 617, 621, 623

R

Radix, 23–24
  conversion, 25–28
Radix complement (RC), 201–203
  diminished, 202–204
Radix mark, 25
Read-Only-Memory (ROM), 182–185, 187, 500–506
  control, 401–405
REDUCE, 681
Redundancy, 231–237, 683–701
  array connection, 646–648
Redundant input variable, 699–701
Reed-Muller expansions, 118–120
Reflected codes, 226
REgister declarations, 123
Register file, 476–479
Registers, 8
  address extension, 483
  addressing, 481
  advanced applications, 475–488